he Unite

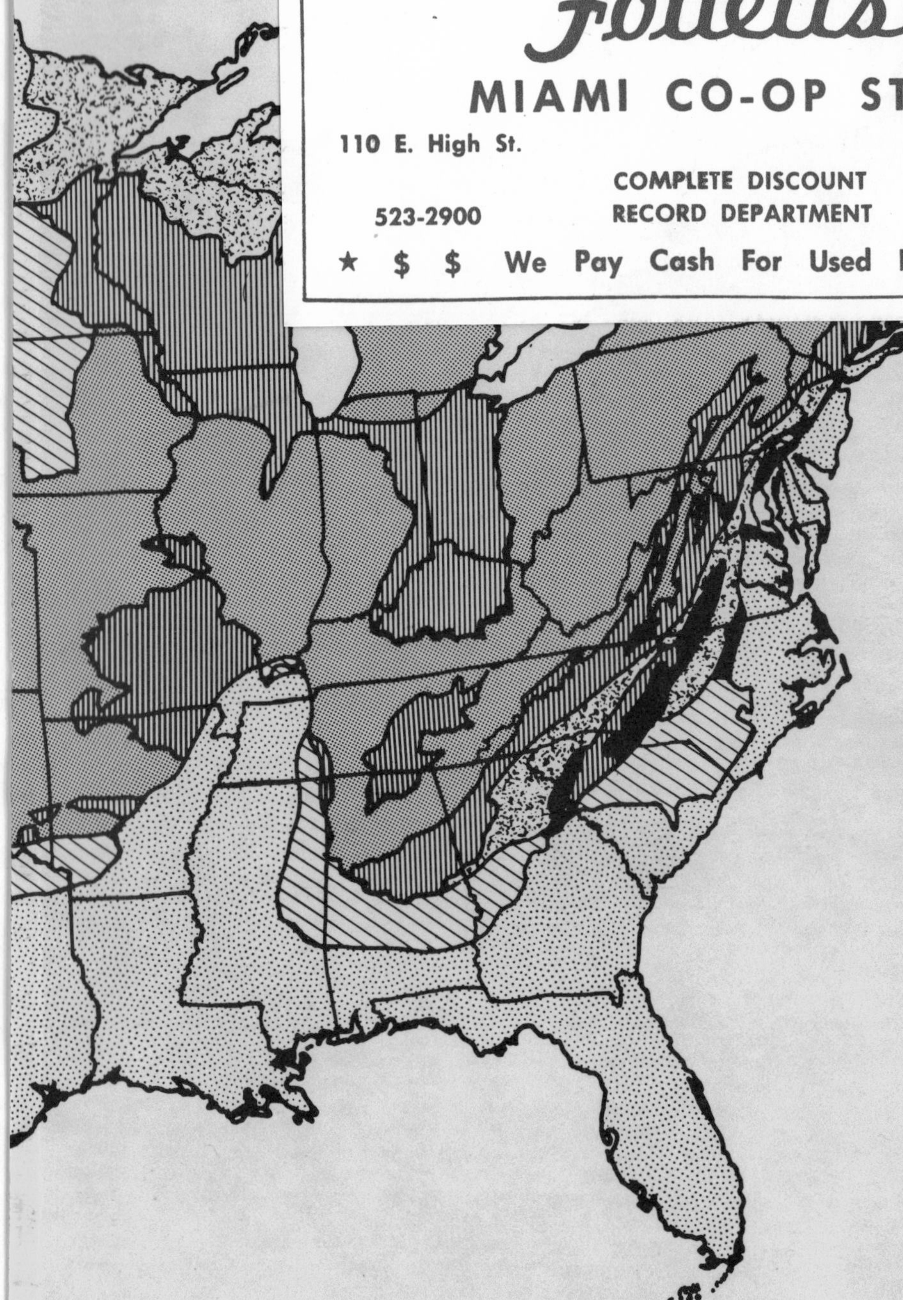

IGNEOUS ROCKS

Historical G

The Science of a D

Historical Geology

The Science of a Dynamic Earth

2nd Edition

Leigh W. Mintz
California State University, Hayward

Charles E. Merrill Publishing Company
A Bell & Howell Company
Columbus, Ohio

Cover Photos Top: *Jasper National Park.* (Courtesy of Alberta Travel Bureau, Canada.)
Bottom: *Middle Eocene Mammals, Reptiles, and Plants.* Mural by J.H. Matternes. (Courtesy of Smithsonian Institution, photo no. 62-1446.)

Published by
Charles E. Merrill Publishing Company
A Bell & Howell Company
Columbus, Ohio 43216

This book was set in Times Roman.
The production editor was Ann Mirels.
The cover was prepared by Will Chenoweth.

International Standard Book Number: 0-675-08603-5
Library of Congress Catalog Card Number: 77-76777

3-4-5-6-7-8-9-10—86-85-84-83-82-81-80-79-78

Printed in the United States of America

to the friends of the earth

Preface

When the first edition of this text appeared in 1972, a great revolution in the earth sciences, engendered by the discovery of plate tectonics, had just culminated. The majority of the geologic community, long defenders of continental and oceanic stability, had been won over to the idea of a mobile lithosphere and its momentous implications for the physical and biological history of the earth. As a recent convert to the new theories, I had attempted at that time to write the first of a new generation of historical geology texts in which earth history is interwoven in the context of shifting continents and ocean basins. Only the broader outlines were clear then, and many educated (and some uneducated) guesses had to be made in interpreting geologic history. In the intervening years the majority of the interpretations have been further substantiated, although some have had to be modified or discarded. But that is the essence of science and what makes it so intellectually stimulating. This new edition may lack some of the earlier revolutionary fervor, but it rests on a firmer foundation of new information gathered by legions of geologists using the paradigm of plate tectonics. As noted throughout the book, there are still many unknowns and ample speculation to stimulate the inquisitive mind.

The reception of the first edition was gratifying in its magnitude. Several suggestions and corrections that were made by its users have been incorporated into the second edition. Although the basic format and approach have been retained, the physically intimidating size of the first edition has been drastically reduced through the skilled production editing of the Merrill staff. All chapters have been extensively rewritten; those involving the earth's physical history in terms of plate tectonics are virtually new. As in the first edition, I have attempted to strike a balance between the physical and biological aspects of earth history. Many new illustrations have been added, some outdated ones removed, and the quality of still others, substantially improved.

I would like to thank the following individuals for providing much useful input: Professors Thomas Roberts, Peter Whaley, O.B. Nye, Jr., Rodney Griffin, Herbert Tischler, John Martin, Robert Gernant, and Raman

Singh. Many of their comments have been heeded, others have not, and all responsibility for the interpretations made in the text rests with the author.

I would like to thank my wife, Carol, for her invaluable assistance in the tedious job of indexing both editions and Judith Moreland and Mary Harrington for typing the manuscript of the second edition. And, finally, my thanks to the Merrill staff for their work in improving both the layout of the book and its content.

Preface to the First Edition

This book has been written with the missionary purpose of corporating the exciting discoveries about global tectonics made in the last ten years into the fabric of an account of earth history. Until now, essentially all historical geology texts discussed global tectonics in the vacuum of a separate chapter, but incorporated little of the subject in their recounting of the earth's evolution. The great impact of plate tectonics on the physical and biological development of the earth has thus been lost.

On the other hand, much of the traditional approach to historical geology, particularly the basic principles of stratigraphy and paleontology, are as important as ever in the process of unravelling the planet's biography. There is a conscious attempt in this book not to submerge the significance these time-proved principles to the nearly overwhelming new discoveries about earth history made in recent years. Both are vital to an understanding of how a geologist interprets earth history and what he has learned.

This book makes no attempt to provide detailed regional stratigraphic histories, but instead concentrates on the broader aspects of plate motions and collisions and the major trends in the fossil record. Rarely are local formation names used and fossil genera are listed only to assist the instructor in providing suitable laboratory material from North American supply houses. The intent is to provide a useful book for students in an introductory one-quarter or one-semester course. Too long a book with too much detail is simply incomprehensible to students in such situations.

Geology is a science that is growing so rapidly that it is hard for any individual to keep abreast of all current hypotheses, theories, and facts. Parts of this book will undoubtedly be superseded even before it is published, but the attempt to be general rather than detailed in the discussion of the earth's evolution should alleviate some of this problem. Several colleagues have read the manuscript and contributed helpful suggestions and criticisms, some of which have been heeded and some have not. All responsibility for factual errors or over enthusiastic interpretations rests with the author, however. The following individuals have read all or most of the manuscript: William B.N. Berry, Robert J. Foster, James Gilluly, Norman J. Silberling, Gary D. Webster, and A.O. Woodford. My colleague at Cal State, Hayward, Alexis Moiseyev, has been an ongoing stimulus for my interest in plate tectonics. As a paleontologist by training, I find it impossible to acknowledge in a limited space all who have stimulated my study of fossils, but Robert V. Kesling and J. Wyatt Durham come most readily to mind. My father provided many of the photographs and my wife, the patience which made this book possible. Finally, I wish to thank Mrs. Anna D. Watson, Mrs. Alice Gould, and Miss Jean McMillan for their many hours of typing and retyping the manuscript.

Contents

Part 1 Principles

Part 2 Earth History

Part 1

Principles

1

The Present Is the Key to the Past: Uniformitarianism

1

The Scientific Method

Basic to the scientific way of viewing the world is the so-called scientific method. The scientific method accepts as a fundamental postulate the existence of a real or tangible world that we can perceive with our senses. Debates about the validity of this assumption are in the realm of philosophy.

To the scientist, the universe consists of two parts, this real world and the conceptual world of the human mind (Fig. 1.1). The boundary

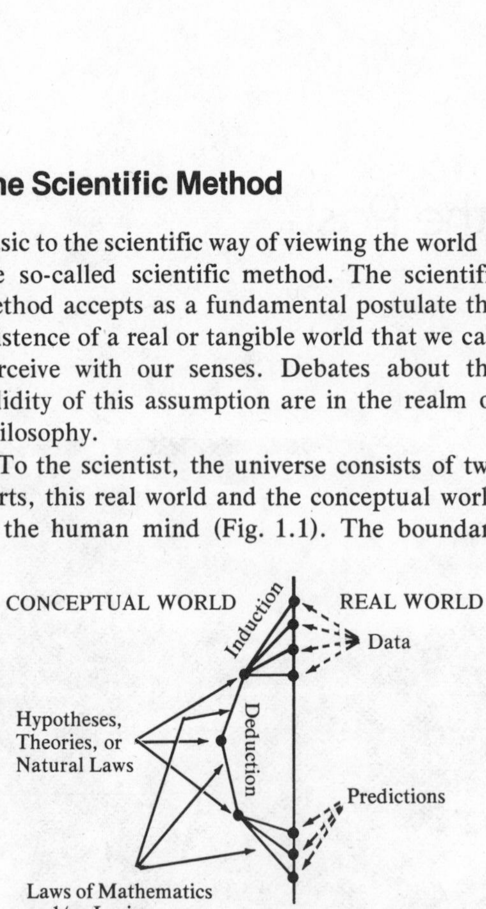

Figure 1.1. *The Scientific Method.* This diagram shows the relationships between the real and conceptual worlds as they appear to the scientist.

between the two is formed by empirical observations. Empirical observations are important to the scientific method at the two ends of the process. First, there is the gathering of data and the formulation of **hypotheses** to explain the data. This process is called **induction.** Using the "laws" of logic and mathematics, the mind discovers that new mental constructs follow from the original hypotheses. These constructs enable the mind to predict observations about the real world. This process is called **deduction**. Those hypotheses whose predictions are verified by observing the real world attain the status of **theories**. Notice that to a scientist, the word *theory* is very strong and implies a high degree of validity, unlike popular usage of the term. After a theory has been tested over a long period of time and found to always make correct predictions within its circumscribed limits of applicability, it becomes a **scientific "law," "principle,"** or **"truth."** A prevailing principle that helps define the entire structure of a particular science is called a **paradigm.**

It is very important to note that no scientific paradigm, law, principle, or truth is ever considered absolutely true in the sense of the laws of logic and mathematics where truth is inherent in the symbols themselves. Although a law has always explained and agreed with observations in the past, there is no guarantee that tomorrow someone will not make an observation or perform an experiment that contradicts it. This then means either that the limits of applicability of the law have to be revised or that the law may have to be discarded.

Science advances by successive replacement of paradigms by new and better ones. Because the discovery of new paradigms alters the world view of a science, it involves strokes of brilliant genius and intuition quite different from the normal progress of science described earlier. The proponents of new paradigms appear revolutionary to the majority of scientists and frequently encounter bitter opposition within the scientific community.

The nonabsolutism of science is no doubt surprising to many people who believe that science deals only with facts or absolute certainties. It is, however, the construction of theories and laws that truly distinguishes the scientific endeavor and not just the collecting of facts. A telephone book is a collection of facts, but it is not scientific. You should also note that the scientific method and, hence, science can be applied to any area of human activity and are not restricted to the pursuits usually considered "scientific." You undoubtedly used the scientific method in learning and performing most of your daily activities.

In the first half of this text, we will explore the development of those laws and principles which enable geologists to unravel the history of the earth's past. Many will be nothing more than what most of us in this scientific age consider to be "common sense," but the long histories of their implementation testify that this has not always been so. Indeed, to many of the readers of this book who are not familiar with those areas commonly called "sciences," these laws and principles may be anything but intuitively obvious.

Uniformitarianism

Fundamental to all geologic thinking is the paradigm, or principle, usually called **uniformitarianism,** but sometimes called **actualism.** The principle states that the physical and biological features of the earth (both past and present) were produced by the same processes that are acting today. This is another way of saying that supernatural causes need not be invoked to explain the real world. To the earliest human intelligence, all of the real world other than one's own actions probably seemed inexplicable and, hence, supernatural. Gradually, there must have dawned the notion that some events are "natural" or explicable in terms of the properties of substances and cause-and-effect relationships. This conflict about the supernatural versus the natural has gone on throughout human existence; however, as people's knowledge has expanded, the realm of the supernatural has been constantly decreasing.

Viewed in this context, uniformitarianism ceases to be a principle unique to geology, but becomes the essence of science itself. Thus, uniformitarianism can be defined by the statement that the physical and chemical laws of the universe have always been the same. This concept is implicit in all scientific endeavors, though the extension of it to the vast immensity of geologic time with opportunity for virtually unlimited interactions to occur is truly unique to geology. Without uniformitarianism, history would be impossible for there would be nothing left today that could reliably explain the past. If sandstones or river valleys had not originated in the past as they do now, there would be no way they could tell us anything about the past. To talk of earth history, then, is to imply uniformitarianism.

Historical Recognition

The earliest written records of the application of uniformitarianism to geological concerns are those of the classic *Greeks* of the sixth and fifth centuries B.C.—*Xenophanes* (about 600 B.C.), *Xanthus* (about 400 B.C.), and *Herodotus* (about 400 B.C.). All observed fossil seashells in the rocks of mountains high above present sea level. They interpreted the fossil seashells as the remains of once-living marine organisms, and their presence in areas well away from and above the level of the present sea indicated that these areas must have been submerged at some time in the past. Similar observations about a changing or dynamic earth were made by the later Greeks Heraclitus, Pythagoras, Empedocles, Democritus, and Aristotle and by the Romans Lucretius, Strabo, and Pliny. These ideas were lost in the early Christian era as supernatural explanations of nature dominated the Dark and Middle Ages.

In the *Renaissance,* the scientific outlook began to reappear. Around 1500, *Leonardo da Vinci* (Fig. 1.2A), that most complete of Renaissance men, revived uniformitarian notions by insisting that fossils were the remains of once-living creatures and that the presence of seashells in the mountains proved that the distribution of lands and seas had changed through time. Unfortunately, Leonardo exerted little influence on his contemporaries. Perhaps because of fear of suffering the fate of others who were imprisoned or killed for similar statements contrary to church decrees, Leonardo wrote all of his notes backwards by using a mirror. As a result, no one could decipher them for several hundred years. From the 1500s through the 1700s, only occasional scholars such as the Englishmen *Hooke* and *Burnet*; the Frenchmen *Buffon, Guettard,* and *Desmarest*; and the German *Füchsel* enunciated uniformitarian thoughts.

The modern demonstration of the validity of uniformitarianism is usually attributed to the Scotsman *James Hutton* (Fig. 1.2B) in the late 1700s and the Englishman *Charles Lyell* (Fig. 1.2C) in the early to middle 1800s. Both painstakingly gathered numerous careful observations about the rocks and landforms of Western Europe, showing that they could be produced by the same processes that act today if they were allowed plenty of time. Time is the crux of the whole matter. Only tiny amounts of erosion, deposition, uplift, and settling, all apparently without observable effects in the life span of an individual or of a few generations, can be observed today. These can, however, accomplish the wearing away of mountains, the building of

Figure 1.2. A. *Leonardo da Vinci* (1452–1519), **B.** *James Hutton* (1726–1797), **C.** *Charles Lyell* (1797–1875), **D.** *Georges Cuvier* (1769–1832).

deltas, and the shift in position of land and seas if only they are given enough time. Once the uniformitarians realized this, it immediately became obvious to them from the detailed rock record that the earth had a very long and eventful history. As Hutton stated, he saw "no vestige of a beginning—no prospect of an end."

Uniformitarianism and Actualism

Lyell went a bit farther in his discussion of uniformitarianism, however. He not only stated that the processes were the same in the past as they are today, but that they have always operated at the same rate and in the same patterns or configurations. Also, there were no additive effects of the same or different processes. In other words, the physical conditions on the earth were never much different from what they now are. The positions of lands and seas, mountains and valleys had changed, but the overall average condition had always been similar. Many geologists of his day who accepted the first part of his thesis on uniformitarianism could not follow Lyell in this second aspect of the principle. One of these men, *Constant Prevost,* coined the term **actualism** for the principle that processes have always been the same, while recognizing that their rates and configurations change through time because of additive effects and chance juxtapositions. Most modern geologists also accept only this part of Lyell's uniformitarianism and because of this, some of them prefer *actualism* to *uniformitarianism*. There is evidence that in his later years Lyell also came around to this view of uniformitarianism as evidenced by his acceptance of the Pleistocene Ice Age as a unique event in earth history and of evolution as a process which produced ever more complex organisms.

Uniformitarianism versus Catastrophism

In the early 1800s, uniformitarianism was still not accepted by most geologists because it was in competition with a very attractive hypothesis, more acceptable to theologians, known as **catastrophism.** As propounded by its leading exponent, the Frenchman *Baron Georges Cuvier* (Fig. 1.2D), this paradigm treated the earth's history as a series of catastrophies. Each catastrophe exterminated the existing life, after which a new influx of organisms was either created or migrated in from elsewhere. There were apparently six major catastrophies, and these could be inferred to represent the six "days" of biblical creation if one interpreted "days" to mean "eras" in a loose reading of scripture. Also, the last catastrophe could conveniently be synonymized with Noah's Flood. This made the whole notion very inviting to theologians and to those geologists with theological training or leanings who were numerous in the 1800s.

Eventually, however, catastrophism passed from the scientific scene as normal deposits in other areas were shown to coincide with the depositional breaks or gaps representing the "catastrophies" in the Paris Basin record studied by Cuvier. Also, some species supposedly exterminated by the catastrophies began to turn up in the rocks overlying these breaks. By midcentury, catastrophism in its original sense was essentially dead in scientific circles, though it persists widely today in fundamentalist religious sects.

Biological Uniformitarianism

Although the uniformity of process in the physical world was largely accepted by scientists in 1850, the notion that uniformitarianism also could be applied to biology was not widely recognized at that time. The idea that the slow changes which some scientists saw occurring in organisms today could also produce great change if given enough time seems to follow directly from Lyell's uniformitarianism. It appears remarkable, therefore, that he did not accept this idea until late in life. If one remembers, however, that Lyell's uniformitarianism was essentially cyclical and nonprogressive, whereas evolution is clearly progressive, his reticence is not so surprising.

It remained for another man, *Charles Darwin* (Fig. 4.6), in 1859, to convince the scientific community that the aspect of biologic uniformitarianism called **evolution** also existed. Darwin did for biology precisely what Hutton and Lyell did for geology. He gathered data about the changes in organisms today and indicated what changes in life they could produce if given great lengths of time. Thus, the paradigm of evolution replaced the one of species constancy. The topic of evolution is very important in earth history and will be discussed at greater length in Chapter 4.

Uniformitarianism Today

"The present is the key to the past" is a widely stated axiom in geology. As we have seen, it has meant different things to different scientists at different times. Today, to most geologists it means only uniformity of process, physical, chemical, and biologic, and not uniformity of rate. The rates, the times and places of occurrences, the additive effects of occurrences, and the joint occurrences of two or more processes are all variable. They have given rise to the ever-changing pageant of earth history we now interpret and not the one of cyclic recurrences envisioned by Hutton and Lyell. Sometimes in the earth's history, continents were higher than at others; hence, rates of erosion were greater. At other times, when temperatures, precipitation, and geography were appropriate, glaciers formed. We have also seen that, in a sense, uniformitarianism is just another way of stating the basic scientific assumption that the physical-chemical-biological laws of the universe are invariable. Geology has shown us how we can use this principle to unravel the past.

REVIEW

1. What are the differences between hypotheses, theories, and scientific laws or principles?
2. What is a **paradigm?** Give an example from a science other than geology.
3. What is the difference between mathematical truth and scientific truth?
4. Cite an example from your everyday life of the application of the scientific method.
5. What is the modern definition of **uniformitarianism?** How does it differ from that of Lyell?
6. How does **catastrophism** differ from uniformitarianism?
7. Does uniformitarianism encompass the possibility that an event could occur which has never before been witnessed?

2

The Time Element: Superposition, Cross-Cutting Relationships, and Faunal Succession

2

To the first people who looked at the rocks of the earth's crust, there probably appeared no way of reading a history from the apparent disarray of rock types and structures. Even today, a non-geologist is often mystified by how a geologist can interpret a time sequence of events by viewing a hillside or roadcut. Although many of the principles involved in making these interpretations can be deduced by simple logic, we have no written record that anyone recognized them until the second half of the seventeenth century.

Superposition

Nicolaus Steno (Fig. 2.1), a physician in Florence, Italy, first formulated the fundamental principle called **superposition.** Simply stated,

Figure 2.1. *Nicholas Steno* (1638–1686).

superposition means that in any sequence of layered rocks undeformed by folding or faulting, each layer was formed after the one below it and before the one above it (Fig. 2.2). It may seem obvious that a layer could not be deposited without anything beneath it for support, but it was not until 1669 that Steno clearly stated this principle. Notice that superposition does not refer to the age of the particles composing the layer, but to the time of formation of the layer.

In Steno's principle is implied another one, often called the principle of original horizontality. When one observes the formation of layers of sediments at a particular place, it usually appears that they form in more or less horizontal bands. Obvious exceptions include foreset beds in deltas, dunes, and alluvial fans, although even here each older layer is beneath each younger layer despite the inclination. The overly zealous application of original horizontality, however, has led to errors in interpreting earth history. Most serious of these is the assumption that layers were truly formed absolutely horizontally and, hence, any widespread layer of the same composition must have formed at the same time everywhere it is found. The serious consequences of this error will be more fully explored in Chapter 5.

One might logically ask how a geologist deals with rocks deformed by folding and faulting which can overturn the sequence or repeat parts of it. Through the years, geologists have devised several methods of telling the tops from the bottoms of beds to determine if they are in correct sequence or upside down. These include the distinctive top and bottom shapes of mud cracks, raindrop impressions, oscillation ripple marks, and cross-beds; the gradation from coarse at the bottom to fine at the top of a bed, a feature called graded bedding; and the formation of scour marks and load casts on the bottom of sand layers that made or filled depressions in

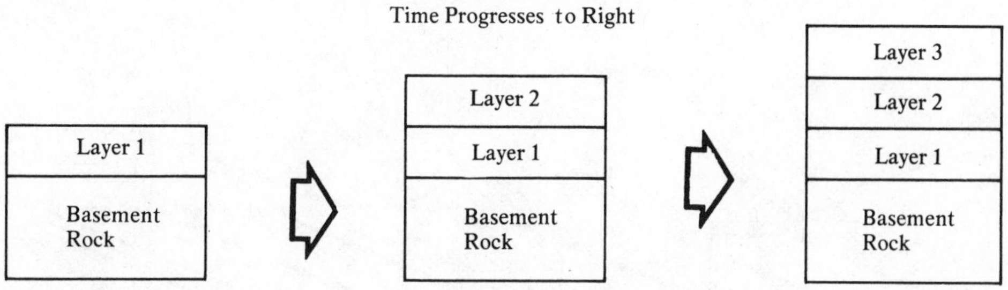

Figure 2.2. *Superposition.* This diagram shows the simple, logical reasoning behind the principle of superposition. Each layer was deposited before the one above it and after the one below it.

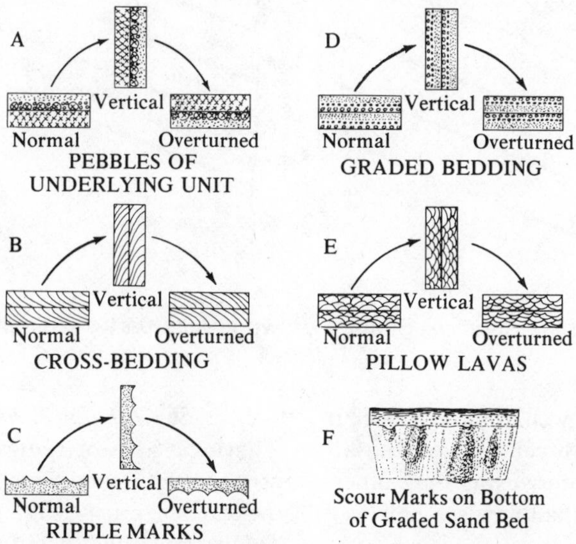

Figure 2.3. *Tops and Bottoms.* Usage of primary sedimentary structures and pillow lavas in distinguishing tops from bottoms of rock layers. Each structure has a distinctive shape or orientation relative to the top and/or bottom bedding plane. **A.** *Pebbles from older bed,* **B.** *Cross bedding,* **C.** *Ripple marks,* **D.** *Graded bedding,* **E.** *Pillows of lava,* **F.** *Linear sole marks.*

underlying mud layers. Many of these features and some additional ones are shown in Figure 2.3. Sequences inverted or repeated by faulting can often be recognized by the features that occur along fault planes such as gouge and the truncation of strata which lie above a surface of discontinuity (Fig. 2.4). Where these features are not present and where beds have been repeated, superposition alone is usually insufficient to tell the correct age sequence unless the layers can be traced to an area where they are undeformed and in the correct sequence.

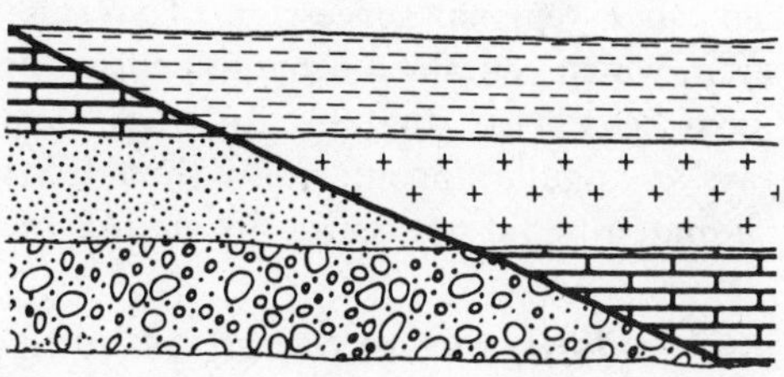

Figure 2.4. *Recognition of a Fault.* Base of units above surface of discontinuity is abruptly truncated. Normally, this does not occur along an unconformity as the rock unit overlying the surface of discontinuity parallels the unconformity.

Superposition is obviously a very useful tool for telling the *relative* geologic ages of strata (i.e., whether a particular layer was older or younger than another layer in the sequence) in one area. However, for superposition to indicate correct relative ages over great distances, the

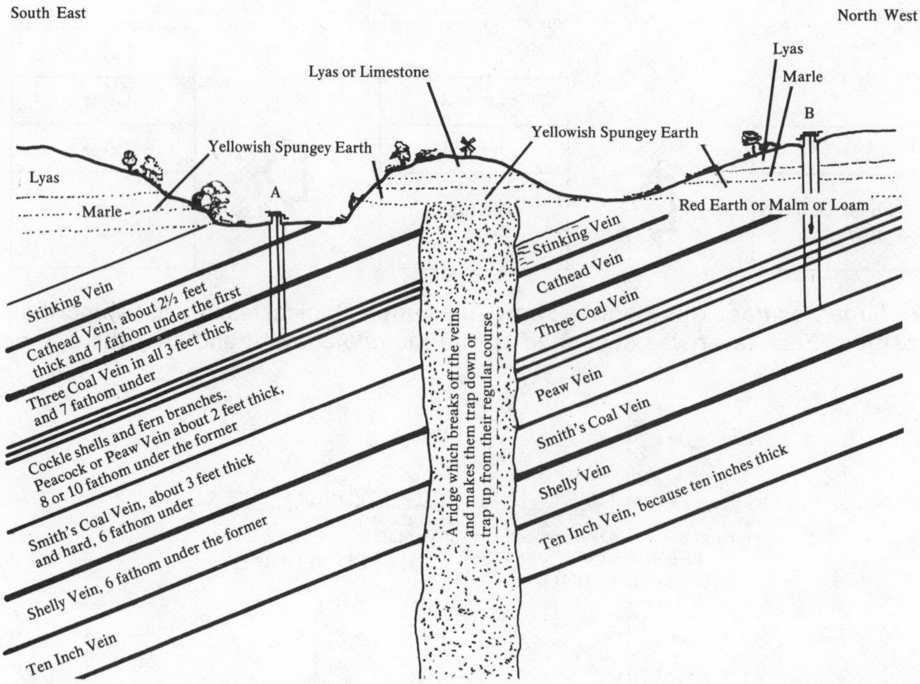

Figure 2.5. *Strachey's Cross Section.* Section drawn across part of Somerset, south of Bristol, England, in 1719.

same sequence of rocks would have to occur worldwide. The eighteenth century saw the attempt to unravel earth history using only superposition as a guide. We shall now see how well this effort fared.

An Englishman, *John Strachey,* in 1719 and 1725 described and illustrated a cross section of the strata in Somerset, England (Fig. 2.5). Strachey, using superpositional relationships, worked out a regional succession of strata for this area, which contained economically important coal beds. Note that his diagram clearly portrays an angular unconformity, though Strachey apparently did not realize its significance.

A Time Scale Based on Superposition

Geologists then turned their attention to establishing units of wider scope and greater magnitude. In 1740 the Italian *Lazzaro Moro* noted that generally the stratified sedimentary rocks overlie unstratified crystalline rocks and proposed a twofold subdivision of the rocks of the earth's crust into *Primary* and *Secondary* types.

In 1756, *Johann Lehmann,* working in the Thuringia area of central Germany (Fig. 2.6), proposed a threefold subdivision of the rocks of the earth's crust (Fig. 2.7). He called them "Mountains" for their topographic expression. In the cores of the highest mountains and passing superpositionally beneath the younger mountains as one moved outward in either direction are the oldest crystalline (igneous and metamorphic) rocks. He called these the Primitive or Ore "Mountains" (*Gang-Geburge*) because they were the source of many important metallic ores. They were inferred to have formed at the time of the earth's origin. Outward from the Ore Mountains and superpositionally above them are lower ranges called the Secondary or Stratified "Mountains" (*Flötz-Geburge*), composed of layered sedimentary rocks with fossils. These limestones, sandstones, and shales were supposed to have formed at the time of the biblical Deluge. Finally, the youngest rocks, forming the lowest foothills, are the "Alluvial Mountains" (*Angeschwemmt-Geburge*). These consist of unconsolidated alluvium, chiefly sands and

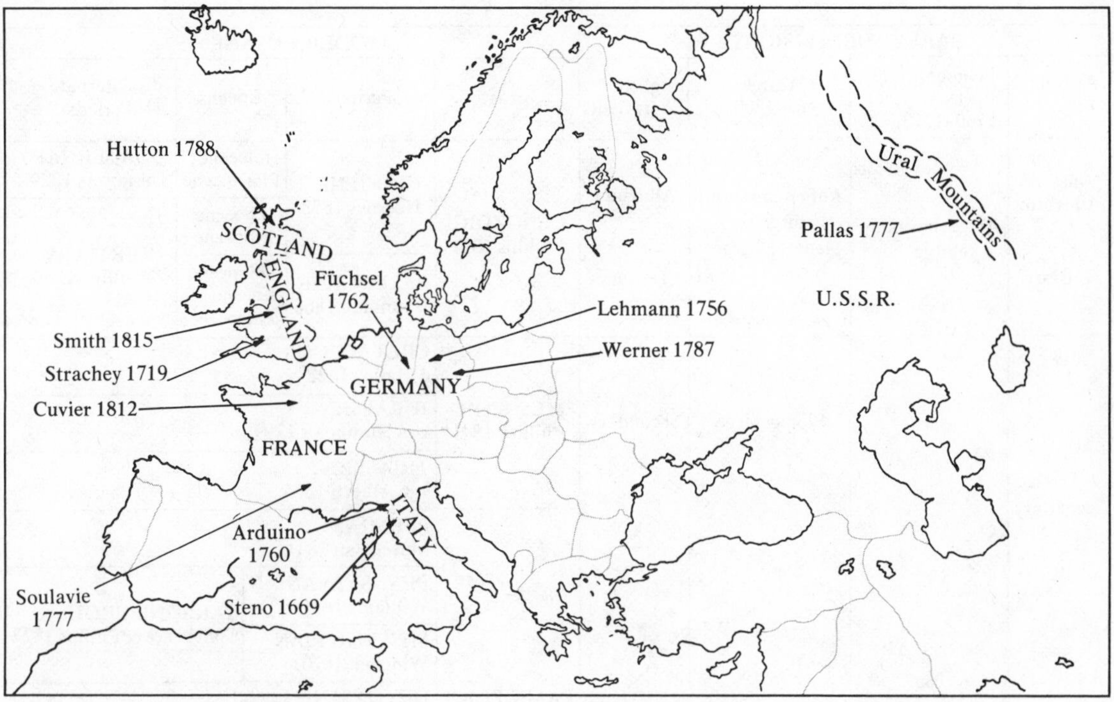

Figure 2.6. *European Localities of Pioneer Geologic Work.* Dates of major publications are indicated.

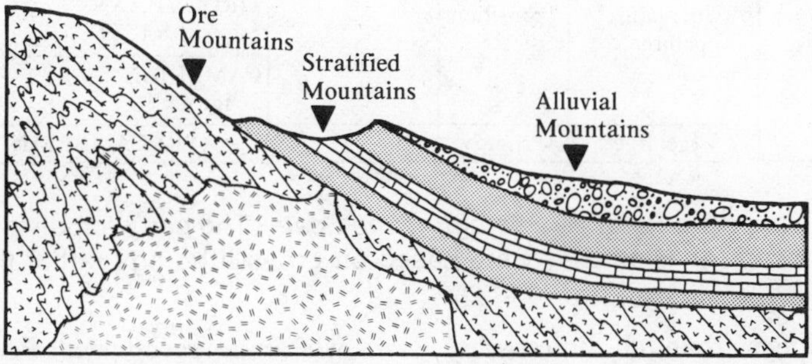

Figure 2.7. *Lehmann's Subdivisions.* A general cross section showing the superpositional relationships of the various types of rocks in central Germany.

gravels, and volcanic rocks. They were supposed to have formed since the Flood by minor events such as local floods, earthquakes, and volcanic eruptions. To determine the age of a rock (i.e., its time of formation), all one had to do was determine what type of rock it was.

In 1760, *Giovanni Arduino,* working in the same general area of northern Italy as Steno, proposed a similar sequence of mountains and their rocks. Arduino's sequence (Fig. 2.8), from oldest to youngest, was: *Primitive* or *Primary Mountains* composed of crystalline rocks containing metallic ores; *Secondary Mountains* composed of marbles, limestones, and clays which contain fossils; *Tertiary Mountains* consisting of gravels, sands, clays, and marls; *Alluvium* of the plains consisting of unconsolidated debris; and finally, *Volcanic Mountains* consisting of recent volcanic materials. Arduino, in contrast to Lehmann, believed that this sequence was only a generalization and not universally true. In other words, not all gravels and sands

EARLY SUBDIVISIONS				MODERN USAGE			
Arduino 1760	Lehmann 1756 Füchsel 1760–1773	Werner ca. 1800	English Equivalents	Eras	Periods	Epochs	Alternate Periods
Volcanic Alluvium	Angeschwemmt-gebirge	Aufgeschwemmt-gebirge	Alluvium	CENOZOIC Phillips 1841	NEOGENE Hoernes 1853	Holocene Pleistocene	QUATERNARY Desnoyers 1829
						Pliocene Miocene	TERTIARY Arduino 1760
Tertiary		Neues Flötzge-birge	Tertiary		PALEOGENE Naumann 1866	Oligocene Eocene Paleocene	
Secondary	Flotzgeburge	Flötzgebirge	Secondary	MESOZOIC Phillips 1841	CRETACEOUS d'Halloy 1822		
					JURASSIC von Humboldt 1799		
					TRIASSIC von Alberti 1834		
				PALEOZOIC Sedgwick 1838	PERMIAN Murchison 1841		
					PENNSYLVANIAN Williams 1891		CARBONIFEROUS Conybeare & Philips 1822
					MISSISSIPPIAN Winchell 1870		
					DEVONIAN Murchison, Sedgwick 1839		
Primitive	Ganggeburge	Übergangs-gebirge	Transition		SILURIAN Murchison 1835		
					ORDOVICIAN Lapworth 1879		
					CAMBRIAN Sedgwick 1835		
		Urgebirge	Primary	PRECAMBRIAN			

Figure 2.8. *The Growth of the Relative Geologic Time Scale.* The early subdivisions only roughly correspond to the modern ones as they were based on gross lithology rather than fossils.

were necessarily younger than all limestones. Geologists had to determine their own local sequences by superposition.

Another German geologist working in Thuringia, *George Füchsel,* studied Lehmann's Stratified Mountains in some detail. In 1762 and 1773 he subdivided these mountains into nine units he called series, each of which was further subdivided into statumina. He also was the first to erect time units corresponding to the times of formation of these units of rock. In Chapter 5 we will explore the significance of this dichotomy.

In 1777, another German geologist, *Peter Pallas,* extended the ideas of these men to Russia where he recognized Primary, Secondary, and Tertiary subdivisions. His Primary "Mountains" consist of granite and schists; the succeeding Secondary "Mountains," of fossil-bearing limestones; and the overlying Tertiary "Mountains," of fossiliferous clays and marls.

Werner

The foremost individual associated with these attempts to determine the earth's history from superposition alone and to tell the age of strata by the type of rock composing them was the German *Abraham Werner* (Fig. 2.9). Werner was the world's leading geologist in the late

Figure 2.9. *Abraham Gottlob Werner* (1749–1817).

1700s, and he called his scheme of historical geology the science of geognosy. To Werner, all rocks were the precipitates of a worldwide ocean which gradually disappeared somewhere. All the layers precipitated from this ocean were thus the same age everywhere, and each layer had its own unique composition. At the base were the crystalline Primitive "Mountains" (*Urgebirge*) consisting of granites and gneisses. Above these were the sparsely fossiliferous Transition "Mountains" (*Übergangsgebirge*) consisting of hard graywackes and limestones. Overlying these were the fossiliferous Stratified "Mountains" (*Flötzgebirge*) consisting of typical sedimentary rocks and above these were the Alluvial "Mountains" (*Aufgeschwemmtgebirge*) consisting of surficial sands and gravels. Volcanic rocks were at the very top. The earth's crust was like the layers of an onion-skin, except that the outer layers were not continuous because, as the universal sea disappeared, it left islands of older rock which were not overlapped by the younger deposits. Werner published only a brief summary of his work in 1787, but he was a great teacher and produced many outstanding students who propagated his work. Werner never left the region, in what is now East Germany (Fig. 2.6), where he taught and, therefore, never checked to see if his sequence was indeed universal or if all rocks formed as precipitates. He and his followers were known as **neptunists** because of this belief about the marine origin of rocks, which brought them into confrontation with those who insisted that some rocks formed by heat.

The **plutonists,** as the latter geologists were called, proclaimed that such rocks as basalt could clearly be seen to be volcanic products and not oceanic precipitates in regions such as the Auvergne of central France. The evidence was so clear that even some of Werner's own students came away from the Auvergne convinced he was wrong. Even more telling a blow to Werner's hypotheses, however, were the works of numerous European geologists who showed that rock units were not worldwide and that each region had its own sequence of strata. Though in a very general, worldwide sense Werner's sequence has some truth to it, it turned out that just about any rock type could occur anyplace in the superpositional sequence depending on the region.

Thus, it resulted that superposition is a very useful tool in determining the local sequence of geological events, but it cannot be used for matching strata from one region to another where the sequence is different. If rock units were worldwide, superposition alone would be sufficient to unravel earth history. But they are not, and clearly some other principles are needed.

Cross-Cutting Relationships

Hutton developed some useful lithologic tools. He first clearly pointed out another almost intuitively obvious principle, which is often called the **law of cross-cutting relationships.** He stated that any body of rock that cut across the boundaries of other units of rock must be younger than those it cuts (Fig. 2.10). How else could it cut them if they were not already there in the first place? This principle led Hutton to recognize the difference between what we today call intrusive contacts (where bodies of igneous rock cut across the boundaries of pre-existing rocks) and the structures he called **unconformities.** In defining an unconformity, Hutton noted that in

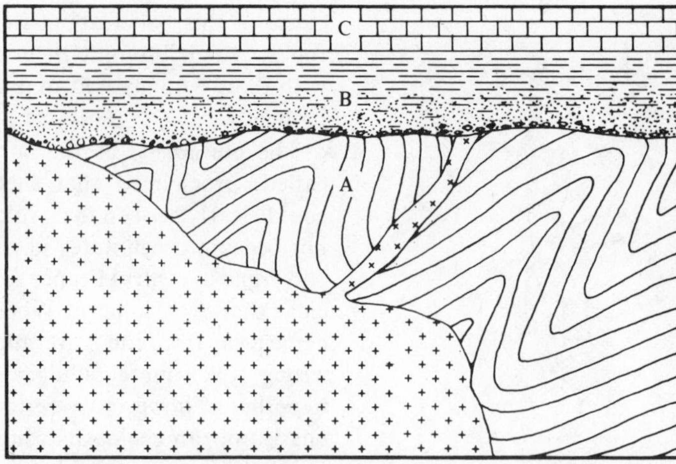

Figure 2.10. *Cross-Cutting Relationships.* The granite intrusion cuts Formation A and thus is younger. Formation B, which has an angular unconformity at its base, cuts both the granite and Formation A and, hence, is younger than both. (Formation C is younger than Formation B by superposition.)

Scotland the upturned edges of the folded Transition rocks were truncated by the essentially flat-lying and overlying Secondary rocks (Fig. 2.11). (This today would be called an **angular unconformity**, only one type of unconformity.) Hutton reasoned that the Transition rocks must have originally been deposited more or less horizontally as are most rocks and then later deformed. Then there must have been a long period of erosion, represented by the unconformable surface, during which time the folded mountains were reduced to a nearly horizontal surface so that when the overlying Secondary rocks were deposited, these younger strata were also essentially flat-lying.

Inclusions

A similar principle first clearly formulated by Hutton is the so-called **law of inclusions.** This principle states that a rock body containing

A

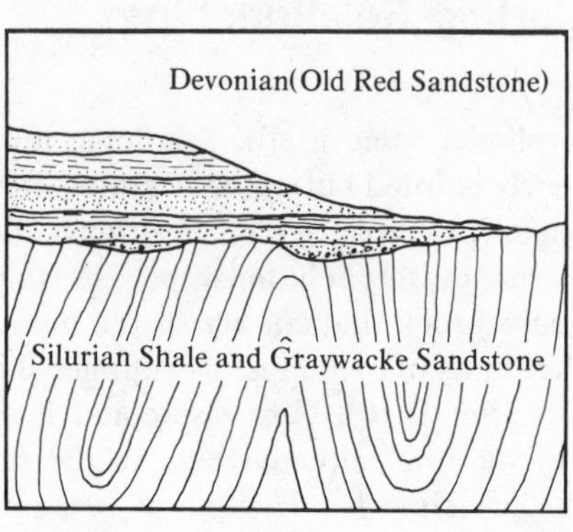

B

Figure 2.11. *The Angular Unconformity at Siccar Point, Scotland.* **A.** *Diagrammatic cross section of the unconformity,* **B.** *View of the unconformity in the field.* (Crown Copyright. Institute of Geological Sciences photograph. Reproduced by permission of the Controller, Her Britannic Majesty's Stationery Office, London.)

fragments of another body of rock must be younger than the unit whose fragments it contains (Fig. 2.12). Again, it seems untuitively

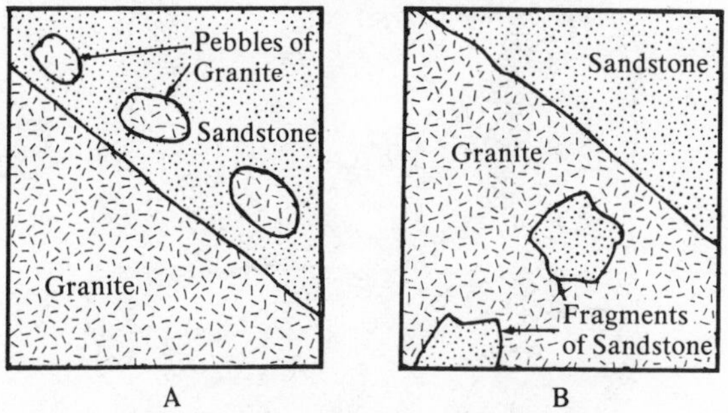

Figure 2.12. *The Law of Inclusions.* The fragments of the older rock occur in the younger rock. **A.** *An erosional surface, a nonconformity.* Pebbles eroded from the granite are deposited near the base of the sandstone. **B.** *An intrusive contact between the granite and the intruded rock.* The granite engulfed the sandstone, but some unmelted fragments of the latter remain within the intrusion.

obvious that a rock unit could not contain fragments of another rock body unless that other body existed first. But it remained for Hutton's sharp eyes and mind to bring the obvious to the attention of other geologists. Hutton's principles made it possible to determine the proper time of formation of new units or sequences that appeared in an area, but they still did not solve the problem of determining the relative ages of rocks in different areas with different sequences. A much more powerful tool was needed.

The Significance of Fossils

As early as 1777, the churchman *Giraud de Soulavie* had worked out a fivefold subdivision of the rocks in the Vivarais district of southern France and noted that, although the types of rocks were repeated in the succession, the types of fossils were not. There was a definite change in the fossils as one went up the sequence. Fossils more like present forms of life gradually replaced the older, more primitive types. Soulavie suggested, therefore, that fossils and not lithology might be the key to determining the relative age of a rock. This was a momentous discovery, but Soulavie was far ahead of his time and it was not until many years later that the principle he suggested became widely accepted. Before pursuing the matter of using fossils to date rocks, however, we must backtrack a bit to explore the history of people's opinions concerning fossils.

The Origin of Fossils

Fossils are usually defined as the remains or traces of prehistoric organisms. (Although the word originally meant anything dug from the earth, through the centuries, it has gradually been restricted to its present meaning.) Fossils were noted even by prehistoric people who buried them with their dead, perhaps attaching mystical significance to them. The classical *Greeks* again provide our first written account of opinions concerning fossils. Clearly Xenophanes, Xanthus, and Herodotus recognized that the seashells they found in the mountains of the eastern Mediterranean area were the remains of once-living creatures. *Aristotle,* though he felt some fossils such as fish bones were the remains of organisms, generally believed that fossils formed in rocks under the influence of celestial bodies.

This sort of geological astrology enjoyed wide popularity in the Dark and Middle Ages when the Christian Church accepted Aristotle's words as dogma. Other explanations arose in the Middle Ages to compete with the idea of celestial influences forming fossils. Some theologians felt fossils grew in the rocks from seeds, solutions, or vapors. Others thought they were unsuccessful creations cast aside by the Creator or were tricks of the devil to deceive people about the earth's true history. Another popular notion was that fossils were "sports of nature," creations of Mother Nature in a playful mood. One of the major reasons that fossils could not be accepted as the remains of organisms was because some were clearly unlike living types, and the benevolent God who had saved all creatures from the Flood in the Ark could not have allowed any species to perish. All of the preceding ideas persisted in learned circles even until the 1700s;

and, of course, many exist widely today in the nonscientific community.

In the Renaissance, some original thinkers such as the Italians Leonardo, Fracastoro, and Cesalpino revived the idea of an organic origin of fossils. Their ideas did not enjoy wide circulation, however, because of fear of the persecution that sent others to the dungeon or the stake for heresy.

By the late 1600s and early 1700s, however, it was apparent that fossils were too much like living things to be inorganic. Most scientists, at the urging of *Hooke* in England, *Vallisnieri* in Italy, and *Scheuchzer* in Germany, had come around to accepting fossils as the remains of organisms. Once the scientific tide had swung in this direction, however, the theologians came forward with a new explanation. They then argued that fossils were the remains of creatures killed by the Flood. (No one yet would admit the possibility of extinct organisms, however.) Scheuchzer, in particular, championed this view and passed off a fossil giant salamander as the remains of a poor human sinner who had perished in the flood (Fig. 2.13). He also debated with others about whether the Deluge occurred in the spring or the fall because of reproductive structures found on fossil plants.

One of the last opponents of the organic origin of fossils was a German professor, *Johannes Beringer.* Beringer was deceived into believing that a series of carvings made on rocks by another professor were "unique manifestations of nature" (Fig. 2.14). These included not only the images of animals and plants, but also Hebrew writing and celestial objects. Beringer described and figured these artifacts in 1726 and only later discovered the hoax, much to his chagrin.

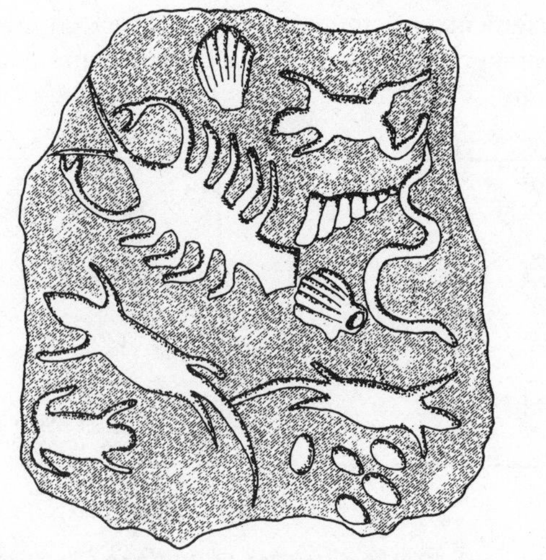

Figure 2.14. *Beringer's Pseudofossils.*

By the time of Soulavie, then, the organic nature of fossils (an aspect of biologic uniformitarianism) was generally accepted in educated circles, although the dispute continued about whether or not they were embedded in the rocks by the biblical Flood. Still, except for isolated individuals such as Soulavie, no one had yet noted anything significant about the pattern of occurrence of fossils in the rocks of earth's crust. The man who finally brought to the attention of the scientific world that fossils did not occur haphazardly but in a definite sequence was *William Smith* (Fig. 2.15).

Smith was an English canal-builder in the late 1700s and early 1800s who traveled widely about Great Britain in his work. He carefully collected fossils and noted the rock units from which they came (Fig. 2.16). As a result, he discovered that types of fossil organisms occur in the rocks of the earth's crust in a definite and determinable order and, hence, the rocks

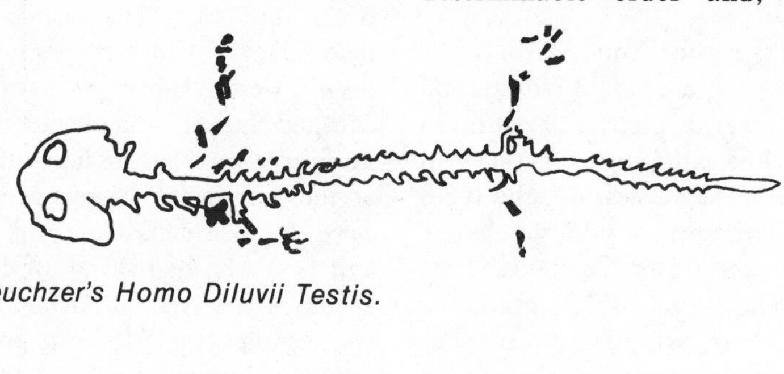

Figure 2.13. *Scheuchzer's Homo Diluvii Testis.*

Figure 2.15. *William Smith* (1769–1839).

formed during a particular interval of geologic time can be recognized by their characteristic fossil content. This most significant geologic discovery has become known as the **principle** or **law of faunal succession.** It is the key that has opened the door to our present understanding of earth history, for here at last, in fossils, are things in the rocks that do not occur in a random fashion but in a regular and irrevocable order. Although a half-century would elapse before Darwin, with his discovery of evolution, would provide the reason for this sequence, the application of faunal succession was immediate.

Strata	Thickness	Springs	Fossils Petrifactions, etc., etc.	Descriptive Characters & Situations.
1. Chalk	300	Intermitting on the Downs	Echinites, Pyrites, Mytilites, Dentalia, Funnel-shaped Corals & Madrepores, Nautilites, Strombites, Cochliae, Ostreae, Serpulae	Strata of Silex imbedded
2. Sand	70			The fertile vales intersecting Salisbury Plain & the Downs.
3. Clay	30	Between the Black Dog and Berkley		
4. Sand & Stone	30	Hinton, Norton, Woolverton, Bradford Leigh.		Imbedded in a thin Stratum of calcareous Grit. The Stone flat smooth and rounded at the Edges.
5. Clay	15			
6. Forest Marble	10		A Mass of Anomiae and High-waved Cockles with calcareous Cement	The Cover of the upper Bed of Freestone or Oolite.
7	60		Scarcely any fossils besides the Coral	Oolyte resting on a thin bed of Coral - Prior Park, Southstoke, Twiney, Winsley, Farley Castle, Westwood, Berfield, Conkwell, Monkton Farley, Colerne, Marshfield, Coldashton.
8. Blue Clay	6	Above Bath		
9. Yellow Clay	8			
10. Fullers Earth	6			Visible at a distance by the Slips on the Declivities of the Hills round Bath.
11. Bastard ditto and Sundries	80		Striated Cardia, Mytilites, Anomia, Pectinites and Duck - Muscles.	
12. Freestone	30		Top covering Anomiae with Calcareous Cement, Strombites, Ammonites, Nautilites, Cochliae, Hippocephaloides, Fibrous Shell resembling Amianth, Cardia, prickly Cockle, Mytilites, lower Stratum of Coral. Large Scollop, Nidus of the Muscle with its Cables	Lincomb, Devonshire Buildings, English combe, English batch, Wilmerton, Dunkerton, Coomhay, Monkton Combe, Wellow, Mitford, Stoke, Freshford, Claverton, Bathford, Batheaston, and Hampton, Charlcombe, Swainswick, Tadwick, Langridge.
13. Sand	30		Ammonites, Belemnites	Sand Burrs.
14. Marle Blue	40	Round Bath	Pectinites, Belemnites, Gryphites, High-waved Cockles.	Cross Bath?, Mineral Springs of Lincombe, Middle Hill, Cheltenham.
15. Lias Blue	25		Same as the Marl with Nautilites, Ammonites, Dentalia, & fragments of the Enchrinni.	The fertile Marl Land of Somersetshire, Twerton, Newton, Preston, Sutton, Stanton Prior, Timsbury, Paulton, Markesbury, Farmborough, Corston, Hunstreet, Burnet, Keynsham, Whitchurch, Salford, Kelston, Weston, Pucklechurch, Queencharlton, Norton-Malreward, Knowle, Charlton, Kilmersdon, Babington.
16. Ditto White				
17. Marl stone, Indigo and black Marle	15		Pyrites and Ochre	A rich Manure
18. Red-ground	180		No fossil known	Beds of Ruddle beneath this bed no fossil, shells or animal remains are found: above it are vegetable Impressions. The waters of this Stratum petrify in the trunks in which they are conveyed, so as to fill them in about 15 years, with red Matter, which takes a fine polish. Highlittleton.
19. Millstone				
20. Pennant Stone			Impressions of unknown plants resembling Equisetum	
21. Grays				Fragments of Coal & Iron Nodules. Hanham, Bislington, Mangotsfield, Brunend, Winterbourn, Forest of Dean.

Figure 2.16. *William Smith's Table of Strata* near Bath, England dictated to Reverend Townsend (1799).

In the 1800s, in relatively rapid succession, the large-scale subdivisions of relative geologic time we now call **periods** and **eras** were established. Many of these were in turn subdivided into still smaller units, all by utilization of this principle.

As so often happens in science, when the general level of knowledge reaches a certain point, several workers independently discover great principles. The discovery of the principle of faunal succession was no exception. At the same time that Smith was making his momentous finding known in Great Britain, *Cuvier* and *Brongniart,* working in the Paris Basin of France, came to exactly the same conclusion about the time value of fossils. Cuvier, you may recall, turned faunal succession into a line of support for catastrophism (Fig. 2.17). This idea eventually died when it was discovered that, in

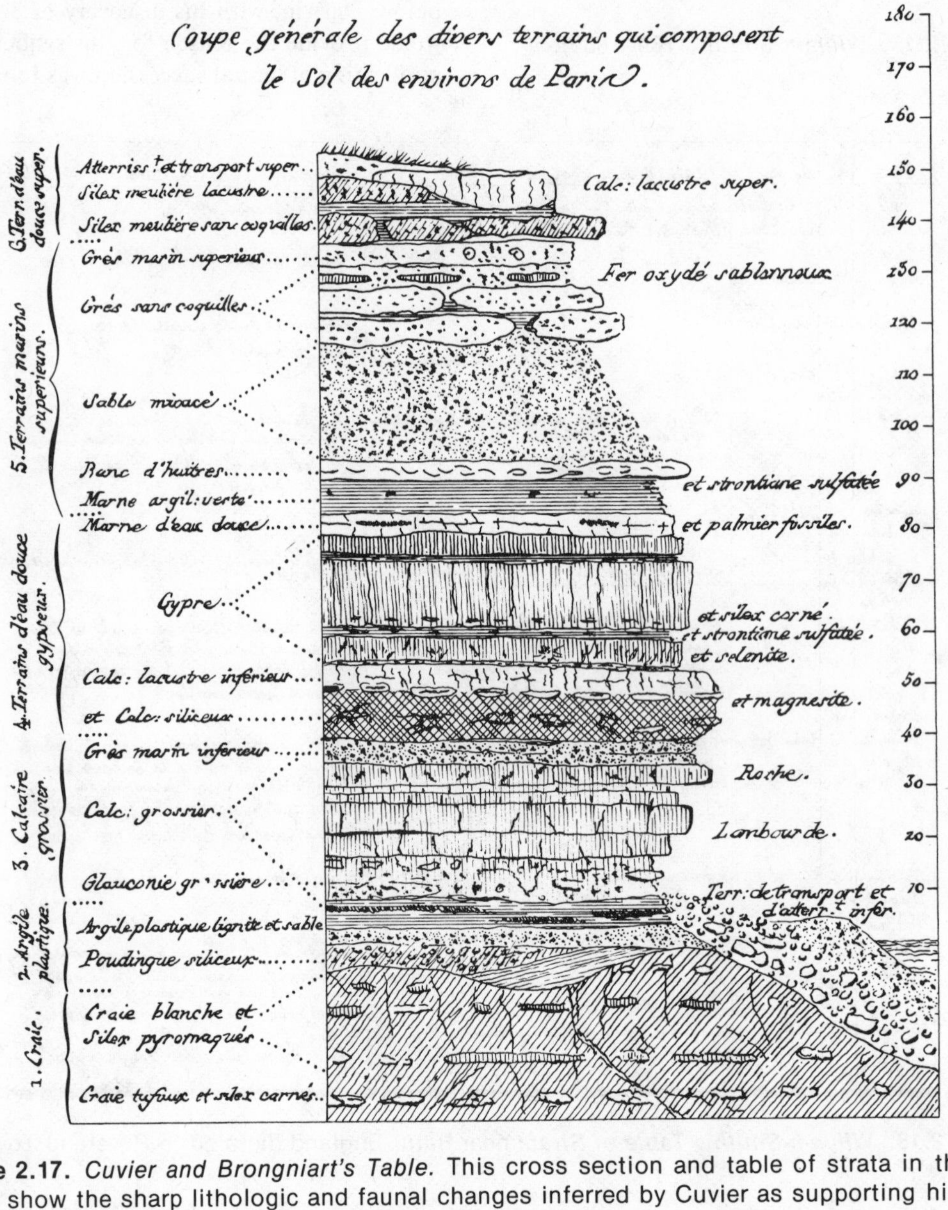

Figure 2.17. *Cuvier and Brongniart's Table.* This cross section and table of strata in the Paris Basin show the sharp lithologic and faunal changes inferred by Cuvier as supporting his catastrophist theories. (The fossils are shown by tiny sketches, except where the layer is too thin to permit.)

complete sequences, changes in life were gradual and not abrupt. As we shall see later, abrupt faunal changes do occur in the rock sequence of a particular region, but these are caused either by unconformities or breaks in the geologic record of the area, or by a regional change in environment carrying with it different organisms adapted to the new ecology.

The Relative Geologic Time Scale

We shall now chronologically explore the naming and establishment of the geologic periods (Fig. 2.8) and eras that occurred as soon as faunal succession became well established. (Those established prior to about 1830 received their original names before the discovery of faunal succession, but the present usage follows the implementation of the principle.)

The Periods

The oldest period name that comes down to us today is the **Tertiary,** which dates from *Arduino's* subdivisions of 1760 and thus antedates faunal succession. The Tertiary "Mountains," however, were described sheerly on a lithologic basis (gravels, sands, and limestones) and a superpositional basis (between the Secondary and the Alluvial "Mountains"). The type area, or region where the unit was first described, is in the Apennine Mountains of northern Italy (Tuscany). The usage of the word in its present sense, based on fossils, dates from the work of *Lyell* and the French paleontologist (one who studies fossils) *Deshayes* in the 1820s and '30s.

Likewise, the **Jurassic** Period derived its name from a prominent lithologic unit. The German geographer *Alexander von Humboldt* in 1799 named the great cliff-forming carbonates of the upper Stratified "Mountains" in the Jura Mountains of northwest Switzerland the Jura Limestone. The present usage of *Jurassic,* based on its fossil content, dates from the work of the German geologist *Leopold von Buch* in the 1830s.

In 1822, *Conybeare* and *Phillips,* two British geologists, published a textbook of British geology in which they named a **Carboniferous** "Order." This subdivision of rocks was near the boundary between Werner's Transition and Stratified "Mountains." It consisted of the well-known coal-bearing rocks of the British Isles and some underlying sandstones and a limestone. Some mention of the unique fossils found in the unit was made, but it was several years before the detailed studies of fossils both in Britain and Western Europe were made that gave the period its present meaning and allowed rocks of that age to be recognized even where there are no coal beds.

Also in 1822, a Belgian geologist, *D'Omalius d'Halloy,* working along the English Channel in northern France, applied the name **Cretaceous** "Terrain" to the prominent chalk beds and underlying greensands, ironsands, and marls exposed there. (*Creta* is the Latin word for *chalk.*) These rocks were at the top of the Secondary or Stratified "Mountains" of Werner and other workers. The Cretaceous chalk forms the famous White Cliffs of Dover on the English side of the channel. Again, the original usage was based on rock type and superposition, but application of the principle of faunal succession by many British and French geologists led to the description of the large and distinctive Cretaceous assemblage of fossils which could be used to recognize rocks of this age whether or not chalk was present. The most famous of these workers was *Alcide d'Orbigny,* who published (1840–1855) a monumental work on Cretaceous fossils. In this work he subdivided the Cretaceous into numerous smaller units which could also be recognized by their fossils.

In 1829, a French geologist, *Paul Desnoyers,* used the term **Quaternary** for the unconsolidated alluvium on the top of the rock record in northern France. The term was proposed as an additional complement to the older Primary, Secondary, and Tertiary units. Quaternary rocks of the type area contain few fossils which could be used to recognize them elsewhere, so the basis for its recognition again was lithologic and superpositional. However, with the discovery that the Quaternary at least partly coincided with the Pleistocene described by Lyell (see p. 22) and based on fossils, and with the discovery of the unique ice age that occurred in the period, its recognition became reliable.

Lyell, in addition to his other great achievements in geology, described and named four major subdivisions of the Tertiary, which he called periods. Here for the first time, we see fossils used as an essential part of the definition of units of geologic time. In 1833, Lyell proposed the **Eocene** (Dawn of the Recent), **Miocene** (Less Recent), and **Older** and **Newer Pliocene** (More Recent) periods in order of decreasing antiquity. These subdivisions were based on the percentage of species of living mollusks found in the rocks of the different ages. All identifications were made by the French paleontologist *Deshayes* to ensure consistency. The Eocene, which Lyell admitted might be worthy of further subdivision, contained 3.5 percent living species and had its type area in the Paris and London basins. The Miocene contained 18 percent of species which survived to the present and had its type area in western France. The Older Pliocene contained 33 to 50 percent living species and had its type area in northern Italy, and the Newer Pliocene contained 90 percent living species and had its type area in southern Italy. Lyell also had a **Recent** Period for the time since the appearance of human beings. In 1839, Lyell followed his original instincts and separated the Newer Pliocene as a distinct period which he called the **Pleistocene** (Most Recent). The Older Pliocene became simply Pliocene. As time went on, most geologists came to feel that Lyell's periods of the Tertiary were of much smaller magnitude than the other periods of geologic time, and they were reduced to smaller-scale subdivisions of periods and called **epochs.** In the most-used scheme, the Eocene, Miocene, and Pliocene are epochs of the Tertiary Period; the Pleistocene and Recent are epochs of the Quaternary Period. With the discovery of the ice age that occurred in the Pleistocene, many geologists ignored the original definition of Pleistocene based on fossils and began to consider it synonymous with the glaciation. This switch in definitions, which even Lyell eventually accepted, has caused no end of difficulty in attempting to establish the boundaries of the Pleistocene.

In 1834, a German geologist, *Friedrich von Alberti,* described and named the **Triassic** "Formation," a series of terrestrial redbeds separated by a fossiliferous marine limestone, which is prominent in central Germany. Despite the lithologic name, the Triassic was defined on the basis of its fossils which allow strata of this age to be recognized in areas where there is no tripartite sequence such as the Alps which contain some of the finest Triassic sequences known.

The naming of the next three periods described is inextricably interwoven with the lives of two famous English geologists, *Adam Sedgwick* and *Roderick Murchison* (Fig. 2.18). They

Figure 2.18. A. *Roderick Impey Murchison* (1792–1871), **B.** *Adam Sedgwick* (1785–1873).

set out in the 1830s to study the essentially unknown Transition rocks which were known to be well-exposed in Wales. Sedgwick went to the base of the section in the northwest, where the rocks are highly deformed and sparsely fossiliferous. Murchison started along the Welsh-English boundary at the top of the Transition (the base of the well-known Old Red Sandstone unit, then placed in the Carboniferous), where the rocks were less deformed and more fossiliferous. In 1835, the two men proposed two new periods of geologic time, the **Cambrian,** named by Sedgwick for the ancient Roman name for Wales (Cambria), and the **Silurian,** named by Murchison for the Silures, an ancient tribe that lived in the type area. The Cambrian "System" was based solely on lithology and superposition while the Silurian "System" was based on its fossil content, which enabled it to be recognized where lithologies were different. Because the two men began at opposite ends of the sequence, it turned out that each claimed some of the strata in the middle for his unit. This led to a great feud between the two that affected British geology for decades. Eventually Murchison got the upper hand because he alone had described fossils for his period, and so every fossil that was found in the lower Transition rocks gradually became Silurian. With no faunal basis for his period, Sedgwick saw his Cambrian swallowed up by the Silurian, particularly after Murchison became director of the Geological Survey of Great Britain. It was not until British geologists finally found distinctive fossils in the type Cambrian and a fine fossil assemblage was described from Czechoslovakia by *Barrand* in the mid-1850s that the Cambrian was salvaged. Final resurrection, however, awaited the resolution of the original problem of overlap, settled in 1879 when these strata were set aside as the **Ordovician** "System" (see below).

While Sedgwick and Murchison were still friends, they studied the geology of Devonshire and Cornwall in southwestern England. Here are a series of contorted graywackes and limestones similar to the type Cambrian. Both men were ready to call the rocks Cambrian until a local paleontologist, *Lonsdale,* pointed out that the fossils (particularly the corals) were intermediate in structure between those of the Silurian and the Carboniferous. Eventually convinced that fossils were more important than lithology in recognizing time, the two men proposed the **Devonian** "System" in 1839 for the rocks deposited during this interval. They also noted that in the rest of Britain, the rock unit in this interval between the Silurian and the Carboniferous is the Old Red Sandstone, quite different lithologically and paleontologically from the marine rocks of Devonshire. The recognition that different rocks and fossils can be formed at the same time in different environments was an important geologic discovery. Such units of the same age, but from different environments, are now called **facies** and will be discussed in Chapter 3.

In 1840–1841, Murchison, by then one of the premier geologists of the world, was invited along with several continental geologists to Russia by the czar to study the geology of that country. While in Russia, Murchison discovered a flat-lying, fossiliferous sequence of marine rocks between the Carboniferous and Triassic, an interval that was represented in Western Europe only by a thick sequence of poorly fossiliferous terrestrial redbeds. These marine rocks are well exposed on the western side of the Ural Mountains in the province of the town of Perm. In 1841, Murchison applied the name **Permian** "System" to the rocks of this age as recognized by their abundant fauna.

In 1853, a German geologist named *R. Hoernes* proposed the term **Neogene** "System" for the rocks of Miocene and Pliocene age. Subsequently, the Pleistocene and Recent epochs were also included.

Heinrich von Beyrich, a German geologist, took Lyell at his word that the Eocene Period probably could be split and in 1854 erected the **Oligocene** (Little Recent) for the upper third of the Eocene. The type area of the Oligocene is in northern Germany and contains numerous fossils by which it can be recognized there and elsewhere.

In 1866, another German geologist named *C.F. Naumann* proposed that the Eocene and Oligocene epochs be grouped together in a **Paleogene** Period. Other Europeans prefer calling the interval the **Nummulitic** after a characteristic type of fossil foram (see Chapter 7) found in it.

Philipp Schimper, a Frenchman, in 1874 completed the process of breaking up Lyell's Eocene by splitting off the lower third as the **Paleocene** (Early Dawn of the Recent). The type area in northern France contains some beds with marine fossils, but consists chiefly of terrestrial beds with plant fossils intermediate in structure between those of the Cretaceous and the Eocene, the basis of its definition. Mammal fossils from these terrestrial beds have been the chief tools for its recognition elsewhere, however.

In 1879, *Charles Lapworth,* a Scotsman, finally resolved the long-standing dispute over the overlap of the Silurian and the Cambrian. Lapworth found that these disputed strata contained an assemblage of fossils different from both the Silurian and Cambrian. He proposed the name **Ordovician** for the rocks, after the Ordovices, an ancient tribe inhabiting the type area in Wales. Although the Ordovician "System" has been widely recognized since Lapworth's time, many Europeans still refer to these rocks as Lower Silurian.

In 1885, *Paul Gervais* proposed the term **Holocene** for Lyell's Recent Epoch at the Third International Geologic Congress. The term gave a certain sense of symmetry to the names of the Tertiary-Quaternary epochs, making them all end in the suffix *-cene.* Not all workers accept Holocene and Recent as synonymous, however, reserving Holocene for the time since the retreat of the last ice sheet and Recent for that part of the Holocene since the beginning of written history.

By 1900, the standard geologic time scale reached the form it now has in most of the world (Fig. 2.8). At that time, most geologists believed that the periods were subdivisions of earth history whose boundaries were marked by worldwide mountain-building disturbances that produced the unconformities which bounded most of the systems in their type areas (areas where originally described, see Fig. 2.19). Once again, as in Wernerian times, geologists extrapolated from what they saw in a limited area, Europe, to the whole world; and once again they encountered difficulties, as we shall see later.

However, the geologists of the late 1800s and early 1900s operated on this principle; and when they left Europe and studied the geology of other continents, they found major unconformities at other positions in the rock sequences of those regions. Reasoning that these also must represent worldwide system boundaries that had been overlooked in Europe, they began the wholesale erection of new periods. North America was particularly fruitful in this regard as the patriotic Americans jumped into the contest with new periods such as Ozarkian, Canadian, and Co-

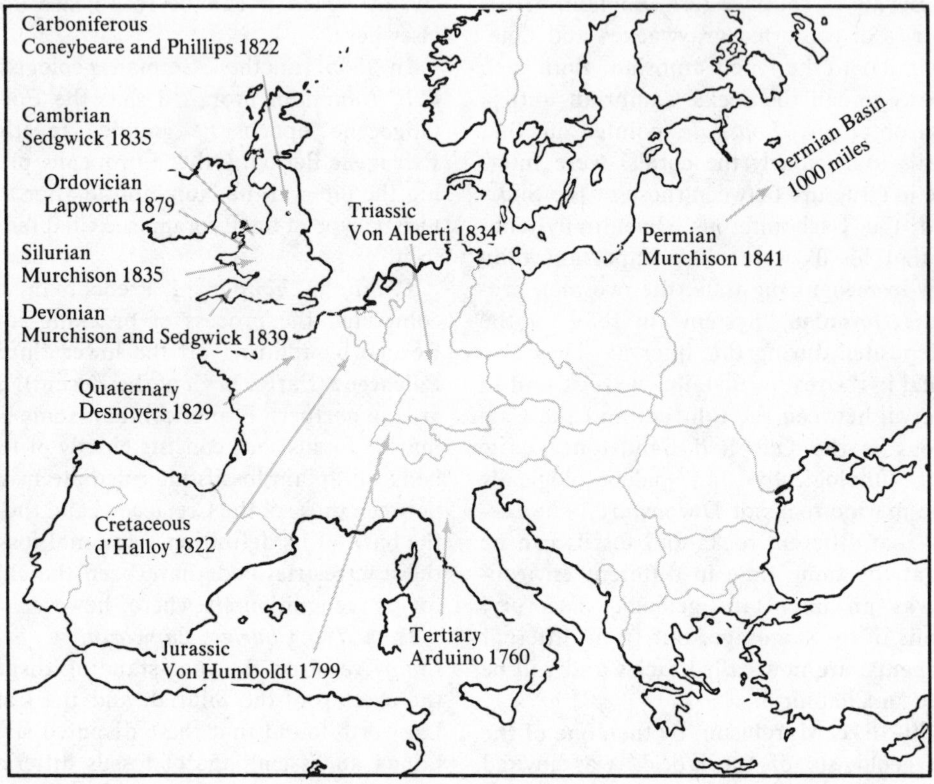

Figure 2.19. *Type Localities of the Geologic Systems.*

manchean. Only two, however, ever enjoyed any widespread acceptance. Those are the Mississippian and Pennsylvanian, which roughly correspond to the Early and Late Carboniferous of the remainder of the earth. Only in North America and places where North American geologists have worked extensively are these terms widely used.

In recent decades, many geologists have come to believe that periods are not units based on the occurrence of worldwide physical events, but instead that their boundaries are arbitrarily fixed at certain events in the evolution of life. Thus, the tendency to propose new periods has largely ceased. Nevertheless, some physical events, such as the worldwide Late Cretaceous flooding and regression of seas from the continents, are apparently synchronous and worldwide. These may be related to plate tectonics (see Chapter 10), which can have worldwide significance. Hence, at least some physical events are of sufficient magnitude to have provided the original period boundaries.

In 1891, *H.S. Williams* of the U.S. Geological Survey took the lithologic term **Mississippian,** proposed by *Alexander Winchell* in 1870 for the Lower Carboniferous limestones of the Upper Mississippi River Valley, and his own term, **Pennsylvanian,** proposed for the overlying coal measures of the state of Pennsylvania, and elevated them to the status of epochs or major subdivisions of the Carboniferous Period. In 1906, *T.C. Chamberlin* and *R.S. Salisbury* published a geology textbook that was to serve as an American standard for decades. In the book, the authors raised Mississippian and Pennsylvanian to the status of periods, and the widespread use of their book has made this terminology nearly universal in North America. Recently, with the widespread acceptance of the biological rather than physical basis of periods, there has been some tendency to revert to the traditional European usage so that everyone has a common terminology.

Before leaving the discussion of periods, one recent development should be noted. For the vast thickness of apparently unfossiliferous and typically highly deformed strata beneath the Cambrian, the name **Precambrian** has been in long usage. In recent years, many microfossils (fossils so small they must be studied with a microscope) have been found throughout the Precambrian sequence, and in the youngest rocks some large or megafossils have been found. These megafossils are much different from those of the following Cambrian, and it has been suggested that a new period, the **Eocambrian**, be established for the period of time in which these organisms lived. Many workers have adopted this usage.

The Eras

The periods of geologic time are grouped into large units called **eras.** The names most widely used for these are **Paleozoic** (Ancient Life) for the Cambrian through the Permian periods, **Mesozoic** (Middle Life) for the Triassic through the Cretaceous periods, and **Cenozoic** (Recent Life) for the Tertiary and Quaternary or Paleogene and Neogene periods. These terms were given their present definition by *John Phillips* in 1841. Phillips did not coin the word *Paleozoic* as that term had been proposed earlier by Sedgwick (1838) for the Cambrian-Silurian rocks. Phillips originally spelled the youngest era *Kainozoic* or *Cainozoic,* as is still often done in Britain. Those geologists who use the Eocambrian Period generally include it in the Paleozoic Era. Several names have been proposed for eras in the Precambrian such as Archeozoic, Archean, Proterozoic, and Algonkian; but none have been widely accepted as they were based chiefly on units of rock and not time.

Eons

To avoid the awkward term "Postprecambrian" in talking about the rocks of the Cambrian (or now Eocambrian) onward in which megafossils occur, *Chadwick* in 1930 proposed to group the eras into two larger units called **eons.** The **Cryptozoic Eon** (Hidden Life) consists of the Precambrian; and the **Phanerozoic Eon** (Evident Life) contains the Paleozoic, Mesozoic, and Cenozoic eras. The terms have enjoyed some degree of acceptance.

Boundaries

A significant point to note about the original development of the geologic time scale is that the major subdivisions were erected on the basis of major sequences of strata which contain unique fossil assemblages and which are bounded by unconformities representing much missing time. The early geologists assumed that these natural geologic breaks which bounded most of their subdivisions were worldwide. However, as they explored other regions of the world, they found strata with fossils which represent the time missing in the type areas. Thus, it became necessary to establish arbitrary boundaries for the subdivisions of geologic time. The problem has always been complicated by the fact that the strata that are missing in the type area but found elsewhere are not part of the original definition of any of the units; hence, great arguments have arisen as to the time period to which they ought to be assigned. To this day, geologists are still involved in the process of fixing the period boundaries. If exact boundaries are to be set for the subdivisions of the relative geologic time scale, some methodology must be established.

If natural catastrophies had occurred, as workers such as d'Orbigny thought, the task would be easy, but the evidence showed they had not. In the 1850s, a German geologist, *Friedrich Quenstedt,* suggested that the probable key to the whole problem would be the determination of the precise stratigraphic range of each fossil species. In other words, geologists had to record, on a centimeter-by-centimeter basis, the exact thickness of strata through which a single species of fossil ranged. Quenstedt was convinced that analysis of these precise ranges could provide units with boundaries.

The man who brought this scheme to fruition was another German, *Albert Oppel,* who studied the precise ranges of Jurassic fossils, particularly the ammonoids (relatives of today's pearly nautilus and squid). Using the overlapping ranges of species of fossils, Oppel was able to delineate the small-scale units called **zones** which were characterized by the joint occurrence of species not found together above or below their zone (Fig. 2.20). Some species began at the

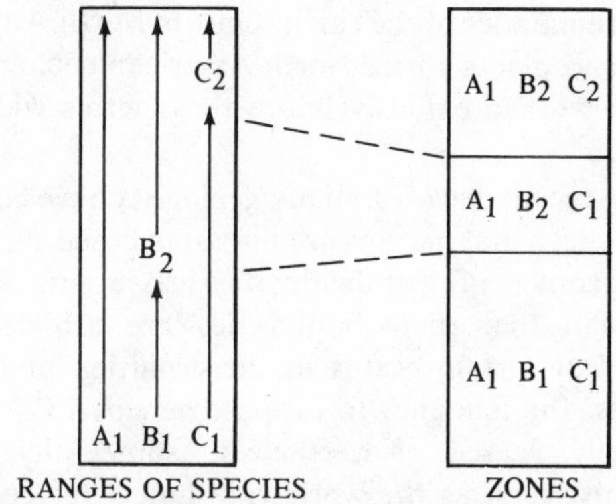

Figure 2.20. *Oppelian Zones.* Each zone is characterized by a unique fossil assemblage that does not occur above or below. Each species in this hypothetical example is designated by a number.

base of the zone; others ended at its top; still others ranged throughout the zone. Here at last are units with boundaries. Ultimately, the boundaries between units of rock assigned to one period and those assigned to the succeeding period are also the boundaries between successive zones. In other words, the Jurassic-Cretaceous boundary is also the boundary between the highest Jurassic zone and the lowest Cretaceous zone. The principle of zonation based on the overlapping ranges of species of fossils has been successfully applied in this country in this century by *Robert Kleinpell* and others. Oppelian zones are often called **concurrent range zones** to differentiate them from other types of zones established by geologists.

Recently, some geologists have noted that there is no compelling biologic reason why several species should begin and end at exactly the same time. Hence, even zonal boundaries are "fuzzy." By this we mean that the boundary of two zones marked by the beginning of two species and the end of two others is in itself a very narrow zone and not a line. Some newer, and apparently more precise, statistical tools have been developed by geologists such as *Shaw,* who are working at such extremely fine levels (Chapter 9).

REVIEW

1. What is the **law of superposition** and who formulated it?
2. How can oscillation ripple marks be used to tell the tops from the bottoms of beds?
3. What was the basic rock sequence determined by Lehmann, Arduino, and Werner?
4. What was the fundamental conflict between **neptunism** and **plutonism?**
5. If a dike cross-cuts a sedimentary layer, which rock unit is older?
6. Why do you suppose medieval people did not recognize the organic nature of fossils?
7. What was the major contribution of William Smith to geology?
8. Name the two geologists who named the Cambrian and Silurian **Periods** and then disputed their boundaries.
9. What have been the major ideas about the actual basis of the relative geologic time scale?
10. What are the eras and periods of the relative geologic time scale?
11. Why have geologists had endless debates about system boundaries?
12. How did Oppel define a **zone?**

3

The Perversity of Sedimentation and Organisms: Facies and Biogeography

3

We have seen how geologists originally believed the earth's crust to resemble an onion-skin, composed of several worldwide layers each of uniform composition, and how this belief proved to be an illusion. One of the most striking demonstrations of this fact came in the late 1830s when Sedgwick and Murchison discovered that the marine limestones and graywackes of the type Devonian area, with their coral and brachiopod fossils, occupied the same level in the rock sequence of Great Britain as the terrestrial Old Red Sandstone, with its fish fossils, found to the north and east.

Facies

Also in the 1830s, *Amanz Gressley,* a Swiss geologist working in the Jura Mountains, traced several Triassic and Jurassic rock units laterally along the mountains and discovered that both the types of rock and the types of fossils changed. For example, in the hypothetical illustration shown in Figure 3.1, shale with its characteristic assemblage of fossils gradually changes laterally into a limestone with another characteristic assemblage of fossils. To such lateral variations in both the rocks and fossils of the same age, Gressley applied the term **facies** in 1838. Gressley recognized that these differences were caused by differences in the environment in which the rocks were deposited and the organisms lived, just as there are different environments on the earth's surface today each of which produces its own sediment type and is inhabited by an assemblage of organisms adapted to its conditions (Fig. 3.2).

Gressley and many other geologists experienced a great disillusionment, however, with the realization that species of fossils were not worldwide and, therefore, fossils are not perfect tools for dating rocks. Some geologists despaired that unless one could trace strata laterally to determine their time equivalency, there might be no infallible way to date rocks in new areas. The French geologist *Constant Prevost* stated that similar lithologies and fossil assemblages indicate only equivalent environments and not equivalent ages.

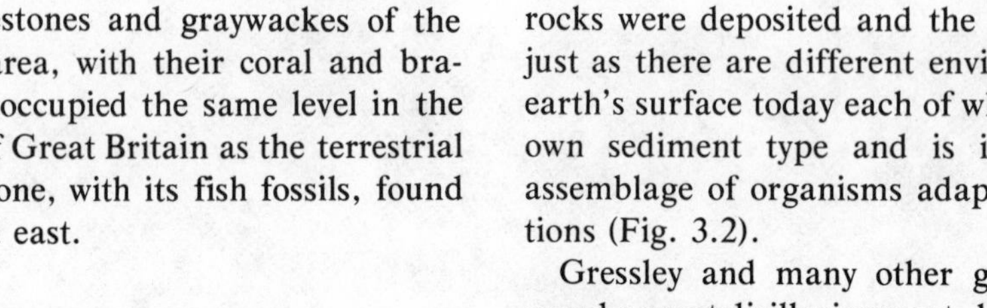

Figure 3.1. *Facies.* This cross section of a hypothetical region illustrates Gressly's original concept of facies as lateral differences among rocks of the same age. Unfortunately, the word has been used in many other ways in stratigraphy since his time.

Not all geologists despaired this much, however, for many noted that certain kinds of fossils are found in rocks formed in many different environments. Most notable among these are the ammonites, typically coiled shells with all the loops in one plane and subdivided by internal partitions with crinkly edges (Fig. 7.67). These fossils are very abundant in the widely exposed Mesozoic rocks of Western Europe. In 1842, the French geologist *Alcide d'Orbigny* proposed to group together all rocks deposited during the same interval of time, regardless of their lithology and general fossil content, into units called **stages.** Most stages were defined on the basis of widespread ammonite species, though some were based on species of other organisms with wide environmental tolerances. D'Orbigny established a sequence of stages for both Jurassic and Cretaceous rocks. In d'Orbigny's work, we see the first formal separation of the terms applied to rocks on the basis of their composition and those applied on the basis of the time of their formation. In Chapter 5 we will explore this most important development in some detail.

D'Orbigny's work was refined in the 1850s when *Albert Oppel* established, for the Jurassic, **zones**, still smaller subdivisions of rocks based on their time of formation as defined by species of wide-ranging fossils such as ammonites. Thus by the middle of the last century, Europeans widely recognized the existence of facies; that is, the fact that different types of rocks and fossils could form at the same time and that some separate sort of terminology was needed to differentiate units of rocks based on their lithology and units based on their time of formation.

The concept of facies did not achieve widespread acceptance in North America until much later, however. In the late 1800s and early 1900s in North America, an almost Wernerian concept of geologic history prevailed. In this theory of geology, whose leading exponents were *Chamberlin, Ulrich,* and *Schuchert,* the periods were natural subdivisions of time based on alternate advances and withdrawals of the seas from the oceans onto the continents. At the beginning of a period, all the continents were high and the seas were marginal; in the middle of the period, the continent was flooded; and at the close, the sea withdrew. The thin, widespread rock units of the continental interior were easily visualized as the contemporaneous deposits of a sea in full flood. Each rock unit, such as the Ordovician St. Peter Sandstone (Fig. 3.3A) or the Devonian-Carboniferous Chattanooga Shale (Fig. 3.3B), was considered a deposit of essentially the same age throughout its vast lateral extent. Most geol-

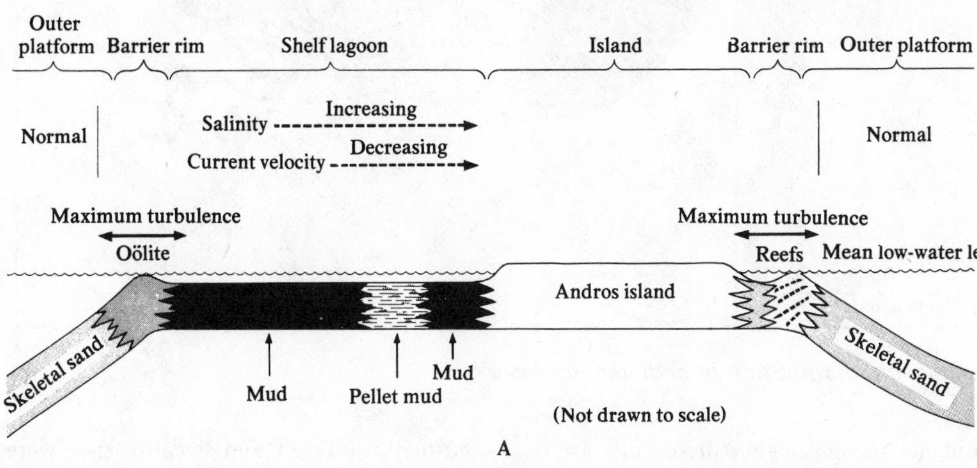

Figure 3.2. *The Bahama Banks.* The distribution of the different sedimentary and biological environments of the Bahama Banks today and some of the factors influencing these patterns. **A.** *Cross section showing environments, environmental factors, and sediment types.*

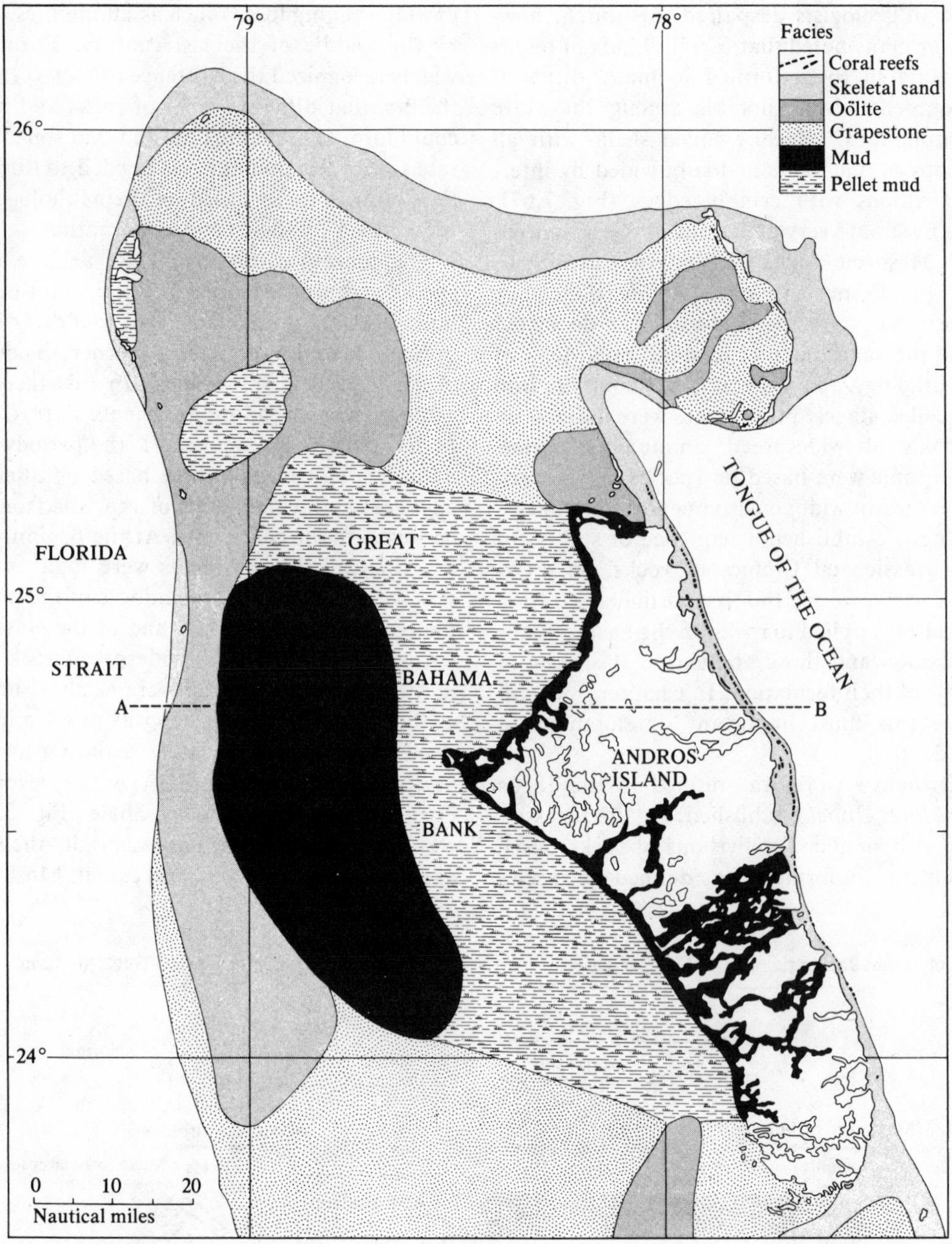

B. *Map showing distribution of sedimentary facies.*

ogists did not recognize that these units are of different ages in different places (Fig. 3.4), depending on the geographic location of that environment at various times (see also Chapter 8). Because these deposits cover virtually the whole continental interior and because they were assumed to be the same age throughout, it was natural that no one ever found the lateral facies equivalents of these units. Therefore, most American geologists rejected facies out of hand.

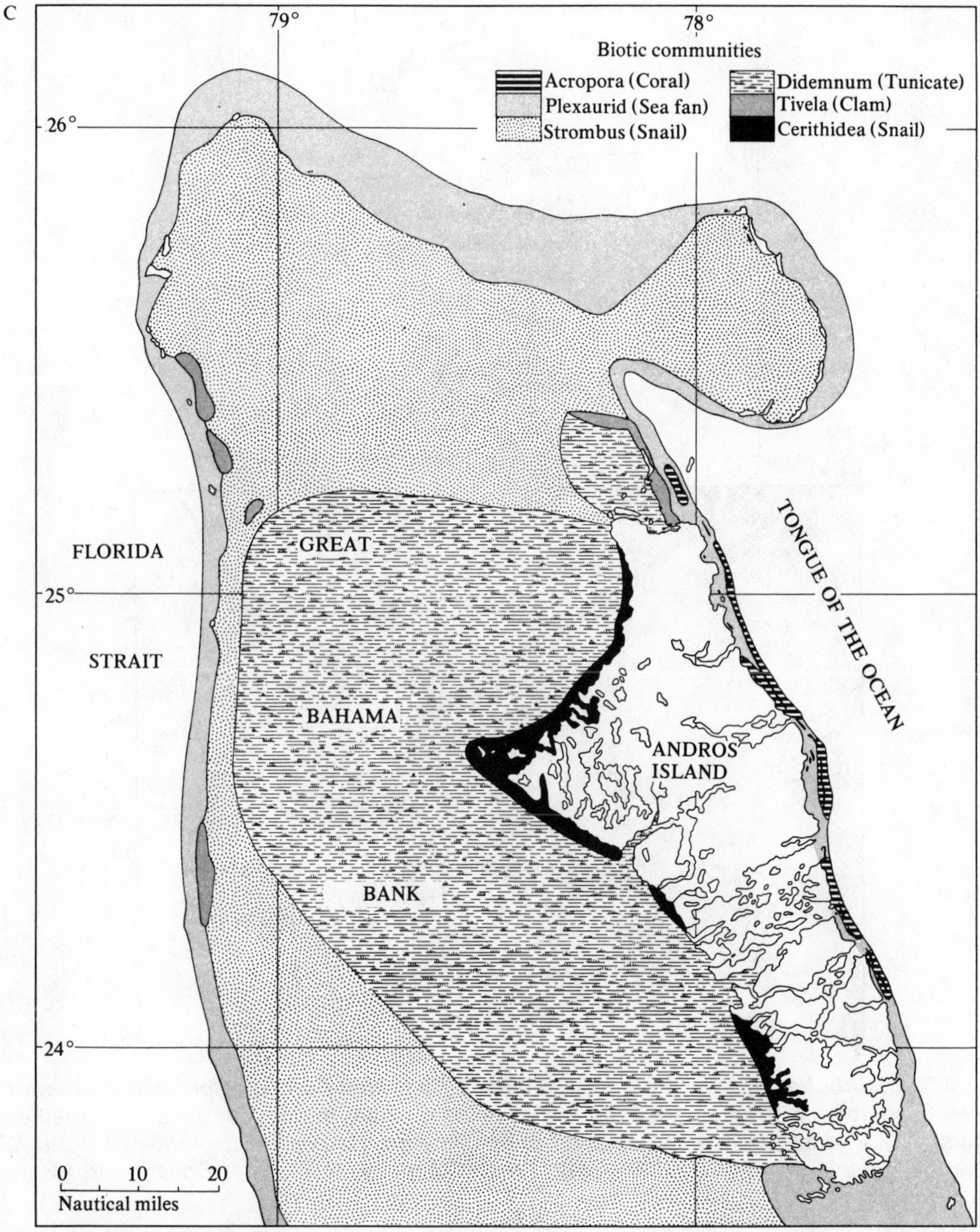

C. *Map showing distribution of biofacies.* (Parts A-C from Leo F. Laporte, *Ancient Environments,* © 1968 Prentice-Hall, Inc., Englewood Cliffs, N.J.)

E.O. Ulrich, one of the most important geologists in America in the first decades of this century, was completely convinced of these ideas. His great influence kept in the background the few people who recognized this conceptual error and supported the idea of facies. *Amadeus Grabau,* a leading proponent of facies, actually left the country to direct the Geological Survey of China. Beginning in the 1930s, however, the concept of facies enjoyed a

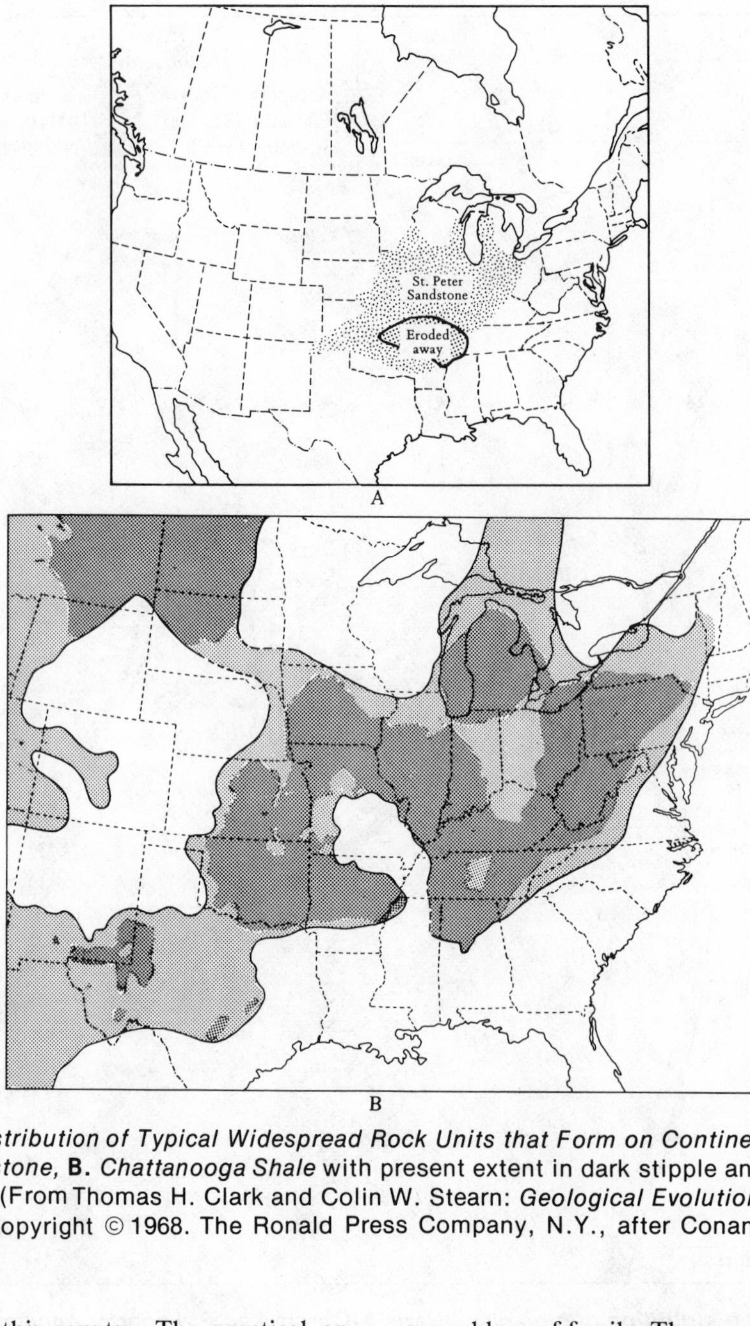

Figure 3.3. *Distribution of Typical Widespread Rock Units that Form on Continental Interiors.* **A.** *St. Peter Sandstone,* **B.** *Chattanooga Shale* with present extent in dark stipple and inferred extent in light stipple.(From Thomas H. Clark and Colin W. Stearn: *Geological Evolution of North America,* 2nd ed., Copyright © 1968. The Ronald Press Company, N.Y., after Conant and Swanson, U.S.G.S.)

renaissance in this country. The practical application of the facies concept still lags far behind its theoretical recognition.

As we noted earlier, geologists generally recognize two major subdivisions of facies, those based on the type of rock and those based on the assemblage of fossils. These are known, respectively, as **lithofacies** and **biofacies.** The study of the causes of the lateral variations in rocks formed at the same time is the province of the science of **sedimentology** and is usually discussed in detail in physical geology texts. Only a brief

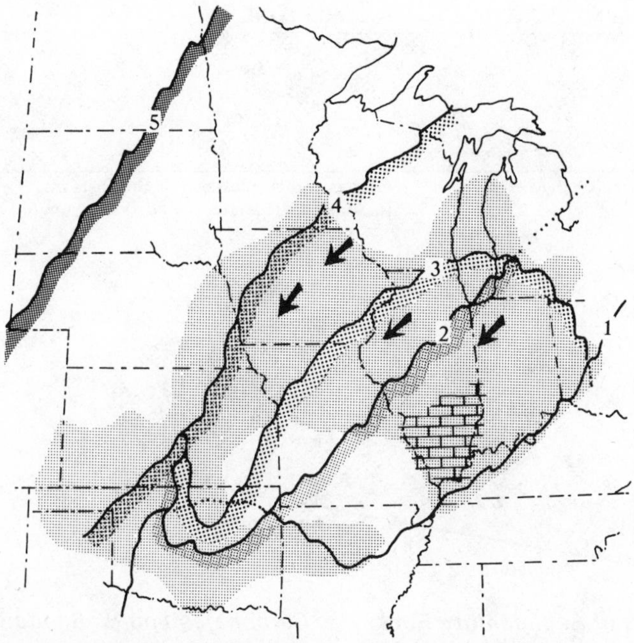

Figure 3.4. *Different Ages of St. Peter Sandstone.* This map shows several successive positions of the shoreline during the deposition of the St. Peter Sandstone. Thus, the rock unit is of different ages in different places. (From Edgar Winston Spencer, *Basic Concepts of Historical Geology,* Copyright © 1962, Thomas Y. Crowell Company, Inc. Reprinted with permission of the publisher.)

survey is provided here. The study of the causes of lateral variations in assemblages of organisms living at the same time is the science of **biogeography.** This topic is not covered in physical geology courses and will be discussed in more detail.

Sedimentation

The major clue to the environment of formation of most sedimentary rocks is their **texture,** the way the grains fit together and their sizes. The two basic textures (Fig. 3.5) of sedimentary rocks are **clastic,** composed of individual fragments, and **crystalline,** composed of interlocking crystals that have grown in place.

The major clastic (or detrital) rocks, in order of decreasing particle size, are **conglomerate, sandstone,** and **shale** or **mudstone** (Fig. 3.6). In addition, there are also many **clastic limestones.** In clastic rocks, the smaller the particles and the more rounded they are, the longer and generally farther they have been transported. Larger particles cannot be moved much or far without attrition and rounding. Smaller particles, once suspended in a fluid, take longer to settle out and, hence, are carried farther from the source. A particle of any size will become more rounded the more chances it has to collide with other particles.

Another aspect of clasic texture is size **sorting** (Fig. 3.5B), which refers to the degree to which the sediment consists of particles of uniform size. A well-sorted sediment consists of particles of about the same size; a poorly sorted one, of particles of numerous sizes. Sorting is also a measure of time of transportation; the poorly sorted or immature sediments having been formed rapidly with little chance for currents to winnow out the finer grains, while well-sorted sediments have undergone a selective removal of these smaller particles. Many clastic limestones

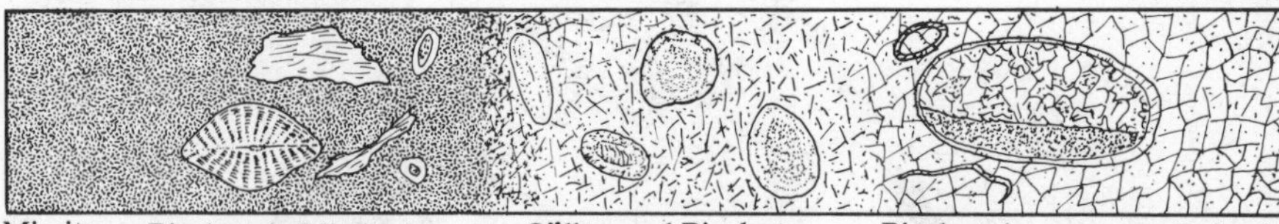

A

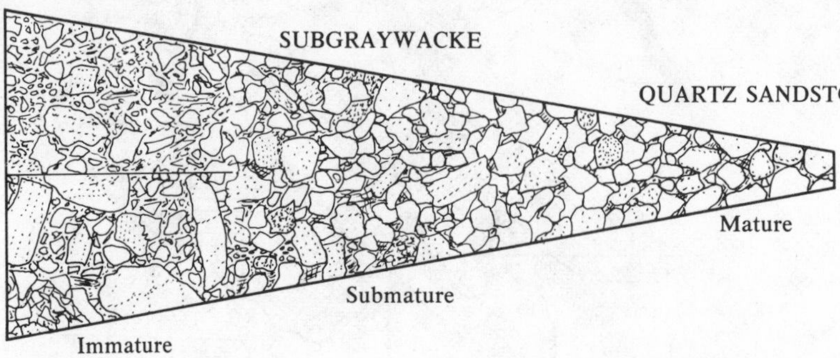

B

Figure 3.5. *Textures of Sedimentary Rocks.* **A.** *Carbonates* and **B.** *Sandstones.* Note crystalline texture of some carbonates and clastic texture of remaining carbonates and of the sandstones. (After Brice and Levin, 1969.)

(Fig. 3.5A) consist of whole or partial fossils. the former indicating either quiet or agitated water depending on the type of fossil and the latter typically indicating agitated water. Other clastic limestones are composed of the tiny spherical, inorganically formed grains called **oolites** characteristic of agitated waters. Clastic limestones consisting of limey mud, called **micrite** when lithified, are characteristic of quiet waters.

Crystalline textures (Fig. 3.5A) are most commonly found in carbonate rocks and are usually secondary; that is, they developed after the original formation of the rock by solution and recrystallization producing crystal growth. Crystalline textures also occur in evaporites such as gypsum and halite which formed under conditions of restricted circulation, high salinity, and high evaporation. Many limestones also formed in this manner.

Primary sedimentary structures (Figs. 2.3 and 3.7), structures that develop in sediments as they are originally deposited, also furnish evidence of the environment of formation of sedimentary rocks. For example, **cross-bedding** indicates that currents dominated the depositional environment; **ripple marks** commonly indicate shallow-water deposition (although deep-water ripple marks also occur); **mud cracks** indicate alternating wet and dry conditions; and **graded bedding** and **sole marks** indicate deposition by turbidity currents originating on steep submarine slopes.

Color commonly offers environmental clues with reds, yellows, and browns typifying terrestrial rocks; gray and green, marine rocks; and black, stagnant water sediments. There are, however, many exceptions.

Associations (Fig. 3.8) of sedimentary rocks also furnish evidence of regional environments. Shallow-water or shelf marine conditions typically produce widespread thin carbonates and well-sorted quartz sandstones; rapidly subsiding marine troughs with adjacent highlands typically produce narrow but thick sequences of graywacke sandstones with interbedded shales; and rapidly subsiding continental basins adja-

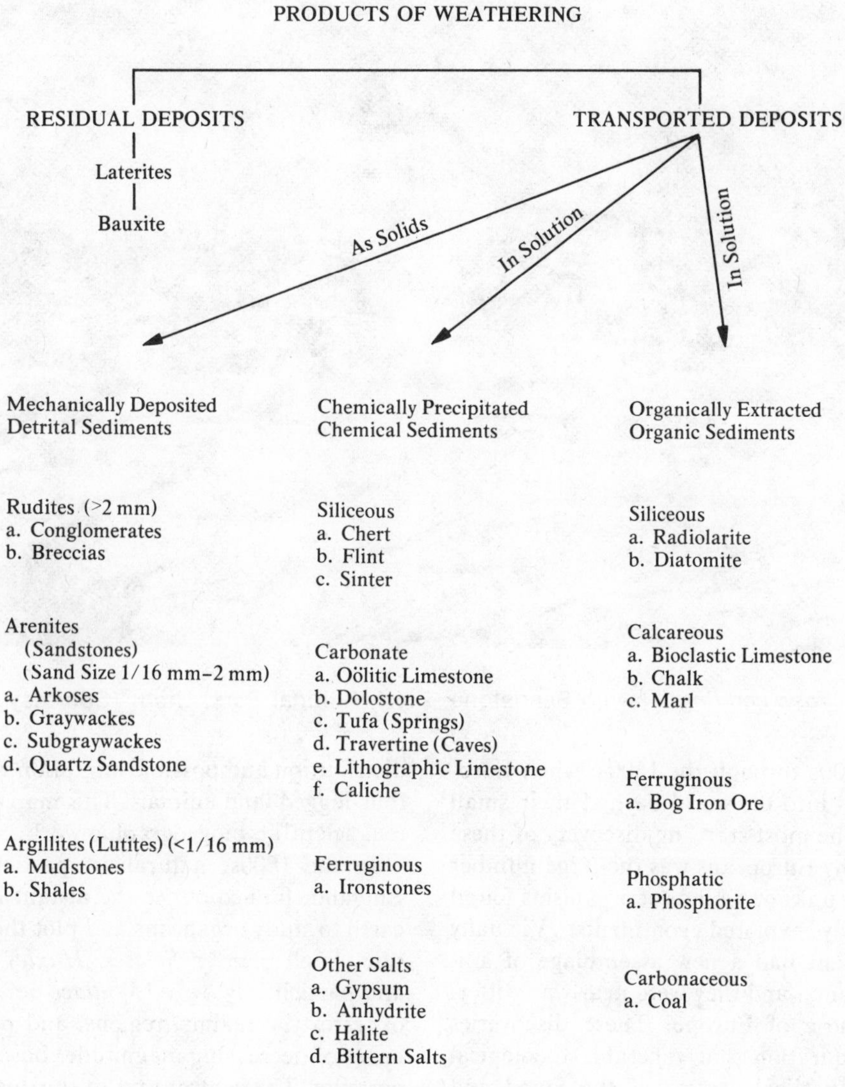

Figure 3.6. *A Classification of Sedimentary Rocks.* (After Spencer, 1966.)

cent to highlands typically produce thick accumulations of red arkose sandstone and conglomerate. Sediments also offer climatic evidence. Evaporites today form only in a band extending 20° north and south of the equator. Red sediments also form today only in the tropics. Coal, the compacted and carbonized remains of plants, presently forms in two humid settings, equatorial regions and a high-latitude zone in both northern and southern hemispheres. Numerous other associations also occur. More will be said about patterns of sedimentation in Chapters 8 and 10.

Biogeography

The science of biogeography can be said to have begun with the Ages of Discovery of Western

Figure 3.7. *Cross-bedding.* Navajo Sandstone, Zion National Park, Utah. (Courtesy of William M. Mintz.)

Man, the 1400s through the 1600s, when Europeans moved into the world beyond their small continent. The most startling discovery of these trips for many Europeans was the large number of previously unknown kinds of organisms found on the newly explored continents. Virtually every continent had a new assemblage of animals and plants, and they were nearly all different from those of Europe. These discoveries called into question that venerable theological dogma, the biblical version of the Flood and Noah's Ark. It was painfully obvious that no boat could hold two individuals of all the vast number of species of land organisms. As if this were not bad enough, the distribution of land animals and plants did not fit with the story that the center of their present distribution should be Mt. Ararat where the Ark supposedly came aground. Some of the new animals, such as those of Australia, were found only on their own isolated continents and did not occur anywhere even remotely close to Mt. Ararat.

In 1766, the great French naturalist, *Comte de Buffon*, prepared a world map showing the distribution and possible migration routes of the four-legged land animals. This map was the first real scientific biogeographic work.

In the 1800s, naturalists, particularly from England, fanned out to the distant lands of the earth to study organisms and plot their distribution. Such men as *Sclater, Huxley, Lydekker,* and particularly *Alfred Wallace* defined a series of terrestrial realms, regions, and provinces (in order of decreasing magnitude) based on biogeography. The central unit in this hierarchy, the region, is the most significant. The earth's terrestrial regions (Fig. 3.9) are the **Palearctic** (Europe, Africa north of the Sahara, and Asia north of the Himalayas); the **Nearctic** (North America north of central Mexico); the **Oriental** (India and Southeast Asia); the **Ethiopian** (sub-Sahara Africa); the **Neotropical** (Central and South America); and the **Australian** (Australia and the South Pacific islands).

The determination of the distribution of organisms in the sea began later, in the late 1800s, with the work of *Woodward* and has reached fruition with the research of *Sven Ekman* and

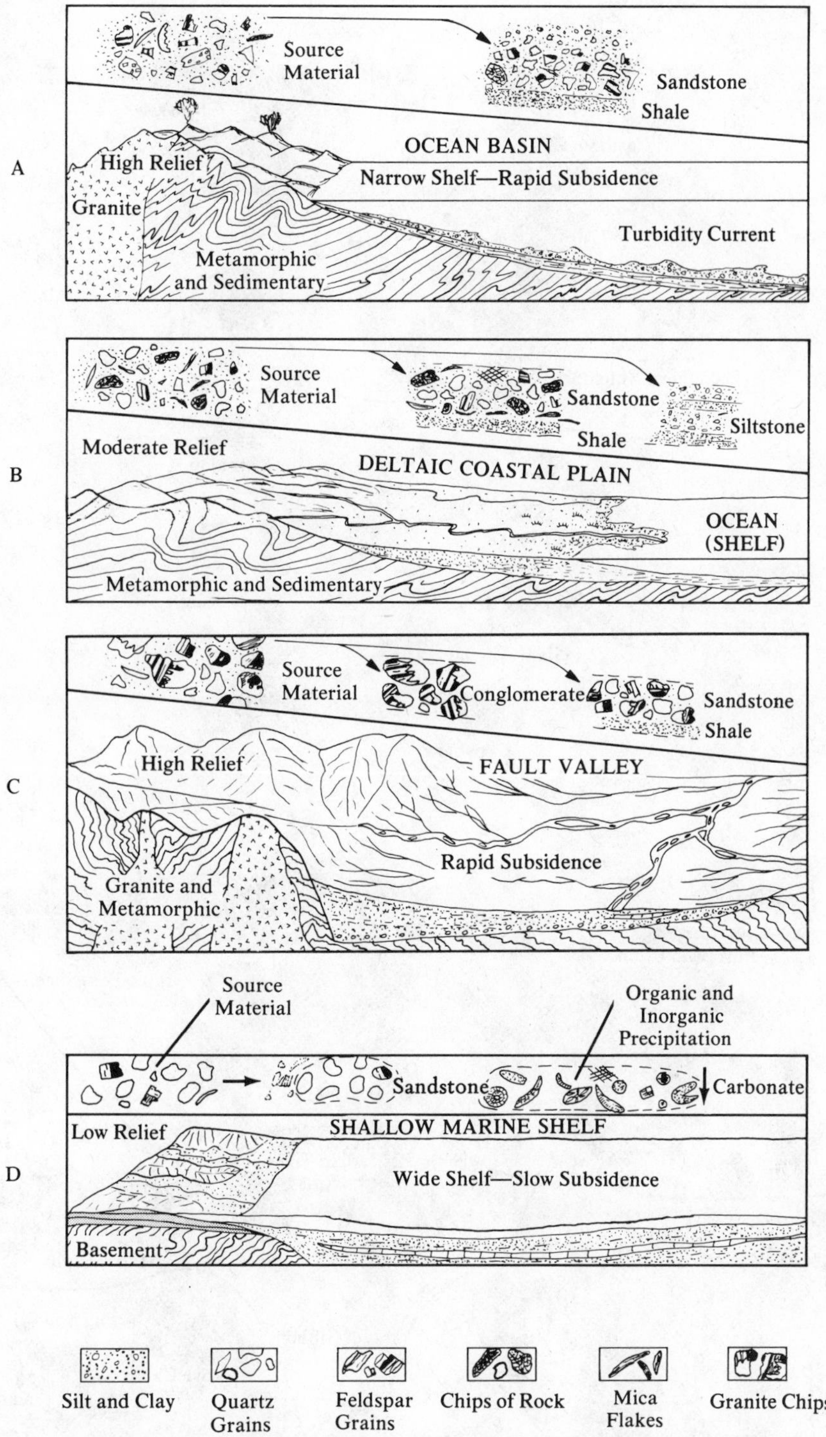

Figure 3.8. *Environments of Deposition and Associations of Sedimentary Rocks.* **A.** *Graywacke Sandstone,* **B.** *Subgraywacke Sandstone,* **C.** *Arkose Sandstone,* **D.** *Quartz Sandstone and Carbonates.* Source areas and materials shown on left side; depositional areas and materials, to right. (After Brice, 1960.)

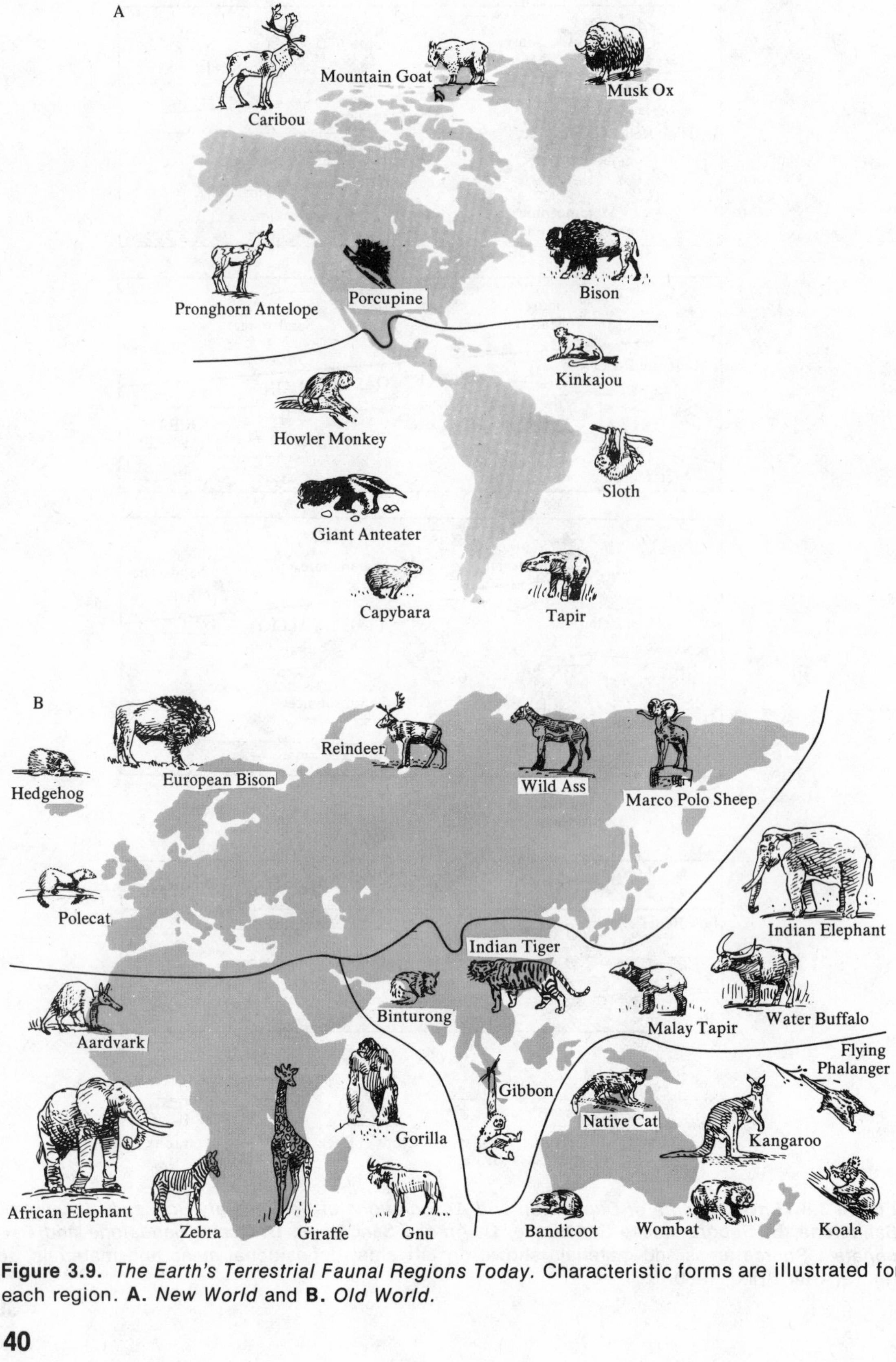

Figure 3.9. *The Earth's Terrestrial Faunal Regions Today.* Characteristic forms are illustrated for each region. **A.** *New World* and **B.** *Old World.*

others in this century. Much more, however, needs to be done as most of the current data are only on the organisms of the continental shelves (Fig. 3.10). The gathering of data does not a science make, however, and the real scientific effort in biogeography has developed as people have searched for the causes of the patterns of distribution.

From studies of both terrestrial and marine organisms, it has become apparent that three basic factors influence the distribution of organisms. They are the local **environment** (ecology), the presence of **barriers** (geography), and the **geologic history** of the region.

Marine Environments

Most fossils are those of marine organisms because the oceans are major sites of deposition, while the land is chiefly a region of erosion. For this reason, we will concentrate on marine environments. The major ecologic factors that influence the distribution of marine organisms are temperature, dissolved gases (chiefly O_2 and CO_2), sunlight, salinity, water turbulence, substrate, and feeding relationships of organisms. Typically, all of these and many other lesser factors interact with one another to produce the local environment (Fig. 3.11).

Temperature is among the most significant factors as it affects the rate of metabolism of organisms both directly (Fig. 3.12) and indirectly by affecting the solubility of essential gases and salts and establishing convective water movements. In a cross section of an ocean (Fig. 3.13), the most diverse and abundant life occurs in the shallower waters above 200 meters where warmer temperatures favor more efficient metabolism. Sunlight is also a vital factor as it provides the energy for photosynthesis, the process whereby

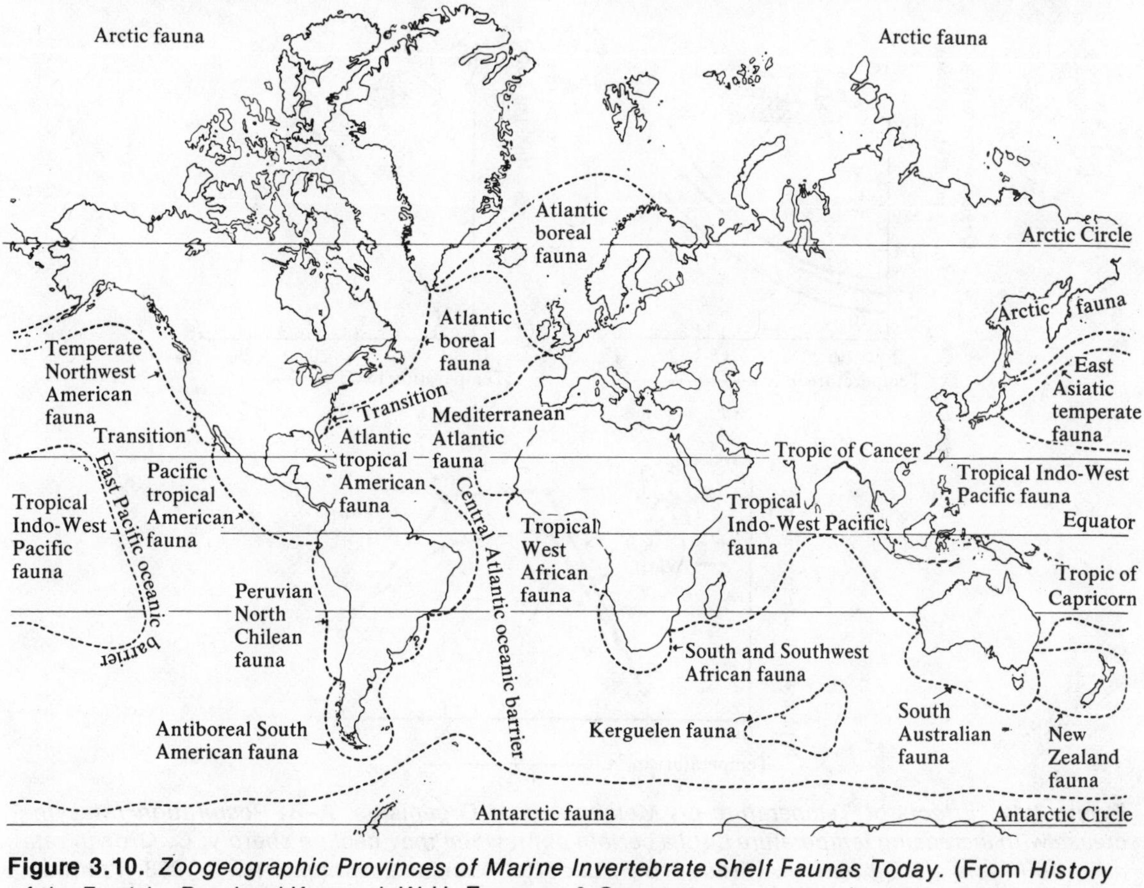

Figure 3.10. *Zoogeographic Provinces of Marine Invertebrate Shelf Faunas Today.* (From *History of the Earth* by Bernhard Kummel, W.H. Freeman & Company, Copyright © 1970, 2nd ed.)

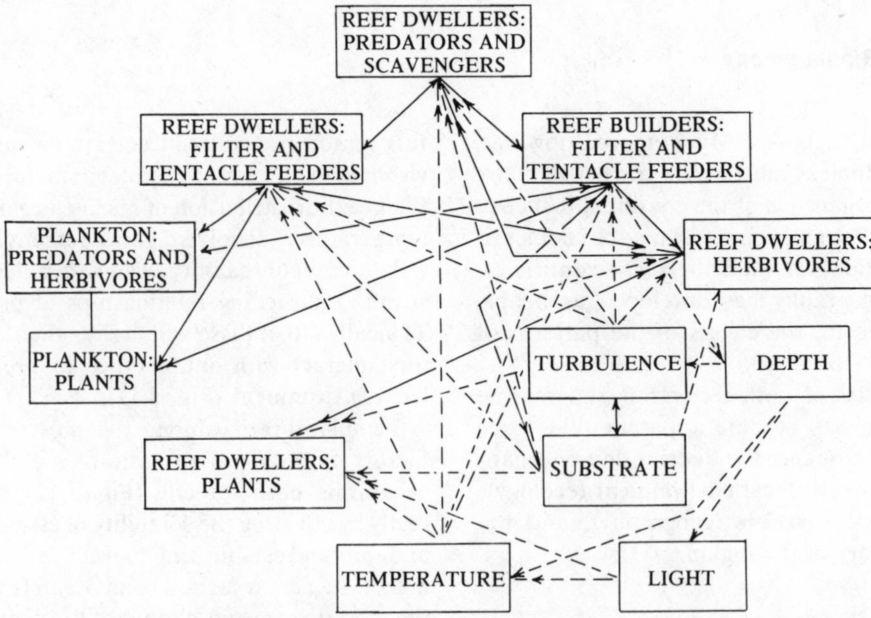

Figure 3.11. *Complex Interaction of Factors Producing a Silurian Reef Environment.* Arrows indicate direction of action. Most significant interactions indicated by solid lines. (From James R. Beerbower, *Search for the Past: An Introduction to Paleontology*, 2nd ed., © 1968. Reprinted by permission of Prentice-Hall, Inc., Englewood Cliffs, N.J.)

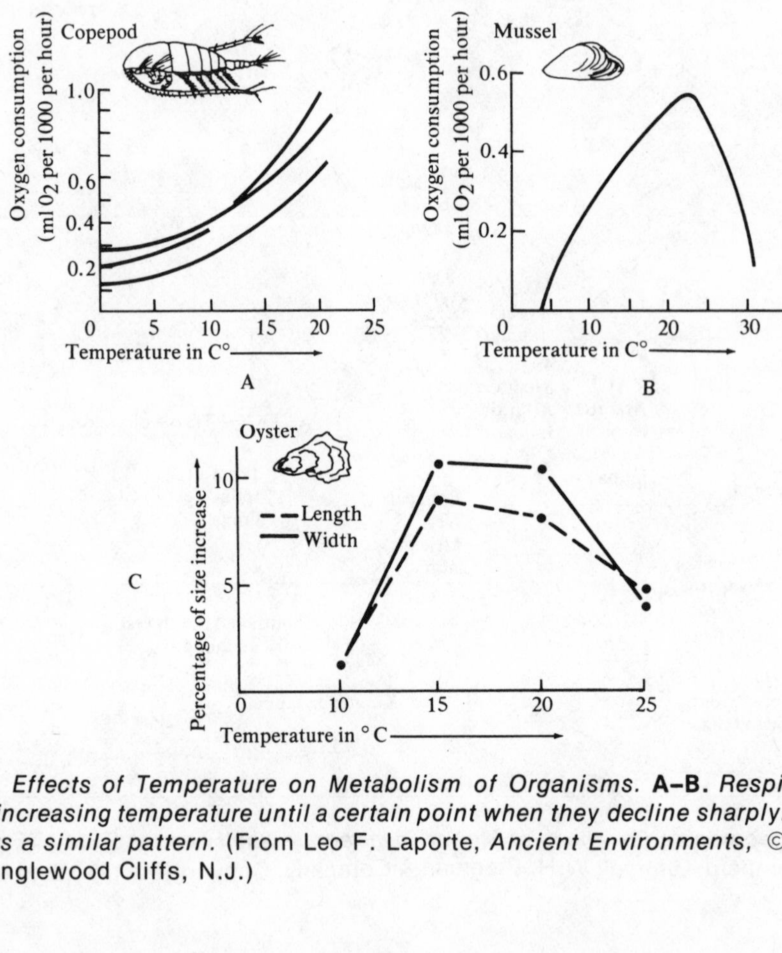

Figure 3.12. *Effects of Temperature on Metabolism of Organisms.* **A–B.** *Respiration rates increase with increasing temperature until a certain point when they decline sharply.* **C.** *Growth rate curve follows a similar pattern.* (From Leo F. Laporte, *Ancient Environments*, © 1968, Prentice-Hall, Inc., Englewood Cliffs, N.J.)

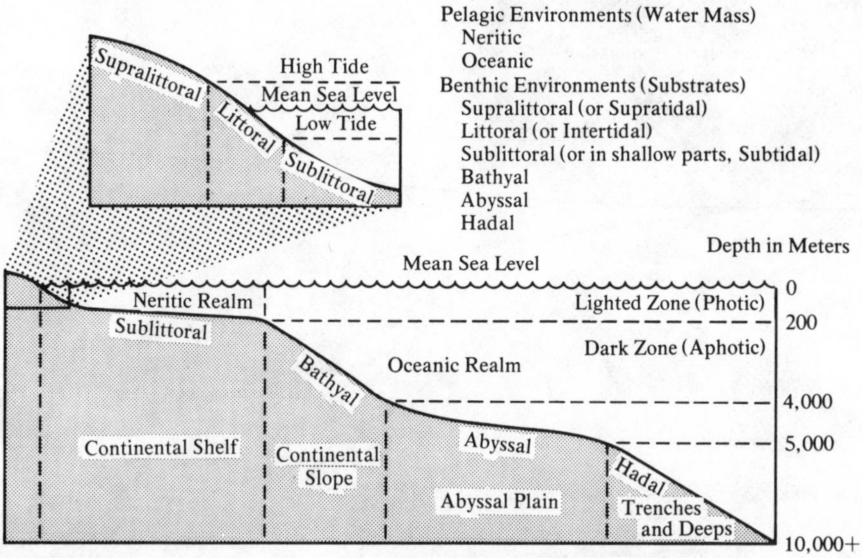

Figure 3.13. *Partial Cross Section of an Ocean Showing Major Marine Environments.* Generally these correspond with major geomorphic zones indicated in diagram.

chlorophyll-bearing organisms ("plants" in the loose sense) convert inorganic substances to the complex organic substances used for metabolism and structure. These organisms form the basis for all life as they are fed upon either directly by herbivores or indirectly through carnivores eating herbivores or other carnivores. Sunlight typically penetrates only the upper 200 meters or so of the ocean; hence, it is only there that one finds plants or plantlike microorganisms. This is another reason why the shallower seas have the most diversified life. However, this does not mean that the depths are devoid of life. In fact, oceanic depths often contain numerous individuals of a few species.

In our cross section of an ocean (Fig. 3.13), you will note that there are two basic categories of environments—the **benthic** which embraces the surface of the bottom and its substrate, and the **pelagic** which includes all of the ocean waters above the bottom. The benthic environment is subdivided into a series of depth zones which roughly correspond to the physiographic subdivisions listed after their names in parentheses: **littoral** (intertidal), **sublittoral** (continental shelf), **bathyal** (continental slope), **abyssal** (abyssal plains), and **hadal** (trenches). The littoral and sublittoral zones not only have sunlight and more favorable temperatures than the deeper zones, but also have a greater variation in substrates, turbulences, and nutrients which accounts for the plethora of life. A large percentage of the marine rocks on the continents today were deposited in ancient sublittoral environments and, fortunately for geologists, contain the most abundant benthic fossils. Representative benthic organisms, or benthos, are shown in Figure 3.14.

The pelagic environment is generally subdivided into a thin upper, lighted **photic** zone in which all the living photosynthesizers and many animals are found and a huge lower **aphotic** zone inhabited only by animal carnivores and scavengers, and by the bacteria that break down dead organisms and return basic mineral salts to the water to be later reused by the photosynthesizers. Pelagic organisms are either floaters, the **plankton**, or swimmers, the **nekton** (Fig. 3.14).

Marine organisms can also be ecologically classified on the basis of their modes of nutrition (Fig. 3.15). Those organisms which can manufacture complex organic molecules from simpler inorganic ones, chiefly the photosynthesizers, are called **producers** or **autotrophs.** These are the true plants and plant-like microorganisms. Those organisms that require an external supply of organic molecules are called **consumers, het-**

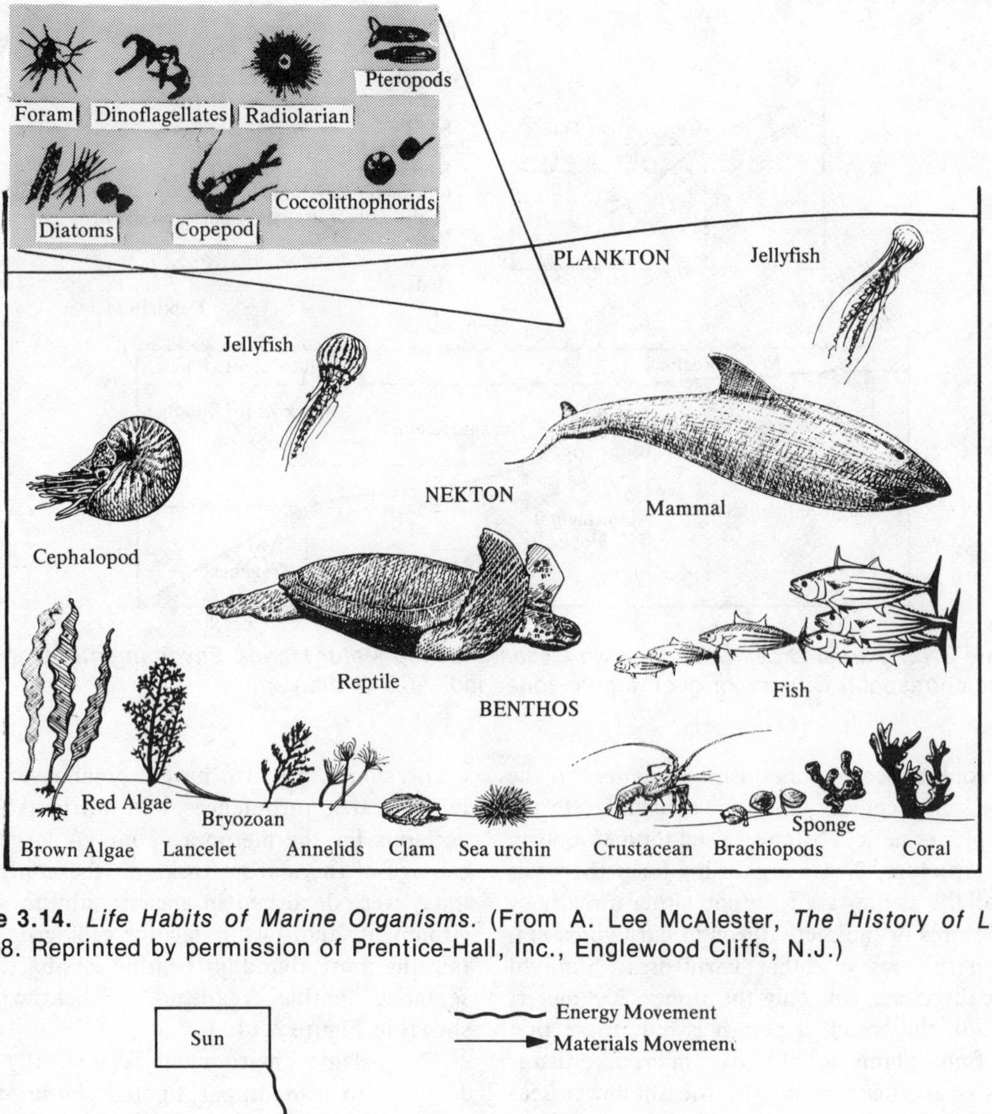

Figure 3.14. *Life Habits of Marine Organisms.* (From A. Lee McAlester, *The History of Life,* ©1968. Reprinted by permission of Prentice-Hall, Inc., Englewood Cliffs, N.J.)

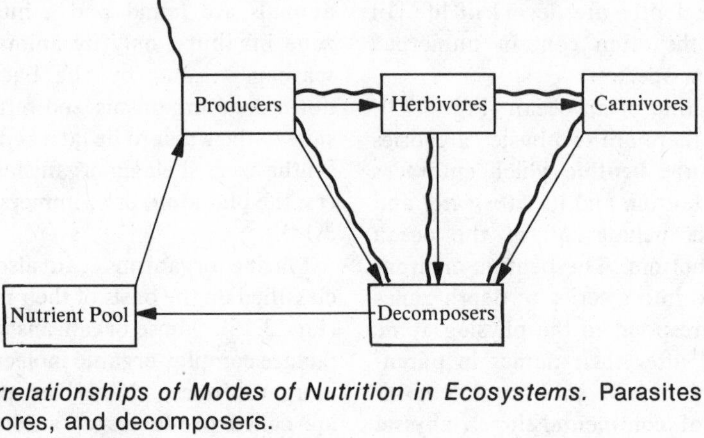

Figure 3.15. *Interrelationships of Modes of Nutrition in Ecosystems.* Parasites occur on and in producers, herbivores, and decomposers.

erotrophs, or **allotrophs.** These fall into four subtypes, the **herbivores,** which eat producers; the **predators** or **carnivores**, which eat other consumers; **parasites**, which absorb nutrients from other living organisms; and the decomposers or **saprophytes** and **scavengers,** which obtain their nutrients from dead organic matter. Animals and animal-like microorganisms are found in all the consumer subtypes; the bacteria and fungi are also major contributors to the last two categories.

The interrelationships of organisms through these modes of feeding produce **food chains** and **food webs** (Fig. 3.16) which compose the marine ecosystem (Fig. 3.17). In the marine ecosystem, the producers are small, floating, plant-like microorganisms, the phytoplankton (Figs. 3.14, 3.22). Attached larger algae are significant only in shallow waters where light reaches bottom. Herbivores include the tiny, floating, animal-like microorganisms and true animals, the zooplankton (Figs. 3.14, 3.22), which actively devour the phytoplankton. Also included are the groups of larger benthic animals (Fig. 3.18B) which feed on the larger attached algae of shallow waters. The predators come in several levels of successively larger animals (Figs. 3.14, 3.18C). Most of these animal groups also contain scavengers. In the sea there are many important fixed, or sessile, animals that collect masses of plankton by filtering the water (Fig. 3.18A). In addition, there are numerous mobile, or vagrant, benthonic animals (Fig. 3.18B) that selectively or nonselectively remove organic detritus from the bottom. The coelenterates are a unique group of attached carnivores (Fig. 3.18C). Parasites enter the marine food chains at all levels. The bacteria are the **transformers** which break down organic detritus and return the simple inorganic compounds to the water where they can be reused by the producers.

A cross section of the oceans gives only part of the ecologic story, however, as there are also the differences which affect the basic environmental factors as one goes from the equator toward the poles. For example, proceeding northward along the North Atlantic coast of North America, the shallow-water temperatures decrease and the organic populations change (Fig. 3.19). Some organisms, such as reef-building corals (Fig. 3.20), can live only in waters of tropical temperature and shallow depth because the algae that live in their tissues and support their metabolism require sunlight. If reef corals have not changed their environmental requirements throughout geologic history, they should serve as valuable guides in reconstructing ancient environments.

We have seen that all organisms are influenced by their environments; therefore, if we can compare a fossil organism with a living relative, we may have a key to help us understand past environments. This is not the complete picture, however, for many areas have the same ecologic conditions and yet are inhabited by quite different kinds of organisms. For example, the shallow-water, marine molluscan faunas on the two sides of the Atlantic, where ecologies are basically the same, are different (Fig. 3.10). Likewise, the mollusks on the east coast and the west coast of North America are different where the environments are the same. Here we see the

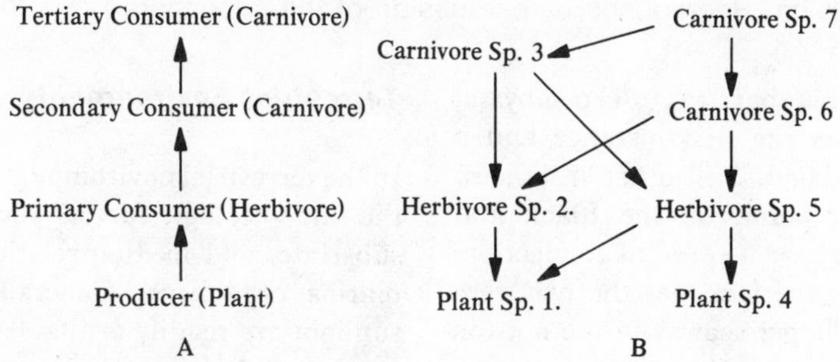

Figure 3.16. A. *Food chain* and **B.** *Food web.*

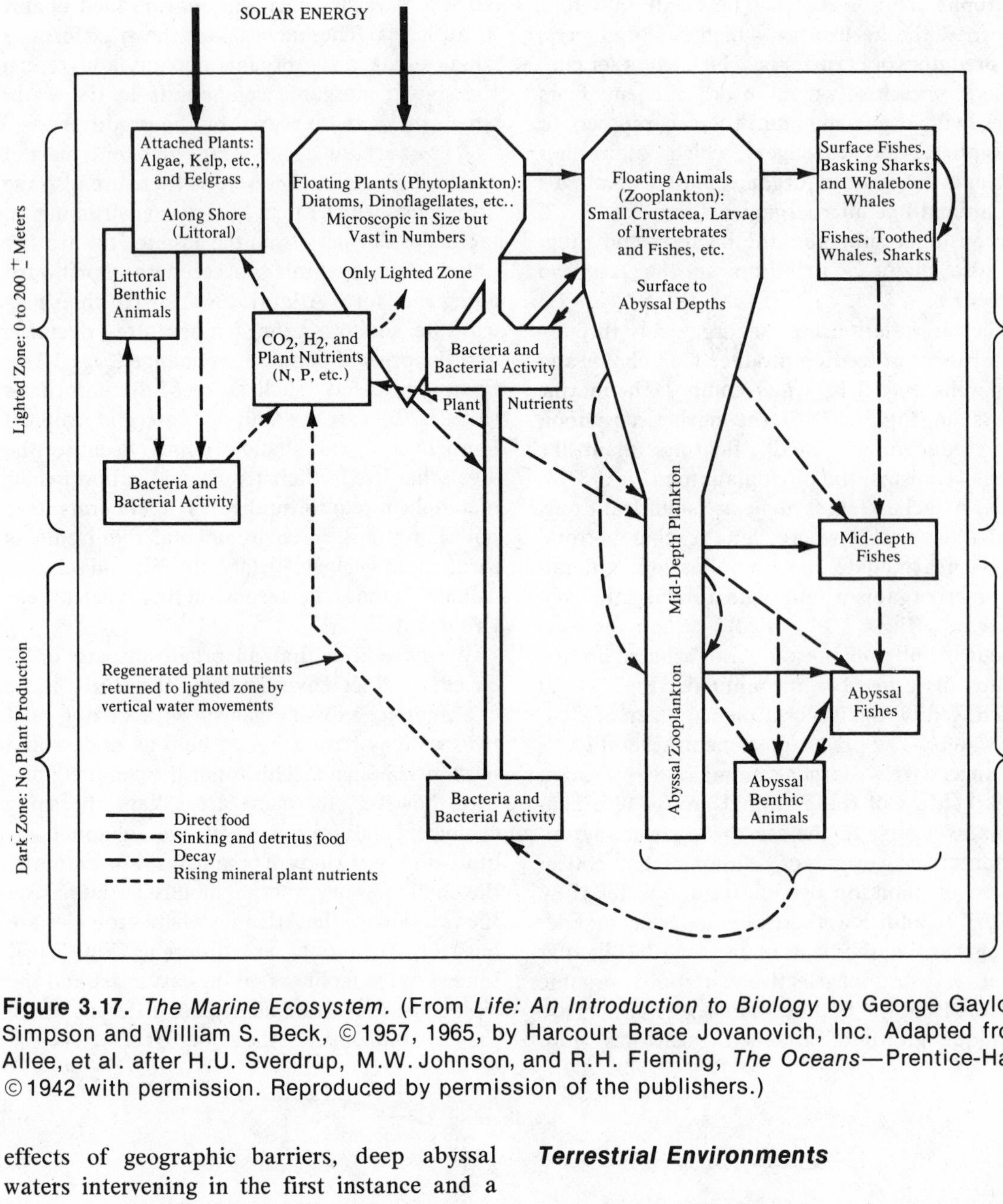

Figure 3.17. *The Marine Ecosystem.* (From *Life: An Introduction to Biology* by George Gaylord Simpson and William S. Beck, ©1957, 1965, by Harcourt Brace Jovanovich, Inc. Adapted from Allee, et al. after H.U. Sverdrup, M.W. Johnson, and R.H. Fleming, *The Oceans*—Prentice-Hall, ©1942 with permission. Reproduced by permission of the publishers.)

effects of geographic barriers, deep abyssal waters intervening in the first instance and a land mass in the second. Still other instances, such as the similar faunas of the Black and Caspian seas, which have no present connection, can be explained by the fact that the two were united as part of a larger seaway in the not-too-distant geologic past.

Terrestrial Environments

In the terrestrial environment, there are some of the same ecologic factors such as temperature, substrate, and feeding relationships that affect marine organisms. Generally, the gases and sunlight are readily available and have little influence. Salinity and water turbulence are not

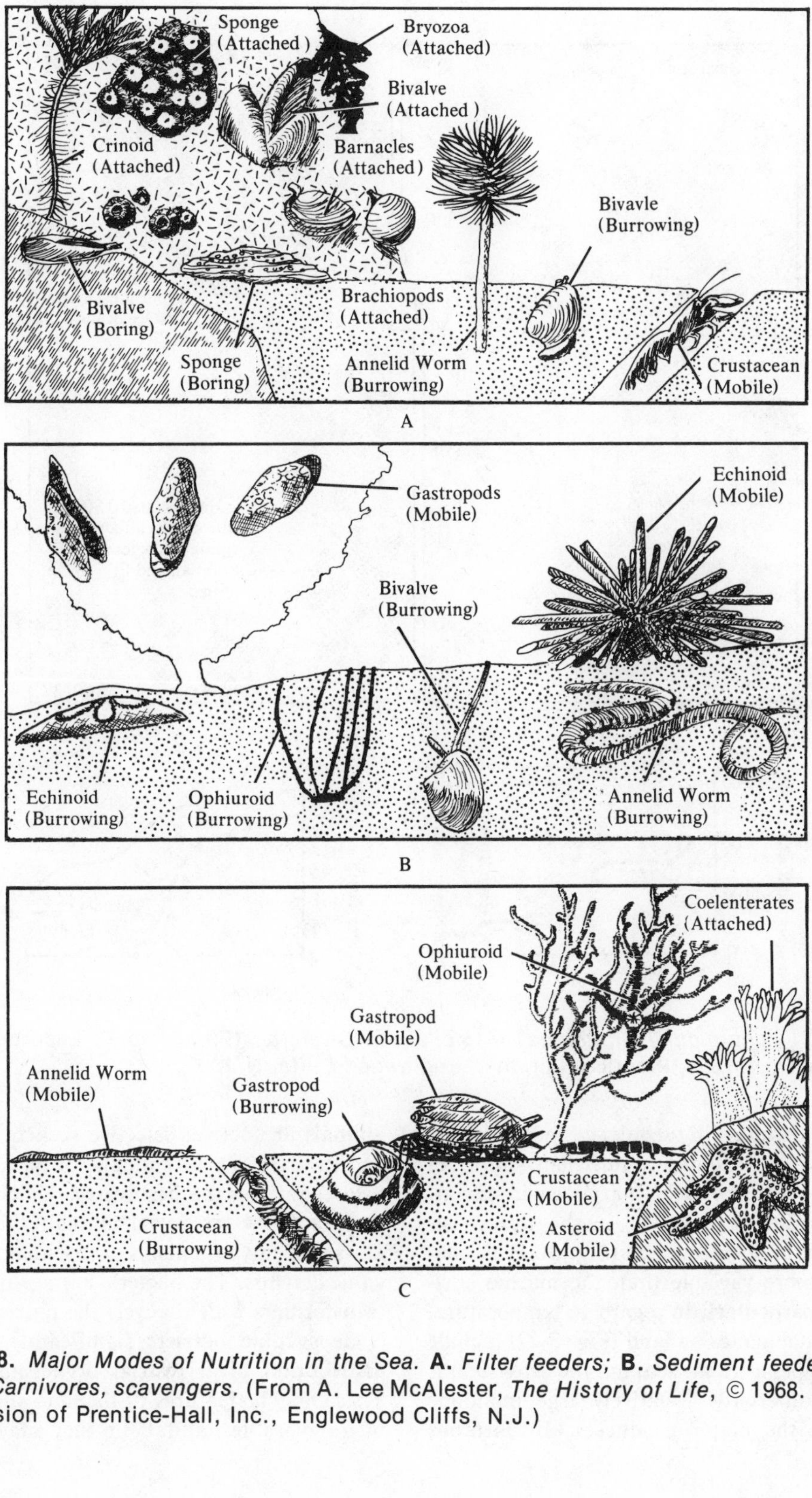

Figure 3.18. *Major Modes of Nutrition in the Sea.* **A.** *Filter feeders;* **B.** *Sediment feeders, herbivores;* **C.** *Carnivores, scavengers.* (From A. Lee McAlester, *The History of Life*, © 1968. Reprinted by permission of Prentice-Hall, Inc., Englewood Cliffs, N.J.)

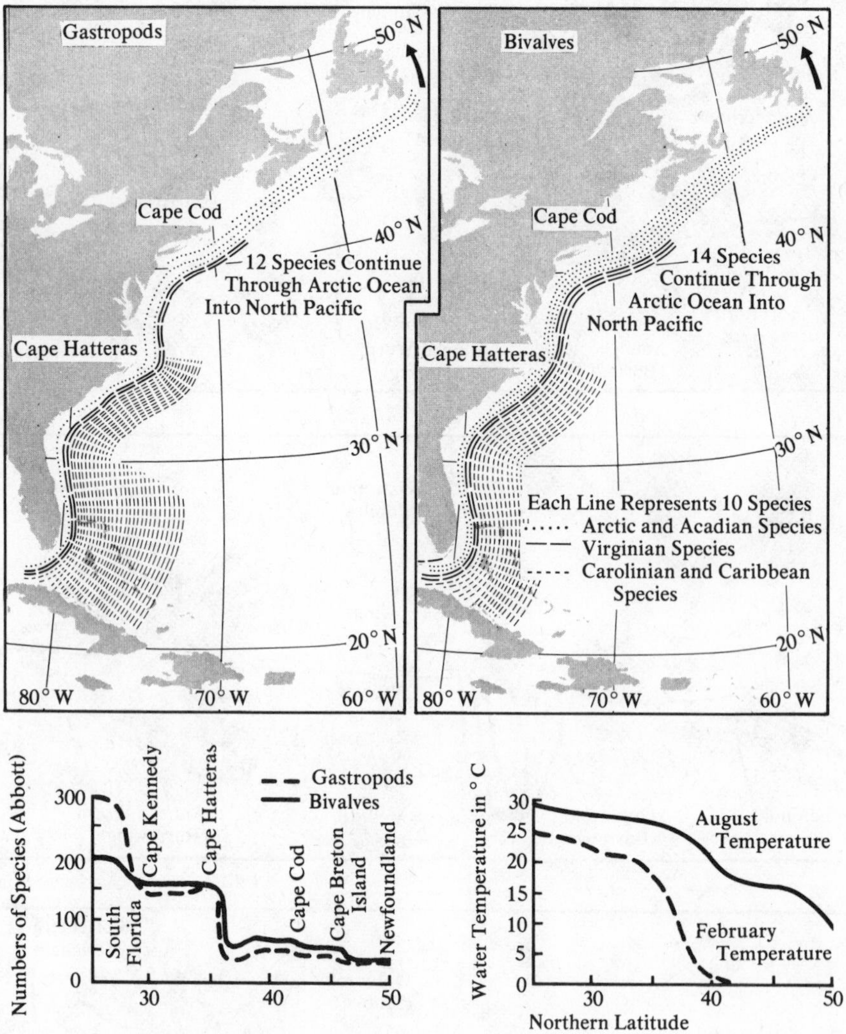

Figure 3.19. *Longitudinal Differences in Molluscan Diversity.* (From Leo F. Laporte, *Ancient Environments,* © 1968, Prentice-Hall, Inc., Englewood Cliffs, N.J.)

pertinent, though air turbulence may be. Obtaining and maintaining a sufficient supply of water, of no concern in the marine environment, is of overriding importance in the terrestrial. The terrestrial environment (Fig. 3.21) is considerably more variable than the marine environment, particularly in regard to temperature.

Nutritional modes on land (Fig. 3.22) include the same types as in the sea, but the ecosystem is organized differently. Relatively large, immobile plants are the major producers. Herbivorous animals are active, selective seekers of plants; and the carnivorous animals, in turn, are selective feeders on the herbivores. Unlike the sea, there are few terrestrial animals that feed on masses of tiny organisms or undifferentiated organic detritus. The bacteria are again the major transformers which recycle the nutrient salts.

Geographic barriers significantly affect the distribution of terrestrial organisms as well. These may be features of the land such as belts of unfavorable habitat, or they may be water

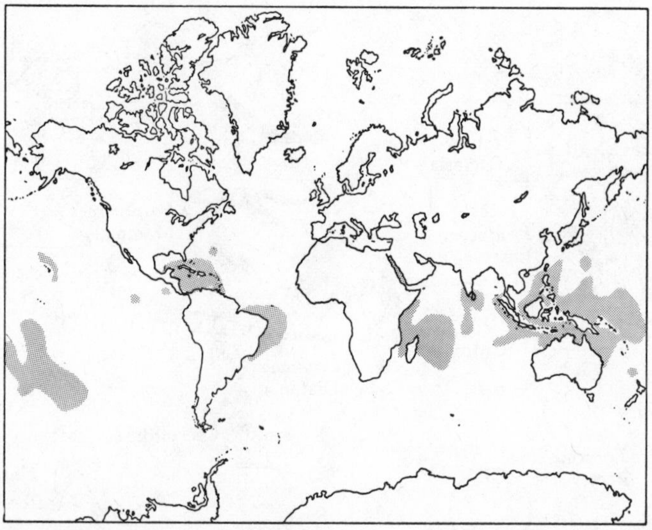

Figure 3.20. *Distribution of Modern Reef Corals.*

bodies that separate land masses. Geologic history is important, too. For example, South America, which is today in contact with the northern continents through Central America, has numerous local organisms not found elsewhere on earth (Fig. 3.9). These are explained by the long isolation of South America throughout most of the Cenozoic.

All these factors have, of course, also influenced the past distribution of organisms. Thus, fossils are guides to the conditions of the past, though until recently there appeared to be many peculiar occurrences because tropical organisms were found in polar regions and polar organisms in the tropics. The discovery of continental drift and sea-floor spreading, discussed in Chapter 10, has explained these anomalous occurrences.

Circumventing the Facies Problem

This brings us back to the original question of how to circumvent the problem of facies if we wish to determine whether two bodies of rock, deposited in different environments, were deposited at the same time. Our review of life in the sea suggests some answers. You will recall that floating (planktonic) and swimming (nek-

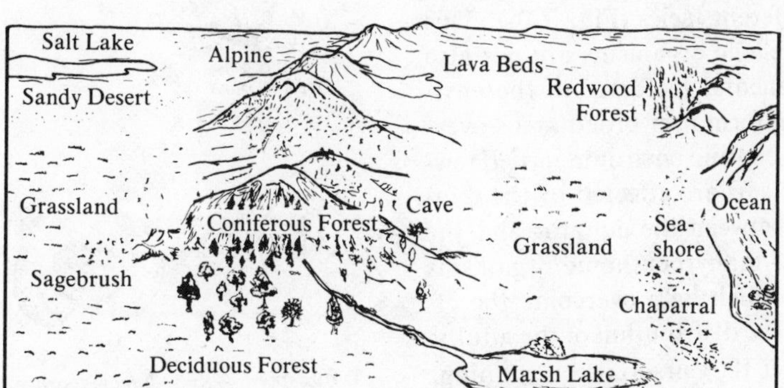

Figure 3.21. *Representative Terrestrial Environments in the Middle Latitudes of North America.*

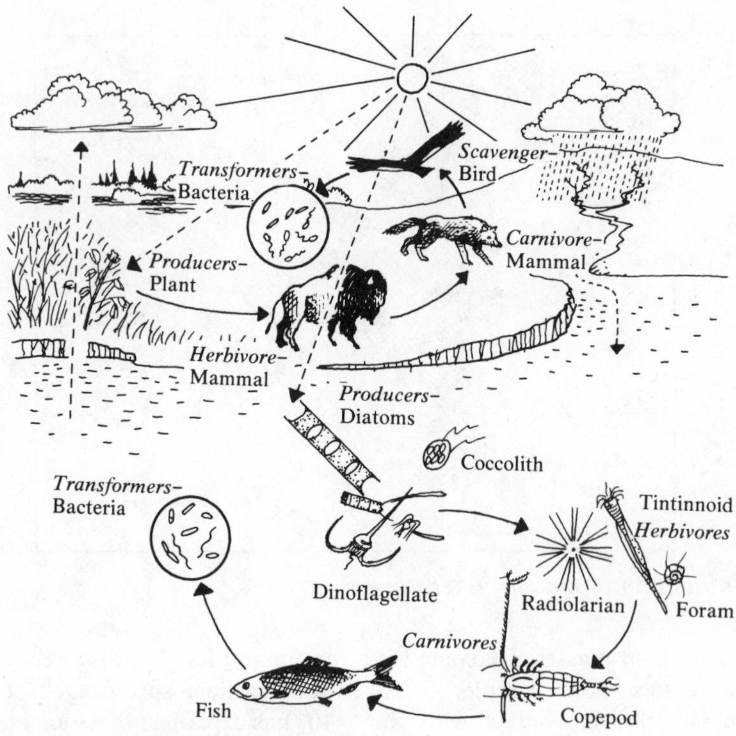

Figure 3.22. *Terrestrial and Marine Ecosystems.*

tonic) organisms of the pelagic environment are found in the waters over several different benthic zones (Figs. 3.14, 3.22). When they die, their remains may settle out in several different environments where the benthic organisms and sediment types are different. Indeed, it is the shelled planktonic microorganisms and the nektonic cephalopods (particularly the ammonoids) which are the most useful in determining the ages of rocks or different facies (Fig. 3.14). This is not to say that pelagic organisms are not also limited by environmental factors, but the environment may be uniform over broad areas overlying a diversity of benthic environments. (However, pelagic organisms are affected by the temperature gradient between the equator and the poles.) In addition, many benthonic organisms have pelagic larva which help overcome the effects of barriers on the distribution of the adults.

One might expect the cursorial, or running, animals and the flyers to help circumvent the facies problem in terrestrial deposits. Cursorial animals often do have wide distribution as long as they have interconnected favorable environments. Flying creatures such as insects and birds are generally too delicate to be preserved commonly enough to be widely used for dating. The pollen and spores of land plants (Fig. 3.23),

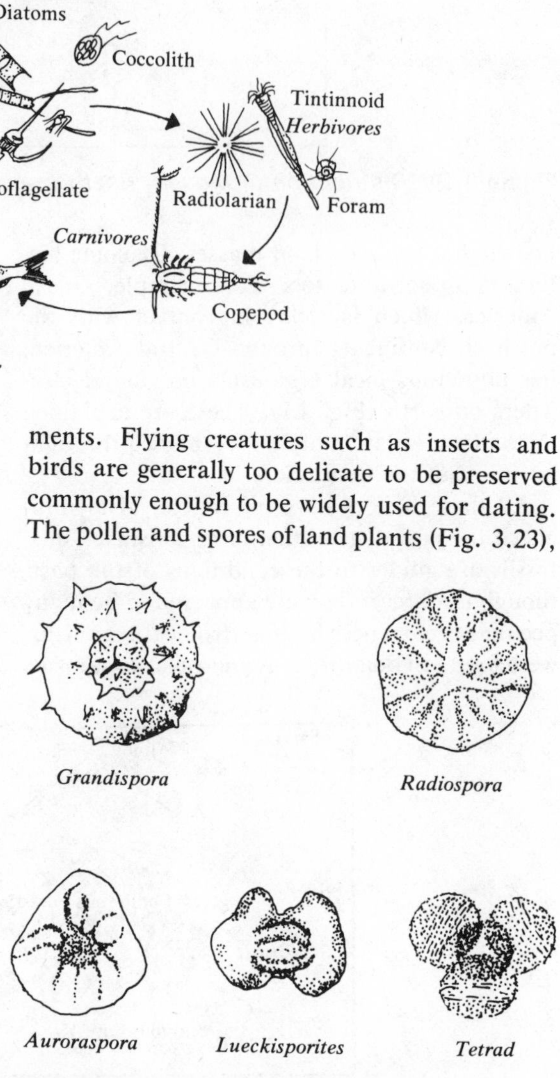

Figure 3.23. *Fossil Spores and Pollen.* X 300. The study of these fossils is called palynology. (After Jones.)

however, are highly resistant and widely disseminated, making them the most useful of terrestrial fossils for determining ages.

Before leaving the subject of facies, we should briefly note that the word has been used in several other senses in historical geology since its original definition. Some geologists use *facies* to refer to the lithogic or biologic record of a particular environment regardless of its age. This is unfortunate, not only because it distorts a useful concept, but also because there are already words in common geologic usage for those records. The rock record of an environment is called a **formation**, and a fossil record of an environment is called an **assemblage zone** or a **zonule.** Other geologists use *facies* to refer to the lateral variations within formations. In Chapter 5 we will discuss the various types of geologic units and the problems encountered by such misdefinitions. *Facies* is best left with its original definition of the lateral lithogic and fossil variations within a unit of rocks of the same age.

REVIEW

1. What was the original definition of **facies** and who proposed it?
2. Why did the concept of facies encounter so much opposition in North America?
3. What is the environmental significance of size sorting in sedimentary rocks?
4. Which of the major sedimentary rock associations would you expect to dominate in your state?
5. What are the six major terrestrial faunal regions today?
6. List the three major factors that influence **biogeography** and give an example of each.
7. What is the difference between the **benthic** and the **pelagic** environments?
8. What are the major producers in the sea? On land?
9. Give an example of an organism in each of the following categories: marine benthic filter feeder, terrestrial scavenger, marine mobile carnivore, marine mobile herbivore.
10. Why are the marine shallow-water faunas off east Africa different from those off west Africa?
11. What types of organisms are best suited ecologically to circumvent facies problems and date strata?

4

The Key to Faunal Succession and Biofacies: Evolution

4

Just as it took people centuries to accept successive changes or evolution of the physical world, the recognition of the evolution of life was a long and difficult process. Both concepts came into widespread use in the mid 1800s as people shifted their notion of a static universe to that of a dynamic one. Biological evolution, in simplest terms, means the change of one kind or species of organism into another, which is another way of saying that all creatures are interrelated. It explains why fossils occur in a definite sequence and why geologic history is important to biogeography.

Historical Development

Once again, it was the classical *Greeks* who left us the first written records of ideas of the transmutation of species. *Anaximander,* who lived in the sixth century B.C., stated that life arose from a mixture of water and mud warmed by the sun and that because of this and the succession of fossils, people had ultimately come from fish. *Empedocles,* who wrote a century later, also spoke of a succession of living things. It is uncertain whether these men had original ideas or derived them from earlier Egyptian scholars.

Aristotle, the greatest Greek scholar, and others developed the idea of a great "Chain of Being" which meant that there is a complete gradation in the organic world between the simplest and the most complex organisms. In this scheme, however, it appears that these scholars thought of each species as static on its rung of the ladder and not capable of evolving into anything else. This concept carried on for centuries, even into the time of Linnaeus in the 1700s.

The Hebrew creation story, adopted from still earlier legends of other Near Eastern peoples and enshrined in the Book of Genesis, was the dominent western view of nature through the Dark and Middle Ages. In this legend, all species were created at once for people's use, and each reproduced its own kind without any changes through time. In addition, all species had been saved from the great Flood by the benevolent Creator; hence, none had perished.

Organic Diversity and Similarity

With the *Renaissance* and the accompanying Voyages of Discovery in the 1400s and 1500s came a renewed interest in observing nature. Disturbing things for the theological viewpoint came to light. We have already spoken of the space problems on the Ark with all the new species and of the fact that the geographical distribution of organisms was not a concentric series of rings around Mt. Ararat. In addition, many creatures were discovered that were intermediate in structure between known species. This meant many new rungs on the "Chain of Being." The biggest surprise, of course, was the discovery of the great apes, creatures that are in many ways intermediate between other animals and people (Fig. 4.1).

Extinct Species

The recognition of *fossils* as organic remains also led to problems for the theological viewpoint. It soon became apparent to forward-

Figure 4.1. *Early Drawing of Chimpanzee.* (After Buffon, 1766.)

thinking men such as *Palissy* (1580), *Hooke* (late 1600s), *Jussieu* (1718), and *Sutton* (1749) that certain species were not alive today, particularly the strange Carboniferous coal-age plants. Theologians were able to counter this argument for many years by stating the organisms were living in some as yet unexplored region of the earth. By the close of the eighteenth century, however, it was obvious that certain large species of fossil organisms could no longer remain undiscovered if still alive. In 1784, *Suckow* stated that the coal age plant *Calamites* was extinct; and, in 1796, *Cuvier* successfully demonstrated the existence of several species of extinct animals, including mastodons and ground sloths (Fig. 4.2).

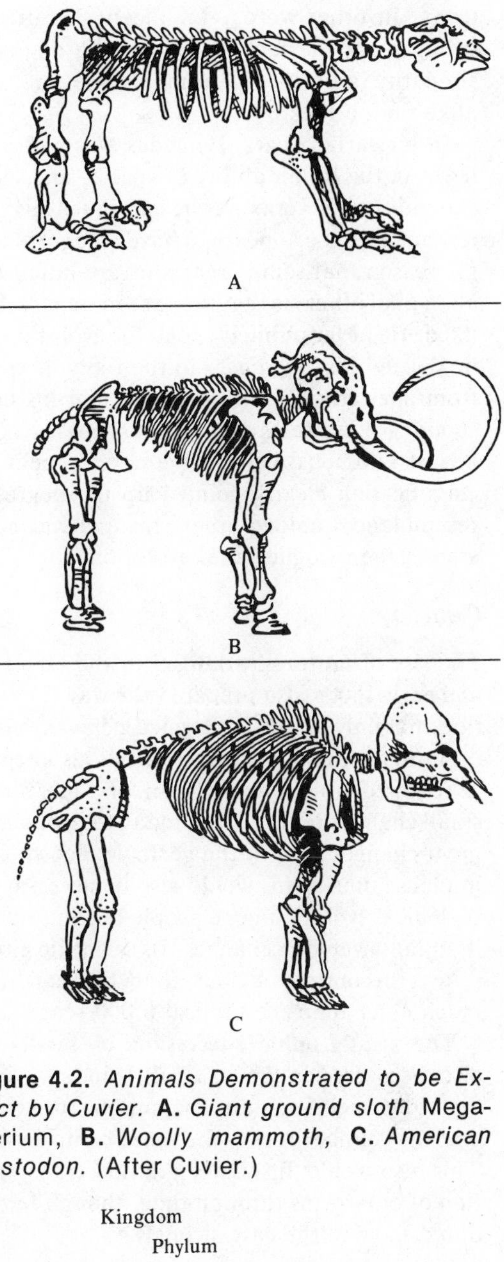

Figure 4.2. *Animals Demonstrated to be Extinct by Cuvier.* **A.** *Giant ground sloth* Megatherium, **B.** *Woolly mammoth,* **C.** *American mastodon.* (After Cuvier.)

Kingdom
Phylum
Class
Order
Family
Genus
Species

Figure 4.3. *The Linnaean Hierarchy of Nomenclature.* Linnaeus himself did not establish all of the units now used.

Classification of Organisms

Another line of scientific advance that paved the way for the recognition of evolution was the development of a formal *classification* of organisms by the great Swedish naturalist, *Carolus Linnaeus* in the 1700s. Linnaeus grouped organisms together in his classification (Fig. 4.3) on the basis of the number of characteristics they have in common. Those with numerous characteristics in common are grouped in the lowest and least inclusive categories. As one goes up the hierarchy of classification, each level encompasses more creatures with less in common until one reaches the highest and most inclusive level (Kingdom) in which the organisms share few traits. In other words, Linnaean classification brought out the fact that there are degrees of similarity among organisms; some are more alike than others.

In his earlier years, Linnaeus was a firm believer in the immutability of species and clearly stated it in his works. As he continued his studies, however, he appears to have recognized that the reason that some organisms are much more alike than others is that they are more closely related. He never publicly spoke for evolution; but he deleted his references to the fixity of species from later editions of his works, and his correspondence indicates a shift in his ideas. Regardless of Linnaeus' beliefs, the development of a classification clearly pointed up the degrees of resemblances among organisms and was a vital step in the recognition of evolution.

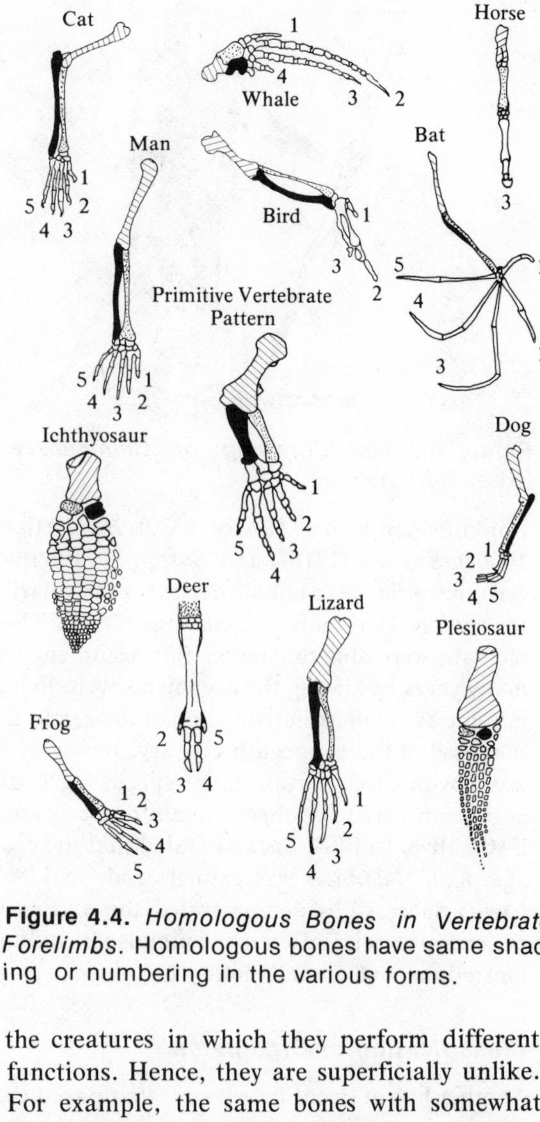

Figure 4.4. *Homologous Bones in Vertebrate Forelimbs.* Homologous bones have same shading or numbering in the various forms.

Geology

The rise of **uniformitarianism** in the late 1700s and early 1800s also prepared the way for evolution, not only because it argued persuasively for evolution in the physical world, but also because it required an immense span of time for the small changes taking place today to produce the great changes seen in the geologic record. Such immense time spans would also be necessary for biological evolution once people began to notice its small, everyday changes. In scientific circles, it was becoming obvious that the earth was much older than the biblical 6,000 years.

The **stratigraphic succession of fossils** was also discovered in this interval. It indicated that the older rocks contain the simpler organisms and the younger rocks, the more complex ones. This also would fit with a gradual transformation of organisms through time, though few still dared to state the case strongly.

Comparative Anatomy

The rise of **comparative anatomy** with *Cuvier, Lorenz Oken,* and *Geoffrey Saint-Hilaire* was another significant event at this time. These men noted that certain structures (Fig. 4.4), which have the same position in the structural organization of dissimilar organisms (and, hence, are really the same structure), are modified differently by the different modes of life of the creatures in which they perform different functions. Hence, they are superficially unlike. For example, the same bones with somewhat different proportions and shapes are found in the same positions in a person's arm, a bat's wing, and a horse's leg. Again, the solution seems obvious that these similarities, called **homologies,** were due to a common ancestry, but most scientists did not, at least publicly, recognize it.

Early Evolutionists

Gradually, through the 1700s and early 1800s, it occurred to the more original thinkers in scientific circles that there had been an evolution of the biological, as well as the physical, world. *Benoit de Maillet* (1656–1738) wrote a book discussing the evolution of life from the sea onto the land and anticipated the more advanced ideas of mutations and natural selection which will be discussed below. *Georges Buffon* (1707–1788) suggested evolutionary changes in organisms, but was forced to recant by the Church. *Erasmus Darwin* (1731–1802), Charles' grandfather, noted **vestigial structures,** non-functional structures of an organism apparently inherited from an ancestor in which they did function (Fig. 4.5). He called these "the wounds of evolution."

Lamarckism

Finally, around 1800, the French paleontologist *Chevalier de Lamarck* advanced the first coherent theory to explain the evolution of life. To Lamarck, the impulse for evolution came from within an organism. It evolved a structure or improved a structure it needed to survive. And once it had made such an improvement, this advance was passed on to its offspring. If structures were no longer useful, they decayed from disuse.

Although Lamarck's theory sounds attractive, particularly from the social viewpoint which has ensured its survival in many minds down to the present, much experimental and observational evidence has been advanced to refute it. No one has ever seen any organism evolve new structures that were not in its hereditary potential, and it is clear that acquired characteristics are not inherited. The son of a steelworker with big biceps will not have his father's big biceps if he becomes an accountant sitting at a desk all day. Despite his erroneous notion of the causes of evolution, Lamarck prepared the way for Darwin by exposing many people to the realtiy of evolution. Much of the theological opposition to evolution was drained off by Lamarck's work.

Other workers, such as *William Wells, Patrick Matthew,* and *Robert Chambers,* advanced theories of evolution in the first half of the nineteenth century. Although their works were often a curious admixture of Lamarckian and Darwinian concepts, they did point to lines of evidence supporting evolution such as the modification of species produced by artificial breeding, the reality of natural variation among individuals of the same species, the changes in the fossil record, and the struggle for existence between the members of a species. These men were not taken too seriously as this was the heyday of catastrophism, that brief marriage of theology and geology.

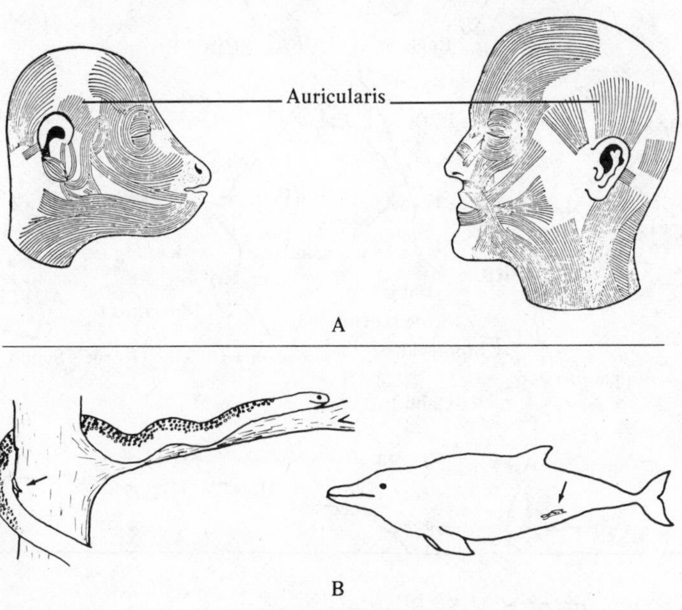

Figure 4.5. *Vestigial Structures.* **A.** *Auricularis muscles, functional in manipulating ears of monkeys, but not man;* **B.** *Posterior appendages in snakes and whales.* (After Elliott, 1957.)

Darwin

Charles Darwin (Fig. 4.6) was the real giant of science who at last succeeded in winning over the scientific community to the reality of evolution by propounding a consistent theory explaining its causes and supporting it with evidence from many areas of biology and geology. In the 1830s, Darwin traveled around the world on the scientific voyage of the ship *Beagle* (Fig. 4.7). The observations he made on this voyage transformed him from a believer in species fixity to a doubter at the end of the voyage. He spent the remainder of his life following up ideas suggested to him by the voyage, which succeeded in convincing him of the reality of evolution. Still, he might never have found time to publish his ideas unless *Alfred Wallace* had not independently reached the same conclusions in the 1850s, an event which stimulated Darwin to action.

Figure 4.6. *Charles Darwin* (1809–1882).

On the *Beagle,* Darwin read Lyell's *Principles of Geology,* which spoke consistently of the slow evolutionary changes in the physical world that had produced great cumulative effects. No doubt, this prepared Darwin psychologically to

Figure 4.7. *Five-Year Voyage of H.M.S. Beagle.*

see the evidence of biological change which he would encounter.

During stopovers in South America, Darwin found the fossils of several extinct Cenozoic mammals (Fig. 4.8). To him, they appeared to be forerunners of the mammals inhabiting that continent today. They were certainly unlike any fossil or recent Old World mammals. This suggested to Darwin that perhaps the present South American animals did not migrate in from Mt. Ararat but somehow had descended from ancestors who had lived in South America for a long time.

When he reached the equatorial **Galapagos Islands**, a volcanic archipelago many hundreds of miles west of South America in the Pacific, Darwin saw many things that really triggered ideas about evolution in his mind. This distant group of islands has no free interchange of species with South America. Instead, only a few species had fortunately survived the great journey because there is little diversity of species. There are, however, several groups of very similar species, most notable of which are the finches (Fig. 4.9). Here are numerous species of birds, all very similar in structure, but with slight differences because they are adapted to different modes of life. Surprisingly, most of these species make a living in ways like those of unrelated birds on the mainland and unlike those of the mainland finch species they resemble. One has even taken up the mode of life of a woodpecker. Could it be, Darwin guessed, that the mainland finch species had gained a

Figure 4.8. *Extinct South American Mammals.* Darwin found the remains of some of these forms. (After Marshall Kay, Edwin H. Colbert, *Stratigraphy and Life History,* ©1965, by permission of John Wiley & Sons, Inc.)

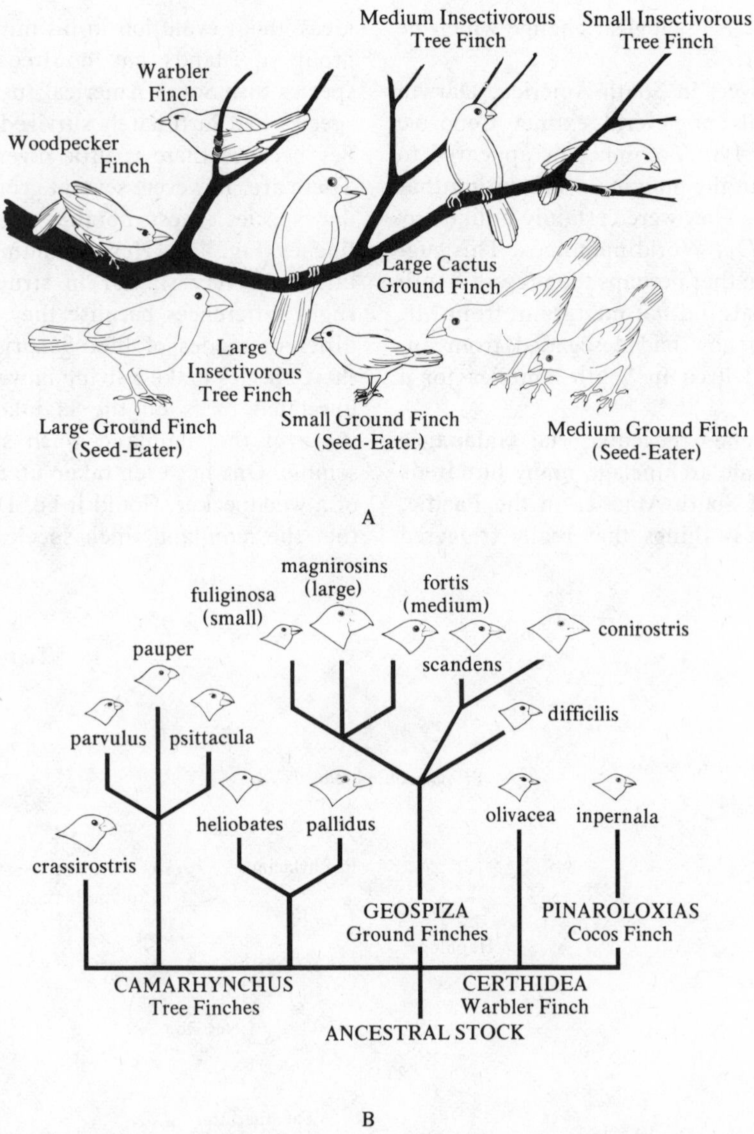

Figure 4.9. *Darwin's Finches.* **A.** *Life habits and* **B.** *Evolutionary relationships* (Adapted from E. Peter Volpe, *Understanding Evolution,* © 1970, William C. Brown Co., Publishers, Dubuque, Iowa.)

foothold here and, in the absence of the competition of other land birds, had evolved to occupy the niches normally filled by other kinds of birds? Certainly it seemed more likely than for the Creator to make all these separate, but similar, Galapagos species when well-adapted, different types of birds on the mainland had already been made.

Another observation Darwin made was that some kinds of animals, particularly the land tortoises, have different species on just about every island. Why should the Creator design all these slightly different separate species, reasoned Darwin? Wouldn't it be more logical to imagine that an occasional tortoise or two got from one island to another and that each tortoise had a slightly different hereditary makeup (it was obvious at this time that no two individuals of any species are ever identical) from the remainder of its original population so that its offspring only contained and perpetuated its peculiarities? Eventually, its isolated descendants would be so different from the original population that they would form a separate species.

Darwin's Theory

These observations and many others set Darwin to exploring further the possibility of evolution upon his return to England. Darwin gathered data from every conceivable source relative to the problem, including information on comparative anatomy, embryology and later developmental stages, vestigial structures, artificial breeding, biogeography, paleontology, and geology. Finally, in 1859, appeared his monumental work, *The Origin of Species.* Darwin not only demonstrated successfully that evolution had occurred, but also offered an explanation.

Darwin's basic argument was as follows. All species of organisms have very high reproductive rates, yet populations appear to remain about the same size in the long run. Therefore, there must be a competition for survival among the offspring because most do not survive. There is variation among the members of a species population; no two individuals have identical characteristics. Therefore, those that survive must have characteristics that give them an edge in the struggle for survival. In Darwin's theory, the organisms do not evolve as Lamarck envisioned, but populations change in composition with time because those individuals with better

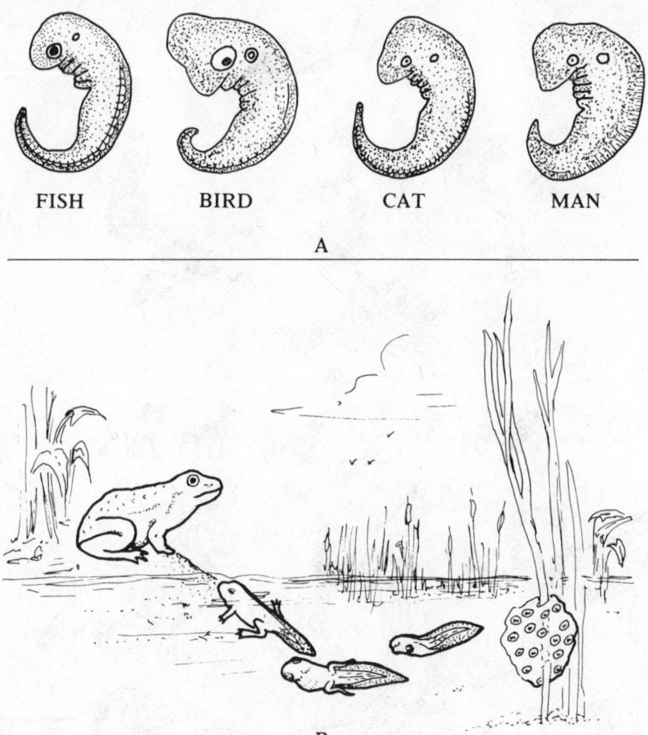

Figure 4.10. *The Biogenetic Law.* **A.** *The early stages of all vertebrates greatly resemble one another.* Note the presence of gill slits even in the nonaquatic forms. **B.** *The development (ontogeny) of a frog duplicates its evolutionary history (phylogeny).* The tadpole larval stage lives essentially as a fish.

adaptations stand a better chance of surviving and reproducing. In a way, then, Darwinism is the reverse of Lamarckism. The organisms are, in a sense, passive because all they do is vary, and the impulse for evolution comes from the environment which does the selecting. Darwin called his theory of the cause of evolution the **theory of natural selection.** In the main, his theory is the basis of our own modern one, although it is modified because Darwin did not understand why organisms vary or how characteristics are transmitted from parents to offspring.

Darwin's Evidence

We will only briefly summarize Darwin's evidence here. In **comparative anatomy,** he pointed to homologous structures (Fig. 4.4) which occupy the same position in the organization of

A

B

Figure 4.11. *Selective Breeding.* **A.** *Some of the many varieties of pigeons produced by selective breeding.* **B.** *Several of the many varieties of dogs produced by selective breeding.*

different animals, yet are modified differently to perform different functions. As noted earlier, the best explanation for such structures is that they were inherited from a common ancestry.

In embryology and later **ontogenetic** (developmental) stages, Darwin noted that many creatures pass through stages structurally similar to more primitive organisms. For example, the embryo of a human being (Fig. 4.10) contains gill slits like those of fish, yet they are never used and are soon lost. Darwin stated that the only logical reason for this recapitulation is that people had had fish ancestors at one time and had not yet lost all their traits.

Vestigial structures (Fig. 4.5) inherited from ancestors where they were functional could also be best explained by evolution. Their existence was in direct opposition to the theological notion that a Creator made each creature perfectly adapted to its environment, for here were useless structures.

Darwin pointed to the great changes produced by **selective breeding** of domestic animals such as pigeons, dogs, and sheep (Fig. 4.11). By breeding only individuals with certain desirable traits, people were able to produce structural modifications in an original species as great as those between a Chihuahua and a Great Dane. Just imagine what nature could do over a much longer period even if it were much less selective!

Biogeography also supported evolution, for how else could one explain the fact that in isolated regions such as Australia one group of animals, in this case the pouched marsupials (Fig. 4.12), could fill most of the niches occupied by the typical placental mammals elsewhere. If the biblical story were true, marsupials should have the center of their distribution at Mr. Ararat and should not be isolated on one continent.

Darwin pointed out that the **fossil record** (Fig. 4.13) showed a progressive change from simpler to more complex organisms as one went up the geologic time scale. That was what one would expect if organisms had evolved, but would not occur if all organisms had been made at once. Furthermore, the appearance of the different kinds of animals in the geologic record did not match the six-day biblical sequence. Darwin was pessimistic about the chances of ever

Figure 4.12. *The Marsupial Mammals of Australia.* A representative sampling of the unique mammalian fauna of the isolated continent.

finding complete sequences of fossils showing the small changes from one species to the next because of the basically fluctuating nature of deposition and erosion in an area (Fig. 4.14). He drew attention to a second type of unconformity, later called the **disconformity** by Grabau in 1905 (Fig. 4.15), in which the strata above and below are parallel, but a gap in the sequence of fossils indicates much missing time. Darwin felt that these gaps are abundant in the geologic record and precluded the finding of complete evolutionary lineages. As we shall see in a moment, his reasoning about gaps is good, but he was overly pessimistic because several good evolutionary lineages are now known.

Lastly, Darwin pointed to the great length of geologic time, indicated by the principle of **uniformitarianism,** as being vital for the slow transformations he envisioned. Although temporarily stymied on this point by the physicists, he was later vindicated by the discovery of radiometric dating which proved the earth's great antiquity as we shall see in Chapter 6.

Darwin's Problems

The great stumbling blocks for Darwin were that he did not understand why organisms vary and how characteristics are transmitted

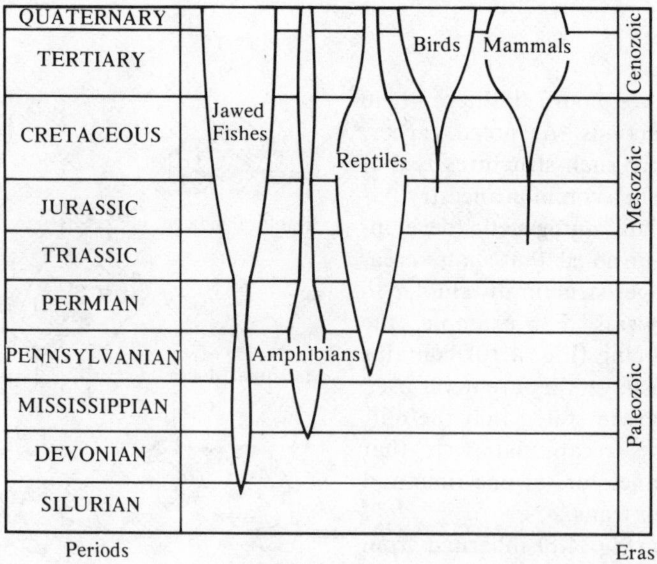

Figure 4.13. *The Fossil Record of the Vertebrates* shows the appearance of more complex types in younger rocks, a fact known even in Darwin's time. The width of the bar is proportional to the known abundance of the group.

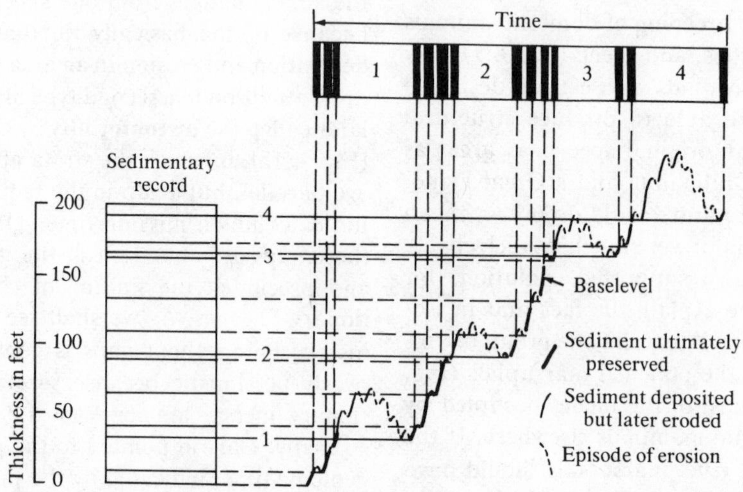

Figure 4.14. *The Basically Discontinuous Nature of Rock Deposition.* The baselevel (line above which erosion occurs and below which deposition occurs) in a sedimentary basin continually fluctuates. The major trend is one of rising baselevel or aggradation, but second- and third-order cycles create a discontinuous record. Where the baselevel rises, sediment accumulates; where it falls, erosion occurs. Solid portions of the curve are ultimately preserved in the sedimentary record. Total amount of time represented in the stratigraphic record at left is shown in black bars at top. Major breaks in deposition are numbered. (From Don L. Eicher, *Geologic Time,* ©1968. Reprinted by permission of Prentice-Hall, Inc., Englewood Cliffs, N.J.)

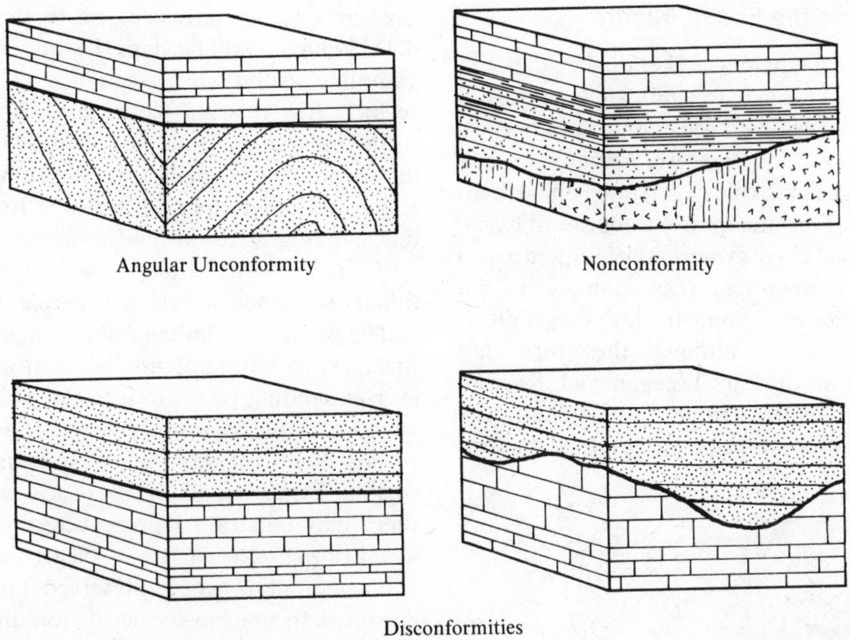

Figure 4.15. *The Three Major Varieties of Unconformity.* Darwin recognized the fact that disconformities exist and are numerous. In angular unconformities, the beds below the unconformity have been deformed prior to deposition of the overlying strata. In disconformities they have not, and the strata above and below are parallel, though the unconformity itself may be irregular. In a nonconformity, the underlying eroded beds are crystalline igenous and/or metamorphic rocks.

from generation to generation. He, like nearly all his contemporaries, envisioned heredity as a blending of parental traits. His critics argued that if this were true, any favorable variation would be swamped out by breeding with those individuals which did not have it. Unless one could envision numerous hereditary changes or **mutations** all in the same direction, no variation could survive. This argument troubled Darwin until his death. At the same time as Darwin's work, an obscure monk, *Gregor Mendel* (Fig. 4.16), had discovered the basic laws of **heredity** by breeding pea plants in a Bohemian monastery. Mendel discovered that the factors which caused individual traits are inherited as separate entities and, hence, retain their identities and are not swamped out. But Mendel's work remained unknown until 1900 when its rediscovery vindicated Darwin. In this century, the science of heredity, **genetics,** has undergone tremendous advances, including the discovery of how spontaneous mutations occur and the chemical basis of inheritance. How far people will go with the control of their own destiny remains to be seen now that they possess this knowledge.

Figure 4.16. *Gregor Mendel* (1822–1864).

Evolution and the Fossil Record

As mentioned, Darwin was skeptical of the ability of historical geology to vindicate his thesis of slow change from one species into another. In the years after 1859, however, numerous unbroken lineages of fossils began to be unearthed. In 1869, the German paleontologists *Waagen* and *Karpinsky* cited evolutionary sequences in lineages of **ammonites.** (An example of an ammonite lineage demonstrating evolution is shown in Figure 4.17, although these forms are not the ones studied by Waagen and Karpinsky.) In the 1870s, *Huxley* in England and *Marsh* in North America discovered the unbroken Eocene to Recent **horse** lineage (Fig. 4.18), which is still perhaps the longest and most complete evolutionary sequence in the fossil record. An Austrian paleontologist, *Neumayr,* in 1875 traced an evolutionary lineage of smooth into ornamented **snails** (Fig. 4.19). In Britain, *Rowe* and others found a series of **heart urchin** lineages (Fig. 4.20) in the Cretaceous chalk. By 1910, *Carruthers* displayed a complete evolutionary sequence of Paleozoic **corals** (Fig. 4.21).

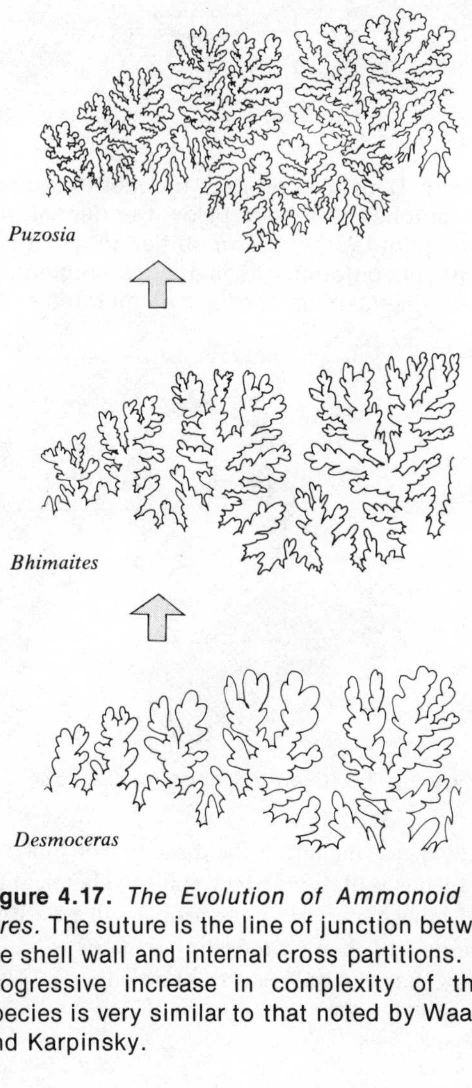

Figure 4.17. *The Evolution of Ammonoid Sutures.* The suture is the line of junction between the shell wall and internal cross partitions. The progressive increase in complexity of these species is very similar to that noted by Waagen and Karpinsky.

The process of finding unbroken evolutionary lineages, including plants, has continued at an ever-expanding pace down to the present. Darwin was still basically correct. The geologic record is full of gaps of all sizes in any one region. Even though in the world as a whole there may be strata representing every year of Cambrian to Recent time, clearly every individual cannot have been preserved if for no other reason than that no species is ubiquitous. As a result, we will probably never know the complete evolutionary lineages of all organisms or even of those likely to be preserved. But enough examples have been found to establish beyond question that the process of evolution occurs.

The Species Problem

One result of the discovery that evolution is indeed recorded in the geologic record by a small series of transformations is that it makes defining a species impossible. Biologists define a **species** as a population or group of populations whose members are capable of interbreeding with one another, but not with members of other species. Such a definition breaks down when the time factor is introduced because then there is no isolation of populations (Fig. 4.22). They are all interconnected through the evolutionary pathway of life. Only when one takes a slice of geologic time, such as the biologist is doing with the present, do populations appear isolated.

Paleontologists have tried to solve the problem by allowing no more variation among the members of a species in a "vertical" time sense than exists among the members of a "horizontal" contemporaneous population. Even this solution fails when one considers that a popula-

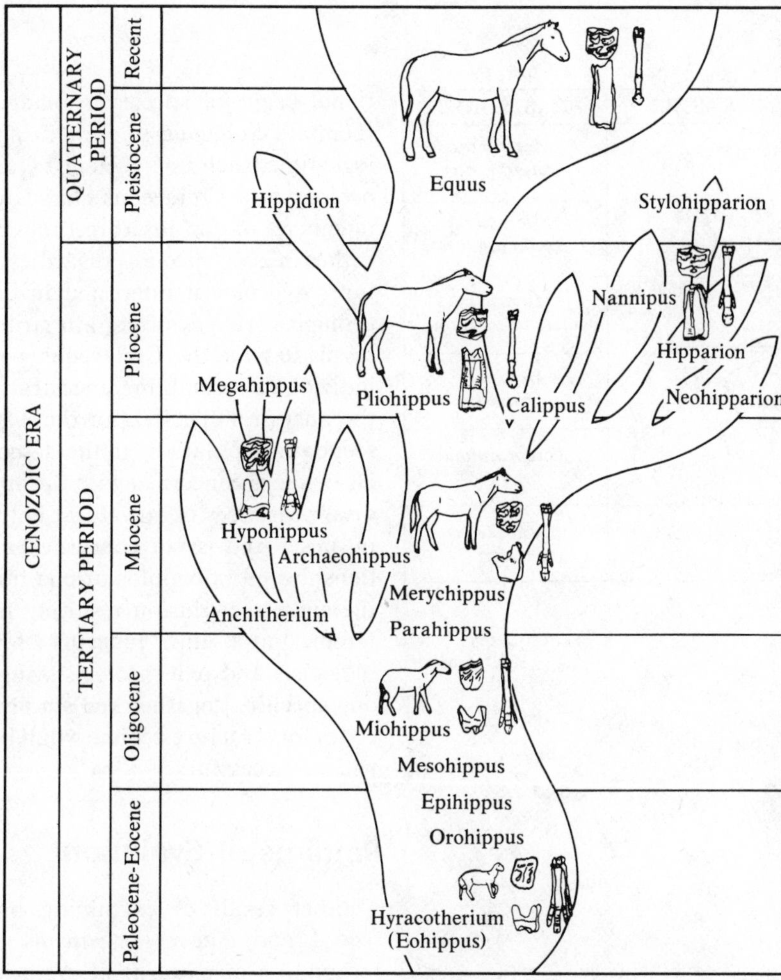

Figure 4.18. *Evolution of the Horse Family.* Limbs, teeth (top and side views), and general appearance of selected genera are illustrated. (After E. Peter Volpe, *Understanding Evolution,* ©1970, William C. Brown Co., Dubuque, Iowa, and Bernhard Kummel, *History of the Earth,* W.H. Freeman & Co., San Francisco, Calif.)

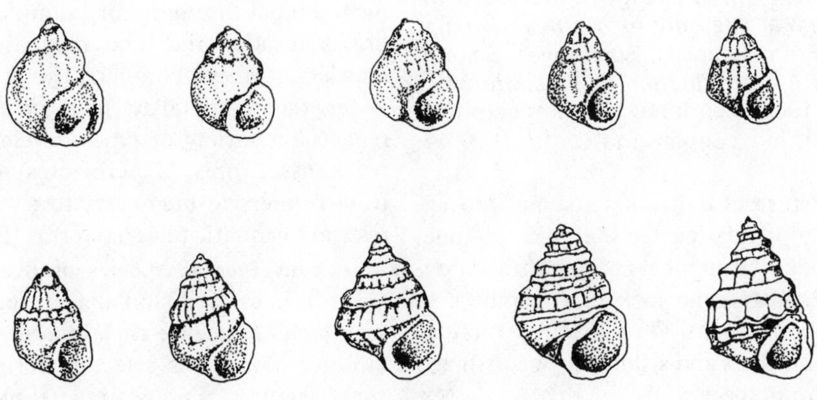

Figure 4.19. *Evolution of the Fossil Snail Paludina.* This trend from smooth to ornamented was noted by Neumayr in 1875.

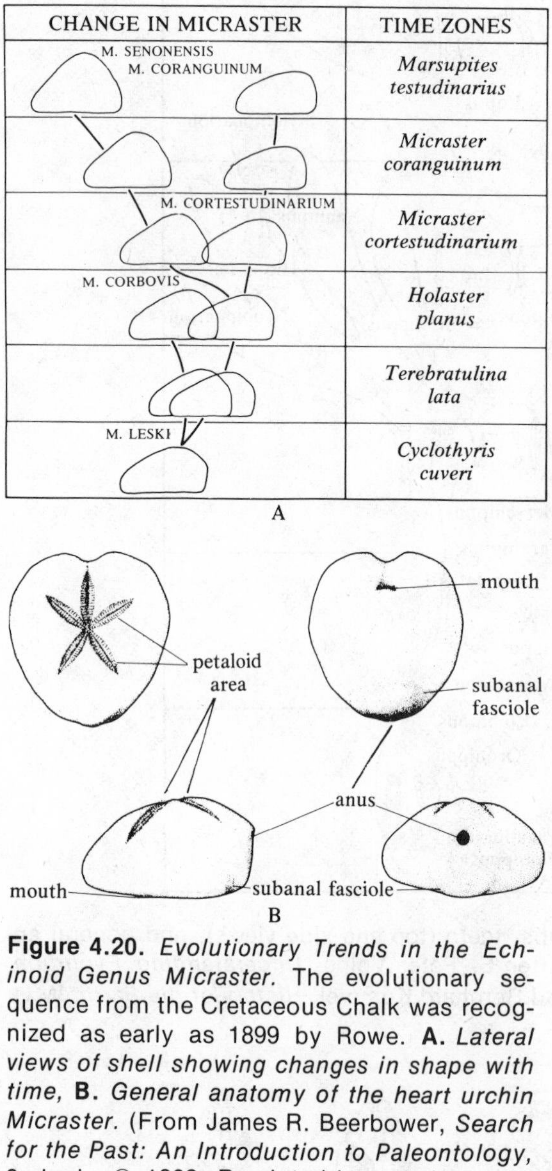

Figure 4.20. *Evolutionary Trends in the Echinoid Genus Micraster.* The evolutionary sequence from the Cretaceous Chalk was recognized as early as 1899 by Rowe. **A.** *Lateral views of shell showing changes in shape with time,* **B.** *General anatomy of the heart urchin Micraster.* (From James R. Beerbower, *Search for the Past: An Introduction to Paleontology,* 2nd ed., © 1968. Reprinted by permission of Prentice-Hall, Inc., Englewood Cliffs, N.J.)

tion is a mixture of individuals, some who are typical of the population for that slice of time, some who look like members of ancestral populations, and others who look like members of descendant populations. On a strictly structural basis, then, brothers and sisters may be assigned to three separate species.

Clearly there are several ways of looking at fossil populations (Fig. 4.23), and the traditional usage of species is inadequate to cover them all. Someone interested in a slice of geologic time, such as a geologist studying the paleocology of a Cretaceous chalk, would lump all members of an interbreeding population together in one "species" regardless of their structure. A geologist interested in establishing the geological ranges of certain structural types of fossils to zone the Cretaceous would assign the individuals resembling ancestral populations to the ancestral species and those resembling descendant populations to the descendant species. He or she is concerned with the appearances and disappearances of structural entities for dating purposes and is not concerned about the relationships of individuals at one time. Because of the endless confusion resulting from these different, but useful, concepts of species, many biologists and paleontologists suggest we abandon species altogether and simply use computer codes for structures. Time will tell if their efforts will be successful.

Patterns of Evolution

Another result of examining both the fossil record and Recent organisms has been the recognition of patterns of evolution. The more significant ones are adaptive radiation, convergence, parallelism, and extinction.

Adaptive radiation is the process of diversification or multiple-branching of a lineage which develops a significant structural and/or environmental breakthrough. For example, when vertebrates invaded the land they diversified into crawlers, runners, climbers, flyers, and so forth. When the mammalian level of evolution was reached, a variety of types developed, ranging from insectivores to herbivores to carnivores, from runners to burrowers to flyers, from terrestrial to aquatic to aerial forms (Fig. 4.24).

In **convergence,** species of distant relationships have evolved similar features and, hence, look similar because they inhabit the same environment which is selecting for generally the same features. For example (Fig. 4.25), fish, whales (mammals), penguins (birds), and the extinct icthyosaurs (reptiles) resemble one an-

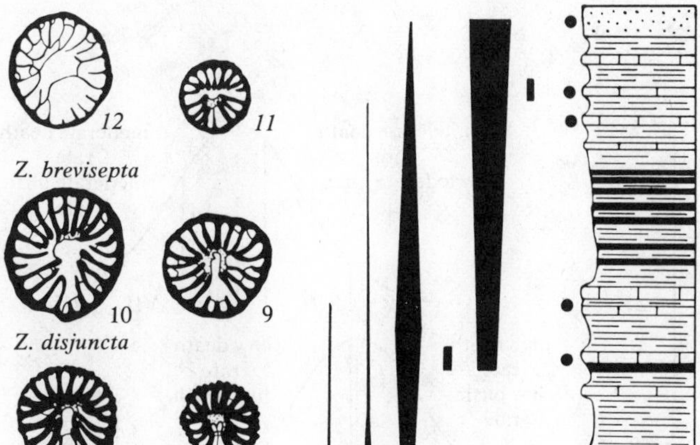

Figure 4.21. *Evolution and Recapitulation in a Lineage of Fossil Corals.* Cross sections of the coral skeletons are shown in 1–12. The even-numbered sections are adult stages showing evolution of internal structures. The odd-numbered sections are the immature stages. Note their resemblances to the adults of the preceding species. On the right are shown the stratigraphic ranges of the various species. Dots are collecting levels. (From *Invertebrate Paleontology,* by William H. Easton, © 1960, Harper & Row Publishers, Inc., Fig. 5.6, "Evolution of *Zaphrentites delanouei,*" after Carruthers, 1910, *London Quarterly Journal.*)

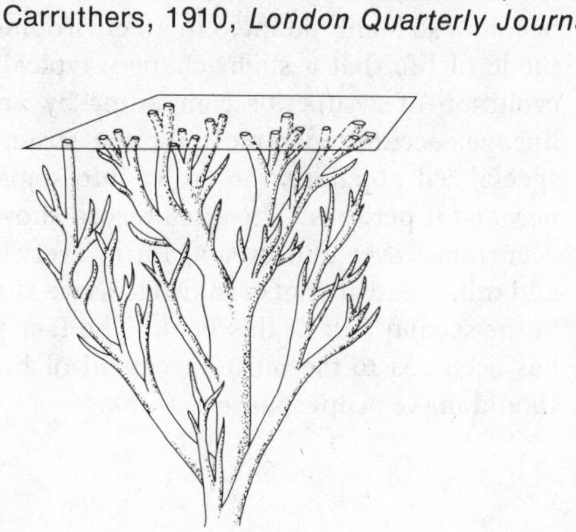

Figure 4.22. *A Portion of the "Bush of Life"* showing the lack of discontinuity in the organic world when viewed from the geologic perspective. The biologist deals only with a slice of time, the present (top of diagram); and species appear distinct. In the geologic past, however, all intermediates existed and may be preserved.

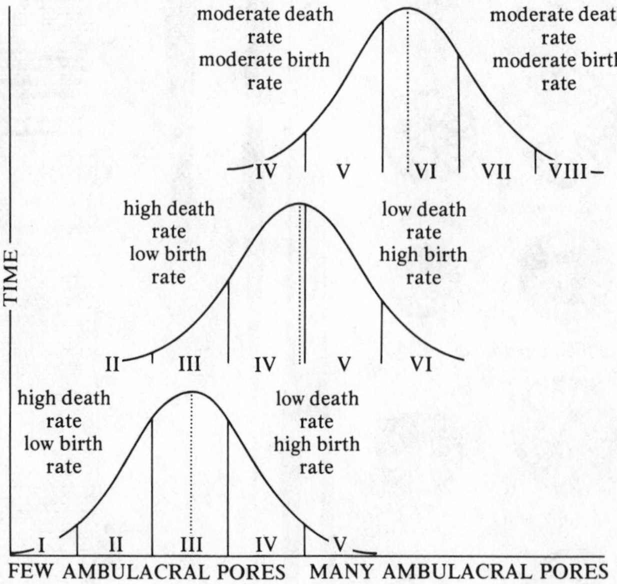

Figure 4.23. *The Species Problem in Paleontology.* Each Roman numeral represents one variation of a particular trait, the number of pores for tube feet in a petal of the heart urchin *Micraster.* Each successive population has a different mix of individuals and a different average. Each population could be defined as a separate species, but individuals with IV and V pores occur in all three populations. Because successive populations overlap in morphology, assignment to a species is purely arbitrary and based on the purposes of a worker as noted in the text. (From James R. Beerbower, *Search for the Past: An Introduction to Paleontology,* 2nd ed., © 1968. Reprinted by permission of Prentice-Hall, Inc., Englewood Cliffs, N.J.)

other as do birds, bats (mammals), and pterosaurs (reptiles); yet all are only distantly related.

Parallelism is similar to convergence in that species of different lineages tend to look alike. But in this case, the lineages are closely related. For example, the horses and the extinct palaeotheres are members of the same order of mammals and have a close common ancestor (Fig. 4.26). Both lived in the same environment and, hence, went through a series of similar evolutionary changes to better adapt to that environment.

Finally, **extinction**, or the final termination of a lineage, is the lot of most evolutionary lines sooner or later. (We will exclude from extinction the disappearance of a species because it evolved into another species.) Sooner or later, a lineage becomes so highly adapted to an environment or mode of life that a slight change, typically the evolution of a superior competitor by another lineage, occurs too rapidly for the structurally specialized population to evolve into something new and it perishes. The fossil record shows this occurrence over and over again in every major and minor category of organisms, as we shall see in the second half of this book. The fact that it has occurred to the most successful of lineages should make people pause to think.

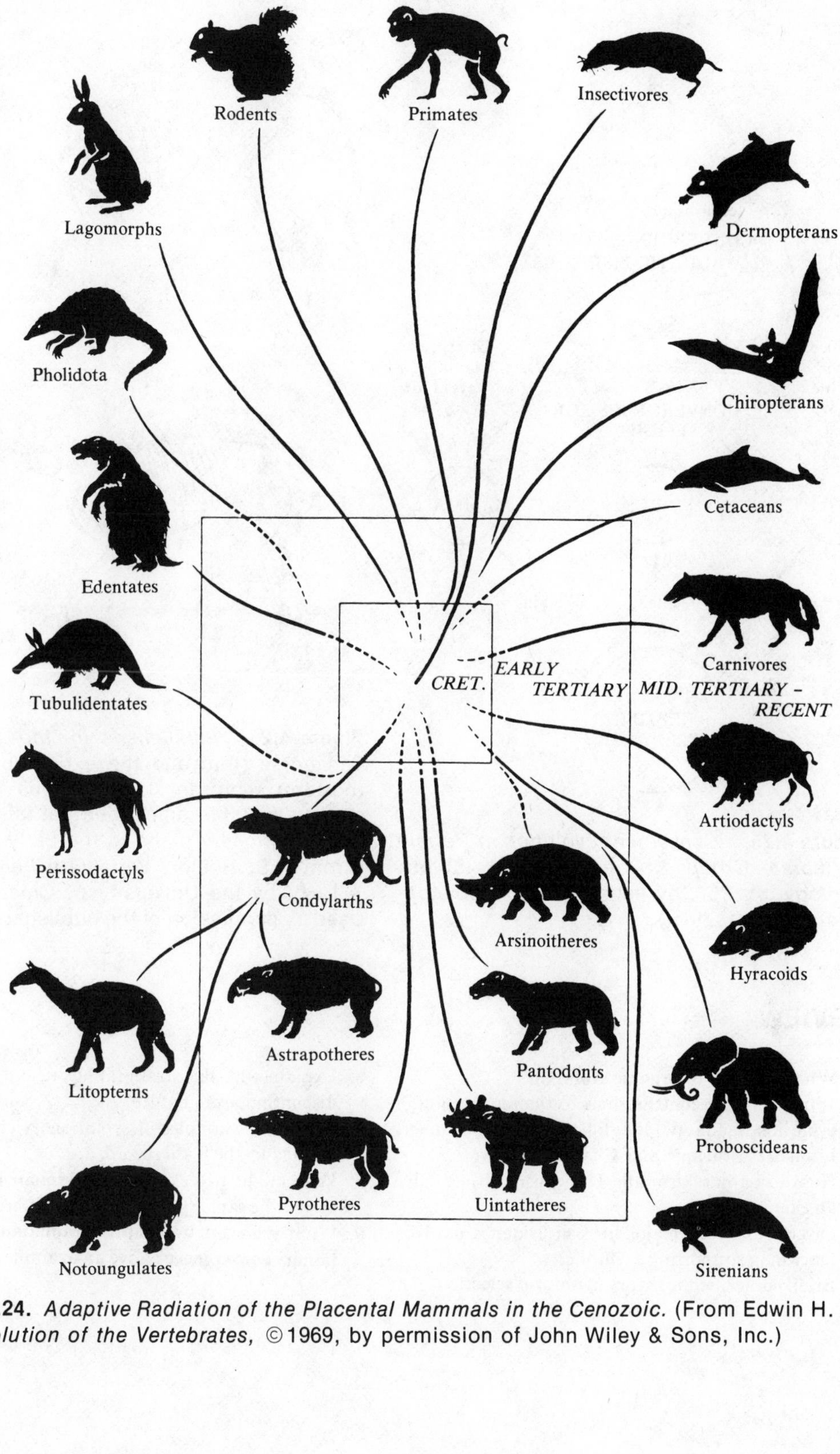

Figure 4.24. *Adaptive Radiation of the Placental Mammals in the Cenozoic.* (From Edwin H. Colbert, *Evolution of the Vertebrates,* ©1969, by permission of John Wiley & Sons, Inc.)

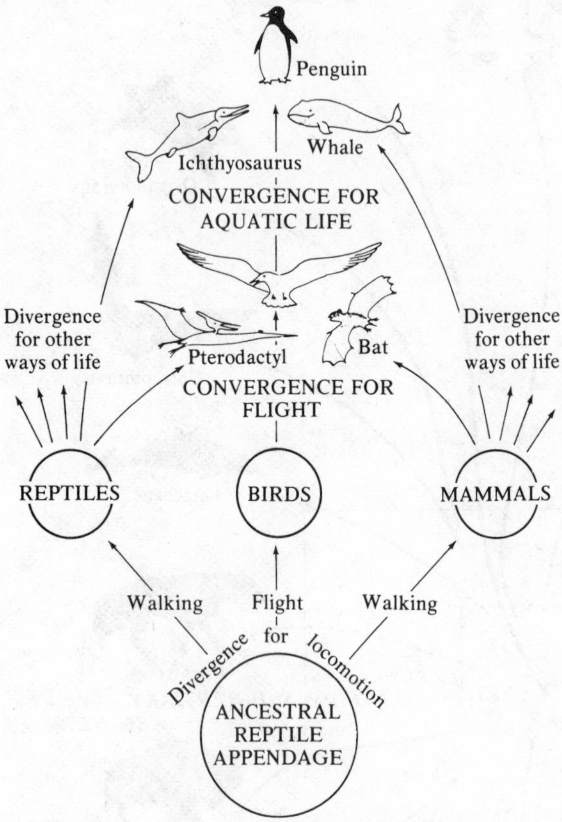

Figure 4.25. *Convergent Evolution in Certain Vertebrate Groups.* (After Alfred M. Elliott, *Zoology,* © 1957, by permission of Appleton-Century-Crofts, Inc.)

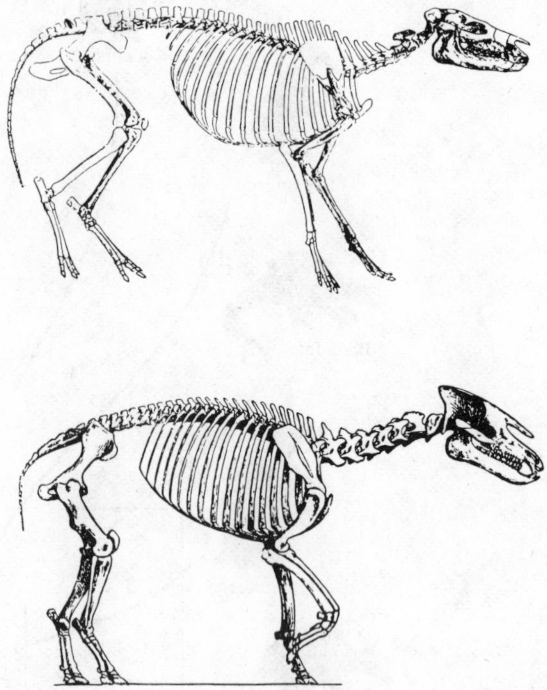

Figure 4.26. *Parallelism in the Mammals.* The horse (top) and the extinct palaeothere (bottom) represent two lineages of hoofed mammals of the same order that followed similar evolutionary paths in the Early Cenozoic. (From A.S. Romer, *Vertebrate Paleontology,* © 1966, by the University of Chicago Press. Used by permission of the publisher.)

REVIEW

1. What is meant by **organic evolution?**
2. What were the contributions to the recognition of evolution made by the following men: Cuvier, Linnaeus, Erasmus and Charles Darwin?
3. How do Lamarckism and Darwinism differ? Give an example.
4. Discuss any four major lines of evidence used by Darwin in supporting evolution.
5. Briefly describe the **theory of natural selection.**
6. Explain why the stratigraphic record is typically discontinuous in nature.
7. Cite some examples of evolutionary change documented by the fossil record.
8. Why might two experts on Permian brachiopods assign the same specimen to two different species?
9. What is meant by **adaptive radiation** and **evolutionary convergence?** Give an example of each.

5

The Dichotomy between Rocks and Time: Formations versus Stages

5

In Chapter 3, we noted that it took many centuries for geologists to recognize there are at least two major ways of subdividing the geologic record. One is on the basis of the types of rocks (lithology) which are present, and the other is on the basis of the times of formation of the various rocks (chronology). At first, it was not apparent that there is any difference between the two. In Wernerian geognosy, for example, each type of rock formed at one particular time in the past over the entire world; hence, rock type and time of formation coincided. Geologists soon demonstrated, however, that rock layers are not worldwide. With the recognition of facies by Gressly and others, it became apparent that several varieties of rock could form at the same time. Hence, lithology was primarily a guide to environment and not age.

Stages

In the early 1800s, a multiplicity of rock units were named and described on the basis of their lithologies. The great French geologist *Alcide d'Orbigny* bemoaned this endless proliferation of terminology and suggested that a universally applicable term was needed to encompass all rocks deposted during the same time interval, regardless of their lithologies. D'Orbigny proposed the term **stage** for this purpose in the 1840s and defined a series of stages for the rocks of Jurassic and Cretaceous ages. These stages were based chiefly on wide-ranging species of ammonoids that were found in numerous facies. D'Orbigny was a catastrophist and believed that the stages were bounded by natural breaks in the record.

Zones

The German geologist *Albert Oppel* was not a catastrophist, however, and he noted that by careful fossil collecting from numerous facies over a wide area, he could piece together a more or less complete time sequence. Species did not all appear and disappear as abruptly as d'Orbigny thought, but instead there was a continuous replacement of old species by new. By utilizing the overlapping ranges of fossil species (Fig. 2.20), Oppel was able to define a continuous sequence of small intervals of strata each deposited at the same time and characterized by a unique assemblage of fossils which did not occur earlier or later in the record. He called these units of rock based on their time of formation, **zones.** Several successive zones that appeared to have much in common paleontologically were assembled into stages.

Rocks versus Time

In examining a simple cross section, such as Figure 5.1, showing one type of rock grading laterally into another, one cannot comprehend the great difficulties that still awaited geologists in separating these two aspects of rocks, their lithology and their time of formation as indicated by fossils. In areas such as the great inter-

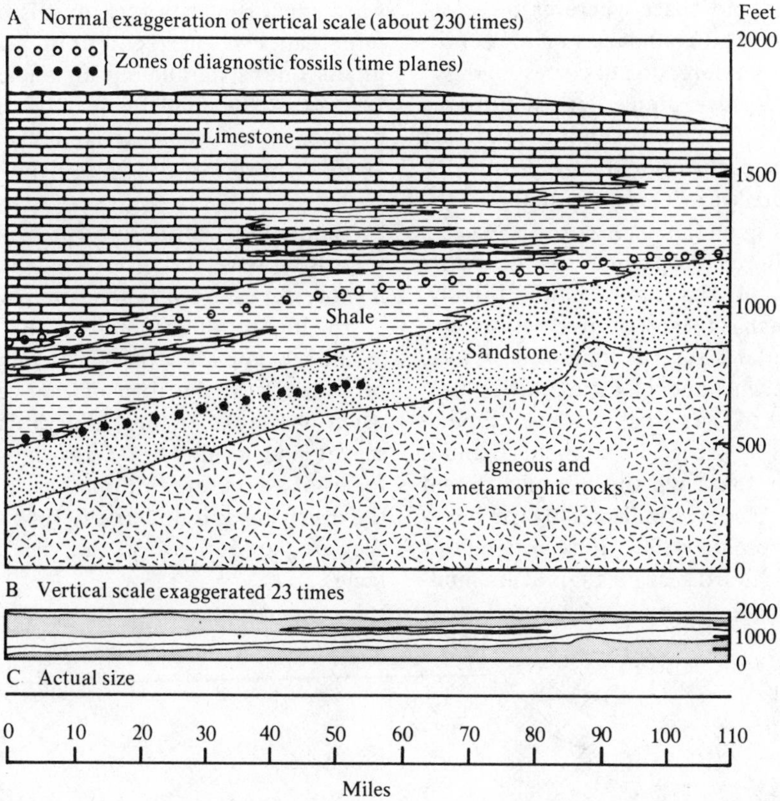

Figure 5.1. *Effect of Vertical Exaggeration.* Cross sections illustrating the difficulties encountered by geologists in separating lithologic and time aspects of rocks in areas of widespread, nearly flat-lying strata. (From Don L. Eicher, *Geologic Time,* © 1968. Reprinted by permission of Prentice-Hall, Inc., Englewood Cliffs, N.J.)

ior plains of North America, which at times in the geologic past were covered by immense epicontinental or epeiric seas (much like gigantic continental shelves), there are widespread rock units which appear essentially horizontal as one sees them at any one locality. It was very easy for early geologists to envision each of these layers as an essentially contemporaneous deposit. The accurate positions of fossils within these rocks were generally not noted in detail as done by Oppel in Europe. It was typically not reported whether diagnostic fossil species occurred lower in the rock unit near the margins of the deposit and higher near the center of the continent; the whole rock unit was one integral entity entirely deposited in the same time span. This hardly seems logical when one considers that the seas must have spread from the edges toward the center of the continent, and this must have taken some sizable amount of time. Thus, any type of sediment, such as a near-shore beach sand, must have occupied only a relatively narrow band at any one time. But until recently, the majority of North American geologists operated with an almost Wernerian approach to geology. To them, the boundaries of rock units were also the boundaries of time units, for each deposit was contemporaneous throughout.

Proceeding on both observational and logical grounds, *H.S. Williams* pointed out in 1894 that lithologies were nothing more than the record of a particular geologic environment and migrated in time and space wherever these environments went. The boundaries of rock units and time units, therefore, do not correspond.

As a hypothetical example, envision a sea transgressing over a gently sloping landmass furnishing clastic particles of all sizes (Fig. 5.2). Nearshore, where wave energy is highest, the coarsest particles, in this case pebbles, occur. Farther from shore, the particle size decreases through bands of sand and mud. Finally, beyond the reach of land-derived clasts are chemical precipitates such as limestone. As the sea advances, each band of facies moves progressively inland so that eventually each type of sediment leaves a sheet of sediment behind. Notice, however, that each sheet is of different ages in different places as indicated by the heavy time lines on the illustration. It is appreciably younger on its seaward margin than at its landward edge. Hence, a rock unit is of slightly different ages as one goes from place to place, unless one is paralleling an ancient shoreline. Instead of the sand being all formed contemporaneously, various parts of it are the same age as adjacent rock types. Note also that with time, each deposit of a transgressing sea is overlapped by the deposits of more offshore sediments. In regressing seas, the reverse is true as the more inshore deposits overlap the more offshore ones.

An important generalization, often called **Walther's law** after the German geologist who proposed it at the turn of this century, arises when one looks at these diagrams. In order to see at one locality what the lateral lithologic facies of a rock are, all one has to do is look at the overlying and underlying rock types because they are part of the very same rock units as those facies.

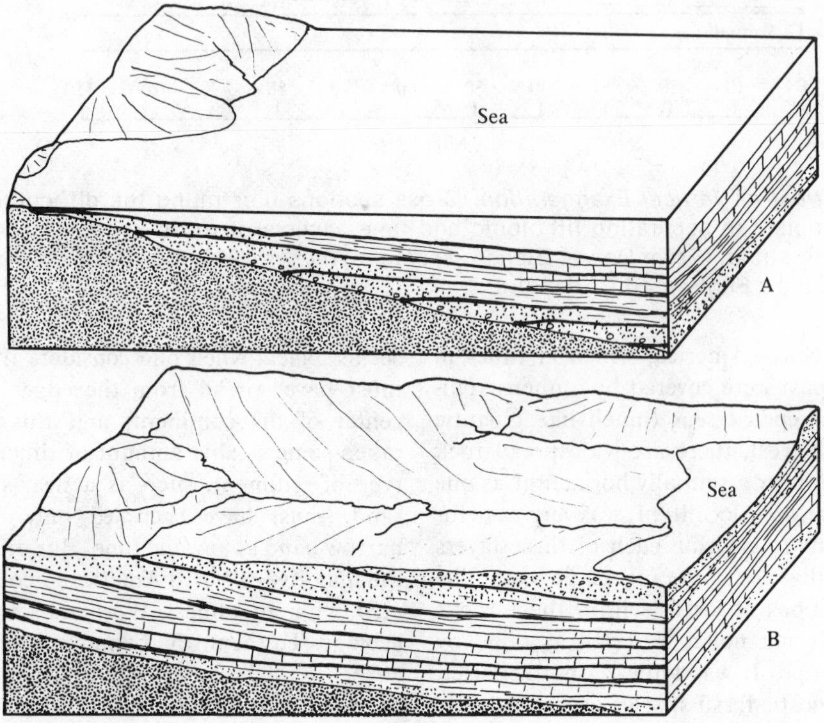

Figure 5.2. *The Diachronous (Time-Transgressive) Nature of Rock Units.* Heavy lines are time lines. Rock units are indicated by symbols: circles for conglomerate, dots for sandstone, dashes for shale, and boxes for limestone. **A.** *Onlap pattern produced by advancing sea,* **B.** *Offlap pattern produced by retreating sea.* Note time lines cross all rock units.

A contemporary of Williams, the noted geologist and Colorado River explorer, *John Wesley Powell,* who then directed the United Geological Survey, proposed usage of a separate lithologic term, **formation.** This term was to be used to clearly distinguish these units of variable age but constant environment from stages and other rock subdivisions based on times of formation. As we have seen in an earlier chapter, however, most leading North American geologists of the time would have none of this reasoning. To them, the alternate rise and fall of the seas were worldwide and essentially instantaneous. Thus, every rock unit was the same age everywhere, a throwback to Werner. Once a fossil giving a certain age was found in a rock unit, the age of that entire unit was fixed everywhere. Different lithologic units always meant different ages; they could not have formed contemporaneously in the same region.

Codes of Nomenclature

Both international and American geological commissions were established in the late 1800s and early 1900s to bring some order out of this chaos of terminology. The international commission (1900) came up with a dual hierarchy of time terms (era, period, epoch, age, and phase in descending order of magnitude) and rock terms (group, system, series, stage, and zone in order of decreasing magnitude) based on the time formation of the rocks as defined by their fossils. Thus, there was an exact one-to-one correspondence between the two categories of terms. One set of units was for pure time; the other applied to rocks deposited during that time. For example, the Devonian System encompasses all rocks deposited during the Devonian Period. There was no formal terminology for lithologic terms, so chaotic usage continued.

The American Commission (1933) came up with a dual set of terms also, but there was an important difference. On one hand there were units of pure time (era, period, and several levels of epochs); and on the other there were rock units (system, series, group, formation, member, and bed) based on the time of formation of the rocks, which was determined by their lithology. In other words, a formation was a lithologic unit that was all deposited during a second-order epoch. Lithology and time were firmly wedded to each other; and most American geologists went their own way, ignoring Gressly, Williams, Powell, and Grabau.

Such obvious contradictions could not endure forever, and beginning with the 1930s and continuing to the present, the trend has reversed itself. American geologists once again talk of stages and formations as different things although theory and practice are often still at variance. The movement back toward recognizing the fundamental differences between units of rock based on time and those based on lithology had as its leaders *Hubert Schenck, Siemon Muller, H.D. Hedberg,* and *Robert Kleinpell.* In a series of papers in the 1930s and culminating in the 1941 work of Schenck and Muller, they argued convincinly that at least three separate sets of terms are needed (Fig. 5.3). First

Geologic-Time Units	Time-Stratigraphic Units	Rock-Stratigraphic Units
Era Period Epoch Age	System Series Stage Zone	Group Formation Member, etc. Bed, etc.

Figure 5.3. *Table of Stratigraphic Units Proposed by Schenck and Muller* (1941).

there are the units of pure time, such as period, epoch, and age; and corresponding to these exactly are units of rock defined on the basis of their time of formation, such as system, series, and stage. (Of course, the time-rock units must be recognized first; they are the material basis on which units of time are established.) Written at right angles to these parallel sets of terms are the lithologic units, group, formation, member, and bed, in an effort to clearly denote the fact that these sets of units in no way correspond to the time-based units, but instead transgress time in virtually all instances. In 1961, the American Commission adopted a new code of units embodying these principles and some additional ones. The International Code of the same year and its revisions are similar.

Modern Stratigraphic Terminology

The new code recognizes four major categories of terms used by those geologists who studied layered or stratified rocks, the stratigraphers. These are the **lithostratigraphic (rock-stratigraphic) units** based solely on lithology and having no necessary time relationship, the **chronostratigraphic (time-stratigraphic) units** based solely on the time of formation of the included rocks, the **geochronologic (geologic time) units** which are subdivisions of pure time recognized on the basis of chronostratigraphic units, and finally, a new set of terms called biostratigraphic units.

Biostratigraphic units are units of rock defined solely on their fossil content without any necessary time connotations. This may seem strange at first, when we have been saying all along that fossils are the basis of chronostratigraphic units. There is an important difference, however. The presence of fossils in rocks is an objective thing, just as the presence of certain lithologies defines lithostratigraphic units. Time, however, is an interpretation made from the rocks; it cannot be seen or touched. We have already seen that lithologic identity or continuity has no time significance except for certain rock bodies such as ash beds and turbidite beds that represent a single depositional event. In other words, the major significance of lithology is in recognizing environments and not time.

We have seen that fossils have both environmental and time significance. A certain assemblage of fossil organisms that lived together in an interdependent fossil community has chiefly ecologic significance at first glance. It, like sediment type, will shift about wherever the proper environment is situated for the moment. However, we must not forget the important factor of evolution. No group of populations stays static for very long. Before much time passes, at least one and typically several species in a community evolve into something else or become extinct. Or a new species may migrate in from elsewhere when a distant barrier disappears, thus altering the composition of the community. As a result, there can never be exactly identical communities of organisms living at separate times in the geologic past.

Also, it is a statistically impossible event for a species to evolve twice, considering the vast number of coincidental hereditary changes that would have to occur. Therefore, the time of existence of a species is a unique biological event in time. Migrations into an area and extinctions are other time-significant and nonrecurrent biologic events that characterize the appearance and disappearance of fossils, but not rock types.

Thus, the simple occurrence of fossils is an objective thing. The interpretive processes of stripping away the ecologic and other factors that influence the distribution of fossils to get at their time significance produces chronostratigraphic units from this raw data. For example, the occurrence of a species of brachiopod that ranges through two and a half geologic periods has little more than general time significance and just records the presence of the well-adapted species whenever the environment was right. On the other hand, a species of ammonite belonging to a rapidly evolving lineage may delineate a span of time as short as one sixty-second portion of a period. As another example, we can consider the appearance or disappearance of a species at a formation boundary as being related to a change in the environment of a region and of no time significance. On the other hand, the extinction of a species or its evolution into another species within the same environment is an event of great time significance (Fig. 5.4).

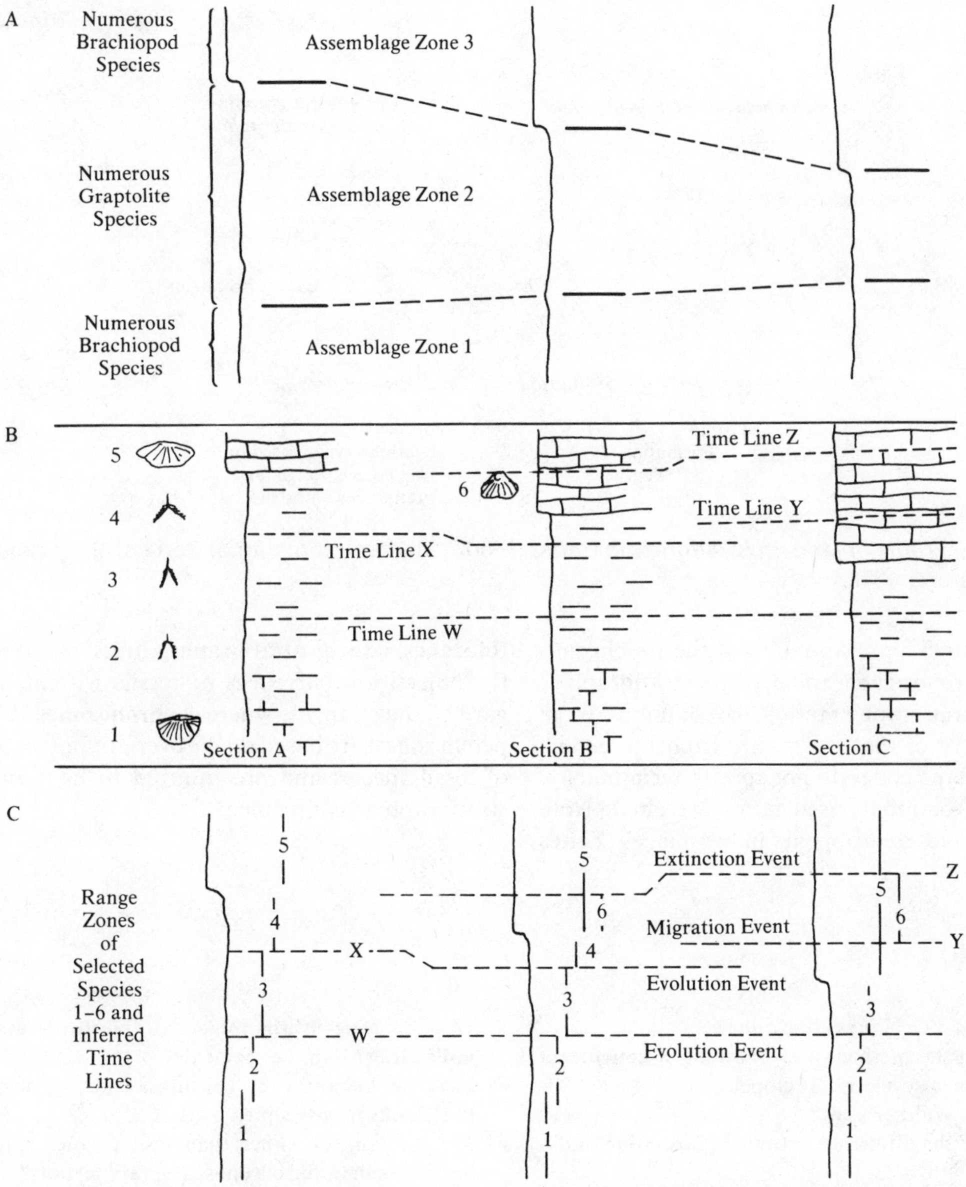

Figure 5.4. *Diagram Illustrating the Relationships between Objective Biostratigraphic and Rock-Stratigraphic Terminology and Inferential Time-Significant Events.* **A.** *Biostratigraphic assemblage zones defined by three separate communities.* Note correspondence with rock types shown in Part B. **B.** *Four time lines based on beginnings or ends of fossil species illustrated at left margin.* Note that only appearances or disappearances within rock units have time value. The others are controlled by the shift in environment recorded by the formation boundaries. **C.** *Ranges of the five species with interpretations of the origins of the time lines.* Lines W and X represent the evolution of one graptolite species into another. Line Y represents the migration into the area of a species from elsewhere. Line Z represents the extinction of a brachiopod species. (From Don L. Eicher, *Geologic Time,* © 1968. Reprinted by permission of Prentice-Hall, Inc., Englewood Cliffs, N.J.)

Geochronologic or Geologic Time	*Chronostratigraphic or Time-Stratigraphic*
Eon	Eonothem
Era	Erathem
Period	System
Epoch	Series
Age	Stage
Chron	Chronozone

Lithostratigraphic or Rock Stratigraphic	*Biostratigraphic*
Group	Biozones
Formation	(various types based on
Member	occurrence, range, and
Bed	abundance of fossils)

Figure 5.5. *Table of Modern Stratigraphic Units.* (Modified after International Code of Stratigraphic Nomenclature, 1972.)

In Figure 5.5 are summarized the geochronologic, chronostratigraphic, lithostratigraphic, and biostratigraphic terms in common usage today. Most of these terms are from the codes, but where the codes do not specify terminology, the most commonly used terms are cited. Note that the word *zone* appears in two places. **Zones** (**biozones**) are biostratigraphic units based on the objective occurrences of fossils without regard to their ranges, whereas **chronozones** (Oppelian zones) are based on the overlapping ranges of fossil species and are inferred to have time stratigraphic significance.

REVIEW

1. How did d'Orbigny define a **stage?**
2. Explain the method of establishing a **concurrent range zone** developed by Oppel.
3. What is **Walther's law?**
4. What is the difference between a **formation** and a **stage?**
5. What was the major contribution of Schenck and Muller?
6. How do **time-stratigraphic** and **biostratigraphic units** differ? Can they coincide?
7. List the hierarchy of **geochronologic** (geologic-time) **units** now accepted by the codes.
8. Are there means other than fossils which might be used in defining chronostratigraphic units?

6

"No Trace of a Beginning—No Prospect of an End": Absolute Time in Geology

6

With superposition, cross-cutting relationships, inclusions, and faunal succession, geologists had the tools to establish the relative geologic time scale of Precambrian through Recent. They could determine the proper sequence of rocks and the order of events in early history. This task still absorbs the work of many geologists worldwide, not only in improving the resolution of detail in known regions, but also in deciphering the history of new regions and incorporating their history into the world pageant. Indeed, for many geologic purposes, this type of dating is all that is necessary. At the same time, however, geologists have also been concerned about absolute time; that is, how many years ago did a particular event occur? Just when did the earth begin? How long was the Devonian Period? How many years does it take one species to evolve into another or for a foot of sandstone to form? How does one date unfossiliferous strata? And so on.

Early Ideas

The early writings of many religions ascribe various lengths to earth history, ranging from relatively very short to rather immense spans of time. The Christian religion is intimately bound with the history of Western Man, in whose culture the modern science of geology evolved. To many scholars, the emphasis of Christianity that the earth was made for people to master through knowledge is responsible for the development of that unique Western contribution to civilization we call science.

It is natural then, that Western people first turned to their theologians to answer questions about the antiquity of the earth. In the 1600s, biblical scholars reasoned that if they worked backward through the genealogies in the Bible, which generally give the life spans of individuals and sometimes the birth years of their children, they could, with reasonable extrapolations for the gaps, arrive at the time of the creation of Adam and Eve and of the world. In 1644 Bishop Lightfoot fixed the time at 9:00 A.M. on September 17, 3928 B.C. Archbishop Ussher, in 1658, determined a slightly different date of October 23, 4004 B.C. In general, the dates were in close agreement and indicated that the earth was less than 6,000 years old. Marginal notations of the dates were made in many editions of the Bible, making them literally the "gospel truth." Opposition to them was considered heresy.

In the mid 1700s, the great French scientist *Buffon* suggested that the earth must have initially been molten because of the presence of volcanoes and the fact that temperatures increased in deep mines. If so, using the then-known data on the earth's volume, its present temperature, and its inferred molten temperature, Buffon reasoned that it must have taken the earth 75,000 years to cool to its present state. Buffon had to recant, and although we would tend to discount his postulates today, he did open the door to consideration that the earth was much older than the biblical estimate.

The Application of Uniformitarianism

With the rise of uniformitarianism because of Hutton and Lyell's work and the validation of biological evolution by Darwin, it became apparent to scientists that vast lengths of time were involved in the geologic past. But how much, no one yet knew. It remained for someone to measure the rates of present processes and then extrapolate these to the history that was recorded in the earth's rock record.

Three classic approaches to the problem were tried in the second half of the nineteenth century, and all disagreed for various reasons. One method involved the salinity of the oceans. The oceans were assumed to originally have been fresh water. Scientists reasoned that if they could correctly estimate the amount of salt being carried to the oceans each year by rivers and the total amount of salt now in the oceans, they could determine the earth's age by dividing the former figure into the latter. In 1899, *John Joly* made this calculation and came up with an estimate of 90 million years. That was certainly a large figure and lent credence to the uniformitarian's contention of vast amounts of time. It was, however, less than Darwin had thought, and we now know it is far too low because of many unconsidered factors. Joly had no reliable notion of the vast amounts salts lost from the oceans by evaporation of sea water, winds blowing salt inland, the formation of saline sediments, and the uplift of saline rocks to form land again. It also seems possible that the amount of salt presently being added to the oceans may be slightly higher than normal for the past because the continents may be higher than in much of the past; hence, erosion rates are probably greater. It now appears that the salinity of the oceans has been essentially constant throughout much of geologic history once the factors (Fig. 6.1) reached equilibrium with one another.

Other geologists reasoned that if they could determine the average rate of accumulation of a foot of sediment and then measure the total thicknesses of the geologic record, they could determine the age of the earth by dividing the former figure into the latter. This notion, actually suggested by the ancient Greek Heredotus, was fraught with even greater uncertainties. Geologists recognized that each type of sediment accumulates at different rates, that even one type of sediment accumulates at different rates under different conditions, that the geologic column in any area is full of gaps which may represent as much time as the rocks themselves, and that there was some uncertain amount of compaction which occurred in the conversion of sediment to sedimentary rock. In addition, geologists kept discovering newer and thicker columns in other regions; so estimates constantly had to be revised. Between 1860 and 1910, many great geologists made estimates of the length of earth history based on this method (Fig. 6.2). They ranged from a low of 3 million years to a high of a billion and a half years, but most were between 80 and 100 million years. Again, these figures indicated great lengths for geologic time, but not as much as some uniformitarians had guessed.

It was from the realm of physics, however, that the most authoritative estimates of the length of geologic time came in the late 1800s. Physics has always been considered the most basic and precise of the sciences and, therefore, the opinions of its practitioners have always carried heavy weight in scientific circles. *Lord Kelvin,* the foremost physicist of this time, arrived at his estimates of the earth's age using two approaches. One was the old assumption of Buffon that the earth had originally been molten and cooled to its present condition. Kelvin realized the limitations of his data and calculations much better than did Buffon of his, but it was still obvious that the maximum amount of time was less than 100 million years and probably as low as 25 million years. This was much less time than evolutionists and uniformitarians guessed and also indicated that even less time was available for biologic evolution because the crust was too hot early in its history to support life.

The other method of attacking the problem was based on the assumption that the source of the sun's energy was of a conventional sort; the sun was either a burning incandescent body or released energy by the gravitational contraction of its immense mass. Either way, it could not continue to produce energy at its present fantastically high rate for very long, for its heat would soon be exhausted. The sun, then, could only have illuminated the earth for a few tens of millions of years. In the past, it was much hotter, and in the future it would get much cooler. Therefore, Kelvin estimated that at most the earth was inhabitable for organisms between 20 and 40 million years. Evolutionists found it virtually impossible to accept these figures, but all

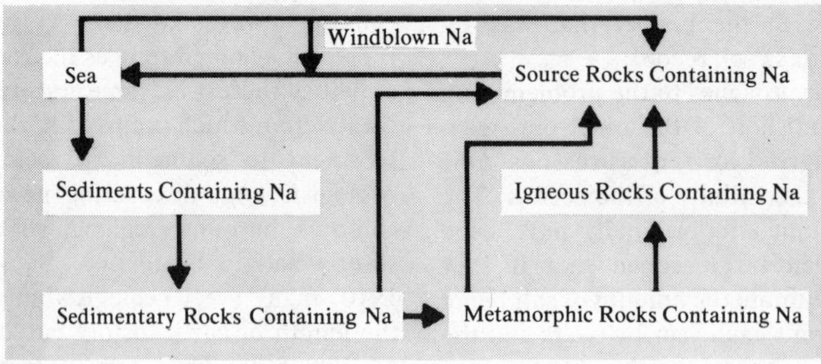

Figure 6.1. *The Sodium Cycle.* This cycle probably attained equilibrium long ago, thus invalidating the saltiness-of-the-sea method of dating the earth.

AUTHOR	MAXIMUM THICKNESS (In Feet)	RATE OF DEPOSIT (Years for 1 Foot)	DATE	TIME (In Millions of Years)
Phillips	72,000	1332	1860	96
Huxley	100,000	1000	1869	100
Haughton	177,200	8616	1871	1526
Haughton	177,200	?	1878	200
Winchell	—	—	1883	3
Croll	12,000 [1]	6000 [2]	1889	72
de Lapparent	150,000	600	1890	90
Wallace	177,200	158	1892	28
Geikie	100,000	730–6800	1892	73–680
McGee	264,000	6000	1893	1584
Upham	264,000	316	1893	100
Walcott	—	—	1893	45–70
Reade	31,680 [1]	3000 [2]	1893	95
Sollas	164,000	100	1895	17
Sederholm	—	—	1897	35–40
Geikie	—	—	1899	100
Sollas	265,000	100	1900	26.5
Joly	265,000	300	1908	80
Sollas	335,800	100	1909	80

After Arthur Holmes, 1913.
[1]Spread evenly over the land areas.
[2]Rate of denudation.

Figure 6.2. *Estimates of the Earth's Age Based on Rates of Sedimentation.* (From Don L. Eicher, *Geologic Time,* © 1968. Reprinted by permission of Prentice-Hall, Inc., Englewood Cliffs, N.J.)

they had were educated guesses in the face of Kelvin's potent mathematics. Darwin and others compromised their original theories in their later years in an effort to reconcile evolution and uniformitarianism with the physicists' estimates. Eventually, however, they were vindicated.

Radioactivity

In all fairness to the physicists, it was one of them, *Henri Becquerel,* who in 1896 made the startling discovery that eventually demolished Kelvin's theories and proved the geologists and biologists right. Becquerel discovered radioactivity, the spontaneous decay of certain unstable elements or forms of elements (isotopes), which produces large amounts of energy by the transformation of a tiny amount of mass. Hence, there was no reason to suppose that the earth's crust had been any hotter in the past than now or that it was going to get appreciably cooler in the foreseeable future. The natural decay of the radioactive elements in the earth's crust was chiefly responsible for its heat, not some residual heat left over from an originally molten stage. These calculations, which proved that the heat generated by the amount of radioactive elements in the earth's crust was more than sufficient to provide its heat flow, were also made by a physicist, *R.J. Strutt.*

Another significant discovery about radioactivity was that it was apparently the source of the sun's energy. The sun, by the loss of infinitesimal amounts of mass, could produce immense amounts of energy by nuclear reactions and, hence, could keep up its present rate of energy output for very great lengths of time. The sun was not just a big bonfire burning in the sky as Kelvin and others thought, but instead was fueled by this miraculous new energy source, radioactivity.

The most significant of all the wonders that flowed from the discovery of radioactivity, however, was that it at long last provided a reliable means of determining the absolute ages of those rocks which contain radioactive substances. It was found that in the process of radioactive decay, elements are transformed from one to another until a stable form is finally reached. In some cases, an element changes directly from a radioactive to a stable element; in others, the transformations are long stepwise processes. In the first decade of this century, an American chemist, *B.B. Boltwood*, found that uranium, a radioactive element, eventually decays to stable lead and that the ratio of lead to uranium is consistently greater the older the rock. The rate of decay appears to be constant; half of the original uranium would decay in a certain amount of time, half of the remaining amount would decay after another equal amount of time, and so on.

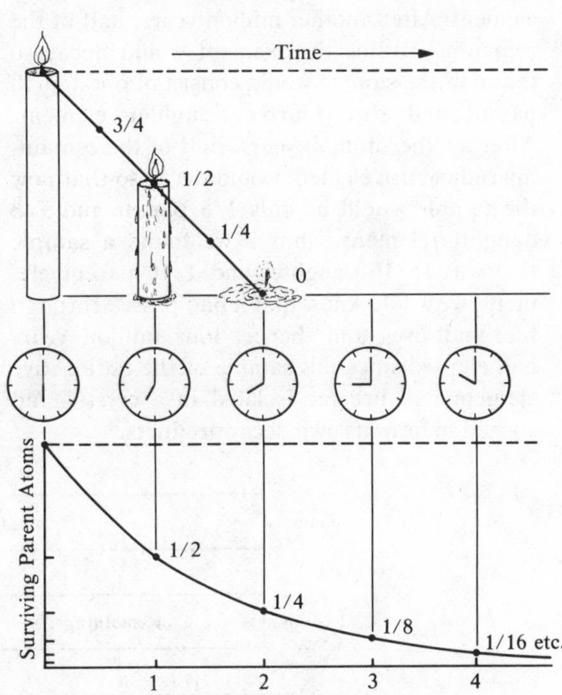

Figure 6.3. *Diagram Illustrating the Exponential Rate of Radioactive Decay* in contrast to normal uniform depletion rates. In a certain period of time, half the candle melts. After another equal period, it has all melted. In contrast, half of the remaining amount of a radioactive substance has decayed after each half-life. The process continues almost indefinitely until the last atom has decayed.

No known physical or chemical process substantially alters this exponential decay rate. By measuring the amounts of lead and uranium and knowing this so-called half-life of the radioactive elements in this transformation, Boltwood reasoned that one could calculate the amount of time that had elapsed since the uranium was first trapped in the crystals of a rock (Fig. 6.3).

To illustrate this method, let us consider a hypothetical example (Fig. 6.4) in which a radioactive element has a half-life of one million years and decays directly to a nonradioactive element. After a million years, then, one half of the original sample consists of the parent radioactive element and one half consists of the daughter element. After another million years, half of the remaining radioactive element would decay so that now the sample would consist of one-fourth parent and three-fourths daughter element. After another million years, half of the remaining radioactive element would decay so that now the sample would be only 1/8 parent and 7/8 daughter element. Thus if we found a sample that was 15/16 daughter and 1/16 parent element, we would know that it had passed through four half-lives and, hence, four million years had elapsed since this sample of the radioactive element had become isolated in a crystal and started to form its own decay products.

Using this method with the uranium-to-lead decay series and a crude estimate of the half-lives involved in this multiple-step decay series, Boltwood arrived at the truly astounding figures of hundreds of millions of years ago for the Paleozoic periods to well over a billion and a half years of age for some Precambrian rocks. The uniformitarians and evolutionists had been right after all. The refinement of using radioactive decay to date rocks, the process called **radiometric dating**, has continued with ever-increasing success down to the present. It is now possible to state with some degree of confidence dates of the beginning and ending of the Paleozoic periods to within a few tens of millions of years, of the Mesozoic periods within a few million years, and the Cenozoic epochs within a fraction of a million years. With the advent of radiometric dating, we have been able to roughly calibrate our relative time scale with an absolute one based on years (Fig. 6.5).

Radiometric Dating Techniques

The major radioactive decay series used in radiometric age determinations are Uranium 238 → Lead 206, Uranium 235 → Lead 207, Potassium 40 → Argon 40, and Rubidium 87 → Strontium

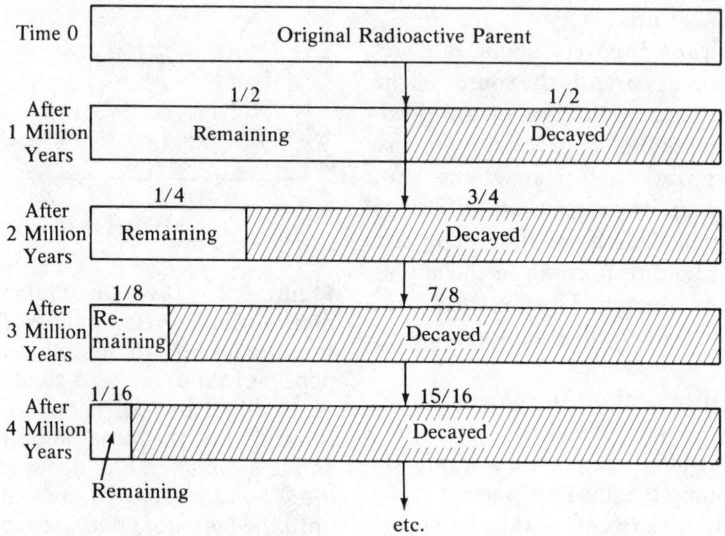

Figure 6.4. *Diagram of the Radioactive Decay of a Hypothetical Element with a Half-Life of 1 Million Years.*

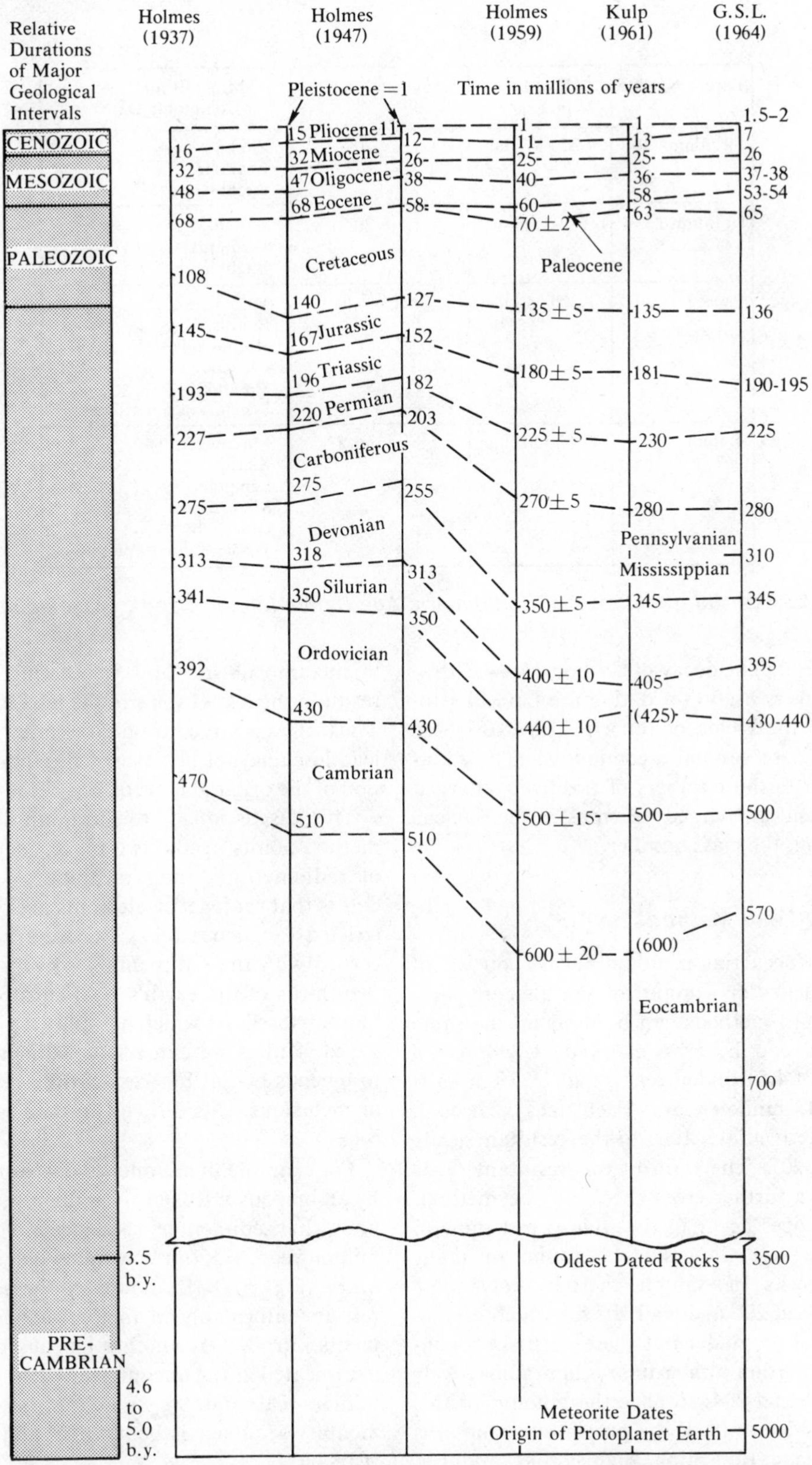

Figure 6.5. *Calibration of Absolute and Relative Geologic Time Scales.* G.S.L. is an abbreviation for the Geologic Society of London. In recent years only slight refinements have been made in the G.S.L. time scale, indicating that the dates are fairly accurate.

Parent Nuclide	Half-Life in Years	Daughter Nuclide	Minerals and Rocks Commonly Dated
Uranium-238	4,510 Million	Lead-206	Zircon Uraninite Pitchblende
Uranium-235	713 Million	Lead-207	Zircon Uraninite Pitchblende
Potassium-40	1,300 Million	Argon-40	Muscovite Biotite Hornblende Glauconite Sanidine Whole Volcanic Rock
Rubidium-87	47,000 Million	Strontium-87	Muscovite Biotite Lepidolite Microcline Glauconite Whole Metamorphic Rock

Figure 6.6. *The Major Methods of Radiometric Age Determination Used by Geologists.*

87 (Fig. 6.6). The decay of Carbon 14 → Nitrogen 14 is very useful for dating the Late Pleistocene and the Holocene epochs. Each of these methods deserves some comments. (The numbers refer to the numbers of positively charged protons and uncharged neutrons in the nucleus of an atom, the mass number.)

Uranium-Lead Methods

Naturally occurring uranium always consists of both varieties or isotopes of the element and, hence, both methods can be used on the same rock and can be cross-checked. U238 has a half-life of 4.51 billion years; and U235, a half-life of 713 million years. Each decays through many intermediate steps to the resultant leads 206 and 207. The ratio of the resultant leads provides a further cross-check on the method. Both isotopes occur in the widespread mineral zircon (a zirconium silicate) found in many igneous rocks. In a sample, there is always some original lead 206 and lead 207 for which corrections must be made, but these occur in a constant proportion with ordinary, nonradioactively produced lead 204; so once the amount of this isotope is determined, the proper subtraction is readily made. Reheating, such as that produced in metamorphism, and weathering commonly result in the loss of some of the lead in a sample. Thus, the ages record only the latest episode of lead loss and not necessarily the time of formation of the original igneous body.

This discussion of the uranium-lead dating method points up the two biggest shortcomings of radiometric dating techniques in geology. One is that radioactive elements are typically restricted to igneous rocks; yet we are more concerned with the sedimentary rocks which cover two-thirds of the earth's land surface and contain the fossils on which our relative time scale is based. Unless we can relate sedimentary rocks to igneous bodies by cross-cutting relationships or inclusions, it is difficult to date sedimentary rocks.

For example, if a sequence of sediments is cut by an igneous intrusion 10 million years old, we know that sedimentary sequence is older than 10 million years old, but how much older we do not know (Fig. 6.7). If, however, these sediments rest unconformably on another sequence of sediments intruded by another igneous body which is truncated at the unconformity and which is 20 million years old, we know that our first sedimentary sequence is between 10 and 20 million years old.

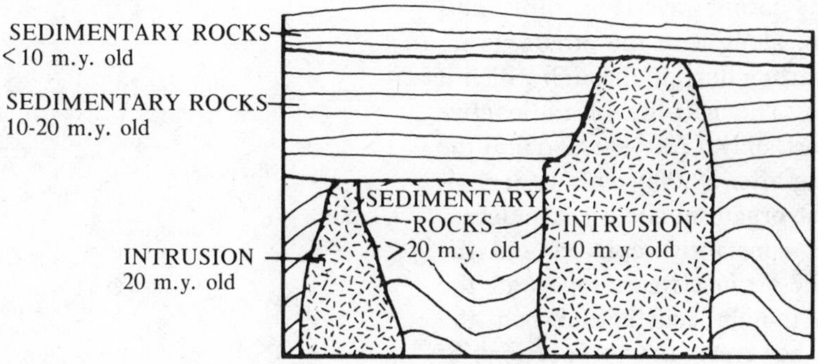

Figure 6.7. *Use of Cross-Cutting Relationships to Date Sedimentary Rocks in Relation to Igneous Rocks Dated by Radiometric Means.*

The other difficulty with radiometric dating is that metamorphism, weathering, and other processes often result in the loss of parent and/or daughter products. Unless this is recognized, erroneous dates can be determined. Enough points on the relative time scale are now calibrated, however, so that such an anomalous date is usually immediately apparent.

Many sedimentary sequences contain extrusive igneous rocks interlayered with sediments. These rocks do not produce the first difficulty, but are susceptible to the second. Such strata are more readily dated by the succeeding two methods, however.

Rubidium-Strontium Method

The rubidium-strontium method is typically applied to micas and K-feldspars of igneous or metamorphic rocks or to whole metamorphic rocks. Because radioactive potassium also occurs in these minerals, the K-Ar method can commonly be used as a cross-check. The amount of nonradioactively produced strontium 87 in a sample can be determined by its constant ratio to the totally nonradiogenic strontium 86, and a simple subtraction made. Rubidium 87 has a half-life of 47 billion years and, like uranium, is most useful in dating ancient rocks because of its long half-life, but less useful for young rocks because the amounts of strontium generated in a few million years are too small to be measured.

Potassium-Argon Method

The one-step decay of potassium 40 to argon 40 is the most useful method of radiometrically dating rocks because potassium is such an abundant, widespread crustal element, and all natural potassium contains a small amount of the radioactive isotope. Argon, the daughter product, is a gas and easily escapes under conditions of low heat or stress. Thus, the technique often yields low ages. Unlike the methods described earlier, there is no way to correct for the amount of nonradioactively produced argon in a crystal. Fortunately, however, there is hardly ever any nonradiogenic argon in the minerals hornblende, biotite, and muscovite, and these are very common minerals in igneous rocks. The K-Ar method can also be used for dating the sedimentary silicate mineral glauconite, providing the enclosing rocks have not been deeply buried or highly weathered. K-Ar dating can also be applied to dating whole volcanic rocks.

The half-life of potassium 40 is 1.3 billion years, and recent analytic techniques have made it possible to detect the tiny amounts of argon even in rocks as young as 50,000 years, thus making the K-Ar method applicable to rocks of nearly all ages. It is a good method for the most ancient rocks down to the point where the Carbon-14 method becomes effective.

Carbon-14 Method

Carbon 14 is generated continuously in the earth's upper atmosphere by the bombardment of Nitrogen 14 by cosmic rays (Fig. 6.8). This isotope of carbon, along with the others, is incorporated into carbon dioxide through which it enters organisms. The half-life of radioactive carbon is very short, only 5,730 years, so that the amount remaining after 50,000 years is negligible. As long as the organism is alive, the decaying Carbon 14 is constantly replaced and the proportions of the radioactive isotope and the normal isotopes remain constant. When the organism dies, however, the proportion of Carbon 14 gradually decreases as it decays to Nitrogen 14. Radiocarbon dates are determined, then, by comparing the proportions of radioactive and normal carbon in a sample. A necessary condition of valid Carbon-14 dating is that the production of C14 in the upper atmosphere has been constant over the last 50,000 years. The dates of known historic events that have been checked by radiocarbon dating clearly indicate that this assumption is valid for historic time though there is presently some reason to believe there were earlier changes of three to five percent. Only Late Pleistocene and Holocene events can be dated by this method because of the short half-life of C14. This is unfortunate

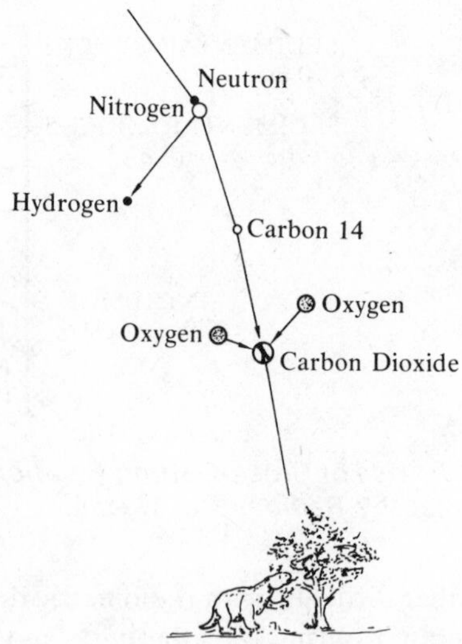

Figure 6.8. *The Formation of Carbon 14 and Its Uptake by Organisms.*

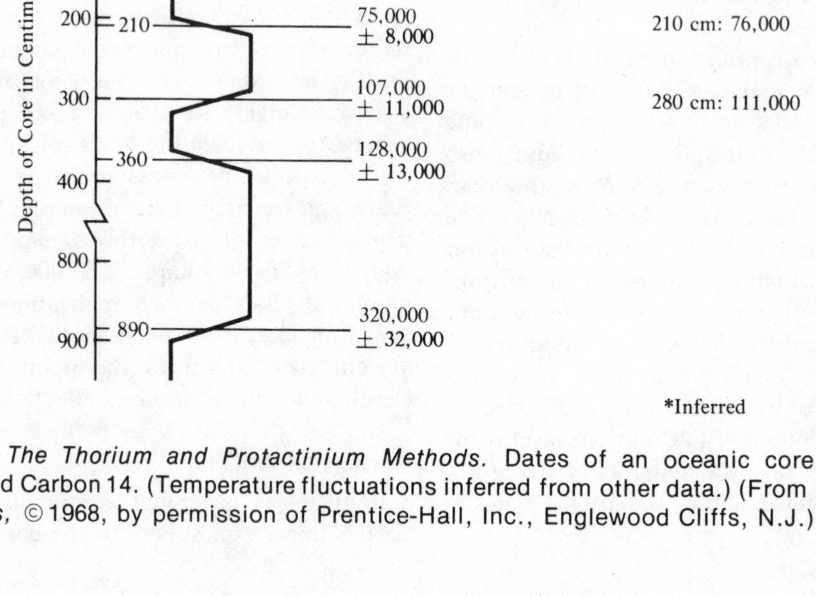

Figure 6.9. *The Thorium and Protactinium Methods.* Dates of an oceanic core using these methods and Carbon 14. (Temperature fluctuations inferred from other data.) (From Karl K. Turelian, *Oceans,* © 1968, by permission of Prentice-Hall, Inc., Englewood Cliffs, N.J.)

because this method is the most useful one for sedimentary rocks.

Thorium, Protactinium Methods

In recent years, some other methods of radiometric dating have proved useful for dating the Upper Pleistocene and Holocene sediment cores brought up from the ocean floors (Fig. 6.9). Thorium 230, one of the by-products of U238 decay, forms insoluble salts in the ocean and is quickly precipitated. It decays with a half-life of 75,000 years; and the amount decreases with the depth of the core—in other words, as one gets down into the older layers. The method is useful for the past several hundred thousand years. Protactinium 231, a by-product of U235 decay, also produces insoluble salts in the sea. Its rate of decay differs from that of Thorium 230; so the ratio of the two can be used to date different layers in cores.

Application of Techniques

For absolute dating of the rocks deposited during most of geologic time, then, geologists must deal with the uranium-lead, rubidium-strontium, and potassium-argon methods. As stated previously, only the K-Ar method can be used to directly date sedimentary rocks as it occurs in the complex silicate of sedimentary origin, glauconite. There is almost always some argon loss, however; and such dates are usually minimum ages unless other methods can be used to cross-check them. Interlayed volcanic rocks found in sedimentary sequences can be dated by either the K-Ar or Rb-Sr methods. In volcanic areas, these methods have provided numerous points on the radiometric time scale. Finally, if sedimentary rocks are closely bracketed by igneous intrusives, the U-Pb and Rb-Sr methods can be used to provide reference points. Though the radiometric time scale is far from perfect, we now know with the correct order of magnitude the duration of the eras and periods of the relative time scale.

It is another matter, however, when we consider dating an individual rock sample by radiometric means. First of all, it is not often possible to ensure that the sample being dated has remained a closed system; in other words, that some of the parent or daughter products have not escaped thus giving erroneous ratios and dates. All radiometric dates are followed by an error figure expressed as plus or minus so many million years or a fraction of a million years. This error figure becomes increasingly important when the amount of the original daughter product present is a large fraction of the radiogenic daughter. In old samples, there is so much radiogenic daughter product that unless the amount of the original daughter is very large, the percentage error is small. But in younger rocks, where the radiogenic daughter has not been accumulating for very long and, hence, is not present in large amounts, any significant amount of original, nonradiogenic daughter could produce gross errors in dating. The analysis is, therefore, a significant part of radiometric dating with many factors to be considered and weighed.

Radiometric dating has proved, at least until recently, the only practical means of dating rocks from the vast interval of geologic time called the Precambrian. Radiometric dating has shown the Precambrian interval was much longer than previously anticipated, probably about 4 billion years long or nearly ninety percent of the time span since the earth formed. The oldest rocks dated are about 3.8 billion years old; but meteorites, which are presumed to be leftover material from the original formation of the solar system, give dates of about 4.6 billion years. Similar dates have been obtained from the moon, which is inferred to have formed at the same time as the other bodies in the solar system (see Chapter 11).

The Paleozoic Era began either .7 or .6 billion years ago, depending on whether one includes the Eocambrian. In Precambrian rocks, excluding the Eocambrian, there are no megafossils (fossils larger than microscopic size) known except for the layered algal mounds called **stromatolites.** Until recently, there did not appear to be much variation in these remains and, thus, no way to apply the principle of faunal succession to Precambrian rocks. Hence, radiometric dating techniques have provided the only practical method of dating the Precambrian. This is why one sees chiefly radiometric dates in Precambrian time scales in place of eras and periods.

In recent years, paleontologists have discovered that the stromatolites are more structurally diverse than thought, and successful attempts to zone the later Precambrian have been made. In addition, geologists using sophisticated new techniques have uncovered a wealth of microfossils in Precambrian rocks, even in the oldest known strata. Powerful new tools, such as the scanning electron microscope, have shown these hitherto uniform-looking fossils also to be diverse and, hence, to have potential for paleontologically subdividing Precambrian time.

Fossils and Absolute Dating

A further development in recent years has been the devising of a method of absolute dating based on the unlikely marriage of paleontology and geophysics, the study of physics applied to the earth. Physicists have long calculated that the earth's rotation rate has been slowing down gradually through geologic time because of tidal friction. The everyday rise and fall of the tides receive their energy from the earth's rotation, the rate of which is thus being slowly reduced. This diminution in the rotation is very tiny; the day is increasing in length by only about two seconds in 100,000 years. The earth's period of revolution around the sun apparently remains the same. Hence, if the days were formerly shorter than now, there must have been more days in the year in the past than now. Physicists calculated that the day was twenty-one hours long at the end of the Cambrian; hence, there were 412 days in the year. Until recently, there was no way to confirm these figures.

However, paleontologists have recently begun to note in many organisms that the annual growth rings are in turn subdivided into smaller units which have turned out to be daily growth increments (Fig. 6.10). By counting these daily

A

B

Figure 6.10. *Daily Growth Lines in Corals. Holophragma,* a Devonian tetracoral. **A.** *Two Thecae,* X 3; **B.** *Enlargement showing fine daily growth bands.* (Specimens provided by John W. Wells; photography by R.M. Eaton. From *Principles of Paleontology* by David M. Raup and Steven M. Stanley. W.H. Freeman & Company, © 1971.)

growth rings in fossil Middle Devonian corals, *Wells* found 385–410 of them. In Pennsylvanian corals, he found 385–390 rings. These figures agree closely with the 400 days predicted for the Middle Devonian year and the slightly more than 390 days calculated for the Pennsylvanian year (Fig. 6.11). Similar observations on other types of fossils of other ages have confirmed these findings. Thus, we may be approaching the day when we can tell the age of a well-preserved fossil by counting its growth rings and comparing this number to the scale of days in a year throughout the geologic past. Similar studies on growth rings representing the lunar (synodic) month have led to the development of a graph showing how its length has decreased through time (Fig. 6.12). Refinement of this diagram could also lead to its usefulness in determining rough absolute dates.

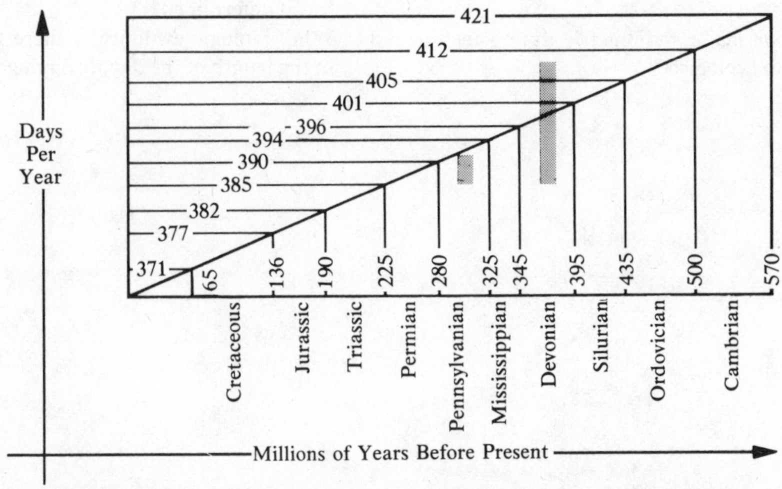

Figure 6.11. *Predicted Number of Days per Year Plotted Against Quantitative Time Scale.* Vertical bars show range of results from daily and annual growth lines on fossil corals. (From J.W. Wells, 1963.)

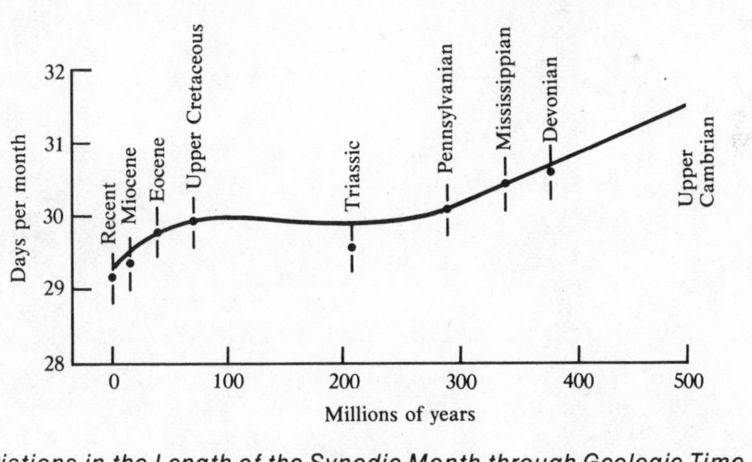

Figure 6.12. *Variations in the Length of the Synodic Month through Geologic Time as Determined from Count of Growth Lines on Fossil Mollusca* (mainly pelecypods). The bars for each point indicate the standard error. (After G. Pannella, C. MacClintock, and M.N. Thompson, *Science,* vol. 162, November 15, 1968, p. 795. Copyright 1968 by the American Association for the Advancement of Science. Used with permission of the authors and AAAS.)

REVIEW

1. What was the rationale behind the **oceanic salinity method** of dating the earth, and why did it fail?
2. Why was there such a large variation in estimates of the earth's age made by the **rate of sedimentation method?**
3. Summarize Lord Kelvin's arguments about the age of the earth. How was he eventually refuted?
4. What is **radioactivity** and how can it be used to date rocks?
5. What are four major radiometric dating techniques useful to geologists?
6. Why is the **Carbon-14 method** not widely applicable in geology?
7. How might an oceanographer determine absolute dates in a sedimentary core?
8. Briefly summarize the major problems with radiometric dating techniques.
9. Why are radiometric dating techniques not widely used in stratigraphy?
10. What is the apparent age of the earth and how was it determined?
11. What biologic evidence is there for the increase in the length of the day through geologic time?

7

The Array of Past Life: Paleontology

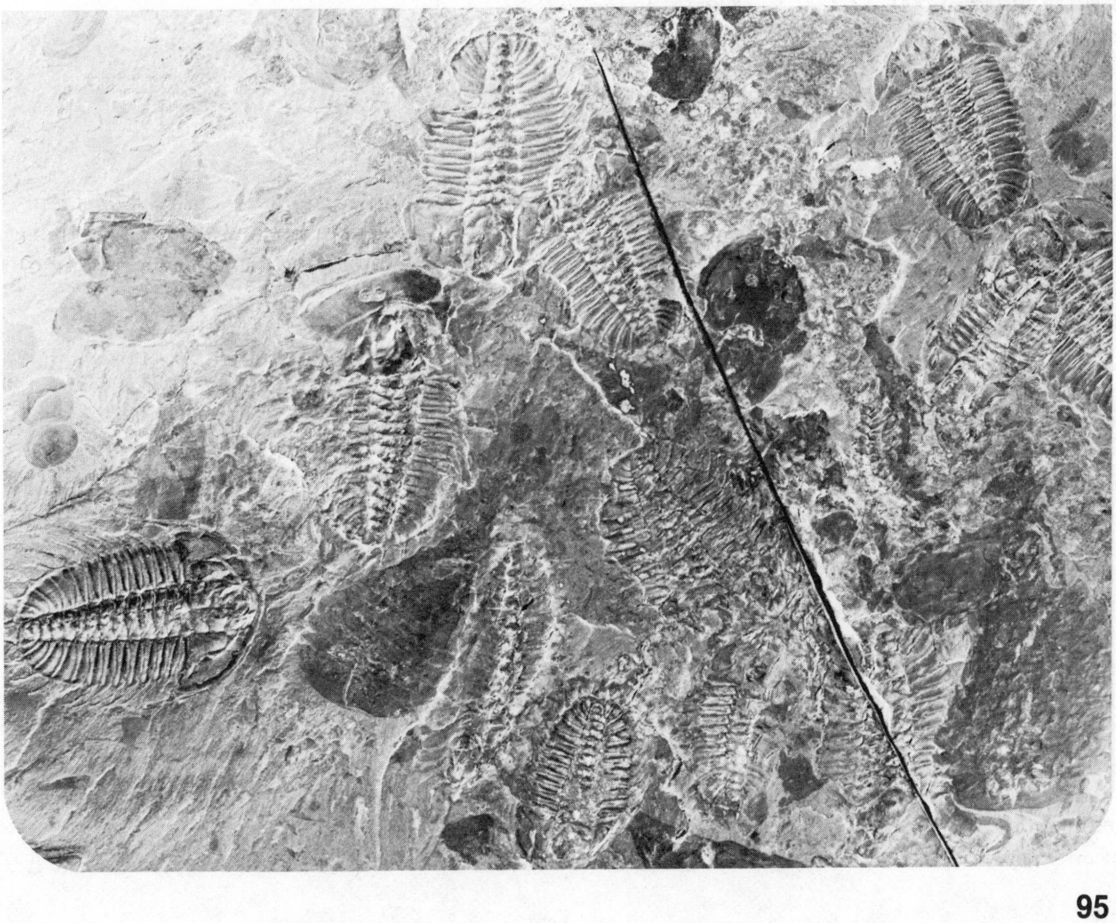

7

In earlier chapters, we defined *fossils* as the remains or traces of prehistoric organisms preserved in the earth's crust and pointed out their geologic significance in the principle of faunal succession and their biologic significance in demonstrating the occurrence of evolution. (By prehistoric, we mean older than the dawn of written history 5,000 to 6,000 years ago, though some geologists exclude any Holocene remains.) We now turn to a study of the different kinds of fossils and how they are preserved.

Fossilization

Paleontology, or paleobiology, is the study of fossils, and its practitioners are called **paleontologists.** Paleontologists still do not understand very well the physical-chemical reactions that operate in the process of fossilization, but they do understand some of its major prerequisites. Generally speaking, an organism must have hard, resistant parts, such as shell or bone, in order to be preserved, for soft, fleshy structures are rapidly destroyed by scavengers and decay bacteria. Furthermore, an organism typically must be quickly buried in some protective medium such as sediment to escape destruction by these agents as well as the geological agents of weathering and erosion, which can eliminate even hard parts. Obviously marine organisms are better candidates for burial than are terrestrial ones; for the oceans are typically sites of sedimentation, while the land is largely a site of erosion.

Organisms with hard parts and which are rapidly buried stand the best chances of being fossilized (Fig. 7.1). For example, most clams

		Hard Parts	
		Yes	No
Rapid Burial	Yes	GOOD clams	POOR jellyfish
	No	POOR birds	VERY POOR butterflies

Figure 7.1. *The Likelihood of Fossilization.* (Modified after Leo F. Laporte, *Ancient Environments,* ©1968. Prentice-Hall, Inc., Englewood Cliffs, N.J.)

have a hard shell and live buried in sand or mud; hence, they have left an excellent fossil record. Organisms of delicate structure which do not live in areas of sedimentation (for example, butterflies) have left a very poor fossil record. Intermediate cases occur when organisms lack hard parts yet live in favorable environments or have hard parts but do not live in good environments for fossilization. Jellyfish illustrate the first case as they are known only from uncommon impressions of the body, while birds illustrate the latter case for their resistant bones only occasionally collect in areas where they can be rapidly buried. The fossil record is, therefore, very imperfect in the sense that most organisms are not fossilized (and this is fortunate as it keeps the elements recycling), but it

A	PRESERVATION WITHOUT ALTERATION	
	1. Organic Compounds	
	a) Soft parts	Frozen; mummified. Such finds are rare and limited largely to Pleistocene deposits.
	b) Skeletal parts	Organic constituents of bone or shell. Chitin in arthropods, scleroprotein in graptolites and some other invertebrates. Cartilage in some vertebrates. Cellulose in plants.
	2. Inorganic Compounds (see also Fig. 7.5)	
	a) Calcium carbonate:	
	Calcite	Fairly stable, found in many invertebrate phyla and algae.
	Aragonite	Moderately stable, rare in rocks older than Mesozoic. Corals and molluscs, some algae.
	b) Tricalcium phosphate	Conularids; some brachiopods, arthropods; vertebrates. Quite stable.
	c) Silica (opaline)	Moderately stable. Rare in rocks older than Cenozoic. Sponges and some protistans.
B	ALTERED IN FOSSILIZATION	
	1. Organic Compounds	
	a) Soft parts	Films of carbon. Rare. Found in fine shales deposited in anerobic environments.
	b) Skeletal parts	Carbonized. Particularly the chitinous skeletons of arthropods, scleroproteins in graptolites, and plant tissues. (Also called distillation.)
	2. Inorganic Compounds	
	a) Permineralized	Deposition of minerals in interstices of skeleton. Commonly $CaCo_3$. Less frequently SiO_2, glauconite, iron compounds, etc.
	b) Recrystallized	Less stable inorganic compounds alter in physical form to more stable state without change in chemical composition, e.g., aragonite to calcite. May be very common mode of preservation, but difficult to distinguish from replacement.
	c) Replacement	Removal of original skeleton material by solution and deposition of new compounds, carbonates, silica, iron compounds, etc., in its place. Very common—intergrades with permineralization and recrystallization.
C	PRESERVATION AS MOLDS OR CASTS	
	1. Organic Compounds	
	a) Soft parts	Imprints in fine-grained shales and lithographic limestones.
	b) Skeletal parts	Imprints or casts.
	2. Inorganic Compounds	
	a) Molds	External and internal molds formed by sediment around or within skeletal parts.
	b) Casts	Filling of mold after skeletal parts are dissolved. Intergrades with replacement.
D	EVIDENCES OF ANIMAL ACTIVITIES	
	1. Tracks	Mode of locomotion. Preserved as molds and casts.
	2. Burrows	Animal habitat and behavior. Mode of burrowing. Preserved as molds and casts.
	3. Coprolites	Fossilized excrement. Diet. Structure of gut. May be preserved in any of ways in this table.
	4. Borings and Tooth Marks	Evidence of predation.
	5. Human Artifacts	Evidence of culture.

Figure 7.2. *Types of Fossil Preservation.* (Modified after James R. Beerbower, *Search for the Past: An Introduction to Paleontology,* 2nd ed., © 1968. Reprinted by permission of Prentice-Hall, Inc., Englewood Cliffs, N.J.)

does provide us with a good sampling of those organisms with hard parts that lived in areas of sedimentation and an occasional glimpse of the rest.

Types of Preservation

Paleontologists recognize four broad and intergrading categories of types of fossilization—unaltered remains, altered remains, molds and casts, and traces of organic activity (Fig. 7.2). In general, the amount of desirable information decreases in this order with unaltered remains furnishing maximum information and trace fossils, only fragmentary information. But paleontologists take what they can find. They will probably ignore molds and casts if they have unaltered or altered remains; but if molds and casts are all that is preserved, they are the only sources of valuable information and must be utilized.

Different types of sedimentary and postdepositional environments tend to favor certain types of preservation. For example, the quiet waters in which muds accumulate are not likely to abrade or break the shells of dead organisms, but the higher energy environments in which sand accumulates often lead to broken and worn fossils. Also, sandstone is much more permeable than shale; hence, groundwater is more likely to dissolve away the shell, leaving only a mold in the sandstone.

Each category of fossilized remains can be further subdivided. **Unaltered remains** can be split into organic, carbon-containing compounds and inorganic, noncarbon-containing compounds. **Organic compounds** are complex mixtures of carbon bonded to hydrogen, oxygen, nitrogen, phosphorus, and sulfur. They occur both in fleshy soft parts and as hard parts of many organisms. **Soft parts** are very rare in the fossil record, as you might expect, for these are very liable to decay and do not survive long under any circumstances. They are restricted to the recent geologic past, Late Pleistocene or Holocene, and to very cold and/or dry conditions. The most famous of these fossils are the not uncommon mammoths, extinct hairy elephants, frozen in the permafrost of Siberia and Alaska (Fig. 7.3). Even parasites and the

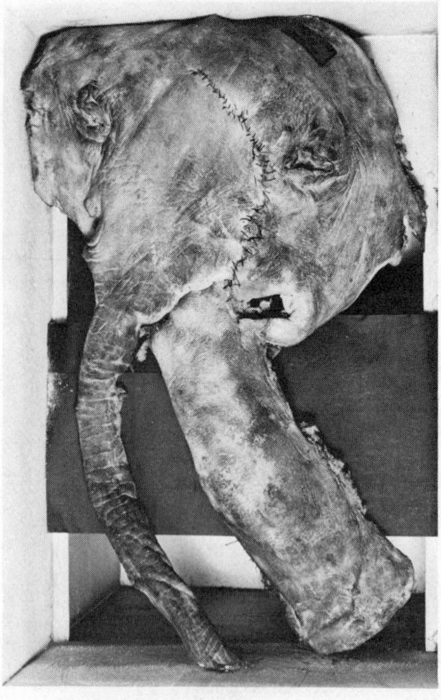

Figure 7.3. *Young Woolly Mammoth* dug from frozen ground in Alaska. (Courtesy of the American Museum of Natural History.)

gut contents are preserved in these remarkable fossils. A rhinoceros pickled in tar is known from Poland; pieces of ground sloth flesh and fur are known from dry desert caves in both North and South America; mummified clams are found in Greenland; and mummified small crustaceans called ostracods occur in Alaska. This is the entire extent of known fossils of this type of preservation, but degraded organic compounds which cannot be ascribed to any particular creature are known far back into the Precambrian.

Organic skeletal remains survive only slightly better in the unaltered state. These include carbohydrates such as chitin (Fig. 7.4) which composes arthropod (crustaceans, insects, etc.) skeletons and worm jaws and spines, and cellulose which forms the supporting structures of plants and the shells of the microscopic dino-

Figure 7.4. *Ant (Sphecomyrma freyi) in Cretaceous Amber* from New Jersey. (Courtesy of Frank M. Carpenter.)

flagellates; scleroproteins which compose the skeletons of certain coral-like groups and the extinct graptolites; and organic phosphates such as the cartilage of sharks and other vertebrates. (If you are unfamiliar with any of these groups of organisms, they are discussed and illustrated later in this chapter for reference.) With the exception of the cellulose composing spores and pollen (Fig. 3.23), which are studied by the paleontologists called **palynologists**, these skeletal structures are much more commonly preserved as altered remains.

Inorganic hard parts form the vast majority of unaltered fossils. These are nearly all composed of one of three compounds (Fig. 7.5)—calcium carbonate ($CaCO_3$), calcium phosphate ($Ca_5(PO_4)_3OH$), and hydrous silicon dioxide ($SiO_2 \cdot nH_2O$). (The carbon in calcium carbonate is part of the carbonate ion and not elemental carbon as in organic compounds, so it is classified here.) **Calcium carbonate** (Fig. 7.5) is by far the most widely used and abundant skeletal material in the organic world. It occurs in two forms, the minerals calcite and aragonite. Aragonite is less stable and recrystallizes to calcite with time; so there are few unaltered aragonite fossils older than Late Mesozoic. Calcite composes or contributes to the hard parts of many blue-green algae, charophytes, many red algae, coccolithophores, most foraminifera, some sponges, stromatoporoids, tetracorals, tabulate corals, many bryozoans, most brachiopods, some snails, many clams, some worm tubes, trilobites, many crustaceans, and echinoderms. Aragonite is found in many green algae, hexacorals, many bryozoans, most snails, many clams, most cephalopods, and some worm tubes.

Calcium phosphate (Fig. 7.5), often in the form of the mineral apatite, is much less common; but it does compose the skeletons and teeth of most vertebrates and is therefore very important from our point of view. Other organisms with phosphatic hard parts are the extinct conularoids, some brachiopods, many trilobites, many crustaceans, and the common microfossils called conodonts.

Silicon dioxide (Fig. 7.5) occurs in organisms in the hydrous, noncrystalline mineral opal. With time, opal gradually crystallizes to quartz;

Phyla		Calcium Carbonate $CaCO_3$: Calcite	Calcium Carbonate $CaCO_3$: Aragonite	Silica SiO_2	Calcium Phosphate $Ca_5(PO_4)_3OH$
Cyanophyta (Blue-Green Algae)		Frequent			
Chlorophyta (Green Algae)			Frequent		
Charophyta (Stone Worts)		Frequent			
Chrysophyta (Coccolithophorids)		Common			
Bacillariophyta (Diatoms)				Common	
Rhodophyta (Red Algae)		Common			
Sarcoda	Foraminiferans	Common			
	Radiolarians			Common	
Porifera (Sponges)		Frequent		Common	
Coelenterata (Corals)		Common	Common		
Bryozoa (Bryozoans)		Common	Common		
Brachiopoda (Brachiopods)		Common			Frequent
Mollusca	Snails	Frequent	Common		
	Clams	Common	Common		
	Cephalopods		Common		
Annelida (Segmented Worms)		Frequent	Frequent		
Arthropoda	Trilobites Crustaceans	Common			Common
	Arachnids Insects				
Echinodermata (Echinoderms)		Common			
Chordata	Acorn Worms Tunicates, Lancelets				
	Vertebrates				Common

Legend: Frequent | Common | Rare or Absent

Figure 7.5. *Distribution of Major Inorganic Skeletal Compounds among Geologically Significant Phyla of Organisms.* (Adapted from A. Lee McAlester, *The History of Life,* © 1968. Reprinted by permission of Prentice-Hall, Inc., Englewood Cliffs, N.J.)

so most older sedimentary rocks contain altered siliceous remains. Silica occurs in three abundant groups of microfossils, the silicoflagellates, diatoms, and radiolaria. It also composes the skeletons of most sponges.

Unaltered remains of all compositions occur in rocks of every age, but become more abundant as one approaches the Recent.

Altered remains are also quite abundant and come in varying degrees of alteration. Organic remains, both hard and soft, are commonly altered by the process called **distillation** or **carbonization.** In this process, the weight and heat created by overlying rocks force out the volatile carbon compounds in an organism, eventually leaving only a black carbon film or residue. Some aquatic vertebrates have had their bodies preserved as sheets of carbon by this process (Fig. 7.6). More common, however, is the distillation of hard parts such as the chitin skeleton of arthropods, the protein skeleton of graptolites, and the cellulose of plants (Fig. 7.7). Indeed, the sedimentary rock coal is a product of the distillation of great thicknesses of plant remains.

Inorganic compounds are commonly altered by three processes. In **permineralization,** porous cavities in hard parts such as bone are filled with chemical precipitates, most commonly calcite or silica, or with the sedimentary matrix of the surrounding rock, thus giving the fossil much more strength. **Replacement** is the process whereby the atoms of original material are individually removed and replaced with other types by percolating solutions. The identical structure is preserved, but the composition is different (Fig. 7.8). The two most common replacing substances are calcium carbonate and silica whose stabilities are mutually opposite, the former replacing ions under basic conditions, the latter, under acidic conditions. If a calcium carbonate fossil in a limestone is replaced by silica, it can often be easily extracted by dissolving the rock in acid. Many delicate fossils have been obtained this way which could not have been freed from the matrix in any other manner (Fig. 7.9).

Recrystallization, or **crystallization,** is the process whereby an unstable form of a substance changes to a more stable form. This involves an internal shuffling of atoms which distorts or expands the crystals of the original material, thus altering the fossil. The most common examples are aragonite recrystallizing to calcite and opal crystallizing to quartz.

Molds and casts are generally less desirable fossils than the above because they preserve only the shape of the surface, either inside or outside,

Figure 7.6. *Distilled or Carbonized Fossil Fish* (*Diplomystus analis*) from Eocene Green River Shale. The bones have also been preserved. (Courtesy American Museum of Natural History.)

A

B

Figure 7.7. *Distilled or Carbonized Fossil Plant Leaves.* **A.** *Carboniferous fern,* **B.** *Cenozoic angiosperm.* (Part A courtesy of T. Delevoryas, Yale Peabody Museum. Part B courtesy of William L. Stokes, University of Utah.)

A B

Figure 7.8. *Replaced Fossil Wood.* **A.** *"Old Faithful" petrified log.* Petrified Forest National Park, Arizona. (Courtesy of William M. Mintz.) **B.** *Transverse section of silicified wood.* (Courtesy of T. Delevoryas, Yale Peabody Museum.)

Figure 7.9. *Silicified Fossil Brachiopods.* Spiny brachiopods of the Order Productoida from the Permian of the Glass Mountains, Texas. (Courtesy of G. Arthur Cooper and U.S. National Museum.)

of a structure. No internal structure of the shell or bone is preserved. However, they are almost always the only remains of soft-bodied creatures or soft parts of any organisms. A **mold** is an impression of a surface. For example (Fig. 7.10), if one were to plunge a fist into a piece of clay and then extract it, the clay would contain a reverse replica of the outside of the hand, an

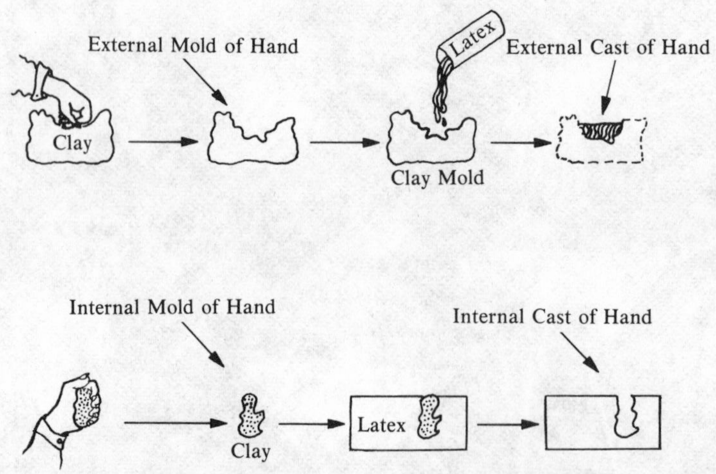

Figure 7.10. *Diagram Illustrating Formation of External and Internal Molds and Casts.*

external mold. If one were to squeeze a ball of clay in the palm of a hand, the clay would contain a reverse replica of the inside or palm of the hand, an **internal mold.** If a mold is later filled either naturally with sediment or by artificial processes, a **cast** results. If the external mold of one's fist were filled with plaster which was allowed to dry and then was removed, the plaster would contain a replica of the surface of the fist, an external cast. If the internal mold of one's palm were covered with plaster which was allowed to dry and then removed from around the mold, a replica of the surface of the palm, an internal cast, would result. Molds and natural casts are common fossils and include both soft and hard, organic and inorganic structures. If the molds and casts are so thin that they cannot be distinguished, as in plant leaves, the word **impression** (Fig. 7.11) is often used.

Figure 7.11. *Fossil Plant Impression.* Pennsylvanian, Mazon Creek, Illinois. (Courtesy of T. Delevoryas, Yale Peabody Museum.)

Trace fossils record only the effects of an organism, usually an animal, on its environment. **Trails,** which are records of the passage of a body, and **tracks** (Fig. 7.12), which record the passage of the hands and/or feet of an animal, are commonly preserved as molds and casts. They give some idea of the method of locomotion, the size, and the shape of the animal that made them. Unless actual remains occur nearby, however, it may be possible to do no more than identify them as "worm trails" or "reptile" tracks." **Burrows**, which are also preserved as molds and casts, give information about the size and habits of an organism, but are even more difficult to assign to a particular kind of animal. **Coprolites** (Fig. 7.13) are fossilized feces and are preserved by any of the above means. They tell something about the diet and the digestive tract structure of their makers. Small, delicate fossils, not found elswere as fossils, often turn up in coprolites. **Borings** in shells and **tooth marks** (Fig. 7.14) on bones and shells offer evidence of who ate whom and how certain organisms fed. Lastly,

Figure 7.12. *Fossil Vertebrate Tracks.* Reptile tracks in Triassic strata of the Connecticut Valley. (Courtesy of Yale Peabody Museum.)

Figure 7.13. *Coprolites.* Fossilized ground sloth dung, Rampart Cave, Lake Mead National Recreation Area, Nevada. (Courtesy of National Park Service.)

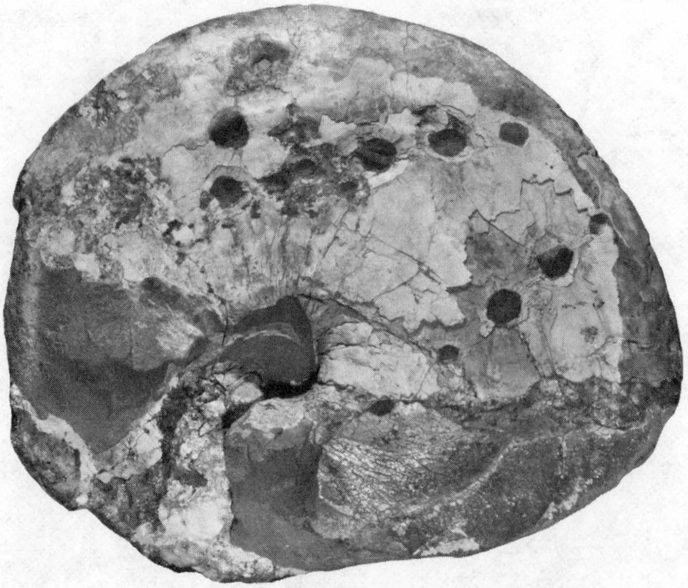

Figure 7.14. *Fossil Toothmarks of Mosasaur on Ammonoid Shell, Cretaceous Age.* (Courtesy Robert V. Kesling and University of Michigan Museum of Paleontology.)

human artifacts made by prehistoric people are trace fossils that offer evidence of the culture of their makers and some evidence of their makers' size and structure.

Geologists and Fossils

In studying a fossil, paleontologists are typically interested in three major tasks: the determination of its position in the evolution of life so that it can be properly assigned in the formal classification, the determination of its ecology so that the environment of the deposit containing it can be ascertained, and the determination of its position in the geologic column so that its age can be known. Fossils can be classified in these three ways or any number of other ways, depending on one's purposes. The ecology of fossils has already been discussed in Chapter 3; the use of fossils in stratigraphy will be developed in Chapter 9. In this chapter we will pursue the first type of classification.

Formal Classification of Fossils

The formal classification used by paleontologists is the one used by biologists and devised by *Linnaeus* in the 1700s. In this scheme, each type of organism or species is assigned a two-part (or binomial) species name that is always italicized. The first word, which is capitalized, is the generic name and is usually shared with several other, very closely related species. The second word, lowercased, is the trivial, or species, name and is unique in the particular genus, although it may be used for species of other genera. In scientific works, the name of the original author who described the species and the date of this work appear after the name. For example, the horse is *Equus caballus* Linnaeus, 1758. The scientific name of a species is in either Greek or Latin form; and, although it seems cumbersome to the uninitiated, its universal use eliminates the problems of language barriers and lack of enough everyday words to name every creature. Each species must have a unique binomial name; the name cannot be used twice nor can two names be applied to the same kind of organism.

Another aspect of the Linnaean system is that it establishes a hierarchy of classification terms; that is, each term is less inclusive than the one above it and more inclusive than the one below it. The terms used in the hierarchy today, most of which were coined by Linnaeus, are

Kingdom
Phylum (Botanists use Division here)
Class
Order
Family
Genus
Species

An order, for example, encompasses several families; and it, in turn, belongs with several other orders to a class. There is more diversity recognized in the organic world today than in Linnaeus' time; hence, biologists have found it necessary to insert several new terms between those of the traditional hierarchy. Most of these have been created by adding the prefixes *super-* and *sub-* to existing categories. For example, a class could be subdivided into several subclasses, each of which could contain several superorders, each of which in turn could consist of several orders. Sometimes, new words such as *cohort* and *tribe* have been devised where even more subdivisions are needed.

You may recall that Linnaeus initially did not accept evolution, but when he devised his classification it became obvious that the degrees of similarity expressed by his classification could be best explained by evolution. Today, we recognize that the evolutionary relationships of organisms form the basis of our classification. In biological jargon, it is **phylogenetic**, or at least we try to make it so, though we are not certain of the relationships of many organisms. A necessary implication of such a classification is that each category has only one ancestral species; otherwise we would be classifying together unrelated organisms. For example, birds, bats, and pterosaurs all fly and have numerous similar structural adaptations for flight, but we classify them in separate classes of vertebrates because they are not closely related. This is an obvious example, but there are numerous others, particularly among the invertebrates (animals without backbones) that go undetected.

A Modern Classification of Organisms

In an introductory course in historical geology, it is rarely necessary to go below the order level in discussing significant fossil groups. In the following pages, we will examine the major groups of organisms which have left a significant fossil record. In some groups, it will be sufficient to discuss the phylum only. In other, more diverse groups we will proceed down to the order level.

Zoologists have never standardized the endings for terms above the family level in animals, though this is clearly desirable from the student's viewpoint. In this text, the endings will be standardized as follows: phylum—*a*, subphylum—*zoa*, class—*ea*, subclass—*ia*, order—*oida*. These are the most commonly used endings, but are not universal; so their application to certain terms, such as the Class Gastropodea, may seem strange to the biologists reading this text. This step is taken, however, in the fervent hope that zoologists will finally take this desirable action.

Where finer subdivisions, such as orders, are discussed in this text, an attempt is made to establish groups whose typical members have about the same amounts of differences among each other. As a result, some common orders based on minor differences are combined, and some important subgroups which differ widely from other members are considered orders. This task of defining groups whose typical members are separated from other groups by roughly equal differences is another important task facing biologists and paleontologists.

Traditional classifications of the organic world have consisted of only two kingdoms, animal and plant. In recent years, however, there has been a decided shift in biology to recognizing one, two, or even three additional kingdoms. There are two major reasons for this. One is that there are many microorganisms which share features of both traditional animals and plants. The microscopic green flagellates such as *Volvox* and *Euglena* (Fig. 7.15) appear as animals in zoology texts and plants in botany texts. Clearly, the traditional kingdoms break down in classifying these creatures.

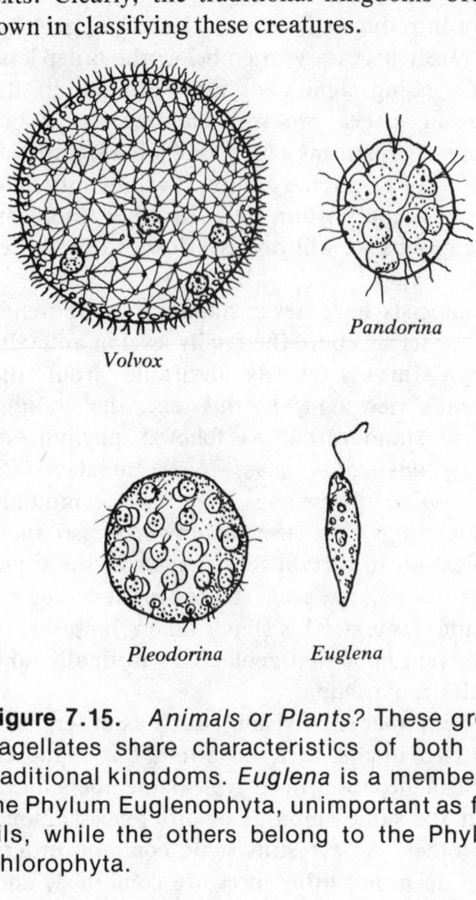

Figure 7.15. *Animals or Plants?* These green flagellates share characteristics of both the traditional kingdoms. *Euglena* is a member of the Phylum Euglenophyta, unimportant as fossils, while the others belong to the Phylum Chlorophyta.

The second reason is that people have traditionally studied the organic world from their anthropocentric viewpoint. It was easy to recognize the major differences among the complex animal groups, but the more distantly an organism is related to humans, the less obvious significant differences appeared. In studying organisms at the one-celled level of organization, this became particularly apparent. Zoologists classed all one-celled "animals" as the Phylum Protozoa; and botanists, the simple "plants" in the Division Thallophyta. We have begun to discover the profound differences among these organisms, and our classification is now beginning to reflect them. Biologists now generally recognize the Moneran, Protistan, Animal, and Plant kingdoms. Some exclude the fungi from the protistans and add a fifth kingdom, but because fungi are relatively rare fossils, we need not concern ourselves with the problem here.

Moneran Kingdom

These organisms are distinguished from all others by their lack of a nucleus; the hereditary material is not concentrated and surrounded by a membrane but is diffuse. The evolution of the nucleus was probably one of the most significant evolutionary events in life history, and some of its steps are shown in the surviving monerans. Organisms without a nucleus are called **procaryotic;** those with one (or more), **eucaryotic.** Monerans are all one-celled or colonies of essentially independent cells. The **bacteria** (Fig. 7.16), a diverse assemblage of abundant microorganisms, are assigned here, but their geologic importance is presently limited to their appearance among the oldest fossils in rocks over three billion years old and their possible geochemical effects. There is only one abundant group of monerans in the fossil record.

Phylum Cyanophyta—The **blue-green algae,** so named because they contain blue and green (chlorophyll) pigments when alive, are autotrophic; that is, they produce their own food by photosynthesis. They have no hard parts; but several of the sheet-like, benthonic, colonial forms entrap and bind layers of calcium carbonate (calcite) among their mucilaginous filaments. These layered mounds, called

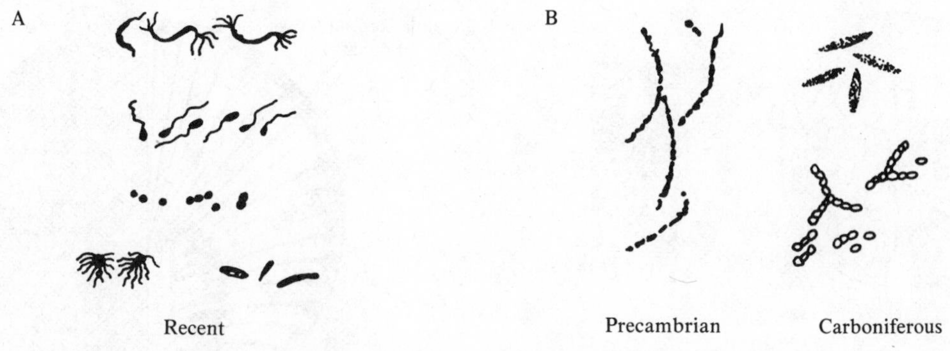

Figure 7.16. *Recent and Fossil Bacteria.* **A.** *Recent forms* and **B.** *Fossil forms.* X 250.

stromatolites (Fig. 7.17), range in size from a few inches to many feet across and often resemble petrified heads of cabbage. They range throughout the geologic record from Precambrian to Recent, but are most notable in Precambrian rocks in which they are the only megafossils. Most living types of blue-green algae live in fresh water, but marine and terrestrial kinds occur. The fossils lived in all three environments, but predominated in the littoral environment of oceans and lakes.

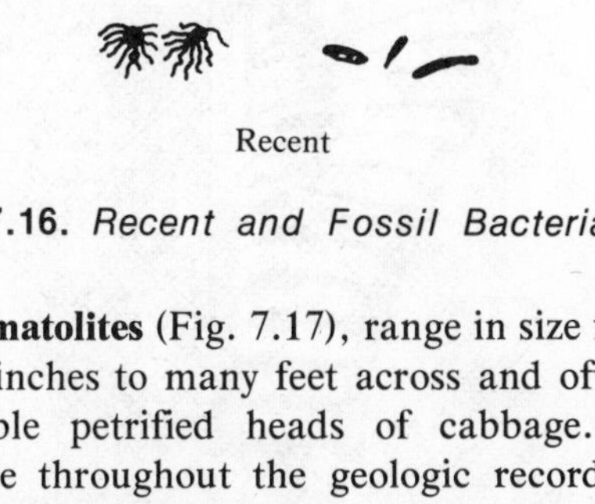

Figure 7.17. *Stromatolites,* the fossil remains of colonial blue-green algae (Phylum Cyanophyta). X 1/5.

Protistan Kingdom

The protistans are the first group of organisms which evolved a nucleus. Most are still one-celled or simple aggregates of cells, but some of the larger algae are multicellular with the cells differentiated. Protistans form their sex cells (gametes) in unicellular structure and do not form embryos. There are nine major fossil groups of protistans: green algae, charophytes, golden algae, diatoms, dinoflagellates, red algae, foraminifera, radiolaria, and tintinnoids.

Phylum Chlorophyta—The **green algae** are so named because they contain the green pigment chlorophyll, which is utilized in photosynthesis. This chlorophyll is the same type as that found in true plants and indicates they probably arose from green algae. The cells of most chlorophytes possess a whip-like flagellum at some time in their life. Green algae range from single cells through colonial cells on up to complex multicellular forms, not all of which may really be closely related. This assemblage, which ranges from Precambrian to Recent, contains only two groups that have left significant numbers of fossils.

The **dasycladaceous** algae (Cambrian-Recent) have a central stalk with radiating side branches (Fig. 7.18). The side branches bear facets and supporting structures of calcium carbonate (argonite and possibly calcite in some fossils). The facets are commonly preserved as fossils, and some groups calcify the central axis and/or the lateral branches enabling them to be preserved. In external molds of the body, the passageways originally occupied by the calcified side branches were misinterpreted as pores by many paleontologists, and the group was classified with the sponges! Large extinct types called receptaculitids (Fig.

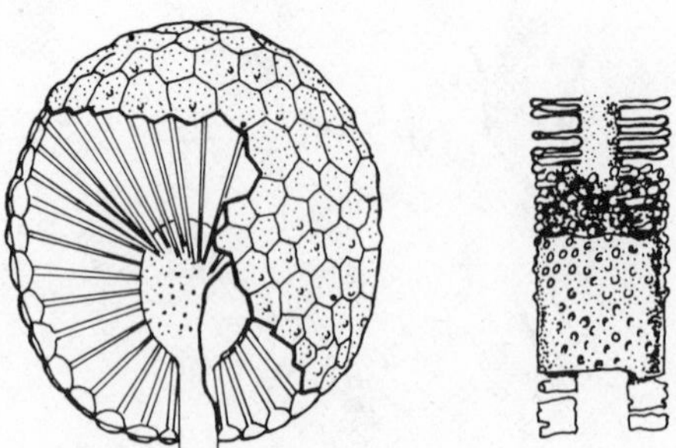

A *Cyclocrinus* Ordovician

B *Gyroporella* Jurassic

C *Triploporella* Jurassic

Figure 7.18. *Dasycladaceous Green Algae. Phylum Chlorophyta.* **A.** *Paleozoic form,* **B–C.** *Mesozoic forms.* Approximately natural size. (After Pia.)

7.19) were common in the Ordovician; more typical types abounded in Eurasia in the Triassic and Jurassic.

Codiaceous algae (Cambrian-Recent) typically have a branching body, or thallus, which is covered by aragonite in many types (Fig. 7.20). The jointed varieties have been major contributors to Cenozoic coral reefs.

Both types of green algae are marine and have a warm-water preference.

Phylum Charophyta—The **stoneworts**, or **charophytes**, are complex, multicellular green algae with structures superficially resembling stems, roots, and leaves (Fig. 7.21A). Some believe they are intermediate between chlorophytes and plants. The leaf-like structures occur only at nodes. The reproductive structures are also found at the nodes. In their metabolism, the benthonic charophytes, which inhabit fresh and brackish water, precipitate calcium carbonate over their thalli, or bodies. When they

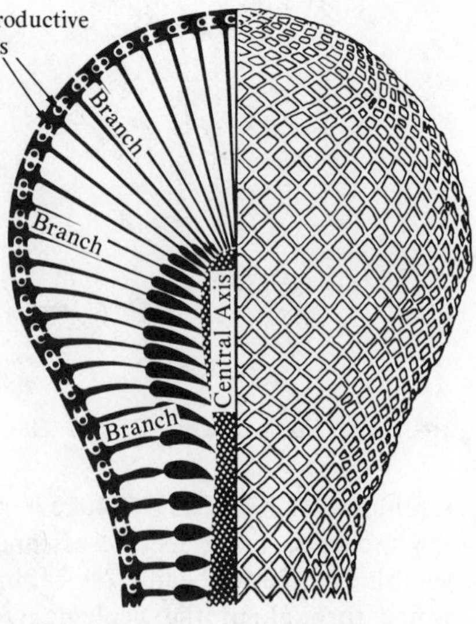

Figure 7.19. *Ischadites, a Receptaculitid.* These Paleozoic fossils were long mistakenly classified with sponges, but appear to form an extinct group of Dasycladaceous Green Algae. X 2. (After Kesling and Graham.)

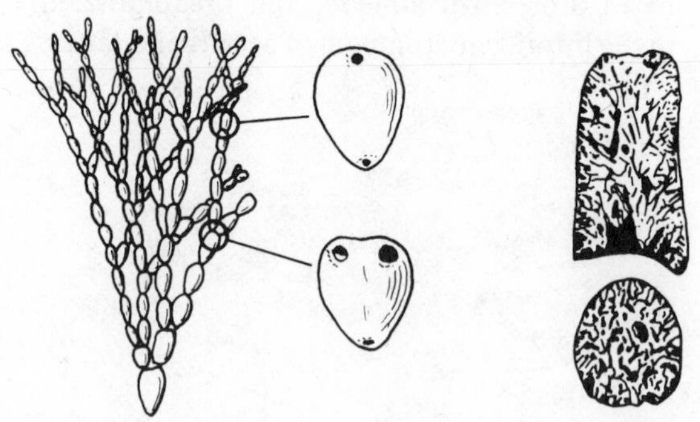

Ovulites Eocene *Codium* Cenozoic

Figure 7.20. *Fossil Codiaceous Green Algae (Phylum Chlorophyta).* **A.** *Form with jointed calcareous plates,* **B.** *Vertical and cross sections through a stem showing internal tubular structure.* Approximately natural size. (After Pia.)

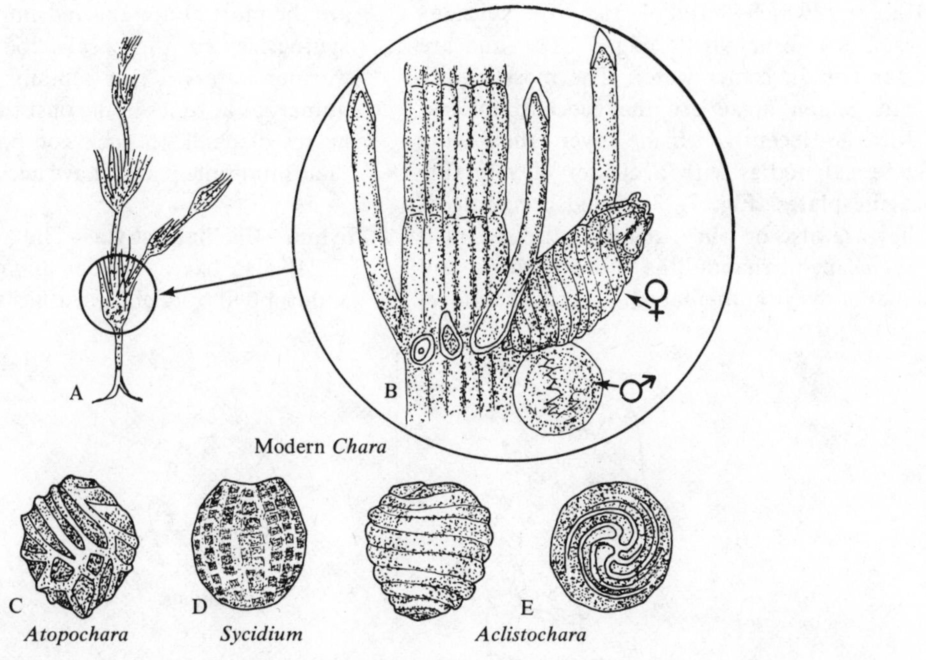

Figure 7.21. *Phylum Charophyta.* **A.** *Recent Chara, approximately natural size;* **B.** *Detail of reproductive structures on stem, X 10;* and **C–E.** *Various fossil oogonia, X 15.* (After Andrews and Peck.)

rot, a spongy mass of limestone called **marl** remains. Rarely is there much structure recognized in these fossils, however.

The female reproductive structures, or **oogonia** (Fig. 7.21B-E), are tiny, subspherical bodies with a spiral or vertical arrangement of cells which are impregnated with calcite. These microfossils are abundant in freshwater deposits of Mesozoic and Cenozoic age, although they are known as far back as the Silurian.

Phylum Chrysophyta—The **golden algae** are so named because in life they contain, in addition to chlorophyll, golden pigments. Most are photosynthesizers, but some must eat other creatures to obtain certain nutrients. Most are one-celled, though some are colonial; and all possess flagella in life. They are all planktonic.

The **chrysomonads** (Cretaceous-Recent) produce siliceous cysts, or resting stages (Fig. 7.22A). They are chiefly fresh water and are occasionally common fossils. More important are the marine **silicoflagellates** (Cretaceous-

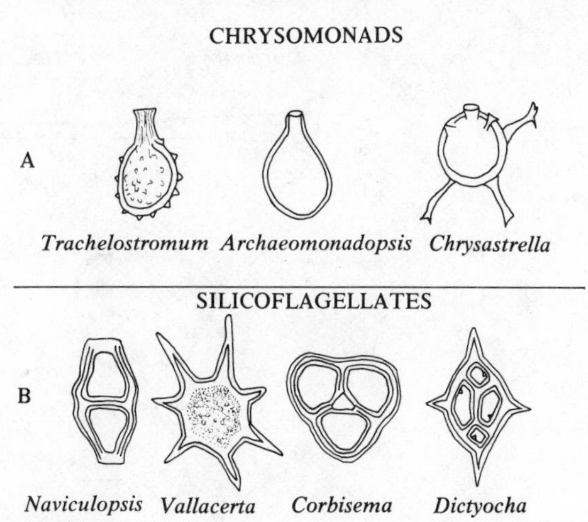

Figure 7.22. A. *Chrysomonads* and **B.** *Silicoflagellates* (*Phylum Chrysophyta*). X 300. (After Piveteau.)

Recent), which build lattice-like siliceous tests, or internal shells (Fig. 7.22B), and are abundant in cooler water. The most significant golden algae are the **coccolithophores** (Jurassic-Recent), which cover their tiny spherical bodies with a cluster of very tiny calcite plates (Fig. 7.23) called coccoliths if they are disc or wheel-shaped, discoasters if star-shaped, rhabdoliths if cylindrical, and lopadoliths if club-shaped. Coccolithophores are the most abundant and important photosynthesizers in the oceans today and prefer warmer waters. Their remains have been so numerous at times in the past that great thicknesses of chalk, a rock composed chiefly of their miniscule plates, have accumulated.

Phylum Bacillariophyta—The **diatoms** (Fig. 7.24) also have a golden pigment and have chlorophyll for photosynthesis. They differ

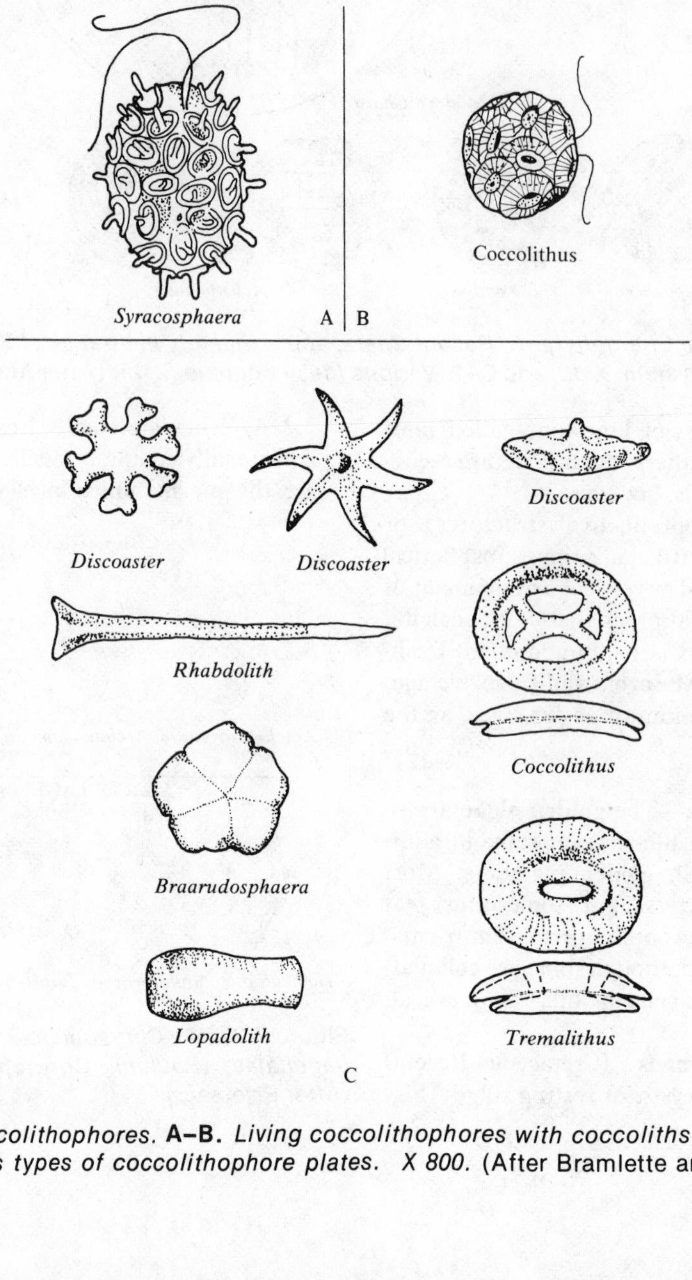

Figure 7.23. *Coccolithophores.* **A–B.** *Living coccolithophores with coccoliths in living position, X 250*; **C.** *Various types of coccolithophore plates. X 800.* (After Bramlette and Riedel.)

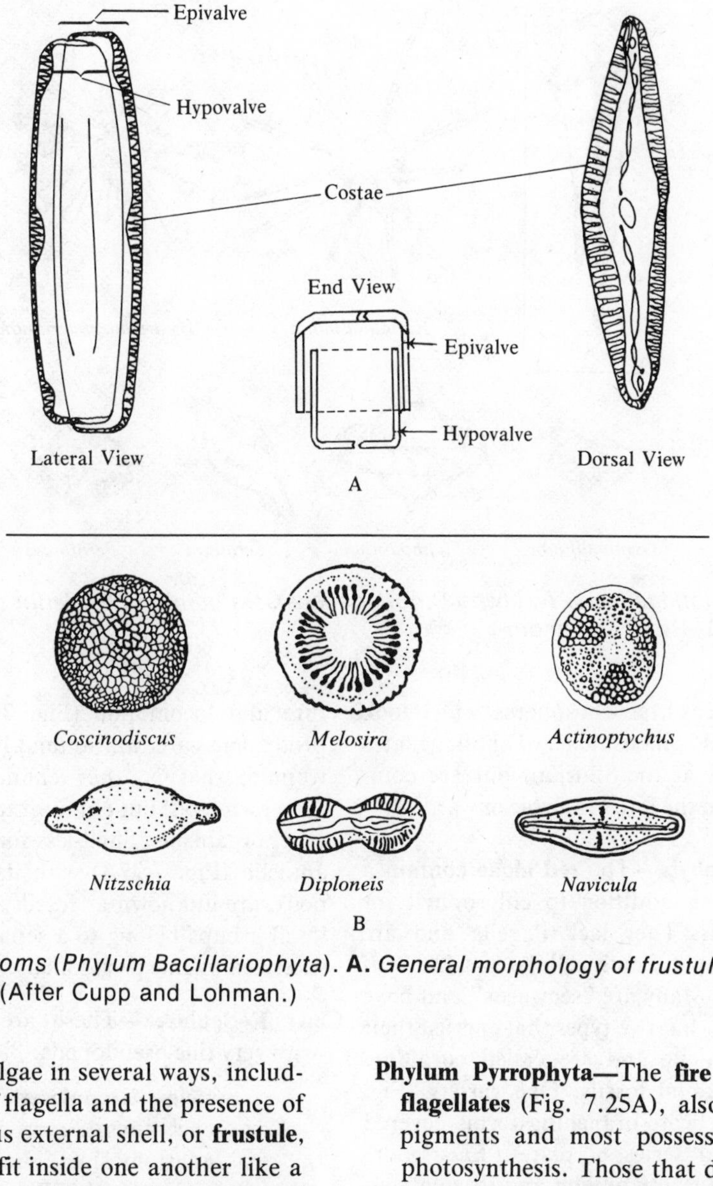

Figure 7.24. *Diatoms (Phylum Bacillariophyta).* **A.** *General morphology of frustule, X 200;* **B.** *Fossil forms, X 100.* (After Cupp and Lohman.)

from the golden algae in several ways, including the absence of flagella and the presence of a bivalved siliceous external shell, or **frustule**, whose two halves fit inside one another like a pillbox. Diatom frustules come in a variety of shapes, but circular disks and spindle-shaped or rod-like forms predominate. Most are single-celled, though many are colonial. Most are marine and pelagic, but several types have inhabited fresh waters since the Miocene and there are several benthonic genera. They range in age from Cretaceous to Recent and have been abundant throughout this interval. Diatoms are among the most important photosynthesizers on earth today.

Phylum Pyrrophyta—The **fire algae,** or **dinoflagellates** (Fig. 7.25A), also contain golden pigments and most possess chlorophyll for photosynthesis. Those that do not, as well as many of the others, are animal-like in that they eat other organisms for nutrition. They have two flagella in grooves, one meridional and one equatorial. Dinoflagellates are pelagic, asymmetrical single cells. The body is encased in a cellulose shell, or lorica (impregnated with silica or calcium carbonate in some). This shell is rarely preserved, but a much thicker cellulose **cyst** (Fig. 7.25B) which forms around the individual when it reproduces is a common fossil, particularly the

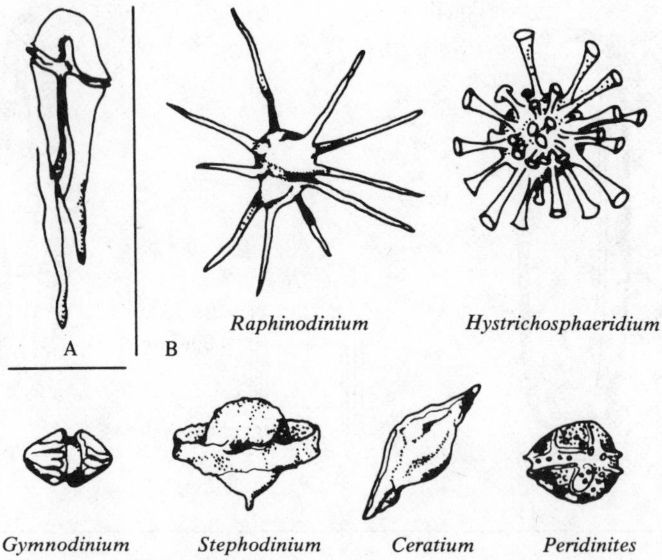

Figure 7.25. *Dinoflagellates.* **A.** *Living Dinoflagellate Ceratodinium,* **B.** *Various fossil dinoflagellate cysts.* X300. (After Deflandre.)

variety known as hystrichospheres, which look like spherical pincushions. Dinoflagellates occur as early as the Silurian, but are common only from the late Mesozoic onward.

Phylum Rhodophyta—The **red algae** contain a red pigment in addition to chlorophyll for photosynthesis. They lack flagella and are nearly all complex, multicellular marine benthonic forms. Many are "seaweeds" and have not left fossils, but two types that encrust their thalli with calcite and are called **corallines** have left common fossils. One variety (Fig. 7.26A) has delicate branching forms covered with a jointed series of plates. Even more common are *Lithothamnion* and its relatives, the nullipores (Fig. 7.26B), which form sheets or lumpy masses that often bind coral reefs together. Red algae are known from the Cambrian to the Recent, but are most abundant in the Cenozoic Era.

Phylum Sarcoda—The protozoans with pseudopods are a diverse lot which may not be a natural group. Pseudopods are fleshy extensions of the one-celled body used in food capture and locomotion (Fig. 7.27). Sarcodine protozoans are all animal-like, or heterotrophic; that is, they cannot manufacture their own food, but obtain it instead by eating other organisms. Shell-less forms, such as the amoeba (Fig. 7.27A) with its thick pseudopods, are unknown as fossils. The significant fossil groups belong to a separate class characterized by fine pseudopods.

Class Reticularea—These are sarcodines with very fine pseudopods.

Subclass Granuloreticulosia—This group encompasses sarcodines with delicate, anastomosing, unstiffened pseudopods (Fig. 7.27B).

Order Foraminiferoida—The foraminifera are probably still the single most important group of protistans in geology. They possess a **test**, or internal shell (Fig. 7.28), usually composed either of agglutinated mineral particles or, much more commonly, of calcium carbonate (calcite in nearly all). The test of most forams is subdivided internally into a series of chambers, though the simpler varieties

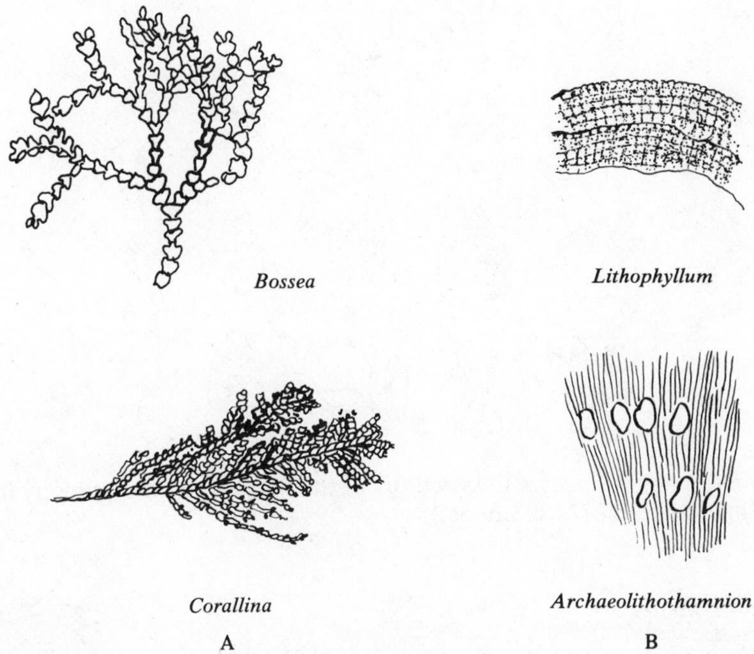

Figure 7.26. *Red Algae (Phylum Rhodophyta).* **A.** *Holocene species of jointed caralline red algae, X 1/2* (after Smith); **B.** *Cross sections of fossil nullipores, X 40* (after Johnson and Ferris).

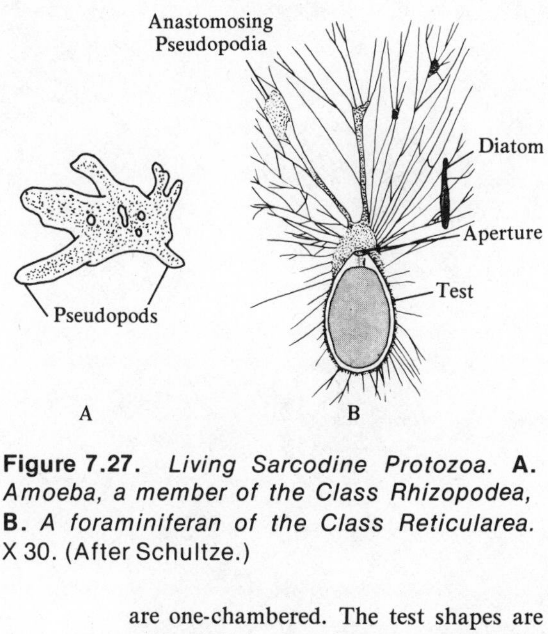

Figure 7.27. *Living Sarcodine Protozoa.* **A.** *Amoeba, a member of the Class Rhizopodea,* **B.** *A foraminiferan of the Class Reticularea.* X 30. (After Schultze.)

are one-chambered. The test shapes are quite diverse, ranging from a simple rod to planispiral and conispiral coils and many others, and the ornamentation is similarly diverse.

Most forams are small fossils barely visible to the naked eye, but some very large varieties have evolved. In the Pennsylvanian and Permian, the large spindle- or grain-shaped **fusulines** (Fig. 7.29A) are abundant; and in the Paleogene, large disk- or coin-shaped **nummulites** and **orbitoids** (Fig. 7.29B) abound. The more typical, smaller forams have been abundant since Jurassic times.

Nearly all forams are marine, and most are benthic. A few very common varieties, the **globigerines** (Fig. 7.30), have become planktonic and are major contributors to oceanic oozes. Forams, being small and abundant, are easily recovered from well cores and cuttings and, hence, have been important tools for petroleum geologists.

Subclass Actinopodia—This group encompasses sarcodines with fine, radiating, stiffened pseudopods called axopods (Fig. 7.31).

Order Radiolaroida—The radiolarians have a spiny, perforate shell or **scleracoma**

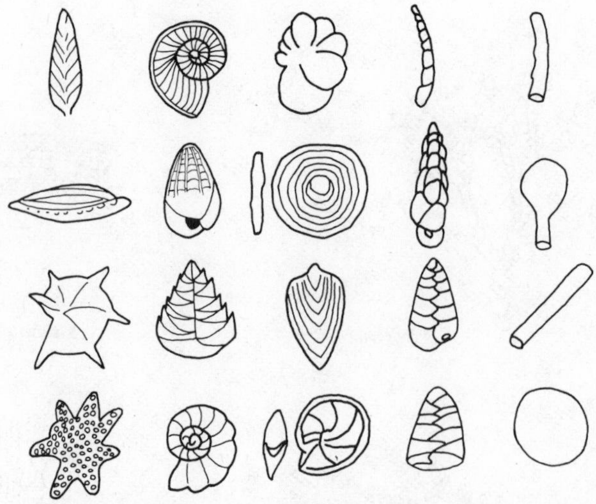

Figure 7.28. *Order Foraminiferoida.* Illustrations of the diversity in morphology found among the forams. Approximately X 25. (After Jones.)

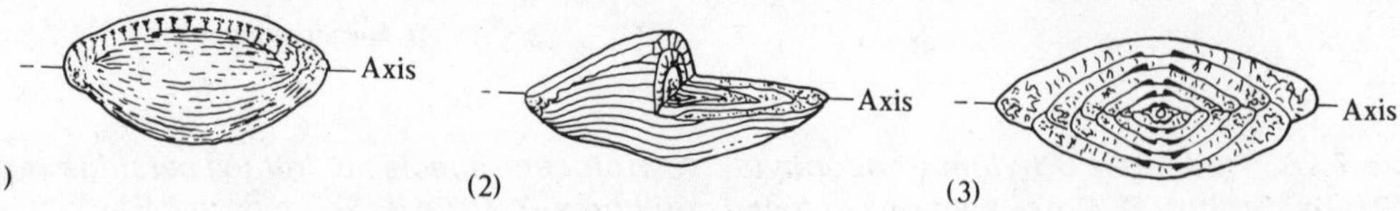

A

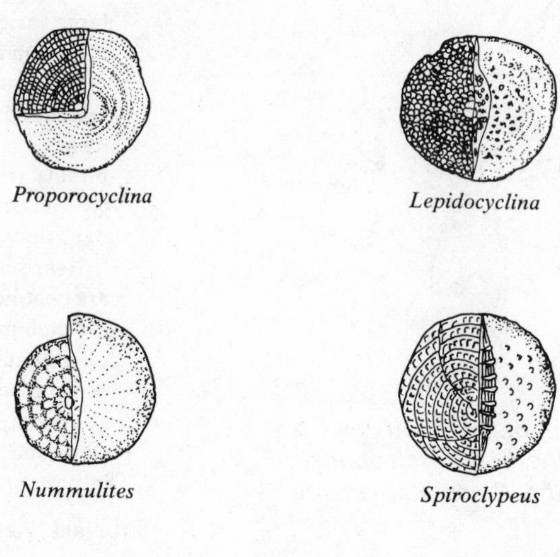

B

Figure 7.29. *Large Forams.* **A.** *Late Paleozoic Fusulines.* 1) External view, 2) partially sectioned shell, 3) cross section showing complex internal anatomy. X 10. (After Henbest.) **B.** *Cretaceous and Cenozoic Nummulites and Orbitoids partially cut away to show internal structures, X 10.* (After Vaughn, Cole, and Cushman.)

(Fig. 7.32) of siliceous composition. The microscopic scleracomae are spherical to conical in shape and occur in marine rocks from Cambrian or Ordovician to

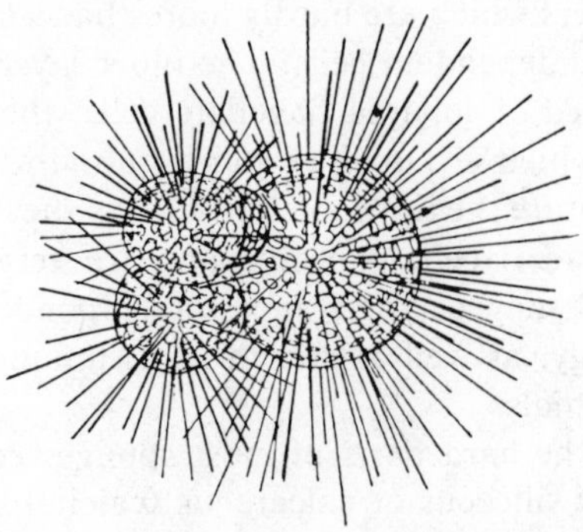

Figure 7.30. *Globigerina, a living planktonic foram.* (From J.M. Crane, Jr., *Introduction to Marine Biology,* © 1973 by Charles E. Merrill Publishing Co.)

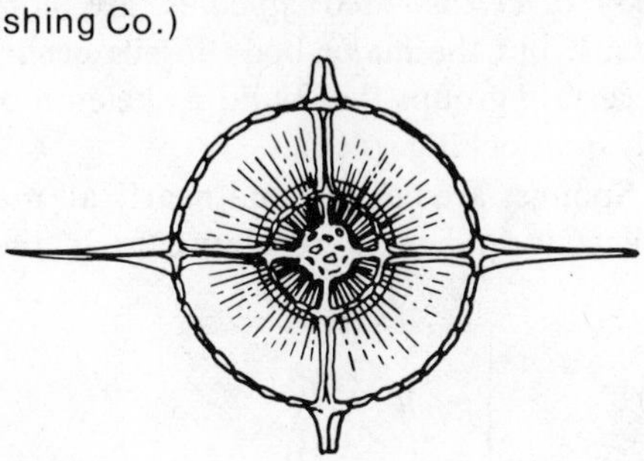

Figure 7.31. *Subclass Actinopodia.* A living radiolarian showing relationship of body and pseudopods to shell or scleracoma. (After Hertwig.)

Recent in age, though they are most abundant from the Late Mesozoic onward. They are all planktonic, and most frequent the photic zone because of photosynthetic algae in their bodies which are necessary to their metabolism.

Phylum Ciliophora—The protozoans with cilia (Fig. 7.33), tiny hair-like projections used chiefly for locomotion, are typically not fossilizable creatures because most lack hard parts. All are one-celled, but very complex; and all are animal-like, eating other creatures for their nutrition. Many familiar ciliates have uniform cilia over their bodies, but the major fossil group does not belong to that class.

Class Spirotrichea—These are ciliates in which the cilia are restricted to the oral region.

Order Tintinnoida—Tintinnoids (Fig. 7.34), which compose nearly half of all living ciliate species and virtually all known fossil types, possess a conical agglutinated or calcareous shell, or **lorica.** Most are marine and planktonic. They range in age from Ordovician to Recent, but most fossils are from the Jurassic and Cretaceous.

Animal Kingdom

Animals, like protistans and plants, have a nucleus. All are multicellular, and the cells are dif-

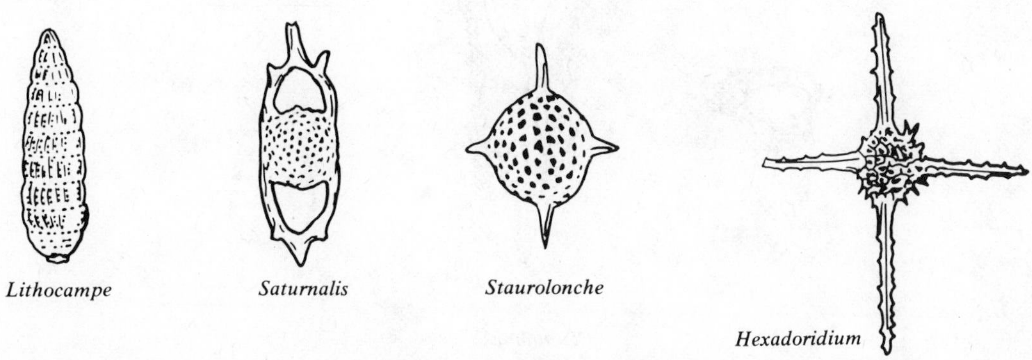

Figure 7.32. *Order Radiolaroida.* Fossil scleracomae. Approximately X 75. (Adapted from *Invertebrate Fossils* by Moore-Lalicker-Fischer, Copyright 1952 by McGraw-Hill, Inc., used by permission of McGraw-Hill Book Company.)

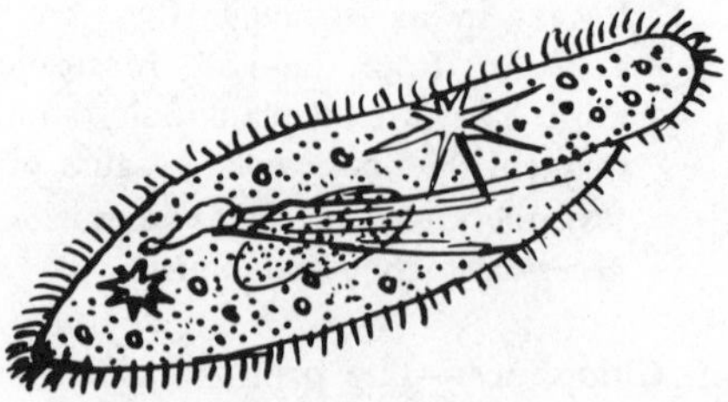

Figure 7.33. *Paramecium, a living ciliate protozoan* of the Class Holotrichea.

ferentiated for different functions. Animals are unable to synthesize their own food from inorganic substances and, hence, must obtain their nutrients from other organisms. Their gametes develop in multicellular structures, and embryos are formed. There are twelve major animal groups: sponges, archaeocyathans, coelenterates, bryozoans, brachiopods, mollusks, annelid worms, arthropods, echinoderms, graptolites, conodonts, and vertebrates.

Phylum Porifera—The **sponges** (Figs. 7.35, 7.36), whose relationship to the other animals is uncertain, have a sac-like body perforated by pores for water intake and a larger opening (osculum) or openings (oscula) for water outflow. The body consists of three weak tissue layers which are hardly more than aggregates of independent cells. The inner layer is composed of collared flagellate cells, the beating of which sets up the water flow and which also capture the tiny food particles as they flow by. Some flagellate protozoans are very similar and may be their ancestors. Sponge embryology, as well as structure, is unique among animals.

The hard parts of most sponges consist of tiny siliceous or calcareous (calcite) elements called **spicules** (Fig. 7.37), which come in a variety of shapes from simple rods to radiating star-shaped patterns and occur in the middle body layer. Isolated spicules are common fossils, but the major body fossils occur only in certain groups that build a skeleton of rigidly interlocking spicules.

Sponges are benthic and nearly all marine. Siliceous sponges were abundant in the De-

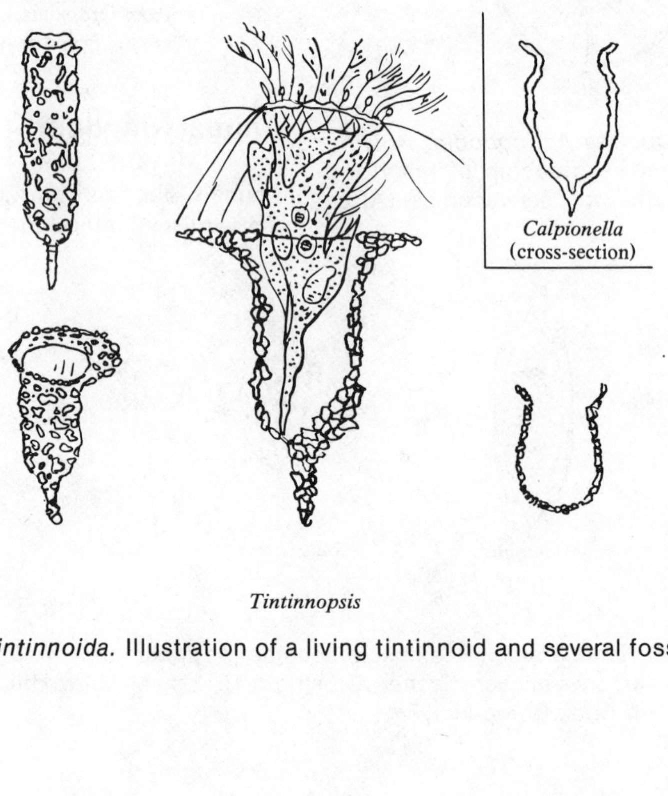

Figure 7.34. *Order Tintinnoida.* Illustration of a living tintinnoid and several fossil loricas. X 150. (After Colom.)

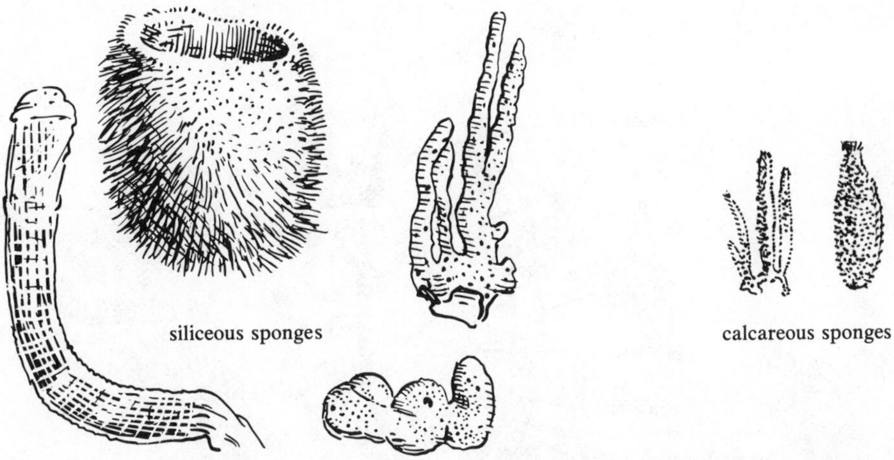

Figure 7.35. *Sponges (Phylum Porifera).* External anatomy of some living sponges. (Adapted from *Invertebrate Fossils* by Moore-Lalicker-Fischer, Copyright 1952 by McGraw-Hill, Inc., used by permission of McGraw-Hill Book Company.)

vonian and again in the Jurassic-Cretaceous interval, while the less common calcareous types reached a peak in the latter two periods.

Some authors include the common Paleozoic fossils called stromatoporoids (Fig. 7.42) in the sponges because of external similarities. The internal anatomy of these extinct creatures is quite different from the sponges and is much closer to the hydrocorals, near which they are placed in this book.

Phylum Archaeocyatha—Archaeocyathans (Fig. 7.38) are one of several common fossil groups that have no living representatives; hence, their systematic position is uncertain. The basic structure of most is a calcareous (calcite), double-walled cone, the **cup,** with the two walls connected by vertical plates called **septa** (singular: septum). All body parts are perforated by pores. Only a few forms are colonial. The general shape and the vertical partitions are like those of corals which, however, lack the inner wall and pores. The perforations suggest sponge affinities, but sponge skeletons consist only of spicules. The general structure of archaeocyathans is also suggestive of dasycladaceous green algae. The consensus of experts is that they are somehow related to sponges. Archaeocyathans (Fig. 7.39) were all marine and commonly built reefs. They are all Cambrian, and most are Lower Cambrian.

Phylum Coelenterata—This large and diverse phylum includes **hydroids, jellyfish, sea anemones,** and **corals.** The body (Fig. 7.40) is saclike with only a single opening into the digestive cavity. This opening, the mouth, is surrounded by tentacles used for capturing prey. Unlike the majority of animals, most coelenterates are fundamentally radially symmetrical; that is, the body parts are disposed in a radial manner about the axis like the rays of the sun, though this symmetry is typically imperfect. Many have two stages in their life cycles (Fig. 7.41), a sessile benthonic polyp attached by its base with tentacles and mouth directed upward, and a pelagic umbrella-shaped medusa, or jellyfish, with these structures directed downward. Polyps asexually give rise to medusas which then reproduce sexually to again form polyps. Most groups lack hard parts and, hence, are rare fossils, but the stony corals and the problematic stromatoporoids and conularoids have hard parts and are common fossils. The fossil

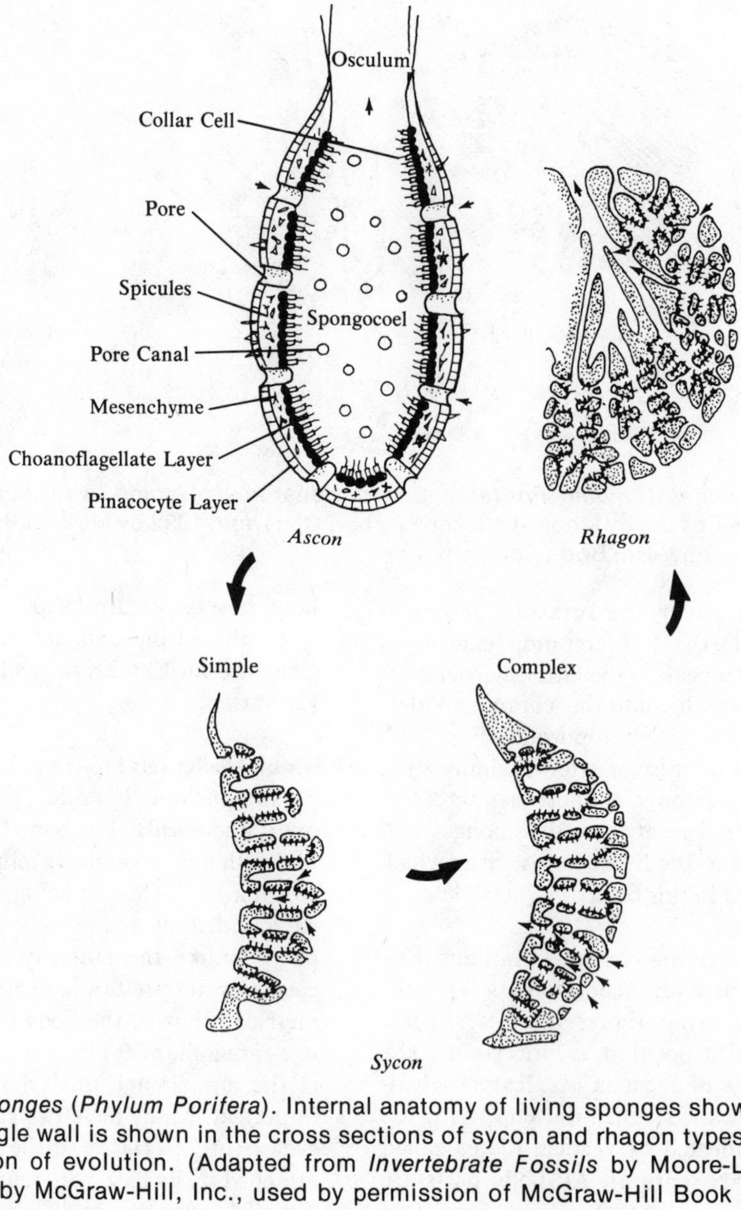

Figure 7.36. *Sponges* (*Phylum Porifera*). Internal anatomy of living sponges shown in cross section. (Only a single wall is shown in the cross sections of sycon and rhagon types.) Small arrows indicate direction of evolution. (Adapted from *Invertebrate Fossils* by Moore-Lalicker-Fischer, Copyright 1952 by McGraw-Hill, Inc., used by permission of McGraw-Hill Book Company.)

record of the phylum goes back to the Eocambrian, where several jellyfish and soft corals occur.

Class Hydrozoea—The polyps, which are typically colonial, are the major stage in the life cycle (Fig. 7.41) of this group, the medusa being small. The skeleton of most living types is a tough organic substance rarely preserved, though a few—the **hydrocorals**—have an aragonite skeleton. The long, sparse fossil record of these forms goes back to the Cambrian, and nearly all are marine.

? Order Stromatoporoida—The stromatoporoids (Fig. 7.42) are another extinct group of uncertain affinities. Their calcite hard

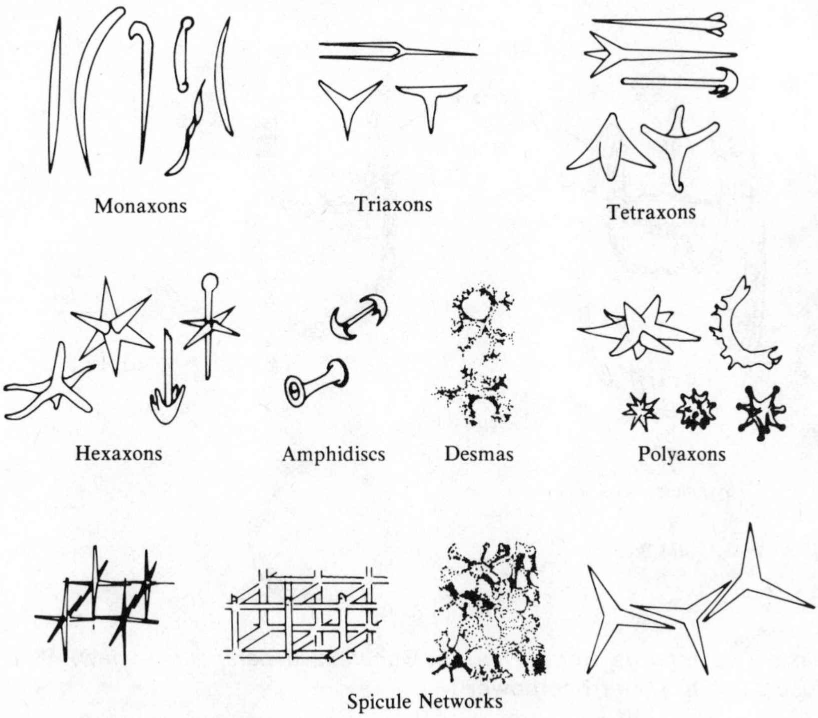

Figure 7.37. *Sponge Spicules.* Various types and their names. Bottom row illustrates networks which form most sponge body fossils.

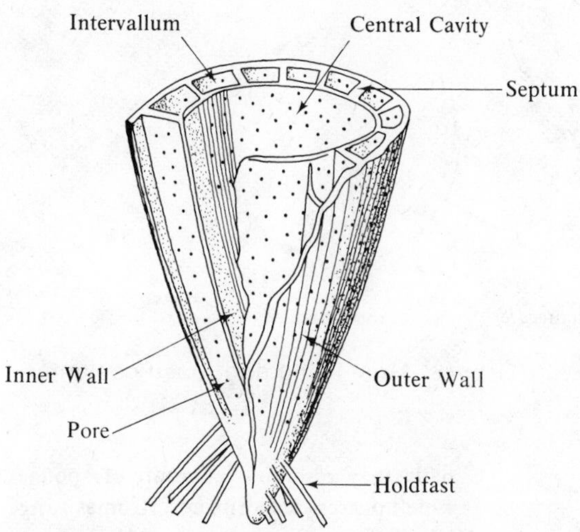

Figure 7.38. *Phylum Archaeocyatha.* Morphology of a typical archaeocyath cup. (Adapted from *Invertebrate Fossils* by Moore-Lalicker-Fischer, Copyright 1952 by McGraw-Hill, Inc., used by permission of McGraw-Hill Book Company.)

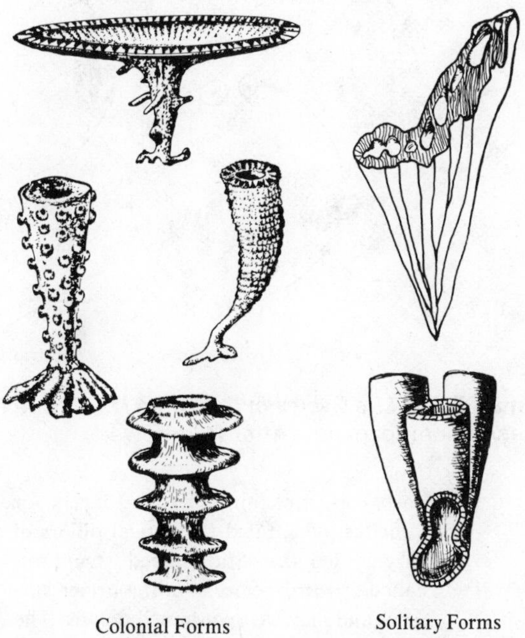

Figure 7.39. *Variation among Archaeocyath Cups.* (From *Treatise on Invertebrate Paleontology,* courtesy of the Geological Society of America and University of Kansas Press.)

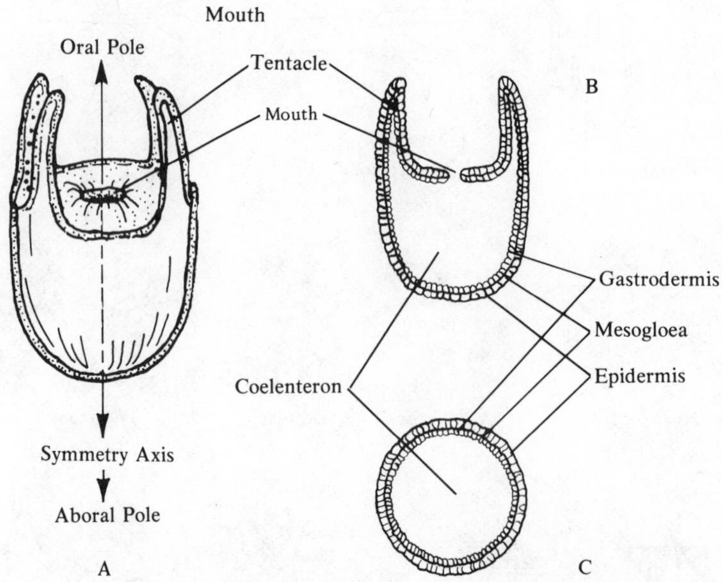

Figure 7.40. *Basic Coelenterate Body Plan.* **A.** *Generalized perspective view,* **B.** *Longitudinal section,* **C.** *Cross section.* (After Beerbower.)

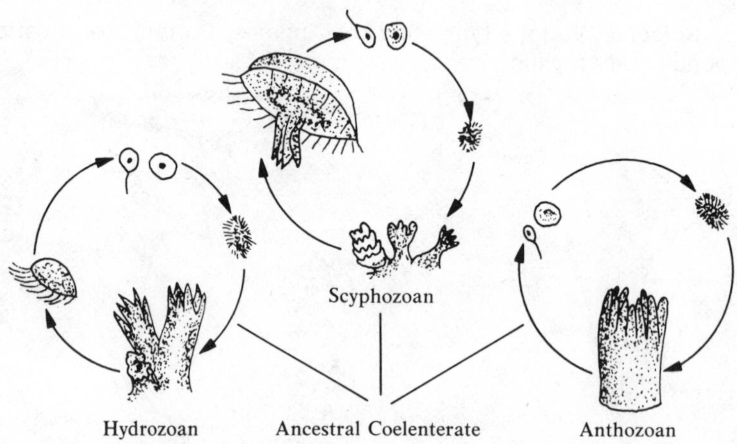

Figure 7.41. *Life Cycles in the Three Major Coelenterate Groups.* Most living and fossil coelenterates are anthozoans. (After Beerbower.)

parts consist of thin horizontal layers, or **lamellae,** penetrated by vertical **pillars** of varying lengths. Star-shaped structures called **astrorhizae** occur in the upper surfaces and may have contained polyps. The shape of the marine benthic colonies is variable, but typically stromes form large, lumpy masses. The internal anatomy is suggestive of some hydorcoral groups, hence, this systematic placement. Externally they resemble a group of sponges which possess astrorhizae. Stromes range in age from Cambrian or Ordovician to Cretaceous, though some experts exclude from the group all post-Devonian types. They were major reef-building elements in Silurian and Devonian times. A related order, the **spongiomorphoids** (Fig. 7.43), is common in Triassic warm-water deposits.

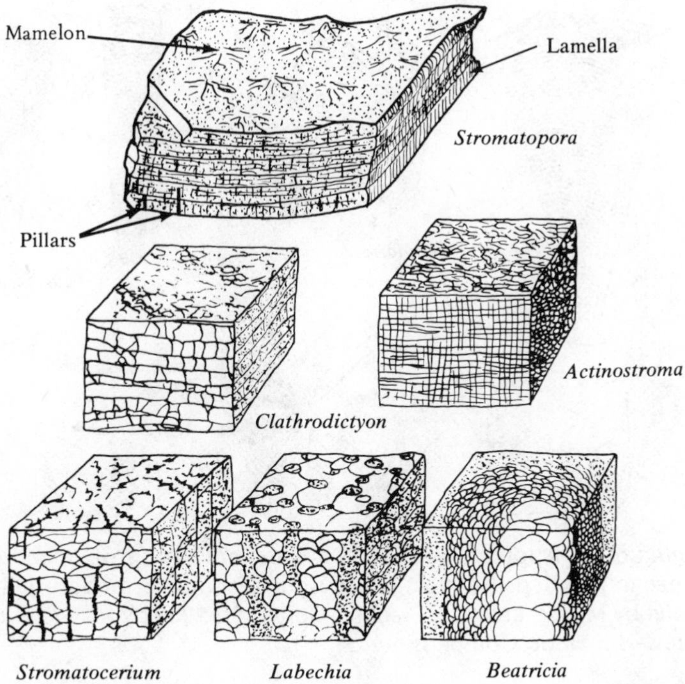

Figure 7.42. *Stromatoporoids.* Members of an extinct group which may belong to the Porifera or Coelenterata. Note astrorhizae on tops of several genera. Illustration shows blocks cut from typically amorphous skeletons. X 1/3. (Adapted from *Invertebrate Fossils* by Moore-Lalicker-Fischer, Copyright 1952 by McGraw-Hill, Inc., used by permission of McGraw-Hill Book Company.)

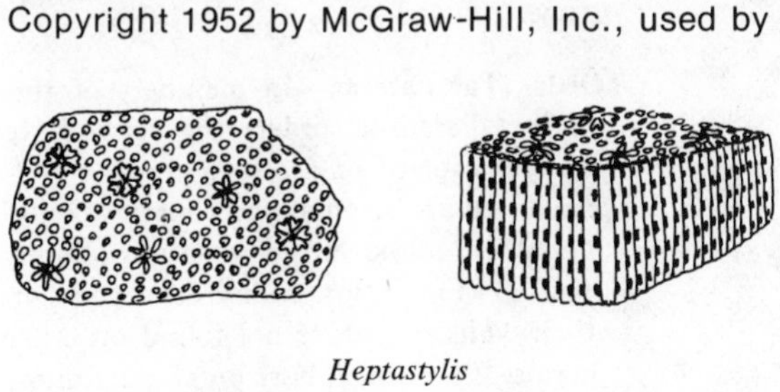

Figure 7.43. *Spongiomorphoid.* Top and side views of a member of this extinct group related to the stromatoporoids. X 2.

Class Scyphozoea—The pelagic medusa stage dominates in this class, the polyps being small (Fig. 7.41). The large jellyfish have a fourfold symmetry, are all marine, and are rare fossils because they lack hard parts.

? Order Conularoida—These flexible, four-sided, conical, chitinophosphatic shells (Fig. 7.44) are representatives of a totally extinct group. At first glance they appear most unlike jellyfish, but they have four internal partitions and tentacles like scyphomedusae; and one form is known with a low domical shape similar to that of jellyfish. They are all marine and range in age from Cambrian through Triassic, reaching a maximum in the Upper Paleozoic.

Class Anthozoea—Anthozoans are polyps only; the medusa stage is lacking. The polyps are complex and subdivided internally by vertical partitions, or mesenteries (Fig. 7.45). All are marine and benthic. The sea anemones and soft corals lack hard parts and are rare as fossils. The **stony corals** have a conical or cylindrical skeleton, the **theca**, subdivided, in most cases, by radiating internal partitions, the **septa**.

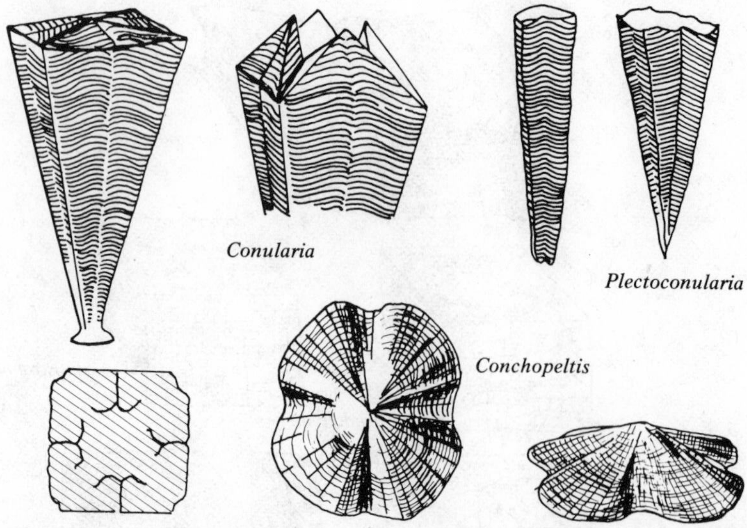

Figure 7.44. *Conularoids.* Approximately X 2. An extinct group of uncertain affinites, perhaps related to scyphozoan jellyfish. Form at lower right resembles modern jellyfish. (Adapted from *Invertebrate Fossils* by Moore-Lalicker-Fischer, Copyright 1952 by McGraw-Hill, Inc., used by permission of McGraw-Hill Book Company.)

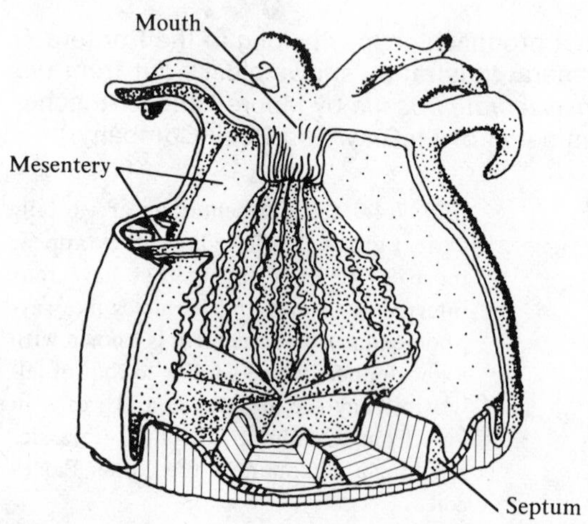

Figure 7.45. *A Living Coral Polyp,* magnified and sectioned to show internal anatomy and its relationship to the hard parts. (Adapted from *Invertebrate Fossils* by Moore-Lalicker-Fischer, Copyright 1952 by McGraw-Hill, Inc., used by permission of McGraw-Hill Book Company.)

Subclass Zoantharia—This group includes the stony corals and most sea anemones. The theca, when present, begins with six major septa, and the orders are based on the positions where new septa are added as the animal grows upward.

? Order Tabulatoida—In members of this extinct Paleozoic order, the septa have not yet evolved or are very poorly developed. All are colonial and consist of polygonal or round tubes generally closely packed (Fig. 7.46). Many have pores in their walls, a feature not found on other corals. Prominent horizontal partitions, or tabulae, supported the base of the animal as it grew upward as in other corals. Tabulates range in age from Late Cambrian through the Permian (a few may have survived later) but were important chiefly in the Lower Paleozoic, especially the Silurian, when they built large reefs. Some workers believe they are really sponges; others believe that the group is artificial and includes a variety of animals of different relationships.

Order Rugosoida—Another extinct group, these corals are often called tetracorals (Fig. 7.47) because they insert their new septa in only four of the six spaces be-

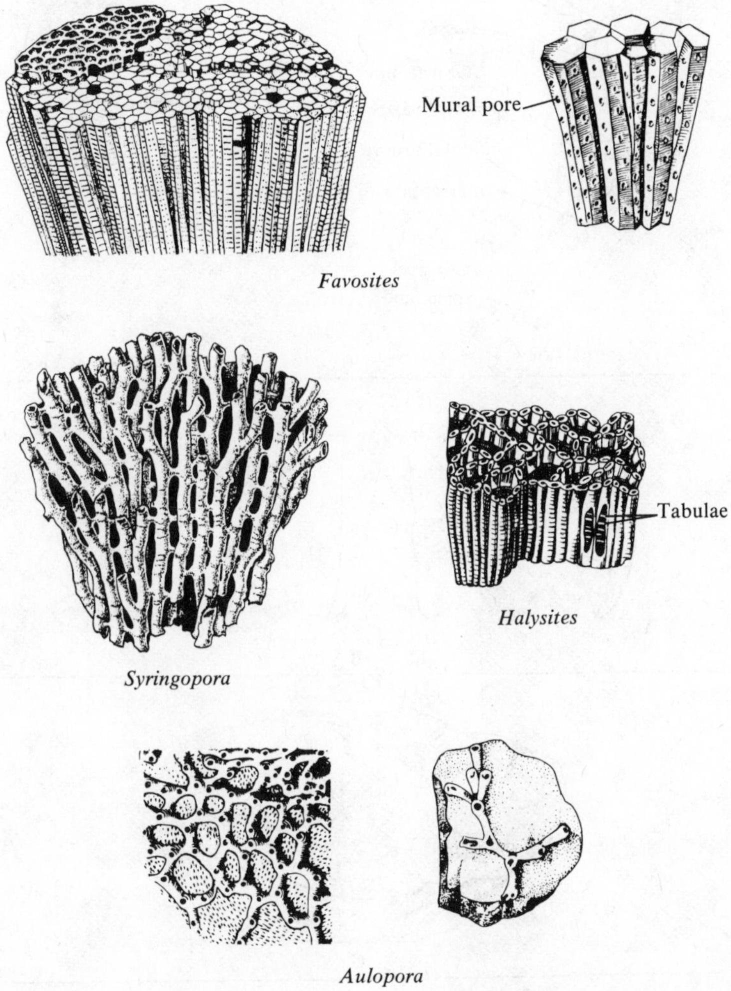

Figure 7.46. *Morphology of Tabulate Corals.* Four representative genera. (From *Treatise on Invertebrate Paleontology*, courtesy of The Geological Society of America and University of Kansas.)

tween the initial septa (Fig. 7.48). They are either solitary or colonial, the former type dominating and having a typically curved theca; hence, the name horn corals. The top usually contains a depression, or **calyx,** for the body. Rugose corals were the major reef builders of the Late Paleozoic, particularly the Devonian. They are known from the Ordovician through the very earliest Triassic.

Order Scleractinoida—These are usually known as hexacorals (Fig. 7.49) because they add new septa in all six spaces between the original six (Fig. 7.48). They are both solitary and colonial, with the latter type dominating. The solitary forms are typically uncurved and lack a distinct outer wall; hence, there is usually no calyx. Hexacorals, which range from Triassic to Recent, include the major reef-building corals of Mesozoic and Cenozoic times. The reef builders are tropical and restricted to shallow water because of photosynthetic algae in their tissues which are necessary for their metabolism.

Phylum Bryozoa—The bryozoans, or moss animals, appeared in the Cambrian and have

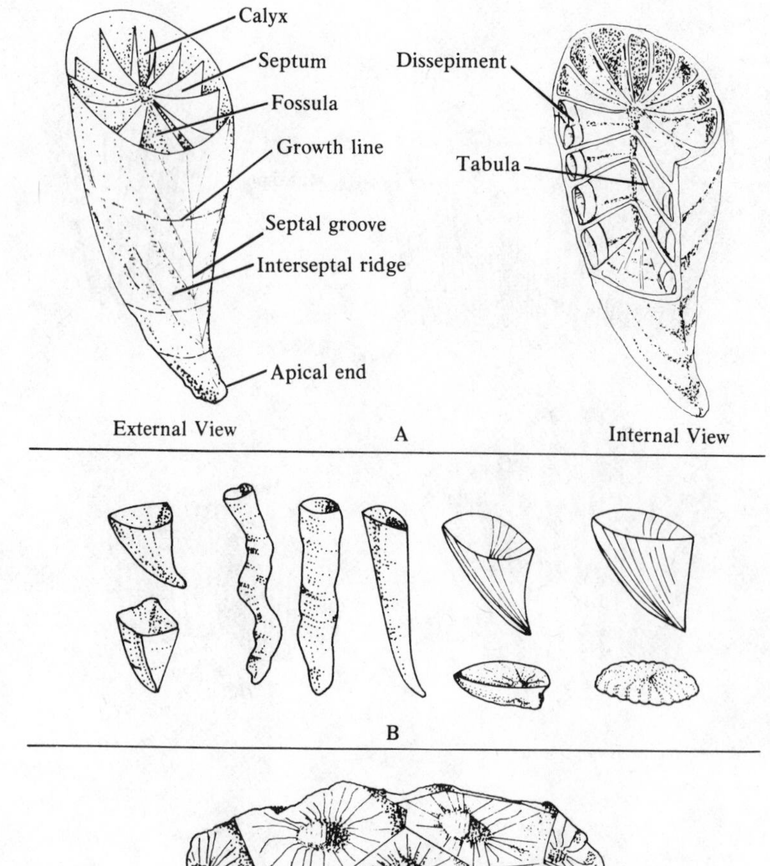

Figure 7.47. *Morphology and Diversity of Rugose Corals* (*Tetracorals*). **A.** *Theca of solitary coral,* **B.** *Variation among solitary coral thecae,* **C.** *Colonial coral,* **D.** *Variation among compound corals.* Parts A and C approximately natural size. (Adapted from *Invertebrate Fossils* by Moore-Lalicker-Fischer,Copyright 1952 by McGraw-Hill, Inc.,used by permission of McGraw-Hill Book Company.)

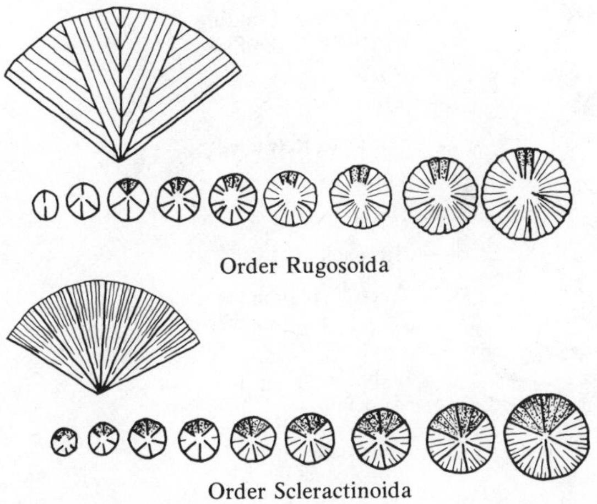

Figure 7.48. *Patterns of Septal insertion in Tetracorals and Hexacorals.* The two dorsal interspaces between initial septa are shaded for reference in the cross sections. The two lateral views of unrolled thecae show how the pattern of septal insertion is recognized externally.

been very abundant from the Ordovician onward. Individually, they are microscopic, but all are colonial and the colonies often are many inches across. The tiny bodies, or zooids. (Fig. 7.50) of individual bryozoans contain a retractible portion, the polypide, which has a U-shaped gut and a ciliated tentacle-bearing ridge, or lophophore, used in capturing microscopic food. The soft parts of most are encased in a calcareous (mixture of calcite and aragonite) skeleton, or **zooecium,** collectively known as the **zooarium** of the colony. Bryozoans are all benthic and nearly all marine.

Class Stenolaematea—These have tubular zooecia with a terminal aperture.

Order Cyclostomoida—Cyclostome bryozoa (Fig. 7.51) either lack the internal horizontal partitions of other tubular bryozoans or have less numerous ones. Zooecial

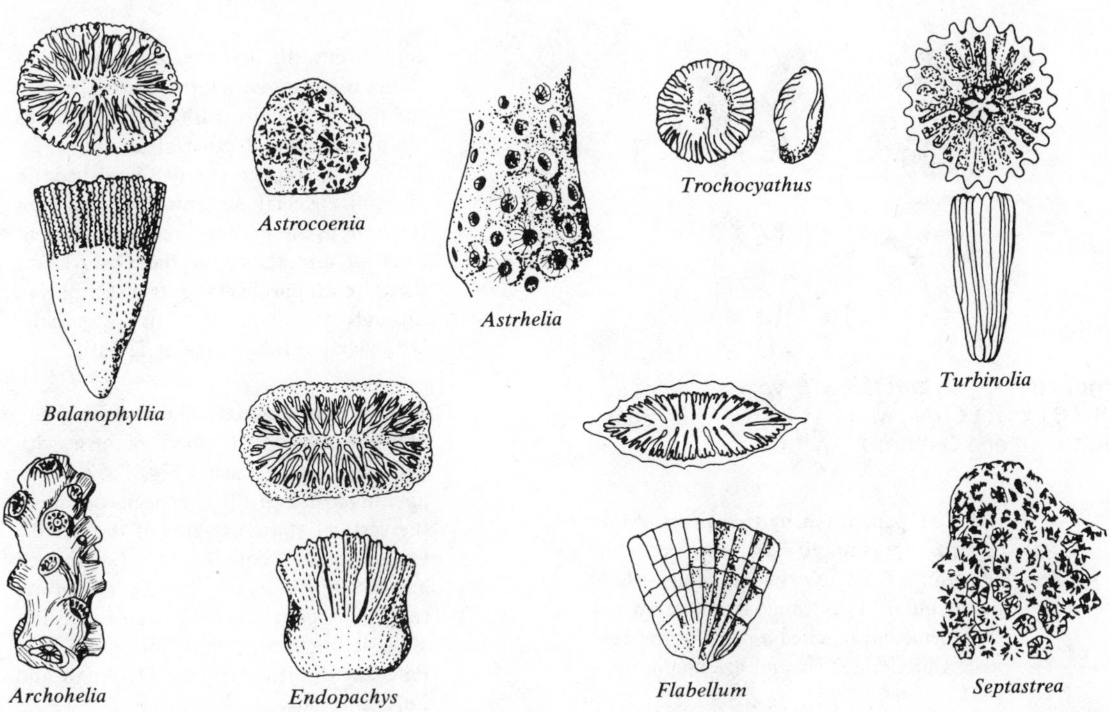

Figure 7.49. *Diversity of Fossil Scleractinian Corals (Hexacorals).* Solitary and colonial types shown. Approximately natural size. (Adapted from *Invertebrate Fossils* by Moore-Lalicker-Fischer, Copyright 1952 by McGraw-Hill, Inc., used by permission of McGraw-Hill Book Company.)

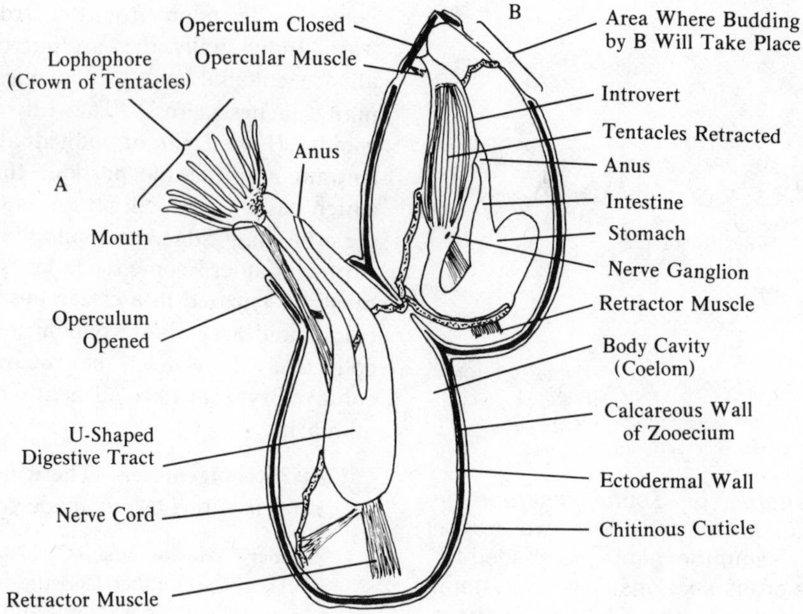

Figure 7.50. *Structure of a Living Bryozoan* **A.** *with polypide extruded,* **B.** *with polypide retracted.* Highly magnified. (Adapted from Delage and Hérouard.)

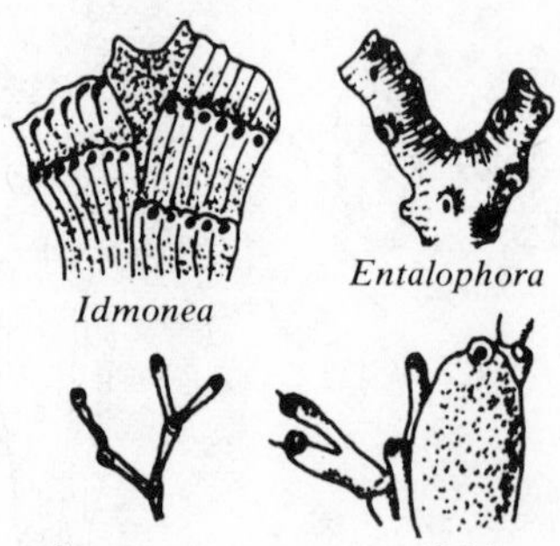

Figure 7.51. *Cyclostome Bryozoan Fossils.* All X 8 except *Crisia* which is X 15. (After Bassler, Canu, and Osburn.)

walls are penetrated by tiny pores. Most zooaria are smooth and twiglike, but some form delicate networks while others are globular. They range from Cambrian to Recent, but reached a peak in the Jurassic and Cretaceous and are common in the Cenozoic.

Orders Cystoporoida and **Trepostomoida—** These extinct stony bryozoa (Fig. 7.52) have typically twig-shaped zooaria with their surfaces covered by mounds and depressions, though the latter order includes many massive forms also. They are complex internally with cross-partitions (diaphragms) and other structures (Fig. 7.52). The orders are distinguished on internal zooecial features and need not concern us here. Both reached a great peak of abundance in the Ordovician, though cystoporoids range from Cambrian through Permian; and trepostomoids, Ordovician through earliest Triassic.

Order Cryptostomoida—This extinct group of bryozoans consists chiefly of forms with delicate, lacy zooaria (Fig. 7.53). The openings of individual zooecia occur on the vertical and cross-bars of the fronds. One common genus has a screw-shaped axis to support its lacy fronds. The group ranges in age from Ordovician to earliest Triassic but is most common in the Late Paleozoic, particularly the Devonian and Carboniferous.

Class Gymnolaematea—These typically have box-like zooecia with lateral openings. (Fig. 7.54).

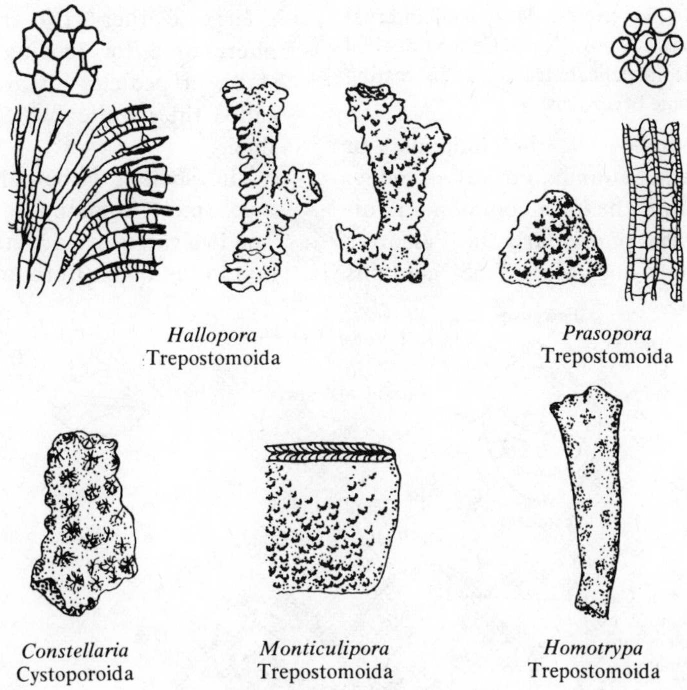

Figure 7.52. *External and Internal Morphology of Trepostome and Cystoporoid Bryozoans.* Cross sections of *Hallopora* and *Prasopora* show complex internal anatomy. All X 1 except cross sections, which are X 10. (After Bassler, Cummings, and Ulrich.)

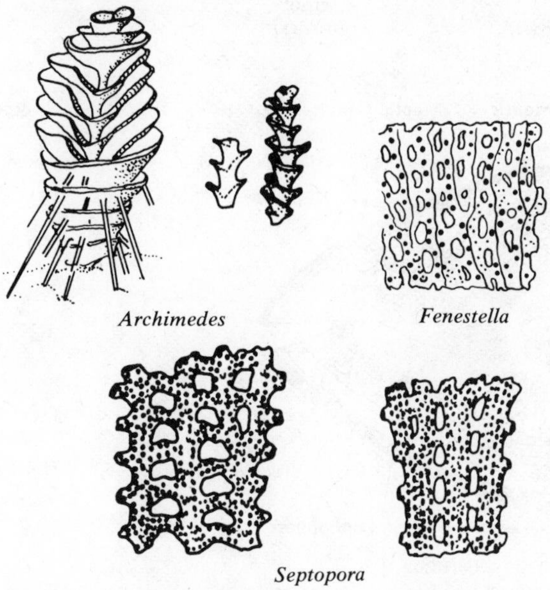

Figure 7.53. *External Morphology of Cryptostome Bryozoans. Archimedes* X 1, others X 5. (After Easton and Ulrich.)

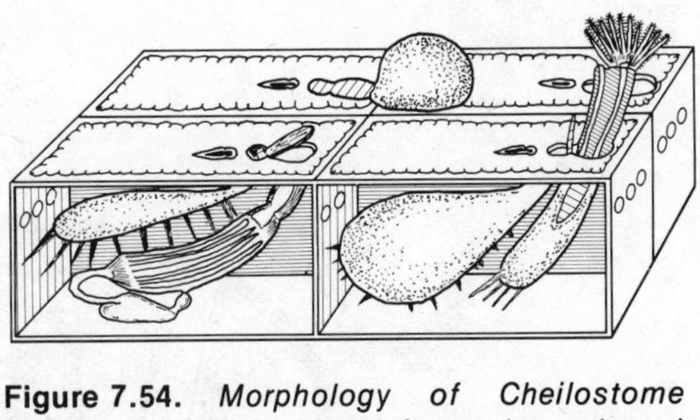

Figure 7.54. *Morphology of Cheilostome Bryozoans.* Individual at left rear has a brood pouch. Individual at left front is retracted, while individual at right front is extruded by pressure of expanded compensation sac. X 100. (After Bassler.)

Order Cheilostomoida—This, the largest group of living bryozoans, appeared in the Late Jurassic and has dominated from the Cretaceous onward (Figs. 19.15 and 21.11). The zooaria are very diverse, but are always very thin, commonly forming sheet-like incrustations, delicate twigs, or blade-like fronds. Most lack internal partitions. Almost any Cenozoic shell contains the delicate tracery of encrusting cheilosome bryozoans.

Phylum Brachiopoda—The brachiopods, or lamp shells, are not abundant or varied today, but they are among the most common and diverse megafossils, particularly in Paleozoic rocks. The simple body (Fig. 7.55) contains a large, feathery, two-armed anterior lophophore used for feeding; a fleshy posterior stalk, or pedicle, for attachment; and flaps called the mantle, which secrete a bivalved shell.

The **shell** is either chitinophosphatic or, much more commonly, calcareous (calcite). The two valves are dorsal and ventral in relation to the body. Both are bilaterally symme-

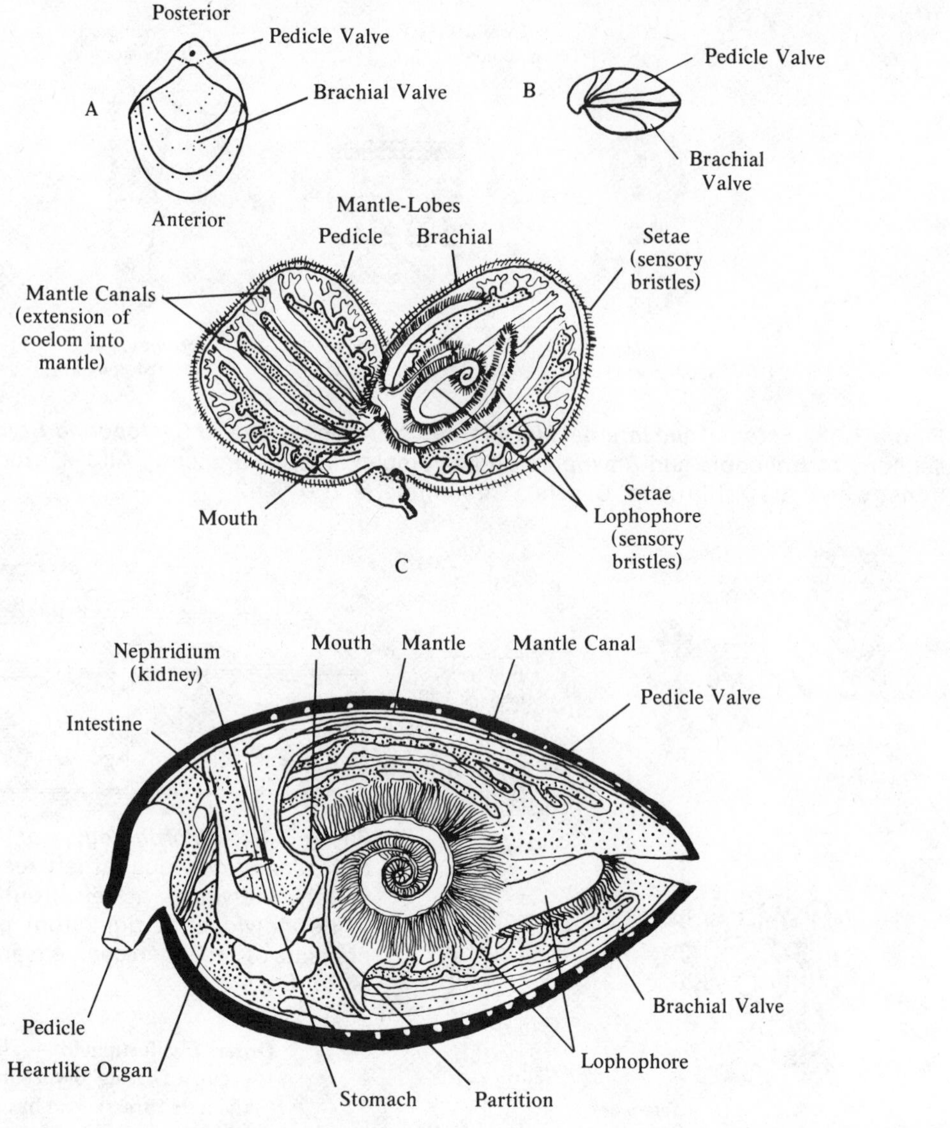

Figure 7.55. *Phylum Brachiopoda.* The shells and soft parts of a Recent Brachiopod. **A.** *Ventral view,* **B.** *Lateral view,* **C.** *Body with mantle lobes separated,* **D.** *Section among symmetry plane.* (Adapted from Davidson and Parker and Haswell.)

trical with one, the valve housing the pedicle and called the pedicle valve, typically larger than the other, called the brachial valve because it supports the lophophore arms. The posterior prolongations of the valves, the beaks, are typically much larger on the pedicle valve because they enclose the pedicle. Beginners often confuse brachiopods with clams. The latter belong to the much more complex phylum Mollusca and are also bivalved; however, in clams, the valves are typically right and left in relation to the body, and the plane of symmetry passes between the valves, each valve not being symmetrical about its midline (Fig. 7.56).

Brachiopods are all marine and benthic. They range in age from Cambrian to Recent, but are most abundant in Paleozoic and some Mesozoic rocks. The major types of this most significant fossil group are as follows:

Class Inarticulatea—In the inarticulate brachiopods (Fig. 7.57), the two valves are not hinged and are typically chitinophosphatic in composition. The inarticulates have a long geologic record from Cambrian to Recent, but were most abundant in the Cambrian, when they were the major brachiopod group, and in the Ordovician.

Class Articulatea—Members of this much larger group of brachiopods have hinged, calcareous shells. On the inside of the brachial valve are supports for the lophophore.

Order Orthoida—Orthoids (Fig. 7.58A) have squarish-to-round, biconvex shells. The posterior hinge line is typically wide. The valves are covered with fine radial ornamentation, or costae, in most. Orthoids range from Cambrian through earliest Triassic, reaching an Ordovician peak.

Order Strophomenoida — Strophomenoid shells (Fig. 7.58B) are similar to orthoids, but one valve is concave. The pedicle opening is sealed in adults, indicating loss of the pedicle. The order ranges from Ordovician through Carboniferous, but has an Ordovician peak.

Order Chonetoida—Chonetoids (Fig. 7.58C) are like strophomenoids, but have a row of spines along the posterior margin of

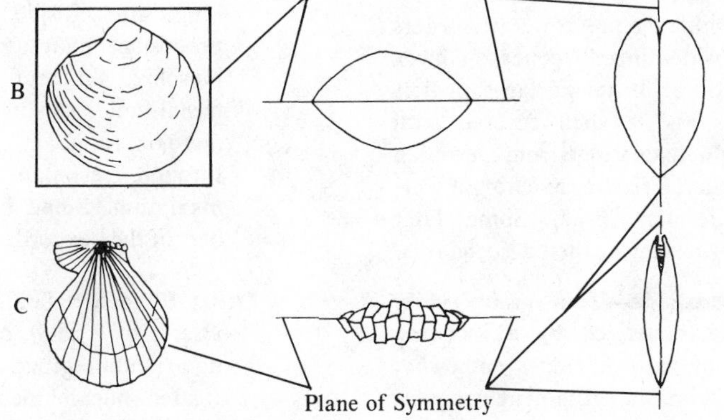

Figure 7.56. *The Morphology of Brachiopod and Bivalve Shells Contrasted.* **A.** *Typical brachiopods,* **B.** *Typical bivalve,* **C.** *Scallop, a bivalve with symmetry resembling a brachiopod.*

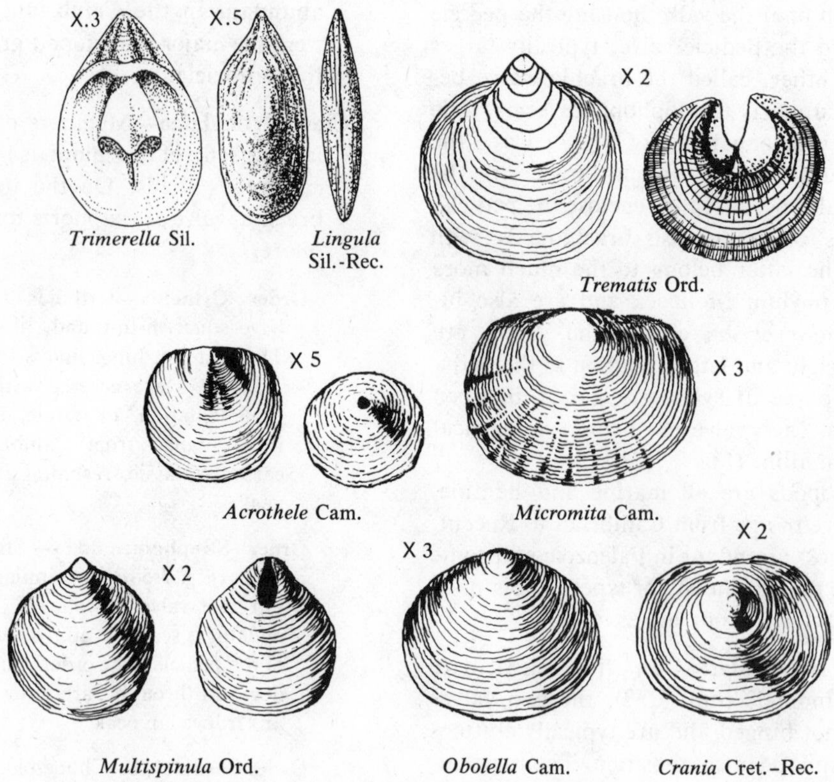

Figure 7.57. *Class Inarticulatea.* Representative fossil inarticulate brachiopods. Magnifications indicated.

their concave-convex shell. They range from Ordovician through Permian, reaching a Devonian-Carboniferous peak. Some peculiar Mesozoic forms may be their descendants.

Order Productoida — Productoids (Fig. 7.58D) resemble the preceding two orders in having wide-hinged, concave-convex shells, but differ in having large pedicle valve beaks and a shell covered with spines. Productoids range from Devonian through earliest Triassic, reaching a Carboniferous-Permian peak. Some later Mesozoic forms may be their descendants.

Order Davidsonioida—This group (Fig. 7.58E) is characterized by a biconvex shell with one valve having a somewhat conical shape and a prominent irregular beak. Davidsonioids range from Ordovician through Permian and are common in Carboniferous and Permian rocks. Some Triassic forms may belong here.

Order Pentameroida—Pentameroids (Fig. 7.58F) have biconvex shells with short hinge lines. Inside the valves they have prominent plates which form areas of muscle attachment. These appear on internal molds as slits (Fig. 7.58F). Pentameroids range in age from Cambrian through Devonian, reaching a Silurian maximum. Some later Paleozoic members of the next order may belong here.

Order Rhynchonelloida—Members of this order (Fig. 7.58G), one of the two surviving articulate groups, possess highly corrugated (plicate) biconvex shells with very narrow hinges and sharp beaks. They

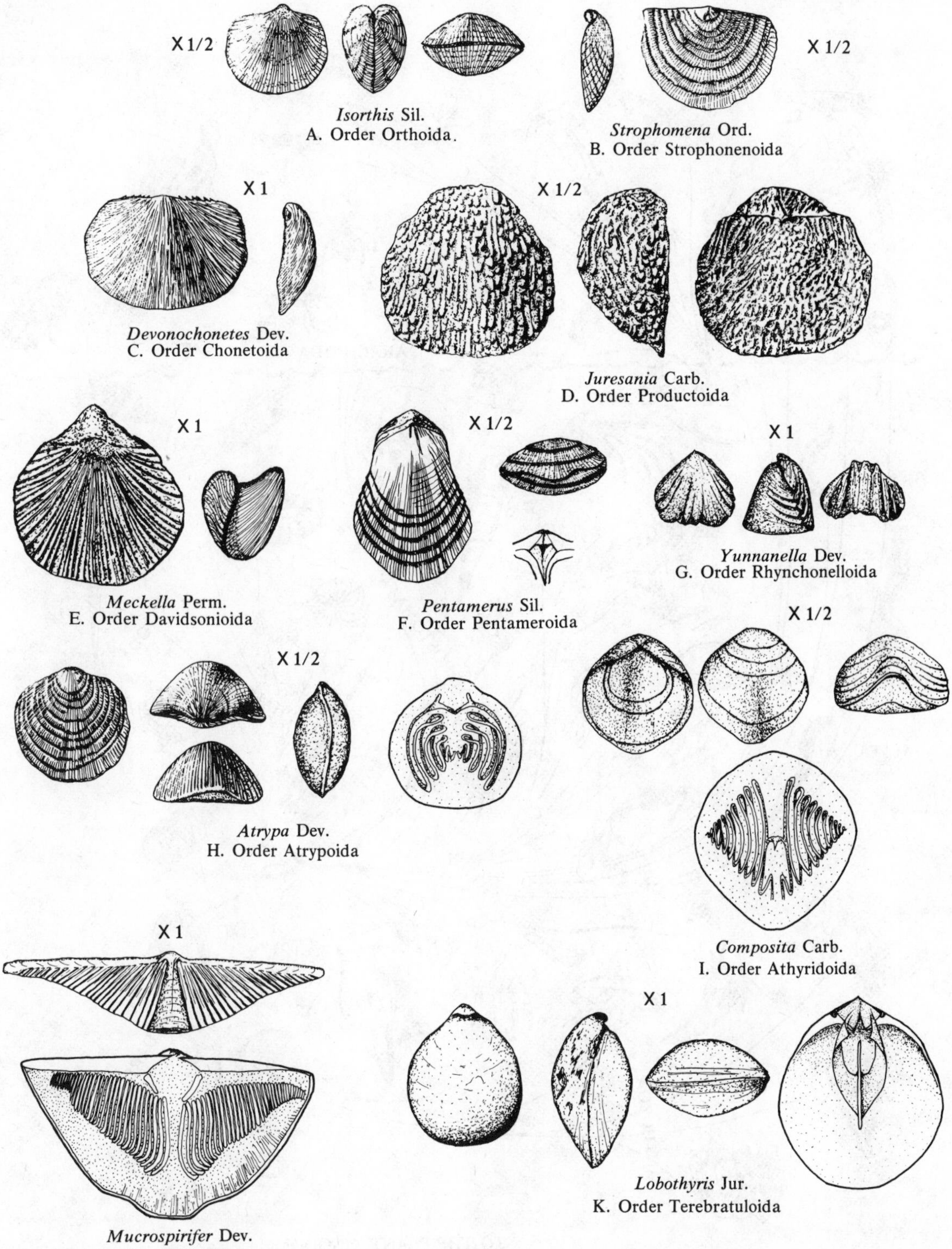

Figure 7.58. *Class Articulatea.* Representatives of the major articulate brachiopod orders. Magnifications indicated. (After *Treatise on Invertebrate Paleontology*.)

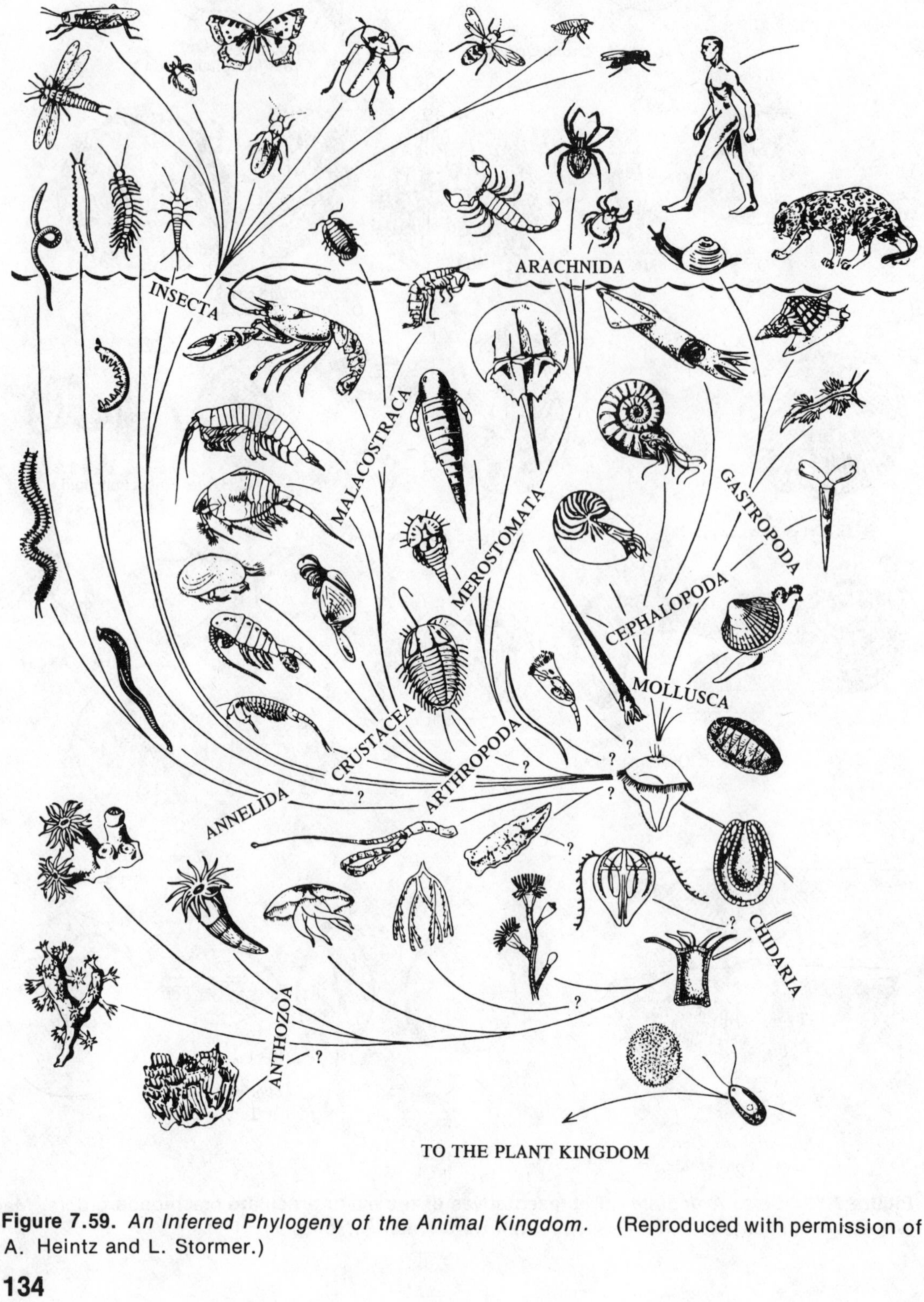

Figure 7.59. *An Inferred Phylogeny of the Animal Kingdom.* (Reproduced with permission of A. Heintz and L. Stormer.)

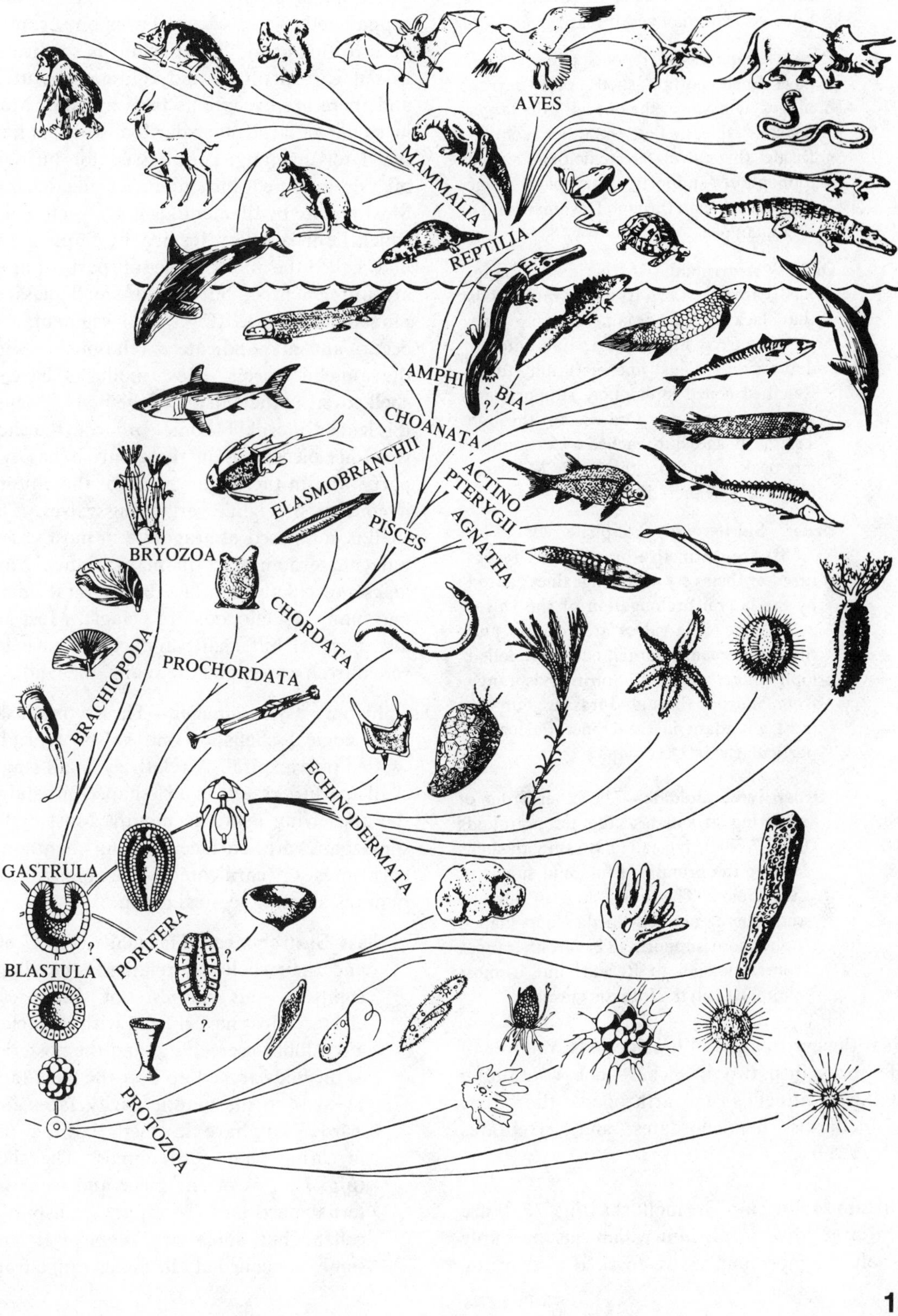
AVES
MAMMALIA
REPTILIA
AMPHI BIA
?
CHOANATA
ELASMOBRANCHII
ACTINO PTERYGII
PISCES
AGNATHA
BRYOZOA
CHORDATA
BRACHIOPODA
PROCHORDATA
ECHINODERMATA
GASTRULA
?
PORIFERA
?
BLASTULA
?
PROTOZOA

range from Ordovician to Recent, with Silurian-Devonian and Jurassic peaks.

Order Atrypoida—Atrypoids (Fig. 7.58H) have finely costate shells with narrow hinges. One valve, the one with the larger beak, is typically flat; the other, convex. Inside the valves are calcified, coiled lophophore supports. Atrypoids range from Ordovician through Devonian with a Devonian peak.

Order Athyridoida—Arthyridoids (Fig. 7.58I) are one of two types of brachiopods that lack radial ornamentation. They have a narrow hinge line and an anterior depression (sulcus) and elevation (fold). Calcified coiled lophophore supports are present. They are Ordovician through Triassic in age, but reached a Carboniferous peak. Some peculiar later Mesozoic forms may be their descendants.

Order Spiriferoida — Spiriferoids (Fig. 7.58J) are distinctive brachiopods recognized by their very wide hinge lines caused by the lateral prolongation of the valves called alae. The valves are costate, and there are broad, internal, calcified, coiled lophophore supports. Spiriferoids range from Silurian through Jurassic, but are most abundant in the Upper Paleozoic, particularly the Devonian.

Order Terebratuloida—The other order of surviving articulates, the terebratuloids (Fig. 7.58K), typically have smooth shells lacking the prominent fold and sulcus of athyridoids. The hinge line is narrow, and there are calcified, loop-shaped lophophore supports. The group ranges from Devonian to Recent, but is most common from the Jurassic onward.

Near the bryozoans and brachiopods, the tree of life forks into two branches—one containing mollusks, annelids, and arthropods; the other, echinoderms, hemichordates, and vertebrates (Fig. 7.59).

Phylum Mollusca—The mollusks (Fig. 7.60) are a large, diverse phylum which has probably left the most impressive fossil record of all. Three major groups—the snails, clams, and cephalopods—and several lesser ones contribute to this array. The phylum is so diverse that it is difficult to find common features, and there are exceptions to every rule. Mollusks have a large muscular foot which is flattened for creeping, compressed for burrowing, or modified into tentacles for seizing. Most respire by filamentous gills, or ctenidia, which lie in a cavity formed by flaps of the body called the mantle. The soft parts of most are unsegmented, but in the small class of monoplacophorans (Fig. 7.61) segmentation occurs and may indicate a relationship with the annelid worms. Most mollusks have a **shell** covering the outside of the body, though it is lost or internal in some. In most, the shell is of one piece; but in the clams it has two parts, and in the minor group of the chitons it consists of eight overlapping valves. The shell is composed of aragonite in most cases, but calcite dominates in many clams. Mollusks are chiefly benthic, but several snails and many cephalopods are pelagic. Most are also marine, but snails and clams have invaded fresh water; and the snails, the land.

Subphylum Cyrtosomazoa—These mollusks have conical shells of one valve, typically twisted into a spiral. A relatively small single shell aperture results in a U-shaped gut and in the anus lying near the mouth. Most cyrtosomazoans are surface-dwelling benthonic herbivores or carnivores. The monoplacophorans are the ancestral group.

Class Gastropodea—The **snails** (Fig. 7.60) and their shell-less relatives, the slugs, compose this largest of molluscan classes. Most have a head with tentacles, a flat foot for creeping, and the posterior of the body arched up over the back (and twisted) so the mantle cavity faces forward. Snails have tiny teeth that they use in eating plants or animals. The shell (Fig. 7.62) is of one piece and lacks internal partitions. Most are conispirally coiled, but some are planispiral and some, not coiled at all. Snails range from

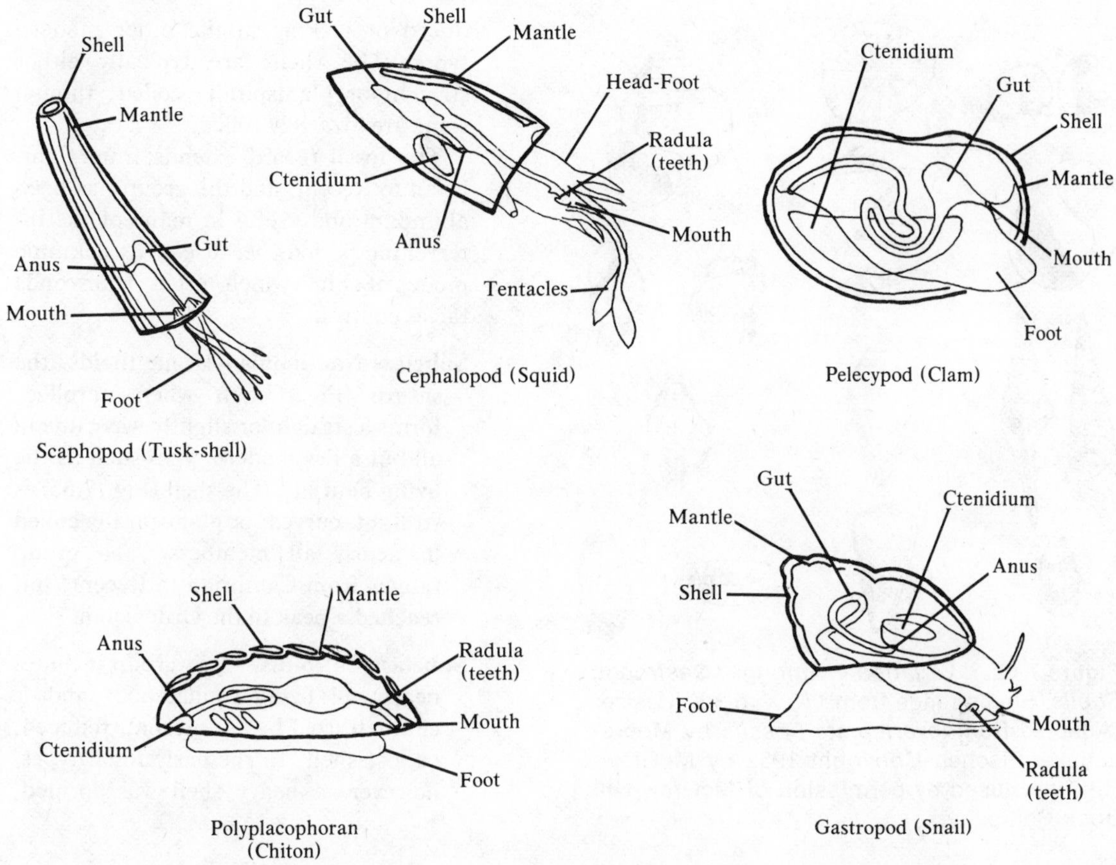

Figure 7.60. *Phylum Mollusca.* The major classes showing the shell and its relationships to the soft parts.

Figure 7.61. *Pilina, a Monoplacophoran Mollusk.* Exterior and interior views of fossil shell. Note multiple muscle scars on interior. X 1/2.

Cambrian to Recent and have been at their peak from the Cretaceous onward.

Class Cephalopodea—This group includes the living squids (Fig. 7.60), cuttlefish, octopuses, and the pearly nautilus which is more like the host of fossil types. The cephalopod foot is modified to tentacles for seizing prey and a funnel used for jet-propelled locomotion. Cephalopods are chiefly active, pelagic mollusks and have highly developed nervous and circulatory systems.

Most have a one-piece shell divided into a series of internal chambers by cross-partitions called **septa** (Fig. 7.63). The lines of junction between septa and shell are called **sutures,** and their structure is the basis for classifying most cephalopods. The shell was external in most fossils, as it is in the modern pearly nautilus, but it has become internal and is re-

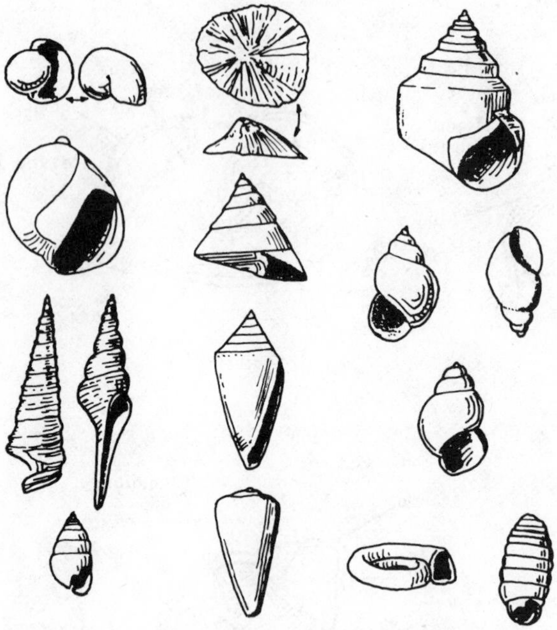

Figure 7.62. *Variation Among Gastropod Shells.* Shells range from 1/3 X to 3 X lifesize. (Adapted from *Invertebrate Fossils* by Moore-Lalicker-Fischer. Copyright 1952 by McGraw-Hill, Inc., used by permission of McGraw-Hill Book Company.)

duced or lacking in the other modern types. The shells are typically either straight or planispirally coiled, though some are bizarrely coiled.

The fossil record extends from Cambrian to Recent, and the group has been abundant and useful in many of the intervening periods because of its nektonic mode of life which often transcends facies control.

Subclass Nautiloidia—In nautiloids, the suture (Fig. 7.64), when unrolled, forms a straight or slightly wavy line in all but a few modern types such as the living nautilus. The shell (Fig. 7.65) is straight, curved, or planispirally coiled in nearly all members. The group ranges from Cambrian to Recent, but reached a peak in the Ordovician.

Subclass Coleoidia—This group includes nearly all living cephalopods and is characterized by an internal, reduced, or lost shell. In the early fossil types, however, a heavy shell was formed.

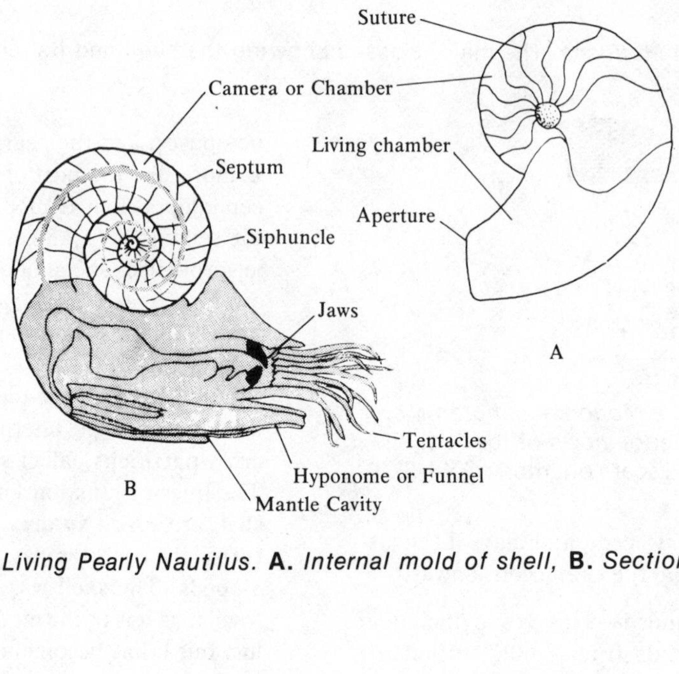

Figure 7.63. *The Living Pearly Nautilus.* **A.** *Internal mold of shell,* **B.** *Section through living organism.* X 1/3.

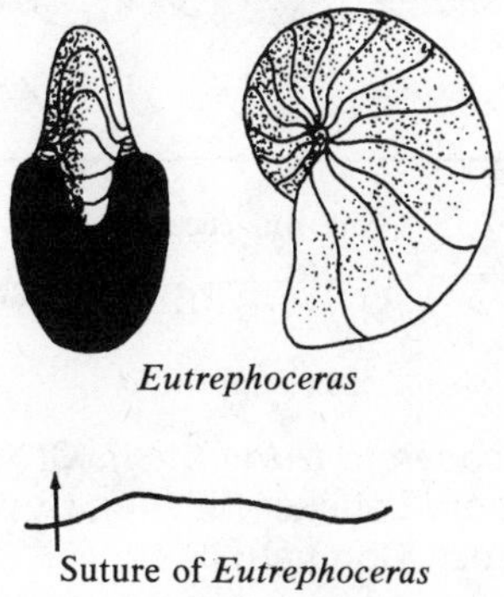

Figure 7.64. *Coiled Nautiloid Shell and Nautiloid Suture.* X 1/3. (After Miller.)

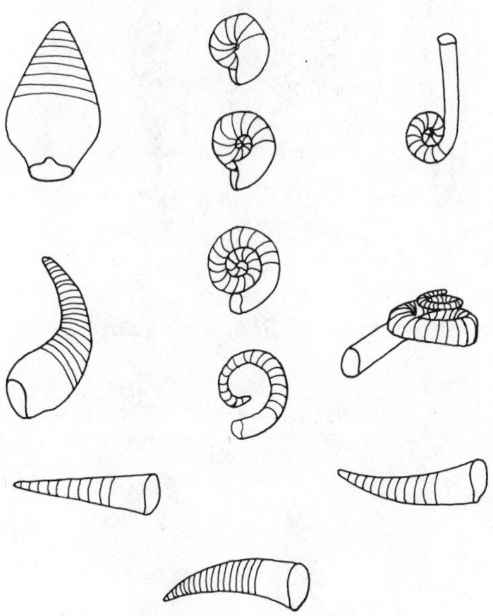

Figure 7.65. *Variation Among Fossil Nautiloid Shells.* Shells range from X 1/10 to X 1. (After Flower.)

Coleoids were derived from straight nautiloids.

Order Belemnitoida—In belemnoids (Fig. 7.66), the chambered part of the shell is reduced, but the outer walls are tremendously thickened to form a heavy cigar-shaped guard, or rostrum. Belemnites range from Carboniferous through Paleogene, but are abundant only in the Jurassic and Cretaceous. Members of a similar and closely related group, the **aulacoceroids,** are common in the Triassic.

Subclass Ammonoidia—The extinct ammonoids are the most geologically significant cephalopods. They have complex sutures and chiefly planispiral shells (Fig. 7.67). They range from Devonian through Cretaceous.

Order Goniatitoida—These chiefly have goniatite sutures (Fig. 7.68); that is, the flexures toward the aperture (saddles) and those away from the aperture (lobes) are not crenulate. They range from Devonian through earliest Triassic with a Carboniferous and Permian peak.

Order Ceratitoida—These chiefly have ceratite sutures (Fig. 7.68); that is, the lobes, but not the saddles, are crenulate. They range from Devonian through Triassic, though ceratite sutures do not appear until the Mississippian. Most are Triassic in age.

Order Ammonitoida—These chiefly have ammonite sutures (Fig. 7.68); that is, both lobes and saddles are crenulate. Most are highly ornamented. Some, called **lytocerines,** have very bizarre patterns of coiling or uncoil (Fig. 7.69). They range from Triassic through Cretaceous and have a Jurassic-Cretaceous peak. (Ammonitic sutures appear in the Permian in advanced members of the other orders.)

Subphylum Diasomazoa—These mollusks have a shell that is primitively open at both ends, a relatively straight gut, and are typically burrowing filter feeders or deposit feeders.

Class Bivalvea or **Pelecypodea**—The **clams** (Fig. 7.60) have a laterally compressed, headless body and foot. The mantle cavity is large and extends forward from the posterior along the sides of the body. It contains the large gills which are the

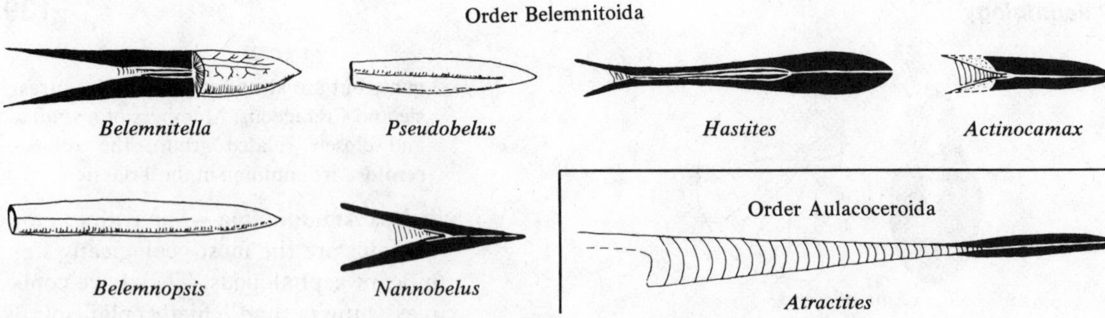

Figure 7.66. *Shells of Fossil Belemnoids and a Related Aulacoceroid* (*Atractites*). Cross sections show in black. X1/2. (Adapted from *Invertebrate Fossils* by Moore-Lalicker-Fischer, Copyright 1952 by McGraw-Hill, Inc., used by permission of McGraw-Hill Book Company.)

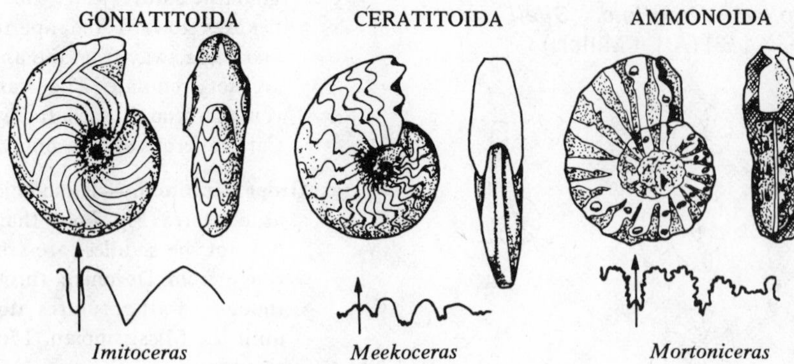

Figure 7.67. *Ammonoids of Each of the Three Orders Showing Shell Shape and Sutures.* X 1/3. (After Smith and Reeside.)

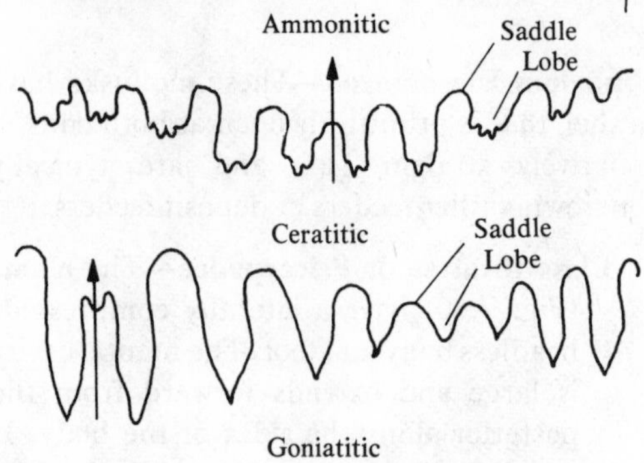

Figure 7.68. *Types of Ammonoid Sutures.* (After Easton.)

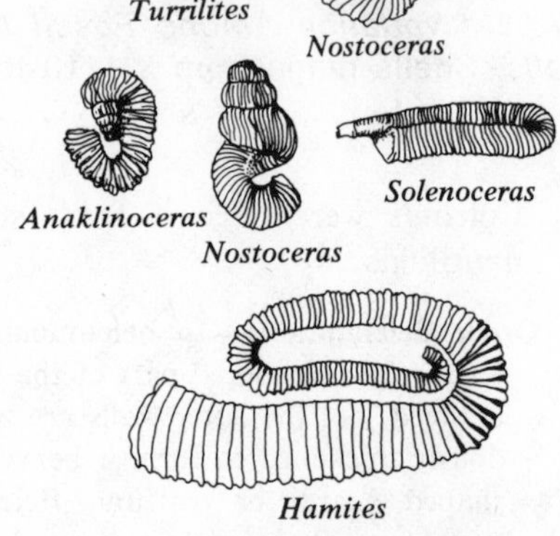

Figure 7.69. *Lytocerine Ammonoid Shells.* X 1/3. (After Griepenkerl, Reeside, and Stephenson.)

chief feeding organs, straining out microscopic food particles from water which enters and leaves by a pair of posterior siphons in most clams. Clams are typically burrowers, though many are attached or free on the surface. The shell (Fig. 7.70) consists of two valves which hang down on either side of the body and which are mirror images of each other except in those that lie on one side like the scallops, oysters, and the extinct coral-shaped rudists. Clams range from Cambrian to Recent, but, like snails, they dominated Cretaceous and Cenozoic aquatic environments.

Class Scaphopodea—The curved, conical shells (Fig. 7.71) of this small group, the **tusk** or **tooth shells,** are open at both ends. The animal lives buried in the bottom, with only the narrower top protruding. Water currents enter and leave this opening; tiny tentacles gather food at the other end. Scaphopods range from Ordovician to Recent and are common elements in Cenozoic faunas.

Subphylum Uncertain

?Class Hyolithea—This is another extinct group of uncertain relationships. Hyolithans have a three-sided, conical, calcite shell (Fig. 7.72). They range from Cambrian through Permian in age but are most abundant in the Cambrian.

?Class Tentaculitea—Still another extinct group, the tentaculites have conical ringed shells (Fig. 7.73) of circular cross section. They range in age from Ordovician through Devonian and are common throughout this iterval. They may be related to annelid worms.

Phylum Annulata—The segmented worms, or annelids (Fig. 7.74), have a body characterized by the repetition of similar parts in a series. Only the marine group, the polychaetes, have left much of a record. The mobile forms have few hard parts except for their chitinous jaws or **scolecodonts** (Fig. 7.75). Consequently, their fossil record consists of some body impressions and occasionally common scolecodonts. Many of the attached or burrowing forms build tubes (Fig. 7.76), which are typically agglutinated sand grains; but some build calcite or aragonite tubes, and these are common fossils. Annelids range from Eocambrian to Recent with identifiable tubes appearing in the Cambrian and scolecodonts, in the Ordovician.

Phylum Arthropoda—This is the largest phylum of living organisms. The body is segmented, suggesting an annelid relationship, but there

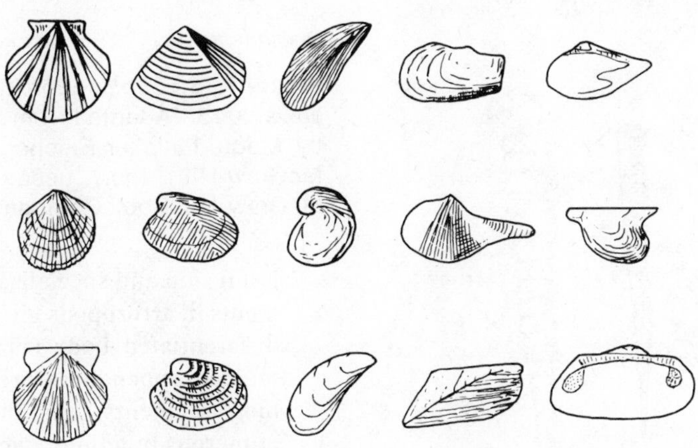

Figure 7.70. *Variation among Bivalve Shells.* Shells range from X 1/10 to X 2 lifesize. (Adapted from *Invertebrate Fossils* by Moore-Lalicker-Fischer, Copyright 1952 by McGraw-Hill, Inc., used by permission of McGraw-Hill Book Company.)

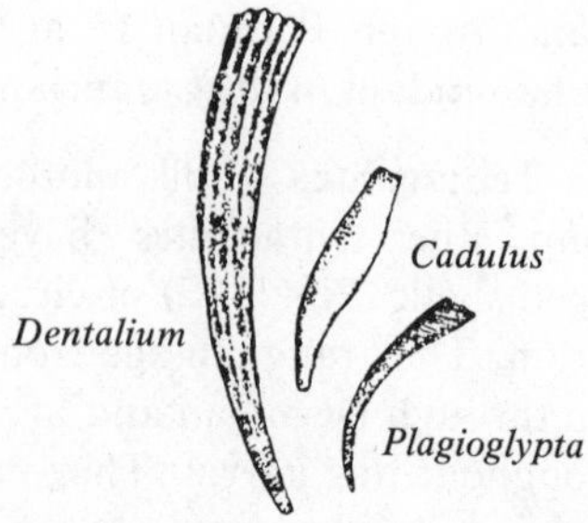

Figure 7.71. *Class Scaphopodea.* Representative fossil tusk or tooth shells. X3/4. (From H.L. Levin, *Life Through Time,* © 1975 by Wm. C. Brown Co.)

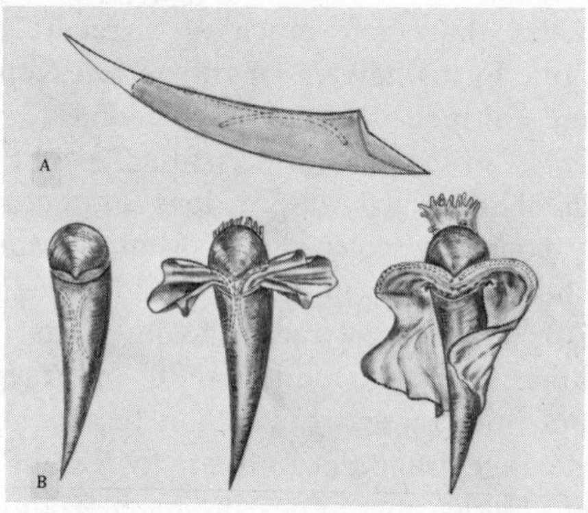

Figure 7.72. *Class Hyolithea.* **A.** *Fossil shell,* **B.** *Restorations.* X1/2. (From *Treatise on Invertebrate Paleontology,* courtesy of The Geological Society of America and University of Kansas.)

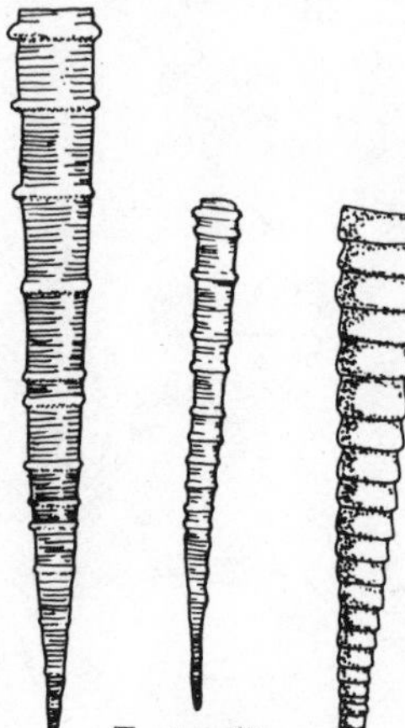

Figure 7.73. *Class Tentaculitea. Tentaculites.* X10.

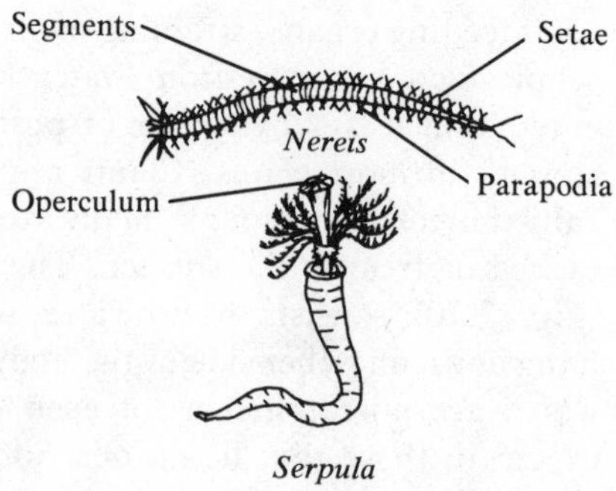

Figure 7.74. *Living Polychaete Annelids.* A mobile (X 1/3) and an attached form (X 1) are shown.

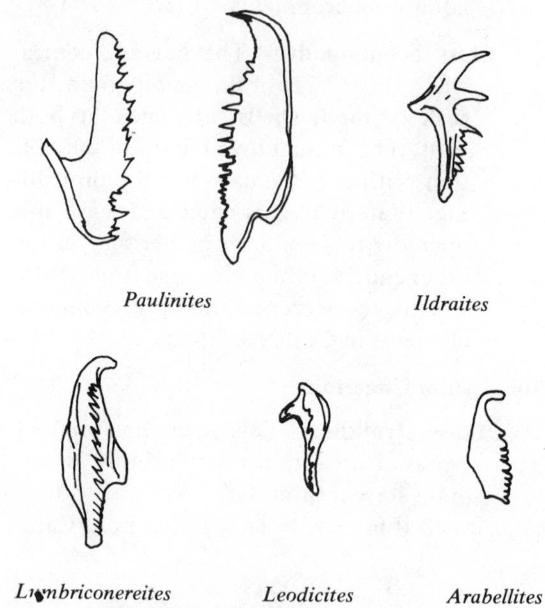

Figure 7.75. *Scolecodonts or Fossil Worm Jaws.* X 15. (Adapted from *Invertebrate Fossils* by Moore-Lalicker-Fischer, Copyright 1952 by McGraw-Hill, Inc., used by permission of McGraw-Hill Book Company.)

is a fusion and specialization of groups of segments in arthropods giving rise to a regionally differentiated body (Fig. 7.77). Paired, segmented appendages are typically present on most segments; and these are modified for numerous functions, such as sensing, feeding, walking, and reproducing. They are also specialized on the various regions of the body. Being active creatures, arthropods have a

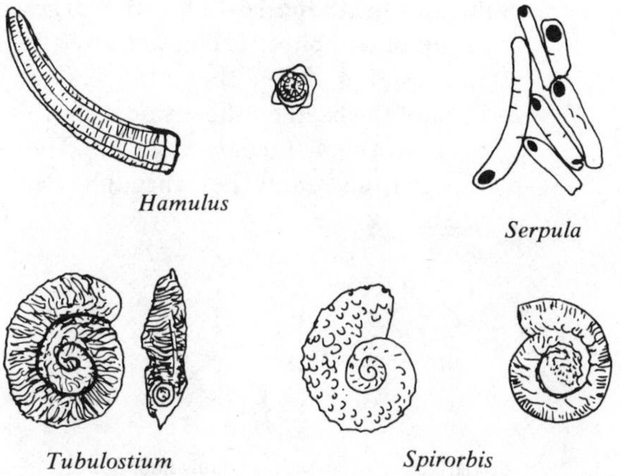

Figure 7.76. *Fossil Annelid Worm Tubes.* X 3. (Adapted from *Invertebrate Fossils* by Moore-Lalicker-Fischer, Copyright 1952 by McGraw-Hill, Inc., used by permission of McGraw-Hill Book Company.)

well-developed nervous system. There is a chitin **exoskeleton** which covers the outside of the body and the appendages. In some groups, the exoskeleton is mineralized with calcite and/or calcium phosphate which greatly improves its chances of fossilization. Because an arthropod has an outside skeleton, the organism must shed it periodically in order to grow; hence, an individual may leave several fossils. The fossil record ranges from Cambrian (possibly Eocambrian) to Recent with trilobites dominating the Lower Paleozoic and crustaceans taking over thereafter, with insects spasmodically common after their Devonian appearance. Arthropods dwell in marine, fresh-water, and terrestrial environments with the greatest numbers living in the last, but the fossil record is biased toward the first.

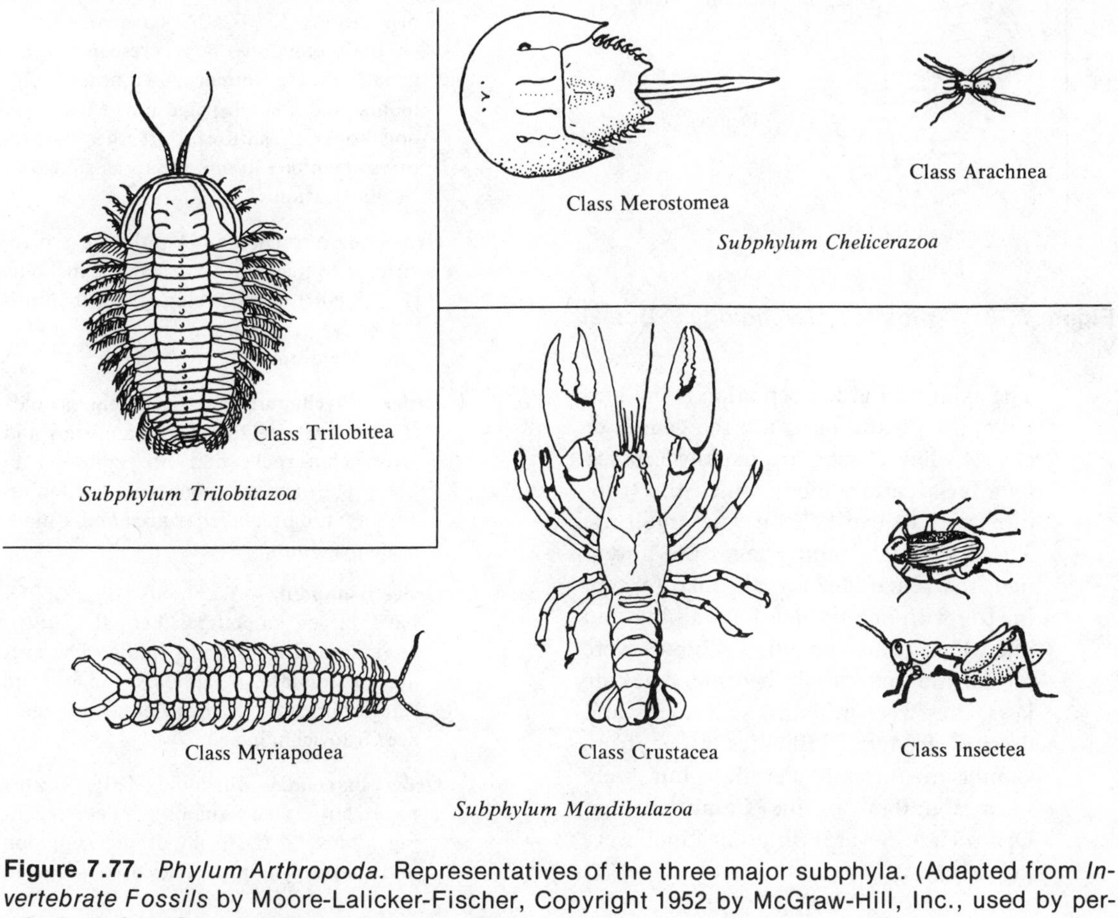

Figure 7.77. *Phylum Arthropoda.* Representatives of the three major subphyla. (Adapted from *Invertebrate Fossils* by Moore-Lalicker-Fischer, Copyright 1952 by McGraw-Hill, Inc., used by permission of McGraw-Hill Book Company.)

Subphylum Trilobitazoa—Trilobites and some related forms are placed here.

Class Trilobitea—Trilobites (Fig. 7.78) are an extinct group of arthropods whose members are longitudinally trilobed, with an axis and two pleurae on either side. Their body is also transversely specialized into a **cephalon,** or head; a **thorax,** or body; and a **pygidium,** or abdomen. The first and last consist of fused segments and are the commonest fossils.

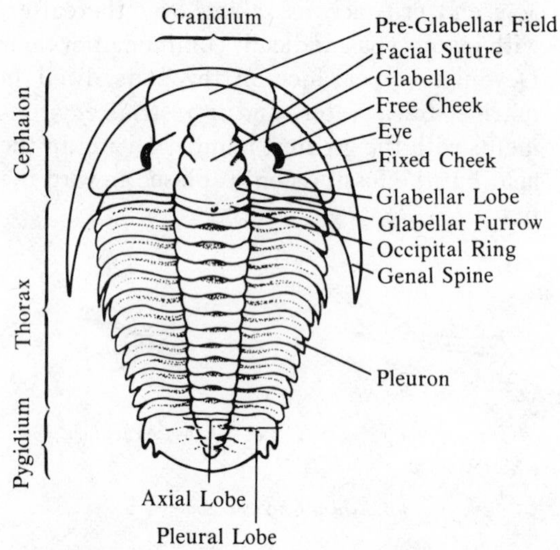

Figure 7.78. *Trilobite Morphology.* Dorsal view.

The axial part of the cephalon is the **glabella;** the pleural parts are the **genae,** or cheeks. The cheeks are usually crossed by a **facial suture** along which the trilobite splits its exoskeleton when molting. The first pair of appendages are antennae; the remainder are all alike (except in size) with one branch for walking and one which bears the gill. Trilobites were all marine and chiefly benthic. Lacking jaws, they were probably scavengers and detritus feeders. Trilobites range from Cambrian through Permian but were most abundant in the Cambrian and Ordovician. Several different kinds were important.

Subclass Opisthoparia—This, the largest group of trilobites, is characterized by sutures (Fig. 7.79) that intersect the back of the head on the inside from the back corners (genal angles). They range from Cambrian through Permian.

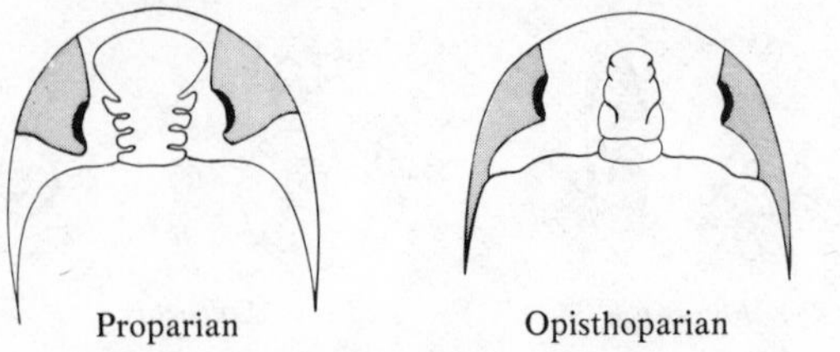

Figure 7.79. *Trilobite Facial Sutures.* The two major varieties. Free cheeks are stippled. (After Tasch, *Paleobiology of the Invertebrates,* © 1973 by John Wiley & Sons, Inc.)

Order Redlichioida—These strictly Cambrian forms (Fig. 7.80A) have large semicircular cephalons, large crescentic eyes, genal spines, numerous thoracic segments, and a small pygidium. Most have opisthoparian sutures, but these structures atrophied in one group as shown in the illustration.

Order Corynexochoida—Another group restricted to the Cambrian, corynexochoids (Fig. 7.80B) have an elongate, subelliptical body, genal spines (in most), and a large pygidium.

Order Ptychoparioida — Ptychoparioids (Fig. 7.80C) are found in Cambrian and Ordovician rocks and are common in both. They are very generalized trilobites with a large preglabellar area and a moderate pygidium.

Order Asaphoida—Asaphoids (Fig. 7.80D) have large, smooth, subequal, subtriangular cephalons and pygidia. The eyes are tuberculate. Asaphoids range from Cambrian through Ordovician, but most are Ordovician in age.

Order Illaenoida—Illaenoids (Fig. 7.80E) have a large, nearly smooth glabella reaching almost to the front of the cephalon and a large pygidium. They are Cambrian

CLASS TRILOBITEA

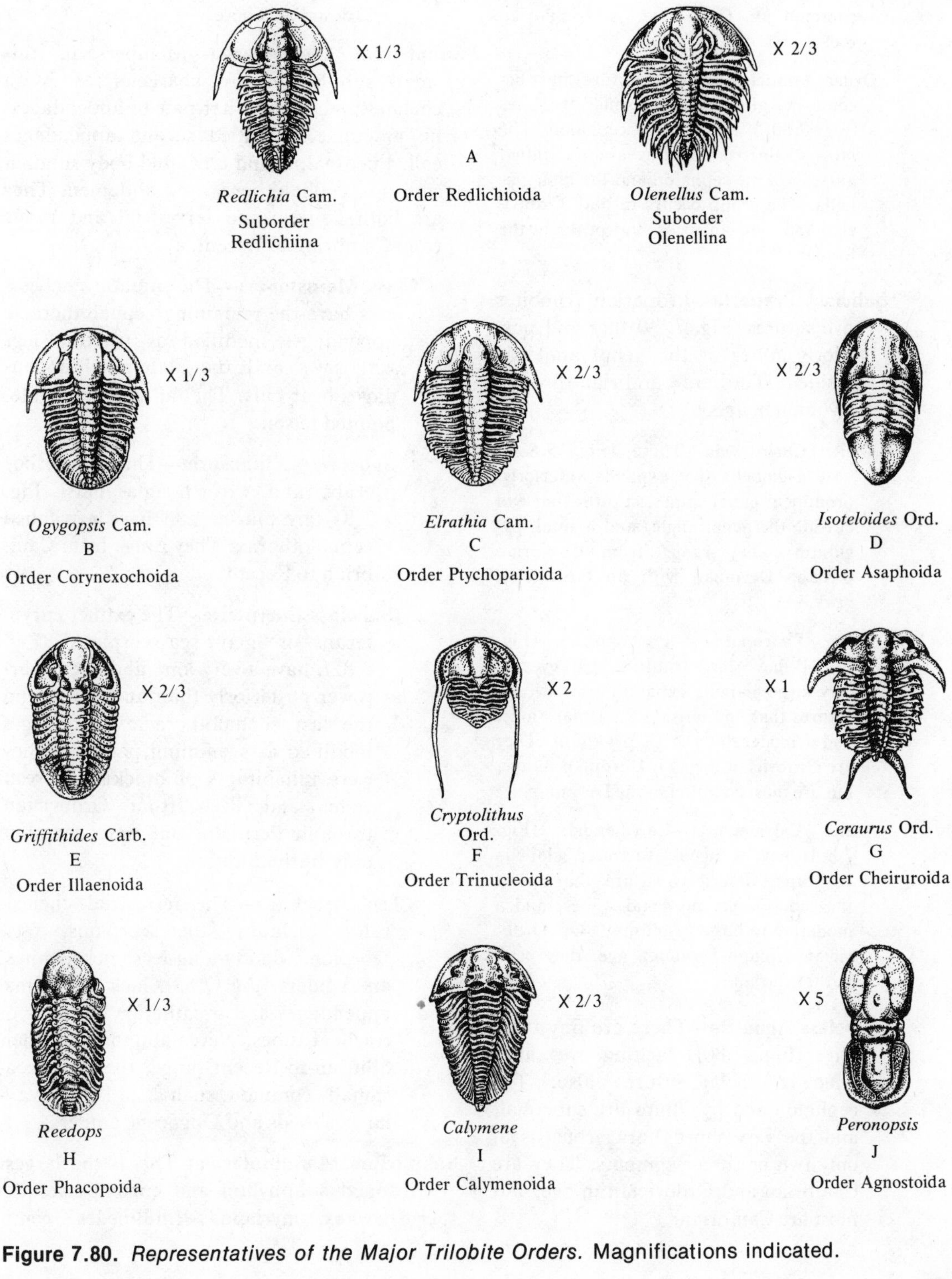

Figure 7.80. *Representatives of the Major Trilobite Orders.* Magnifications indicated.

through Permian in age, but are most common in Ordovician and Silurian rocks.

Order Trinucleoida—The suture has become marginal in these small trilobites (Fig. 7.80F) with a large cephalon. The most distinctive features are the pitted anterior of the cephalon and the high glabella. The group occurs in both Ordovician and Silurian rocks, but peaks in the former.

Subclass Proparia—Proparian trilobites have sutures (Fig. 7.79) that end posteriorly either at the genal angle or outside it. They are Cambrian through Devonian in age.

Order Cheiruroida—These (Fig. 7.80G) have a glabella that expands anteriorly, prominent genal spines, sutures that end outside the genal angle, and a small pygidium. They range from Cambrian through Devonian with an Ordovician peak.

Order Phacopoida — Phacopoids (Fig. 7.80H) have large multifaceted eyes, a very large glabella expanding anteriorly, sutures that end outside the genal angle, and a moderate-to-large pygidium. They are Ordovician through Devonian in age, but are most common in the Devonian.

Order Calymenoida—Calymenoids (Fig. 7.80I) have a deeply furrowed glabella that tapers anteriorly, sutures that end at the genal angle, no genal spines, and a moderate-to-large pygidium. Of Ordovician through Devonian age, they peak in the Silurian.

Subclass Agnostia—These are tiny trilobites (Fig. 7.80J) lacking eyes and, typically, facial sutures also. The cephalon and pygidium are subequal, and the very short thorax consists of only two or three segments. They are Cambrian and Ordovician in age, but most are Cambrian.

Order Agnostoida—These are typical forms as described above.

Subphylum Chelicerazoa—Members of this small subphylum are characterized by a chelate (pincer-like) first pair of appendages, no antennae, modified second appendages called pedipalps, and a twofold body subdivision into cephalothorax and abdomen. They are both aquatic and terrestrial and range from Cambrian to Recent.

Class Merostomea—The aquatic chelicerates have the remaining cephalothoracic appendages modified as walking legs with jaws and the abdominal appendages bear gills. The tail is an elongate, pointed telson.

Subclass Xiphosuria—The horseshoe crabs and extinct aglaspoids (Fig. 7.81) are marine and have a trilobed cephalothorax. They range from Cambrian to Recent.

Subclass Eurypteria—The extinct eurypterans, or "giant sea scorpions" (Fig. 7.82), have a very long abdomen, narrower posteriorly than anteriorly, and the last cephalothoracic appendages modified as swimming paddles. They were inhabitants of brackish-to-fresh water and lived from Ordovician through Permian, but were common only in the Silurian.

Class Arachnea—The terrestrial chelicerates, including the scorpions, false scorpions, daddy-long-legs, ticks, mites, and spiders (Fig. 7.83), lack abdominal appendages and breathe air by lungs or tracheal tubes. Never abundant in their Silurian-to-Recent range, they are occasionally common, such as in Pennsylvanian coal beds and Oligocene ambers.

Subphylum Mandibulazoa—This is the largest arthropod subphylum and encompasses the crustaceans, myriapods (millipedes, centi-

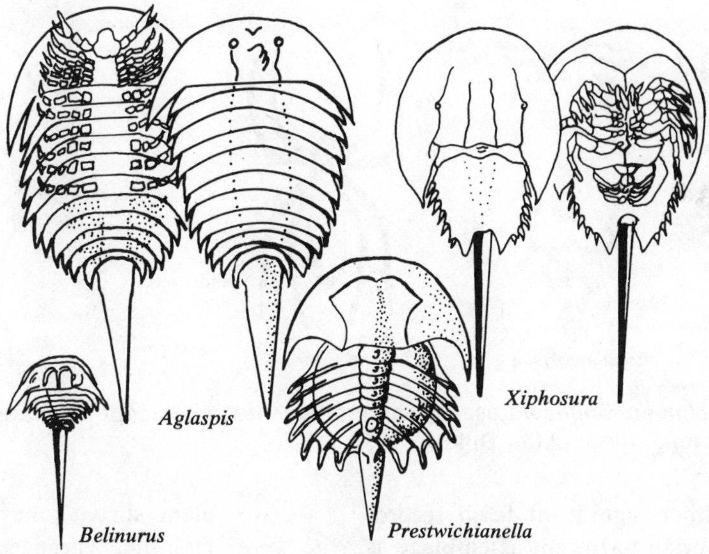

Figure 7.81. *Fossil Xiphosurian Merostomes.* All figures X 1 except *Xiphosura*, which is X 1/3. (Adapted from *Invertebrate Fossils* by Moore-Lalicker-Fischer, Copyright 1952 by McGraw-Hill, Inc., used by permission of McGraw-Hill Book Company.)

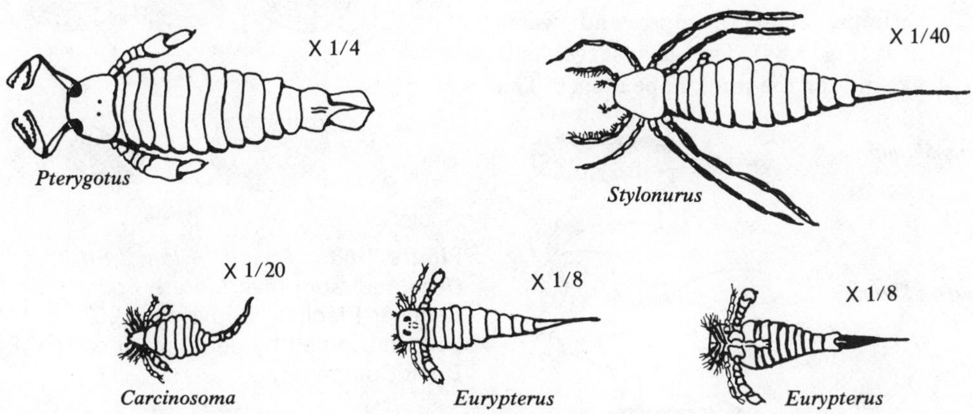

Figure 7.82. *Eurypterans.* Magnifications indicated. (Adapted from *Invertebrate Fossils* by Moore-Lalicker-Fischer, Copyright 1952 by McGraw-Hill, Inc., used by permission of McGraw-Hill Book Company.)

pedes), and insects. The first pair of appendages are sensory antennae; and the third pair are jaws, or mandibles, followed by two pair of accessory feeding appendages, the maxillae. The geologic range is Cambrian to Recent. The group may not be a natural one.

Class Crustacea—Crustaceans are the primarily aquatic mandibulates. They possess two pairs of antennae, and most have a shield, or **carapace,** over much of the body. The carapace is mineralized with calcite and/or phosphate in those

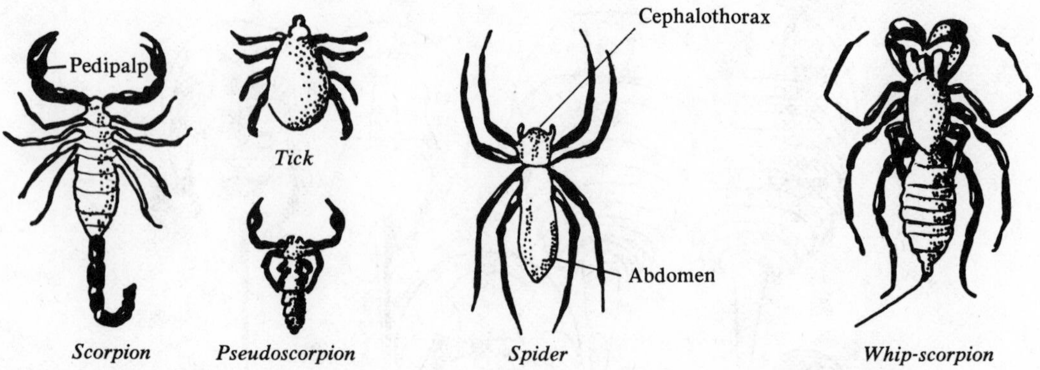

Figure 7.83. *Arachnean Chelicerates.* Approximately lifesize except tick and pseudoscorpion which are highly magnified. (After Dillon.)

groups with a significant fossil record. This Cambrian-to-Recent assemblage is very diverse, but only four groups have left a good fossil record.

Subclass Branchiopodia—This group includes the fairy shrimps, tadpole shrimps, clam shrimps, and water fleas (Fig. 7.84). They have large, leaf-like gills on the trunk appendages. The

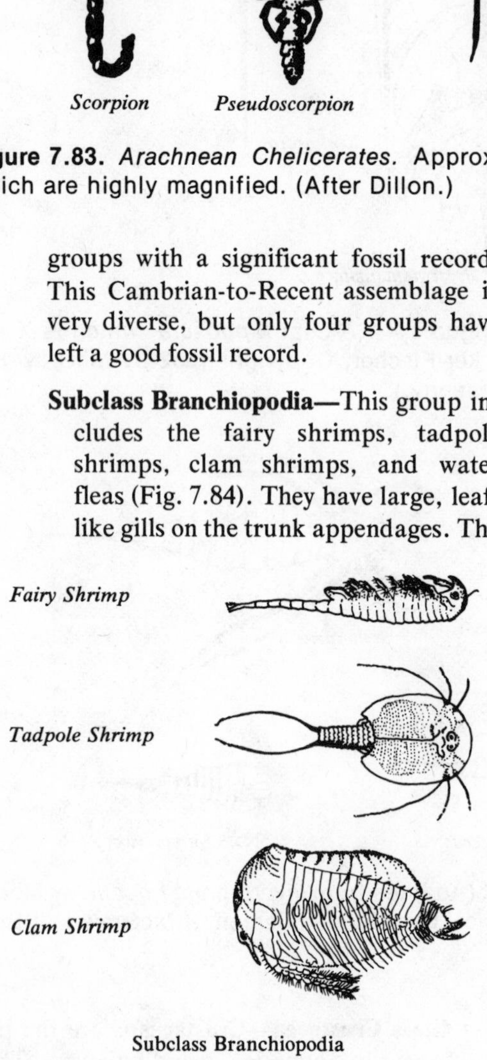

Figure 7.84. *Branchiopod Crustaceans.* X 3. (Adapted from *Invertebrate Fossils* by Moore-Lalicker-Fischer, Copyright 1952 by McGraw-Hill, Inc., used by permission of McGraw-Hill Book Company.

clam shrimps have small, bivalved, clam-like carapaces (Fig. 7.85) with growth lines, a feature lacking in other arthropods that molt their shells. They

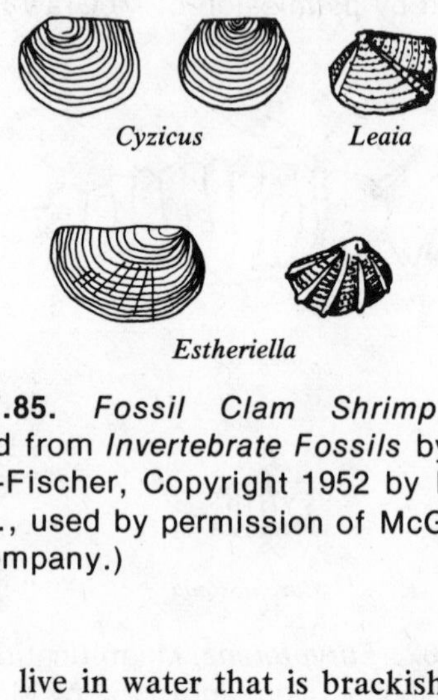

Figure 7.85. *Fossil Clam Shrimp.* X 3. (Adapted from *Invertebrate Fossils* by Moore-Lalicker-Fischer, Copyright 1952 by McGraw-Hill, Inc., used by permission of McGraw-Hill Book Company.)

live in water that is brackish to fresh and range from the Devonian to Recent, though common only in times of abundant nonmarine deposition such as the Carboniferous and Triassic.

Subclass Ostracodia—The most significant crustaceans to the geologist, the **mussel shrimps** (Fig. 7.86), have a highly calcified bivalved carapace. The carapace shape and ornamentation are

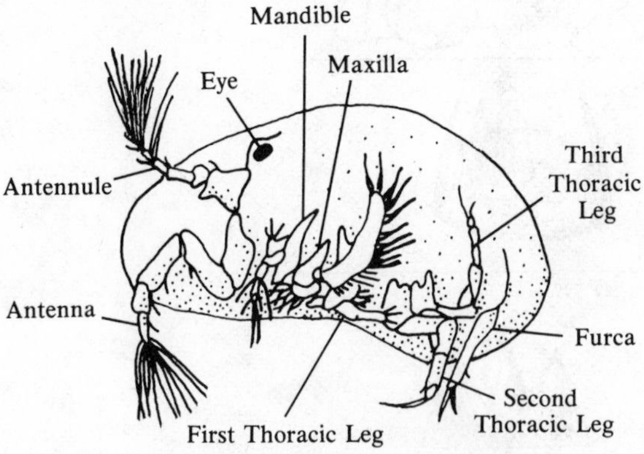

Figure 7.86. *Ostracod Anatomy.* X 10. (After Bassler, Kellett, Kesling.)

very diverse, and there are no growth lines (Fig. 7.87). They are both marine and fresh water. They range from Cambrian to Recent and have been abundant from the Ordovician onward.

Subclass Cirripedia — The **barnacles** (Fig. 7.88) have a shrimp-like body concealed in a cluster of overlapping calcite plates which are not shed and which are fixed to the substrate. They are commonly shaped like a tiny volcano or have a stalk. They range from Silurian to Recent and are particularly large and common in the Neogene.

Subclass Malacostracia—There are the large and familiar crustaceans—the **shrimp, lobsters,** and **crabs**—and the smaller pill bugs and sand fleas (Fig. 7.89). Distinguished by its segmentation pattern, the group as a whole does not have highly mineralized shells, which makes whole fossils uncommon. Fragments of appendages and carapaces are common, particularly in younger rocks. They are both marine and fresh water and range from Cambrian to Recent.

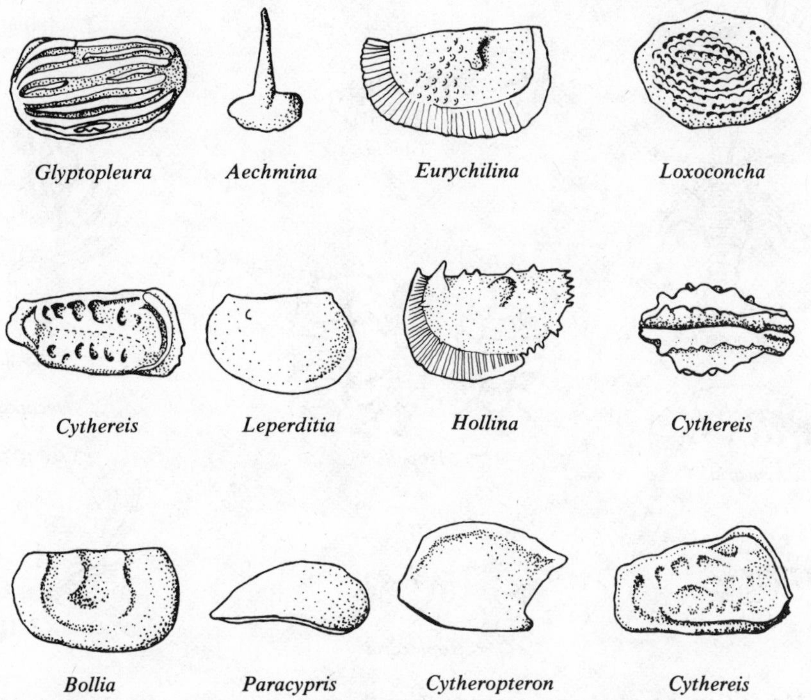

Figure 7.87. *Fossil Ostracod Carapaces.* Approximately X 20. (After Easton.)

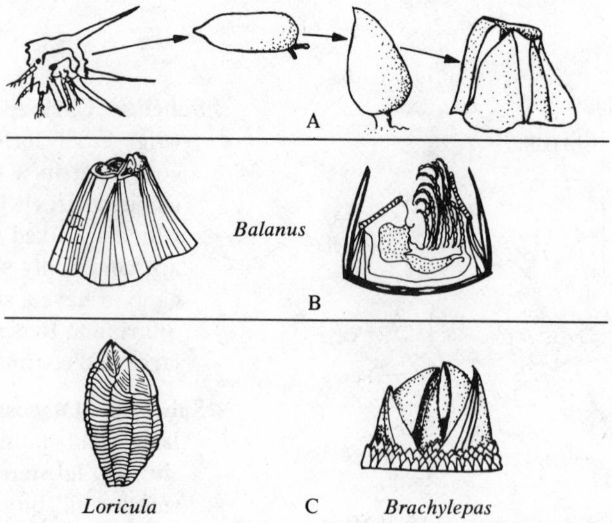

Figure 7.88. *Barnacles.* **A.** *Development of a barnacle,* **B.** *The common genus Balanus with cut-away showing soft parts,* **C.** *Fossil barnacle carapaces.* All X 1 except highly magnified larval stages.

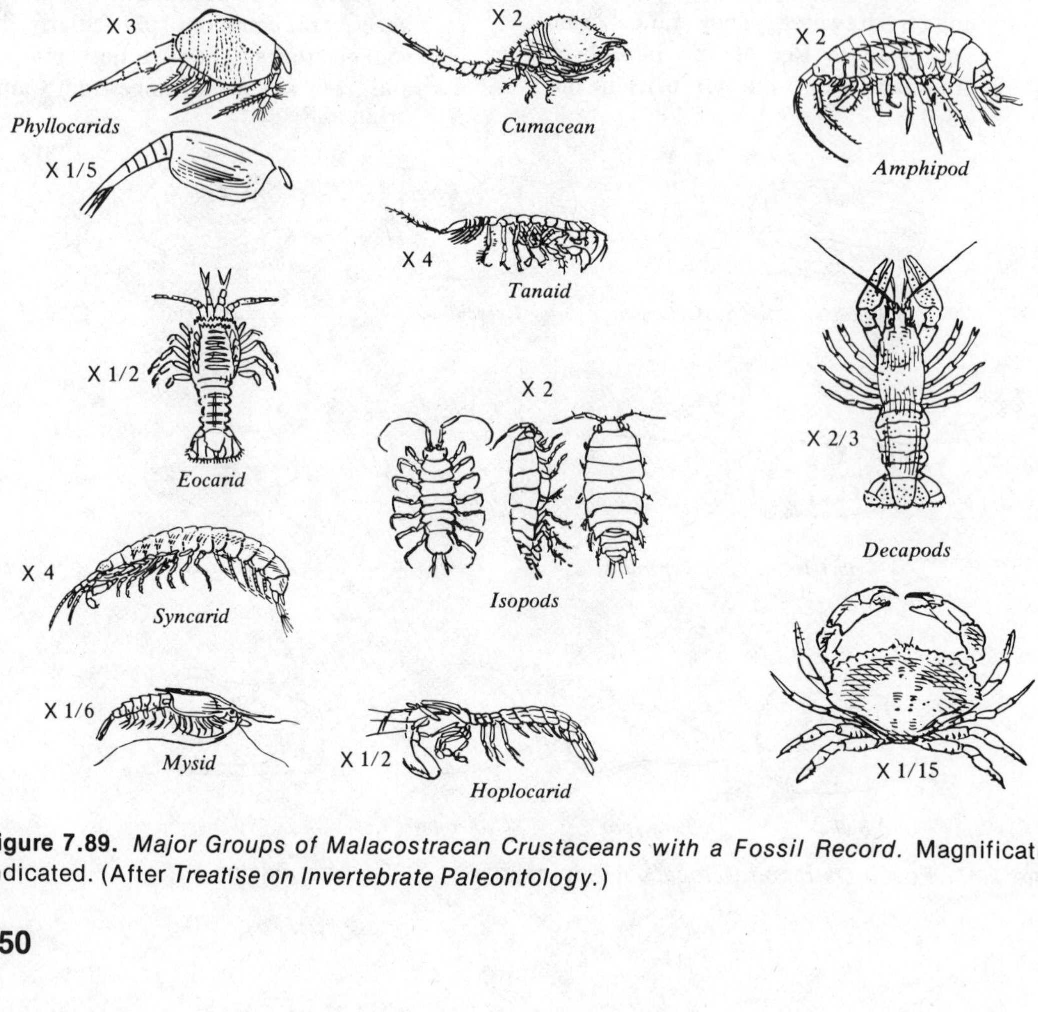

Figure 7.89. *Major Groups of Malacostracan Crustaceans with a Fossil Record.* Magnifications indicated. (After *Treatise on Invertebrate Paleontology.*)

Class Insectea—The insects (Fig. 7.90) are by far the largest living group of arthropods and, although they are also the most numerous fossil arthropods, the fossil record displays only a tiny fraction of the whole group. Insects have a single pair of antennae, no abdominal appendages, three pairs of walking legs on the thorax, and typically two pairs of wings on the thorax. It is because of the aerial habitats and consequently light and poorly mineralized skeleton that the fossil record is chiefly limited to fortuitous circumstances such as swamps,

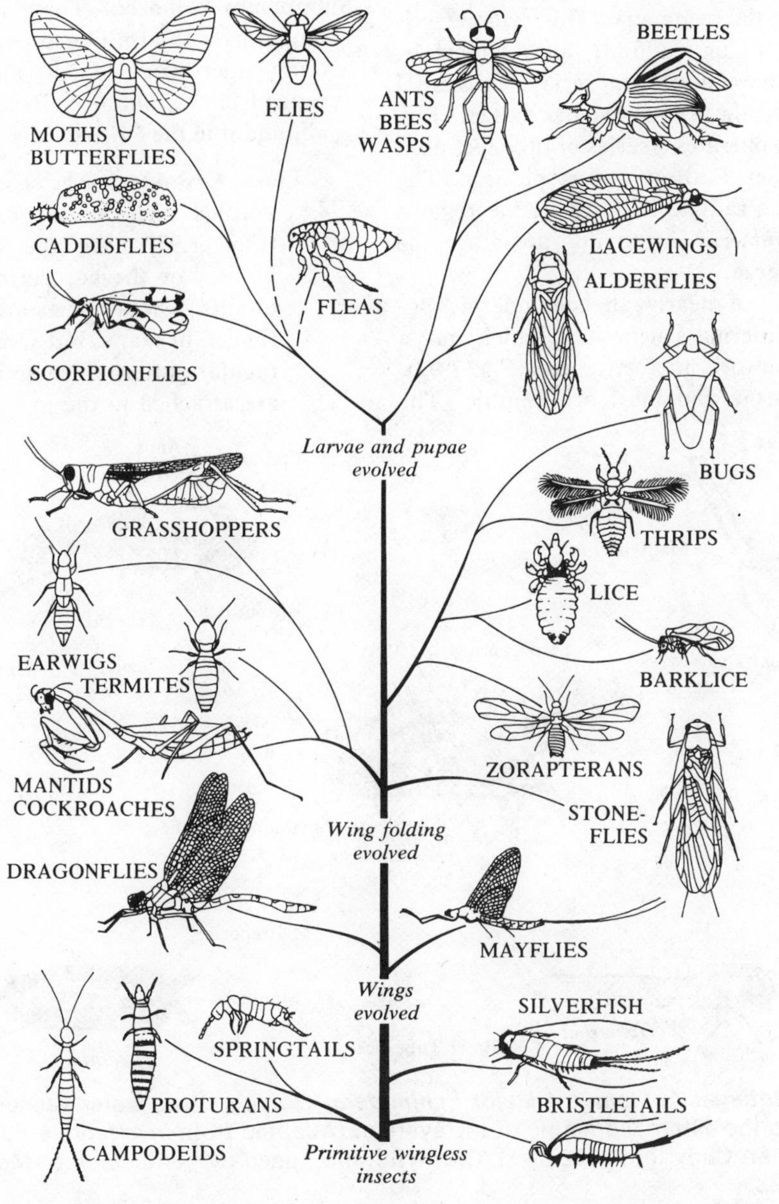

Figure 7.90. *Insects of the Major Living Orders.* (After Illinois Natural History Survey.)

lakes, and amber beds. Insects range from Devonian to Recent, and peaks occur in the Carboniferous, Permian, Cretaceous, and Cenozoic.

Phylum Echinodermata—Echinoderms are unique among the more complex animals in having a radial symmetry. This symmetry is pentamerous, or fivefold, and imperfect. They also have a unique series of internal canals called the **water vascular system** (Fig. 7.91) which operates a series of tube feet used in locomotion, feeding, and respiration. The canals which radiate outward define regions called **ambulacra;** intervening areas are the **interambulacra.**

The body of nearly all echinoderms possesses an internal calcite **test** which has a unique meshwork microstructure (Fig. 7.92). All are marine and most are benthic. The fossil record goes back to the Cambrian (possibly Eocambrian) and embraces a wealth of extinct forms surpassing any other phylum. Many are rare, such as the chiefly bilateral haplozoans and the homalozoans, or carpoids (Fig. 7.93), which may link echinoderms and fish.

Subphylum Crinozoa—These are stalked echinoderms with arm-like appendages for gathering microscopic food (Fig. 7.94). They range from Cambrian to Recent but are most abundant in the Paleozoic.

Class Cystoidea—The eocrinoids, diploporoids, dichoporoids, and fistuliporoids (Fig. 7.95) have the plates of their bodies, or **thecae,** perforated by pores or slits. Thecal plates are irregularly arranged in many, but some have a nearly regular pattern. The ambulacra of most are attached to the top and sides of the

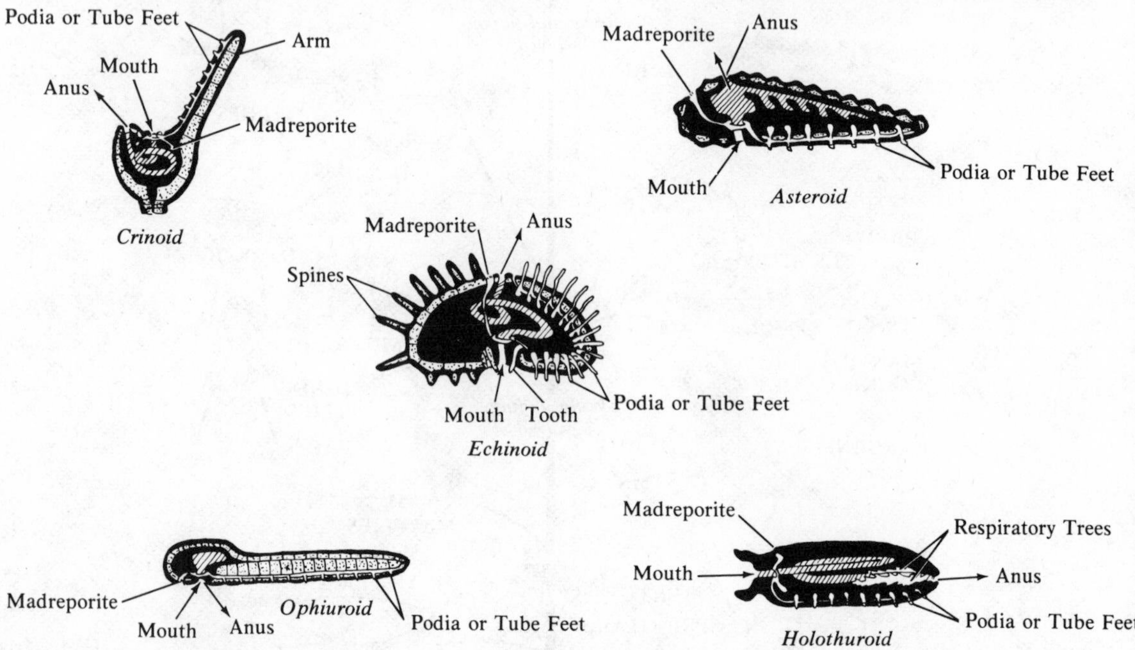

Figure 7.91. *Internal Anatomy of Major Echinoderm Groups.* Note water vascular system. The madreporite is the external opening of the system. (Adapted from *Invertebrate Fossils* by Moore-Lalicker-Fischer, Copyright 1952 by McGraw-Hill, Inc., used by permission of McGraw-Hill Book Company.)

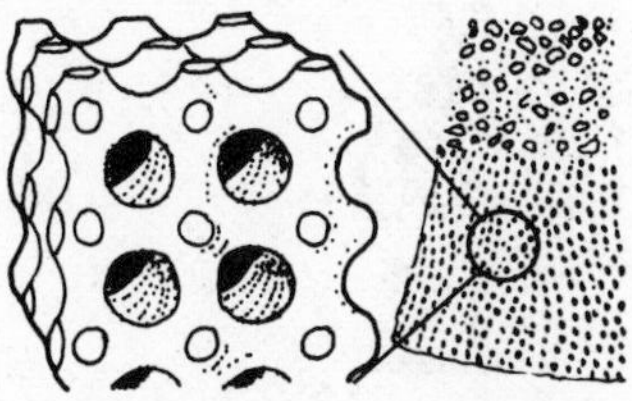

Figure 7.92. *Echinoderm Skeleton Microstructure.* (Adapted from *Invertebrate Fossils* by Moore-Lalicker-Fischer, Copyright 1952 by McGraw-Hill, Inc., used by permission of McGraw-Hill Book Company.)

theca and bear slender, upright, food-gathering **brachioles.** The cystoid group ranges from Cambrian through Devonian and is occasionally common in the Ordovician and Silurian.

Class Blastoidea—Blastoids (Fig. 7.96) resemble a rosebud with their few, regularly arranged plates. The prominent ambulacra bear brachioles as in cystoids, but blastoids have a unique series of folds (**hydrospires**) beneath the ambs which are connected to pores or slits along the sides of the ambs and larger openings at the top. Blastoids range from Silurian through Permian, but reach their maximum in the Mississippian.

Class Crinoidea—Called **sea-lilies** or **feather stars,** the crinoids (Fig. 7.97) are the only modern survivors of the subphylum. Their body, or **calyx,** has regularly arranged plates lacking pores or slits. The ambulacra are not attached (except at the base), but extend upward and are called **arms.** The delicate structures attached to them are called **pinnules** in crinoids. The group ranges from Cambrian to Recent, but was most abundant in the Paleozoic, particularly the Missisippian (Fig. 17.2). Paleozoic types were shallow-water benthos, but later forms gradually became deep-water, or pelagic.

Subphylum Asterozoa—The common **sea stars, brittle stars,** and nearly extinct **somasteroids**

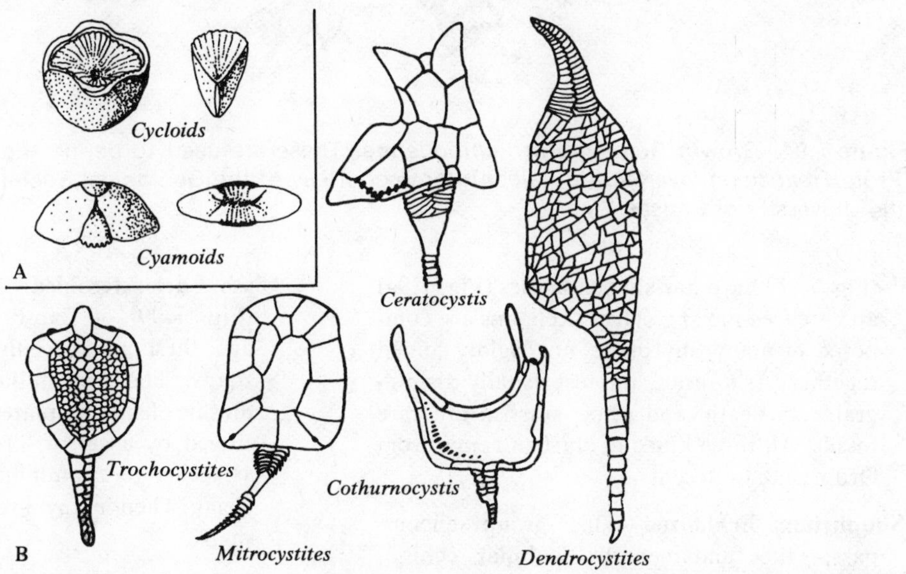

Figure 7.93. *Extinct Echinoderms of Uncertain Relationships.* **A.** *Haplozoans,* **B.** *Carpoids or Homalozoans.* (Part A after Whitehouse and Part B after Barrande, Bather.)

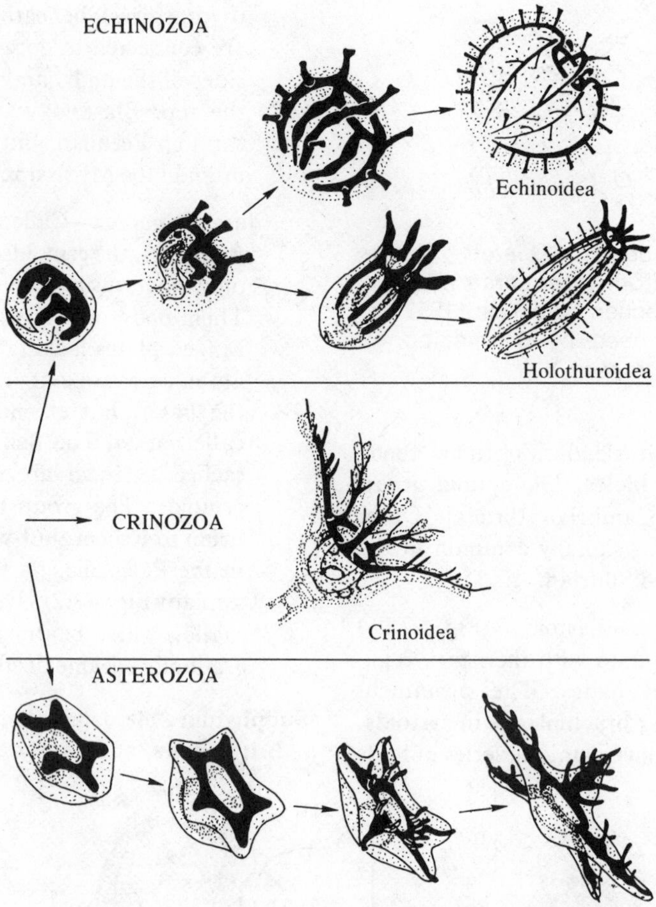

Figure 7.94. *Growth Gradients in Echinoderms.* These are used to define the major subphyla. (From *Treatise on Invertebrate Paleontology,* courtesy of the Geological Society of America and the University of Kansas.)

(Fig. 7.98) have star-shaped bodies (Fig. 7.94) and are free-moving. Their skeletons are composed of many tiny plates not rigidly joined together. As a result, the test usually disintegrates on death, and whole starfish are rare fossils. All three kinds of starfish range from Ordovician to Recent.

Subphylum Echinozoa—This group encompasses the fundamentally globular echinoderms (Fig. 7.94) which are chiefly free-living. They range from Cambrian to Recent.

Class Edrioasteroidea—An extinct group (Fig. 7.99) of Cambrian-Carboniferous age, these are the only attached echinozoans. The five ambulaca are set in a mosaic of smaller plates and are typically ringed by a set of larger plates. Edrioasteroids were common only in the Ordovician when many grew on brachiopod shells.

Class Holothuroidea—The sausage-shaped **sea cucumbers** (Fig. 7.100) have large

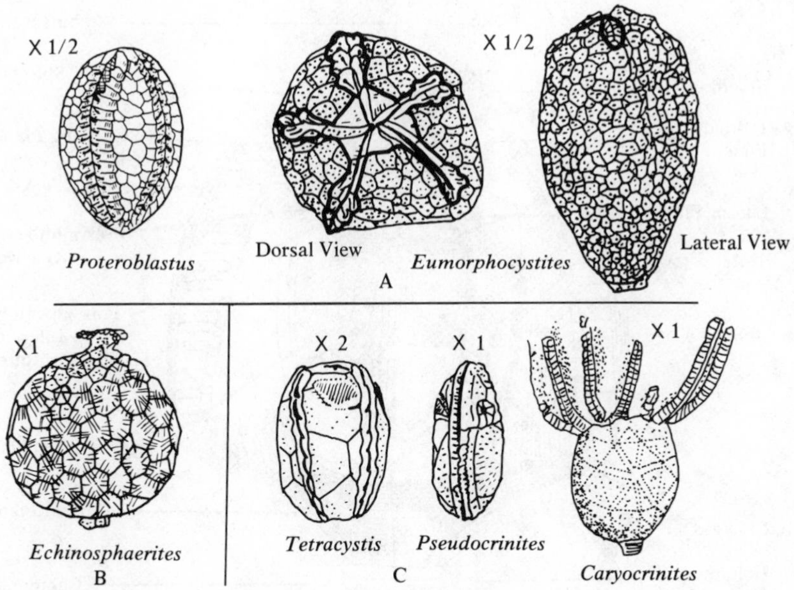

Figure 7.95. *Class Cystoidea.* **A.** *Diploporoid,* **B.** *Fistuliporoid,* and **C.** *Dichoporoid Cystoids.* Magnifications indicated. (After Easton.)

anterior tube feet modified to tentacles, a skeleton reduced to a calcareous ring in the throat area, and isolated plates, or **ossicles.** Only the ossicles are occasionally common fossils.

Class Echinoidea—This is by far the most significant fossil group in the subphylum as it includes the familiar sea urchins, sand dollars, and heart urchins (Fig. 7.101). The plate arrangement is regular with five meridional ambs and five meridional interambs, each consisting of two rows of plates in most echinoids. The mouth is ventral; opposite it is the **apical system** which encircles the anus in **regular** forms and does not in **irregulars** where the anus can be anywhere in the posterior interambulacrum. **Spines** set on **tubercles** cover the test.

The spherical regulars, or **sea urchins,** chiefly inhabit rocks where they use their teeth to eat algae. This environment is not generally favorable for fossilization. Some primitive irregulars, the **holectypoids,** have a sea-urchin-like shape. Most irregulars are bilateral, burrowing forms such as the flattened **sand dollars** so common in the Cenozoic; the **cassiduloids** with their peculiar floscelle around the mouth and peak of abundance in the Paleogene; and the distinctive **heart urchins**, common from the Cretaceous onward.

Phylum Hemichordata—Living members of this phylum (Fig. 7.102) have bilaterally symmetrical bodies divided into three regions—proboscis, collar, and trunk. The trunk bears branchial or gill slits; and there is a small stiffened rod along the back at the anterior end. The acorn worms have left no known fossils, but the deep-water, colonial **pterobranchs** have an irregular or branching scleroprotein coenoecium with a peculiar half-ring, or fusellar, structure. A few fossils extending

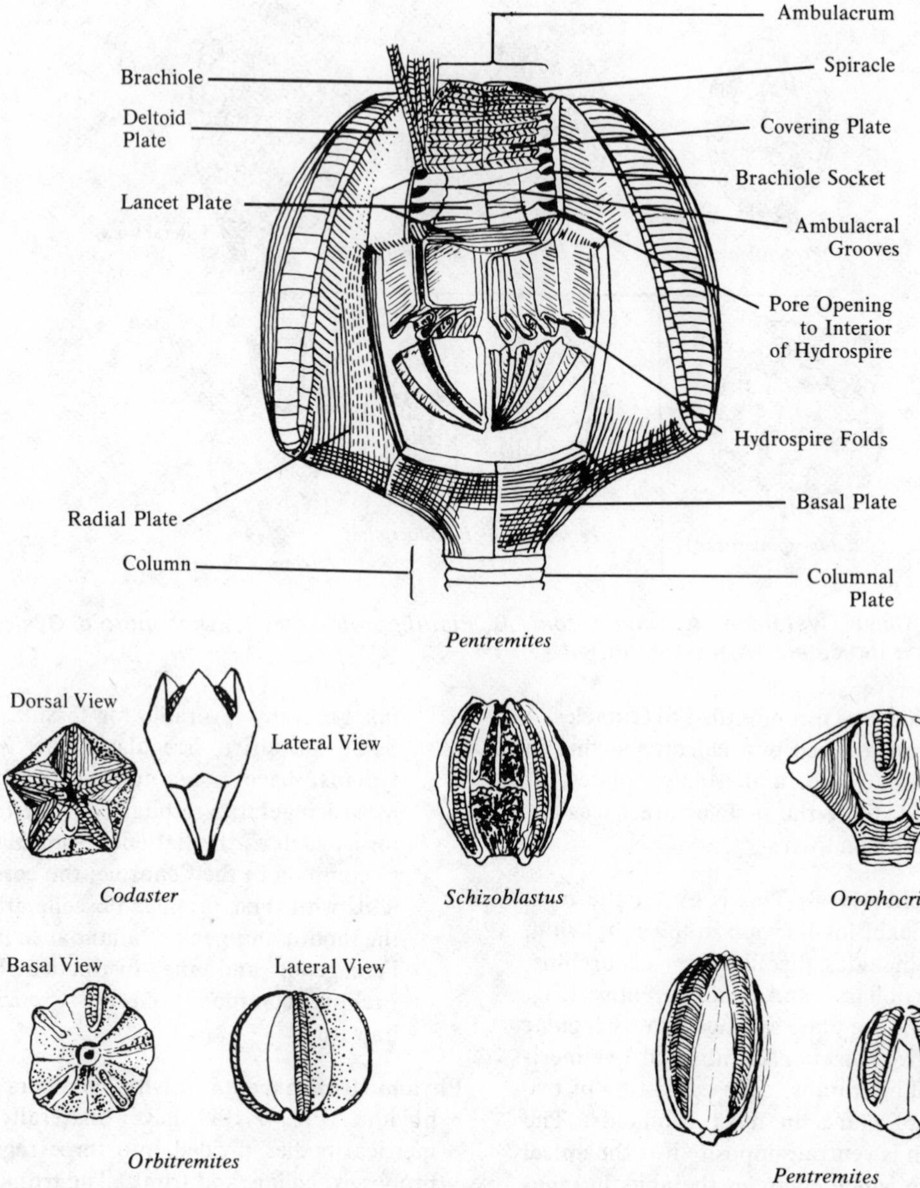

Figure 7.96. *Class Blastoidea.* General morphology and variation in thecal shapes. All natural size except sectioned specimen at top. (Adapted from *Invertebrate Fossils* by Moore-Lalicker-Fischer, Copyright 1952 by McGraw-Hill, Inc., used by permission of McGraw-Hill Book Company.)

back to Cambrian are known, but this group is important chiefly because the structure is virtually identical to that of the extinct graptolites which mystified scientists for so long.

Class Graptolithea — Graptolites (Fig. 7.103) are colonial organisms, each of which inhabited an individual cup, or **theca,** arranged on a branch, or **stipe.** A

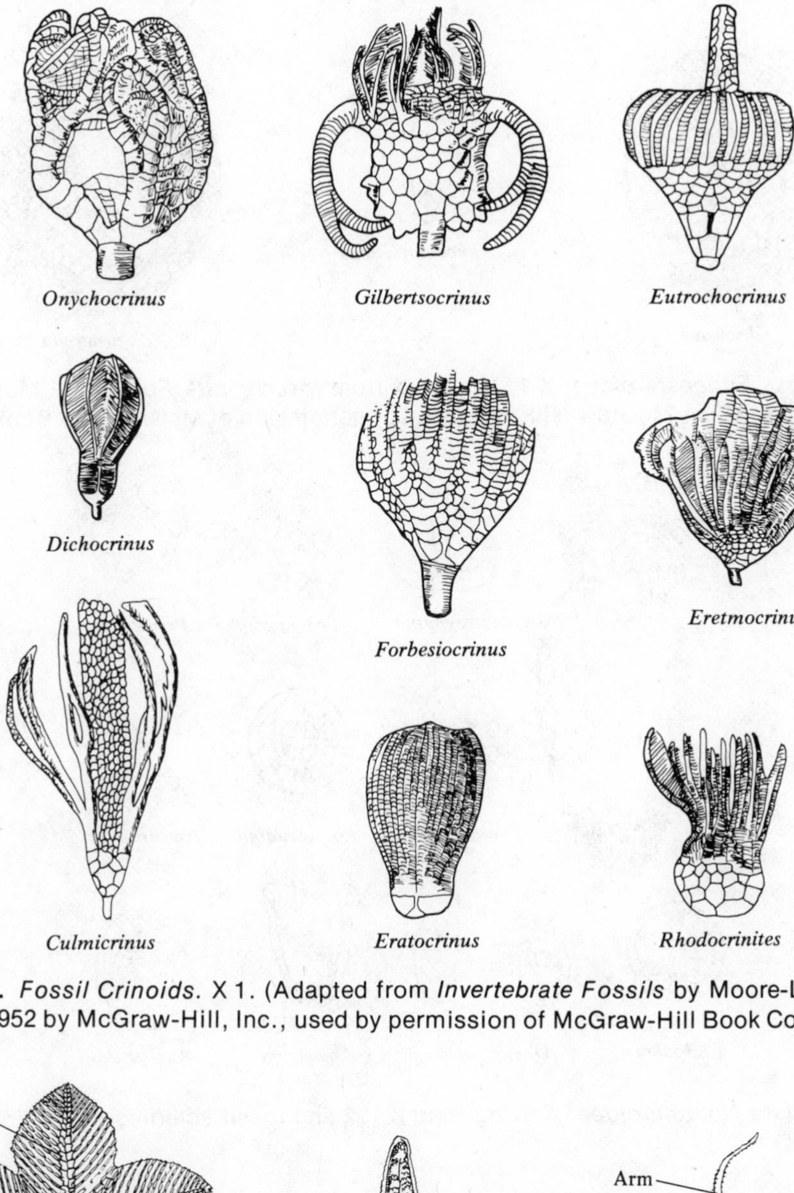

Figure 7.97. *Fossil Crinoids.* X 1. (Adapted from *Invertebrate Fossils* by Moore-Lalicker-Fischer, Copyright 1952 by McGraw-Hill, Inc., used by permission of McGraw-Hill Book Company.)

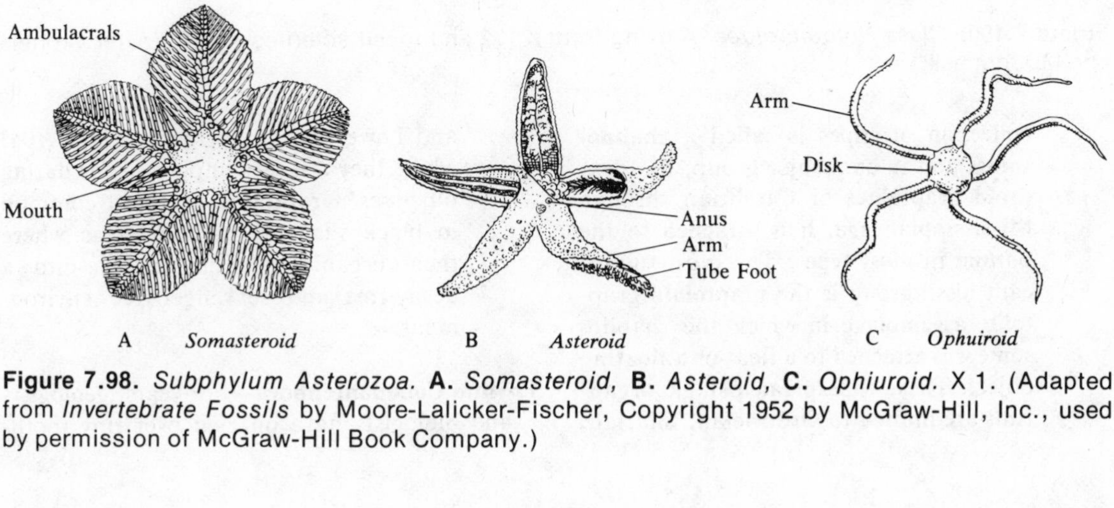

Figure 7.98. *Subphylum Asterozoa.* **A.** *Somasteroid,* **B.** *Asteroid,* **C.** *Ophiuroid.* X 1. (Adapted from *Invertebrate Fossils* by Moore-Lalicker-Fischer, Copyright 1952 by McGraw-Hill, Inc., used by permission of McGraw-Hill Book Company.)

Figure 7.99. *Class Edrioasteroidea.* X 1. (Adapted from *Invertebrate Fossils* by Moore-Lalicker-Fischer, Copyright 1952 by McGraw-Hill, Inc., used by permission of McGraw-Hill Book Company.)

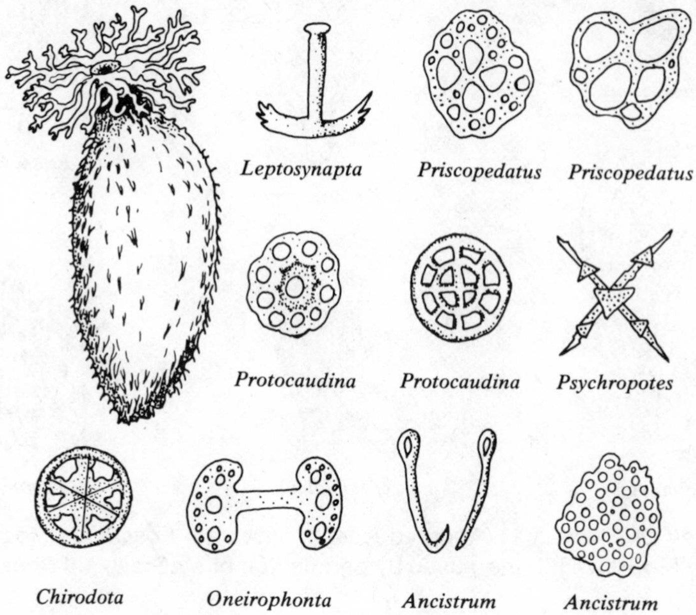

Figure 7.100. *Class Holothuroidea.* A living form X 1/2 and fossil sclerites X 100. (After Cronels and McCormack.)

collection of stipes is called a **rhabdosome;** and in one major group, the **dendroid** graptolites of Cambrian through Mississippian age, it is attached to the bottom in most types. The most significant fossil group is the **graptoloid** graptolite assemblage in which the rhabdosome was attached to a float or a floating object. These widespread pelagic organisms are limited to Ordovician, Silurian, and Lower Devonian rocks (Fig. 7.104) where they are very important for dating purposes. Graptolites are most common in black shales formed in areas where their organic remains settled into a decay-free and scavenger-free environment.

Phylum Conodontophora—For years, geologists and biologists have puzzled over tiny tooth-

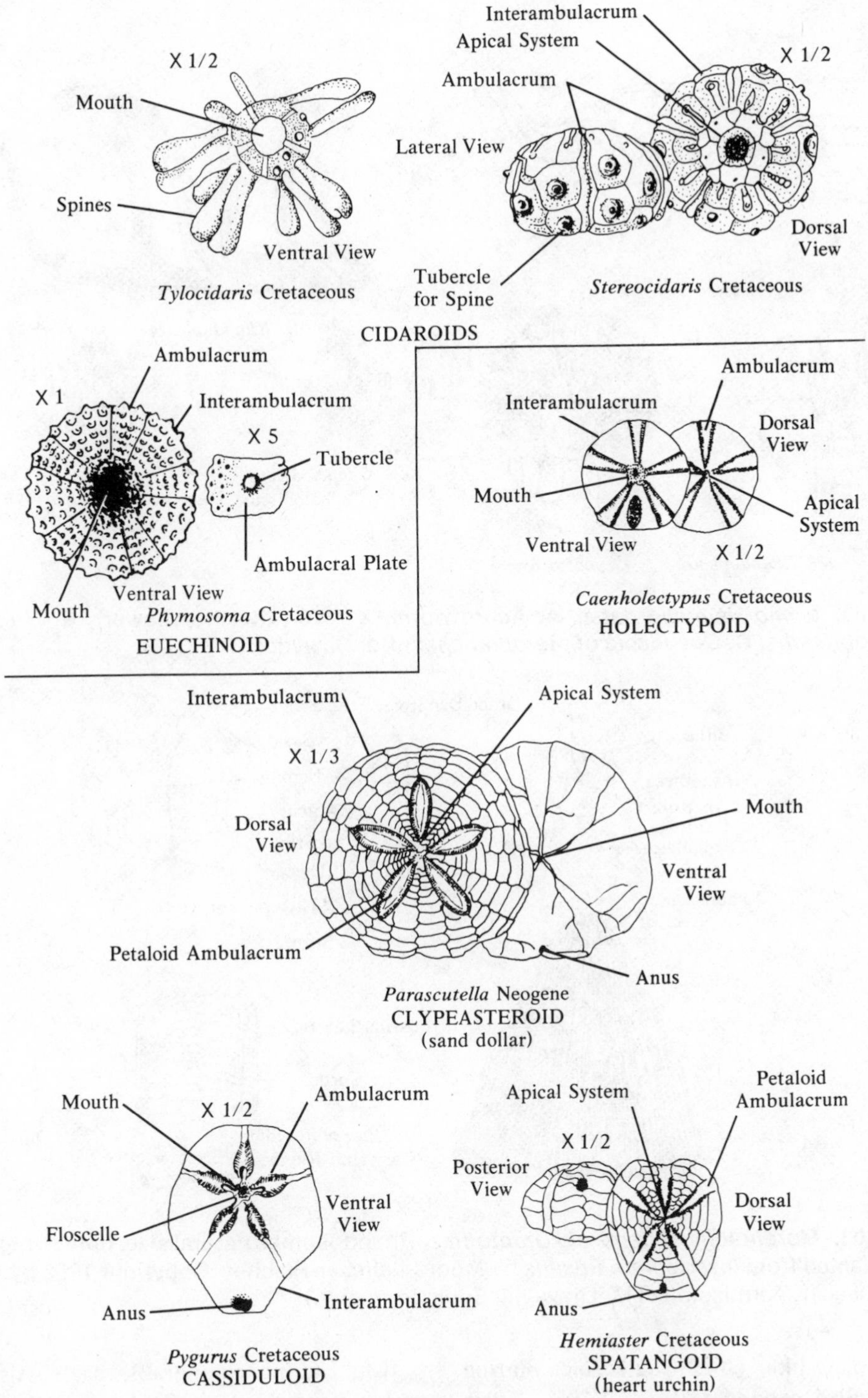

Figure 7.101. *Class Echinoidea.* The major groups and their structural features. Forms above the line are regular, those below are irregulars. (Adapted from *Invertebrate Fossils* by Moore-Lalicker-Fischer, Copyright 1952 by McGraw-Hill, Inc., used by permission of McGraw-Hill Book Company.)

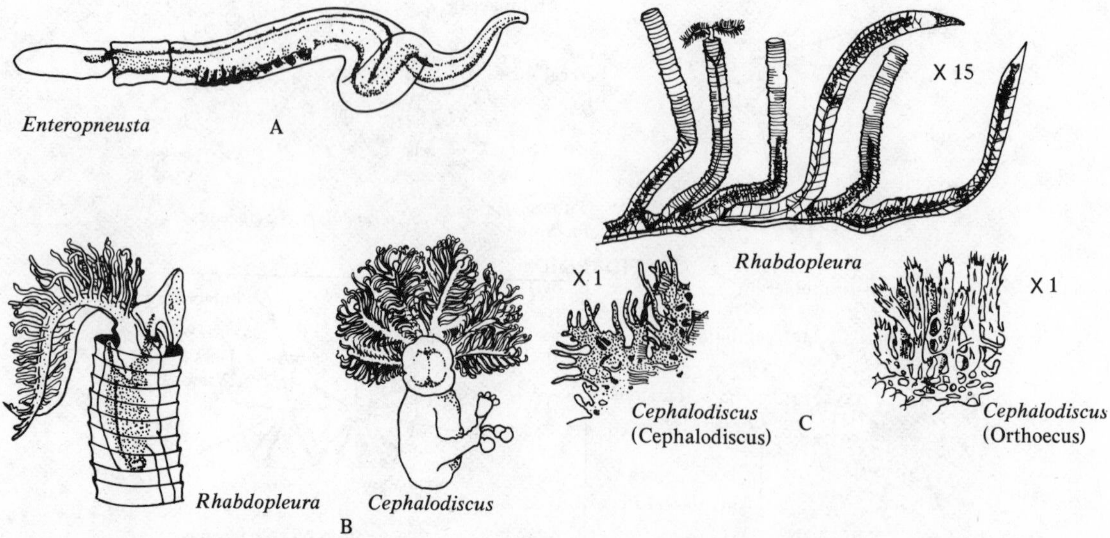

Figure 7.102. *Living Hemichordates.* **A.** *Acorn worm, X 1/10* (after Beerbower); **B.** *Highly magnified pterobranchs;* **C.** *Coenoecia of pterobranchs* (after Dawydoff).

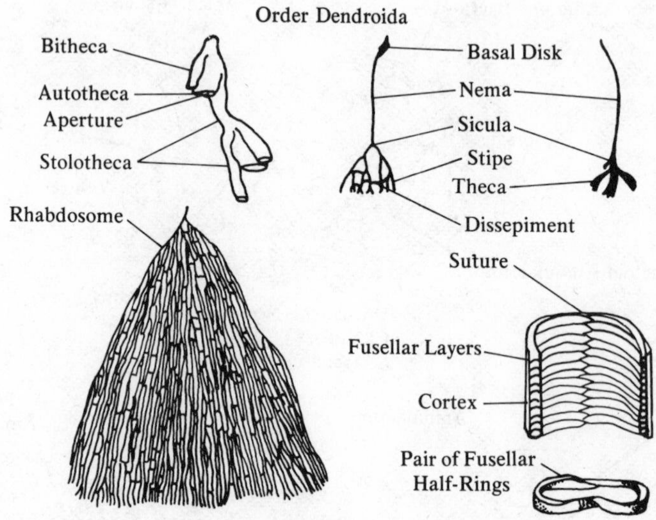

Figure 7.103. *Morphology of Dendroid Graptolites.* Rhabdosomes natural size, details highly magnified. (Adapted from *Invertebrate Fossils* by Moore-Lalicker-Fischer, Copyright 1952 by McGraw-Hill, Inc., used by permission of McGraw-Hill Book Company.)

like or plate-like, chiefly Paleozoic, marine fossils called **conodonts** (Fig. 7.105). Their composition is phosphatic, which indicates some relationship with vertebrates, possibly fish; but no comparable parts exist in any living vertebrate. Furthermore, their layered structure shows that growth occurred only on the outer surfaces, indicating they were not

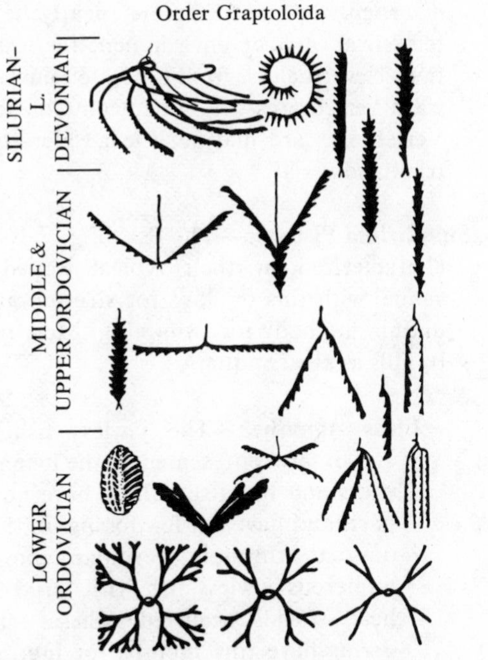

Figure 7.104. *Variation Among Rhabdosomes of Graptoloid Graptolites.* (Adapted from *Invertebrate Fossils* by Moore-Lalicker-Fischer, Copyright 1952 by McGraw-Hill, Inc., used by permission of McGraw-Hill Book Company.)

teeth. Their occurrence in all marine rock types indicates a pelagic creature, and rare assemblages show the creature was bilateral and possessed several types of conodonts (Fig. 7.106). Despite the uncertainty of their origin, conodonts are abundant fossils throughout their Cambrian-through-Triassic range and have become very important fossils for dating Middle to Upper Paleozoic rocks. Recently, what appear to be distillation fossils of the actual animal have been found (Fig. 7.107). The creature is small and resembles a worm with fins. The conodonts are interior structures, apparently part of the digestive system.

Phylum Vertebrata—Most textbooks lump the vertebrates, sea squirts, and amphioxans into a Phylum Chordata based on the fact that all, at some time in their lives, possess a dorsal nerve cord, gill slits, and a dorsal stiffening rod of some type. The last two groups are very different from vertebrates as adults and have left few fossils; so from a paleontologic viewpoint, the vertebrates can best be considered a separate phylum.

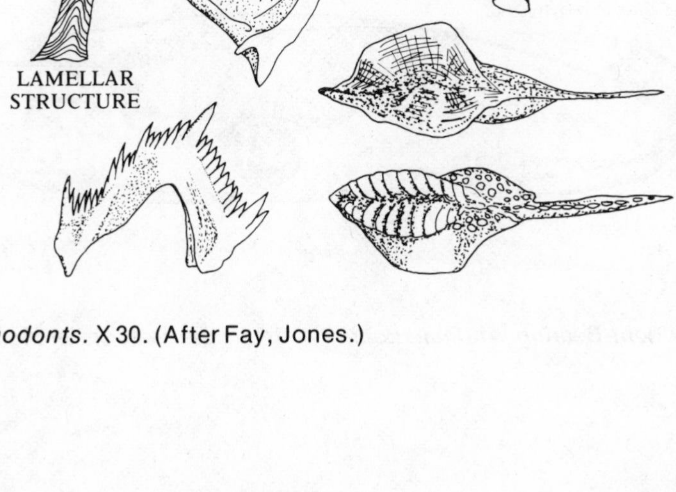

Figure 7.105. *Conodonts.* X30. (After Fay, Jones.)

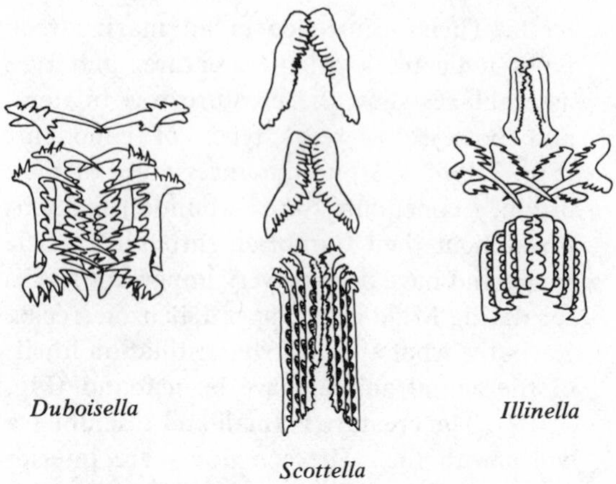

Figure 7.106. *Conodont Assemblages.* X 10. (After Rhodes.)

The vertebrates, except the jawless fish, possess a dorsal supporting structure composed of calcium phosphate (either **bone** or cartilage) and called the **vertebral column.** They all have a phosphatic brain case; hence; their other common name, **Craniata.** Other bones typically present are the thoracic ribs, axes of appendages, and dermal bones. All bones have a meshwork construction much like that of echinoderms. Most have **teeth** of hard enamel.

The fossil record goes back to the Ordovician or possibly the Cambrian, but is not as good as that of the invertebrates. All vertebrates have numerous parts and are easily disarticulated. Furthermore, nearly half are terrestrial, a poor environment for fossilization. Nevertheless, from the Late Silurian onward vertebrates are not uncommon fossils. Vertebrates are marine, fresh water, terrestrial, and aerial.

Subphylum Piscezoa—The **fish** (Fig. 7.108) are characterized by their typical torpedo-like shape, with fins or flaps for steering and an undulating body for swimming. Most respire by gills as all are aquatic.

Class Agnathea—The jawless fish (Fig. 7.108), are represented by the living lampreys and hag-fish, which have no hard parts and have left few fossils. In the Ordovician through Devonian, however, numerous jawless fish with large, bony head shields occurred. These **ostracoderms** have tiny mouths for mud-grubbing, no true fins for stabilization, and large gill areas. Ostracoderms lived in fresh-water, brackish, and marine environments.

Class Placodermea—This is a heterogeneous collection of primitive, armored, jawed fish (Fig. 7.108). The jaws evolved from an anterior gill arch and were not propped securely against the skull (Fig. 7.109). Paired fins of unusual construc-

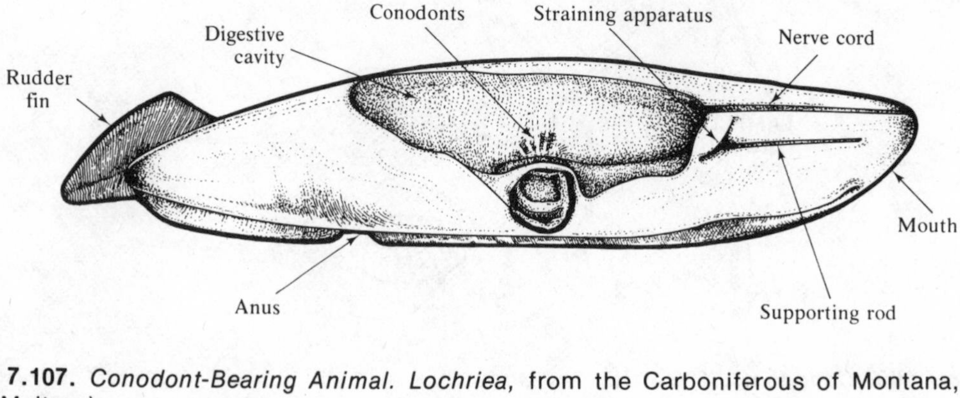

Figure 7.107. *Conodont-Bearing Animal. Lochriea,* from the Carboniferous of Montana, X 15. (After Melton.)

tions were present. Nearly all placoderms are Devonian, but a few Silurian and Mississippian forms are known. Most were marine, including the carnivorous **arthodires.** Some were fresh water, such as the mud-grubbing **antiarchs.**

Class Chondrichthyea—The cartilaginous fish (Fig. 7.108) include the **sharks** and the **skates** and **rays;** the ratfish (**Holocephalea**) appear to be of separate origin. Jaws are present and propped against the skull by the next gill arch, the hyomandibular, whose gill slit is reduced to a hole, or **spiracle** (Fig. 7.109). The gills all open separately to the exterior. The internal skeleton is of cartilage and rarely preserved, but the hard teeth and spines (Fig. 7.110) are common fossils, particularly from the Carboniferous onward. Most are marine.

Class Acanthodea—This group (Fig. 7.108) appears to be ancestral to the bony fish. Its members possess a flap covering the gills (the operculum). They have one external opening as in bony fish. In both groups, the jaw is propped firmly against the skull and there is no spiracle. Both groups also have similar scales. Acanthodians, or spiny sharks, have their fins supported by single large spines unlike those of bony fish, however. They were chiefly fresh water and range from Silurian through Permian with a Devonian peak.

Class Osteichthyea—The **bony fish** (Fig. 7.108) have an internal bony skeleton, a scaly covering, paired fins supported by bony shoulder and hip girdles, jaws propped firmly against the skull with no spiracle present, and an operculum. They are both marine and fresh water and range from Devonian to Recent with a Cretaceous-to-Recent peak. Bony fish are divided into two subclasses (Fig. 7.111).

Subclass Actinopterygia—The **ray-finned fish** have slender, parallel rays of dermal bone supporting their paired fins, lack internal nostrils, possess a swim bladder (early types probably had lungs), have vertebrae entering the upper tail lobe, and have one dorsal fin. They have been by far the most significant bony fish group after the Devonian.

Subclass Sarcopterygia—The **lobe-finned fish** have skeletal bones supporting their paired fins which also bear a large and fleshy fringe, possess internal nostrils and lungs, typically have the vertebrae passing between the tail lobes, and have two dorsal fins. Common only in the Devonian and rare after the Paleozoic, the lobe-finned fish are notable as the ancestors of the tetrapods which their skeletal structure, particularly the fins (Fig. 7.112), skull, and lungs, foreshadow.

Subphylum Tetrapodazoa—These are basically four-limbed, terrestrial, air-breathing vertebrates. Some have secondarily returned to the water, but still breathe with lungs; others have taken up flight, and the front limbs have evolved into wings.

Class Amphibea—Amphibians—the living frogs, toads, and salamanders (Fig. 7.113) and a host of extinct types—may not be a natural group. The limbs are weak in most, indicating that their habitat is still largely aquatic. All but the earliest forms from the Late Devonian have lost the fish-like tail. The eggs are small, shell-less, and lack protective membranes, which necessitates their being laid in water. The young hatch at an immature stage in which they are aquatic, fish-like, and breathe with gills. Only later do they metamorphose to the air-breathing, four-limbed adults. Amphibians, like fish, are "cold-blooded"; that

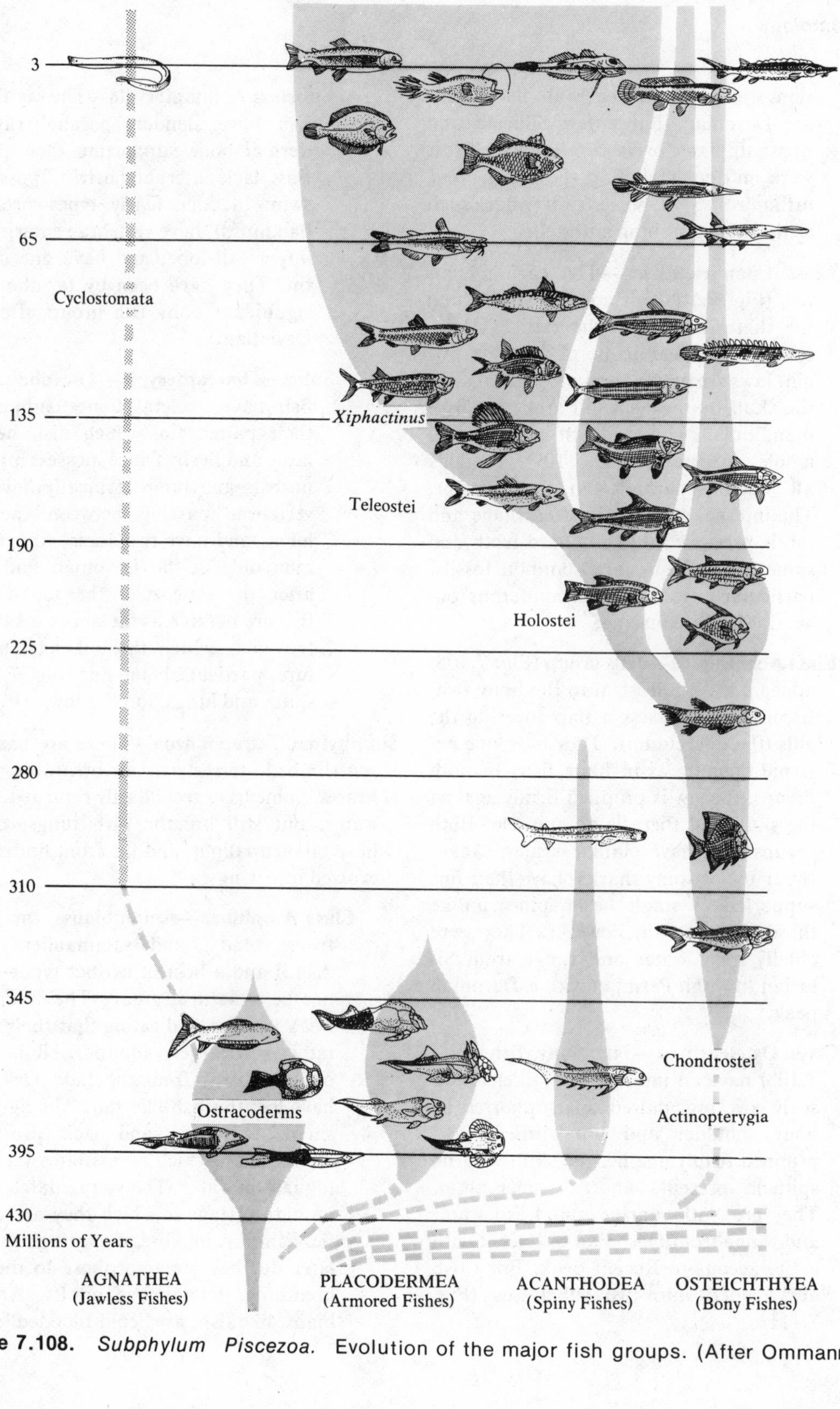

Figure 7.108. *Subphylum Piscezoa.* Evolution of the major fish groups. (After Ommanney.)

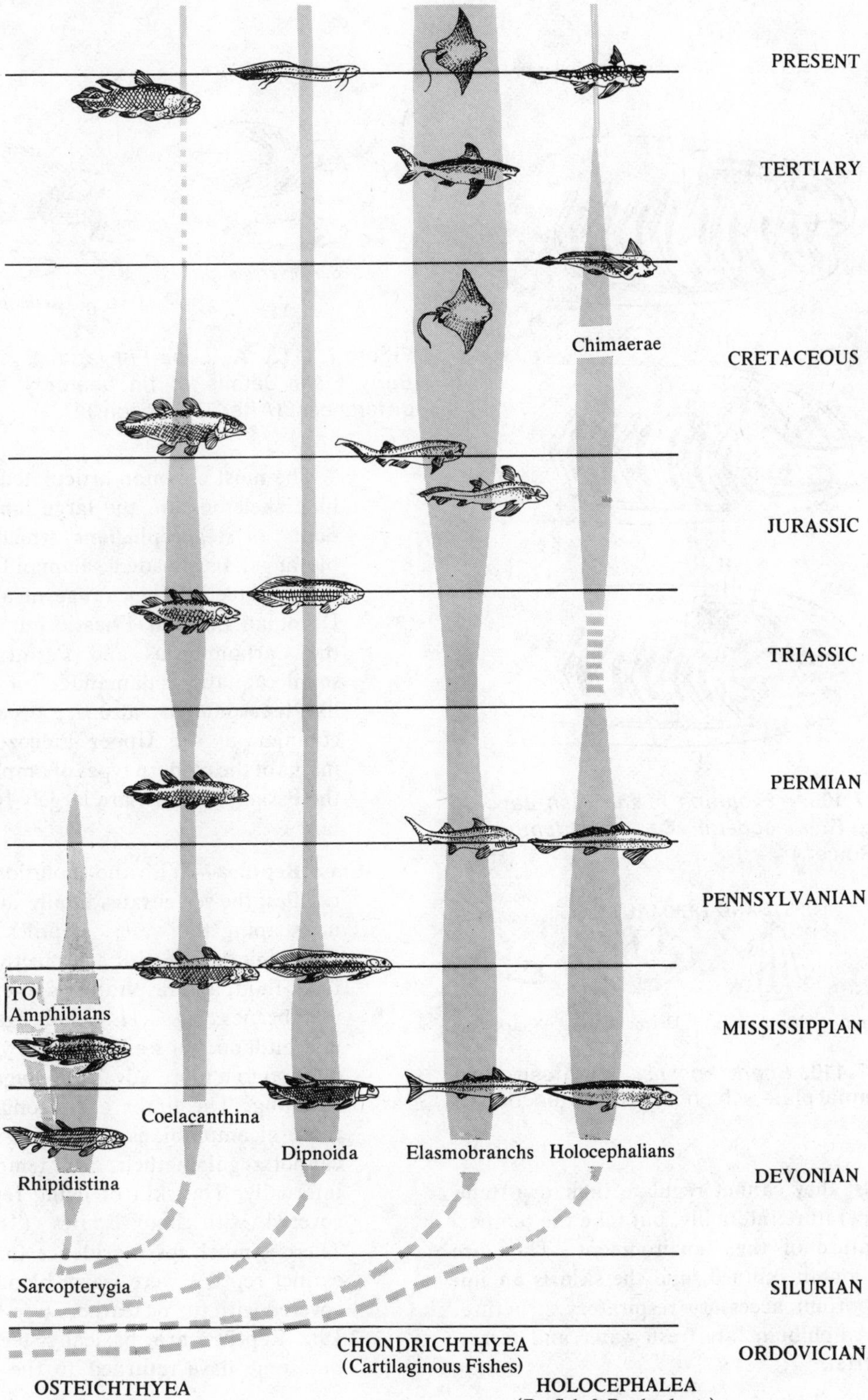

PRESENT
TERTIARY
CRETACEOUS
JURASSIC
TRIASSIC
PERMIAN
PENNSYLVANIAN
MISSISSIPPIAN
DEVONIAN
SILURIAN
ORDOVICIAN
Chimaerae
TO
Amphibians
Coelacanthina
Dipnoida
Elasmobranchs
Holocephalians
Rhipidistina
Sarcopterygia
OSTEICHTHYEA
(Bony Fishes)
CHONDRICHTHYEA
(Cartilaginous Fishes)
HOLOCEPHALEA
(Ratfish & Bradyodonts)

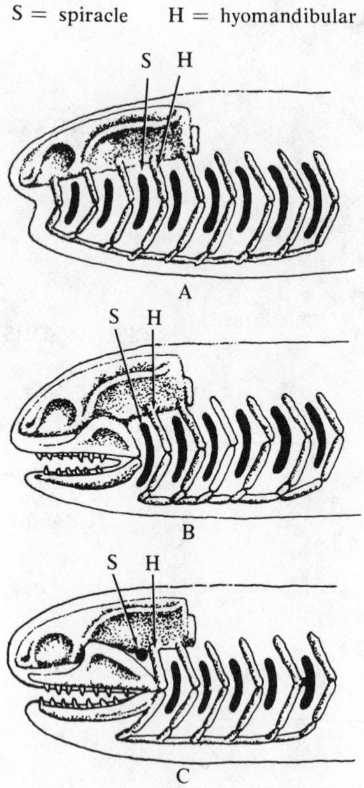

Figure 7.109. *Evolution of the Fish Jaw.* **A.** *Agnath,* **B.** *Placoderm,* **C.** *Chondrichthyean.* (After Romer.)

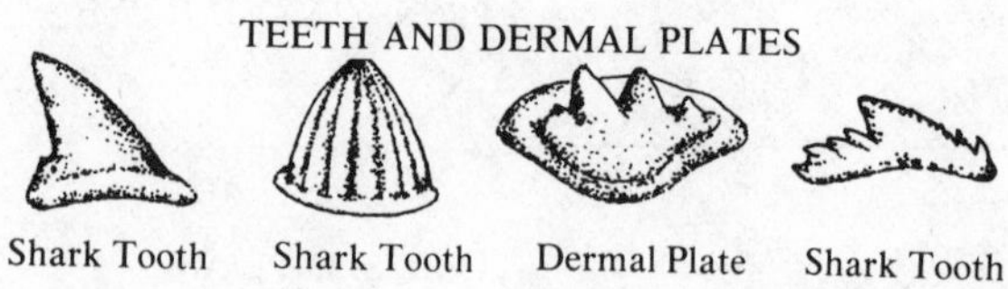

Figure 7.110. *Shark Fossils.* All lifesize except dermal plate, which is X 10. (After Jones.)

is, they cannot regulate their own temperature internally, but take the temperature of their environment. They are smooth-skinned, and the skin is an important accessory respiratory structure. Amphibians are fresh water and terrestrial.

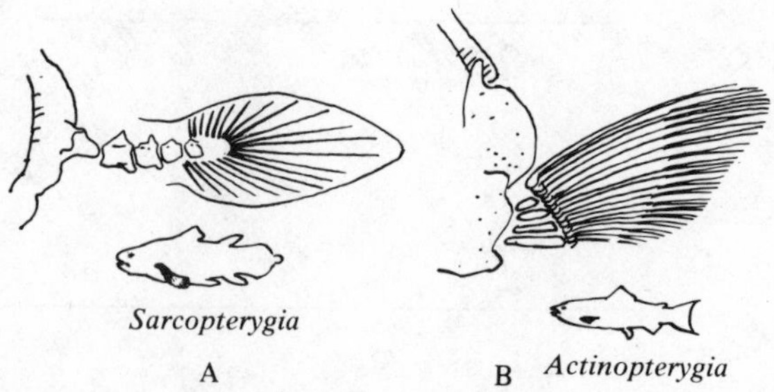

Figure 7.111. **A.** *Lobe-Fin* and **B.** *Ray-Fin Bony Fish.* Details of fin anatomy showing differences. (After Thompson.)

The most common articulated amphibian skeletons are the large **labyrinthodonts,** or **stegocephelians,** which resemble huge, flat-headed salamanders with fish-like teeth. They range in age from Devonian through Triassic but peak in the Carboniferous and Permian. The small aquatic salamander or snake-like **lepospondyls,** also are occasionally common in the Upper Paleozoic. Remains of the modern types of amphibians, the **lissamphibians,** are largely fragmentary.

Class Reptilea—With the evolution of the reptiles, the vertebrates finally succeeded in escaping the water. Reptiles evolved the **amniotic egg** with a protective shell, food, fluid, and respiratory and excretory membranes (Fig. 7.114). This egg can be laid on land, thus allowing the young to develop to a more advanced stage before hatching. The limbs are stronger than those of amphibians, but living reptiles cannot regulate their own temperature internally. The skin of living reptiles is covered with bony scutes ("scales"). There is much fossil evidence that some extinct reptiles were warm-blooded and covered with fur or feathers (see Chapter 19). Reptiles are basically terrestrial, but some have returned to the aquatic

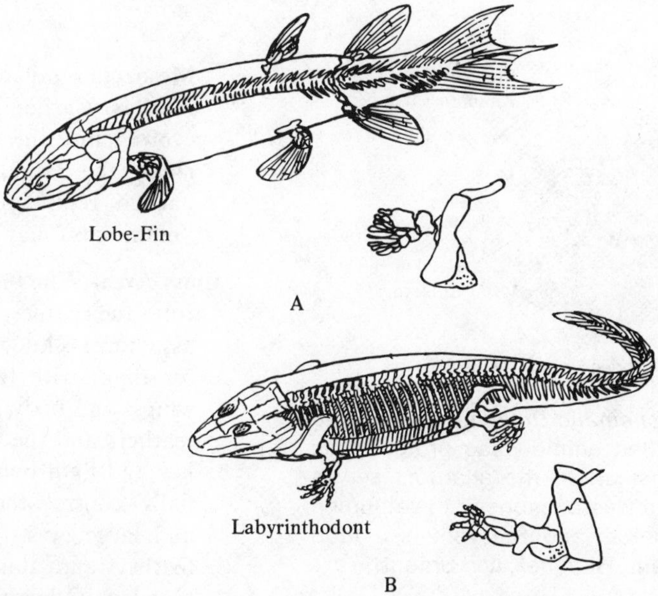

Figure 7.112. *Similarities Between Devonian Lobe-Finned Fish and Amphibians.* **A.** *Lobe-fin fish and its shoulder and forelimb,* **B.** *Early amphibian and its shoulder and forelimb.* (After McAlester, in part.)

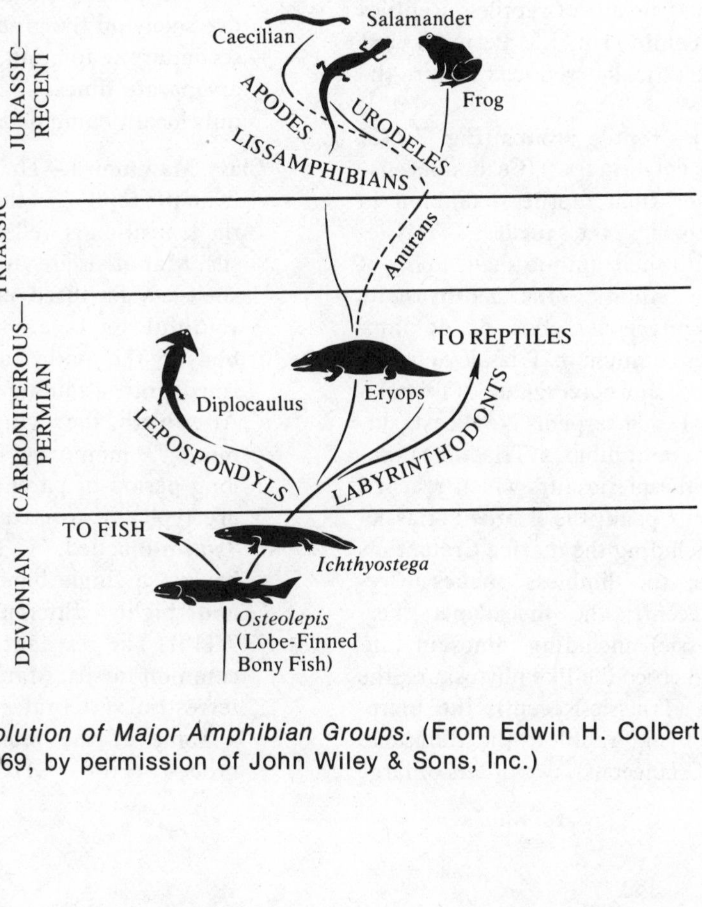

Figure 7.113. *Evolution of Major Amphibian Groups.* (From Edwin H. Colbert, *Evolution of the Vertebrates,* © 1969, by permission of John Wiley & Sons, Inc.)

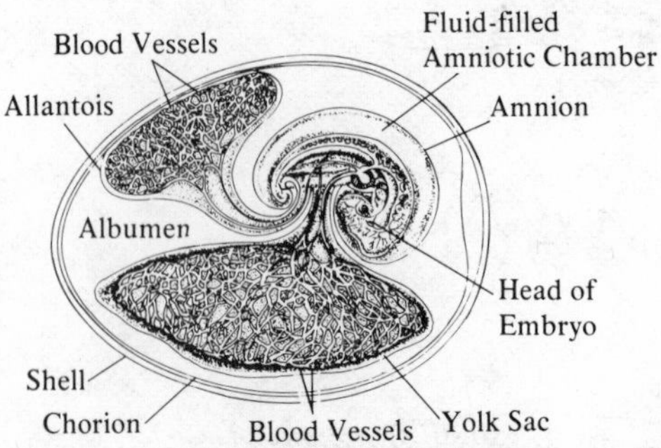

Figure 7.114. *Schematic Drawing of Amniotic Egg.* The fluid-filled amnion sac protects the embryo from dessication; the allantois serves for respiration and waste disposal; the albumen is a moisture source; and the yolk is a food source. (After Oram, Hummer, and Smout.)

environment, both marine and fresh water, and some have developed the ability to fly. Reptiles range in age from Carboniferous to Recent, but the Mesozoic Era was the true age of reptiles. Reptiles were also common in the Permian, but became drastically reduced after the Mesozoic.

The major reptile groups (Fig. 7.115) are the **cotylosaurs** (Carboniferous-Triassic), or stem reptiles, common in Late Paleozoic; the **turtles** (Triassic-Recent) with their unique shell, many of which are aquatic; the **ichthyosaurs** (Triassic-Cretaceous), fish or dolphin-like reptiles common in Triassic and Jurassic; the **sauropterygians** (Triassic-Cretaceous), sea-serpent-like beasts including the amphibious Triassic nothosaurs and the plesiosaurs which reached a Cretaceous peak; the **lizards** (Triassic-Recent) including the marine Cretaceous mosasaurs; the limbless **snakes** (Cretaceous-Recent); the **thecodonts** (Permian-Triassic) including dinosaur ancestors and crocodile-like phytosaurs; the **crocodiles** (Triassic-Recent); the **pterosaurs,** or flying reptiles; the **dinosaurs** (Triassic-Cretaceous), two orders of large Mesozoic reptiles; and the **mammal-like reptiles** (Carboniferous-Jurrasic), which evolved into the mammals and also include the large, Late Paleozoic, finbacked reptiles. The living egg-laying "mammals" may be survivors.

Class Avea—The **birds** (Fig. 7.116) evolved from the reptiles, and a transitional Jurassic form is known. Birds are structured for flight with front limbs modified to wings, and body and wings covered with feathers and the tail made of them, hollow and light bones, the ability to internally control their own temperatures, and large eyes. The Jurassic form had feathers and, thus, was probably warm-blooded. Otherwise, it is structurally a small carnivorous dinosaur. Birds retain the reptilian egg, but all except the Jurassic genus and two Cretaceous forms have lost the teeth. Some large flightless birds live solely on the ground. This may be a secondary adaptation or inheritance from a separate dinosaur ancestry. Birds are only locally common in Cenozoic rocks.

Class Mammalea—The mammals evolved from the reptiles in the Late Triassic, and the transition is well documented by fossils. Mammals are viviparous; that is, the shell-less fertilized egg and succeeding youthful stages are maintained in the body by the mother and internally nourished until attaining an advanced state. After birth, the young are nursed by the mother's mammary glands, ensuring a long period of parental care. Mammals are typically covered with hair and are "warm-blooded." They have large brains, a single bone in the lower jaw, and highly differentiated teeth (Fig. 7.117). The resistant teeth are the most common fossils. Mammals are the major terrestrial vertebrates of the Cenozoic.

Some of the more important fossil groups (Figs. 7.118, 7.119) are the

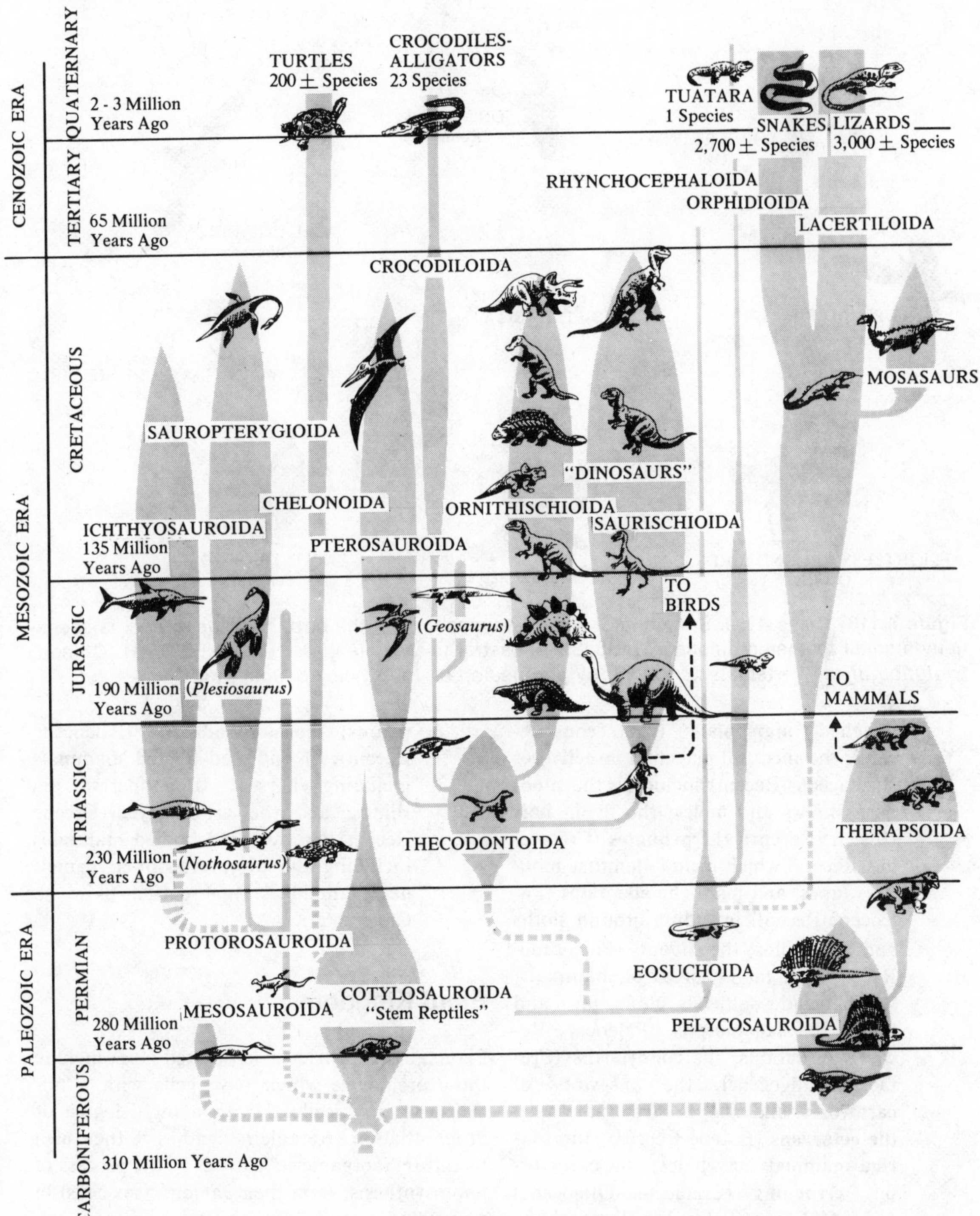

Figure 7.115. *Evolution of the Reptiles.* Note the great diversity of reptiles in the Mesozoic Era. (After Carr.)

Figure 7.116. *Class Avea.* Some major adaptive types among the birds. *Archaeopteryx* is essentially a small feathered dinosaur, probably ancestral to the true birds. (From Edwin H. Colbert, *Evolution of the Vertebrates,* ©1969, by permission of John Wiley & Sons, Inc.)

pouched **marsupials** (Cretaceous-Recent); the ancestral placental **insectivores** (Cretaceous-Recent) including the modern shrews and moles; the flying **bats** (Eocene-Recent); the **primates** (Cretaceous-Recent) which include lemurs, monkeys, apes, and man; the **edentates** (Paleocene-Recent) including ground sloths and armadillos; the **rodents** (Paleocene-Recent), the largest order of mammals, which includes squirrels, guinea pigs, and mice; the **lagomorphs** (Paleocene-Recent), or rabbits; the **condylarths** (Cretaceous-Oligocene), the ancestors of carnivores and the hoofed mammals; the **cetaceans** (Eocene-Recent), the marine mammals or whales; the carnivorous **creodonts** (Cretaceous-Oligocene) and **carnivores** (Paleocene-Recent); the bulky **amblypods** (Paleocene-Oligocene) including the pantodonts and uintatheres; the **proboscideans** (Eocene-Recent) including mastodons and elephants; the **perissodactyls** (Paleocene-Recent), or odd-toed hoofed mammals, including horses, titanopheres, and rhinos; and the **artiodactyls** (Eocene-Recent), or even-toed hoofed mammals, including the pigs, oreodonts, camels, deer, antelopes, and cattle. (Also see Chapter 21).

Plant Kingdom

Plants, in the restricted sense used here, include those organisms which have cells with a nucleus, are multicellular with a high degree of differentiation, are able to synthesize their own food from inorganic substances by the process of photosynthesis, form their gametes (sex cells) in multicellular structures, and form embryos. So restricted, plants consist of two major groups or subkingdoms—the **Bryophyta** comprising the delicate mosses, liverworts, and hornworts, which lack a system of conducting tubes; and

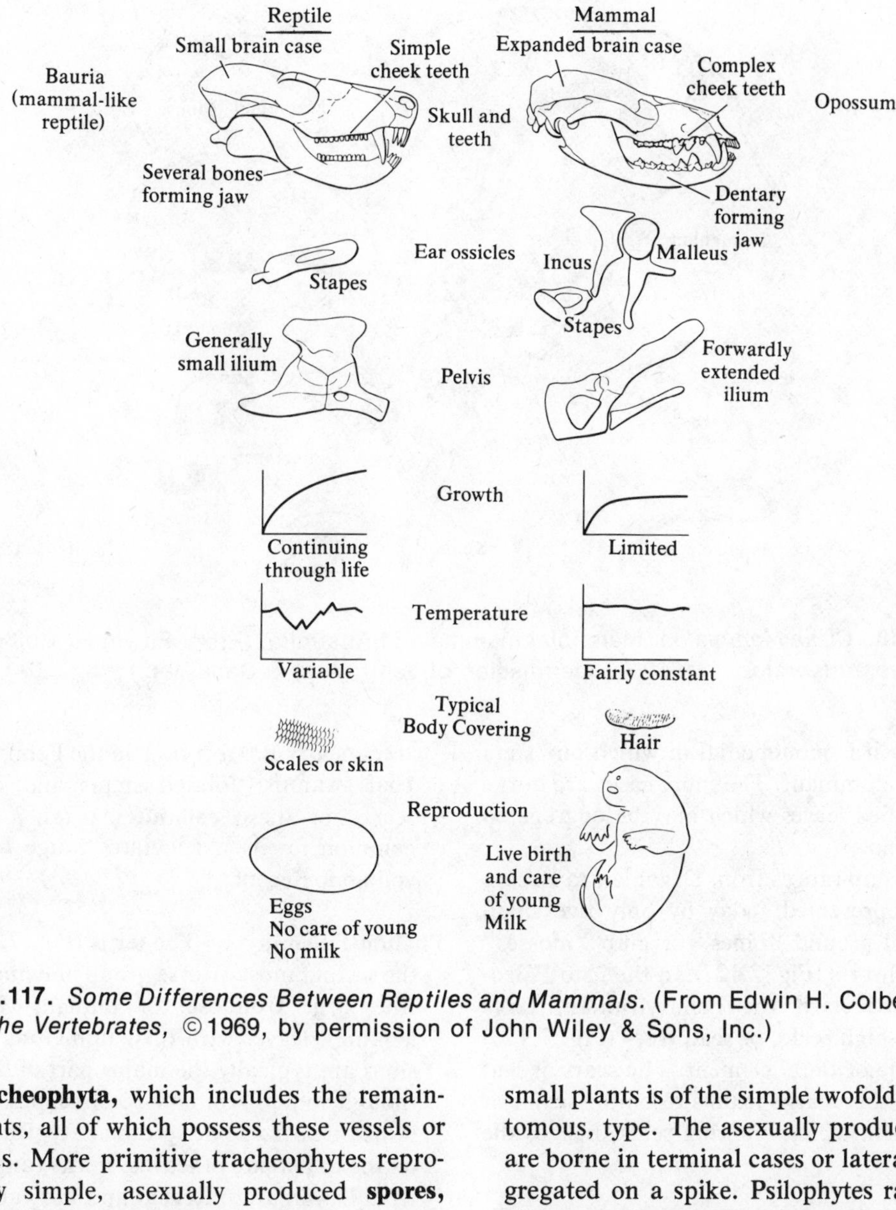

Figure 7.117. *Some Differences Between Reptiles and Mammals.* (From Edwin H. Colbert, *Evolution of the Vertebrates,* © 1969, by permission of John Wiley & Sons, Inc.)

the **Tracheophyta,** which includes the remaining plants, all of which possess these vessels or tracheids. More primitive tracheophytes reproduce by simple, asexually produced **spores,** while advanced groups have developed the more complex, sexually produced bodies called **seeds.**

Subkingdom Tracheophyta

Phylum Psilophyta—The psilophytes, or whisk "ferns" (Fig. 7.120), have no differentiated leaves and roots, just upright stems with chlorophyll, and underground supporting and absorptive stems. The branching in these small plants is of the simple twofold, or dichotomous, type. The asexually produced spores are borne in terminal cases or lateral ones aggregated on a spike. Psilophytes range from Silurian to Recent, but only two genera (Fig. 7.120) apparently belonging to this group survive. The greatest diversity was in the Devonian when several main lineages evolved.

Phylum Lycopodophyta—Lycopods, like the remaining plant groups, have true leaves and roots developed. The leaves are small and spirally arranged, a distinctive feature of this group (Fig. 7.121). Branching is dichoto-

Figure 7.118. *Class Mammalea.* Marsupial mammals of Australia. (From Edwin H. Colbert, *Evolution of the Vertebrates,* ©1969, by permission of John Wiley & Sons, Inc.)

mous, or it is monopodial in which one stem becomes dominant. The spore cases are borne on modified leaves which may be aggregated into a cone.

The group ranges from Devonian to Recent but is represented today by only five small genera of ground "pines" or club "mosses" and quillworts (Fig. 7.122). In the Late Paleozoic, particularly the Pennsylvanian, large 100-foot-high **scale, or seal, trees** (Fig. 7.123) were a major floral element. The scars or leaf bases on these large trunks resemble scales or the imprint of ancient ring seals, hence, the name.

Phylum Arthrophyta—The **articulates,** or **sphenopsids,** have stems and branches which are jointed. The side branches and their small leaves occur in whorls only at the nodes (Fig. 7.124). The spore cases are terminally arrayed on short lateral branches which are typically aggregated into cones. The single living genus (Fig. 7.125)—the small living horsetail, or scouring "rush"—is the last survivor of this group. Articulates reached the size of large trees nearly 100 feet high in the Pennsylvanian coal swamps. Jointed stems and clustered leaves of these **calamites** (Fig. 7.126) are common fossils. Articulates range from Devonian to Recent.

Phylum Filicophyta—The **ferns** (Fig. 7.127) are the second most diverse group of living plants. The large frondose, or pinnate (feathery-looking), leaves with their numerous subdivisions are typically the major part of the body, though some fossil and recent types have attained tree size. The spores are typically borne on the underside of leaves. Ferns range in age from Devonian to Recent and are particularly common in the Mesozoic. Many of the fern-like leaves of the next group were formerly confused with true ferns.

Phylum Pteridospermatophyta—This group is comprised of **seed "ferns"** (Fig. 7.128), which were shrubs or low trees with fern-like foliage and bore seeds or sexually produced reproductive bodies on their leaves. They range in age from Devonian through Cretaceous, but have a

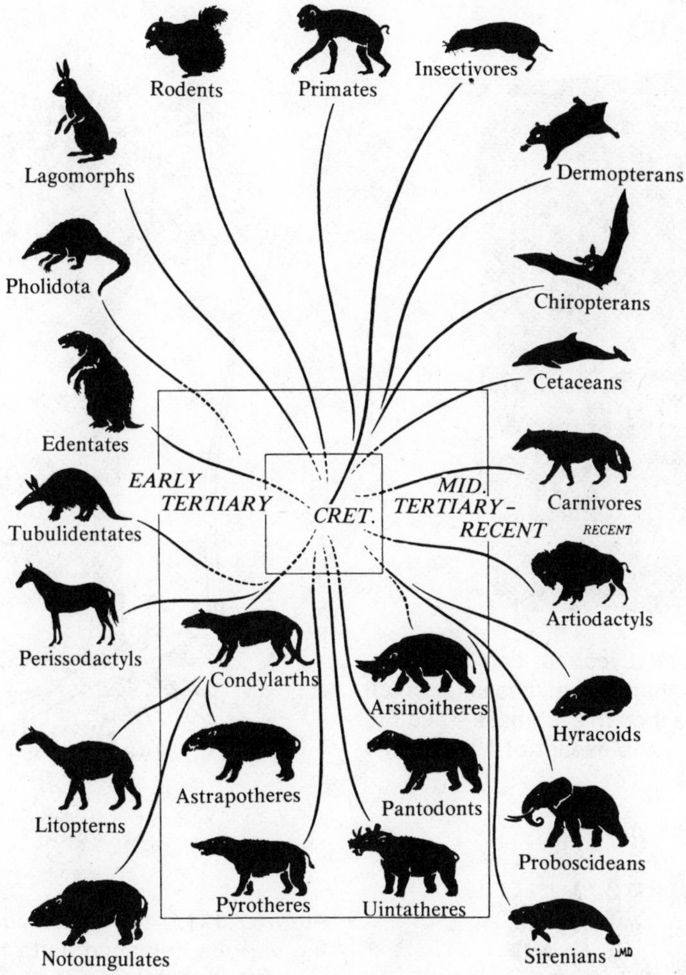

Figure 7.119. *Class Mammalea.* Placental mammals. (From Edwin H. Colbert, *Evolution of the Vertebrates,* © 1969, by permission of John Wiley & Sons, Inc.)

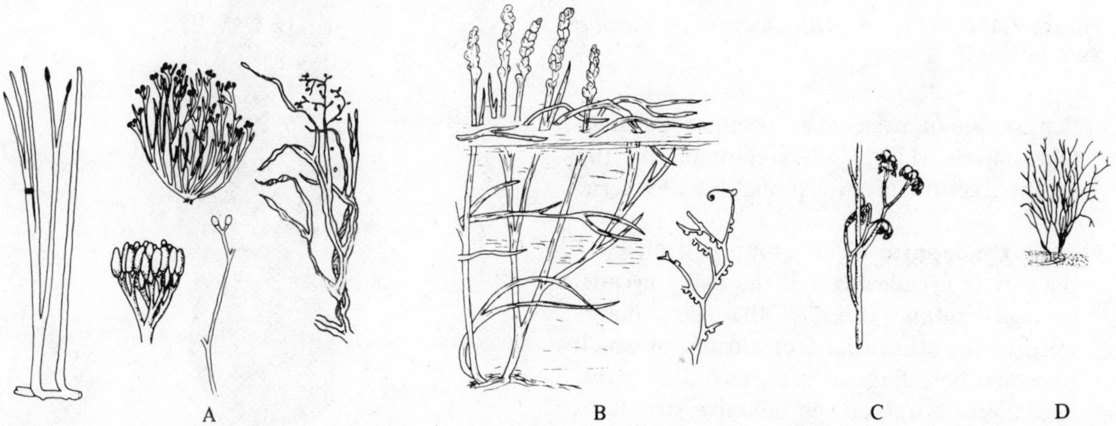

Figure 7.120. *Psilophytes.* **A.** *Rhynia-type,* **B.** *Zosterophyllum-type,* **C.** *Trimerophyton-type,* **D.** *Living Psilotum.* (After Banks.)

Figure 7.121. *Fossil Lycopod Lepidodendron* showing diamond-shaped leaf bases; a small, simple leaf was attached to each base—see Fig. 7.123. (Courtesy Field Museum of Natural History, Chicago.)

Figure 7.122. *Living Herbaceous Lycopod Selaginella.*

Pennsylvanian peak. The peculiar Permian **Glossopteris** (Fig. 7.129), common on the southern continents, was probably a seed fern.

Phylum Cyadophyta—This group encompasses the extinct **cycadeoids** and the living **cycads,** or **sago "palms"** (Fig. 7.130). These plants were of the dimensions of shrubs or small trees and bore large, coarse, fern-like leaves. Seeds were borne on reproductive structures which are specialized leaves and resemble flowers or cones (Fig. 7.130). The group ranges from Carboniferous to Recent, but was dominant in the Mesozoic Era. Only a few

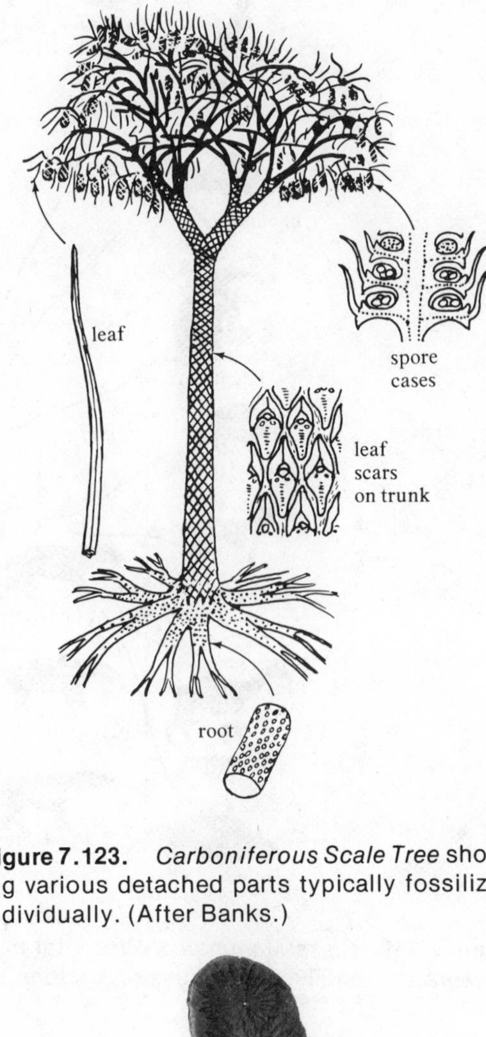

Figure 7.123. *Carboniferous Scale Tree* showing various detached parts typically fossilized individually. (After Banks.)

Figure 7.124. *Arthrophyte Jointing.* Fossil Carboniferous foliage. (Courtesy Field Museum of Natural History, Chicago.)

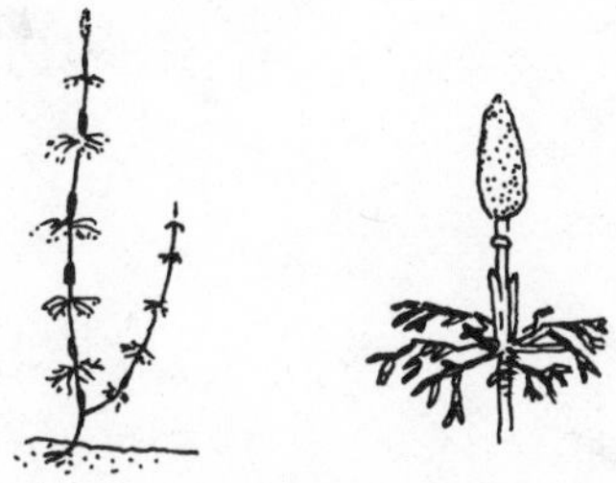

Figure 7.125. *Living Equisetum.* Enlargement shows terminal "cone."

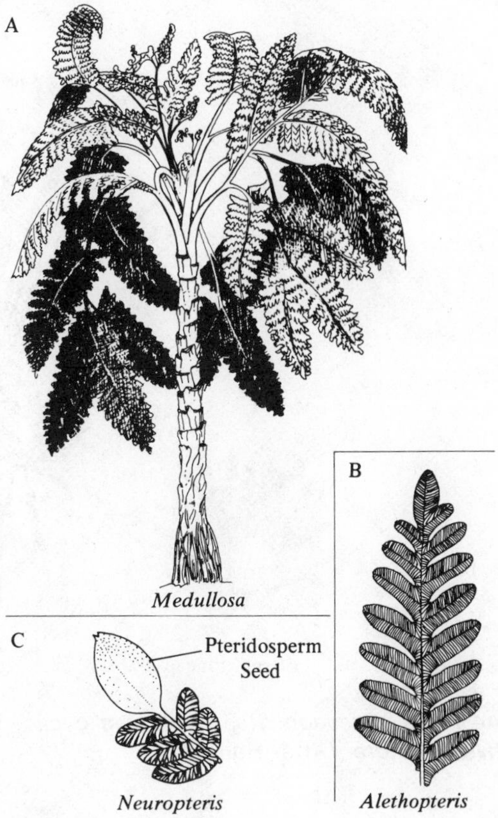

Figure 7.128. *Pteridosperms, or Seed Ferns.* **A.** *Reconstruction of an entire plant,* **B.** *Foliage of another genus,* **C.** *Pteridosperm leaf with seed.* (After Delevoryas.)

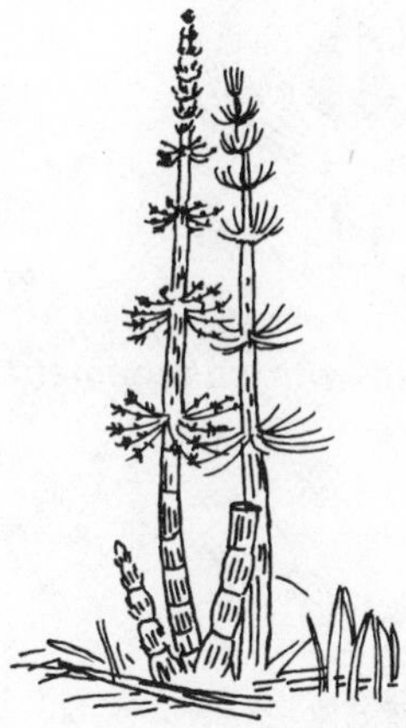

Figure 7.126. *Carboniferous Calamites.*

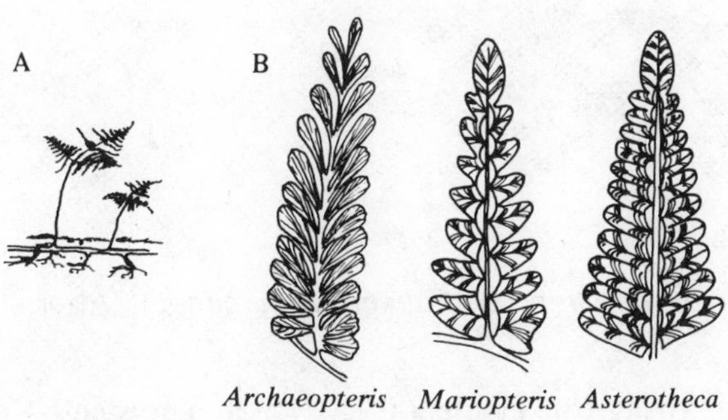

Figure 7.127. *Ferns.* **A.** *Living fern,* **B.** *Fossil fern foliage.*

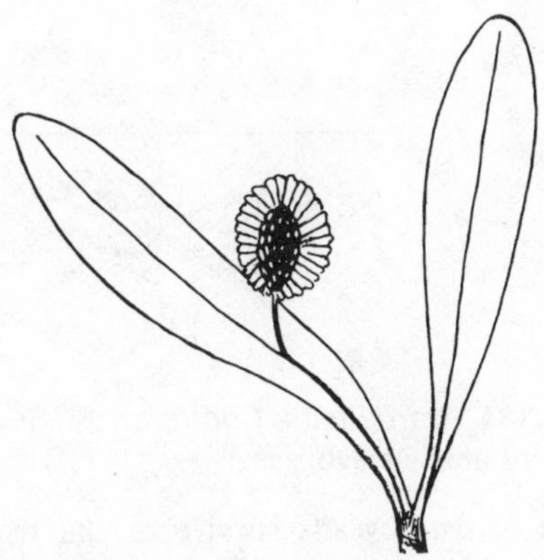

Figure 7.129. *Glossopteris.* Foliage and reproductive structure. (After Plumstead.)

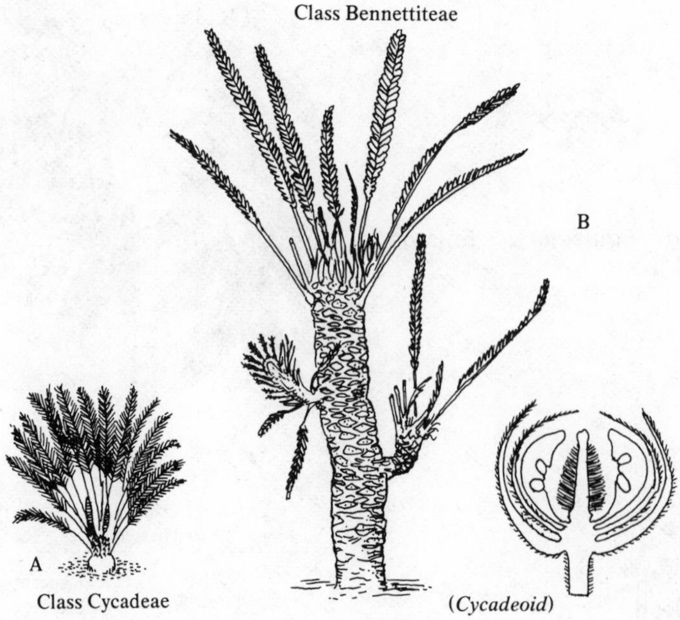

Figure 7.130. *Cycadophytes.* **A.** *Living cycad,* **B.** *Extinct cycadeoid with enlargement of reproductive structure.* (After Harris.)

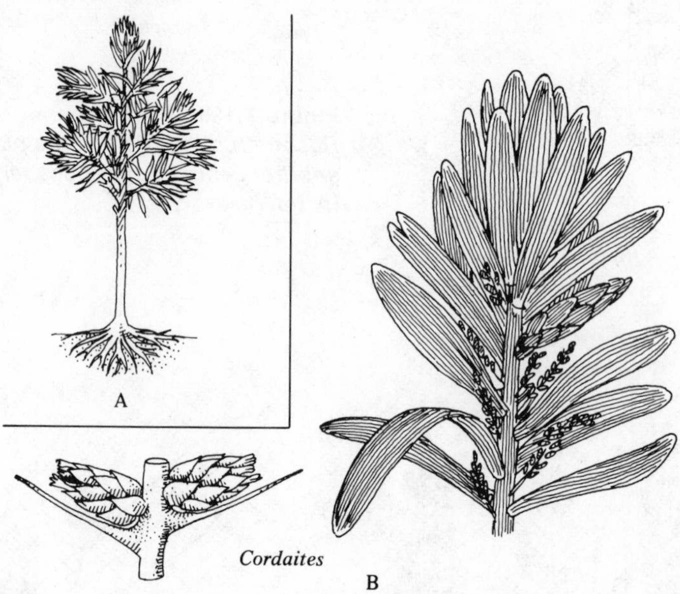

Figure 7.131. *Cardaitean Conifers.* **A.** *Restoration of the tree,* **B.** *Foliage with cones.* (After Grand'Eury and Delevoryas.)

genera of true cycads survive in the tropics and subtropics today.

Phylum Coniferophyta—This group includes the extinct **cordaites** (Fig. 7.131) and the surviving true **conifers** (Fig. 7.132). The stem is well developed as almost all are large trees, some nearly 400 feet tall. The leaves are simple and lack a stalk or petiole (Fig. 7.132). Cordaites leaves were strap-shaped; more

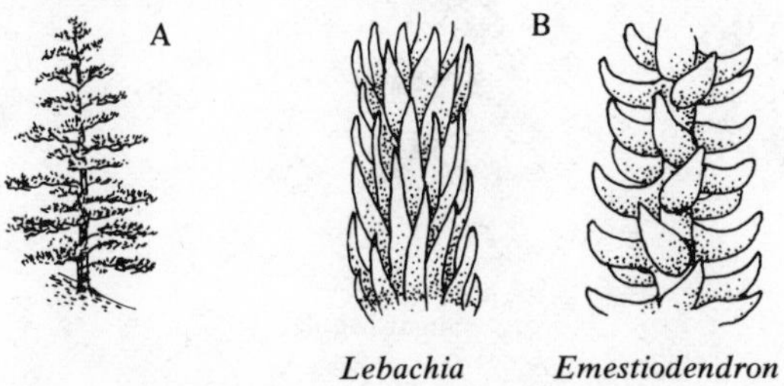

Figure 7.132. *True Conifers.* **A.** *Living pine tree,* **B.** *Permian conifer foliage.* (After Florin.)

modern conifers have needle-like leaves. Seeds occur in reproductive structures called **cones,** which are modified branches. Coniferophytes range from Carboniferous to Recent with cordaites dominant in the Pennsylvanian and the true conifers in the Permian and Mesozoic. Conifers are of low diversity today, but cover vast areas in harsh climates.

Phylum Gingkophyta—The surviving ginko or maiden-hair tree is the sole remaining species of this group of trees which evolved in the Carboniferous and was abundant in the Mesozoic. The gingko has a distinctive, dichotomously branched, petiolate leaf (Fig. 7.133). The seeds are borne in small cones, which are modifed branches similar to conifers.

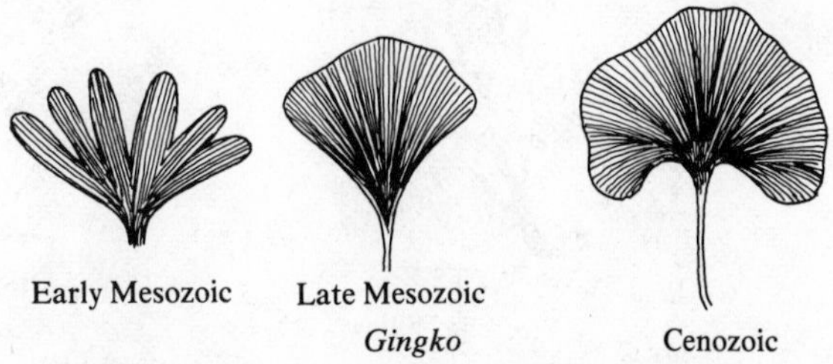

Figure 7.133. *Gingko Foliage.* (After Brown.)

Phylum Anthophyta—The **angiosperms,** or **flowering plants,** are by far the largest group of plants in the world today (Fig. 7.134). They range from huge trees, such as the eucalyptus, to the grasses, which are all leaf; they include all garden flowers, vegetables, shrubs, and many common trees, such as oak, maple, and sycamore. They are characterized by the possession of flowers, showy devices to lure insects which aid the reproductive cycle, and fruits, tasty structures housing the seeds. The edible fruits are eaten by mammals and birds who move about; and by the time the seeds leave their digestive systems, they have been dispersed. Angiosperm leaves are typically petiolate. Flowering plants range from the Cretaceous, or possibly the Jurassic or Triassic, to Recent. They became dominant during the Cretaceous Period.

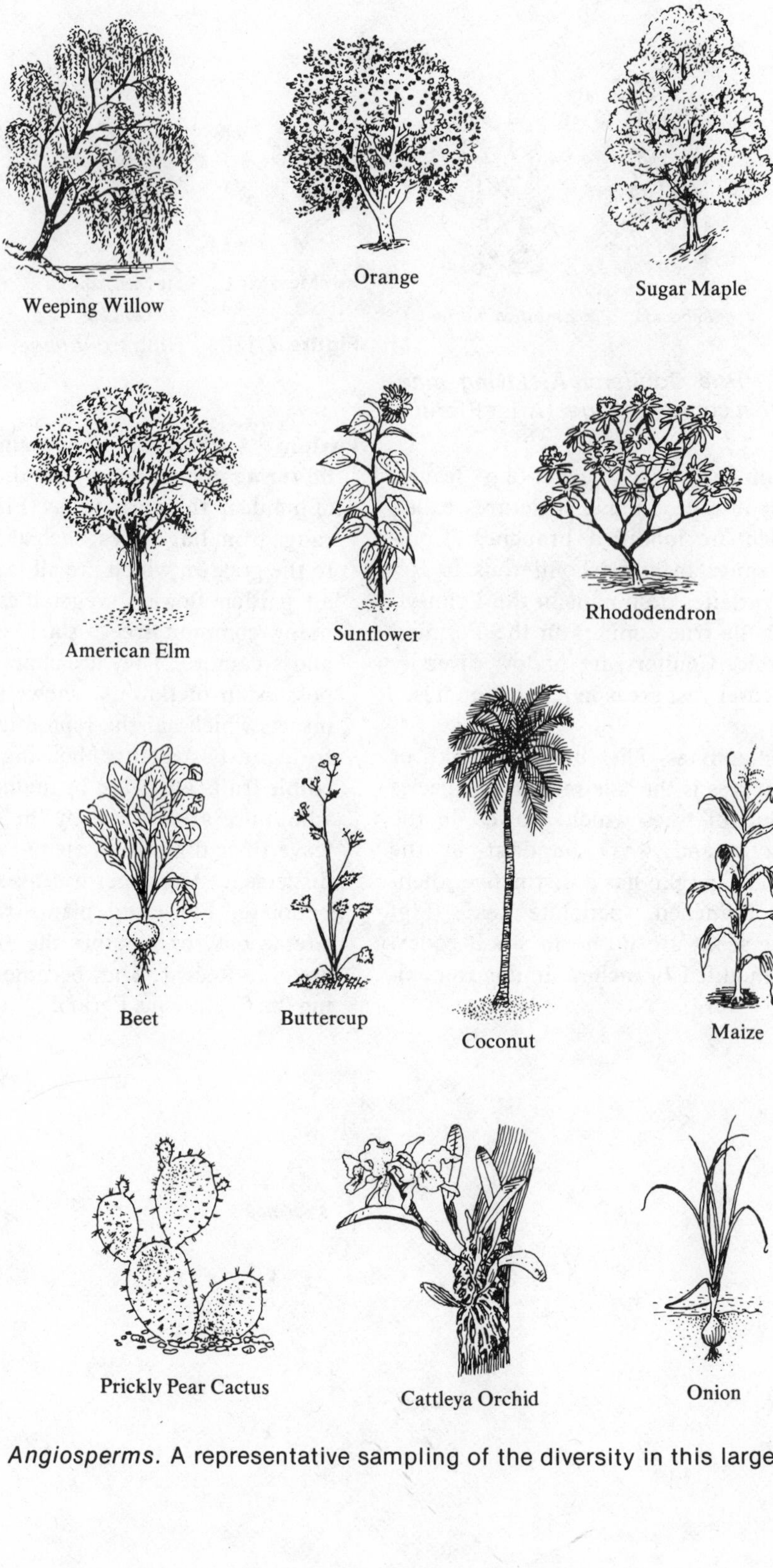

Figure 7.134. *Angiosperms.* A representative sampling of the diversity in this large group. (After Went.)

REVIEW

1. What are the major prerequisites for fossilization?
2. Why do you suppose trilobites have a better fossil record than do spiders?
3. What are four common organic skeletal materials?
4. What are the three major inorganic skeletal compounds? Which is most widespread?
5. What is the difference between **permineralization** and **replacement?** Between a **mold** and a **cast?**
6. What is the value of trace fossils?
7. What is a **phylogenetic classification?**
8. What are the distinguishing characteristics of the four kingdoms of organisms?
9. Which groups of protistans are the major oceanic producers of the latter portion of geologic time?
10. How does a foram test differ from a small snail shell? From a small cephalopod shell?
11. What features of archaeocyaths suggest sponge affinities? Coral affinities?
12. How can tetracorals and hexacorals be distinguished?
13. What are the differences between stromatolites and stromatoporoids?
14. How can one distinguish bryozoans and colonial corals?
15. How can brachiopod shells and clam shells be differentiated?
16. How can one distinguish snail shells from cephalopod shells?
17. How does one differentiate trilobites and xiphosuroid merostomes?
18. How can ostracods be distinguished from branchiopods? From small clams?
19. What are the distinguishing features of echinoderms?
20. Why are graptolites assigned to the hemichordates?
21. What are the distinguishing traits of vertebrates?
22. What are the major differences between reptiles and mammals?
23. How are ferns and seed ferns differentiated?

8

Stability versus Instability: Sedimentation Patterns

8

We saw in Chapter 2 that most early geologists assumed that the rock units they saw in their local areas were worldwide. As knowledge expanded, however, it became apparent that this was not the case and that there were several broad categories of regions, each characterized by particular types of sedimentary environments and, hence, rocks. Analysis of how these patterns form and the time relationships of the rock units deposited within them is still the subject of active research. Much more study of modern analogs is necessary for correctly interpreting ancient environments.

Geologists have recognized at least five of these broad types of regions or environments. These are the terrestrial regions, the continental shelves, the ocean basins, oceanic trenches, and the epeiric seas. Each of these types of regions has a characteristic effect on the patterns in which its sediments are deposited and, hence, profoundly affects the processes of correlation discussed in Chapter 9.

The terrestrial environments (fluvial, desert, lacustrine, glacial) and continental shelf (littoral, sublittoral) environments are the best known today because they are the most accessible. The patterns of sedimentation in these environments are treated in physical geology texts in some detail and will not be duplicated here. In historical geology, we are interested in the time context of the rock record and, in particular, the correspondence or lack of correspondence of time and rock units. Where the rock record is characterized by rapid lateral and vertical shifts in facies, there has been less inclination on the part of geologists to confuse time and rock units because the latter typically have little lateral persistence. Still, Wernerian concepts die hard. Most geologic maps and cross sections still assign rock units a single age or assign two rock units different names if they are of different ages.

In this chapter, we will examine two of the major regions in which there has been the most confusion about rock and time units. These are the epeiric seas on the stable continents and the unstable oceanic-continental interfaces where trenches and mountains occur.

Epeiric Seas

The most widespread pattern whose rock record is accessible on the continents is that of the **epeiric, platform,** or **epicontinental**, seas. These are the widespread, shallow invasions of oceanic waters that have covered most of the stable continental regions (continental platforms or cratons) of the earth at various times in the past. Their dimensions are measured in thousands of miles, and the rock record is very uniform geographically and relatively thin. The exact causes of the advances and retreats of these great seas are still being explored, but their existence has been obvious for a long time. Recently, several geologists have suggested that epeiric seas develop during times of continental fragmentation when the volume of the ocean basins is reduced (see Chapter 10).

One of the great problems in evaluating the epeiric sea environment is apparent by a look at a map of the earth's surface today (Fig. 8.1).

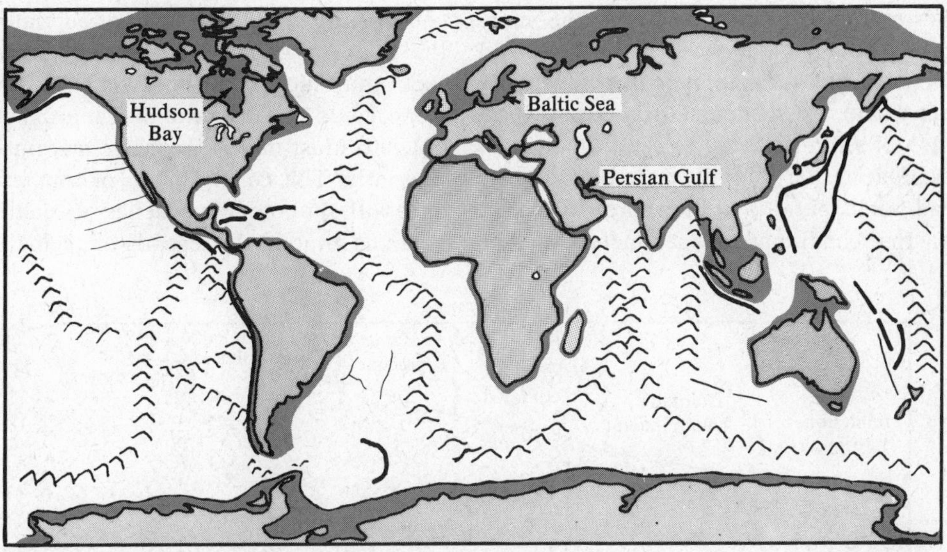

Figure 8.1. *World Map Emphasizing Sea Floor Topography.* Lightly stippled areas are land areas of continents; heavy shading indicates flooded portions of continents; mountain symbols in oceans indicate submarine ridges; and heavy lines are trenches. Note small extent of present epeiric seas. Hudson Bay, the Persian Gulf, and the Baltic Sea are the only bodies of water that are modern analogs of ancient epeiric seas.

With the exceptions of Hudson Bay in North America, the Baltic Sea in Europe, and the Persian Gulf in Asia, there are no widespread marine incursions of this sort on the continents. Even these modern analogs lack the great lateral dimensions of past epeiric seas. Hence, our usage of uniformitarianism is restricted; we cannot observe a number of representative examples of these environments today. We must, therefore, apply known physical and chemical laws to the conditions we can deduce from the rock record. This is a type of uniformitarianism in reverse, but the principle of uniformitarianism is so thoroughly validated now, that such a procedure is acceptable. The following discussion of epeiric sea conditions draws heavily on information developed by *Irwin* and *Shaw*.

We know from gross paleontological dating of the rock record of transgressing epeiric seas that the deposits are older near the continental margins than they are toward the center. In other words, the seas transgressed from the ocean basins inward over the continent over a period of time. Hence, the continental surfaces must have been inclined toward the margins, but very gently because the average slope of their undeformed deposits is typically less than .1 to .5 foot per mile. The types of fossils and rocks indicate a very shallow water depth over immense areas. An important result of these conditions is that the great currents characteristic of today's oceans, both the basins and the shelves, could not exist because the energy would be almost immediately dissipated through bottom friction. (There is much evidence of such agitation in the deposits of epeiric seas.) Furthermore, the normal twice daily rise and fall of the tides could not have occurred, just as it does not today in isolated seas such as the Mediterranean (which is not epeiric) and Florida Bay (which is only a marginal, shallow water shelf).

Chemical Deposits

Because of the lack of the freshening effect of normal tides and currents, epeiric seas would develop a salinity gradient with higher salinities the farther toward the interior of the continent the sea progressed (except very near the shore where the freshening effects of river water could be effective). The development of this salinity gradient would, upon supersaturation and precipitation of sediment, lead to the formation of a series of facies, based on the degree of salinity, in broad bands parallel to the shore (Fig. 8.2). Hence, the conditions of sedimentation for chemical precipitates would not be uniform over the entire sea; and no chemical sedimentary rock unit, such as a thousand-mile-wide limestone, could possibly be of the same age throughout, but must instead be older marginally then interiorly. This conclusion is, of course, at variance with the common practice of equating rock type and time in the legends of geologic maps.

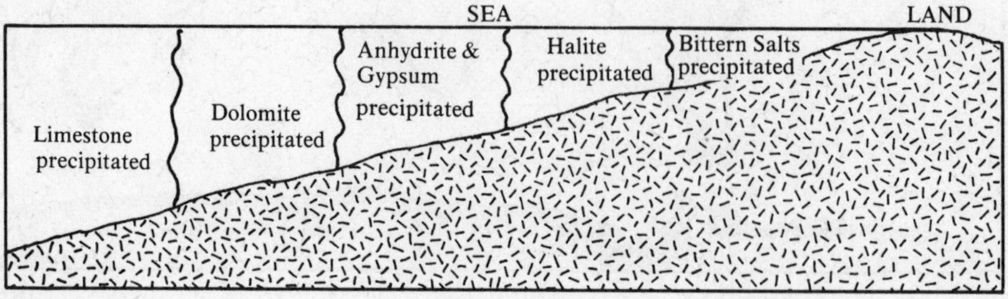

Figure 8.2A. *The Basic Sequence of Chemical Precipitates in an Epeiric Sea.* Vertical scale greatly exaggerated. The effects of normal marine waves and tides would be felt in the limestone range where bioclastic sediments are common.

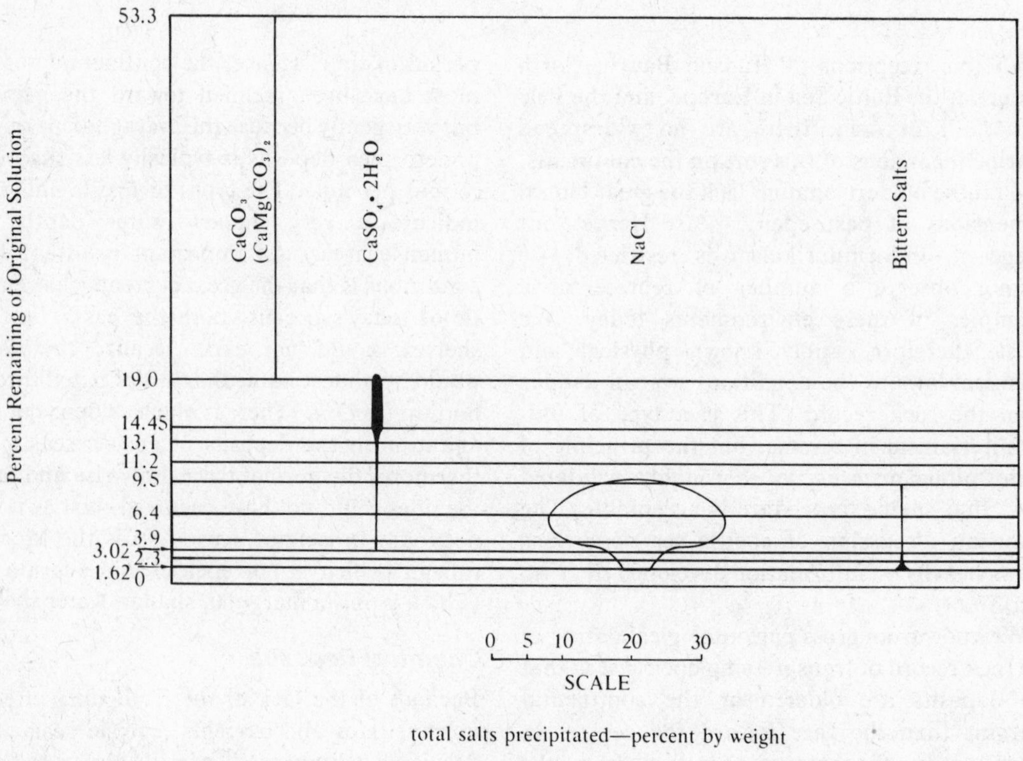

Figure 8.2B. *Theoretical Evaporite Mineral Precipitation Series.* (Adapted from California Division of Mines Bulletin 175.)

The sequence of epeiric sea chemical precipitates (Fig. 8.2), from least restricted conditions at the margins to most restricted conditions toward the continental center, is **limestone, dolomite** (not the type produced by later alteration of limestone), **anhydrite** or **gypsum, halite,** and the **bittern** potassium and magnesium **salts.** (The limestones, lying nearest the active open-sea conditions, will show a progressive variation from high- to low-energy types as one goes inward.) These deposits result from the physical-chemical conditions inherent in epeiric seas, and the sea will always attempt to impress this pattern on its sedimentary record. If there is any appreciable influx of clastics, however, the chemical precipitates will be so completely swamped out that they will compose only an insignificant fraction of the rocks. In a strictly chemical sequence of a transgressing sea, the deposits of more open seas overlie those of more restricted conditions, and the reverse occurs in regressing seas (Fig. 8.3). You should also note from this discussion that the presence of evaporite salts does not require an actual physical barrier to set up restricted conditions, although many evaporites do form in isolated basins. The restriction inherently occurs in epeiric seas because of the lack of normal tides and currents.

Clastic Deposits

If one considers the clastic deposits of epeiric seas, the effects of the restricted conditions are also apparent. Wave action will be confined to narrow bands where local, wind-caused wave systems develop and feel bottom (Fig. 8.4). The energy is rapidly dissipated, and the transportative effects are restricted in scope. Therefore, a widespread clastic rock unit, such as sandstone, which reflects a certain energy environment, cannot be the same age throughout its geographic extent. It too must be older on the continental margins than near the center.

The greatest amounts of energy release will occur in the shallower water where the waves feel bottom, and there will be correspondingly lesser amounts as one moves outward from the shore toward the margins of the epeiric sea. The basic

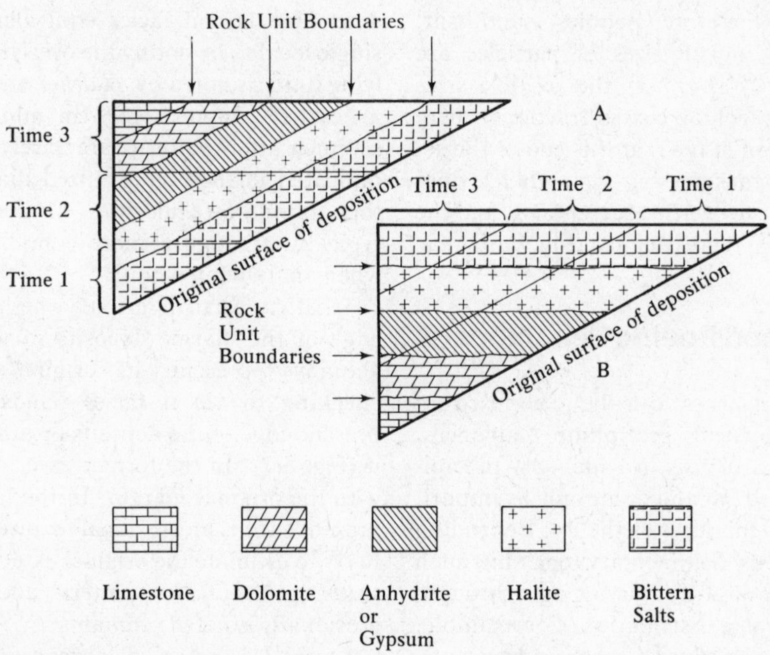

Figure 8.3. *Patterns of* **A** *Transgression and* **B** *Regression in Chemical Precipitate Deposits of Epeiric Seas.* Vertical dimension grossly exaggerated. In transgressive sequences, offshore deposits overlie inshore ones. In regressive sequences, inshore deposits overlie offshore ones.

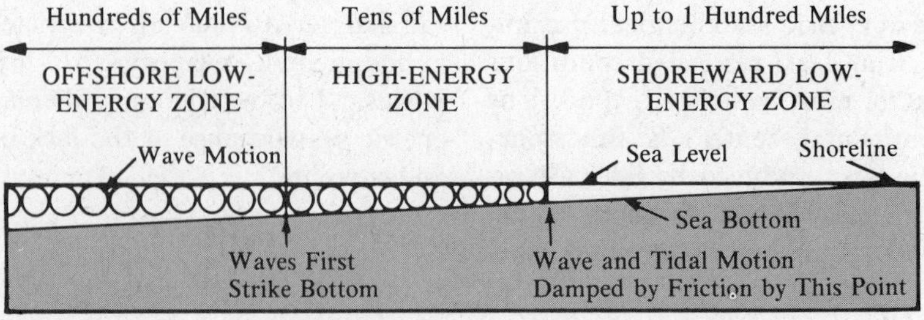

Figure 8.4. *Energy Zones in Epeiric Seas.* (After Irwin.)

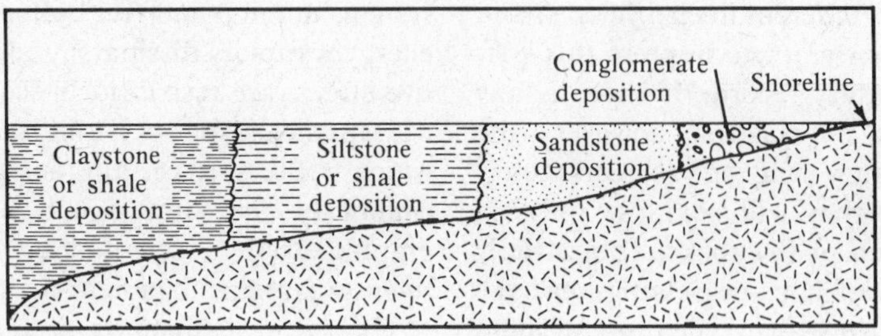

Figure 8.5. *The Basic Sequences of Clastic Sedimentary Rocks Formed in an Epeiric Sea.* This model assumes particles of all sizes are available. Vertical scale greatly exaggerated. The high-energy zone falls toward the right in this diagram.

sequence of clastic rocks (Fig. 8.5), from inshore to offshore, is, therefore, pebbles, sand, silt, and clay, providing all sizes of particles are being provided by rivers. If the particle sizes delivered do not include coarse fractions, there will be a reduction of types at this end of the sequence. In a transgressing sea, then, finer clastics will come to overlie coarser ones, and the reverse occurs in a regressing sea (Fig. 8.6).

Geologists and Epeiric Sea Deposits

Actual rock sequences can be composed of either pure chemical precipitate sequences, "pure" clastic sequences in which the precipitates are swamped out and occur only as impurities, and various mixtures of the two depending on local conditions. Sedimentary rock units such as formations are always of varying ages throughout their geographic distribution. For example, the basal sandstone of the Cambrian transgression is called the Tapeats in Arizona where it is Early Cambrian in age and the Mt. Simon in Minnesota where it is Late Cambrian. One can determine lateral facies equivalents even at a single locality by noting the overlying and underlying units as noted by *Walther* and discussed in an earlier chapter. Keep in mind that actual sedimentation patterns are rarely as simple as indicated in these generalized diagrams. Local topographic irregularities in the sea bottom typically produce a more complicated pattern when analyzed in detail.

Finally, we can determine whether the present edge of the marine deposits of an epeiric sea actually represents its original shoreline by checking to see if facies bands parallel the present edge of the deposits or are truncated by it (Fig. 8.7). In the former case, we are dealing with the original margin. In the latter case, the present edge is an erosional feature, and we have to try to estimate the original extent of the sea by studying the facies pattern and looking for erosionally isolated remnants.

Several series of paleogeographic maps of North America have been prepared by workers who did not accept facies and used the present

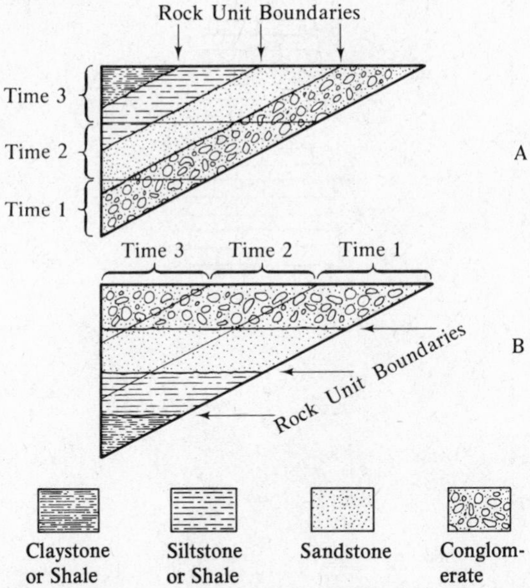

Figure 8.6. *Pattern of* **A** *Transgression and* **B** *Regression in Clastic Deposits of Epeiric Seas.* Vertical dimension grossly exaggerated. In transgressive sequences, finer deposits overlie coarse ones; and in regressive sequences, coarse deposits overlie finer ones.

edge of deposits to mark ancient shorelines. These often-reproduced maps are of diminished value in modern historical geology.

Geologists working with the thin, widespread epeiric sea deposits on the continents have often found it convenient to consider a sedimentary rock unit of uniform age throughout, despite the evidence cited above that it cannot be. As noted above, even the legends of most geologic maps still portray rock- and time-unit boundaries as parallel. Very serious consequences result from this practice, however.

In Figure 8.8 an advancing epeiric sea is transgressing over a continent. At time 1, a band of sand forms nearest shore; a band of mud forms farther out; and finally, beyond the reach of the clastics, a band of lime forms. At time 2, the sea has advanced farther inland and so have the facies bands. There is a band of nearshore sand, more offshore mud, and most offshore lime; only now each of these sediment types is a younger age. At time 3, the sea has moved still farther inland, and the facies bands have shifted correspondingly. A series of maps showing the position of the shoreline and the distribution of sediment types at times 1, 2, and 3 is given on the top of the blocks in Figure 8.8.

If one assumes, however, as many geologists do, that each rock type is of uniform age throughout, note what happens. If the sand, which is lowest superpositionally and, hence, "oldest," is assumed to be the same age throughout, the shoreline is far inland at the time of its formation and the only deposit is sand. If the overlying mud, the next younger unit, is assumed to be the same age everywhere, the shore-

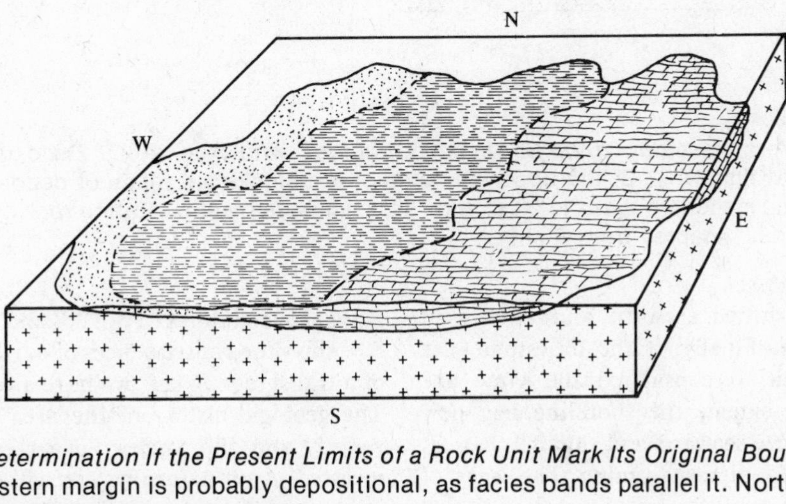

Figure 8.7. *Determination If the Present Limits of a Rock Unit Mark Its Original Boundaries or Are Erosional.* Western margin is probably depositional, as facies bands parallel it. Northern margin is erosional, as facies bands are truncated.

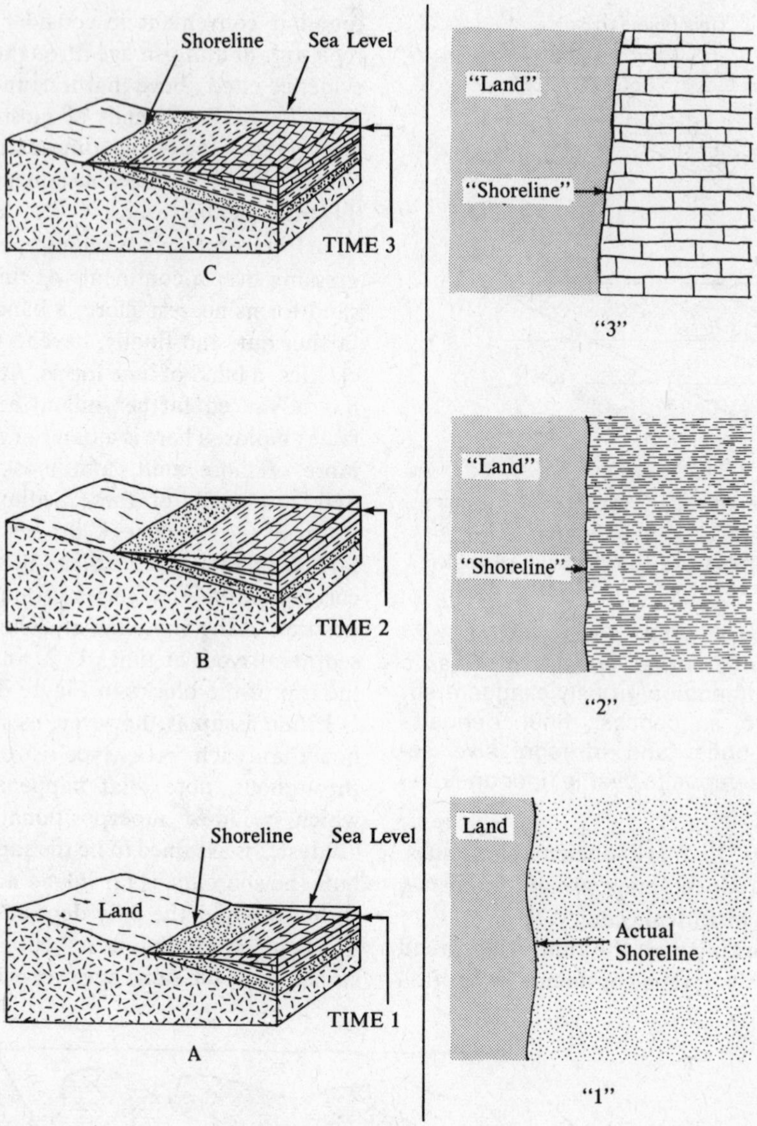

Figure 8.8. *Errors Resulting from Confusing Lithology with Age.* **A–C.** *Deposits of a transgressing sea.* At times 1–2–3, the upper surfaces of the blocks are maps of deposits forming at those times. The maps for times "1," "2," and "3" were made by substituting rock correlations for time correlations. (Adapted from Eicher.)

line has now shifted seaward and only mud occurs offshore. Finally, if the limestone, the "youngest" unit, is considered the same age throughout its extent, the shoreline has now shifted even more seaward and, again, there is only one type of sediment forming. The series of maps, labeled times "1," "2," and "3" in Figure 8.8, show the consequences of our error. Instead of a transgressing sea, we have a regressing sea. The geologic history of the area has been reversed, and the ancient geography has been inverted! And, naturally, one will never find the lateral facies of equivalents of the resulting sandstone, mudstone, and limestone.

What we have just done is to demonstrate again that lithology and time are not synonymous, at least not in the sedimentary deposits of the restricted epeiric sea environment. Let us now analyze the rock record produced at the continent-ocean basin margin.

The Geosynclinal Theory

As early as 1857, the famous American geologist *James Hall* noted that the marine rocks found on the North American continent today occur in two different patterns (Fig. 8.9). One is the thin, widespread, uniform pattern of rocks deposited in epeiric seas that was discussed previously. These sequences are not only thin, but, particularly toward the continental interior, contain many major gaps, which reflect the alternate transgression and regression of the epicontinental seas.

At the continental margins, where the major mountain systems such as the Appalachians and western Cordillera occur today, however, are found very thick, deformed sequences of marine strata in elongate linear or arcuate belts. These rocks represent much more of geologic time than do the platform rocks (Fig. 8.10). Sedimentation on the continental side of the belt occurred in shallow waters as the fossils and rock types indicate. Hence, the region appears to have subsided greatly to receive a several-miles-thick accumulation of sediments, because at any one time, the water was shallow. On the oceanic side, rock types are different and generally have been more difficult to decipher because of the more intense deformation there.

In 1873, *James Dana* applied the term geosynclinal, later changed to **geosyncline,** to linear belts of thick sediments which have been deformed into mountains. The notion grew that a geosyncline was a subsiding region predestined to undergo mountain-building deformation. Eventually, any thick sequence of strata was called a geosyncline, whether or not it had been deformed.

In Europe, the basic concept of a geosyncline underwent some revision because in that area the major geosynclines occurred between the currently stable parts of continents **(cratons)** rather than marginal to them (Fig. 8.11). For example, the great Tethyean Geosyncline, which supposedly spawned the Alps and other Mediterranean-area mountains, lay between Africa and Eurasia; and the Uralian Geosyncline, the supposed parent of the present Ural Mountains, lay between Asia and Europe. The Alpine area, whose deformation was much more recent than that of American geosynclines and, therefore, has a more complete record, became a standard

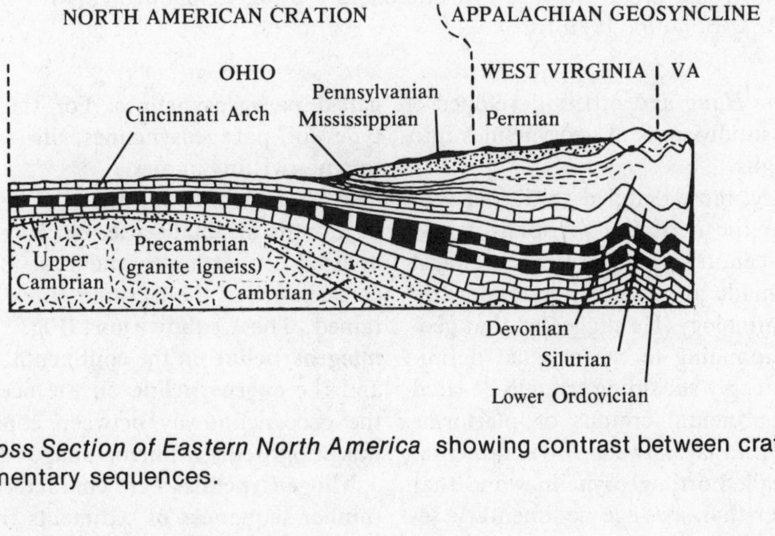

Figure 8.9. *Cross Section of Eastern North America* showing contrast between cratonal and geosynclinal sedimentary sequences.

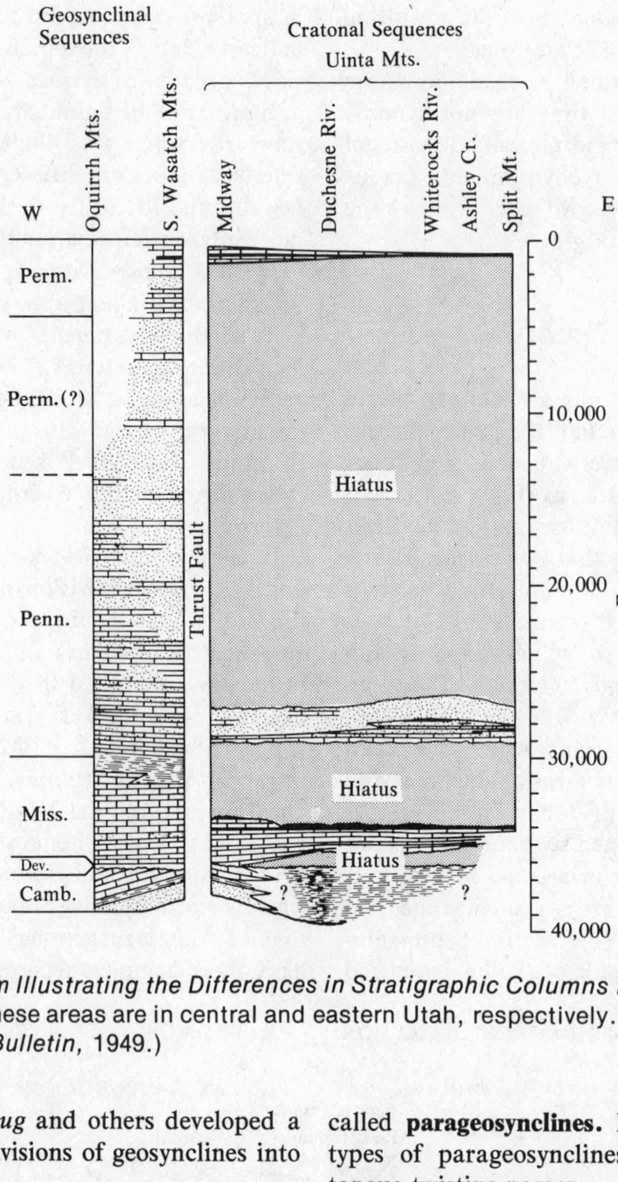

Figure 8.10. *Diagram Illustrating the Differences in Stratigraphic Columns from Geosynclinal and Cratonal Regions.* These areas are in central and eastern Utah, respectively. (After Baker, Huddle, and Kinney, *AAPG Bulletin*, 1949.)

for study. *Emile Haug* and others developed a complex set of subdivisions of geosynclines into ridges and troughs.

In this century, more detailed studies of geosynclines led to the definition of many types. Finally, in mid-century, the American geologist *Marshall Kay* made an attempt to standardize geosyncline terminology. He suggested that geosynclines corresponding to the original definition of linear, deeply subsiding troughs located marginal to continental cratons or platforms (whether or not they lay between two continental platforms) be called **orthogeosynclines** and that areas of thicker-than-average sedimentary sequence on the platforms or their margins be called **parageosynclines.** For the various subtypes of parageosynclines, he coined several tongue-twisting names.

Most geologists continued to use geosyncline only in the original sense, but Kay's formal subdivision of orthogeosynclines, based on the earlier work of the German *Hans Stille,* was retained. These subdivisions (Fig. 8.12) were the **miogeosyncline** on the continental platform side and the **eugeosyncline** on the oceanic side. (If the geosyncline lay between continents, there was supposedly a mirror-image set of both.)

Miogeosynclines were characterized as having thinner sequences of sediments than eugeosynclines, and the sediment types belonged to the

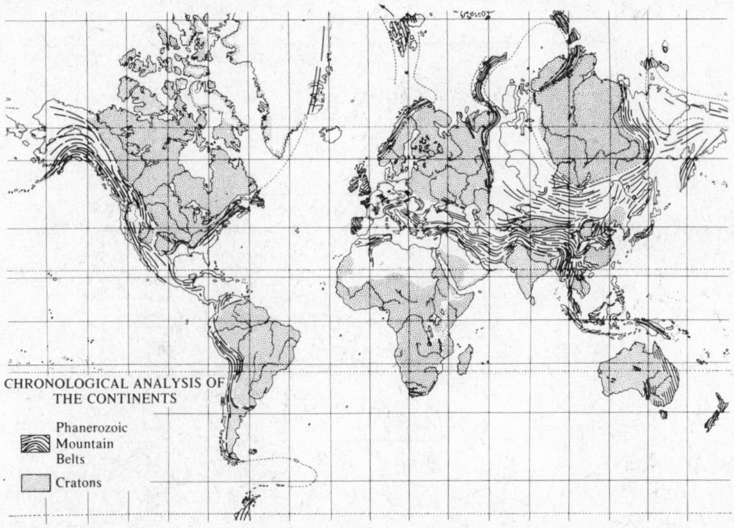

Figure 8.11. *Present Distribution of Cratons and Mountain Belts Whose Deformation Was Late Precambrian to Present in Age.* (Adapted from J.F. Umbgrove, *The Pulse of the Earth,* © 1947 by Martinus Nyhoff, Leiden, The Netherlands.)

quartz sandstone-limestone associations characteristic of the adjacent epeiric seas. Fossils were also those of shallow-water, shelf organisms. The deformation of miogeosynclines was much less intense than eugeosynclines and progressively died out toward the craton. The deformation consisted chiefly of folds and thrust-faults with little metamorphism and no intrusions. After the compressive mountain-building or **orogenic** stage, tension produced normal faulting with some terrestrial sediments accumulating in the down-dropped basins between ranges. The miogeosyncline was supposedly separated from the eugeosyncline by a ridge or **geanticline** during sedimentation.

Eugeosynclines were characterized as having much thicker rock sequences and sediment types of the deeper-water graywacke association. Fossils were uncommon and consisted almost entirely of pelagic organisms. Volcanic rocks were prominent in the sequence and were supposedly derived from a volcanic island arc on the outer side. Deformation began earlier than in miogeosynclines and not only consisted of folding and thrusting, but was followed by the intrusion of coarse-grained granitic rocks and accompanying metamorphism. The post-orogenic, block-faulting stage was usually characterized by the eruption of much lava from fissures and the deposition of thicker terrestrial sequences than in miogeosynclines. Not all geosynclines precisely fit all of these definitions, but in their search for universality, geologists were eager to support a unifying hypothesis such as this so-called geosynclinal theory.

Plate Tectonics

As outlined in Chapter 10, the new discoveries in the ocean basins, which have revived recognition of continental drift, have drastically revised our notions of how mountain building **(orogeny)** proceeds. As this recognition occurred, geologists became aware that the "geosynclinal theory" required drastic alteration, including the discarding of significant portions, and is not the unifying concept it was once thought to be. In its place, a new paradigm of continental and oceanic evolution has arisen.

This theory, plate tectonics, will be more thoroughly discussed in Chapter 10, but for the moment it will suffice to say that the earth's outer surface is chiefly composed of a few large, rigid pieces called **plates,** each consisting of continents and/or ocean basins. The shapes and

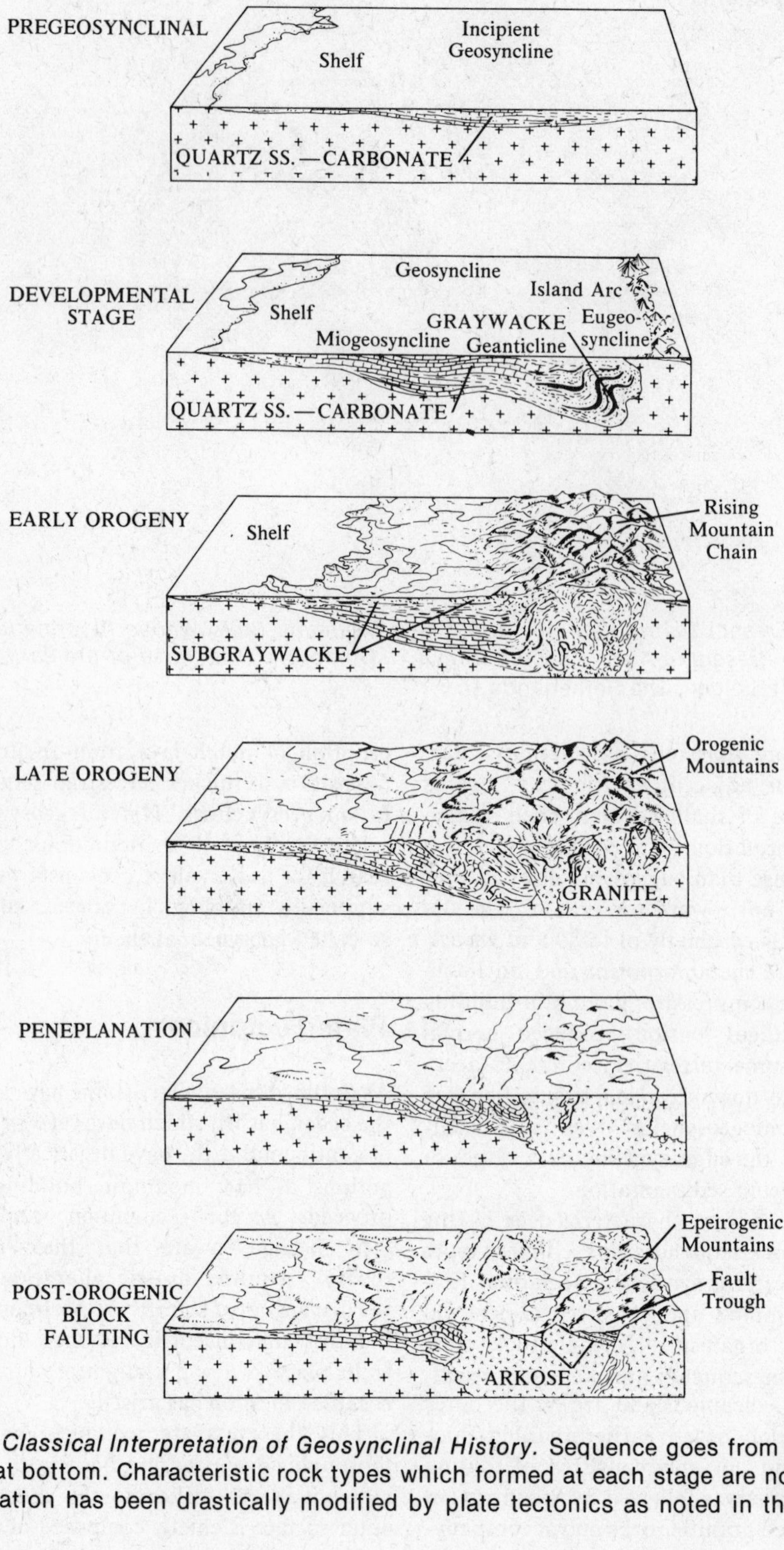

Figure 8.12. *Classical Interpretation of Geosynclinal History.* Sequence goes from oldest at top to youngest at bottom. Characteristic rock types which formed at each stage are noted. Most of this interpretation has been drastically modified by plate tectonics as noted in the text. (After Brice, 1960.)

sizes of these plates continually change as they pull apart, collide with, and/or slide by one another. The collisions and sideswipes produce mountain building.

Very thick accumulations of sediment can occur anywhere there is great subsidence. In some instances, such as where the edge of one plate descends beneath another, mountain building is a part of the process (Figs. 8.14, 8.16, 8.18, 8.19). In other instances, such as the accumulation of land-derived sediments at the base of a continental slope on the foundering edges of separating continents (Fig. 8.13A) or as **clastic wedges** of material on the continent derived from rising mountains (Fig. 8.13B), it does not follow except if, at some time in the distant future, the region becomes a plate boundary.

Three basic types of plate collisions/sideswipes that produce mountain belts have been recognized. They are occurring today, and "geosynclines" whose history fits these patterns have been recognized. These types, defined by *Dewey* and *Bird*, are (a) a trench-continent collision, (b) an island arc-continent collision, and (c) a continent-continent collision.

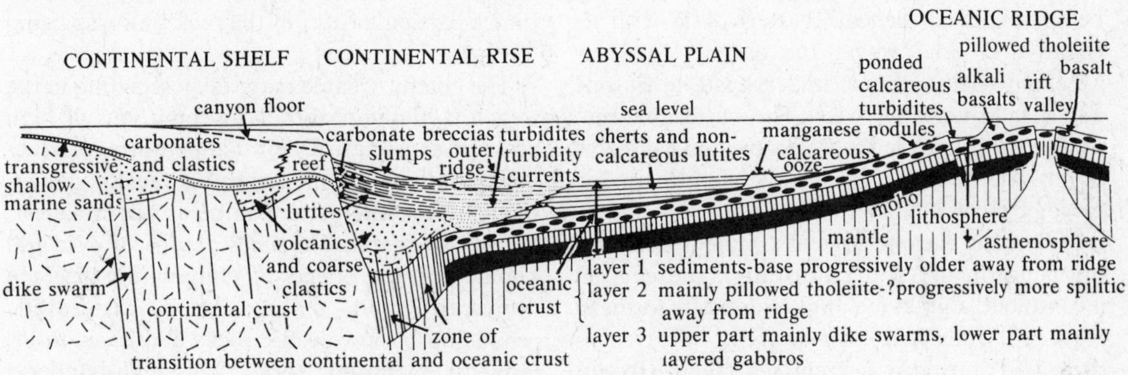

Figure 8.13A. *Thick Sedimentary Accumulation on Foundering Continental Edge.* This is shown on the left side of the diagram, while the right side shows a typical ocean basin floor. The present western Atlantic Ocean off eastern North America closely resembles this idealized diagram. Some of the terms in this diagram are explained in Chapter 10. (From Dewey and Bird, *Journal of Geophysical Research,* vol. 75, no. 14, May 10, 1970. By permission of authors and AGU.)

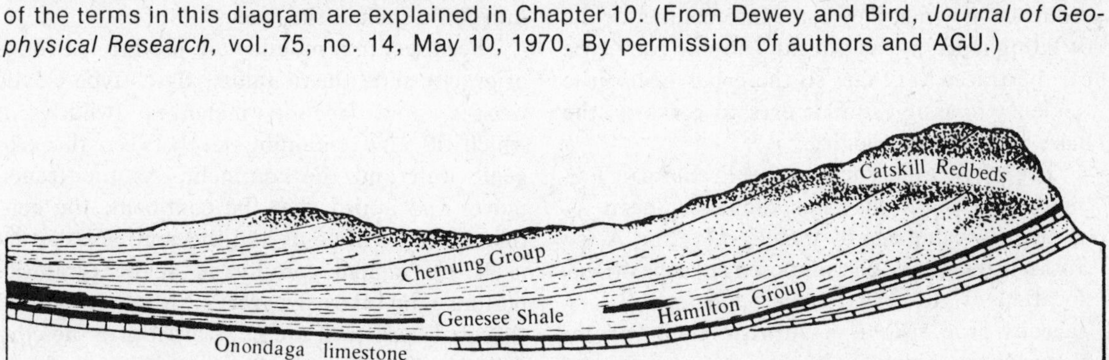

Figure 8.13B. *A Clastic Wedge,* a thick accumulation of sediments adjacent to a mountainous region. The Devonian Catskill Clastic Wedge which today forms low mountains. Original Acadian Mountains were to east. (Adapted from E.W. Spencer, *Basic Concepts of Historical Geology,* Fig. 15-4, p. 255, © 1962 by Harper & Row Publishers, Inc.)

Trench-Continent Collisions

When two plates are in contact and the margin of one is an oceanic trench and the margin of the other is a continent, the results are as indicated in Figure 8.14. In this instance, the denser oceanic slab descends beneath the lighter continental margin. (Because continental margins form by the rupturing and foundering of an original continent, the continental margin is like that shown in Figure 8.13A.) This process, **subduction,** is occurring today where the eastern Pacific is being subducted beneath western South America, where the western Pacific is being subducted beneath eastern Asia with its island arcs, and where the eastern Mediterranean is being thrust under eastern Europe and Asia Minor (Fig. 8.15). The western margin of North America in the Paleozoic and Mesozoic is but one of many excellent examples in the geologic record.

Where the oceanic edge of a plate is descending beneath, or "underthrusting," the edge of a continent, a great melange of deposits forms in a trench. Jumbled, land-derived, but trench-deposited, turbidity current sediments (**flysch**) and pieces of offshore oceanic crust and upper mantle (**ophiolites**), including both pelagic sediments and volcanics, are piled together, not in the time sequence of their deposition, but in the sequence in which they come into contact at the colliding margins. As a result, no one rock type can be traced very far; so there has been little tendency to assign similar ages to rocks on the basis of similar lithologies.

The whole unit, such as the Franciscan Formation of California, has, however, been assigned the same age by many geologists. A few Jurassic fossils and the supposed superpositional location of the Franciscan beneath the Cretaceous gave rise to a Jurassic date for the deposit as a whole. The picture was complicated when several Upper Cretaceous fossils were found in widely scattered places. The deposit then was considered the eugeosynclinal facies of the western geosyncline. Now that its mode of origin is partially understood, the jumbled interrelationships of its components become more intelligible, and its time of formation appears to range up into the Paleogene as demonstrated by the occurrence of Eocene dinoflagellates. The upper pieces are superpositionally the youngest to be incorporated into the Franciscan, but the original time of formation of incorporated rocks is variable and unrelated to the formation of the Franciscan. The fossils only tell us the age of the incorporated blocks, but not the time of formation of the Franciscan, except that they set a maximum age. The age of the Franciscan must be determined by cross-cutting relationships with adjacent strata and possible recognition of those fossils which settled into the trench at the time a particular part of the rock unit was being formed.

The chiefly oceanic material descending in the trench is metamorphosed by conditions of high pressure and relatively low temperature, producing the blueschist metamorphic facies (glaucophane, lawsonite, jadeite mineral assemblages). When this material reaches depths of over 100 kilometers, the effects of temperature become dominant and magmas are generated, producing basic volcanic rocks and basic-to-intermediate plutonic rocks. This high-temperature deformation produces a rising and expanding dome, or orogenic welt, which grows toward the continent. This dome may take the form of an island arc as in east Asia or a volcanic mountain chain such as the Andes.

The spreading, mobile core produces a series of gravity-slide thrust faults, flysch-type clastic wedges, and chaotic mêlanges (wildflysch) which fill any remaining sea between the orogenic welt and the continent. As the trench moves and approaches the continent, the continental rise sediments and the continental crust itself are melted, producing acidic (granitic) plutons and extrusive rocks. Metamorphosed sheets of hard rock are now thrust over the continent, and erosion from these sheets produces the terrestrial sandstones (**molasse**) of the clastic wedges. As the welt expands, it fractures by tension, producing normal faults and a basin-range topography. Note that the thrust faulting in the type of orogenic belt occurs in two distinct phases, one early and caused by subduction of

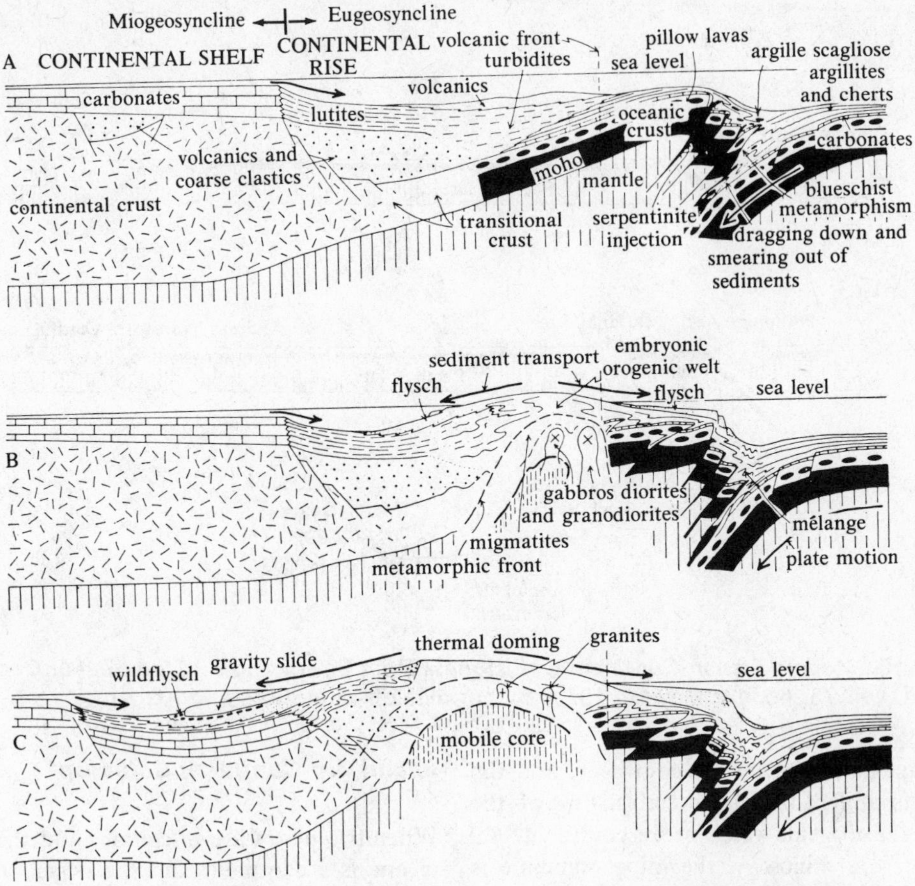

Figure 8.14. *Trench-Continent Collision.* The oldest stage is at top; the youngest, at the bottom. Stages **A** and **B** resemble east Asia today. Stage **E** resembles western South America today. (From Dewey and Bird, *Journal of Geophysical Research,* vol. 75, no. 14, May 10, 1970. By permission of authors and AGU.)

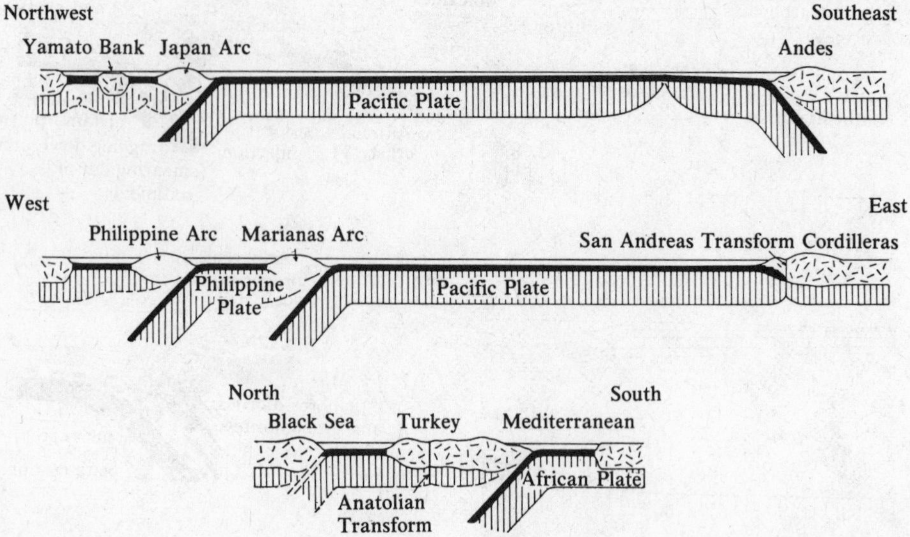

Figure 8.15. *Modern Trench-Continent Collisions.* (From Dewey and Bird, *Journal of Geophysical Research,* vol. 75, no. 14, May 10, 1970. By permission of authors and AGU.)

the ocean floor beneath the continent, and the other late and caused by overthrusting of the rocks of the orogenic welt over the continent.

In Kay's terminology, the miogeosyncline is the area from the edge of the original continental shelf inward to the limit of deformation toward the continent and contains the preorogenic quartz sandstone-limestone shelf association (sometimes called a **miogeocline**) covered by the overthrust flysch-molasse rocks. The eugeosyncline encompasses all the clastic, oceanic, and volcanic rocks involved in deformation around the trench and seaward of the continental margin. (Note that many of the eugeosynclinal rocks did not originate in the area of the present mountain belt, but were transported into the region on the moving ocean floor. This is important in paleogeographic reconstructions.) Thus, a geosyncline is really nothing more than the zone of collision between two plates. Such a collision results in thick, laterally diverse rocks in both age and lithology, the juxtaposition of rocks formed in many places, and mountain building. Because usage of the word "geosyncline" and its attendant terminology leads to confusion of times and places, it is probably best to discard those terms in favor of the new jargon of plate tectonics.

Island Arc-Continent Collision

When two plates are in contact and the margin of one is a continent and the other, an island arc, the results are portrayed in Figure 8.16. This process is occurring today where Southeast Asia is closely approaching the northern Philippine Arc. It occurred in the Miocene during the mountain building that produced New Guinea when the Australian continent collided with the Bismarck Arc. The volcanic arc consists of a flysch wedge and a trench mélange (Fig. 8.17), while the continental margin contains its typical assemblage of sediments shown in Figure 8.13A. The ocean basin caught in the impending crunch (such as the South China Sea today) fills with flysch. But when the continent reaches the trench, it is too light to be thrust under the arc. As a result, the arc rocks are thrust over the continental margin, and molasse from these thrust sheets forms on the interior side of the continent.

If the colliding pressure continues, the oceanic crust behind the island arc may buckle, forming a new trench to take up the crustal material. This new underthrust, however, will have the opposite inclination of the initial trench. A variation of this pattern occurs when there is no island arc and a continent collides with oceanic

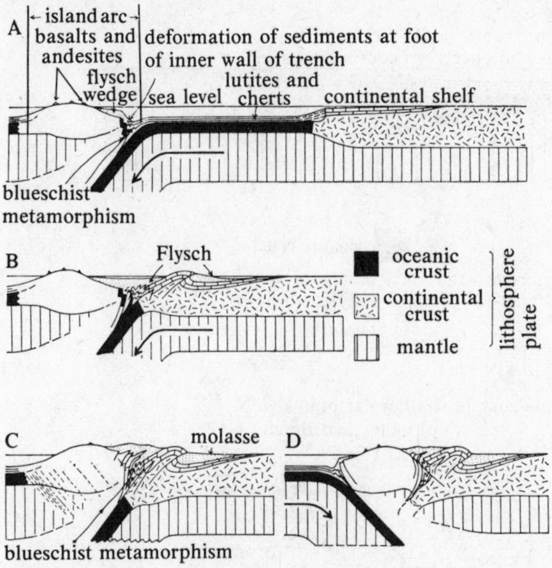

Figure 8.16. *Continent-Island Arc Collision.* The oldest stage, **A**, resembles the South China Sea today; and the youngest stage, **D**, resembles northern New Guinea today. (From Dewey and Bird, *Journal of Geophysical Research,* vol. 75, no. 14, May 10, 1970. By permission of authors and AGU.)

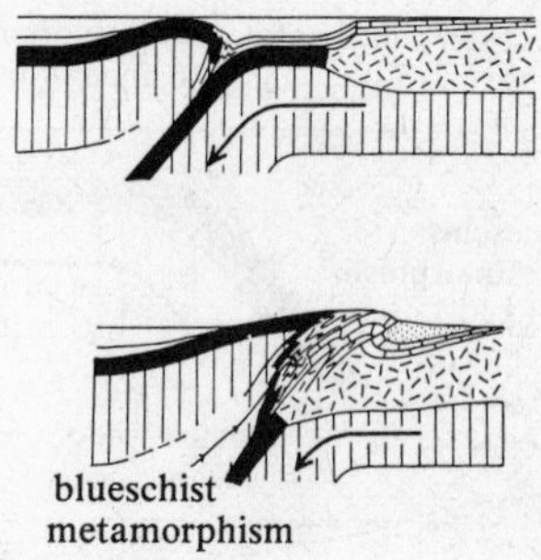

Figure 8.18. *Continent-Ocean Basin Collision.* (From Dewey and Bird, *Journal of Geophysical Research,* vol. 75, no. 14, May 10, 1970. By permission of authors and AGU.)

crust along an intraoceanic trench (Fig. 8.18). In this case, oceanic crust may actually be thrust over continental crust, a process called **obduction.** This has happened and is happening along the boundary between the Australian and Pacific plates adjacent to New Zealand.

Continent-Continent Collision

Finally, when the two edges of colliding plates are both continental, the results are those seen in Figure 8.19. Such processes have occurred and are occurring in the continuing collision of India and Asia. The collision of ancestral North America and Europe to produce the Appalachian-Caledonian Mountains in the Paleozoic is another example. Before the continental margins approach, the situation on one margin is much like that of the trench-continent or continent-arc collisions discussed earlier.

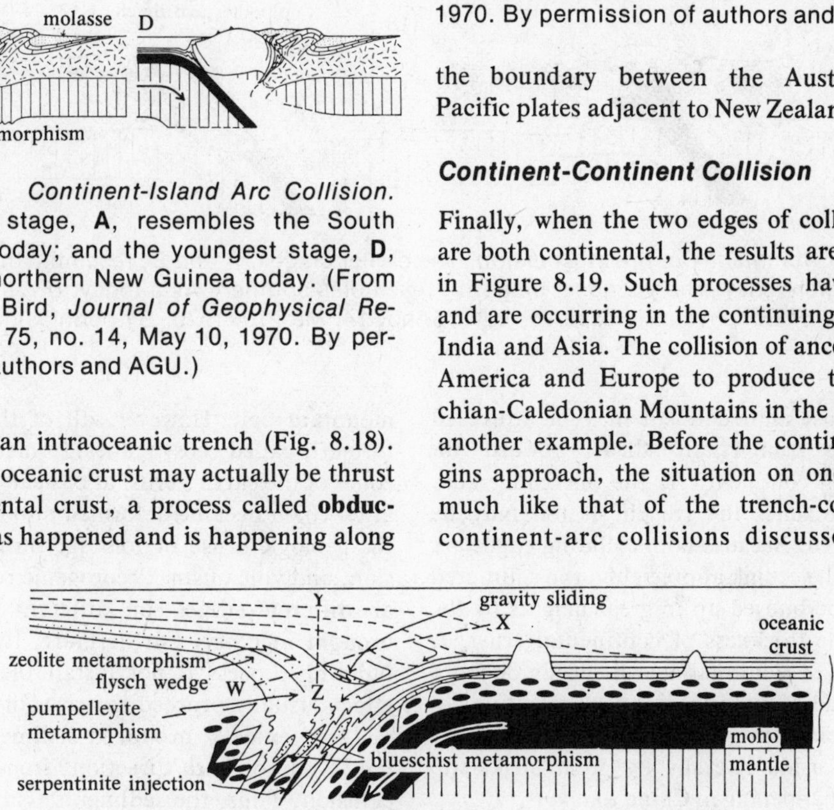

Figure 8.17. *Trench Development.* (From Dewey and Bird, *Journal of Geophysical Research,* vol. 75, no. 14, May 10, 1970. By permission of authors and AGU.)

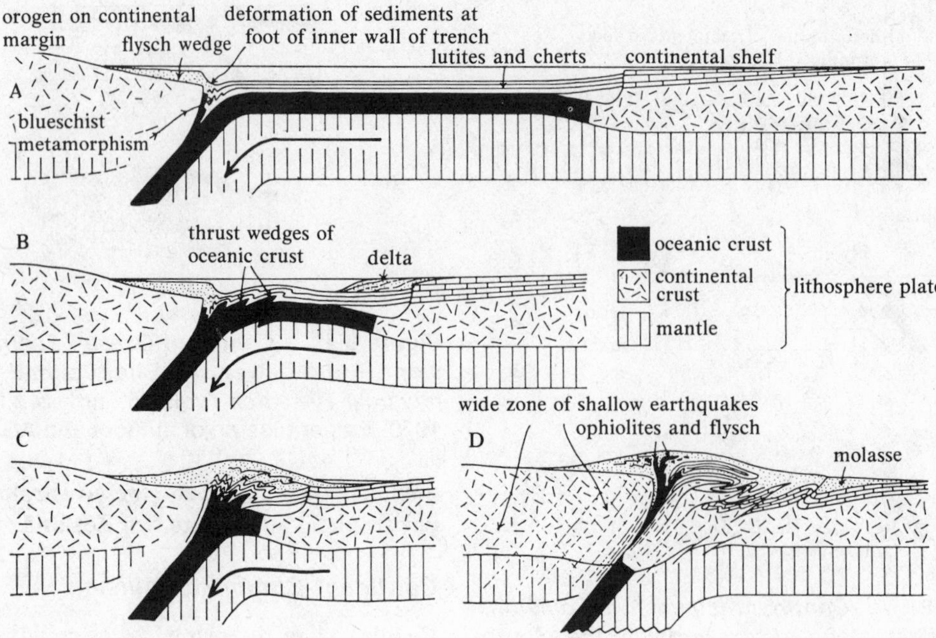

Figure 8.19. *Continent-Continent Collision.* The earlier stages, **A** and **B**, resemble the Mediterranean Sea region today. The youngest stage, **D**, resembles southern Asia today. (From Dewey and Bird, *Journal of Geophysical Research,* vol. 75, no. 14, May 10, 1970. By permission of authors and AGU.)

Oceanic crust and sediment may be squeezed out of the trough and thrust toward the approaching continent. As the oncoming continent approaches the trench, it too may be thrust-faulted. Because both colliding edges are light and will not sink appreciably, the splintered edges will be buoyed up to great heights by the great double thickness of continental crust. A broad zone of deformation will occur on both continents adjacent to the line of their suture. In addition, some pieces caught in the crunch may be pushed aside laterally by great strike-slip faults as is happening in China today.

The Ages of Rocks in Mountain Belts

Much mountain building is infinitely more complex than any of these three generalized cases because mixtures of two or three of them can occur in the historical sequence of any one mountain belt. However, all of these regions, formerly called geosynclines, formed in unstable zones of the earth's crust at or near plate boundaries which became deformed into mountains in the geologic sense of folding, thrusting, intrusion, and vulcanism. Preorogenic rocks, formed at different times in different places, are brought together. Furthermore, the mountain-building process is not instantaneous, and its effects usually proceed in waves of deformation and orogenically produced sediments proceeding in one or both directions from the zone of collision. Thus, the sediments issuing from the mountains (as well as those incorporated within mountains) are of different ages as one crosses a mountain belt (**orogen**).

For example, the famous **flysch** deposits, which consist of land-derived clastics (graywackes, shales) in rhythmic alternation and which form from turbidity currents, are of varying ages as shown in Figures 8.14, 8.16, and 8.19. They are typically older toward the orogenic welt and younger away from it, and older on the original oceanic side than on the continental side. These deposits, which reflect orogeny, typically span many periods or even a whole era. To assume they are the same age throughout because they have the same lithology would cause a gross misinterpretation of earth history. Yet this has been done and only discovered when fossils of vastly different ages have been found in different areas of flysch deposits in one orogen.

Likewise, as the orogenic phase of mountain belt evolution reaches a peak and then passes into the time of isostatic adjustment and normal faulting (which often produces greater relief), another characteristic rock type is formed. This is the **molasse,** a very thick sequence of coarse detrital sediments, chiefly homogeneous calcareous sandstones, which collects in the basins. Molasse is typically a terrestrial deposit but may be marine in basins of very rapid subsidence. It also begins to form earlier near the orogenic welt where the mountains first develop and proceeds toward the continent. It too, then, is **diachronous**, or of different ages in different places.

In summary, the origin of mountains was originally ascribed to a specific geosynclinal cycle. A geosyncline was a subsiding region of thick deposition which eventually was compressed into a mountain belt. Recent knowledge indicates that mountain belts form in various ways by the collision of slabs of crust and upper mantle. The rocks of mountain belts formed at different times and in different places. Thus, as with epeiric sea deposits, rock units of mountain belts do not correspond to time units.

REVIEW

1. Why is it difficult to determine what conditions were like in ancient **epeiric seas?**
2. What is the basic sequence of chemically precipitated deposits in epeiric seas?
3. Why is it unnecessary to have an actual physical barrier in epeiric seas to produce hypersalinity?
4. What is the basic sequence of detrital deposits in epeiric seas having a supply of particles of all sizes?
5. Discuss at least one method a geologist might use in locating an ancient shoreline.
6. What happens to geologic history if lithologic units of epeiric seas are considered time units?
7. What was the traditional concept of a **geosyncline?**
8. In this model, what was the difference between a **miogeosyncline** and a **eugeosyncline?**
9. How has the "geosynclinal theory" been altered by the recent discovery of **plate tectonics?**
10. Discribe the results of a trench-continent collision. An island arc-continent collision. A continent-continent collision.
11. Why is it difficult to date the time of formation of a trench mêlange?
12. Why are **flysch** and **molasse** diachronous?

9

The Maelstrom: Correlation

9

Since the early chapters of this book, we have been talking about dating rocks and determining whether or not two bodies of rock were deposited at the same time. We have seen that this generally cannot be done on the basis of lithology. We have also seen that this is usually done with fossils in sedimentary sequences and by radiometric means in igneous and metamorphic terrains. Most geologists call this process **correlation.**

Unfortunately, this term, like so many in geology, has been used in other senses. For example, many geologists also use the word *correlation* for the process of determining whether two bodies of rock defined lithologically are or were once part of the same lithologic body of rock. They may want to determine if a certain sandstone in one area originally connected with one in another area, and, hence, if they are really the same lithostratigraphic unit (Fig. 9.1). Because of lingering Wernerism, many geologists see little difference between the two very different processes.

In the preceding chapter, we saw that time equivalency and rock equivalency are not the same and that disastrous consequences can result for historical geology if one operates on the assumption that they are. Nevertheless, "correlation" has been so widely used for both processes that it seems unwise to restrict its usage to only one of them. Therefore, an adjective must be applied to "correlation" to assure that there is no confusion.

Accordingly, **time correlation** refers to the process of determining whether two rock bodies were deposited at the same time; and **rock correlation**, to the process of determining whether two rock bodies are (or were) really part of the same lithologic rock body. Only in those instances where a rock body results from a single interval of deposition, such as an ash fall, a lava flow, or a turbidity current deposit, do the two correspond.

Rock Correlation

There are four major methods of **rock** or **lithologic correlation**. The most desirable method is that of tracing the units into each other as this proves unequivocally that the two are one (Fig. 9.2). This technique, called the **physical continuity** method, is only applicable, however, in regions of sparse vegetation and soil and which lack great structural disturbance. Where vegetation and soil cover intervening exposures and/or faults terminate rock units, the geologist must resort to more indirect methods.

If the two rock bodies have absolute or near **lithologic identity**; that is, their mineralogy, texture, primary structures, thickness, etc. correspond, the geologist can be fairly secure in rock correlating the two (Fig. 9.3) even if he or she cannot trace them into each other. A problem arises, however, if there are two or more rock units which have lithologic identity at one or both of the localities.

The next guideline to be used by the geologist in rock correlation of a unit is its **position** in the **stratigraphic sequence** (Fig. 9.3). For example, if a geologist is trying to decide which of two shales at one outcrop to rock correlate with a shale at another, he or she examines their positions in the geologic column. If one shale is underlain by sandstone and overlain by limestone

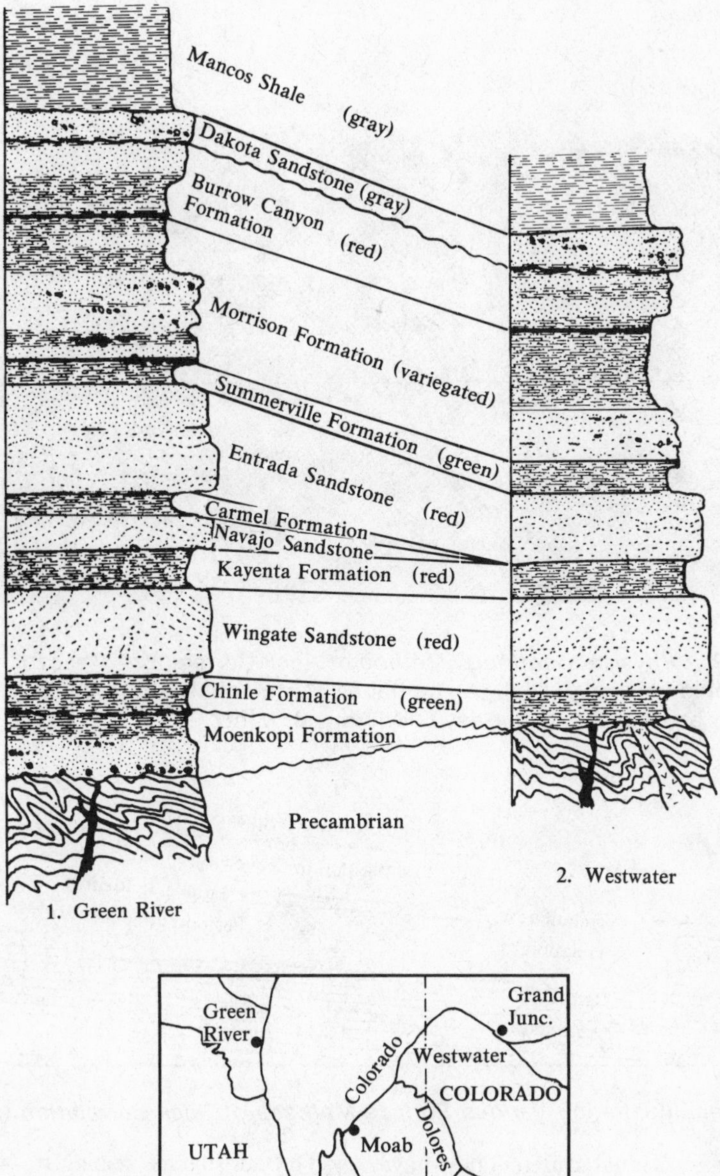

Figure 9.1. *Rock Correlation.* Strata are inferred to have once been continuous between the two localities. Map shows relation of two localities to major rivers and cities in eastern Utah and western Colorado. (After Peterson and Rigby.)

and the other is underlain by limestone and overlain by sandstone, the geologist correlates the first one with the shale at the other outcrop which is also underlain by sandstone and overlain by limestone. Even this method encounters difficulties when one is dealing with rocks that occur in repetitious sequences, such as Pennsylvanian mid-Continent cyclothems (Fig. 9.4) and Cretaceous Great Valley turbidites (Fig. 9.5). In some of these instances, it is virtually impossible

Figure 9.2. *Physical Continuity Method of Rock Correlation.* Mesozoic strata in the Canyonlands area of Utah can be traced continuously along barren mountainsides and canyonsides. (Courtesy of William M. Mintz.)

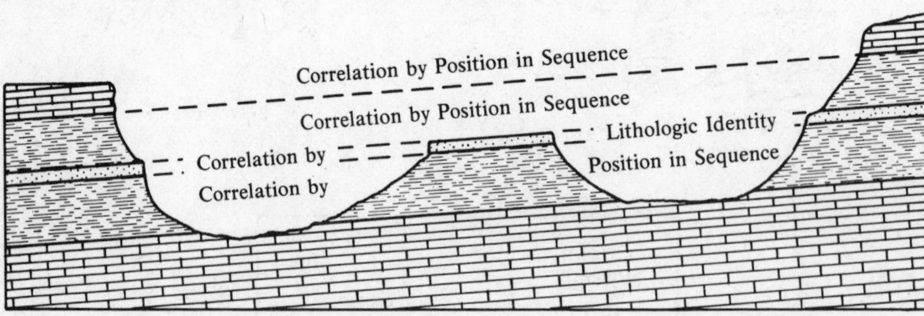

Figure 9.3. *Diagram Illustrating Various Surface Methods of Rock Correlation.* (After Spencer.)

to rock correlate individual units. They have often been time correlated on the basis of fossils. In the case of the epicontinental cyclothems, this only indicates that they cannot be rock correlated unless they are very near one another and, hence, probably still part of the same rock unit before its time lines cross over into another rock unit. Because turbidites result from a single event, fossils may be more reliable lithologic guides, provided several rock units were not deposited in a short interval.

The subsurface geologist, working with the data gathered from drilling wells, has developed a battery of tools for rock correlation that parallel those of his or her surface counterpart. The only exception, of course, is physical continuity, which cannot be used between well holes. If actual cores of sediment are brought to the surface, a geologist can rock correlate by the lithologic-identity or position-in-a-sequence method. The taking of rock cores is a long and expensive process, however, and is only occasionally done.

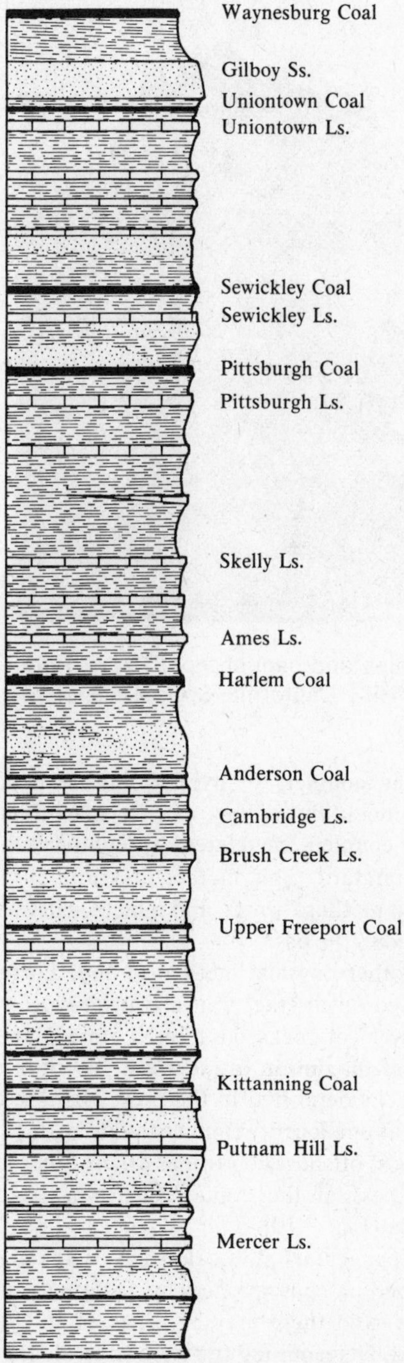

Figure 9.4. *Pennsylvanian Cyclothems.* Position in a sequence and lithologic identity are of little help in rock-correlating these strata. (From Ohio Geological Survey.)

Therefore, the subsurface geologist uses indirect methods of gathering data about rock properties so he or she can apply the identity or position-in-sequence methods. These indirect methods fall under the general heading of **instrumental well logging.**

A very commonly used method is that of **electric logging.** A set of electrodes is lowered into a well by electric cable. They measure both the electromotive force produced in the rocks by the contact of their fluids with the drilling fluids in the drill hole (self-potential), and the resistance they offer to an artificial electromotive force (resistivity). The measurements are carried by the cable to the surface and recorded as a continuous function of depth. The result is a set of curves (Fig. 9.6) whose shape changes with lithology.

Other varieties of logs are used which measure the **natural radioactivity** (such as gamma rays) of the various rock types, the effects of **neutron bombardment** on various rock types (Fig. 9.6), and the velocity of propagation of **sonic waves** in various lithologies (Fig. 9.7). These are all methods of measuring rock properties at a distance and must be interpreted by geologists for lithologic identity and position in a sequence for purposes of rock correlation. (Do not confuse radioactivity well logging, a method of rock correlation, with radiometric dating, a method of time correlation. Both depend on natural radioactivity, but are entirely different techniques for entirely different purposes.)

Unconformities of all varieties can also be used for rock correlations, but, like rock units, they vary in age from place to place. For example, the uppermost Precambrian-lowermost Cambrian interval is represented by an unconformity over most of the North America Craton, but on the margins it embraces only the Eocambrian and lowest Cambrian, while in the center of the continent, the lower two-thirds of the Cambrian is missing. The basal sandstone deposited on this unconformity can be rock correlated across the area, but it is not the same age everywhere, being Eocambrian or Lower Cambrian marginally and Upper Cambrian centrally.

Figure 9.5. *Flysch Turbidites,* interlayed shales and sandstones, Pedro Point near Pacifica, California. (Courtesy of Mary Hill, California State Division of Mines.)

Time Correlation

Time-correlation methods fall into two categories, the physical methods and the biologic or paleontologic methods. The major **physical** method is that of **radiometric dating** discussed in Chapter 6. There, we pointed to a major drawback of radiometric dating—that it is usually not applicable to the sediments that cover sixty-seven percent of the earth's surface.

Another physical method is that of **key** or **marker beds**, which result from single depositional intervals such as volcanic eruptions (Fig. 9.8) and turbidity currents. (Not all so-called key beds have time significance. Many, such as thin conglomerates or reef masses, have only environmental significance.) Recognizing the uniqueness of a particular key bed is often a difficulty, particularly if there are numerous similar flows in the section.

Varves (Fig. 9.9) are types of key beds formed by seasonal climatic changes, such as commonly occur in glacial lakes. They consist of a couplet of strata, and no two couplets are exactly alike because of yearly climatic variations. By matching the couplets of overlapping sequences, a continuous chronology can be assembled and other couplets correlated by comparison with this standard. This method has limited applicability to those times and places where varves formed in the past.

Another physical method of time correlation is based on the fact that a transgressive-regressive cycle of rocks contains a time plane—the **time of maximum inundation.** In other words, the rocks deposited in the most offshore conditions at one locality time correlate with those of the most offshore conditions at any other locality regardless of the actual rock types or water depths (Fig. 9.10).

A particular physical event may trigger a number of consequences in different environments, and these may be reflected in the rock record. Determining the time equivalency of the different lithologic consequences of a common cause is known as **event correlation.** An example is given in Figure 9.11. In the Middle and Upper Devonian sequence of New York, two abrupt sea-level rises are recorded in the stratigraphic record. In the marine environment, rising sea level causes clastic sediments to be trapped in

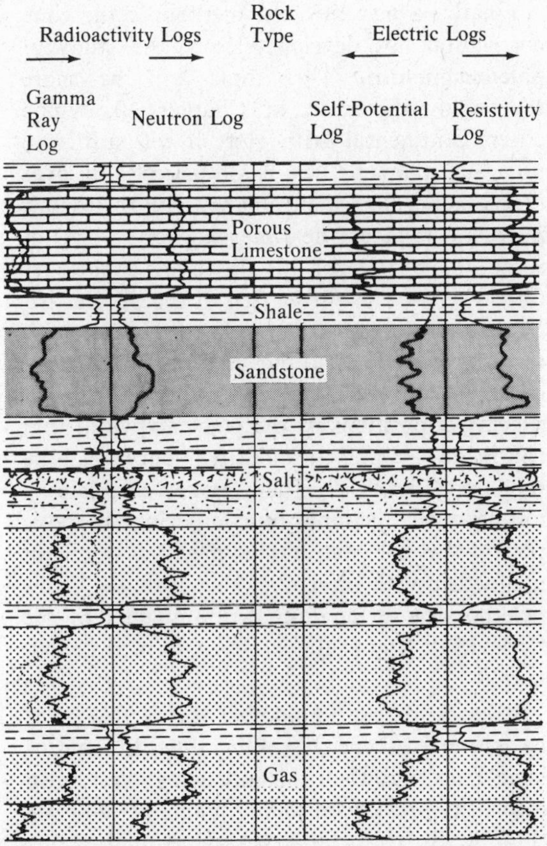

Figure 9.6. *Electric and Radioactivity Logs.* Arrows indicate direction of increase in parameters being measured. (Adapted from Spencer and *Oil and Gas Journal*.)

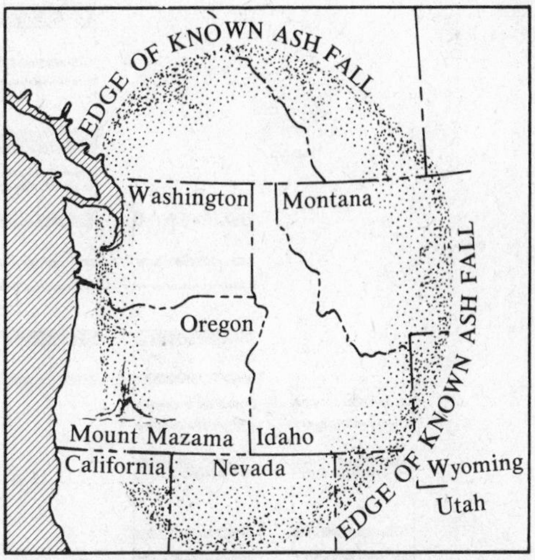

Figure 9.8. *Formation of a Recent Key Bed for Time Correlation.* The Mount Mazama (Crater Lake) eruption of 6,000 years ago scattered a widespread distinctive ash over the West. (After Stokes.)

the heads of drowned river-mouth valleys. This is reflected by a brief period of limestone deposition in the clear waters offshore and in the fluvial environment on land by the accumulation of very thick floodplain deposits. In the rock record, the offshore limestones can be time correlated with the thick terrestrial alluvium because their formation was triggered by the same specific event.

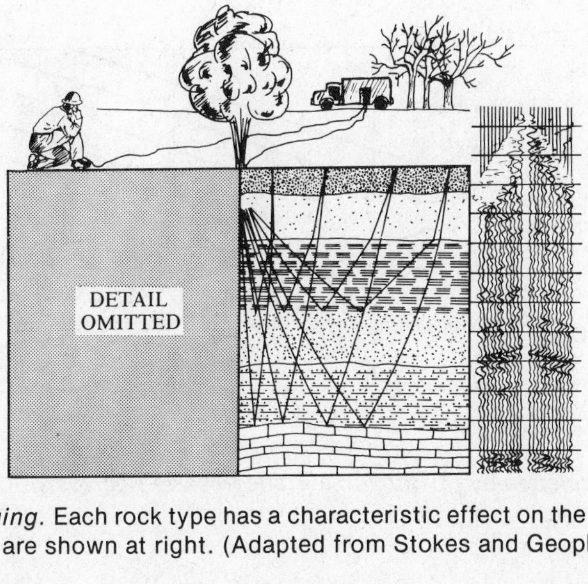

Figure 9.7. *Sonic Logging.* Each rock type has a characteristic effect on the propagation of sound waves. Graphic results are shown at right. (Adapted from Stokes and Geophysical Services, Inc.)

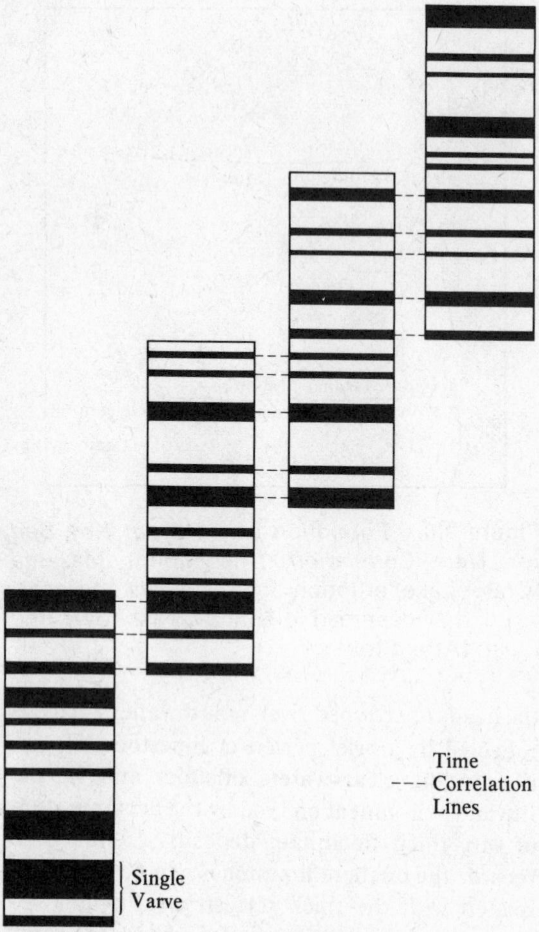

Figure 9.9. *Time Correlation Using Varves.* Similar couplets are matched.

Finally, a new physical method of the time correlation has developed from the study of **paleomagnetism.** This topic will be more thoroughly discussed in Chapter 10, which covers continental drift. Here it will suffice to note that magnetic iron oxide particles in lavas or sediments align their polarity with the earth's magnetic field at the time they crystallize or settle. In older rocks, this alignment of particles differs significantly from that of the present because the continents have been moving with respect to the poles. If one assumes a particular continent was fixed and then determines the successive positions of the pole relative to that continent, a "polar wandering" curve can be prepared (Fig. 9.12). Then one can correlate an unknown rock sample by comparing the position of the pole determined from its remnant magnetism with the curve. For example, if the pole position for an unknown sample fell on the Ordovician part of the curve, it formed in that period.

It has also been found that the earth's poles have reversed themselves many times in the geologic past and that this phenomenon is also recorded in rocks containing magnetic minerals (Fig. 9. 13). Unfortunately, one normal or one reversed time span is like any other (except for its length), so simply finding that a rock has reverse or normal magnetism does not help too much with time correlation. If, however, one is fairly certain by other means of time correla-

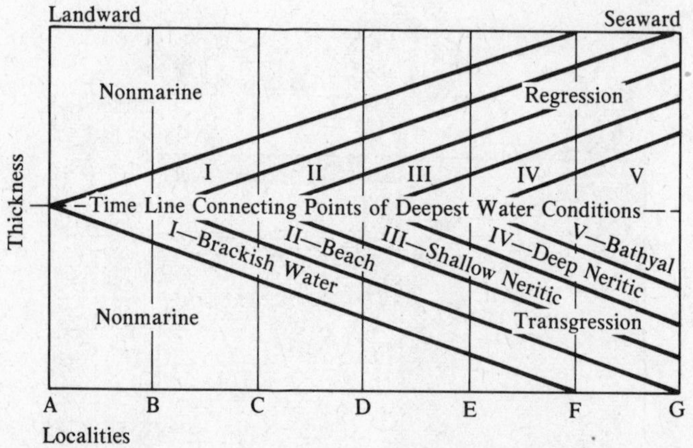

Figure 9.10. *Time Correlation by Position in a Transgressive-Regressive Cycle.* (After Izraelsky.)

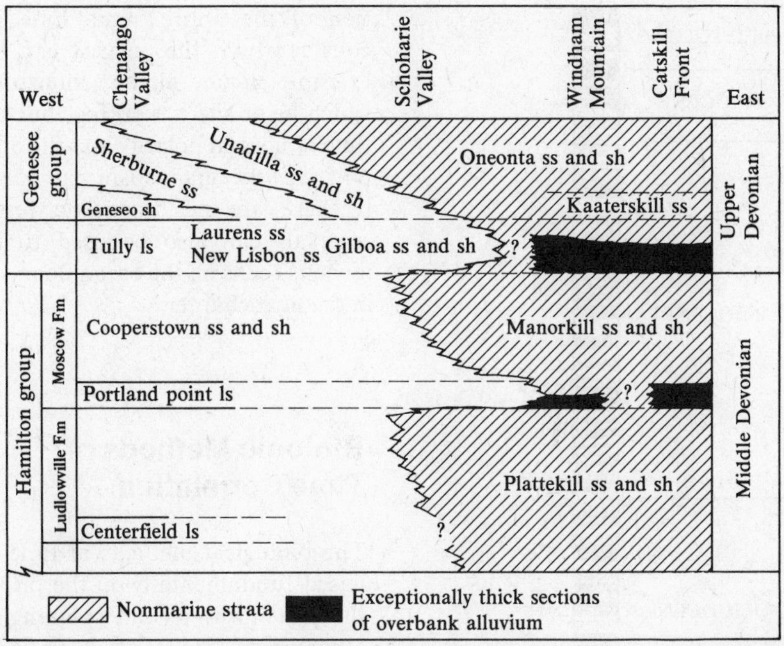

Figure 9.11. *Event Correlation.* Cross section of Devonian strata in New York showing inferred time correlations between marine and nonmarine strata whose formation resulted from the same events. (After I.N. McCave, *AAPG Bulletin,* vol. 53; from Robley K. Matthews, *Dynamic Stratigraphy,* © 1974. Reprinted by permission of Prentice-Hall, Inc., Englewood Cliffs, N.J.)

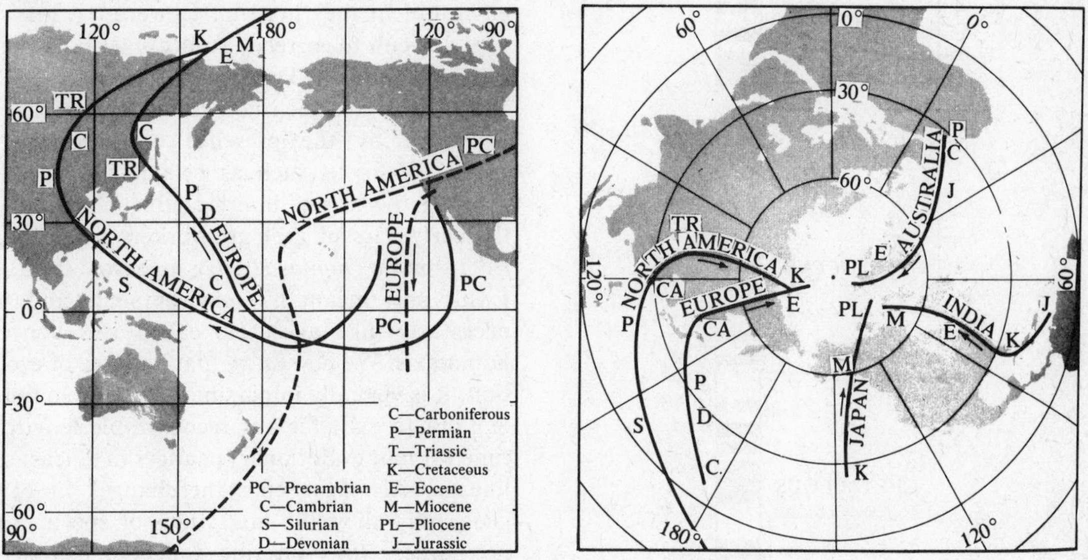

Figure 9.12. *Polar Wandering Curves.* Continent and time periods are labeled. (From Kay and Colbert, *Stratigraphy and Life History,* © 1965, by permission of John Wiley & Sons, Inc.)

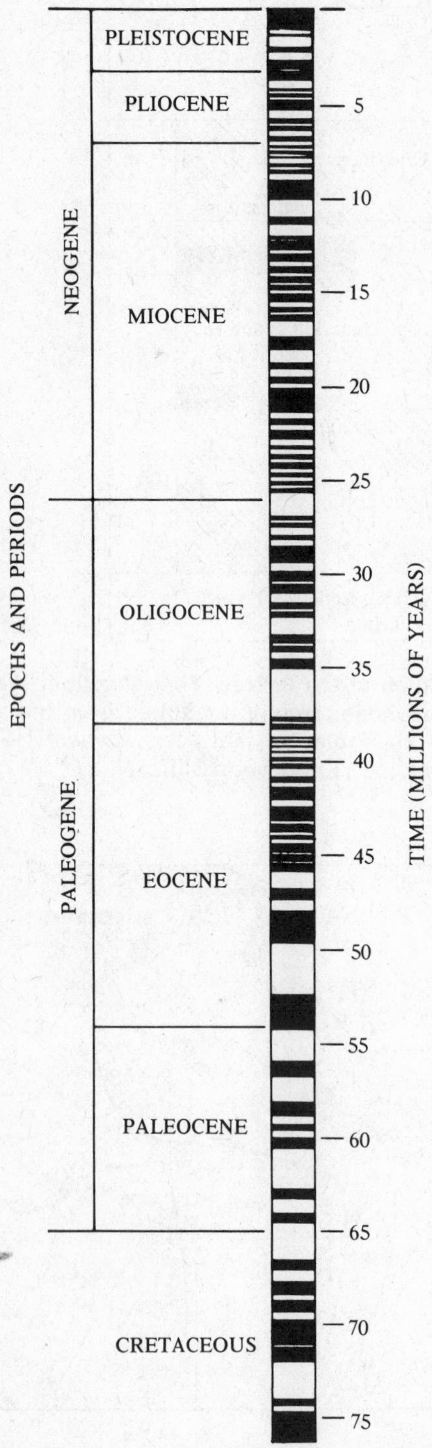

Figure 9.13. *Reversals of the Earth's Magnetic Field Since Late Cretaceous Times.* Black bars are times of normal polarity; white bars, times of reversed polarity. (After Heirtzler, 1968.)

tion of the approximate time interval under consideration, the polarity can be useful. For example, if one already knows the stage with which he or she is working, but one zone was a time of normal polarity and the other of reversed polarity, the magnetism can identify the zone if there are no diagnostic fossils. Magnetic reversals can also be used to date strips of igneous rocks on the ocean floors, as we shall see in the next chapter.

Biologic Methods of Time Correlation

The biological methods of time correlation are based fundamentally on the principle of faunal succession with its underpinning of evolution. In other words, like fossils mean like ages. We have seen the difficulties that facies introduce, particularly at lower levels like stages and zones, though these are less of a problem with systems and erathems. No paleontologist would be likely to confuse Devonian fossils from any environment with Jurassic fossils regardless of what environment they lived in. However, it may be very difficult to correlate sublittoral and abyssal faunas of the same Devonian stage.

We have also seen how this problem can be alleviated by the presence of wide-ranging species of fossils such as pelagic forms or by locating the lateral intertonguing of facies. In the early days of geology, workers such as the Englishman *Thomas Huxley* suggested that like fossils only meant like environments and not necessarily like ages. He called this concept **homotaxis.** We now know that because of evolution, it is virtually impossible for a whole fauna or flora to exist for any recognizable length of time without evolutionary changes in at least one lineage. Also, the constant changes in earth geography allow the migrations of species into new areas, thus altering the composition of faunas and floras. Hence, the concept of nonsynchronous homotaxis of faunas (or floras) need not concern us in time correlation. Individual species, however, may exist at different times in different places as discussed in the next section.

Index Fossils

There are several widely used biologic methods of time correlation. A very popular technique is the so-called **index-fossil** method. After considerable collecting, it was discovered that some lineages of fossil organisms evolved rapidly so that the vertical stratigraphic range of any species is short, spread widely so that the horizontal geographic range of any species is wide, and common enough to be useful to geologists. Species of these lineages have become known as **index fossils.**

One problem that always arises with index fossils is that their ranges are empirically determined; they cannot be known in advance. Therefore, there is no guarantee that an index-fossil species did not survive later in some restricted area than it did elsewhere (Fig. 9.14). For example, the screw-shaped bryozoan *Archimedes* (Fig. 7.53) was long thought to be an index of the Mississippian, but it is now known from Pennsylvanian and Permian rocks in a few places. Furthermore, the appearance of an index fossil in a new faunal province is different from its first appearance in its original province (Fig. 9.14). (The spread of modern exotic organisms indicates that once a new province is invaded, a successful migrant spreads throughout the province instantaneously in geologic terms.) So the time range of any index fossil is always subject to modification based on new discoveries.

Another supposition of the index-fossil method is that only a few types of organisms (Fig. 9.15) evolved rapidly enough and spread widely enough to be useful as index fossils. This double condition of restriction is a limiting feature to be sure, but the development of new techniques and tools has indicated that many groups of organisms, such as the pelagic radiolaria, formerly thought to be too evolutionarily static for correlation purposes, have turned out to be just as rapidly evolving as more widely used organisms.

Fossil Assemblages

Other geologists have reasoned that if a single species is not a perfect correlation tool, an assemblage of organisms should provide better resolution because an assemblage is likely to have a greater geographic range than a single species. All species need not be present to record the presence of an assemblage. Correlation by **fossil assemblages** is based on the similarities of whole faunas rather than on single index species (Fig. 9.16).

The improvement, however, may not be as great as it first appears. Finding ten species in

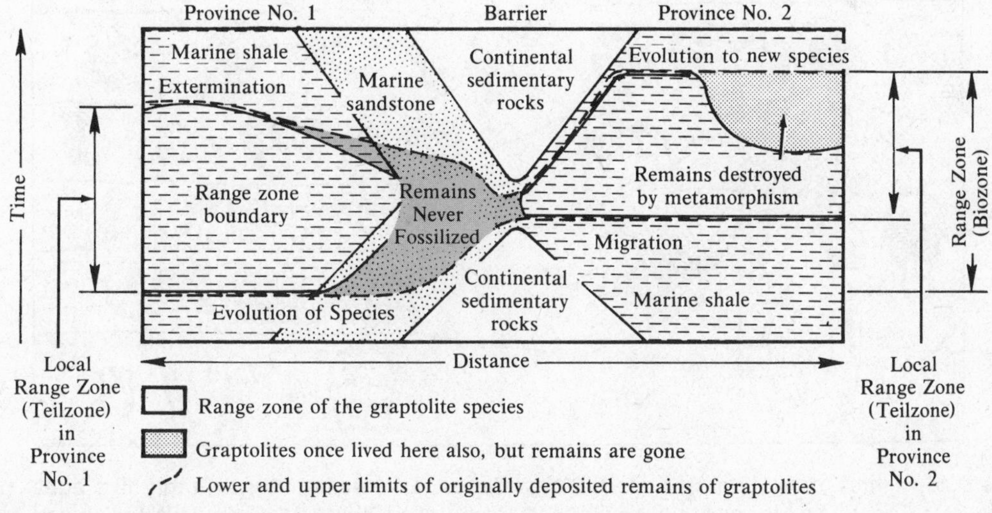

Figure 9.14. *Major Factors Affecting the Range of a Species.* (From Don L. Eicher, *Geologic Time,* © 1968, reprinted by permission of Prentice-Hall, Inc., Englewood Cliffs, N.J.)

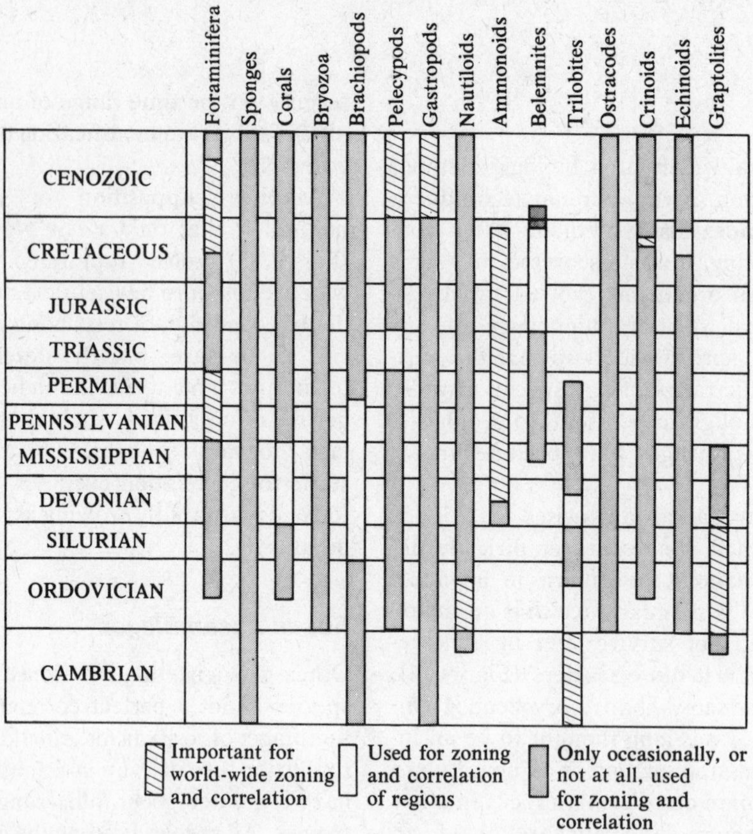

Figure 9.15. *Major Fossil Animal Groups Containing Index Fossils.* Some major groups of index fossils, such as the conodonts, are not included on this chart. (From Don L. Eicher, *Geologic Time,* © 1968, reprinted by permission of Prentice-Hall, Inc., Englewood Cliffs, N.J.)

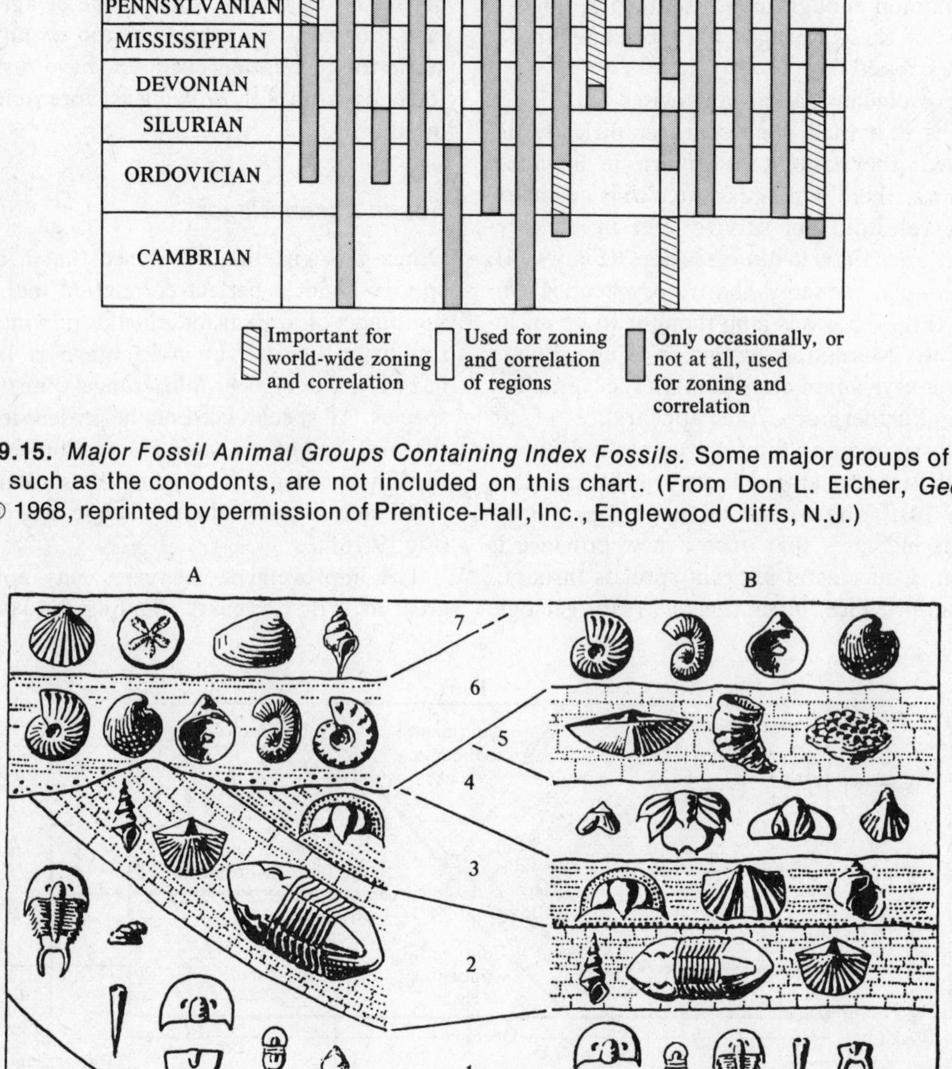

Figure 9.16. *Time Correlation by Fossil Assemblages.* Like assemblages imply like ages. (From *Invertebrate Fossils* by Moore-Lalicker-Fischer, Copyright 1952 by McGraw-Hill, Inc., used by permission of McGraw-Hill Book Company.)

common between two units is not necessarily ten times as good a correlation as one based on a single species if all the species are controlled by the same environmental conditions. Also, the ten species may belong to lineages that did not evolve as rapidly as one index species. Nevertheless, it appears that the usage of more species does improve the accuracy of time correlation because the more that are used, the less the chance that the one changing lineage has been overlooked or that the whole assemblage has changed its stratigraphic range. A late migrant or survivor stands out in an assemblage. This is how erroneous index-fossil correlations were discovered.

Concurrent-Range Zones

A very popular type of fossil-assemblage time correlation is that of using **concurrent-range zones**, or Oppelian zones (Figure 2.20 is a simple example; most are based on many more species). As discussed earlier, these are based on the overlapping stratigraphic ranges of species and are characterized by unique assemblages that do not occur above or below in the geologic record. A characteristic fossil, common in the zone but not necessarily limited to it or found throughout it, is selected as the name-giver and is known as a **guide fossil.**

Keep in mind that any range is empirically determined and that, hence, no range, even one that is used to establish a zonal boundary, can be known with absolute certainty. Formal time-stratigraphic zones or chronozones of the concurrent-range variety are not established until after exhaustive collecting from a wide region in an attempt to overcome the problem, but it can and still does occur. As a result, a zonal boundary established on the basis of the beginning of four species and the ending of four other species eventually becomes a fuzzy zone as slight extensions upward and downward are found (Fig. 9.17). Indeed, the process is inevitable, for there is nothing systematic about the evolution of species by natural selection that could cause the simultaneous appearances or disappearances of several species at once except in those instances where two species cannot survive without each other. If one form were absolutely dependent on another form as its food, then, of course, the two could die out at once, but there is no good reason why two species of forams, clams, or ammonites should die out simultaneously or arise at the same time. Evolution is based on random mutations occurring continuously, most of which are not selected by the environment. Hence, coincident beginnings and endings of ranges are most unlikely. Yet they occur in the geologic record at any one locality. This simply indicates that when such coincidences occur, they represent environmental shifts in the region or gaps in the record there.

Thus, zones can be used for time correlation, but only down to a certain level of refinement. Still more detailed time correlations are becoming desirable, and it has been suggested that statistical methods based on the ranges of individual species be used.

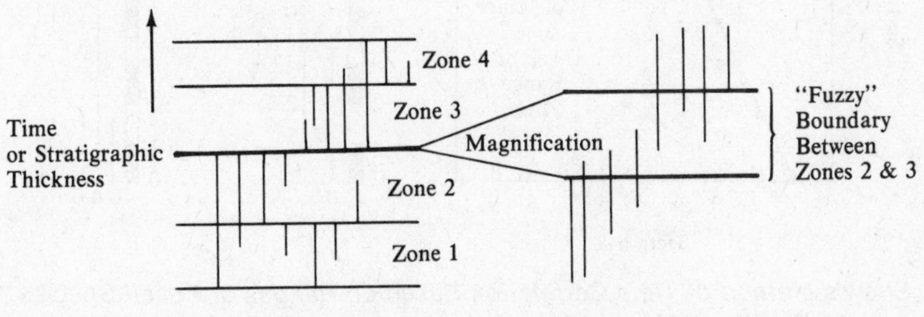

Figure 9.17. *Fuzzy Zonal Boundaries.* Horizontal lines are zonal boundaries; vertical lines are ranges of species. Detailed collecting usually reveals that coincident endings and beginnings of species do not occur. The magnification at the right shows that numbers of species do not have coincident beginnings and endings.

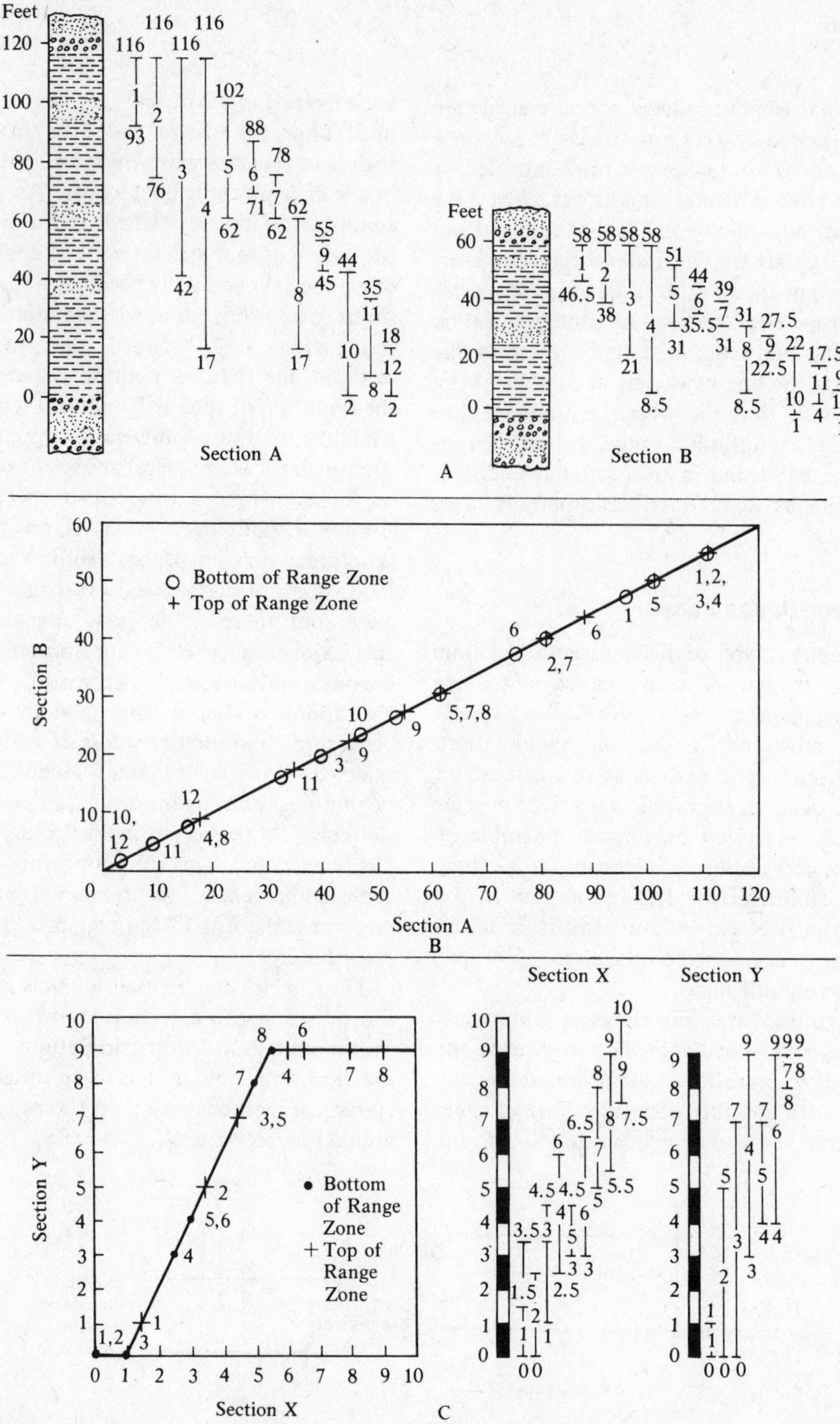

Figure 9.18. *Shaw's Method of Time Correlation Based on Ranges of Fossil Species.* **A.** *Two sections with ranges of fossils graphed in feet above basal datum,* **B.** *Correlation line of sections in Part A,* **C.** *Dog-legs in correlation line produced by cessation of deposition in one section.* (From Don L. Eicher, *Geologic Time,* © 1968, reprinted by permission of Prentice-Hall, Inc., Englewood Cliffs, N.J.)

Statistical Methods

For example, *Shaw* has devised a method based on graphing the ranges of fossils in two geologic sections (Fig. 9.18A). If the two sections correlate, there must be a time-correlation line on the graph because for every point in one column, there is a time-equivalent point on the other (Fig. 9.18B). Where deposition ceased or a gap exists in one section, while sedimentation continued in the other, a dog-leg occurs as the number of feet increases on one axis and stays the same on the other (Fig. 9.18C). Of course, ranges will not precisely coincide from section to section, so a scattering of points rather than an exact line will typically occur. The corresponding line can be visually approximated or statistically determined.

The method can be expanded to correlate one section after another until a composite standard is built up, showing the total stratigraphic distribution of all species in the geographic area. Eventually, if tie points can be established between regions, a standard can be constructed for the whole world. Many geologists object to statistical methods of correlation, arguing that most fossils are facies fossils and only a few forms, such as pelagios, can be used for correlation.

Acmes

From time to time in the past, geologists have attempted to correlate **epiboles,** strata that contain the maximum abundance, or **acme**, of a species. The maximum abundance of a species of fossil is chiefly ecologically controlled, however. It may represent a locally favorable environment or simply a mechanical aggregate of shells. There is no reason why a species should necessarily have a peak of abundance, or if it does, why it should be recorded by layers of abundant specimens. But even if a species did reach a peak and it is recorded, there would be no way to distinguish an abundance of fossils produced this way from one produced mechanically. Hence, the concept of correlating on the basis of the acme of a species is unworkable.

One could assign the same age to strata containing an abundance of specimens of the same species, and they could actually have been formed at different times as a particularly favorable environment shifted about (Fig. 9.19). About the only time value of such an occurrence of fossils is simply that of contributing to the range of the species.

Reconnaissance Methods

There are two other widely used biologic means of time correlation important in the reconnaissance of new regions, but they lack the precision and reliability of the index-fossil, fossil-assemblage, or range-of-fossil-species methods previously described. These three methods should be used in the detailed studies that follow reconnaissance work.

The **stage-of-evolution** method is based upon the thorough knowledge of the history of an evolutionary lineage, such as that of the horse

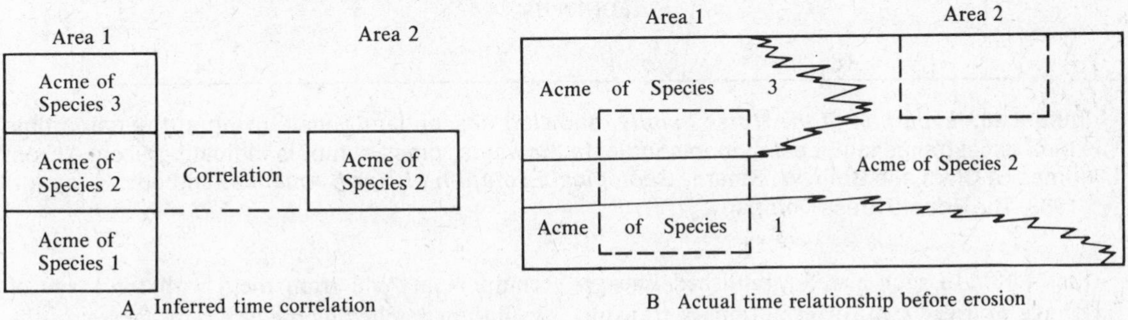

Figure 9.19. *Miscorrelation Using Epiboles.* (Strata containing the acme of a species.) **A.** *Inferred time correlation*, **B.** *Actual time relationships before erosion.* The preserved strata in area 2 equal the youngest strata in area 1, not the middle ones.

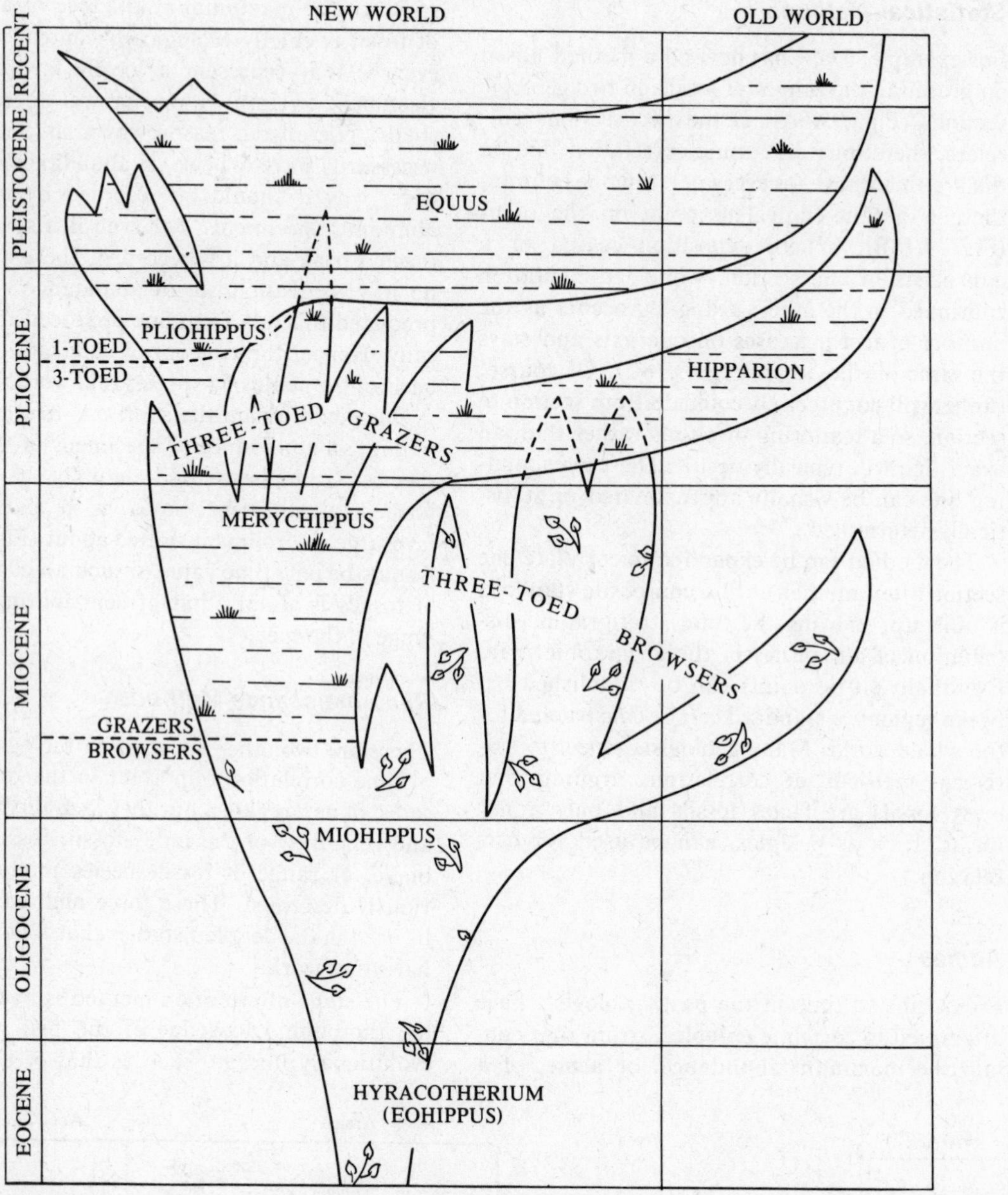

Figure 9.20. *Evolution of the Horse Family,* showing descendant genera living at the same time as their ancestral genera. Leaf symbols indicate browsers; grass symbols indicate grazers. (From Thomas H. Clark and Colin W. Stearn, *Geological Evolution of North America,* 2nd ed., Copyright © 1968, The Ronald Press Company, N.Y.)

(Fig. 4.18). In such a well-established lineage, the age of every step, or evolutionary transformation, in the series is known. To time correlate rocks of unknown age, a geologist would compare a fossil from them with the stage of evolution reached by the group in the rest of the world. For example, a fossil horse tooth from a new locality in Nevada could immediately date the strata containing it because the structure of horse teeth for every stage of the Cenozoic from the Upper Paleocene onward is known.

The method is not infallible, however, because there is no advance guarantee that an ancestral species did not live on at the same time as its descendants or even outlive them. For example (Fig. 9.20), the horse genus *Hipparion* lived on in the Old World at the same time and later than some species of another horse genus *Pliohippus* which evolved after it from the same ancestor. (One could imagine the difficulties an early geologist might have had if he or she knew only the European *Hipparion* and the American *Pliohippus*. One could get the evolutionary history upside down!) The method also has limited application because few lineages are as well known as that of the horse family, though the broad outlines of the evolution of many groups are known.

Finally, we come to Lyell's **percentage-of-living-species** method of time correlation. Lyell, you may recall, established the Cenozoic epochs on the basis of the percentage of surviving species of mollusks they contain (Fig. 9.21). He eliminated the problem of the subjectivity of species by using the identifications of only one man, the French paleontologist *Deshayes.* Lyell's percentages—3.5, 18, 33–50, 90—are discontinuous and, hence, indicate that there was missing time in his sections. We saw earlier that such discontinuities cannot exist biologically and, hence, must have a geologic cause. Furthermore, there is no guarantee that mollusks in environments other than Western Europe and that other types of organisms all evolved at the same rate. Thus, Lyell's percentages serve as useful approximations for time correlating rocks at the series level, but for detailed time correlation, more refined methods are needed.

As noted in Chapter 6, there may be a new biologic method of reconnaissance time correlation coming into use. This method involves comparing the numbers of daily **growth rings** on a fossil with a geophysically determined scale of days in the year for each subdivision of geologic time.

Rock Correlation versus Time Correlation

Before we leave the topic of correlation, it is necessary to emphasize again that rock correlation and time correlation are not the same and, therefore, cannot be substituted for each other. We cannot substitute rock correlations for time correlations because if we do, rocks of different lithologies that formed at the same time will be assigned different ages and rocks of the same lithology which formed at different times will be assigned the same age (Fig. 9.22). If we substitute time correlations for rock correlations, rocks with the same fossils but differing lithologies will be considered parts of the same formation and rocks of the same lithology but different fossils will be considered separate formations (Fig. 9.23).

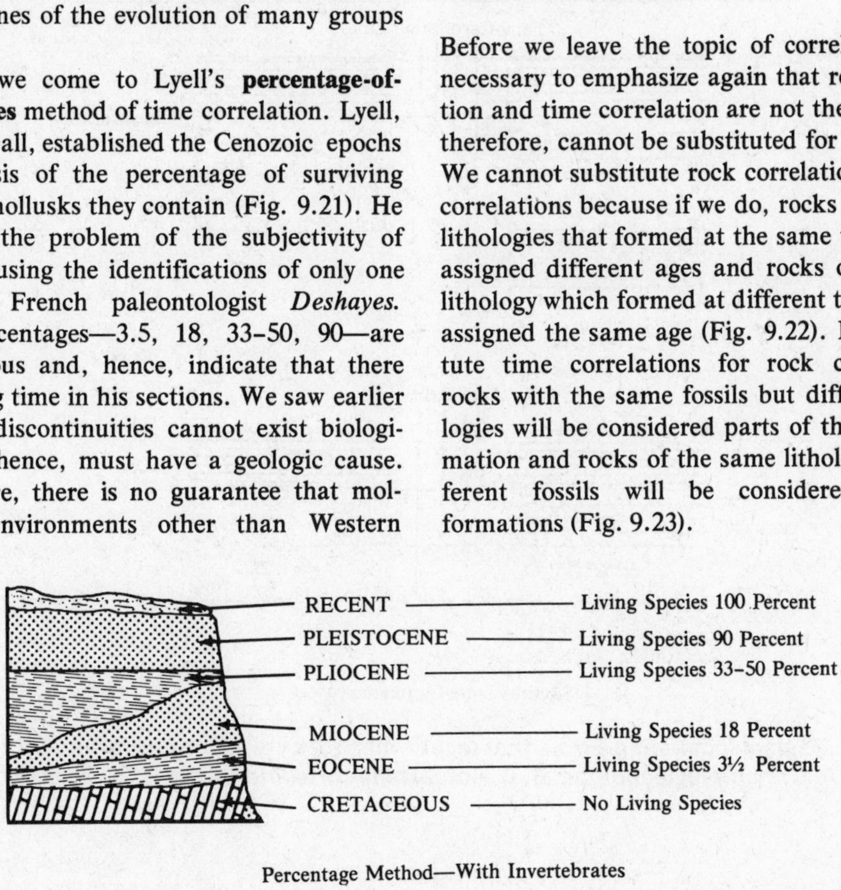

Figure 9.21. *Lyell's Percentage-of-living-Species Method of Time Correlation.* (After Stirton.)

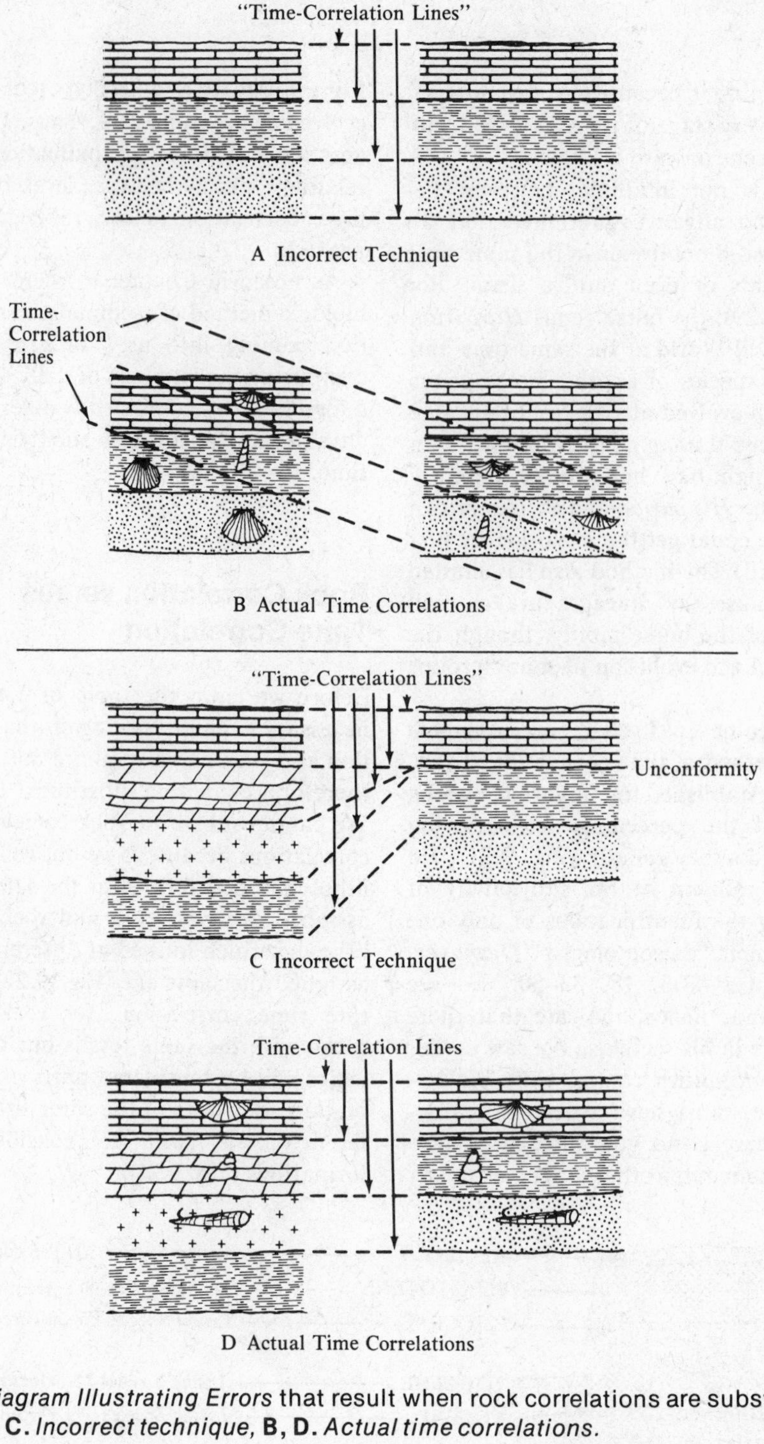

Figure 9.22. *Diagram Illustrating Errors* that result when rock correlations are substituted for time correlations. **A, C.** *Incorrect technique,* **B, D.** *Actual time correlations.*

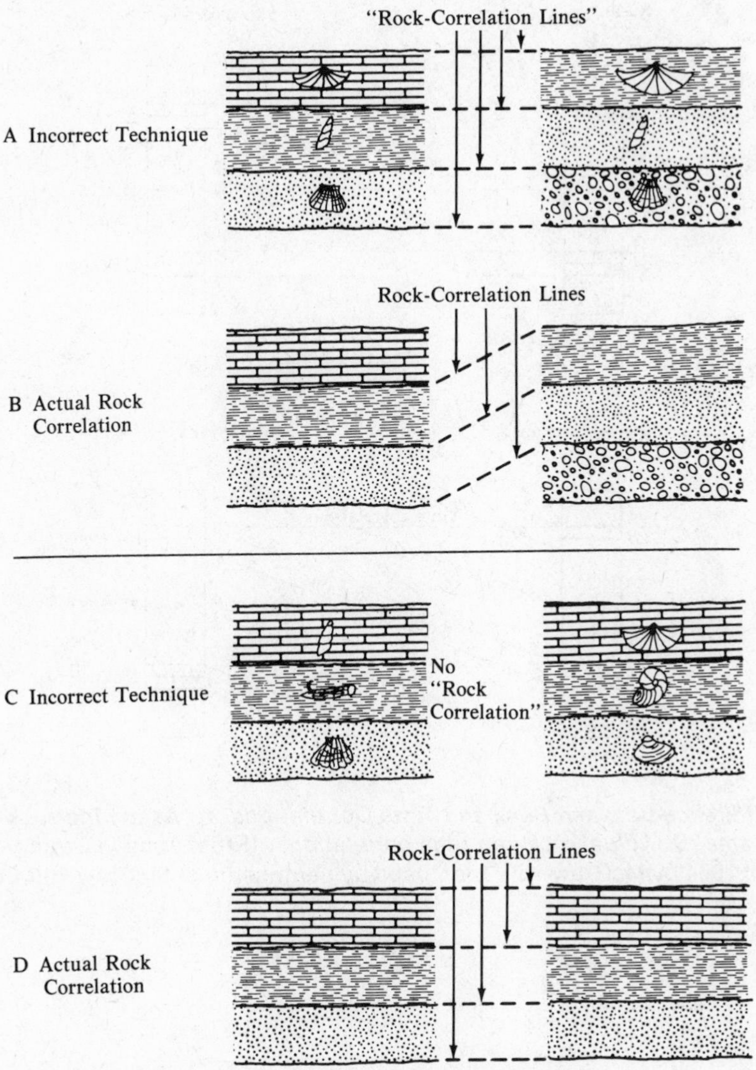

Figure 9.23. *Diagram Illustrating Errors* that result when time correlations are substituted for rock correlations. **A, C.** *Incorrect technique,* **B, D.** *Actual time correlations.*

A particularly troublesome problem can arise with bodies of rock of identical lithology and fossils (Fig. 9.24). The identical fossils indicate contemporaneity and time correlation, but they do not prove lithologic continuity and, hence, rock correlation. They may be parts of entirely different formations forming at the same time as indicated in the lower diagram in Figure 9.24. Therefore, it is always important to distinguish the two types of correlation and not to confuse them or substitute one for the other.

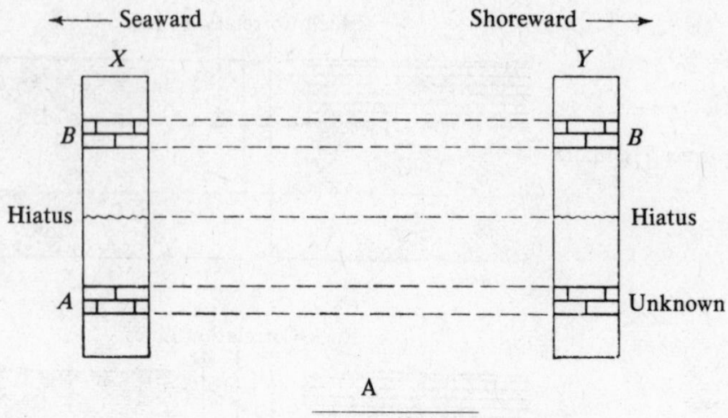

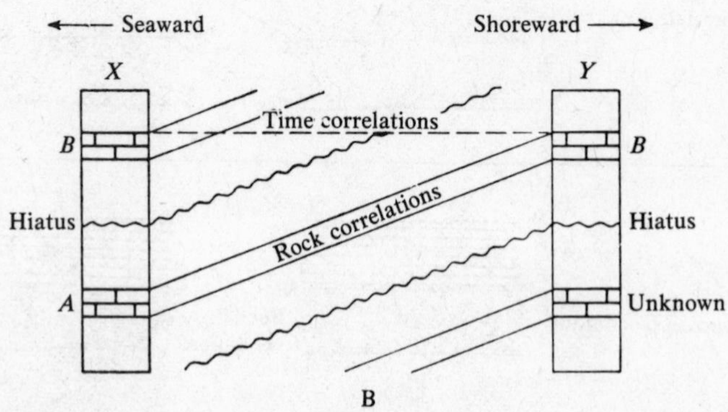

Figure 9.24. *Difference Between Rock and Time Correlations.* **A.** *Assuming rock and time correlations are the same,* **B.** *Actual rock and time correlations.* (From *Time in Stratigraphy* by Alan B. Shaw. Copyright 1964 by McGraw-Hill, Inc., used by permission of McGraw-Hill Book Company.)

REVIEW

1. What is the difference between **rock correlation** and **time correlation?**
2. What are three major techniques of surface rock correlation?
3. What is the purpose of **instrumental well logging?**
4. What are three major physical methods of time correlation?
5. Define **event correlation** and give an example.
6. What is an **index fossil?**
7. Why is nonsynchronous **homotaxis** not considered an obstacle to time correlation with fossils?
8. How does the use of **fossil assemblages** in time correlation improve correlations?
9. Why is the use of **acmes** in time correlation considered of little value?
10. What groups of fossils would be most amenable to the **stage-of-evolution method** of time correlation?
11. What major errors result when time correlations are substituted for rock correlations? When rock correlations are substituted for time correlations?

10

Footloose Continents: Continental Drift and Treadmill Ocean Basins

10

As early as 1620, *Francis Bacon* noted the remarkable similarity in the outlines of the two sides of the Atlantic Ocean (Fig. 8.1). In 1668, *Francois Placet* stated that before the Deluge, the Western Hemisphere continents had been joined to the Old World by islands or perhaps the lost "continent" of Atlantis. *Buffon*, about 1750, suggested that Europe and North America had once been joined because of the similarities of their present fauna and flora. He also suggested continental and polar dislocations to explain anomalous climatic changes in the past. About 1800, *Alexander von Humboldt* also noted the similarity of the South American and African Atlantic coasts and speculated that the Atlantic was a huge valley eroded by the ocean.

In 1858, *Antonio Snider-Pellegrini* drew attention to the great similarity of the Carboniferous coal-age plants on both sides of the North Atlantic. Reasoning that it was most unlikely that the spores of these plants could have blown across the Atlantic Ocean, Snider proposed that all the continents had been joined together in the past. He believed they had split apart and formed the Atlantic Ocean because their initial concentration was an unstable configuration. Near the end of the 1800s, the Austrian geologist *Edward Suess* noted that much of the geologic record of the southern continents of South America, Africa, India, and Australia was nearly identical. He thought that they were all united at once in the past, forming a great supercontinent which he called **Gondwanaland** after a group of rocks which define a geological province of India. (Knowledge of Antarctica, gained at a later date, led to its inclusion in Gondwanaland also.)

The first comprehensive theories about the drifting of continents were independently proposed around 1910 by the American geologists *F.B. Taylor* and *H.H. Baker* and the German meteorologist *Alfred Wegener* (Fig. 10.1). Wegener's work was particularly thorough and detailed, and most geologists today would call him the father of the continental drift concept. Wegener spent many years analyzing geophysical data and matching the geological and paleontological similarities on the two sides of the Atlantic. Wegener noted that geophysical data indicated the possibility of large-scale lateral movements of the crust and the impossibility of lost continents or land bridges in the oceans. He believed that all of the continents had been united in the Late Paleozoic into the vast supercontinent **Pangaea** (Fig. 10.2). (Others noted that connections between the northern and southern continental masses were not complete and preferred

Figure 10.1. *Alfred Wegener* (1880–1930).

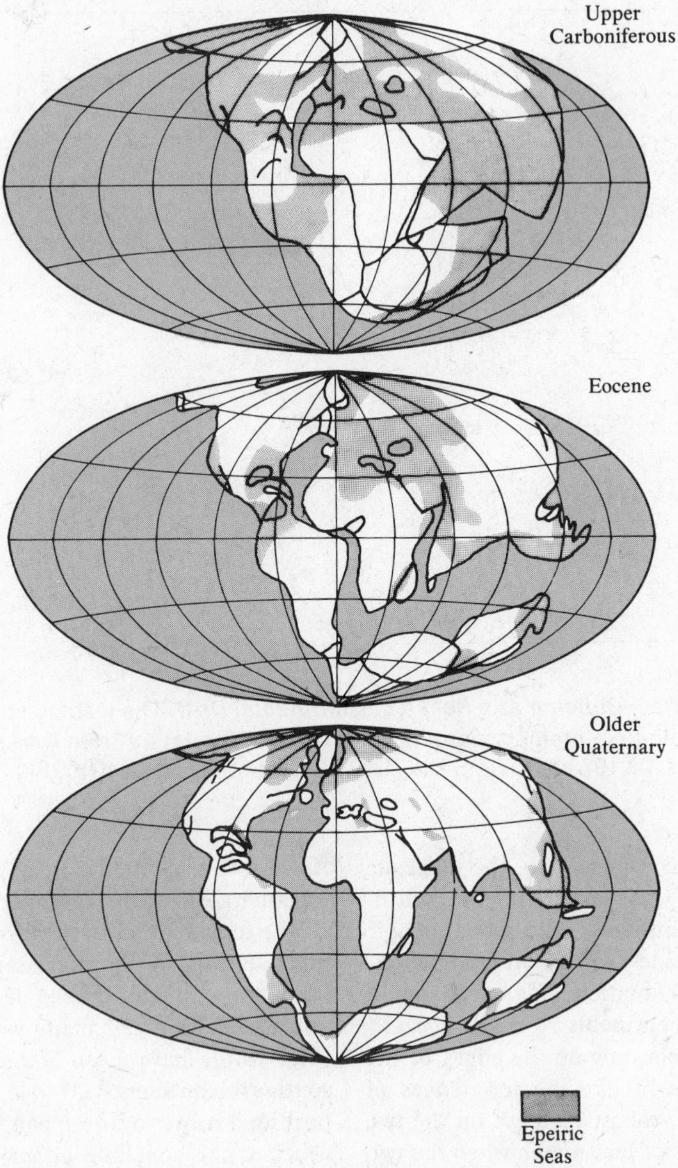

Figure 10.2. *Wegener's Reconstructions of Pangaea and Its Fragmentation.* (After Wegener, 1929.)

Laurasia and **Gondwanaland** respectively.) Wegener suggested that the continents broke up in the Mesozoic and have since drifted to their present positions. He proposed rotational and tidal forces as the driving mechanisms of continental drift and suggested that the long puzzling origin of mountain building could also be explained by his theories. As the continents plowed their way through the oceans, he reasoned that their leading edges would be crumpled, forming geosyclines and mountain belts (Fig. 10.3). The South African geologist *Alexander du Toit* continued and expanded Wegener's work in the 1930s.

Early Evidence

Some of the lines of evidence gathered by these men are the similarity of the geologic rock units

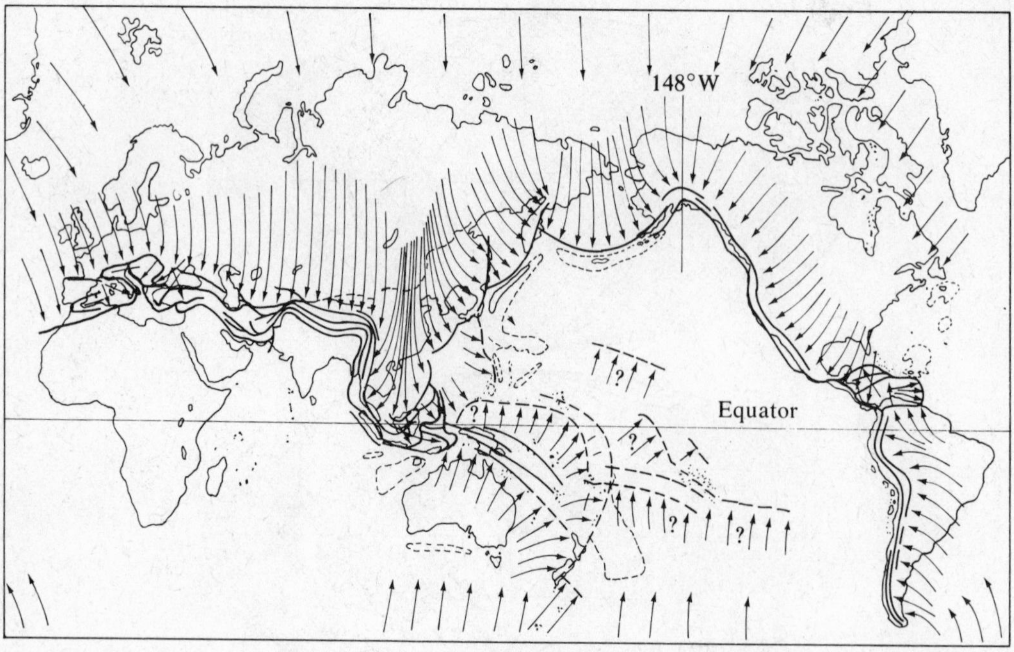

Figure 10.3. *Mountain Building as a Result of Continental Drift.* The leading edges of moving continents buckled to form mountains according to early drift theorists. (From A. Hallam, *A Revolution in the Earth Sciences*, © 1973 by Oxford University Press; after Taylor 1910, Fig. 7.)

and structures of continents on the opposite sides of the oceans (particularly on the Southern Hemisphere continents), the paleontologic similarities of now widely separated landmasses, and the anomalous climatic bands in the past if one assumes static continents.

They noted that not only do the edges of the sundered continents fit like the torn edges of newspaper, but the geology matches on the two sides just as the lines of type match when the two halves of the torn newspaper are put together. Particularly striking is the geologic sequence of the southern continents in the Late Paleozoic and Early Mesozoic (Fig. 10.4). In all these continents, a Permocarboniferous series of glacial deposits is followed by a thick sequence of Permotriassic terrestrial beds, including coal. The sequence is capped by a thick Jurassic sequence of basaltic lava flows erupted from fissures. Geological structures such as the folded Cape Mountains of Africa and the Sierra de la Ventana of Argentina also line up if one puts the continents together along their matching coasts (Fig. 10.5).

The fossils found in these rocks are also strikingly similar. A Permotriassic group of plants with large tongue-shaped leaves, *Glossopteris* and *Gangamopteris* being the most prominent (Fig. 10.6), have been found only on all the southern continents. If one assumes a static position for the southern continents, it would be necessary to conclude either that the *Glossopteris* flora migrated between the southern continents via the more closely connected northern continents or that the seeds of these plants blew across thousands of miles of open ocean. The absence of the *Glossopteris* flora in the north (a few elements of the flora occur along the southern edge of the northern continents in the Mesozoic) refutes the first suggestion; and the second seems most unlikely, particularly in view of the later discoveries of *Glossopteris* in very isolated Antarctica. Also, the tiny fresh-water reptile

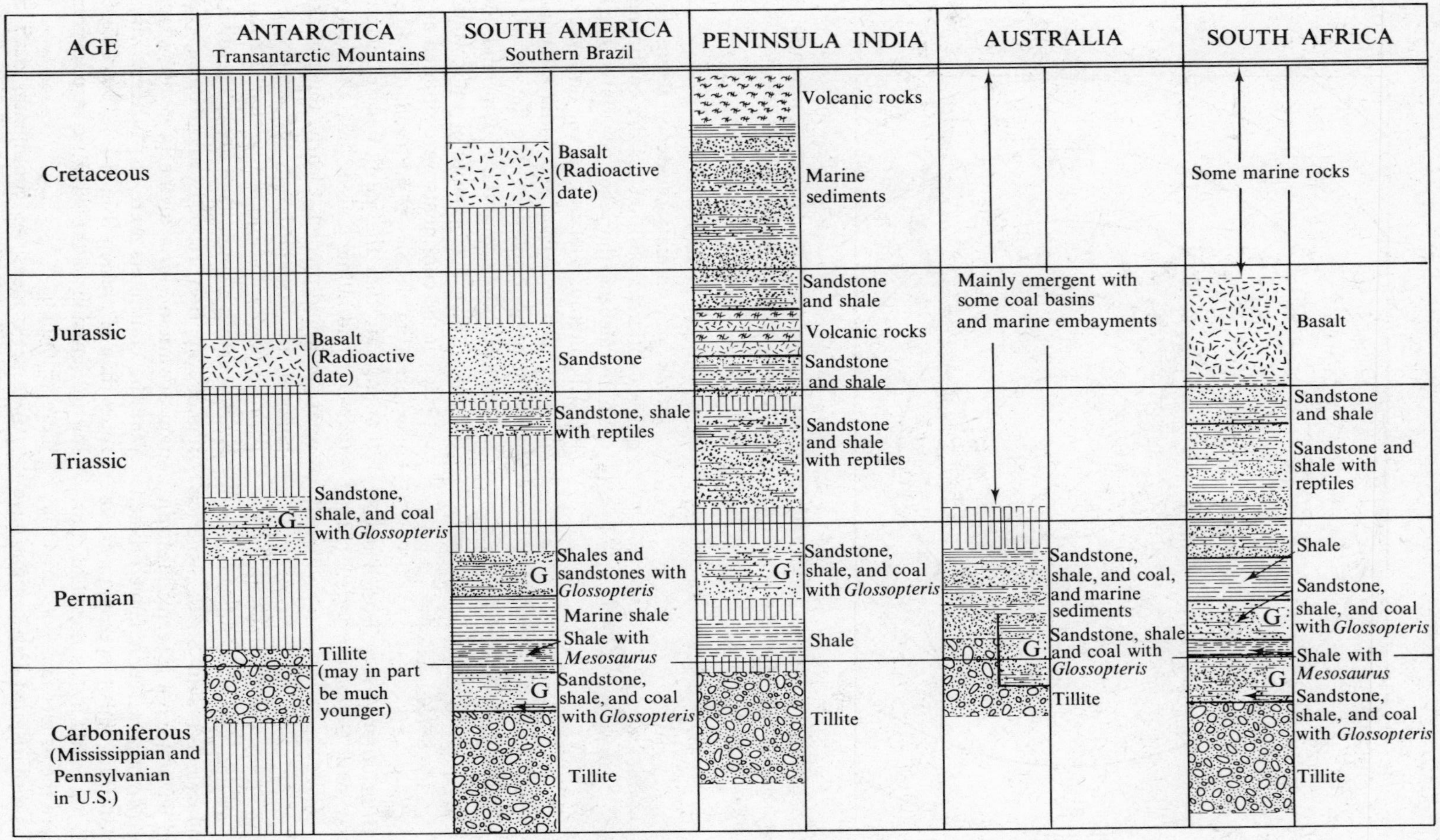

Figure 10.4. *Similarities of the Geologic Records of the Gondwana Continents.* The Carboniferous-Jurassic records of Antartica, South America, Peninsula India, Australia, and South Africa are virtually identical despite their present wide separation. (From R.J. Foster, *General Geology,* 2nd ed., © 1973, 1969 by Charles E. Merrill Publishing Co.)

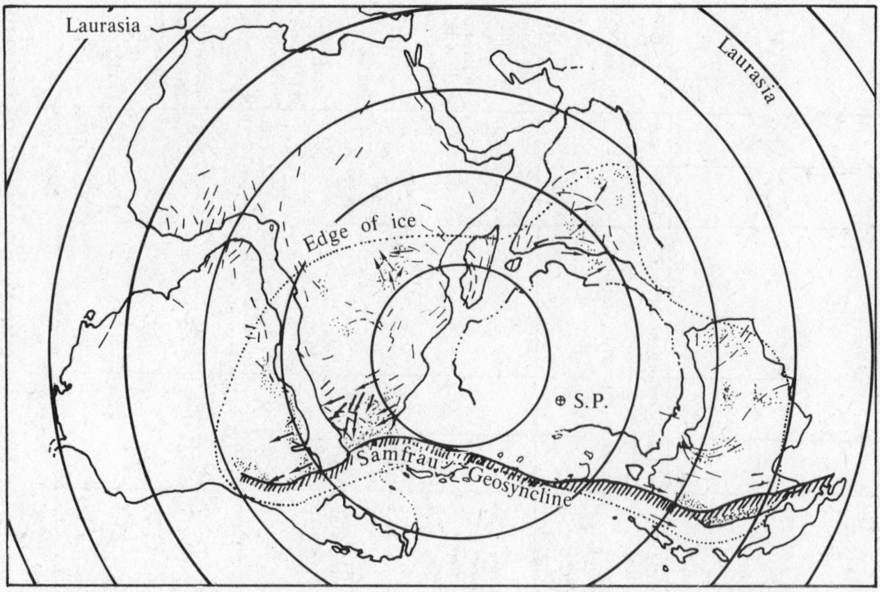

Figure 10.5. *Reconstruction of Gondwanaland in the Late Paleozoic,* showing mountain-belt alignments, glacial deposits, and ice-flow directions. (After A.L. du Toit, *Our Wandering Continents,* © 1937 by Oliver & Boyd, Edinburgh.)

Figure 10.6. *Glossopteris Flora.* Fossil leaves of *Glossopteris* and *Gangamopteris.* (After Hurley.)

Mesosaurus (Fig. 10.7), which is found only in a short interval around the Carboniferous and Permian boundary, occurs in South America and Africa and nowhere else. It seems most unlikely that the *Mesosaurus* could have crossed the present Atlantic in light of its ecologic occurrence, and it is not found in the northern continents even where the proper facies occur. Other fossil organisms with disjunct distributions have also been cited.

Finally, the climatic evidence seems to require a different position for the continents in the Late Paleozoic than they now have. The Permocarboniferous glaciations not only occurred in areas such as Africa and India far removed from the present poles, but the lines of ice flow are presently away from the equator and toward the poles, the exact opposite of what would be expected if the continents were situated as they now are (Fig. 10.8). In addition, the lines of flow indicate that the ice flowed from the sea onto the land in many areas, also the reverse of what one would predict. Clearly, these phenomena can be explained only by assuming the continents were once joined so that the ice was flowing across a single land area. The distribution of evaporites, redbeds, and tropical vegetation at various times in the past also suggests that the continents were in different climatic zones than at present (Fig. 10.9). See Chapter 3 for a discussion of the climatic implications of these sediments.

Wegener's ideas on continental drift were largely rejected by Northern Hemisphere geolo-

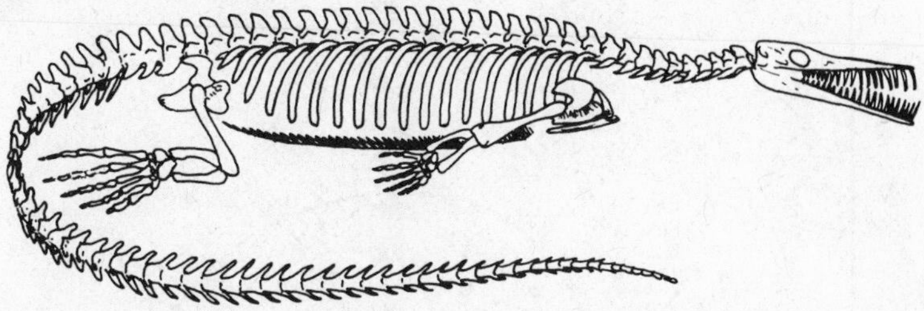

Figure 10.7. *Mesosaurus.* Skeleton of fresh-water reptile known only in South America and Africa. This small aquatic form lived in cool regions and probably hibernated by burrowing into lake bottoms in the winter. (After McGregor.)

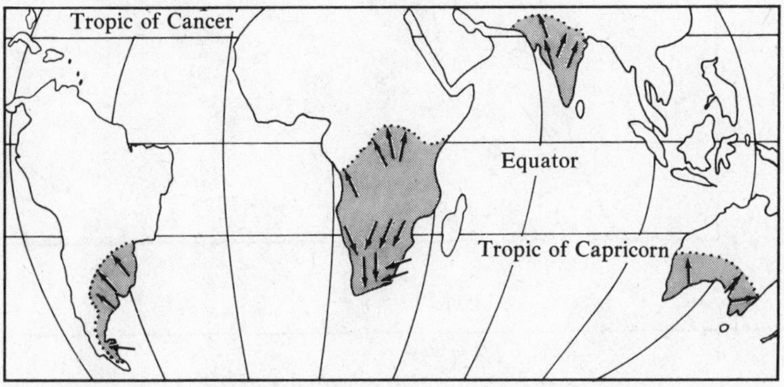

Figure 10.8. *Present Distribution and Flow Directions of Late Paleozoic Continental Glaciers and Glacial Deposits.* Compare with Figure 10.5. (After Holmes.)

gists, though they continued to thrive in the southern continents because of the work of du Toit and others and the more striking evidence there. The reason for the Northern Hemisphere rejection was that the rotational and tidal forces proposed by Wegener appeared too small to move continents, and the physicists could not imagine any larger forces in the earth. In the 1930s and 1940s, there were few Northern Hemisphere geologists who would support continental drift.

Convection Currents

Those who did were the Dutchman *F.A. Vening Meinesz* and the Scotsman *Arthur Holmes.* These men were geophysicists, and they noted that gravity measurements around the oceanic trenches indicate that material there is being held down in opposition to its tendency to attain isostatic equilibrium (Fig. 10.10). It was as if someone beneath a water surface were pulling a cork down against its natural tendency to bob up to a certain level consistent with its height and low density. By this time, geologists were aware that continental crust is lighter and thicker than oceanic crust (Fig. 10.11). The continents are somewhat like corks floating in a sea of denser material.

Vening Meinesz and Holmes hypothesized that the earth's internal heat generates convection currents in a liquid mantle (Fig. 10.12). Hot material rises to the surface, moves outward carrying the lighter continents with it, and cools and descends at the trenches. Here where the cooler material is sinking and its drag effect produces a trench, a piece of the earth's crust is being held down against its own level-seeking tendencies. Trenches were viewed as present-day geosynclines in this theory.

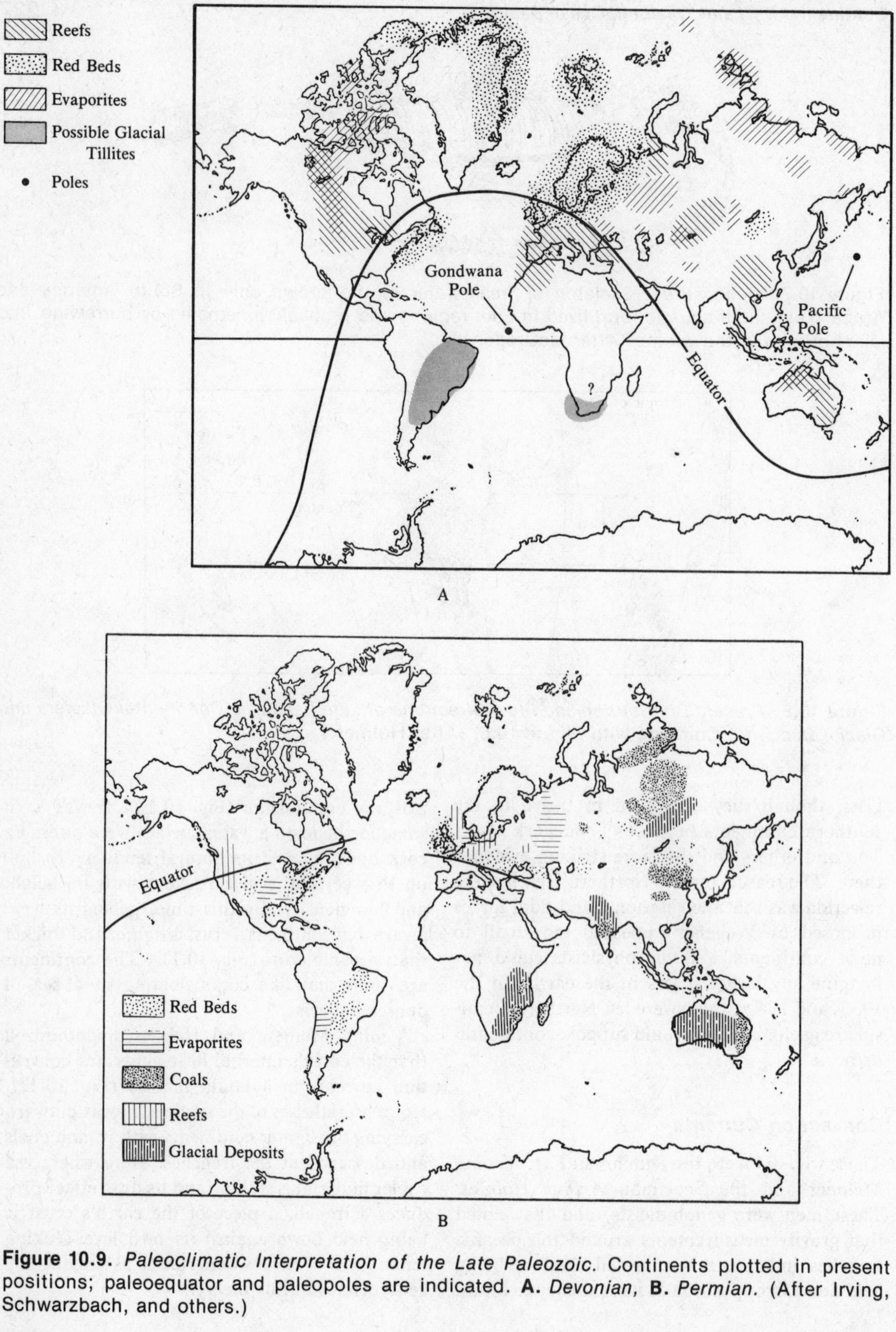

Figure 10.9. *Paleoclimatic Interpretation of the Late Paleozoic.* Continents plotted in present positions; paleoequator and paleopoles are indicated. **A.** *Devonian,* **B.** *Permian.* (After Irving, Schwarzbach, and others.)

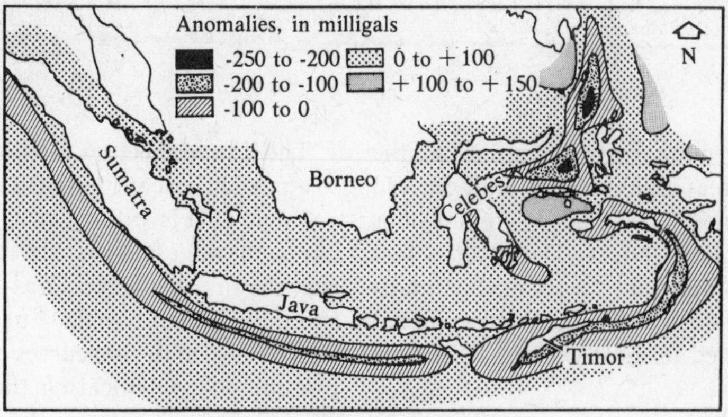

Figure 10.10. *Gravity Anomalies and Oceanic Trenches.* (After Vening-Meinesz.)

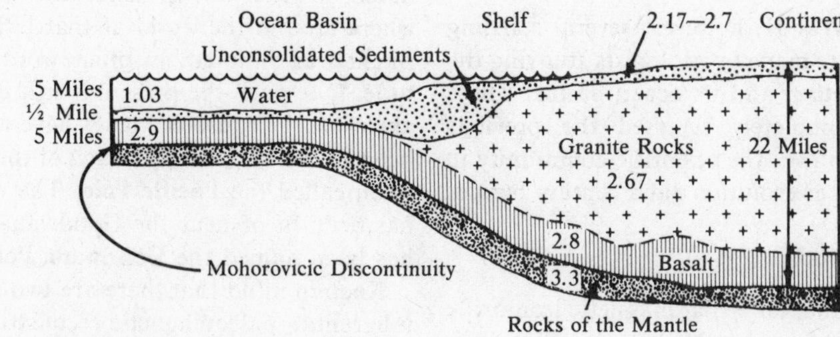

Figure 10.11. *Cross Section of Earth's Crust and Upper Mantle* showing relative densities and thicknesses of constituents.

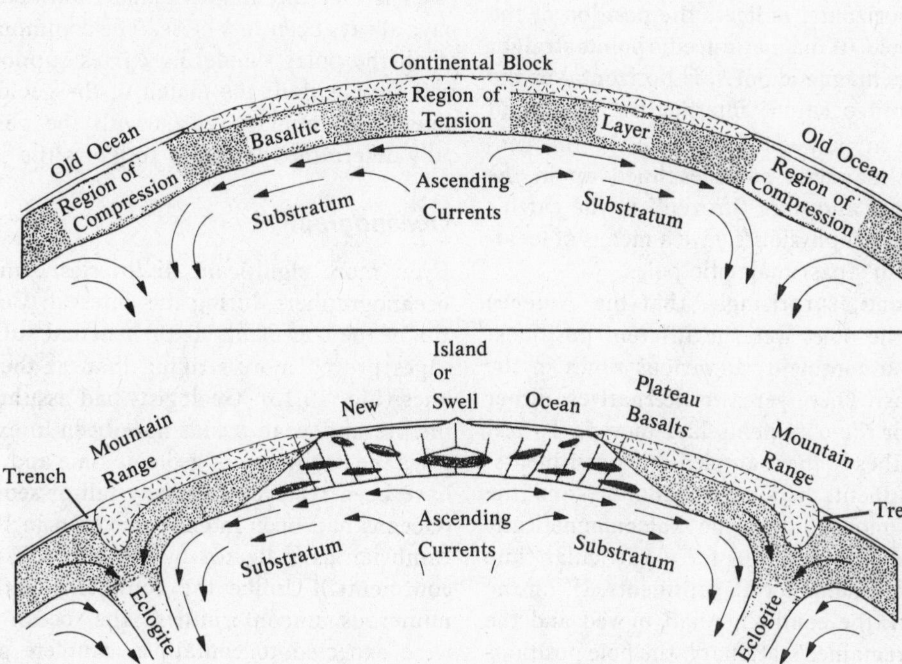

Figure 10.12. *Convection-Current Hypothesis Is Proposed by Arthur Holmes.*

Once the convection stops in an area, the trench with its great thickness of land-derived light sediment is no longer held down and bobs up to form mountain ranges. Despite the fact that the convection-current theory unified many observations in geology with a coherent explanation, it was ridiculed in most geology texts even through the 1950s.

Modern Concepts

After World War II, however, several startling discoveries were made by geologists studying the earth's **magnetism** and its **ocean basins.** These discoveries completely reversed the previous trend and won over the scientific community in the 1960s, just as evolution did a century before.

Paleomagnetism

We noted in Chapter 9 that magnetic iron oxide particles in a cooling lava flow or a settling sediment align themselves with the earth's magnetic field. The direction of alignment and inclination from the horizontal indicate the position of the magnetic pole. (A magnetic needle points straight down at the magnetic poles, is horizontal at the equator, and occupies intermediate angles of inclination in between.) Likewise, in the geologic past, magnetic particles lined up in the same manner and were "frozen" in the earth's crust. Thus, geophysicists have a means of locating the earth's past magnetic poles.

They found, surprisingly, that the evidence indicated the poles were in different positions, relative to a continent, at various times in the geologic past. There were two alternatives: Either the poles or the continents had moved. To test these hypotheses, they sampled other continents. If the continents had been stationary and the poles had moved, then the paleomagnetically determined pole positions for a particular time should be the same for all continents. If, on the other hand, the continents had moved and the poles had remained stationary, the pole positions for a particular time would differ among the continents. The latter case turned out to be the reality (Fig. 9.12).

The line connecting the pole positions relative to a particular continent for various times in the geologic past is still called a **"polar wandering" curve** in reference to the original supposition that the first hypothesis was true. Its form is the same for all continents, but its position is not. The polar wandering curves converge toward the present poles, indicating the movement of the continents to their present positions (Fig. 9.12).

Unfortunately, all paleomagnetism can do is tell us how far from the pole a continent was situated; in other words, its latitude. It cannot tell where around the world at that latitude a continent was situated; in other words, its longitude. One pole, the present North Pole (but not always so, because the poles have reversed) has been in the Pacific area much of the time and is often called the **Pacific Pole.** The other, which has been in or near the Gondwana continents, has been named the **Gondwana Pole.**

Keep in mind that there are two assumptions inherent in paleomagnetic reconstructions. One is that the earth's magnetic field has always been dipolar; that is, with two opposite poles. The other is that the magnetic and rotational poles have always been very close. The common shape of all the polar wandering curves supports the first supposition; the match of the geologically determined climatic zones with the geophysically determined latitudes supports the second.

Oceanography

Even more significant discoveries came from oceanographers during this interval. Computer fits of the continents at their actual submarine edges proved more striking than at the shorelines (Fig. 10.13). Geologists had assumed that the present ocean basins have been in existence since the beginning of geologic time and, hence, have been receiving a steady fall of sediments. (Oceans had been in existence even in the Precambrian as indicated by marine rocks on the continents.) Unlike the continents with their numerous unconformities, the ocean basins were expected to contain a complete geologic record. Surprisingly, however, not only are there unconformities in the oceanic sections, but no rocks older than Jurassic have been encoun-

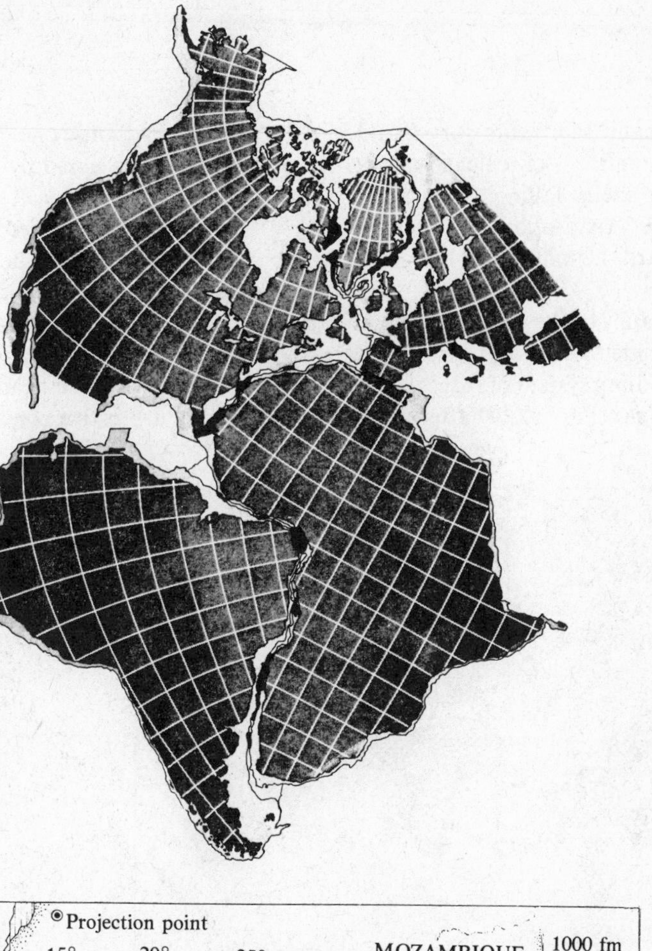

A

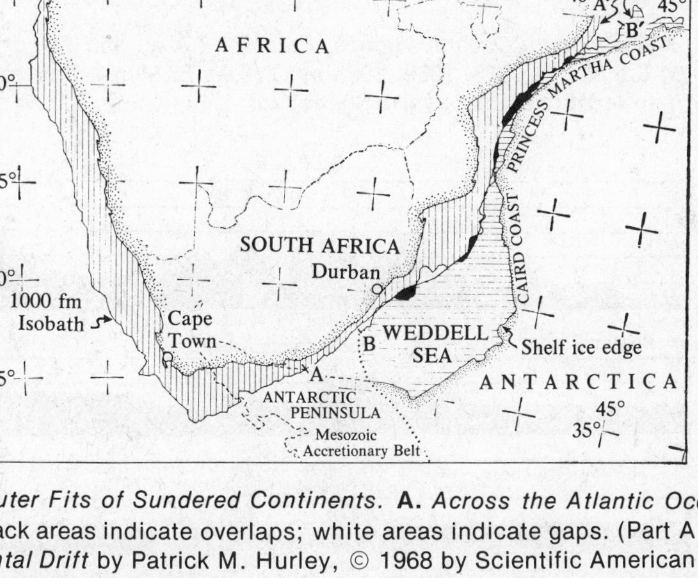

B

Figure 10.13. *Computer Fits of Sundered Continents.* **A.** *Across the Atlantic Ocean,* **B.** *Across the Indian Ocean.* Black areas indicate overlaps; white areas indicate gaps. (Part A from *The Confirmation of Continental Drift* by Patrick M. Hurley, © 1968 by Scientific American, Inc. All rights reserved. Part B from Dietz and Sproll, *Science,* vol. 167, March 1970, p. 1613, © 1970 by AAAS, by permission of the authors and AAAS.)

tered. The oceanic sediments were found to contain great quantities of calcareous debris at depths where shells falling from above today would dissolve. In addition, the rocks regularly are older toward the margins of the ocean basins (Fig. 10.23).

The topography of the ocean floor is not a flat surface from side to side, but instead is crossed by a world-girdling system of ridges, often called **mid-ocean ridges** (Fig. 10.14) after the centrally located Atlantic one. The ridges, whose continuity was first noted by *Maurice Ewing* and *Bruce Heezen,* are generally cracked down the middle by a rift zone characterized by earthquakes, volcanism, and high heat flow (Fig. 10.15). The age of volcanic islands increases outward from the mid-ocean ridges (Fig. 10.16). The ridges go ashore at various places, such as the rift valleys of Africa and the head of the Gulf of California. These are active zones of tension also, and the

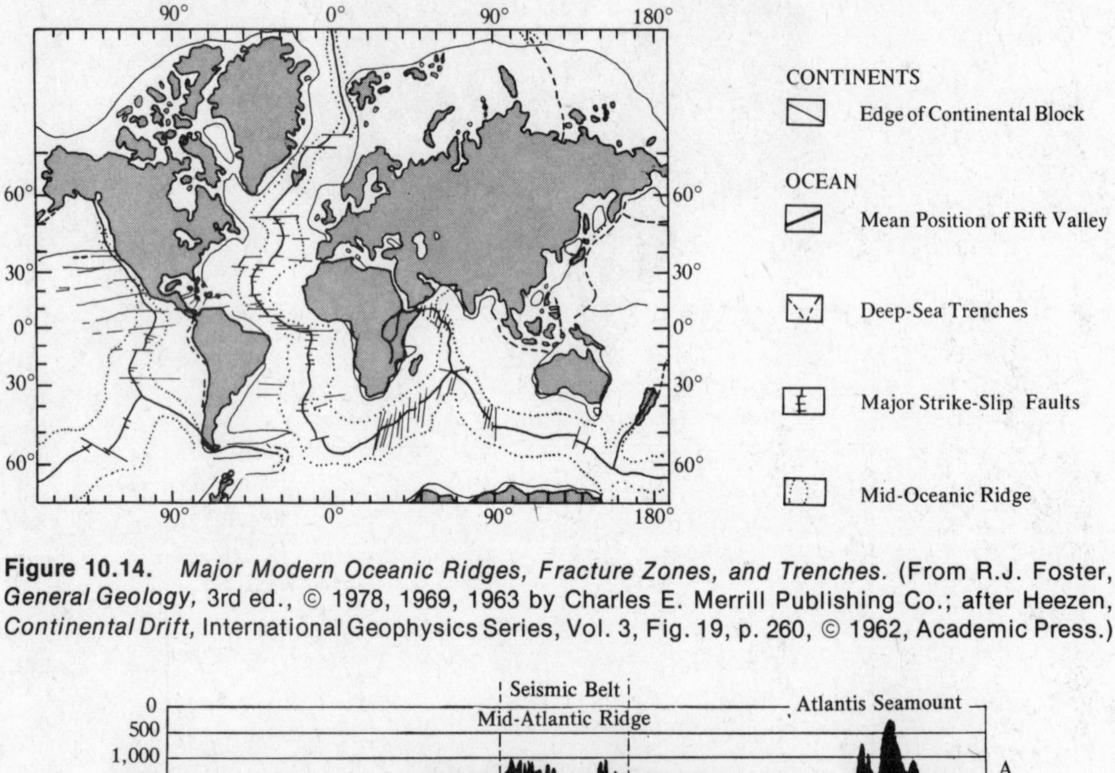

Figure 10.14. *Major Modern Oceanic Ridges, Fracture Zones, and Trenches.* (From R.J. Foster, *General Geology,* 3rd ed., © 1978, 1969, 1963 by Charles E. Merrill Publishing Co.; after Heezen, *Continental Drift,* International Geophysics Series, Vol. 3, Fig. 19, p. 260, © 1962, Academic Press.)

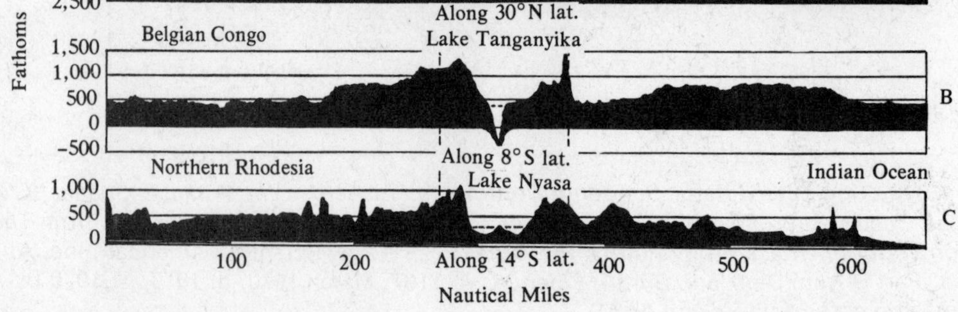

Figure 10.15. *Topographic Profiles Across an Oceanic Ridge and Analogs on the Continents.* **A.** *Atlantic Ocean,* **B–C.** *Rift Valleys of East Africa.* (From Bruce C. Heezen, *The Deep Sea Floor,* © 1962 by Academic Press, N.Y. Used by permission of the author and publisher.)

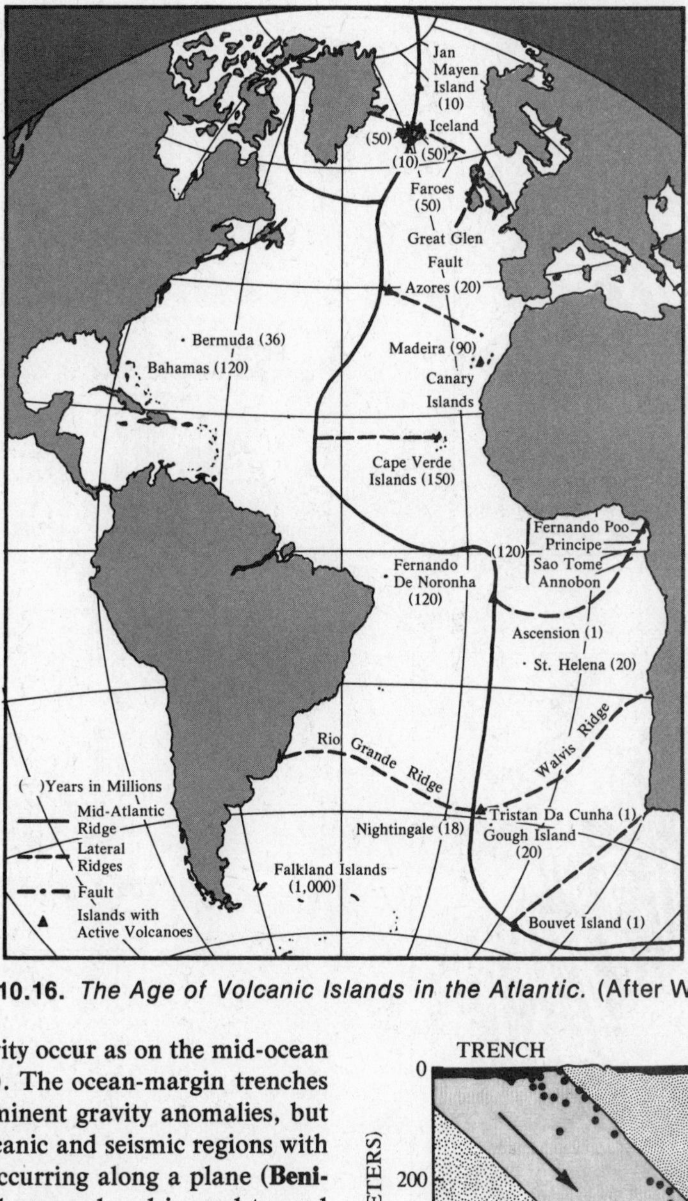

Figure 10.16. *The Age of Volcanic Islands in the Atlantic.* (After Wilson.)

same types of activity occur as on the mid-ocean ridges (Fig. 10.15). The ocean-margin trenches not only have prominent gravity anomalies, but also are active volcanic and seismic regions with the earthquakes occurring along a plane (**Benioff zone**) sloping downward and inward toward the continent (Fig. 10.17).

Another notable discovery was that in the oceans, the intensity of the earth's magnetic field varies. There is a series of parallel bands of alternating intensity (Fig. 10.18). It was discovered that the pattern of these bands is symmetrical about the mid-ocean ridges and the width of each paired set of symmetrical bands is proportional to the duration of the periods of normal or reversed polarity found on the continents.

Finally, the mid-ocean ridges and other ocean basin features are typically offset by large faults;

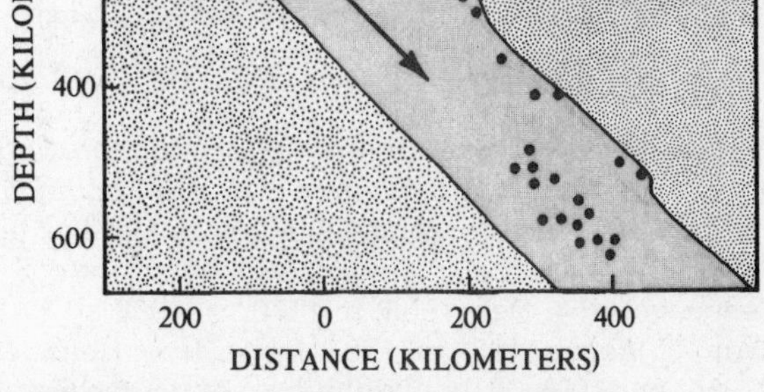

Figure 10.17. *Benioff Zone Beneath a Pacific Trench.* Dots indicate earthquake foci; arrows, direction of movement. (After Heirtzler.)

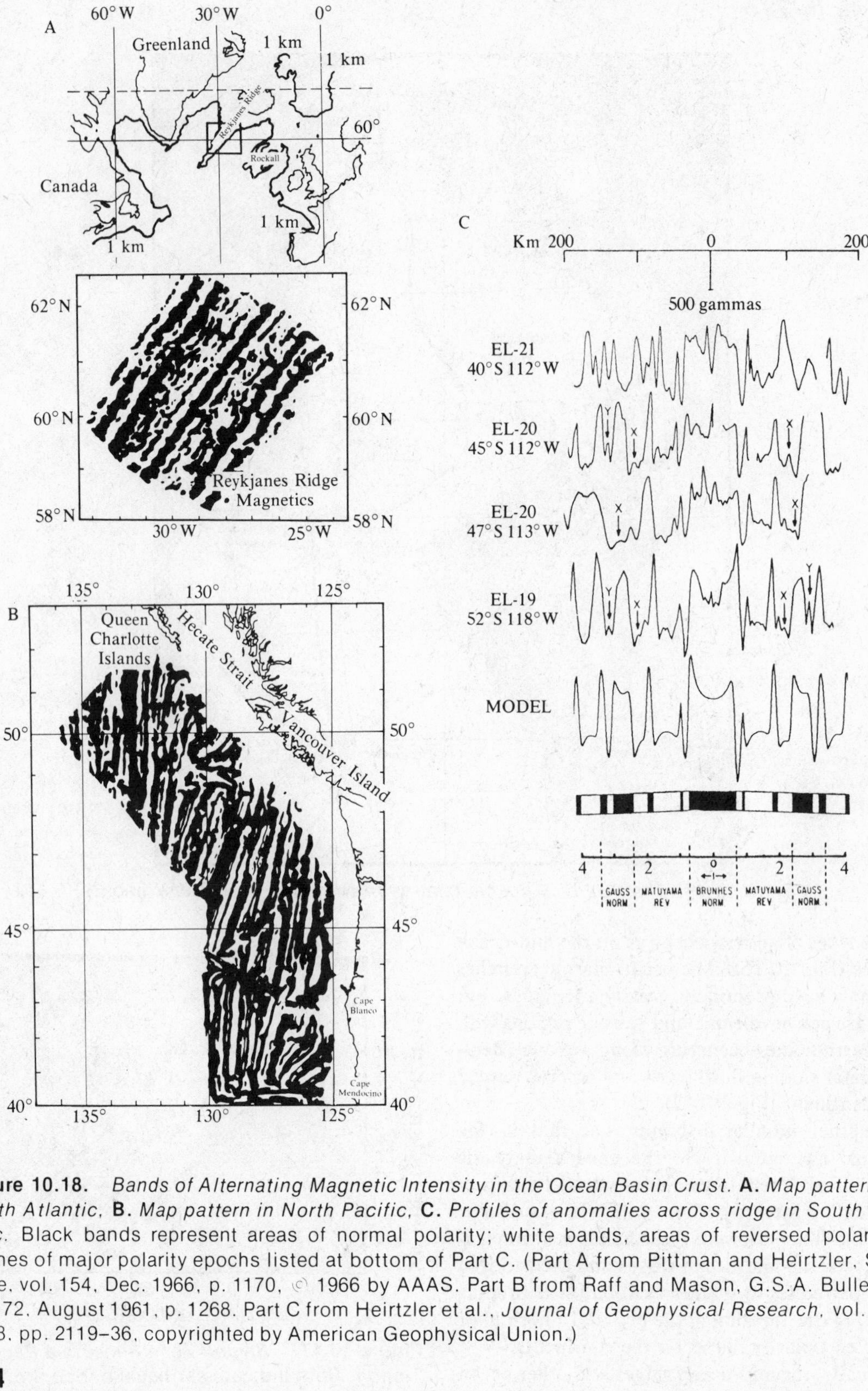

Figure 10.18. *Bands of Alternating Magnetic Intensity in the Ocean Basin Crust.* **A.** *Map pattern in North Atlantic,* **B.** *Map pattern in North Pacific,* **C.** *Profiles of anomalies across ridge in South Pacific.* Black bands represent areas of normal polarity; white bands, areas of reversed polarity. Names of major polarity epochs listed at bottom of Part C. (Part A from Pittman and Heirtzler, Science, vol. 154, Dec. 1966, p. 1170, © 1966 by AAAS. Part B from Raff and Mason, G.S.A. Bulletin, vol. 72, August 1961, p. 1268. Part C from Heirtzler et al., *Journal of Geophysical Research*, vol. 73, 1968, pp. 2119–36, copyrighted by American Geophysical Union.)

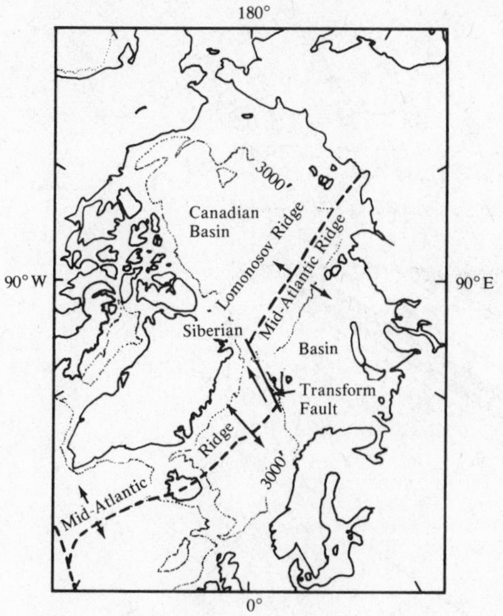

Figure 10.19. *Oceanic "Fracture Zone" Fault.*

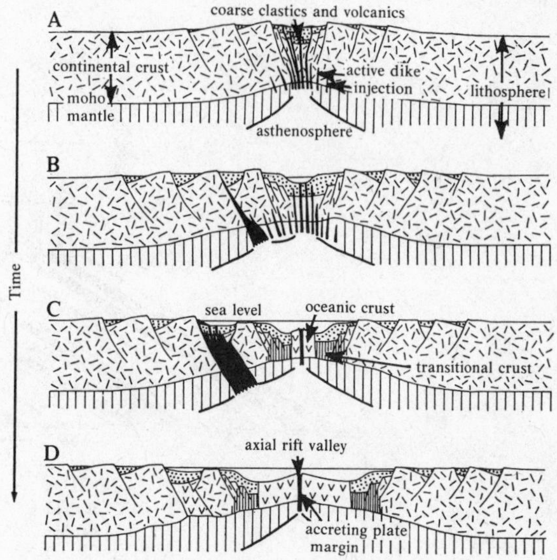

Figure 10.20. *Development of an Oceanic Ridge by the Rupturing of a Continent.* The Moho is the crust-mantle boundary and lies within the lithosphere. See text for explanation of lithosphere and asthenosphere. (From Dewey and Bird, *Journal of Geophysical Research*, vol. 75, no. 14, May 10,1970. By permission of authors and AGU.)

but when the direction of movement of earthquakes on these faults was determined, it was the reverse of the direction of offset (Fig. 10.19).

All these observations have been synthesized in the past ten years into a coherent theory of the evolution of the earth's crust called **plate tectonics** or **global tectonics.**

Plate Tectonics

According to this theory, new crustal material is constantly being generated at the ridge centers where the crust is thin (Fig. 10.20). Lava wells up from below, cools below the so-called Curie point, and the magnetism of the earth's field at the time is frozen in the rock. The constant rising of magma into the tensional rift zones of the ridges is responsible for their vulcanism. This material gradually moves outward from the ridge in both directions. As it does, new magma wells up from below and solidifies, also taking on the magnetism prevailing at the time of its formation. If the earth's two poles have reversed, the rock acquires magnetism in this new direction.

This explains, as *F.J. Vine* and others have shown, the striped pattern of high and low magnetic intensities symmetrical about the ridges (Fig. 10.21). Where the strip is magnetized in the direction of earth's present field, the effect is additive and strong magnetic intensity results. Where the strip is magnetized in the opposite direction, it subtracts from the present magnetic intensity, leaving a low value. The various strips can be matched with the normal and reversed periods of magnetization dated on the continents (Fig. 10.22). (Those older than Late Cenozoic are dated by extrapolation, assuming a constant spreading rate, but radiometric tie-points indicate the assumption is generally valid back into the Cretaceous.) Thus, the age of these strips of the ocean floor can be dated simply by towing a magnetometer behind a ship (Fig. 10.23).

The sediment falling on the new crust near the ridges forms in relatively shallow water. Cal-

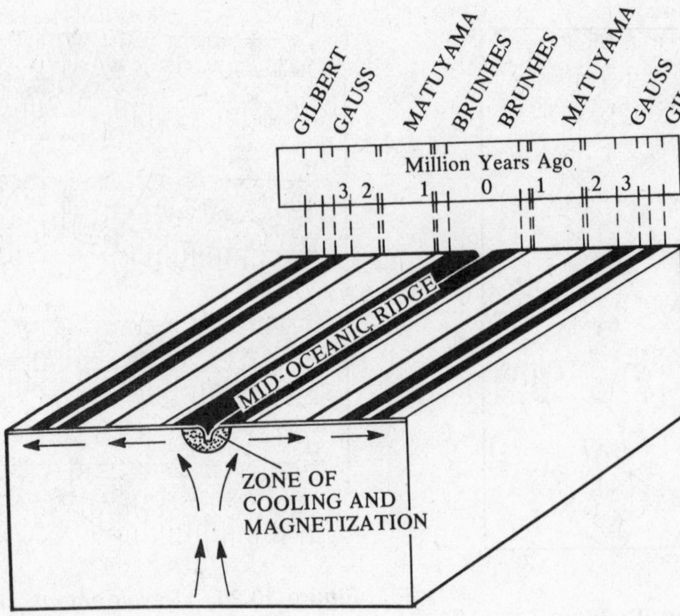

Figure 10.21. *Formation of Magnetic Anomaly Patterns.* Names of polarity epochs and radiometric time scale shown above the block.

cium carbonate shells of pelagic organisms (Fig. 10.24), such as those of coccolithophores, globigerine foraminifera, and small snails called pteropods, accumulate in great numbers. When these move out onto the abyssal floor, they enter deeper waters in which the calcareous shells no longer accumulate, though siliceous ones such as those of radiolaria and diatoms will, if abundant in the surface waters. As a result, the calcium carbonate is slowly dissolved, leaving behind the insoluble residue called brown or red clay that characterizes the more marginal zones of the ocean basins.

The spreading of the sea floor explains why both volcanic and sedimentary rocks are older away from the ridges and why volcanic islands change the composition of their lavas as they migrate across the ocean floors, tapping magmas of different compositions. It also explains the formation of **guyots**, slow-growing volcanic seamounts with tops planed off as they moved outward and, hence, downward away from the ridges (Fig. 10.25). Finally, sea-floor spreading accounts for the thinness of the crust beneath mid-ocean rift zones.

Rifts can open either in ocean basins or under continents. In the latter case, a continent is torn asunder and a new ocean basin gradually develops in between (Fig. 10.20). This is now occurring in Africa (Arabia has already been torn off, creating the Red Sea) and occurred in the past when Gondwanaland ruptured, forming the Atlantic and Indian oceans. Each block of the earth's crust (and some of the upper mantle) moving outward from a ridge is called a **plate,** whether it consists only of an ocean basin or an ocean basin and a continent. All the plates together compose the **lithosphere.** The very viscous fluid mantle material on which they move has been named the **asthenosphere**, or the rheosphere.

Two reasons have been advanced to account for plate motion. Either plates are being pushed outward from the ridges by the rising magma generated by a convection current or rising plume of hot mantle, or they are being dragged into the trenches by the sinking, cooling, heavier margin of the plate. The "push" hypothesis makes it difficult to account for the nearly vertical tension cracks in the mid-ocean rift zones

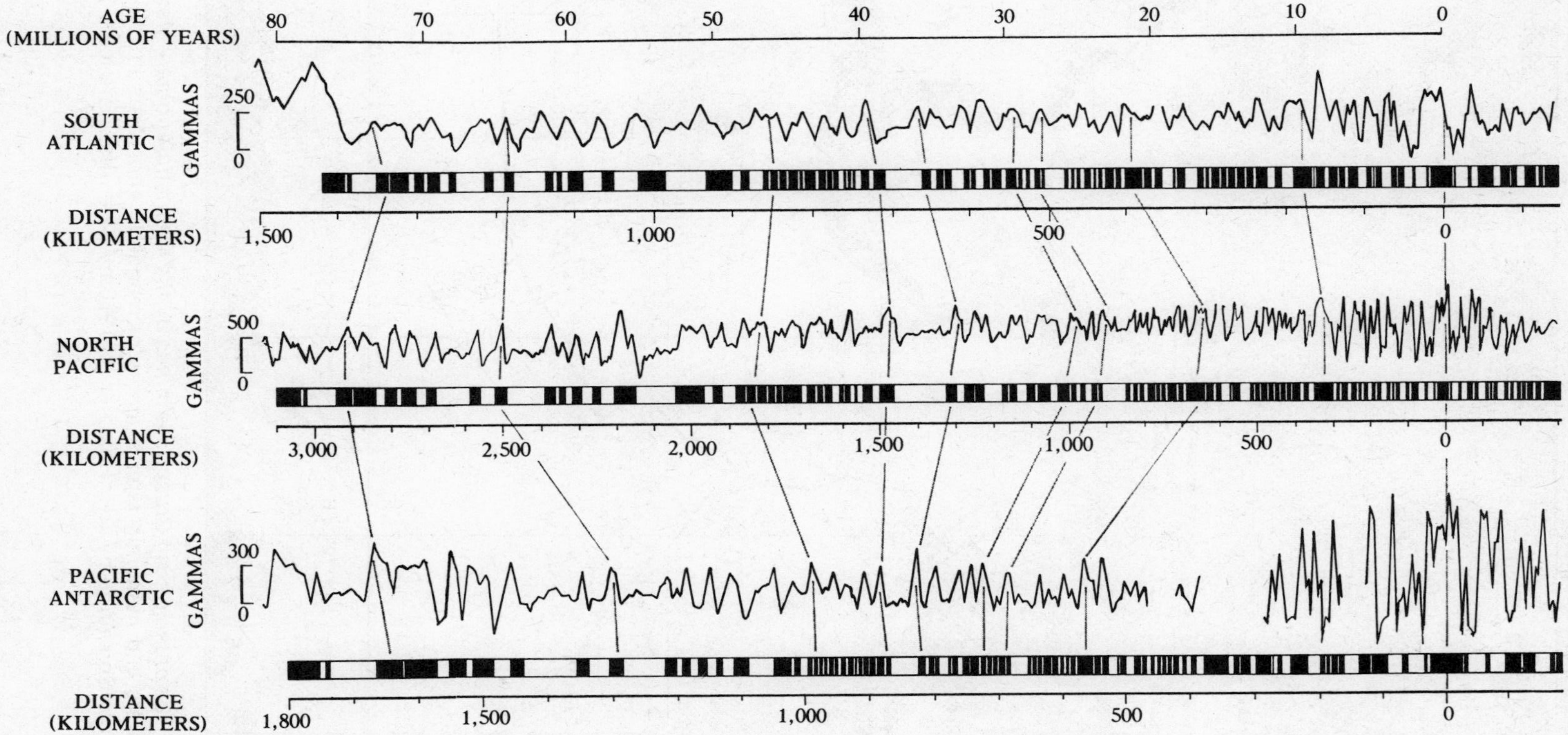

Figure 10.22. *Dating Oceanic Crust by the Magnetic Anomaly Method.* Topographic profiles corresponding to anomaly patterns are shown. Black bars represent normal polarity; white, reversed polarity. Only left half of each pattern shown. (From *Sea-Floor Spreading* by J.R. Heirtzler, © 1968 by Scientific American, Inc. All rights reserved.)

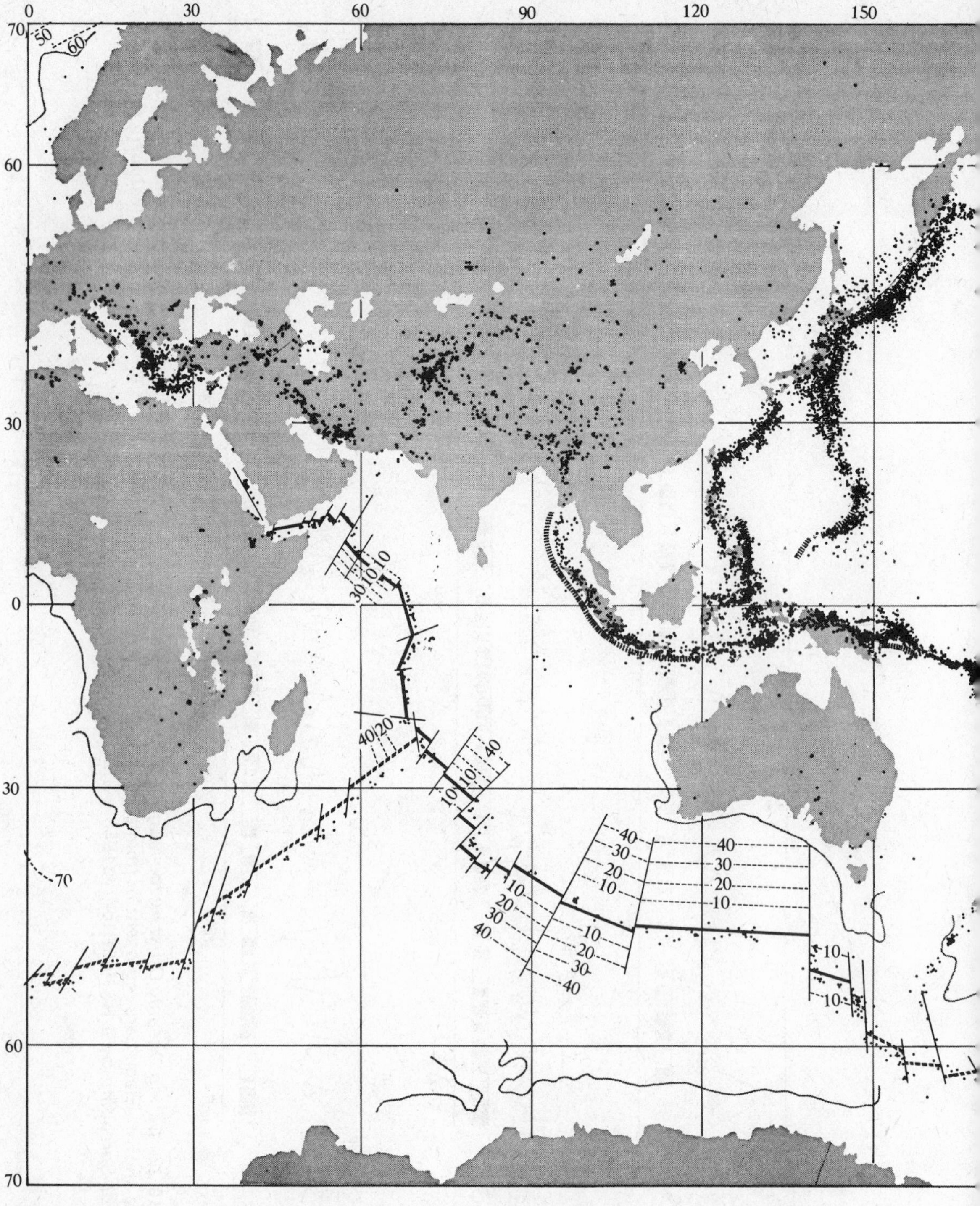

Figure 10.23. *The Age of the Ocean Crust Determined by the Magnetic Anomaly Method.* Dots are earthquake epicenters and numerals refer to dates of magnetic anomalies in millions of years ago. (From *Sea-Floor Spreading* by J.R. Heirtzler,)

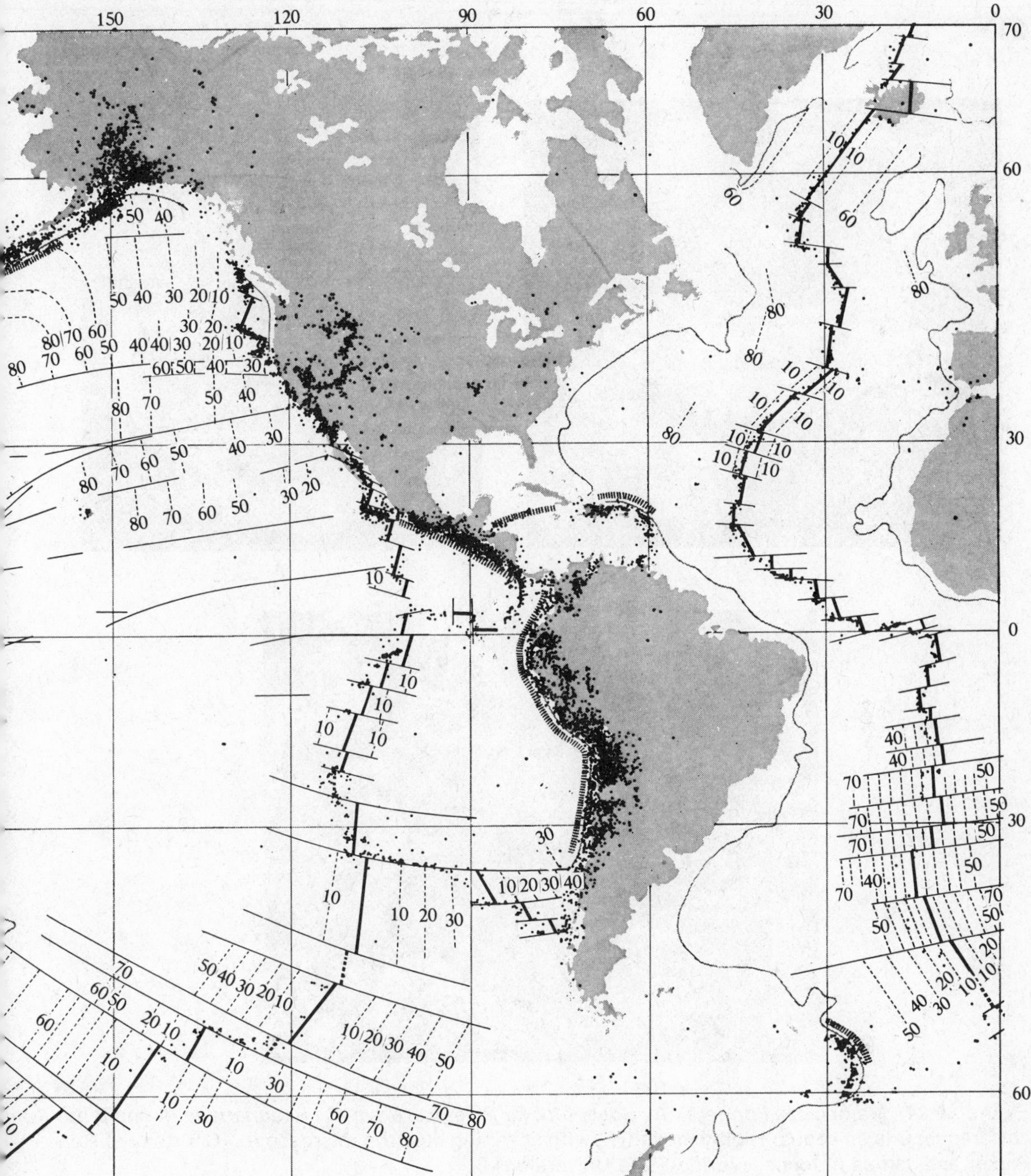
150
120
90
60
30
0
70
60
30
0
30
60
70

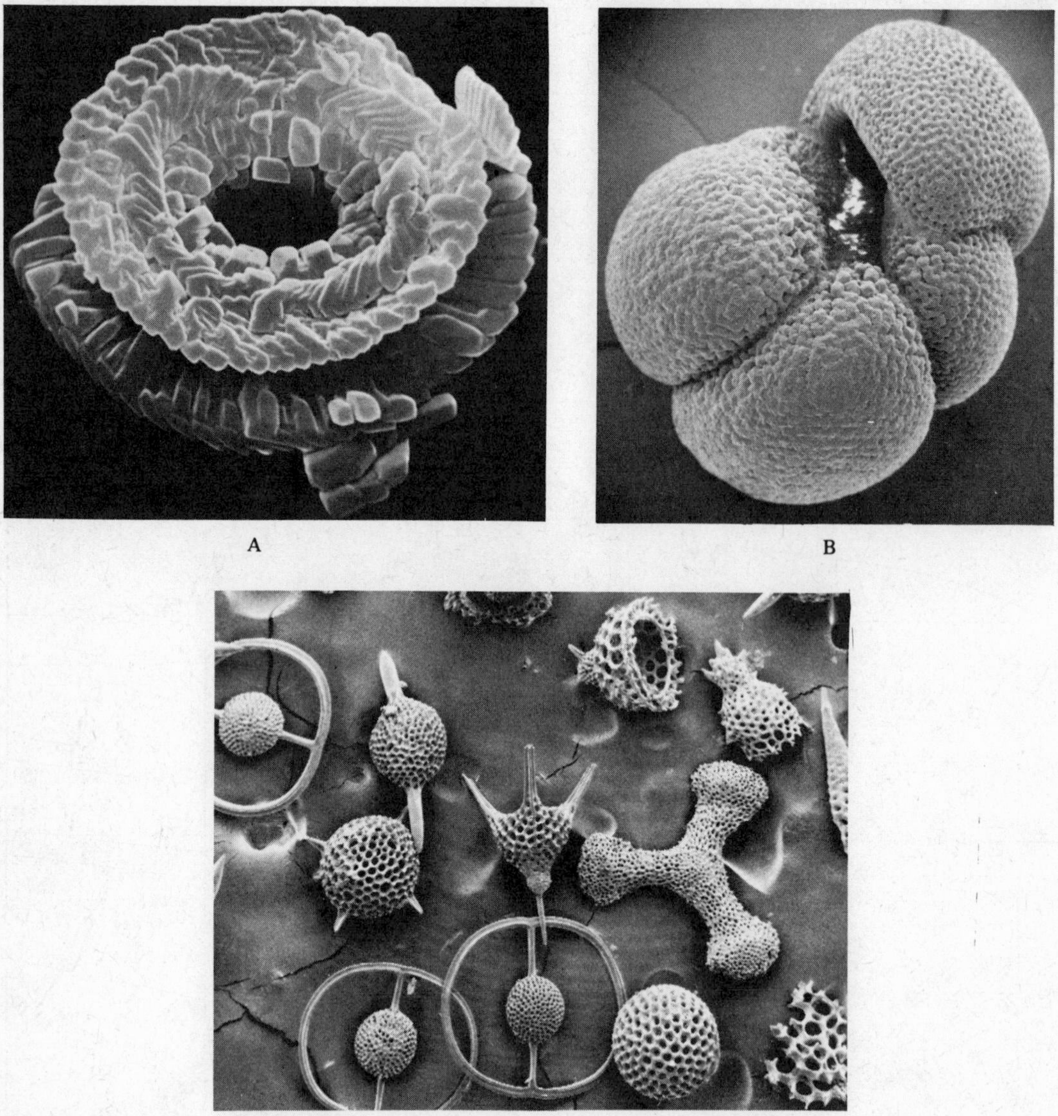

A B

C

Figure 10.24. *Major Ooze Formers:* **A.** *Coccolith*, **B.** *Pelagic foram*, **C.** *Radiolarians*. A and B are calcareous; C is siliceous. Highly magnified with scanning electron microscope. (Courtesy of Ron Parker and James I. Jones, Florida State University.)

(Fig. 10.20). The "pull" hypothesis can account for them and the conditions in the trenches. In a trench, a denser oceanic plate dives under the lighter continental margin of the other plate (Fig. 10.26). This creates the inclined Benioff zone of deep focus earthquakes. As the plate margin sinks, two factors come into play: the increasing pressure which tends to increase the density of the sinking edge or subduction zone and the increasing temperature which tends to decrease its density. The pressure effect should occur more rapidly than that of the temperature, thus increasing the sinking. This would pull the plate away from the ridge and toward

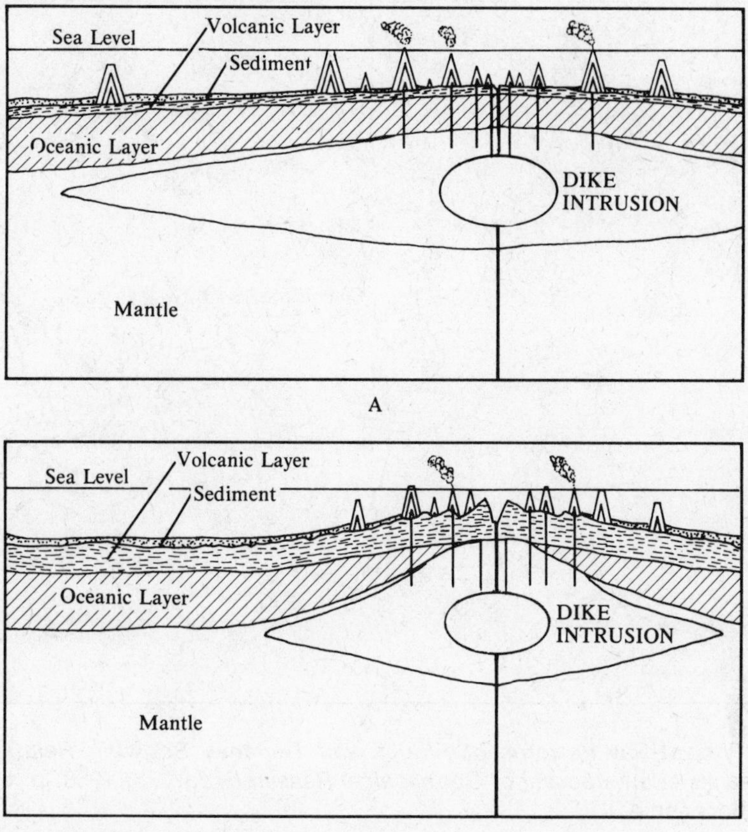

B

Figure 10.25. *The Formation of Guyots on* **A.** *Fast-* and **B.** *Slow-spreading ridges.* Part A represents the current Pacific Ocean and Part B the Atlantic. (After Menard.)

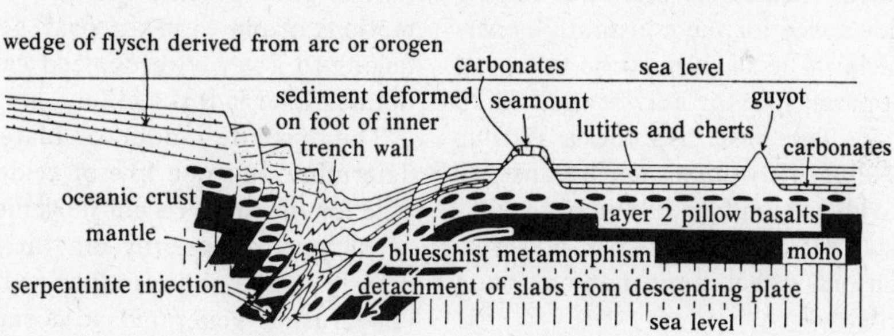

Figure 10.26. *Cross Section of a Modern Trench.* (From Dewey and Bird, *Journal of Geophysical Research,* vol. 75, no. 14, May 10, 1970. Used by permission of authors and AGU.)

the trench. The rapid descent of the cold crust into the high-pressure and temperature zone also metamorphoses and melts the sinking plate edge, as we saw in Chapter 8.

One difficulty with the pull hypothesis is that there are no trenches between some of the earth's ridges. For example (Fig. 10.27), there is no trench between the Mid-Indian and Mid-At-

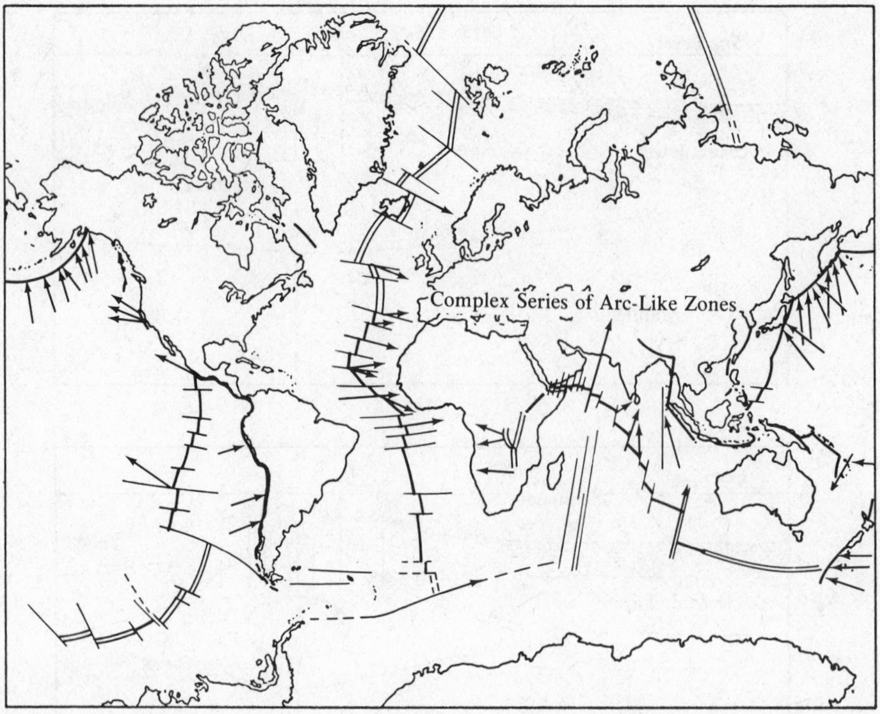

Figure 10.27. *Present-Day Patterns of Ridges and Trenches Showing Relative Plate Motions.* (From Bryan Isacks et al., *Journal of Geophysical Research*, vol. 73, 1968, p. 5861. Used by permission of authors and AGU.)

lantic ridges (in fact, new ridges are forming beneath the African rift valleys). In this instance, the ridges themselves must be moving apart to provide space for the constantly generated new material. The slack must be taken up somewhere, apparently in the Pacific and Indian ocean trenches. The push hypothesis is thus superior in dealing with this issue, because all plates have a ridge boundary but many lack a trench boundary. (It may be possible, however, that one trench may affect more than one plate by its high activity.)

Regardless of whether the surface plate is pushed, pulled, or a combination of both, there must be a compensating reverse flow of mantle material to complete the cycle. The type and depth of this flow are still matters of major debate among geologists. Some view the reverse flow as a surficial phenomenon; others, as a deep-seated process involving much of the mantle.

The side boundaries of plate are termed **megashears,** which are essentially gigantic strike-slip faults. They provide guidelines to the relative motions of plates in the past. In addition, they belong to a newly recognized category of faults defined later in this section.

The absolute motion of plates is difficult to determine, but one line of evidence seems reliable for the more recent movements. Hot spots, or plumes, apparently may arise from a deep mantle source and remain fixed in position. As new crust is generated adjacent to the plume, basaltic volcanic outpourings occur. The trace of these flows leaves a **thread ridge,** or **nematath,** extending from the site of the hotspot outward across the newly formed plates on either side of it (Fig. 10.28). The Walvis-Rio Grande Ridges of the South Atlantic and the Chagos-Laccadive Ridge of the Indian Ocean appear to be nemataths and thus allow the absolute positioning of South America, Africa, and India since the

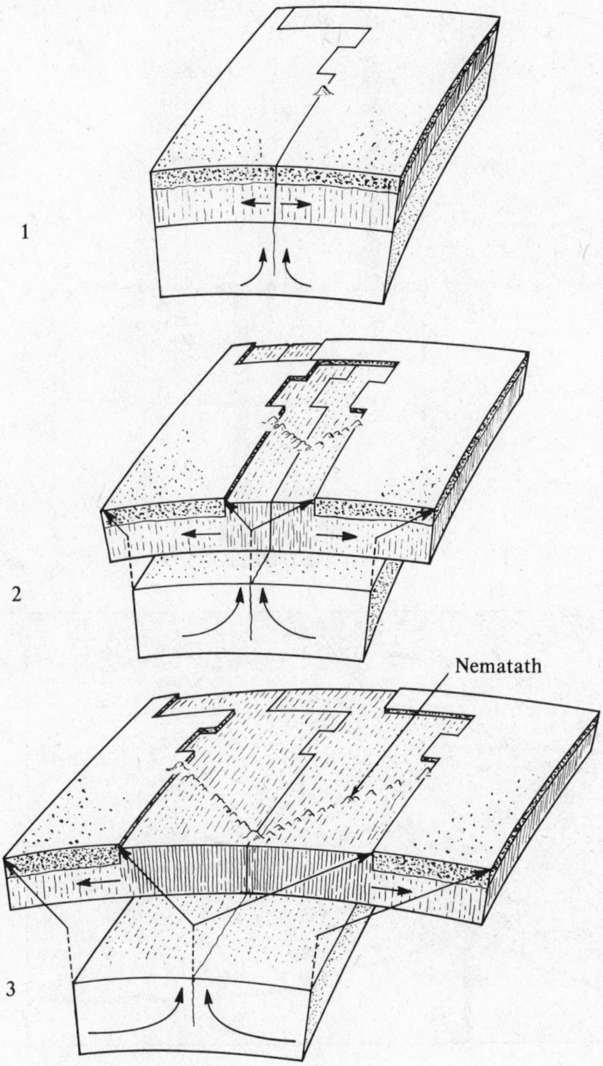

Figure 10.28. *Formation of Nemataths, or Thread Ridges.* The oldest stage is at the top; the youngest, at the bottom. (After Dietz and Holden.)

Jurassic. By coupling these data with that from the magnetic anomalies and megashears, the positions of the other continents can be determined with some accuracy (Fig. 18.4).

The details of the structure and composition of consuming (colliding) and accreting (separating) plate margins were discussed and illustrated in Chapter 8.

As noted, several plate boundaries are zones where the involved plates slide past one another rather than colliding edge-on. In addition, ridges and trenches are offset along their length by similar transverse faults. As they do, where one of these fault zones crosses a ridge, the two offset parts of a ridge continue to form new material at their centers, and this material spreads outward from both (Fig. 10.29). The material on each side of the fault zone moves from the ridge on its side in the direction of the ridge on the opposite side. This produces the anomalous pattern of ridges offset in one sense with the motion of present earthquakes in the opposite sense.

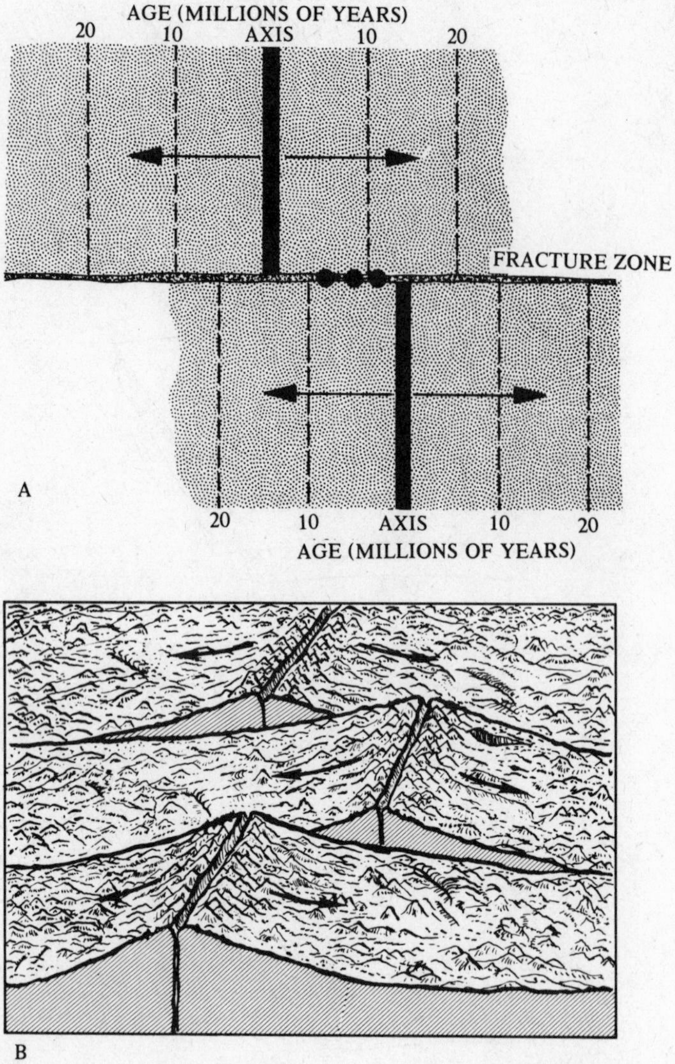

Figure 10.29. *Formation of Transform Faults.* **A.** *Map view with dots representing earthquake epicenters,* **B.** *Oblique view of ocean basin floor with arrows showing direction of movement.* (Part A after Heirtzler and Part B after Menard.)

These peculiar faults have been named **transform faults** by *J. Tuzo Wilson.* They are best known in the ocean basins where they offset most ridges, but also occur on land. For example, the San Andreas Fault connects the ridge in the Gulf of California, Mexico with the Gorda Rise off northern California.

Thus, we can see that the older names of continental drift and sea-floor spreading are not inclusive enough terms for the new global tectonics. Rigid plates with continents and/or ocean basins on them are moving, not just "floating" continents or "conveyor-belt" ocean basins. This is why the name plate tectonics has been widely adopted.

The Plates Today

There are eight major and several minor plates of the earth's crust today (Fig. 10.30). The major

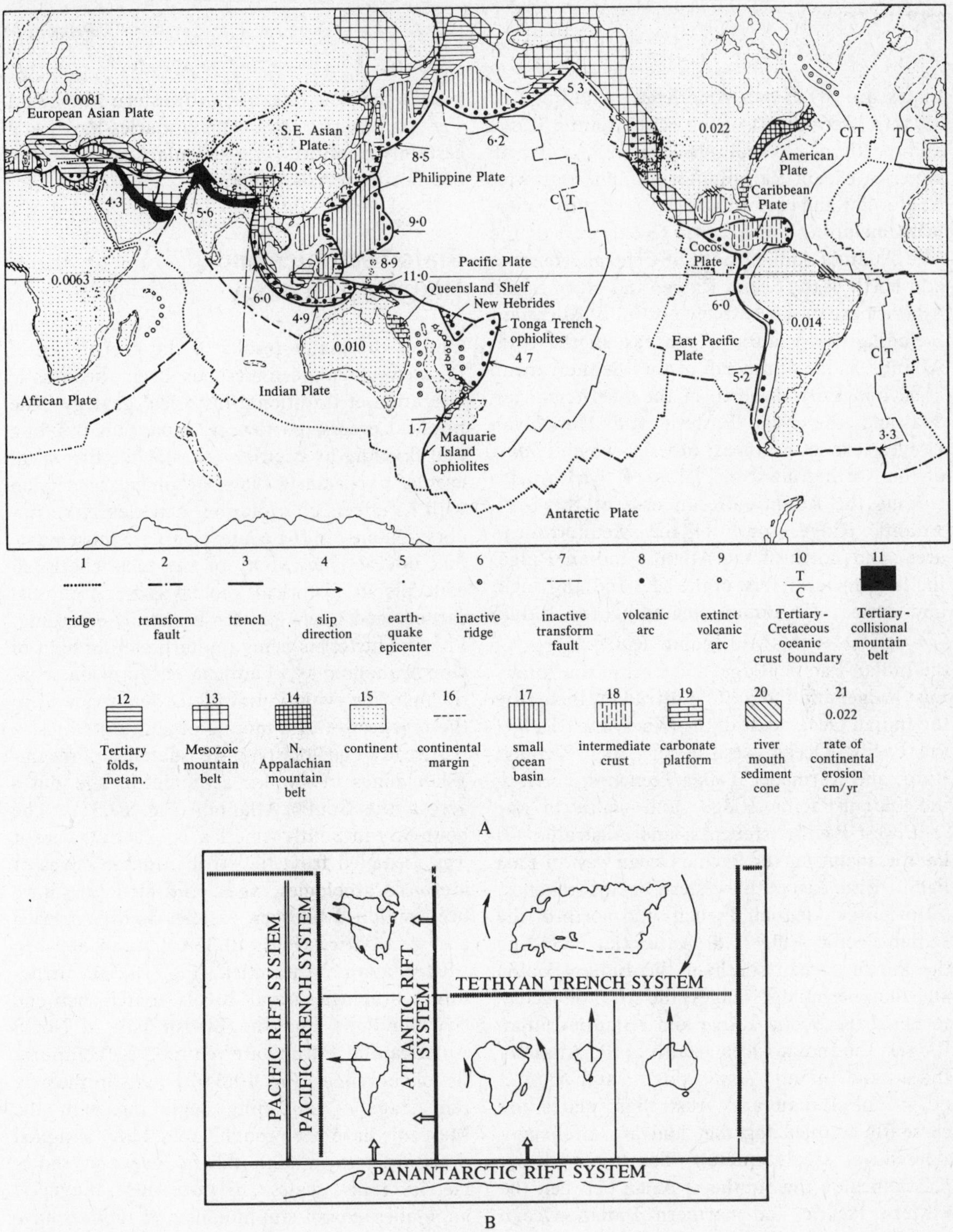

Figure 10.30. *Lithosphere Plates Today.* **A.** *Plates, margins, and their relation to geologic structures;* **B.** *Diagrammatic representation of plates, margins, and movements.* Note that the Indian and Australian plates are assumed to be coupled in Part A. (Part A from Dewey and Bird, *G.S.A. Bulletin*, vol. 75, May 1970; Part B from Deitz and Holden, *Journal of Geophysical Research*, vol. 75, 1970, p. 4954. Used by permission of authors and societies.)

plates are (1) **North American,** including the Atlantic Ocean west of the Mid-Atlantic Ridge and north of Central America; all of North America except Baja California and southwest California; and eastern Siberia; (2) **South American,** including the Atlantic Ocean west of the Mid-Atlantic Ridge, south of Central America, and north of the Scotia Ridge; and all of South America except the extreme north; (3) **Eurasian,** including the Atlantic Ocean east of the Mid-Atlantic Ridge and north of the Mediterranean area; and Eurasia north of the Mediterranean Sea and the Near Eastern and Himalayan ranges (except northeast Siberia, which is part of the North American Plate); (4) **African,** including the Atlantic Ocean east of the Mid-Atlantic Ridge, south of the Mediterranean area, and north of the Atlantic-Indian Ridge; the Indian Ocean west of the Mid-Indian Ridge; and Africa; (5) **Indian,** including the Indian Ocean east of the Mid-Indian Ridge, north of the Indian-Pacific Ridge, and west of the Ninety East Ridge; and India; (6) **Australian,** including the Indian Ocean east of the Ninety East Ridge; the Pacific Ocean west of the New Zealand Ridge and Kermadec-Tonga Trenches, north of the Indian-Pacific Ridge, and south of the southwest Pacific trenches; and Australia; (7) **Pacific,** including the Pacific Ocean west of East Pacific Rise, east of the western Pacific trenches, south of the Aleutian Trench, and north of the Indian-Pacific Ridge; (8) **Antarctic,** including the Pacific Ocean south of the Indian-Pacific and Juan Fernandez ridges; the Atlantic Ocean south of the Scotia Ridge and Atlantic-Indian Ridges; the Indian Ocean south of the Atlantic-Indian and Indian-Pacific ridges; and Antarctica. (The Indian and Australian plates are currently coupled together and are often considered a single plate.) Somewhat lesser plates include the **Southeast Asian** between the western Pacific and northern Indian Ocean trenches and the Himalayan-central Asia mountain chains; the East Pacific (**Nazca**) between the East Pacific Rise, the Peru-Chile Trench, the Galapagos Ridge, and the Juan Fernandez Ridge; the **Caribbean** Plate; the **Arabian** Plate; the **Philippine Sea** Plate; and some smaller ones in the Near East and off western North and Central America. Not all boundaries have been certainly identified yet, particularly in the Arctic Sea-northern Asia region.

Plate Tectonics and Historical Geology

One of the major results of the acceptance of the new global tectonics has been the drastic alteration of traditional historical geology. The physical events, particuarly mountain building and flooding by epeiric seas, and the biological events, particularly biogeographical dispersion with its effects on evolution, can clearly be understood only in the context of plate movements and interactions. Many of the most-cherished concepts in historical geology have been discarded, and a new pattern is rapidly emerging. Physical history is being reinterpreted in light of Gondwanaland and Laurasia and previous continents. The evolution of life is being viewed in the perspective of a mosaic of shifting plates.

Already, geologists have matched Precambrian zones of similar radiometric age dates across the South Atlantic (Fig. 10.31). The boundary in South America is exactly where it was predicted from the African data. Zones of identical lithologies, ages, and structure have been matched between eastern South America and west Africa (Figs. 10.32A–C) and between Australia and Antarctica (Fig. 10.33). Structures, lithologies, and fossils match between Newfoundland and the British Isles if North America and Europe are reunited by computer fits of shorelines (Fig. 10.34). Events in the current stage of spreading beginning with the Mesozoic have been roughly dated and assigned proper places in both the relative and radiometric time scales by utilizing magnetic anomalies. Fossil amphibians and reptiles have been found in Antarctica, a continent to which they could never have migrated with its present isolated geographic situation. Plate tectonics may also explain the epeiric sea inundations of the continents. The parting of plates along a ridge produces a shallower ocean basin than

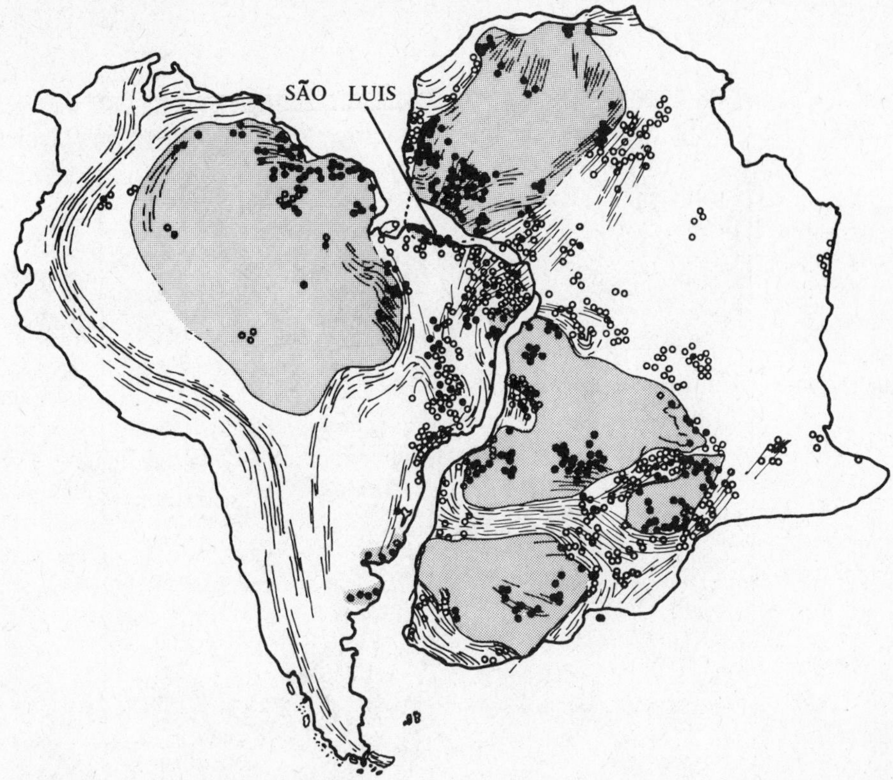

Figure 10.31. *Match of Cratons and Belts of Deformation.* Dark areas are cratons; lineations indicate structural trends in mountain belts. Solid dots are rocks 2,000,000,000 years of age; open dots are younger rocks. Note the correspondence between African and South American age provinces near São Luis, Brazil. (Dates determined radiometrically.) (From *The Confirmation of Continental Drift* by Patrick M. Hurley, © 1968 by Scientific American, Inc. All rights reserved.)

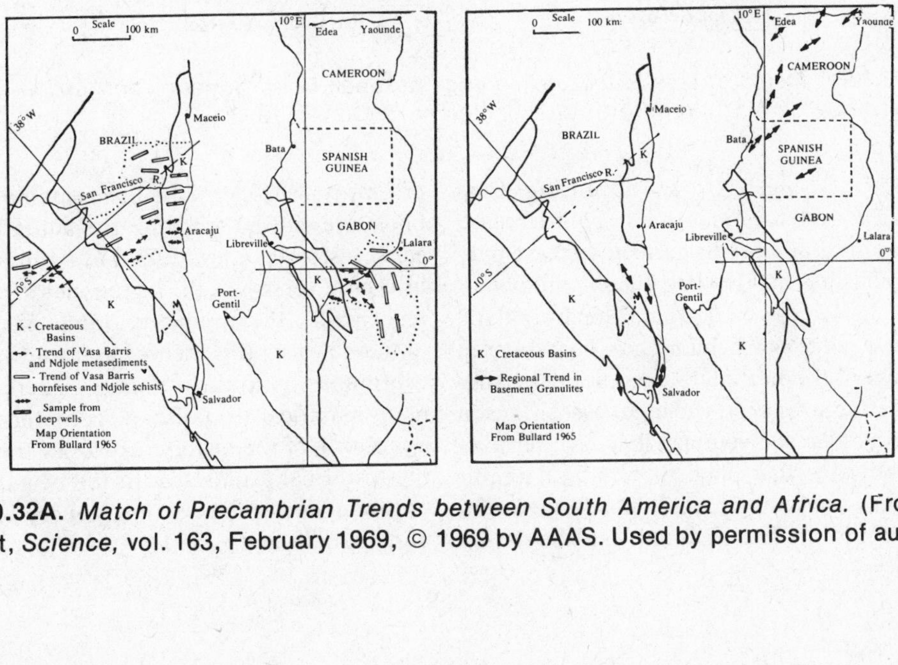

Figure 10.32A. *Match of Precambrian Trends between South America and Africa.* (From Allard and Hurst, *Science,* vol. 163, February 1969, © 1969 by AAAS. Used by permission of authors and AAAS.)

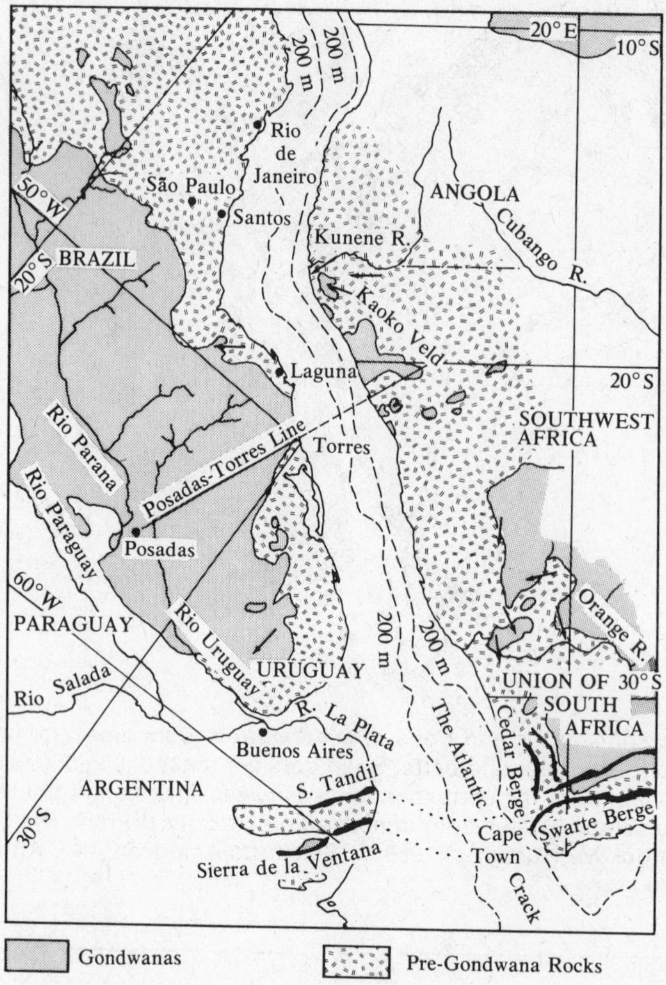

Figure 10.32B. *Match of Late Paleozoic Geology between South America and Africa.* (From *Historical Geology* by A.O. Woodford, W.H. Freeman & Co., © 1965.)

that being consumed by an advancing edge of the split continent (Fig. 10.35). This decrease in oceanic volume may have been at least partially responsible for flooding the continents.

Events before the breakup of Gondwanaland and Laurasia are now beginning to be unraveled in the light of the new plate tectonics. The similarity of structures and lithologic associations in rocks both older and younger than the Permian clearly indicates that plate motion was significant before these supercontinents formed and well back into the Precambrian, but the details are much less clear. In the remainder of this book, we will be studying the actual events in earth history as they unfold. We will attempt to interpret these events in the new framework, recognizing the predictive value of plate tectonics now that this theory has been validated by continuous empirical testing. There are still many significant details not yet understood, and a meshing of the history read from the rocks by the historical geologist with the new discoveries of geophysicists and oceanographers will help us understand them.

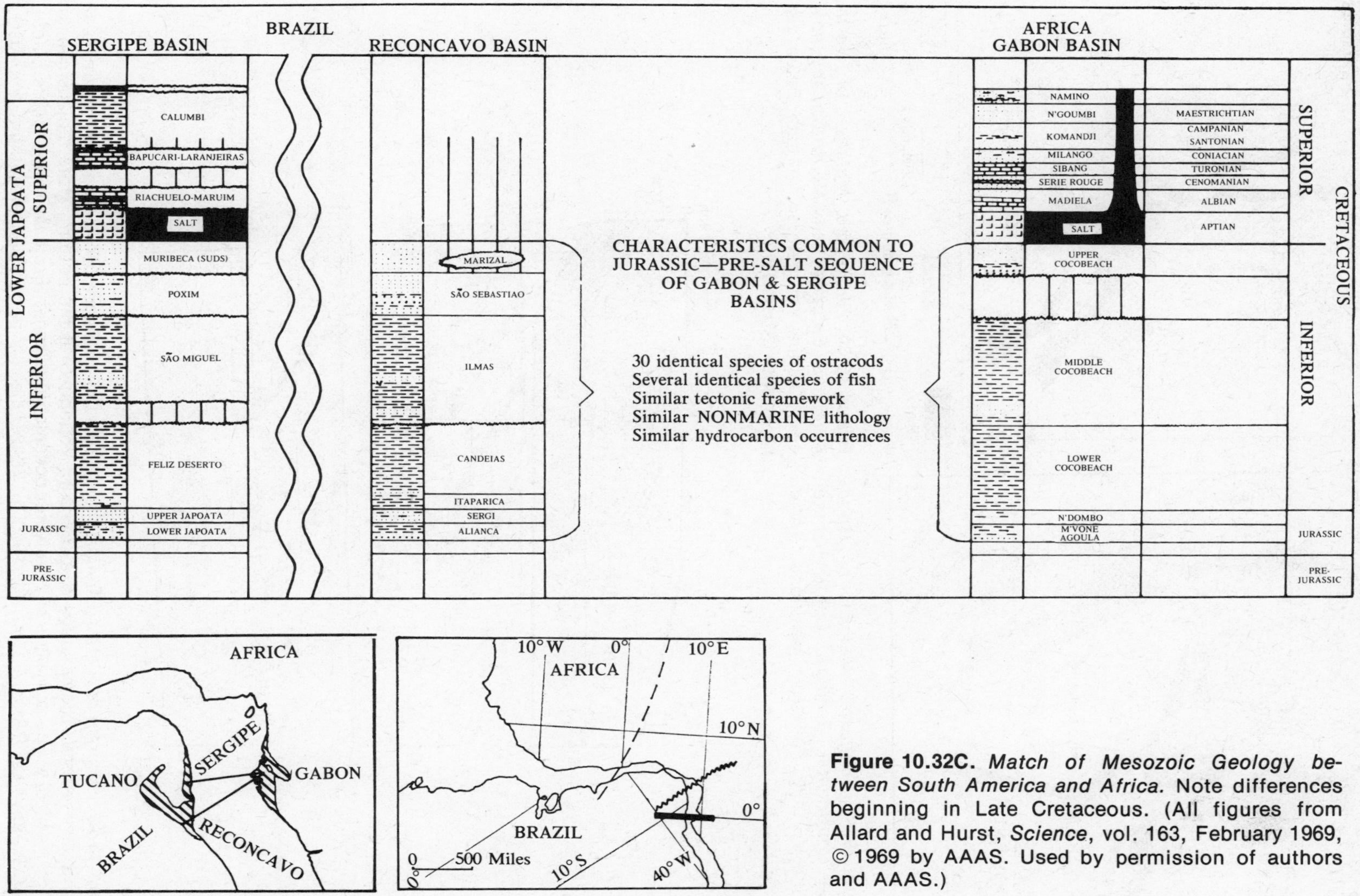

Figure 10.32C. *Match of Mesozoic Geology between South America and Africa.* Note differences beginning in Late Cretaceous. (All figures from Allard and Hurst, *Science*, vol. 163, February 1969, © 1969 by AAAS. Used by permission of authors and AAAS.)

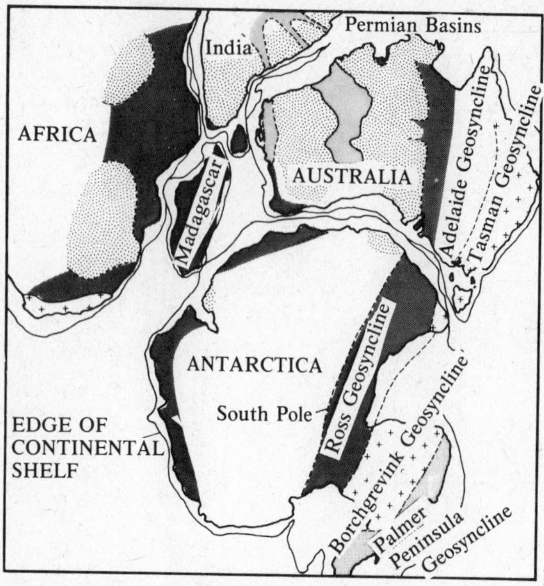

Figure 10.33. *Suggested Matches of Gondwana Continents' Geology.* Continents restored to Late Paleozoic positions. (After Hurley.)

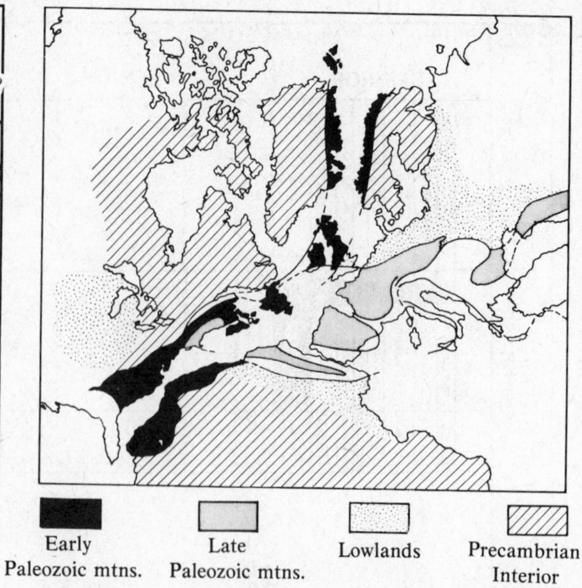

Figure 10.34. *Suggested Matches of North American, European, and North African Geology.* Continents restored to Late Paleozoic positions. (After Hurley.)

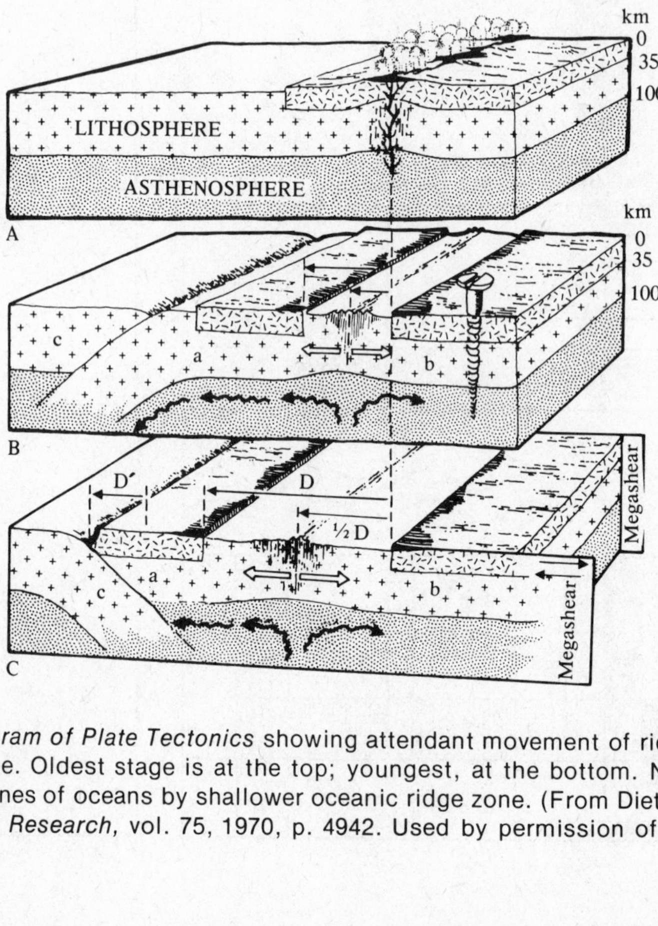

Figure 10.35. *Diagram of Plate Tectonics* showing attendant movement of ridge and decrease in ocean basin volume. Oldest stage is at the top; youngest, at the bottom. Note replacement of deeper marginal zones of oceans by shallower oceanic ridge zone. (From Dietz and Holden, *Journal of Geophysical Research,* vol. 75, 1970, p. 4942. Used by permission of authors and AGU.)

REVIEW

1. What prompted Edward Suess to postulate the existence of **Gondwanaland?**
2. What man is usually considered the father of **continental drift ideas?**
3. Describe the Late Paleozoic-Mesozoic rock sequence of the Gondwana continents.
4. What is the significance of the *Glossopteris* flora?
5. Give at least one example of paleoclimatologic evidence that supports continental dislocations.
6. What is the significance of negative gravity anomalies?
7. How does **paleomagnetism** support the idea that the continents have moved?
8. What are the characteristics of the oceanic ridge systems that suggest they are places where new crust is formed?
9. How have Vine and others explained the alternating bands of magnetic intensity in the ocean floors?
10. What are the strong and weak points of the **pull theory of plate motion?**
11. What is the significance of **nemataths** or thread ridges?
12. What is a **transform fault?** Why do you suppose they form?
13. What are the major plates and their boundaries today?

Part 2

Earth History

11

"In the Beginning": The Origin of the Earth and Life

11

People have speculated about the origin of the earth and living things at least since the dawn of recorded history. All the world's religions have creation myths of some sort, either original or borrowed from their neighbors. The Hebrew creation myth, much of which is derived from other Near Eastern religions and which has become part of the Christian heritage of Western Man, speaks of a world and its inhabitants supernaturally created in six days. By the 1600s and 1700s, however, people were beginning to think in terms of natural processes and longer time spans. In the 1700s, the German philosopher *Emmanuel Kant* speculated that the sun, the earth, and the other planets had condensed from a hot nebular cloud, and the French naturalist *Buffon* suggested that a comet had collided with the sun and spattered out the planets. These two ideas have come down to us in this century as the nebular and encounter hypotheses respectively.

Nebular Hypothesis

The **nebular hypothesis**, as scientifically promulgated by the French mathematician, *P.S. Laplace* in 1820, envisioned a great spherical cloud of hot gases that slowly rotated (Fig. 11.1). As it rotated, it continually speeded up by the law of conservation of momentum and flattened out, producing a thin disk, the ancestral planets, and a still somewhat inflated center, the ancestral sun. When the outward centrifugal force at the rim of the disk exceeded the gravitational attraction of the massive protosun, an outer ring of material detached and the inner mass shrunk. This process repeated itself several times until a ring ancestral to each of the modern planets had formed.

The material in each ring was gradually swept up into a planet, though some remains detached today as meteorites. Each ring, in turn, went through the same process of differentiation, producing satellites. The planets and satellites, being much smaller than the sun, lost their heat more rapidly and developed cold outer crusts. (The continued cooling and shrinking of this originally hot earth was a popular hypothesis to explain mountain building even into this century.) The major problem with the nebular hypothesis was that it indicated the sun should be rotating at a very rapid rate in order to have spun off the planets. Yet ninety-nine percent of the speed of rotation (angular momentum) in the solar system occurs in the rapidly spinning planets and only one percent in the slowly turning sun.

Encounter Hypothesis

The **encounter** or **second-body hypothesis,** given a scientific basis by the geologist *T.C. Chamberlin* and the astronomer *F.R. Moulton* in the early 1900s, dealt with this problem readily. In this hypothesis, another passing star con-

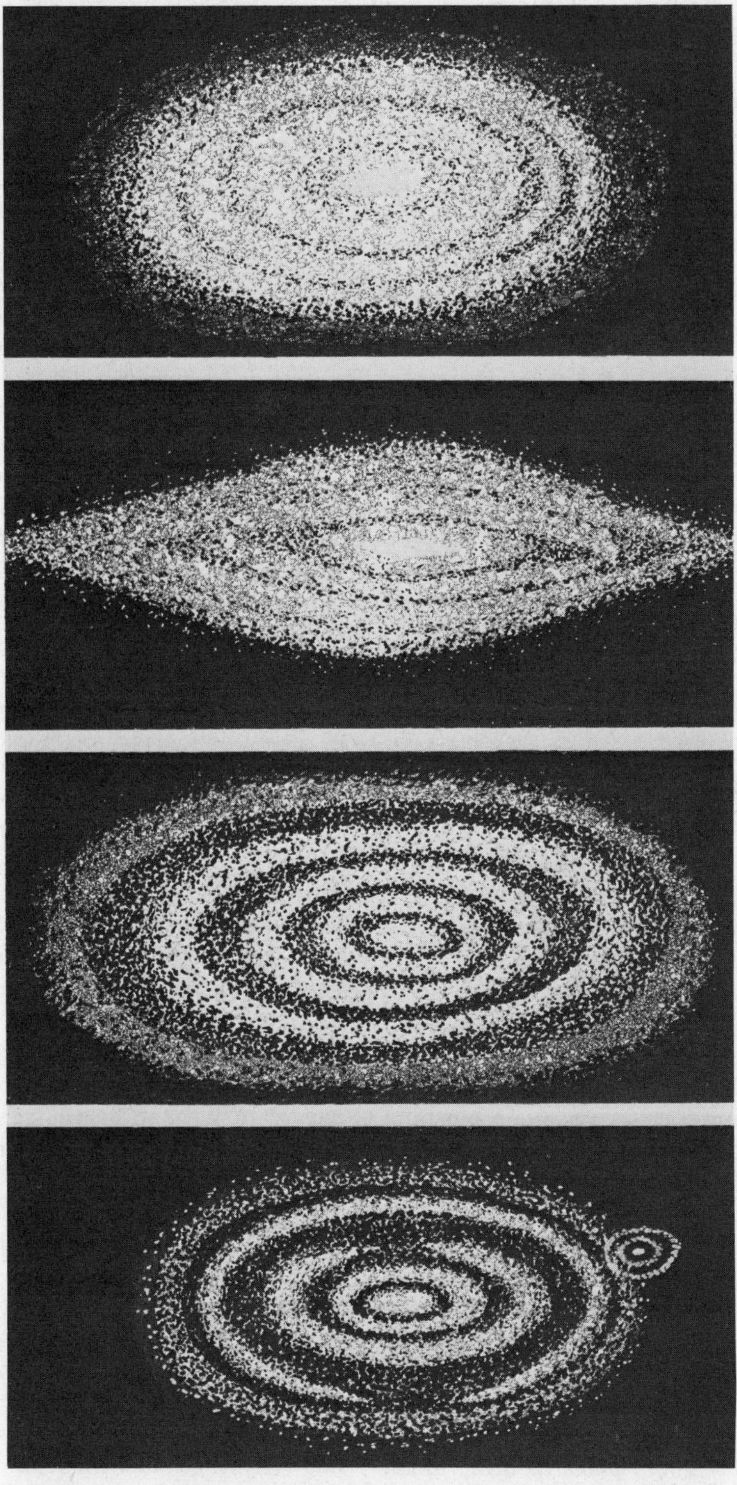

Figure 11.1. *The Nebular Hypothesis of Solar System Origin.* The oldest stage is at the top; the youngest, at the bottom. (After McLaughlin.)

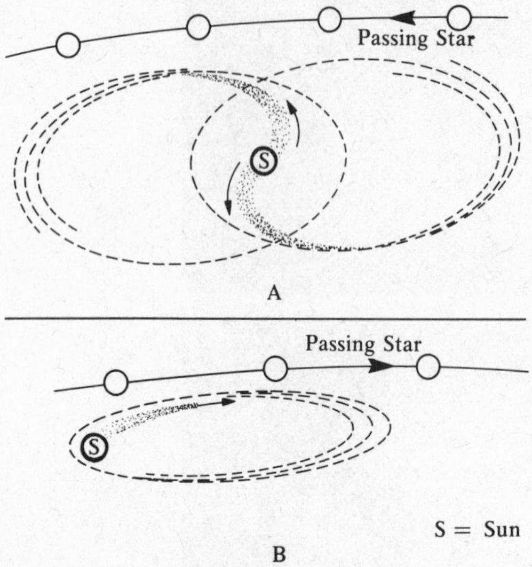

Figure 11.2. *Encounter, or Second-Body, Hypotheses.* **A.** *Original Chamberlin-Moulton version* (1900), **B.** *Modified version of Jeans and Jeffreys* (1925).

tributed greatly to the tidal and eruptive forces of the sun, causing great clouds of gas to be thrown out of the sun's interior into large spirals (Fig. 11.2A). This material was given a great angular momentum and quickly cooled into solids and liquids, which accumulated into planets by collisions. In this hypothesis, then, the planets originally began as cold bodies, in contrast to the nebular hypothesis. Calculations soon indicated the sun has essentially no tides and its eruptions are surface features only. The hypothesis was modified to assume a near collision with the passing star which pulled a long hot filament out of the sun in a very eccentric orbit (Fig. 11.2B). The internal resistance of the cloud eventually produced today's nearly circular orbits. There were several major difficulties with the encounter hypothesis. Calculations indicated that no filament pulled out of the sun this way would ever have enough gravitational cohesion to collect into a planet. Furthermore, the likelihood of any two stars colliding is so statistically remote as to be virtually impossible. Finally, the hypothesis could not adequately explain how the satellites formed.

Modern Theory

The modern theory of the earth's origin, as expounded by *Fred Whipple*, *C.F. von Weizsacker*, and *Gerard Kuiper*, borrows the strong points of both older ideas, but goes further by viewing the earth and solar system in their setting of the galaxy and the universe. The origin of the universe is still widely debated. Opinions fluctuate between the **big-bang hypothesis** of an initial explosion of all matter which is now flying outward from that center and a **continuous-creation hypothesis** in which the processes of forming and destroying matter are continually going on somewhere. Either hypothesis is compatible with the major points of present ideas on solar system formation, and both are probably partly right.

Astronomers claim we can see most stages in the formation of a star and its solar system going on before our eyes in space today. In certain regions of space, interstellar dust particles, composed of frozen nonluminous hydrogen, carbon dioxide, methane, and ammonia, are aggregated together, probably by the pressure of light and other electromagnetic radiation. These **cold diffuse clouds**, probably the initial state of star formation, can be recognized because they shut off the light from stars behind them (Fig. 11.3).

The next stage is represented by small, dark, round nebulae called **globules**, which apparently evolved from the diffuse clouds when the forces of gravitational attraction became effective (Fig. 11.4).

Finally, there are **clouds** of **luminous hot gases** and **dust particles** with bright **stars** scattered in them (Fig. 11.5). These probably evolved from the globules when the amount of aggregated material was large enough for its gravitational attraction to produce the high temperatures and pressures necessary to start the nuclear reactions. These reactions, chiefly the fusion of hydrogen to form helium, are the basis of the energy of stars. At this stage, a star becomes luminous and visible by its own emitted radiation. Stars beyond our own solar system are too distant to determine visually whether they have

Figure 11.3. *Cold Dust Clouds.* Horsehead Nebula in Orion. (Courtesy Hale Observatories.)

Figure 11.4. *Globules.* Rosette Nebula in Monoceras. (Courtesy Hale Observatories.)

Figure 11.5. *Luminous Clouds with Stars.* Great Nebula in Orion. (Courtesy Hale Observatories.)

planets, so the process of planetary formation is based on inference about the collection process of dust clouds into stars.

As now envisioned (Fig. 11.6), a cold nebular cloud that will become a solar system condenses to form a globule. Rotation, a prevalent physical property in the universe, begins thus producing an oblate form. Rotation speeds up, again by the law of the conservation of momentum, and turbulence, another prevalent physical property, develops, forming eddies (Fig. 11.7). The more rapid central spinning is dissipated by communication outward from eddy to eddy. As the cloud continues condensing, a balance between outward centrifugal force and the inward pull of gravity is reached, and cold bodies start to collect in rings around the center. These protoplanets form by collecting debris in their vicinities, including many frozen gases which make them considerably larger than the modern planets. The scars of this sweeping process remain today as great craters on planets with little or no atmospheres (Fig. 11.8). Many uncollected particles still remain as meteorites.

Finally, the gravitational self-compression of the protosun produces the critical temperature and pressure necessary to start the nuclear reactions. A great flood of radiant solar energy, the solar wind, fills the solar system, vaporizing and driving off most of the volatile constituents of the planets. The heat produced by the newly radiant sun influences further condensation, giving rise to the compositional differences among planets. Each ring of cold particles undergoes its own differentiation in the process of planet formation, producing satellites.

The sun's rotation can be further slowed by magnetic braking action. Ions rejected from the sun encounter magnetic lines of force generated by the sun's magnetic field (Fig. 11.9). These

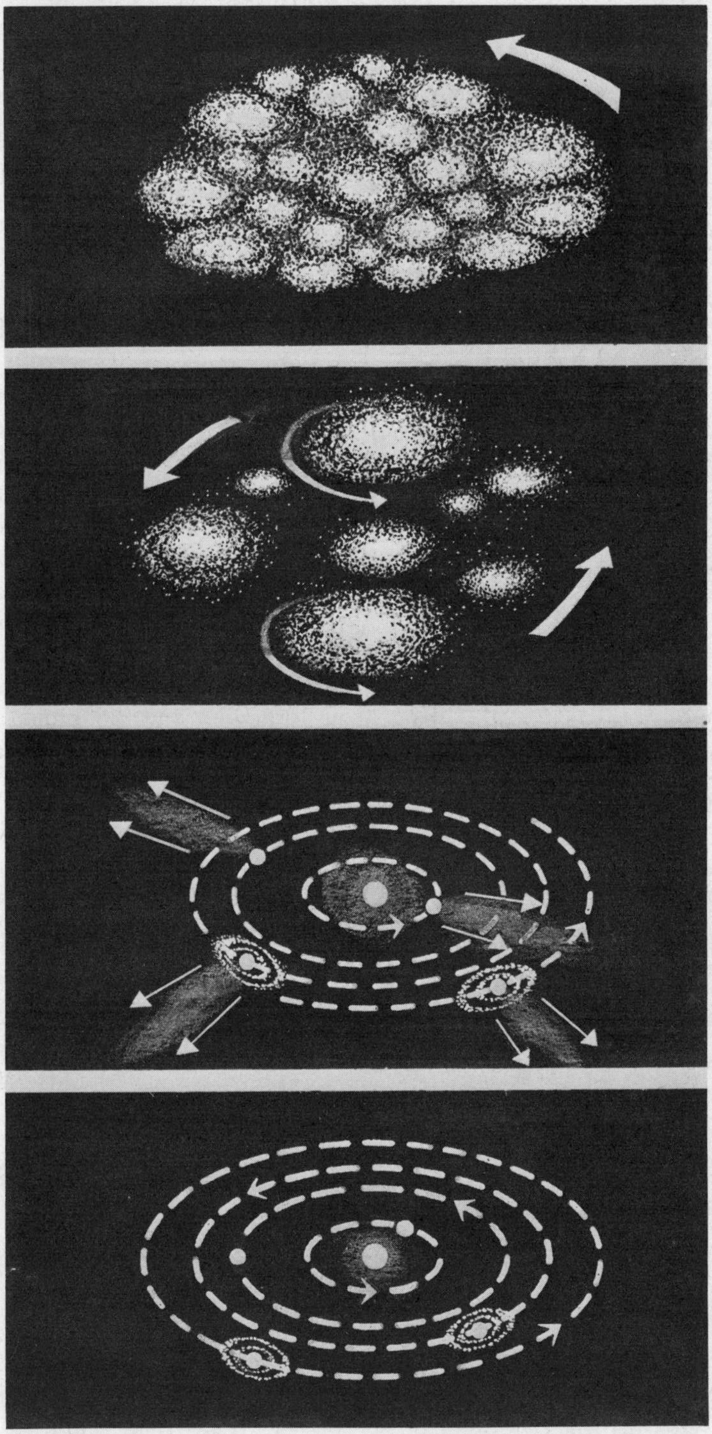

Figure 11.6. *Modern Protoplanet Theory of Solar System Formation.* The oldest stage is at the top; the youngest, at the bottom. (After McLaughlin.)

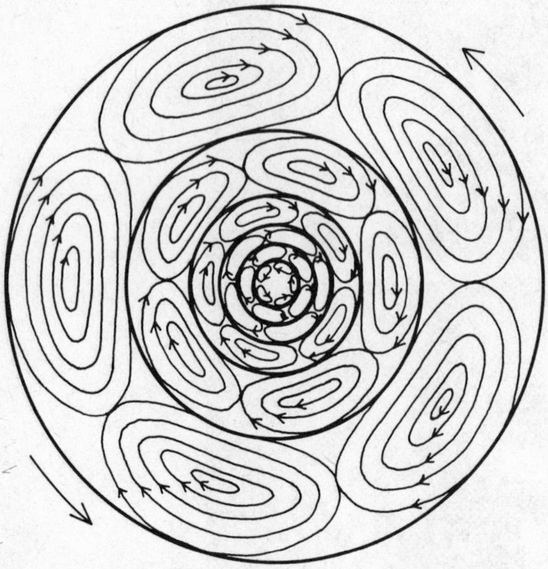

Figure 11.7. *Von Weizsacker's Concept of Turbulent Eddies in Solar Nebulae.* Zones of turbulence create centers around which material collects to form the various bodies in the solar system. (After von Weizsacker.)

propel the ions away from the sun at high speeds, thus carrying off a relatively large amount of the sun's angular momentum for an infinitesimal loss of mass.

Early Development of Earth

Radiometric dating of meteorites and, more recently, of material on the moon's surface, indicates that these events occurred between 4.6 and 5 billion years ago. Extrapolating backwards the ratios of the various isotopes of lead to a time when there was no radioactively produced lead also yields these dates. The oldest known earth rocks are dated at about 3.8 billion years. This leaves a long interval of the earth's history, between 4.6 billion years when the earth completed its formation and 3.8 billion years when permanent crust began to form, unrepresented by direct evidence on the earth's surface today. Most of our theorizing about this interval comes

A

Figure 11.8. A. *Moon* and **B.** *Mars impact craters.* (Photographs courtesy of N.A.S.A.)

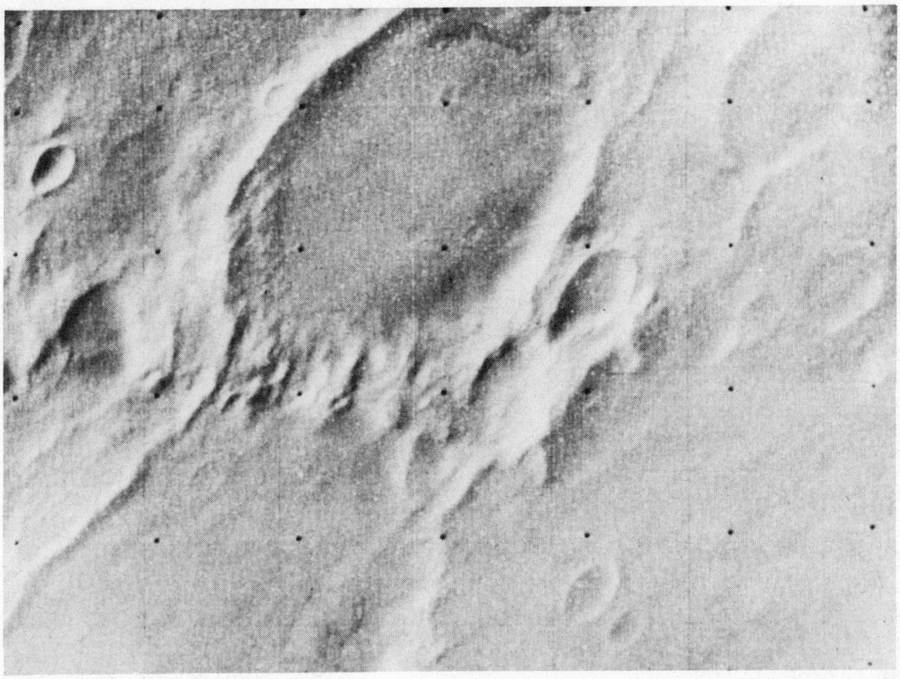

B

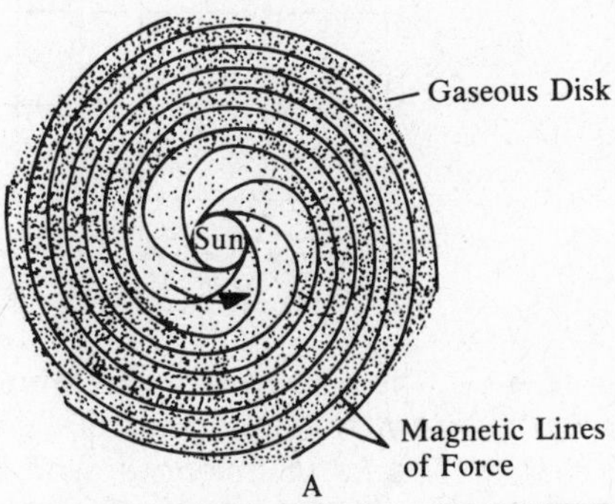

Figure 11.9. *Magnetic Braking Action.* Particles ejected along magnetic lines of force carry away much momentum but little mass. **A.** *Polar view* and **B.** *Equatorial view.* (After Hoyle and Whipple.)

from seismic studies of the earth, the composition of the other planets and their atmospheres, and physical-chemical studies in the laboratory.

The Earth's Interior

The passage of earthquake-generated waves through the earth's interior has disclosed much about its composition and behavior (Fig. 11.10). These waves behave differently, depending on the density and rigidity of the material through which they pass. Seismic studies have demonstrated that the earth's interior is stratified into layers of differing density. These layers form a very dense **core,** a less dense **mantle,** and a light surface **crust** (Fig. 11.11).

The majority of scientists believe the original particles were collected at random, so that some form of segregation has occurred with the heavy material sinking inward to form the core and the lighter material rising to the surface to form the crust. The outer core behaves as if it were a liquid, and the mantle material can also flow plastically under great pressure as demonstrated in

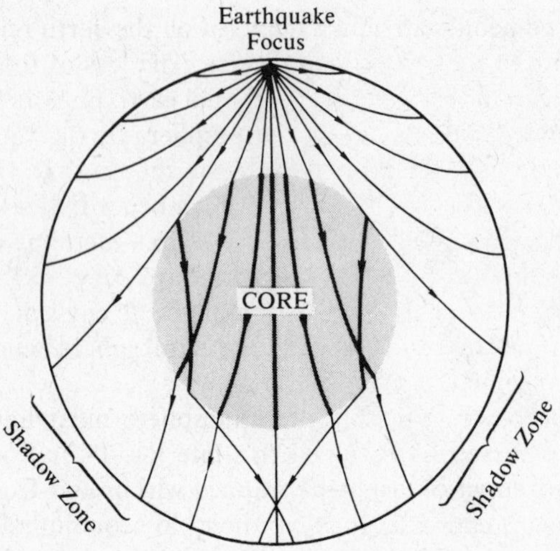

Figure 11.10. *Earthquake Wave Behavior Gives Information about the Earth's Interior.* Note bending of compressional waves by core. This produces a shadow zone where these waves are absent at the surface. (After Stokes.)

laboratory experiments. This suggests that some form of convection currents caused the differentiation. Heat from gravitational compression, from decaying radioactive isotopes trapped in the interior, and/or from tidal friction generated by the moon being much nearer the earth could have been the driving force or forces.

We are uncertain as yet whether the earth's interior ever completely melted or whether the heating is still increasing, decreasing, or about steady. The segregation of the earth's interior does suggest, however, that it has been largely molten. Some geophysicists dispute this contention, however. They reason that diffusion in the solid state was more likely because if the earth were largely molten, it would have lost most of the volatiles it now possesses. The discovery of plate tectonics on the earth's surface also supports the existence of convection currents, though these may be only surface features unrelated to conditions deep in the earth's interior. The speed and timing of the segregation of the earth's interior is another much debated question. We shall explore the evidence from the rock record in the next chapter.

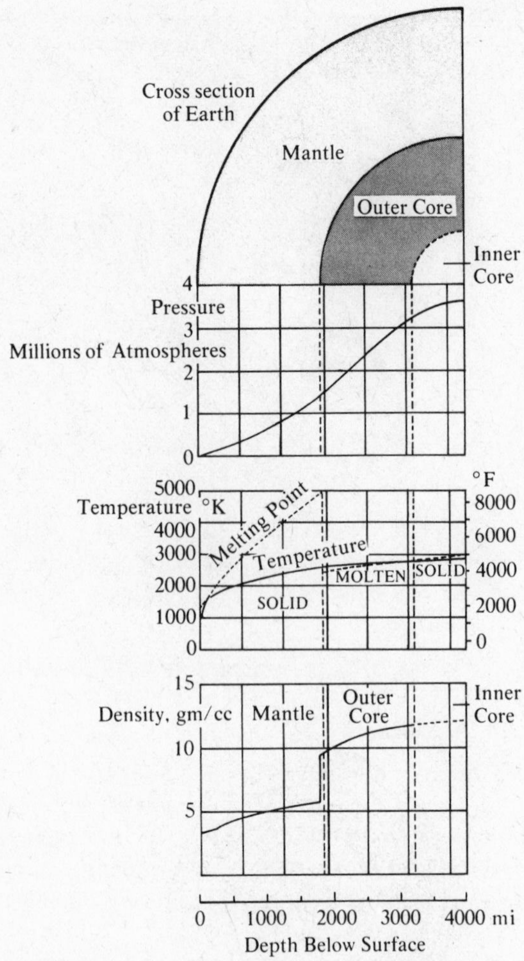

Figure 11.11. *The Structure of the Earth's Interior.* [(Data from J. Verhoogen, 1960 and Bullen, 1963) in *The Earth Sciences,* 2nd ed. by Arthur N. Strahler, © 1963, 1971. By permission of Harper & Row, Publishers, Inc.]

The composition of the earth's interior is presently unknown, but certain ultra-basic (low-silica) igneous rocks, such as peridotite and serpentine, brought up from great depths along plate margins, suggest the upper mantle may be similar in composition. Eclogite, a high-density rock of the same composition as basalt, occurs in these same settings and may indicate that chemical differences are not the only ones between crust and mantle.

The core is very dense and generates the earth's magnetic field (Fig. 11.12), suggesting it is

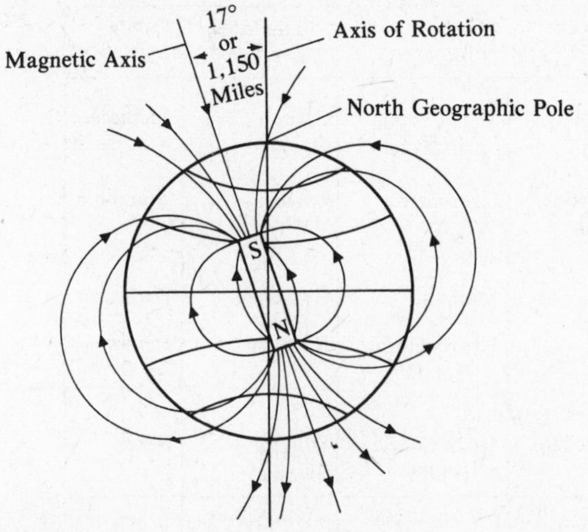

Figure 11.12. *The Earth's Magnetic Field.* Arrows indicate direction of field. The relationship to the rotational axis is shown. (After Elsasser et al.)

composed of iron and related elements. (The so-called transition metals of the chemist's periodic table are the only elements with the correct density characteristics which are common enough to compose the core.) Certain **meteorites,** the **iron-nickel** type, have this composition, and they probably formed in the interior of an asteroid by segregation.

As the original earth, lacking an atmosphere and a magnetic field and swept by the solar wind, underwent heating, segregation, and core formation, a magnetic field developed which shielded its surface from the solar wind.

The Earth's Atmosphere and Oceans

The presence of a primeval atmosphere and ocean at this stage has been much debated. Astronomers, judging from the abundance of various gases present in the universe and particularly in the atmospheres of the outer planets which are less affected by the solar wind, have postulated an original residual atmosphere of **methane, ammonia, water,** and **hydrogen** (Fig. 11.13).

Geochemists, on the other hand, have noted that the inert or noble gases—xenon, krypton, and neon—are much depleted on the earth relative to the rest of the universe. The loss of these gases indicates that the original earth must have been swept clean of an atmosphere by the solar wind. Geochemists have also shown that the rock record of the Early Precambrian indicates no evidence for the presence of free methane or ammonia. Most geologists have now come around to this way of thinking and envision a primeval earth without an atmosphere and, hence, without oceans.

They reason that the atmosphere must have outgassed from the earth's interior, both from frozen gases originally trapped within and from their chemical contribution to compounds. These include clay minerals and other hydrated silicates which contain water, nonbiogenic organic compounds which contain carbon, potassium-bearing minerals with ammonium ions which contain nitrogen, and silicates which contain some chlorine ions as substitutes for oxygen. These compounds occur in **carbonaceous chondrite meteorites,** thus supporting this derivation.

The volatiles that accumulated from outgassing would be chiefly **carbon dioxide** and **monoxide, water vapor, nitrogen,** and **hydrogen** (Fig. 11.14). These compounds could also result if the original atmosphere had contained methane, ammonia, and water (Fig. 11.15). The action of sunlight dissociates water, releasing oxygen which reacts with methane to produce carbon dioxide and with ammonia to produce nitrogen. Because the volume of these gases on earth today vastly exceeds the amount that could have resulted from the weathering of rocks at the earth's surface, they are called "**excess volatiles.**" Oxygen was not present in the earth's early atmosphere as evidenced by the presence of many elements such as iron and uranium in their reduced states, a condition that does not occur in rocks of mid-Precambrian and younger ages.

Geochemists debate whether the outgassing was a relatively rapid event that occurred in the earth's early history or has been a slow, continuous process. The lack of a rock record for the 800 million years and the impossibility of distin-

Earth's present atmosphere			Volcanoes, average	Geysers and fumaroles	Meteorites, average	Jupiter's atmosphere
Gas	Percent	Stability				
Major:						
Nitrogen	(78%)	Stable	Water vapor (73%)	Water vapor (99%)	Carbon dioxide	Methane
Oxygen	(21%)	Unstable (reacts with Fe and C)	Carbon dioxide (12%)	Hydrogen	Carbon monoxide	Ammonia
Argon	(0.9%)	Stable	Sulphur dioxide	Methane	Hydrogen	Hydrogen
Carbon dioxide	(0.03%)	Unstable (reacts with silicates)	Nitrogen	Hydrochloric vapor	Nitrogen	Helium
Water vapor	(variable)	Unstable	Sulphur trioxide	Hydrofluoric vapor	Sulphur dioxide	Neon
Minor: (traces only)			Carbon monoxide	Carbon dioxide	Methane*	
Neon		Stable	Hydrogen	Hydrogen sulfide	Nitrous oxide*	
Helium		Stable	Argon	Ammonia	Carbon disulfide*	
Krypton		Stable	Chlorine	Argon	Benzene*	
Xenon		Stable		Nitrogen	Toluene*	
Hydrogen		Unstable (reacts with oxygen to form water)		Carbon monoxide	Naphthalene*	
					Anthracene*	

*Only present in carbonaceous chondrites.

Figure 11.13. *Comparison of the Earth's Atmosphere with that of an Outer Planet* and gases present in meteorites, volcanic emanations, and geysers.

guishing in present-day volcanic emanations the volatiles derived from groundwater and those coming from outgassing have made a final decision impossible at present. The low boiling point of the volatiles found in carbonaceous chondrite meteorites indicates an early quick release when the earth heated up. On the other hand, even if only one percent of the present volcanic emanations actually represent outgassing, they could have produced the present atmosphere and oceans since the Early Precambrian. The widespread occurrence of marine rocks in Precambrian strata of 3.5 billion years age and younger indicates that the amount of ocean water present then was probably as great as now and argues for the first alternative.

The Origin of Life

Thus, at some time before 3.5 billion years ago there was probably an earth with an atmosphere of CO_2, CO, H_2O, N_2, H_2, oceans of water, and

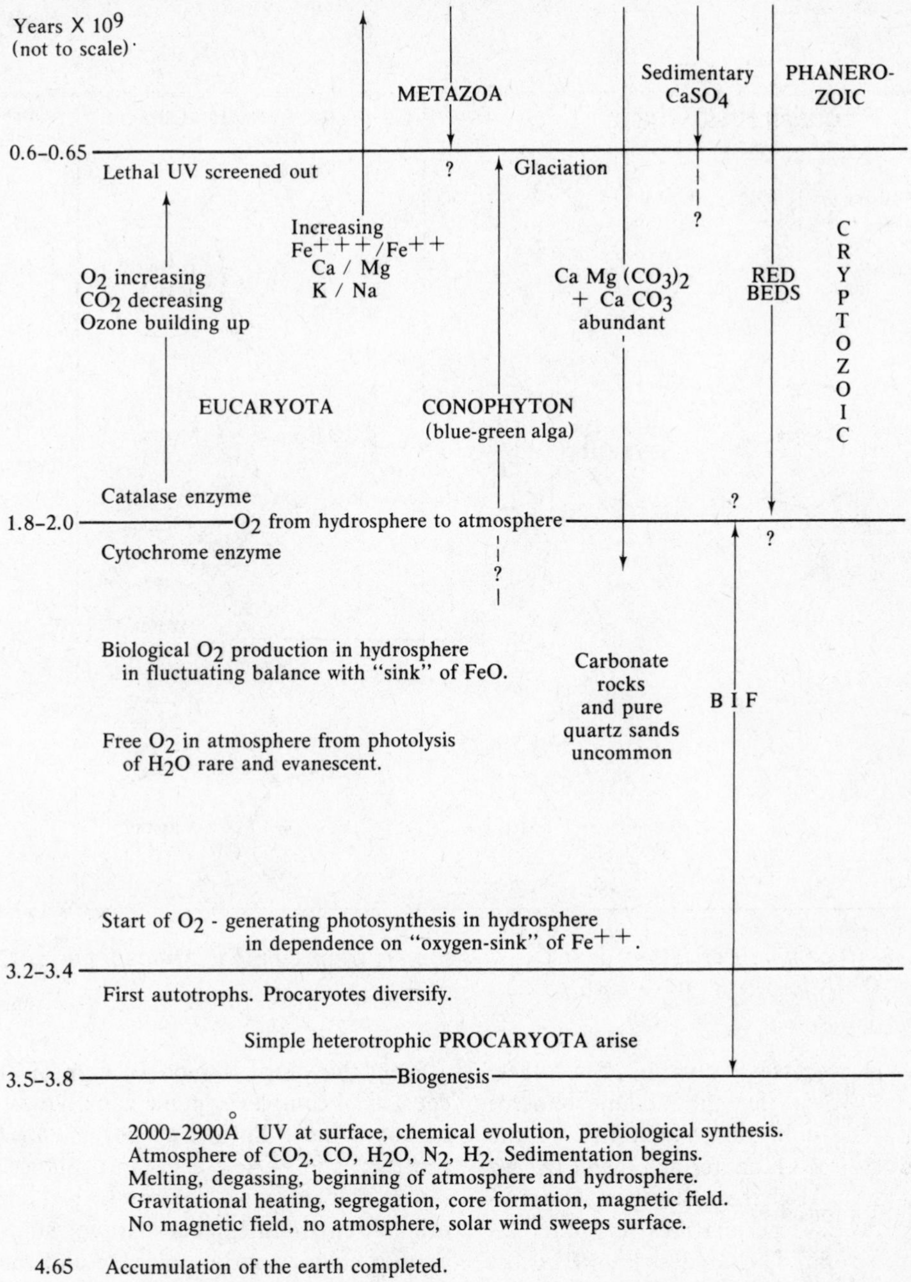

Figure 11.14. *Evolution of the Earth's Early Atmosphere and Life.* (From Preston E. Cloud, Jr., in Drake, *Evolution and Environment,* by permission of Yale University Press.)

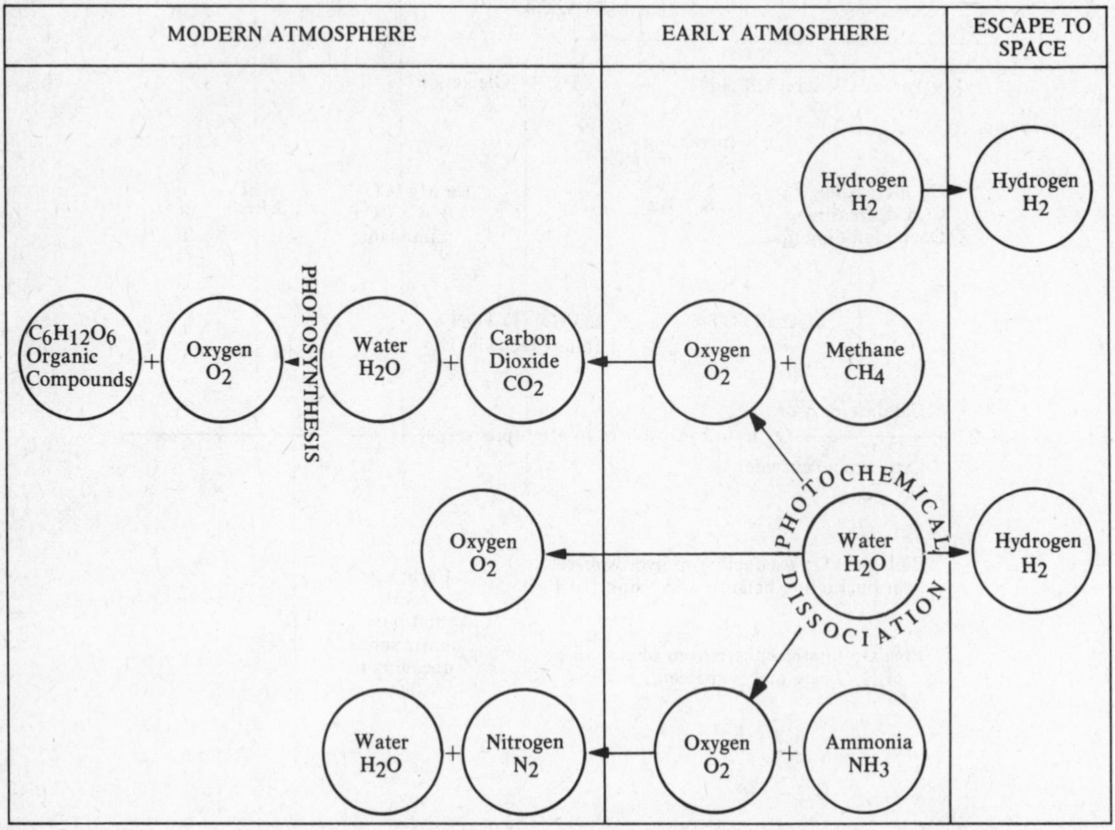

Figure 11.15. *Possible Derivation of CO_2, O_2, and N_2 from Original Atmosphere of CH_4, H_2, NH_3, and H_2O.* (After Clark and Stearn.)

sedimentation processes occurring. The oldest sedimentary rocks, though metamorphosed, clearly show the uniformitarian nature of the sedimentary processes that formed them. (As we shall see in the next chapter, however, Precambrian rocks do have peculiarities not found in Phanerozoic strata.) These oldest known rocks, the Swaziland and Fig Tree groups of South Africa, also contain tiny bodies which appear to be the remains of bacteria and blue-green algae, the simplest of living things. The fact that the oldest rocks contain fossils indicates that the fossil record cannot directly help us determine the steps in the origin of life. We must turn again to the biochemist and geochemist to suggest how life might have originated given the physical conditions just described.

Even the simplest modern type of organism consists of complex organic compounds such as **carbohydrates, lipids, adenosine phosphates, proteins,** and **nucleic acids** in conjunction with water, the universal solvent (Fig. 11.16). Our original environment consisted of simple compounds such as carbon dioxide and monoxide, water, nitrogen, and hydrogen (or possibly methane, ammonia, water, and hydrogen). It is a long way from these to the complexly organized, energy-capturing and expending, self-duplicating thing we call life. Or is it?

In the days (1953) when the original atmosphere was believed to consist chiefly of methane, ammonia, water, and hydrogen, *Miller* and *Urey* circulated such a mixture over an electric spark, analogous to ultraviolet radiation (which

Compounds	Functions	Composition
Water	Universal Solvent	Hydrogen, Oxygen
Carbohydrates	Energy Source	Hydrogen, Oxygen, Carbon
Fats (Lipids)	Energy Storage	Hydrogen, Oxygen, Carbon
Adenosine Phosphates (ADP, ATP)	Energy Transfer	Hydrogen, Oxygen, Carbon, Nitrogen, Phosphorus
Proteins	Structural; Facilitation of Chemical Reactions	Hydrogen, Oxygen, Carbon, Nitrogen, Phosphorus, Sulfur (Organized Into 20 Amino Acid "Building Blocks")
Nucleic Acids (DNA, RNA)	Patterns for Protein Synthesis	Hydrogen, Oxygen, Carbon, Nitrogen, Phosphorus (Organized Into 5 Nucletide "Building Blocks")

Figure 11.16. *Principal Chemical Constituents of Life.*

could penetrate the then oxygenless environment) or lightning (Fig. 11.17). They obtained in this experiment the complex organic substances called **amino acids**, the building blocks of proteins. Since that date, using various possible mixtures of volatiles and energy sources available on the early earth, chemists have succeeded in synthesizing in the laboratory nearly every major constituent of living systems.

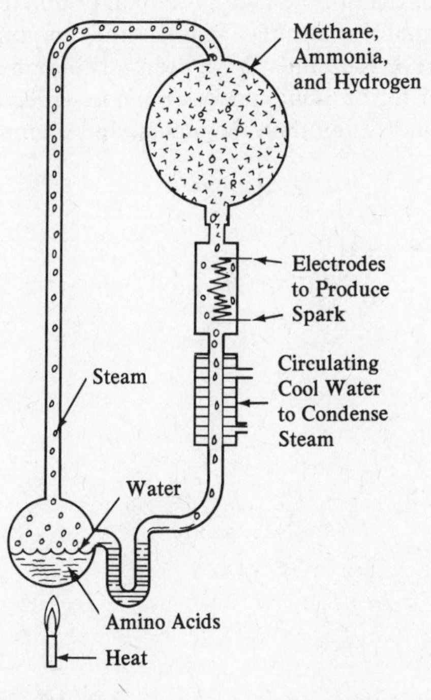

Figure 11.17. *The Miller-Urey Experiment.*

For example, by heating and cooling various amino acid mixtures, *Fox* was able to aggregate them into protein-like chains. *Oro* found the nucleotide-base building blocks of nucleic acids in one of his experiments. *Bar-Nun* found that shock waves from thunderclaps are very efficient producers of amino acids. And, in a very significant breakthrough made in 1968, *Ponnamperuma* and *Hodgson* synthesized chlorophyll, the compound responsible for photosynthesis. Other workers found that even in ordinary mixtures of simple carbon compounds, small amounts of the complex ones are present because those chemical reactions, like all others, are equilibrium reactions. That is, every reaction goes both ways even if only a small amount in one direction under natural conditions. Compounds produced by these reactions in the atmosphere would collect in the seas. So, by an early date in earth history after degassing, the oceans could develop into organic soups of complex molecules. There were no scavengers or any oxygen to destroy the organic compounds, and if they sank below the level of ultraviolet penetration, they were safe from dissociation.

Complex organic compounds by themselves are not organisms, however. The chance fusion of all these compounds into anything even as simple, by biologic standards, as a bacterium is clearly impossible. But chemists began to notice other interesting things about their nonbiogenically produced complex compounds. Fox's "protenoids" produced **microspheres** (Fig. 11.18A). *Oparin* found that the mutual attraction among large molecules produced oriented clumps called **coacervates.** *Salthe* found that molecules soluble only at one end (amphoterics) form structures called **micelles**, which resemble biological membranes in structure and composition (Fig. 11.18B). In other words, much of the order once thought to be found only in living things is derived from the structure of molecules themselves and actually occurs in the nonliving world. (In fact, many of the steps leading to life could probably occur anywhere in space where the right materials and conditions exist.) It was also found that some of the compounds in the soup had greater affinity for the simple building blocks than had others and, hence, became more numerous—a sort of incipient natural selection. Furthermore, some of the complex organic compounds are **catalysts;** that is, they speed up reactions, such as their own formation, that without them would proceed very slowly.

These would include not only this self-catalysis (autocatalysis), but also the step-wise process of continually aggregating more and more complex structures until membranes and other cellular structures resulted, producing the first cell. Even some simple inorganic compounds found in the water inefficiently catalyze organic reactions that today are speeded by more efficient organic compounds, the protein catalysts called **enzymes.** Many of the steps in this chain of catalytic reactions leading from the organic building blocks in the soup to the simplest cells have now been performed in the lab, including the synthesis of the hereditary material itself—**DNA**, or deoxyribonucleic acid. Because the transition from the nonliving to the living is a continuous, evolutionary one, it is hard to decide on a definition of life to be used in deciding if life has been created. The creation of the self-duplicating DNA molecule, however, is certainly on or near the boundary by most definitions.

The chemical processes and structure in all living things are so similar that with the possible exceptions of some groups of bacteria, it seems certain that all life had a common origin. Thus, one type of molecular aggregate in the ancient oceans or soup must have been a superior competitor for building blocks and more efficient in self-duplication than the others, and eliminated

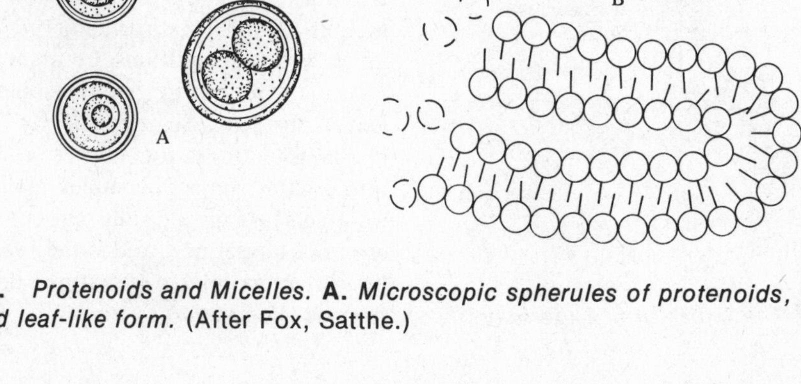

Figure 11.18. *Protenoids and Micelles.* **A.** *Microscopic spherules of protenoids,* **B.** *Micelles of spherical and leaf-like form.* (After Fox, Satthe.)

them. A survivor from this age of subsistence on simple building blocks may be *Kakabekia* (Fig. 13.2), a tiny, umbrella-shaped creature requiring only ammonia as an energy source and known from fossils of two billion years of age.

Fermentation

In the absence of oxygen, the main energy-releasing compound used by cells today, the earliest organisms probably utilized the process of fermentation to release energy. **Fermentation,** still used today by some small organisms like yeast, is the process of splitting organic compounds such as sugars to release energy, carbon dioxide, and alcohol (Fig. 11.19). Clearly, an organism employing fermentation as an energy source depends on an external supply of organic molecules and is heterotrophic.

Photosynthesis

One of the early creatures incorporated into itself the compound chlorophyll which, we have already seen, can be generated abiotically. Chlorophyll allows its possessor to take the carbon dioxide produced by fermentation, water, and solar energy and make its own organic compounds, such as sugar, without depending on an external supply. Oxygen is liberated as a by-product in this process (Fig. 11.19). **Photosynthesis** had appeared. (A few bacteria utilizing a slightly different compound and process can also manufacture their own food along with producing oxygen.) These first simple autotrophs could then make their own organic molecules by photosynthesis and derive energy from them by fermentation.

The free oxygen produced by photosynthesis is a poison to life. Today it is removed by complex organic compounds in the cell which reduce it to nonpoisonous forms. Before these evolved, however, some other device was necessary to get rid of the oxygen. The seas at that time must have had large quantitites of unreduced or ferrous iron derived from the weathering of rocks because there was no free oxygen in the atmosphere or oceans to oxidize it. The oxygen generated by photosynthesis readily combined with this soluble ferrous iron and oxidized it to insoluble oxides and hydroxides which form precipitates.

Here then is the explanation for the large marine iron precipitate deposits of the Precambrian rocks older than two billion years (Fig. 11.20). These strata are called **Banded Iron Formations** (BIF) because they consist of alternating layers of silica (chert) and iron oxides. Such rocks do not occur later in the earth's history because once oxygen became abundant after the spread of photosynthesis, it completed the precipitation of the ferrous iron in the sea and then began to collect in the atmosphere. There it oxidized the ferrous iron before it reached the sea, producing terrestrial redbeds which first appeared around two billion years ago as the marine iron precipitation ceased. Further evidence for the biologic origin of the iron is that the iron occurs in alternating bands with silica, a chemical association that could not occur without organic intervention.

Respiration

If life had remained dependent on inefficient fermentation reactions to obtain energy, the process of photosynthesis would have proceeded much more slowly in supplying the earth with oxygen. However, once the oxygen-reducing compounds evolve in cells, such cells could survive in the presence of free oxygen and eventually use it for a more efficient metabolic process, **respiration** (Fig. 11.19). In cellular respiration, sugars and oxygen combine to produce energy, water, and carbon dioxide. All those by-products

Fermentation: $C_6H_{12}O_6 \longrightarrow 2CO_2 + 2C_2H_5OH + \text{energy}$
Photosynthesis: $6CO_2 + 6H_2O + \text{sunlight} \longrightarrow C_6H_{12}O_6 + 6O_2$
Respiration: $C_6H_{12}O_6 + 6O_2 \longrightarrow 6CO_2 + 6H_2O + \text{energy}$

Figure 11.19. *Major Chemical Equations Important to the Evolution of Life.*

Figure 11.20. *Banded Iron Formation, Michigan.* (Courtesy Michigan Dept. of Natural Resources.)

are harmless to life; in fact, carbon dioxide is a necessary raw material in photosynthesis.

The complementary nature of the photosynthesis-respiration reactions coupled with the much greater efficiency or respiration over fermentation led to the tremendous expansion of life. With it came the gradual evolution of oxygen throughout the oceans and atmosphere. This screened out ultraviolet light, thus removing one of the energy sources which may have started the evolution of living things, but which was also detrimental to near-surface or terrestrial life. Free oxygen also meant the end of the nonoxidizing conditions necessary for the beginning of life and has precluded any additional origins of life.

Organisms then evolved which could obtain their organic compounds by devouring others and using the increasingly abundant oxygen to release energy from these compounds. Animal-like heterotrophs had thus appeared. By mid-Precambrian times, then, the earth had formed, evolved its present atmospheric and oceanic constituents (although the percentages of gases were probably not the same as at present), and was the scene for the unfolding drama of the continuing evolution of life.

REVIEW

1. How does Laplace's **nebular hypothesis** differ from the modern theory of the origin of the earth?
2. What have been the major problems with the **encounter hypothesis?**
3. What is the nature of the earth's interior as disclosed by earthquake waves?
4. What direct evidence do we have about the nature and composition of the mantle?
5. Why do geochemists believe the earth's atmosphere outgassed instead of being a remnant of the protoplanet stage?
6. What are the major compounds found in living things? Of what major elements are they composed?
7. Give some examples of the nonbiogenic generation of complex organic compounds.
8. How might the oriented chemical structures of cells have evolved?
9. What is **fermentation,** and why is it a relatively unimportant process today?
10. Compare and contrast **photosynthesis** and **respiration.**

12

The First Four Billion Years: The Precambrian

12

The **Precambrian** or the **Cryptozoic Eon**—that immensely long interval of geologic time preceding the appearance of abundant megafossils—has not subdivided into a series of units of worldwide application such as those established for later geologic time. Because it covers approximately 7/8 of the earth's entire time of existence, from over 4.6 to .6 or .7 billion years ago, it may seem incongruous to devote only two and one-half chapters in a textbook of earth history to this interval. But because there has been no worldwide tool for time-correlating Precambrian rocks and, hence, for dividing them into universally recognizable units, we simply do not have the detailed knowledge of these strata that fossils give us in later geologic time. The best we can do is assign dates of occurrence to events in terms billions and hundreds of millions of years ago as determined from radiometric dating techniques. But there is still no way to obtain the fine subdivisions possible with fossils. New tools, such as the scanning electron microscope, are enabling paleontologists to uncover new structural details in Precambrian microfossils. Perhaps there will be enough diversity and rapid evolution discovered to provide subdivisions in the future.

In older textbooks, a series of "eras" and "periods" are subdivisions in the Precambrian time scale. However, these units were primarily defined on a lithologic basis. The eras called **Archeozoic,** or **Archean,** and **Proterozoic,** or **Algonkian,** in North America were based on the supposition that all the granites, gneisses, and associated highly deformed and metamorphosed sediments are very old, and that the slightly deformed sedimentary rocks are younger. Once again the illusion that lithology is a guide to age was shattered.

The widespread use of radiometric dating indicated that rocks of both types formed throughout the Precambrian from 3.8 billion years down to the beginning of the Paleozoic Era. As a result, it became obvious that the original usage of these terms was for rock and not time-rock units. (Some geologists have now switched usages, however, and call all rocks older than 2.3–2.4 billion years, regardless of their lithology, Archean, and all younger Precambrian rocks, Proterozoic.)

The incongruity of the switch of definitions is shown dramatically in the Grand Canyon. At the bottom of the Canyon are highly deformed, intensely metamorphosed, intruded, and very "old-looking" rocks originally assigned to the Archean (Fig. 12.1). Radiometric dating, however, has disclosed that they were formed between 1.7 and 1.3 billion years ago, relatively late in the Precambrian. In the new terminology, these Grand Canyon rocks are of Middle Proterozoic age.

In recent years, then, there has been more of a tendency to recognize that most names applied to Precambrian subdivisions have a lithostratigraphic and not a chronostratigraphic basis. Geologists in each area of Precambrian exposures have described the various episodes of sedimentation and deformation, assigning them local era and period names and determining their radiometric ages.

Shields

Precambrian igneous, metamorphic, and sedimentary rocks are found well exposed on all con-

Figure 12.1. *Grand Canyon Precambrian.* Lowest cliff in canyon is composed of dark Vishnu Schist, a highly deformed and metamorphosed group of rocks. Cliff above this on right side of photograph (not present on left side) is composed of younger Precambrian Grand Canyon Group, a series of tilted sedimentary rocks. (Courtesy of William M. Mintz.)

tinents today. In some areas, the Precambrian rocks are widely exposed at the surface over broad, generally low areas called **shields**. The major shields of the earth today (Fig. 12.2) are the **Canadian** in North America, the **Amazonian** in South America, the **Baltic** in Europe, the **African** encompassing most of Africa and recently sundered Arabia, the **Siberian** in Siberia, the **Indian,** the **Chinese,** the **Australian,** and the **Antarctic.** Several small shield areas are also known. In the days when continents were assumed to be static, much genetic significance was attached to shields. They were thought to be stable continental nuclei, or cratons, around which geosynclines developed in a series of concentric rings, each successively plastered to the shield by mountain building. (The word **craton** is applied not only to shields, but to stable regions with a thin Phanerozoic sedimentary cover.)

In North America, where the Phanerozoic mountain belts ring the shield, this concept has been widespread. It has found less support elsewhere, however, because Eurasia and the southern continents do not show any hint of such a

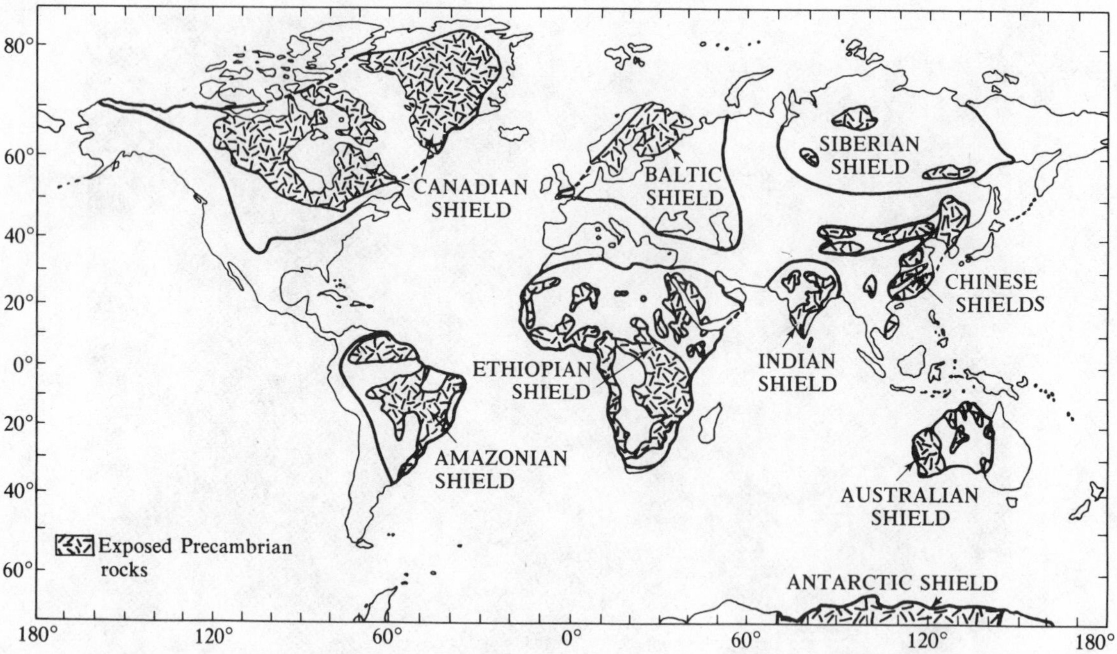

Figure 12.2. *Major Shield Areas of the Earth.* Stippled areas are actual outcrops of Precambrian rocks both in shields and mountains; areas enclosed by lines are total extent of cratons with Precambrian basement.

concentric pattern (Fig. 8.11). For example, a Paleozoic mountain belt runs down the middle of Eurasia, and most margins of Africa are Precambrian. Even a glance at the radiometric dates of the North American shield (Fig. 12.3) shows that there are three or four belts of very old Precambrian rocks; although one is fairly central, the others are not. It now appears that the significance of shields is simply that they are blocks of exposed Precambrian rocks which have not been significantly deformed or destroyed since the Precambrian by plate collisions.

Many have been broken apart by rifting, however. For example, eastern South America, Africa, India, western Australia, and eastern Antarctica are all part of the same shield that has been broken into fragments. The matching boundaries of Precambrian deformed belts can be found on opposite sides of the Atlantic in eastern South America and western Africa (Fig. 10.31). Whatever controls the positions of the margins of the plates, it does not appear systematic as far as the position of continents on the lithosphere is concerned. So, sooner or later, even the existing shields will be near plate margins and be deformed again. This has occurred to many areas of shields in the past; for Precambrian rocks occur in the cores of many of the earth's great mountain ranges, including the Appalachians and western Cordillera of North America.

Precambrian Rocks

The most abundant Precambrian rock types are granites and gneisses (Fig. 12.4). This is not surprising, considering that the vast length of time these rocks have been in existence greatly increased their chances of finding themselves at plate margins and being melted and metamorphosed.

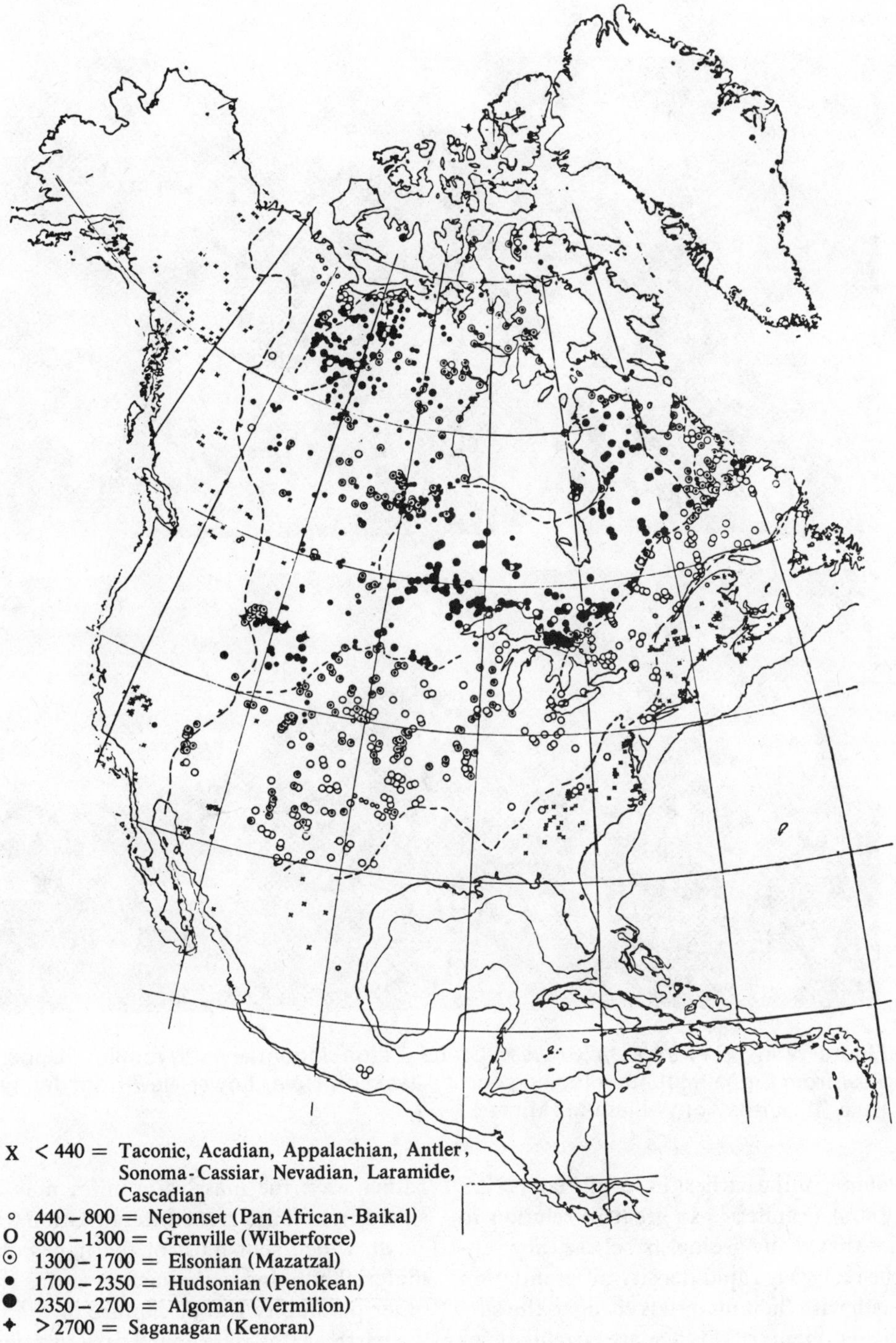

Figure 12.3. *Isotopic Age Dates for North America.* Dashed lines delineate provinces. (From Hurley and Rand, *Science*, vol. 164, June 1969, © AAAS. Used by permission of authors and AAAS.)

Figure 12.4. *Precambrian Gneiss.* Grand Teton Mountains, Wyoming. Upper view from Signal Mountain looking across Jackson Hole. Lower view from Jenny Lake. (Courtesy of William M. Mintz.)

The volume of the earliest Precambrian (>2.3 billion years) granites is so great in relation to later ones that many geologists believe they represent the relatively rapid density differentiation of the earth with light materials rising to the surface. The sediments of this age are oceanic types associated with basic volcanics everywhere they have been found. This may indicate that the original crust was oceanic and was intruded by the massive granites or that the basaltic rifts formed as fissures in an original granitic crust. Either way, the massive granites may have resulted from large-scale differentiation within the earth, which sent most of the light continental material bobbing to the surface at this time. Another possibility is that capture of the moon by the earth at this time could have produced huge tidal forces, bringing up the lighter granite. If these hypotheses are true, granites of Middle Precambrian age and younger may primarily represent reworked sialic material from this one original event. Other geologists believe that granite has continually been added to the continents by differentiation of the lighter fraction from new crust material.

Certain rocks apparently derived from the mantle, such as layered ultramafics (Fe and Mg rich rocks) and anorthsites (rocks made entirely of Ca-plagioclase feldspars), occur in quantity only in the Precambrian. This may indicate that the crust was still thin and in the process of forming.

The Precambrian rocks found in the Phanerozoic mountain belts have been deformed several times, so their history is very difficult to decipher. In the shields, however, where Precambrian rocks have thus far escaped Phanerozoic mountain building, the history is easier to read. Granites and the metamorphic rock gneiss characteristically form only at former plate margins in the cores of mountain ranges in the Phanerozoic. With the possible exception of the earliest granites (Archeozoic or Archean) previously described, we can be fairly certain, therefore, that we are chiefly dealing with the roots of old mountain ranges that formed by plate collisions in what are now the relatively flat shields. These intrusive-metamorphic zones are highly mineralized and are responsible for the great metallic ore deposits found in most shields.

In scattered pods in the shields are occasional remnants of once-overlying sedimentary sequences (Fig. 12.5), now metamorphosed to varying degrees. (These occur within the granites of all ages.) We are fortunate in having even this much left of the sedimentary cover, considering the more than half-billion intervening years during which erosion has occurred. This erosion has not only stripped away most of them, but has also cut down in to the very roots of the Precambrian mountains as evidenced by the abundant migmatites (Fig. 12.6), rocks in which igneous and metamorphosed sedimentary materials are interlayered.

Not all surviving Precambrian rocks originated in the ocean basins, however. The

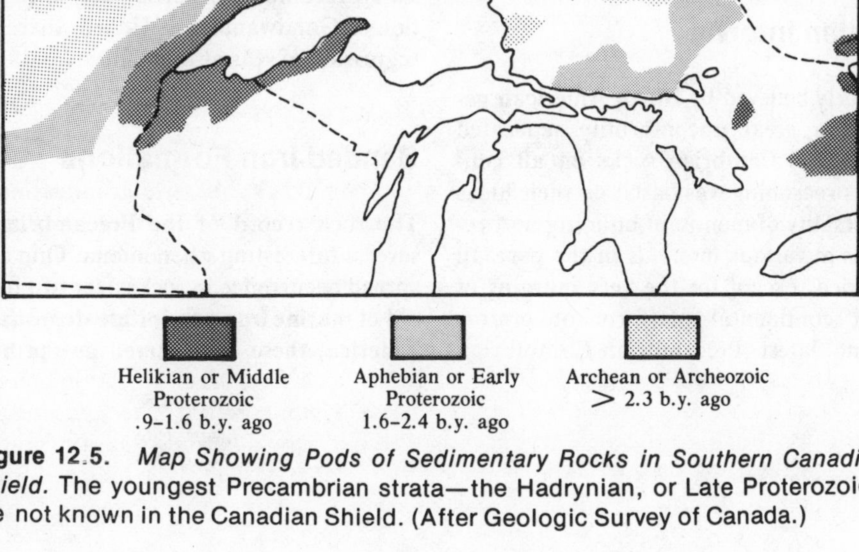

Figure 12.5. *Map Showing Pods of Sedimentary Rocks in Southern Canadian Shield.* The youngest Precambrian strata—the Hadrynian, or Late Proterozoic—are not known in the Canadian Shield. (After Geologic Survey of Canada.)

Figure 12.6. *Migmatite in the Canadian Shield.* (Courtesy of E.R. Brooks, C.S.U.H.)

Middle Precambrian Huronian Group of the southern Canadian Shield (Fig. 12.7) is a shallow sea sequence. The later Precambrian Keewenawan Group of the Canadian Shield (Fig. 12.8) is a great sequence of tilted lava flows and shallow sea sediments, probably representing a rift zone in the area.

The Lipalian Interval

It was formerly believed by North American geologists that a great unconformity separated Precambrian and Cambrian rocks on all continents. This reasoning was based on their ideas of the universality of mountain building and regressing seas at various intervals in the past. In North America, except for the very margins of the present continental platform or craton, there are no latest Precambrian-Eocambrian marine rocks. Hence, there is a craton-wide unconformity in this interval, and geologists assumed it was worldwide.

The notion was also used to explain why there were no megafossils except algae in Precambrian rocks and abundant animal megafossils in the Cambrian rocks. The organisms supposedly evolved in the ocean basins where their remains are not accessible for examination. Then, when the seas reinvaded the craton, they brought with them this influx of creatures whose ancestors were unknown in the platform rocks.

This gap, assumed to be worldwide, was called the **Lipalian Interval.** We now know it is a major unconformity over most of present North America, encompassing increasing amounts of Late Precambrian-Eocambrian time and up to two-thirds of the Cambrian as one moves toward the center of the craton, but nonexistent along the margins. On some other continents and platforms, however, a sequence of shallow marine sediments continues through this interval across the Precambrian-Eocambrian and Eocambrian-Cambrian boundaries. These rocks are called **Riphean** in Russia, **Sparagmitian** in Scandinavia, **Sinian** in Asia, and **Late Proterozoic** in Australia and Antarctica. Strata of this age are also known in ocean basin rocks now incorporated in mountain belts. However, there was considerable continental consolidation in the Late Precambrian, particularly with the formation of Gondwanaland. Hence, there was much regression of seas at this time.

Banded Iron Formations

The rock record of the Precambrian contains several interesting phenomena. One is the widespread occurrence in rocks over two billion years old of **marine iron precipitate deposits.** In North America, these have served as the base of the

Figure 12.7. *Huronian Sediments.* Animikian rocks at Kakabeka Falls, Ontario. (Courtesy of Ontario Dept. of Tourism and Information.)

A

B

C

D

Figure 12.8. *Keewenawan Lavas and Sediments, Michigan.* **A.** *Mott Island, Isle Royale National Park;* **B.** *Monument Rock, Isle Royale National Park;* **C.** *Lake of the Clouds, Porcupine Mountains State Park;* **D.** *Presque Isle River Falls.* (Parts A, B courtesy of National Park Service. Parts C, D courtesy of Michigan Department of Natural Resources.)

iron and steel industry. As noted in the last chapter, these deposits were most probably of biologic origin, precipitated by primitive photosynthesizers as a device for disposing of the poisonous by-product oxygen. In the interlayed bands of silica, numerous types of fossil microorganisms (Fig. 13.2) occur, offering further evidence that their metabolism caused the precipitates.

In rocks younger than two billion years, the banded iron formations do not occur; but, instead, terrestrial redbeds colored by iron oxides are common. This change probably reflects the evolution of oxygen–reducers by primitive organisms, thus allowing them not only to survive in the presence of oxygen, but eventually to utilize it in respiration. Oxygen was then produced in large amounts, which precipitated the remaining ferrous iron in the sea. The gas later accumulated in the ocean and, eventually, the atmosphere. There it could oxidize ferrous iron before it reached the sea, producing iron-oxide stains and coatings on terrestrial clastics.

Precambrian Glaciations

Another notable rock type of the Precambrian is **tillite,** or lithified glacial till, a poorly sorted mixture of clastics of all sizes from boulders to clay. The Lower Proterozoic rocks (about 2.2 billion years old) of the Canadian Shield contain apparent tillites (Fig. 12.9). Whatever the exact cause or causes of glaciation, a topic we will discuss in Chapter 20, it appears certain that continental glaciers originate in polar or subpolar areas. Thus, we have evidence that at this time in the Precambrian, these regions were near a pole, probably the Pacific Pole. Some apparent tillites of this age also occur in South Africa, probably indicating it was near the Gondwana Pole. There is evidence of later Proterozoic glaciations in Africa, Asia, Australia, and North America.

Much more extensive apparent tillites, many asociated with striated, polished pavements, occur in the Upper Precambrian-Eocambrian rocks of the Gondwana continents, Soviet Asia, Scandinavia, China, Scotland, Greenland, Newfoundland, Yukon, and Utah. The paleomagnetic positions determined for the interval indicate, however, that the equator ran through Australia and central Asia very near some of these deposits. It is possible that some of the so-called tillites are fluvial deposits, which are very much like glacial tills but lack a striated pavement. Some, however, do have striated pavements and yet were apparently near the

Figure 12.9. *Gowganda Tillite, Huronian Group.* (Courtesy Geological Survey of Canada, Ottawa.)

equator (Fig. 12.10). Evidence from the fossil record (see Chapter 13) and geophysics indicates that the moon may have been much closer to earth than now and perhaps it triggered large submarine gravity slides which produced the till-like lithology and pavements.

If the deposits are indeed glacial, they indicate the most latitudinally widespread glaciation in earth history. It has been suggested that the onset of glacial conditions was triggered by a decrease in the amount of atmospheric carbon dioxide, a gas which helps the earth retain solar heat by trapping it in the atmosphere. This is called the **greenhouse effect.** The reduction in CO_2 could have been caused by its consumption by phytoplankton colonizing the upper levels of the oceans. The atmosphere probably had sufficiently developed by this time to shut off harmful ultraviolet rays, thus allowing this expansion of photosynthesizers. If less CO_2 means less heat retained at the earth's surface, this may well have started the peculiar Late Precambrian-Eocambrian glaciation. It is unlikely that one could invoke the same argument to explain earlier or later glaciations.

Lack of Carbonates and Sulfates

Notable by their **absence** or **rarity** in the early Precambrian are **carbonates** and **sulfates,** rock types that abound in later geologic time. The lack of shell-forming marine organisms and efficient phosynthesizers (photosynthesis removes CO_2 from solution, thus precipitating $CaCO_3$) in the Early Precambrian probably accounts for the lack of carbonates. Sulfates are absent in the entire Precambrian, probably because the presence of the sulfate ion requires abundant oxygen, not then available, to oxodize reduced sulfur compounds.

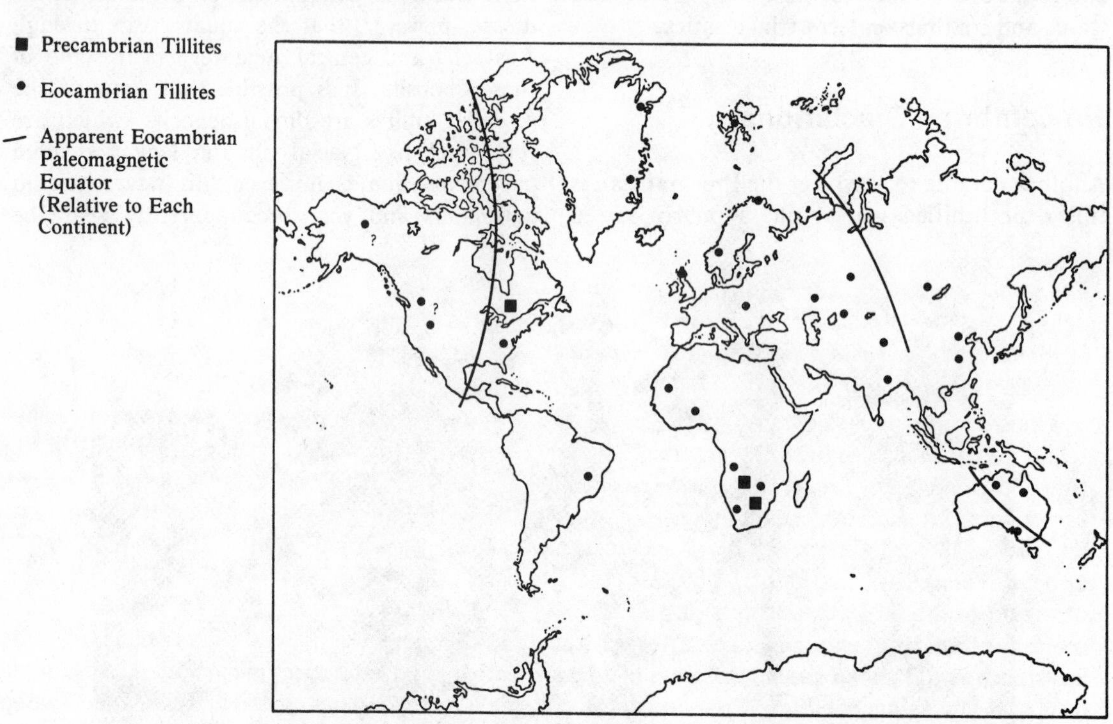

Figure 12.10. *Distribution of Supposed Precambrian and Eocambrian Tillites.* (After Harland and Rudwick.)

Precambrian Physical History

Although the record of various depositional and orogenic episodes is moderately well known, geologists are only now beginning to unravel the Precambrian patterns of shifting plates and, hence, continents and ocean basins. Using the age of deformation of Precambrian mountain belts (Figs. 12.3, 12.17–22), they are beginning to locate some plate boundaries for various times in the Precambrian. And, by combining this with paleomagnetic data, they are able to approximately position the continents at those various times (Fig. 12.16). Therefore, some general pattern of Precambrian plate tectonics is emerging. We will briefly review what is known about the Precambrian history of the present continents and note plate tectonic interpretations where possible.

The Canadian Shield

The Precambrian history of the present North American Craton deserves special comment because it is probably the best known and the most familiar to readers of this book. The dates of deformation cluster around five periods, 3.3, 2.4–2.5, 1.6–1.7, 1.3–1.4, and .9–1 billion years ago (Fig. 12.11). The two oldest orogenies—called the **Saganagan, Kenoran,** or **Laurentian** and the **Algoman,** or **Vermilion,** respectively—deformed a broad east-west trending belt north of the Great Lakes called the **Superior Province** (Fig. 12.12).

The deformed rocks (Fig. 12.13) are assigned to an **Archean** or **Archaeozoic** interval, sometimes divided into lower, or Keewatin, and upper, or Timiskaming, subdivisions. (Keewatin rocks, like all other very old sediments, appear to be oceanic in origin and consist chiefly of metamorphosed lavas and turbidites). These mountain-building episodes also affected the Beartooth and Bighorn mountain areas of Montana and Wyoming, the Slave Province north of Great Slave Lake, and the eastern Nain Province of coastal Labrador. These may have been separate plate edges, or they may have originally been continuous across the intervening regions which now record younger orogenies, the radioactive "clocks" having been reset.

The younger cycles of sedimentation are assigned to the **Proterozoic** interval. The next major disturbance, the **Hudsonian, Penokean,** or **Killarneyan,** chiefly affected a broad geosynclinal band around much of the Superior Province, the **Churchill-Ungava Province.** Most of this province lies to the present north and west of the Superior Province. Its southern part, whose connection with the main body is now buried under epeiric sea deposits, is often called the **Southern Province.** The Hudsonian orogeny culminated about 1.6–1.7 billion years ago. The deformed rocks (Fig. 12.7) are assigned to a subdivision of time called the Early Proterozoic **Aphebian "Era."** In the vicinity of the Great Lakes, these deposits formed in shallow seas; but away from this area, they formed in deeper waters. A subduction zone probably bounded the present northern and northwestern margins of North America in the Aphebian.

The **Elsonian** or **Mazatzal Orogeny,** often lumped with the Hudsonian as a continuation of it, affected the **Central Province,** a belt extending from the present southern edge of the Precambrian shield outcrops across the midcontinental United States to Arizona (Figs. 12.1, 12.12, 20.25A) and adjacent states. The western Nain Province and some of the Churchill Province were also affected by this deformation, which occurred 1.3–1.4 billion years ago.

The final orogeny affecting what is now the Canadian Shield was the **Grenville** or **Wilberforce Orogeny** which occurred about 900 million-1 billion years ago and deformed the areas of the present Appalachian Mountains and southeastern Canadian Shield. Rocks deformed by the orogeny, which include abundant carbonates (Fig. 12.14), are assigned to the **Helikian "Era."** In the northern Rocky Mountains are strata called the **Belt Supergroup** (Fig. 12.15), which yield isotopic dates of this interval also. Though in high mountains today, they are still essentially flat-lying strata in contrast to the deformed shield rocks. The undeformed Kee-

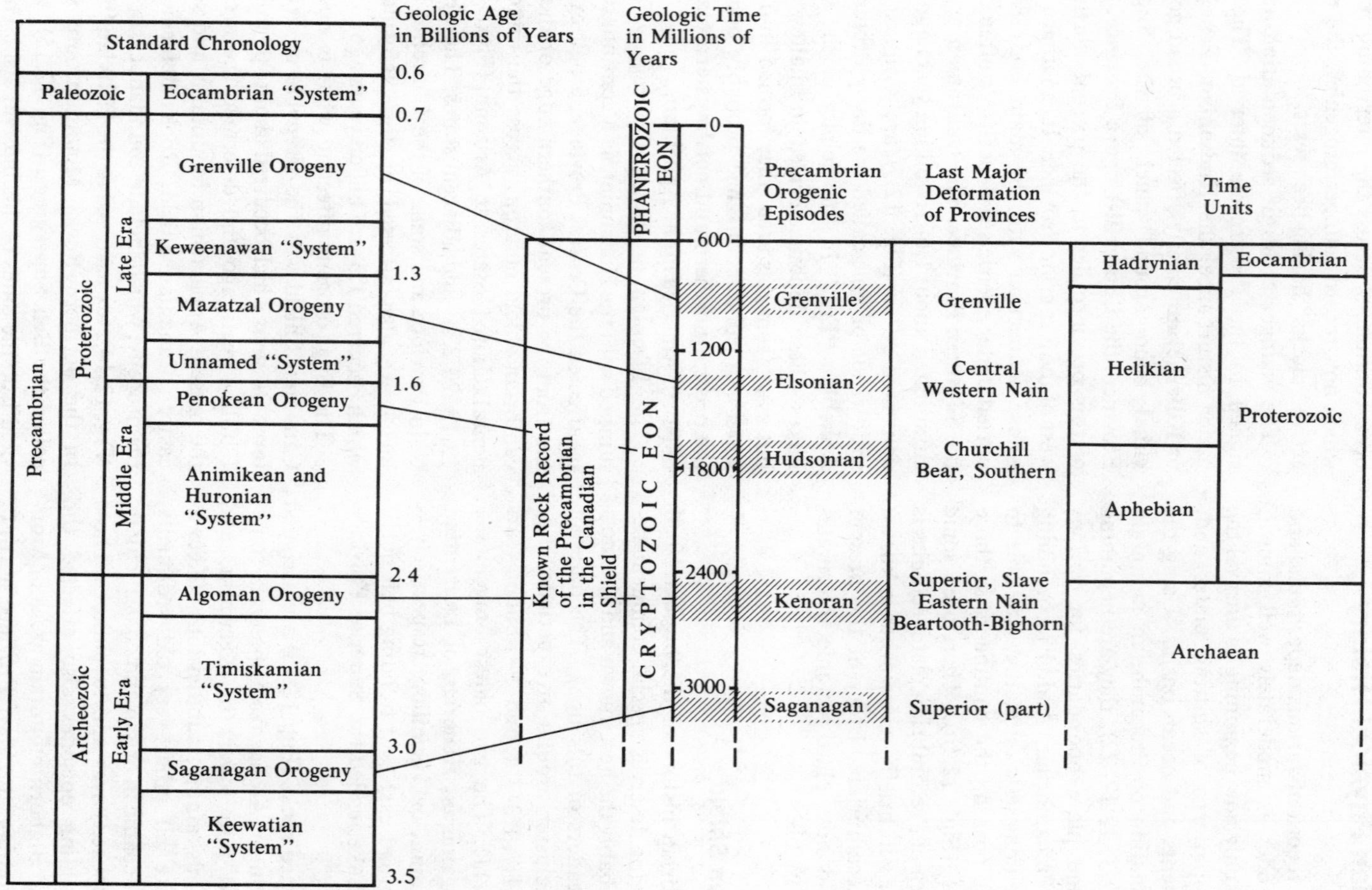

Figure 12.11. *Precambrian History of Canadian Shield.*

Figure 12.12. *Precambrian Structural Provinces of North America.* Compare with Figure 12.3.

Figure 12.13. *Keewatin Volcanic Rocks.* (Courtesy Geological Survey of Canada, Ottawa.)

Figure 12.14. *Grenville Limestone Rocks,* Bancroft area, Ontario. (Courtesy Geological Survey of Canada, Ottawa.)

wenawan rocks (Fig. 12.8) near Lake Superior also date from the Helikian.

The last segment of Precambrian time in North America, represented only in the sedimentary and metamorphic rocks of the Appalachian and Cordilleran mountain belts deformed after the Precambrian, is assigned to the Late Proterozoic or **Hadrynian "Era."**

The geographic location of North America in the Archeozoic is still not known with certainty. In the Early Proterozoic, the continent was essentially upside down in relation to its present position. Its present northern margin lay close to the equator and the tillites of the Great Lakes area formed in high latitudes (Fig. 12.16A). In the Middle to Late Proterozoic, the continent rotated clockwise through about 90°, so that eventually the eastern margin with its abundant carbonates lay close to the equator (Fig. 12.16B). Northern portions of the British Isles were apparently part of North America at this time.

Other Cratons

The **European Craton** has Precambrian rocks exposed primarily in the Baltic Shield, though they are also known in various ancient mountains of northern Europe and just beneath the surface in the Ukrainian part of Russia (Fig. 12.17). Archeozoic strata are gneisses and metamorphosed oceanic sediments as they are in other parts of the world. A group of metamorphosed shallow- and deep-water sediments comprise the **Karelidic "Cycle"**, which ended in deformation 1.8–1.7 billion years ago. A succeeding sequence of metamorphosed sediments and volcanics were deformed 1 billion years ago. Island-arc deposits, including supposed tillites, were formed in the Late Proterozoic and underwent Early Cambrian orogeny.

The **Gondwana** continents, South America, Africa, India, Australia, and Antarctica, were apparently amalgamated into the great supercontinent during the Late Precambrian-Early

Figure 12.15. *Belt Rocks,* Glacier National Park, Montana. (Courtesy William M. Mintz.)

Paleozoic **Pan-African Orogeny** of .7–.5 billion years ago. Prior to that, there were apparently many smaller continents, including four in Africa alone (Figs. 12.18–22). Archeozoic rocks occur on most of them and conform to the general oceanic pattern described above. South Africa was in polar latitudes about 2.7 billion years ago as was East Africa 1.3 billion years ago. Probable tillites are known from both intervals. Mountain building occurred in many parts of present South America, Africa, and India approximately 2 billion years ago. Most of the Gondwana continents also experienced mountain building 1 billion years ago. Widespread tillites occur at the close of the Precambrian and in the Eocambrian of Gondwanaland, particularly those parts comprising present-day Africa and Australia (Fig. 12.10).

The **Chinese Craton** contains Archeozoic metamorphics and intrusives overlain by sedi-

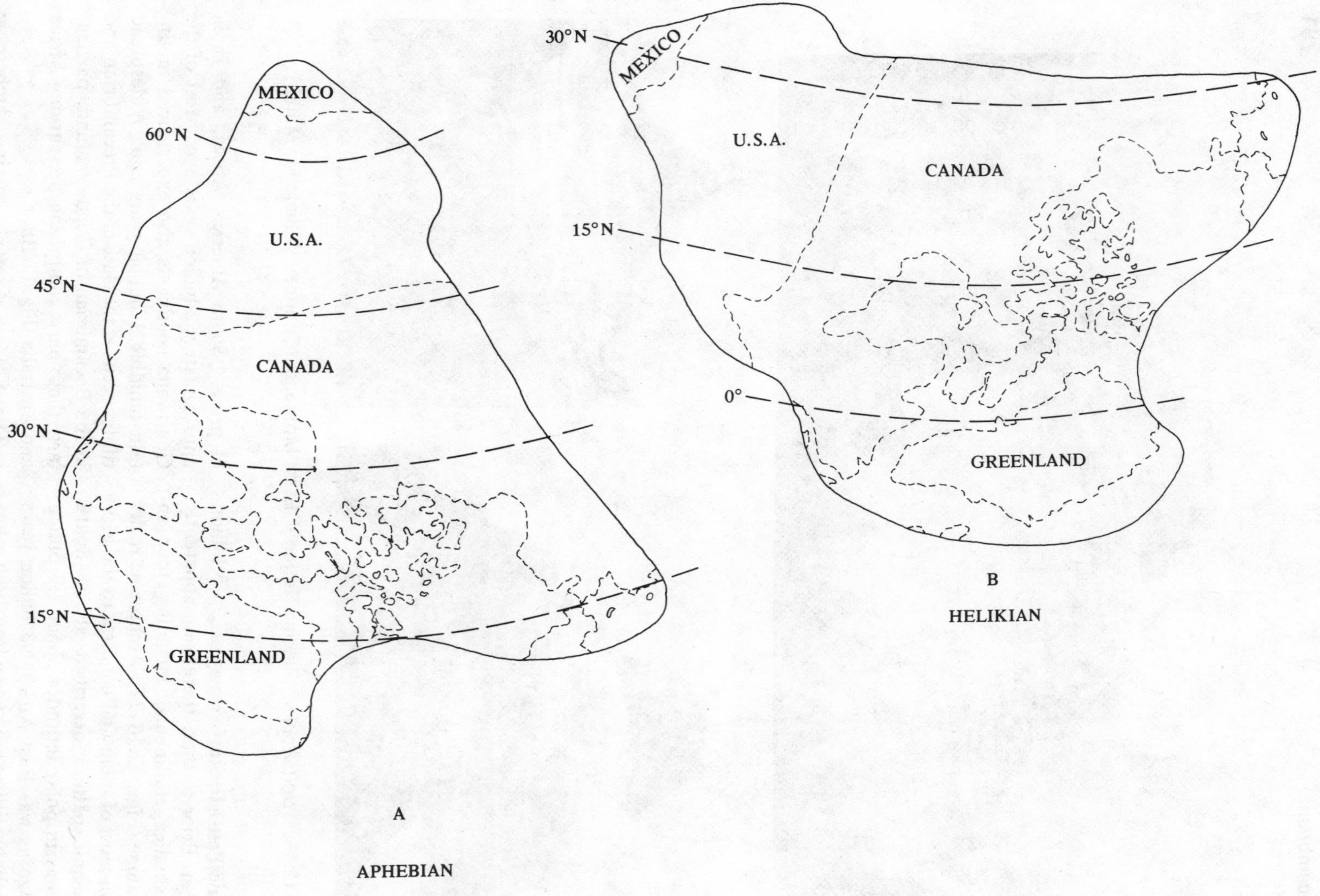

Figure 12.16. *Paleolatitudes for North America in the* **A** *Early and* **B** *Middle Proterozic.* Fine dashed lines show present shorelines and national boundaries.

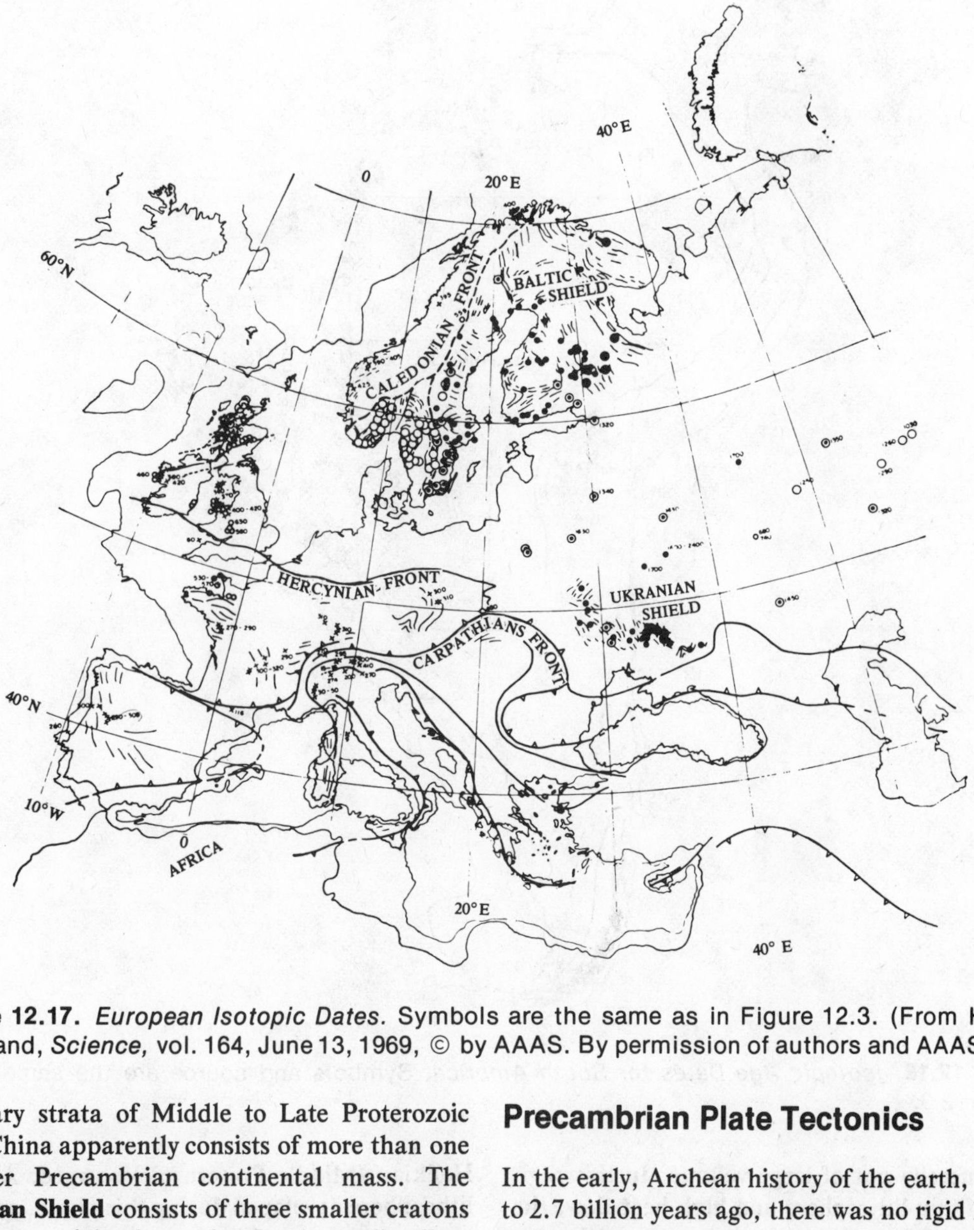

Figure 12.17. *European Isotopic Dates.* Symbols are the same as in Figure 12.3. (From Hurley and Rand, *Science,* vol. 164, June 13, 1969, © by AAAS. By permission of authors and AAAS.)

mentary strata of Middle to Late Proterozoic age. China apparently consists of more than one smaller Precambrian continental mass. The **Siberian Shield** consists of three smaller cratons which were joined together in the earlier Precambrian. Archeozoic gneisses are overlain by metasediments and metavolcanics deformed 1.8 billion years ago. The overlying **Sinian** sediments grade upward from Proterozoic into the Paleozoic. At this time, present western Siberia was in tropical latitudes and limestones were widespread, though possible tillites are also known (Fig. 12.10).

Precambrian Plate Tectonics

In the early, Archean history of the earth, prior to 2.7 billion years ago, there was no rigid lithosphere and, hence, no plate tectonics. In the absence of continents, orogenic regions were broad areas of basaltic lavas and oceanic sediments, such as the Keewatin of present-day North America. These rocks, called **greenstones,** resemble those of modern island arcs but are not confined to narrow belts. Granitic material also differentiated from the mantle, but sedimentary sequences formed from these areas

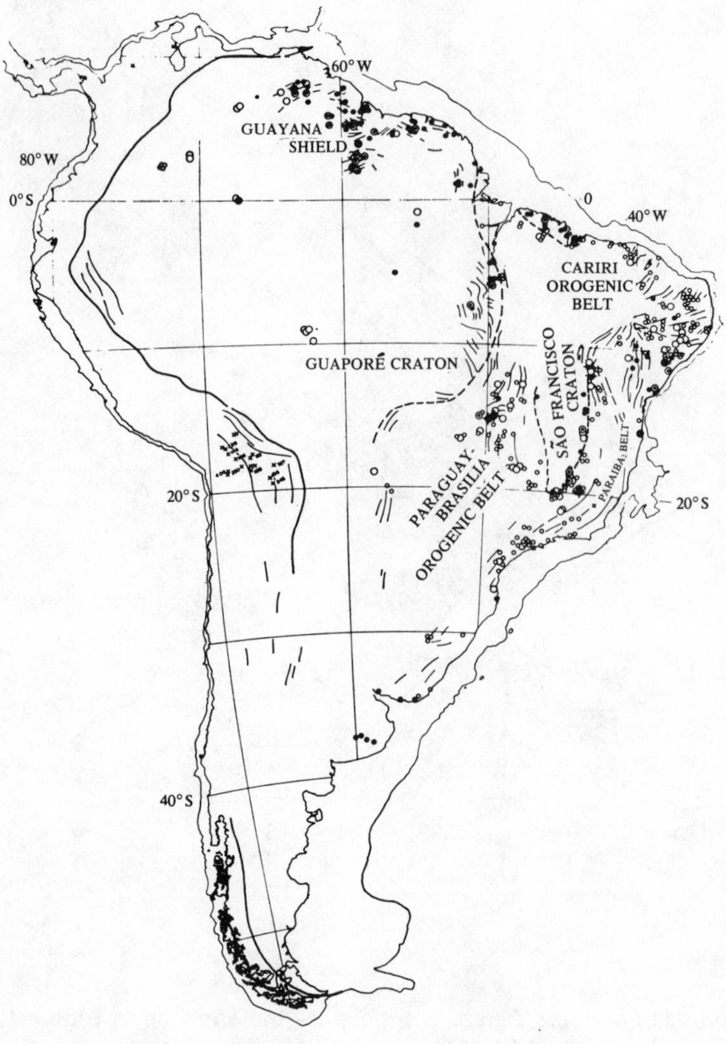

Figure 12.18. *Isotopic Age Dates for South America.* Symbols and source are the same as in Figure 12.3.

are generally not of large volume. In this regimen of thin lithosphere and high heat flow, the basaltic and granitic terrains were subject to continued reactivation. As noted earlier, there are wide differences of opinion among geologists as to whether the basaltic or granitic material represents the prime, original differentiate of the mantle.

During the Aphebian (Early Proterozoic) there appears to have been a transitional phase from this earlier permobile phase to the plate tectonic regime which has dominated from the Helikian (Middle Proterozoic) onward. A rigid lithosphere developed during this phase, and the older rock associations and structures declined in importance. Stable cratons began to form, resulting in thicker sedimentary sequences and the increasing localization of mobile belts. Aphebian cratons contain numerous narrow, failed oceanic rifts, pointing up their instability.

By the Helikian, the presently dominant plate tectonic phase had become established. In the Helikian, however, there were less sharply defined plate margins and a thinner lithosphere

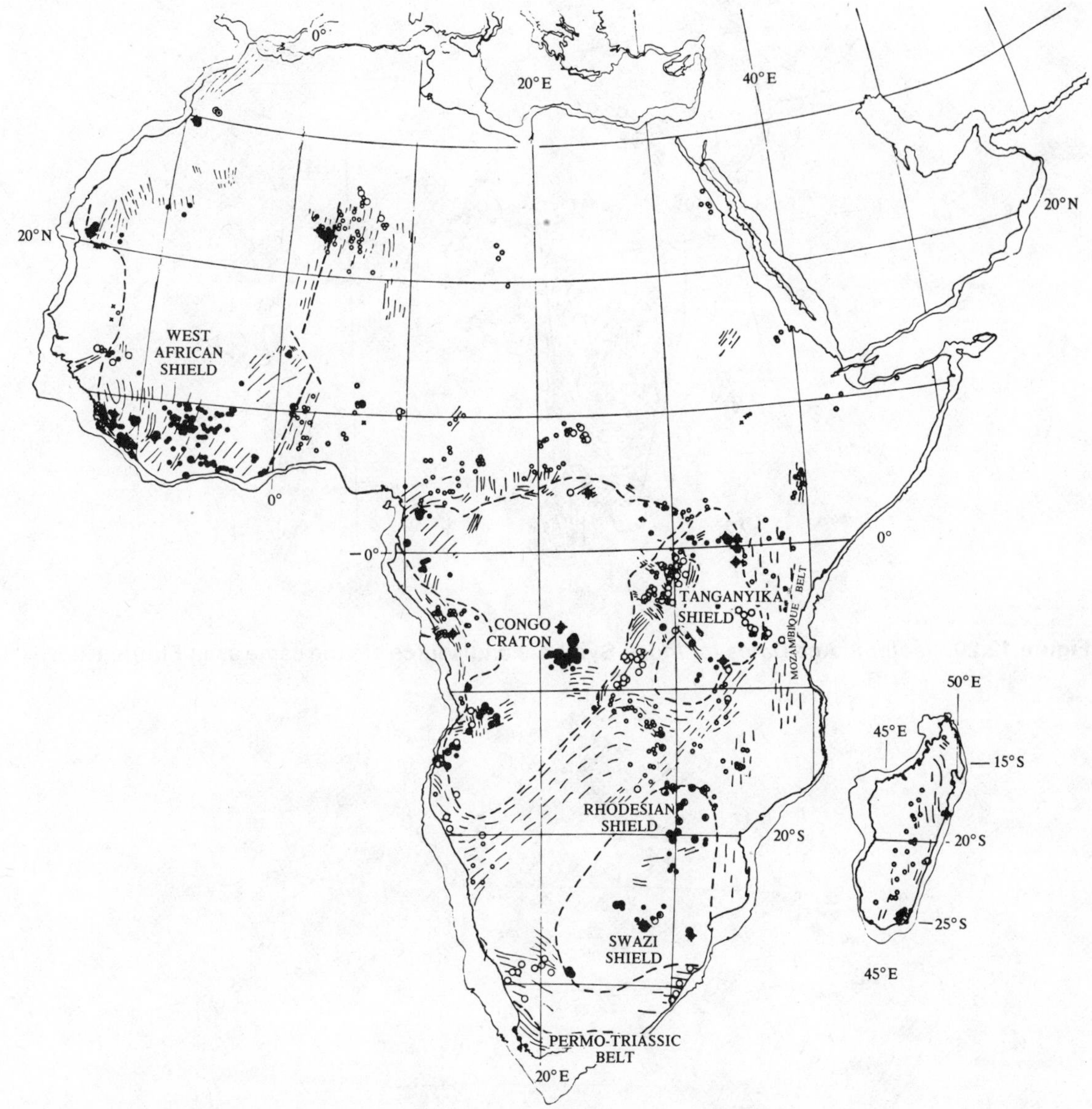

Figure 12.19. *Isotopic Age Dates for Africa.* Symbols and source are the same as in Figure 12.3.

than at present. It appears that continent-continent collisions dominate the plate tectonics phase of the Precambrian because the records of trench-continent and island arc-continent collisions have been altered by subsequent Phanerozoic orogenies in the same areas. Evidences of continent-continent collisions are recognizable between several of the Precambrian African cratons and in the Canadian Shield between the Churchill and Superior Provinces. As geologists continue their studies of Precambrian paleomagnetism, mountain belts, and paleoclimatology, the plate motions of the Proterozoic will become better known, resulting in the paleogeographic reconstructions such as those presently made for the Phanerozoic periods.

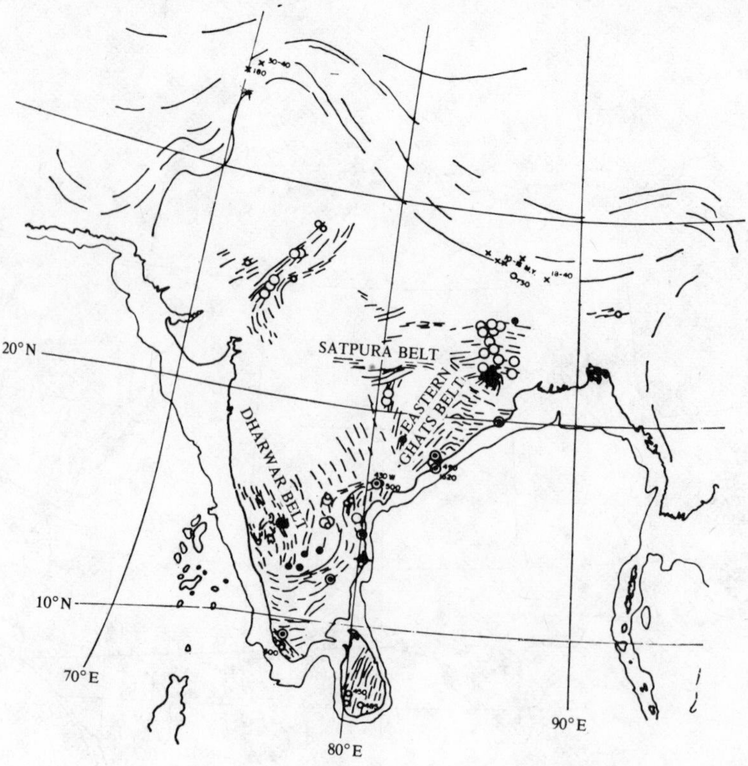

Figure 12.20. *Isotopic Age Dates for India.* Symbols and source are the same as in Figure 12.3.

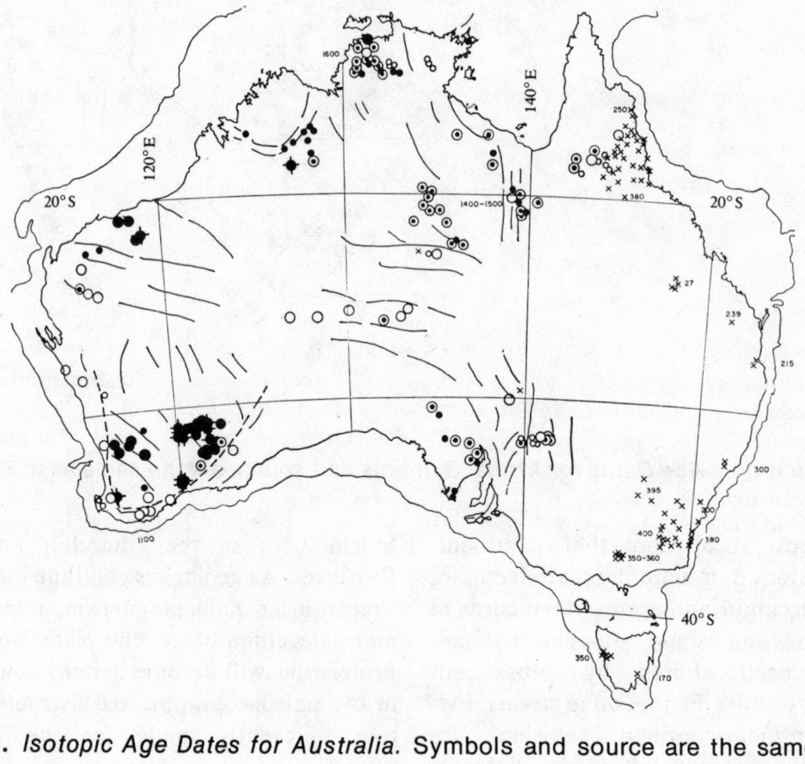

Figure 12.21. *Isotopic Age Dates for Australia.* Symbols and source are the same as in Figure 12.3.

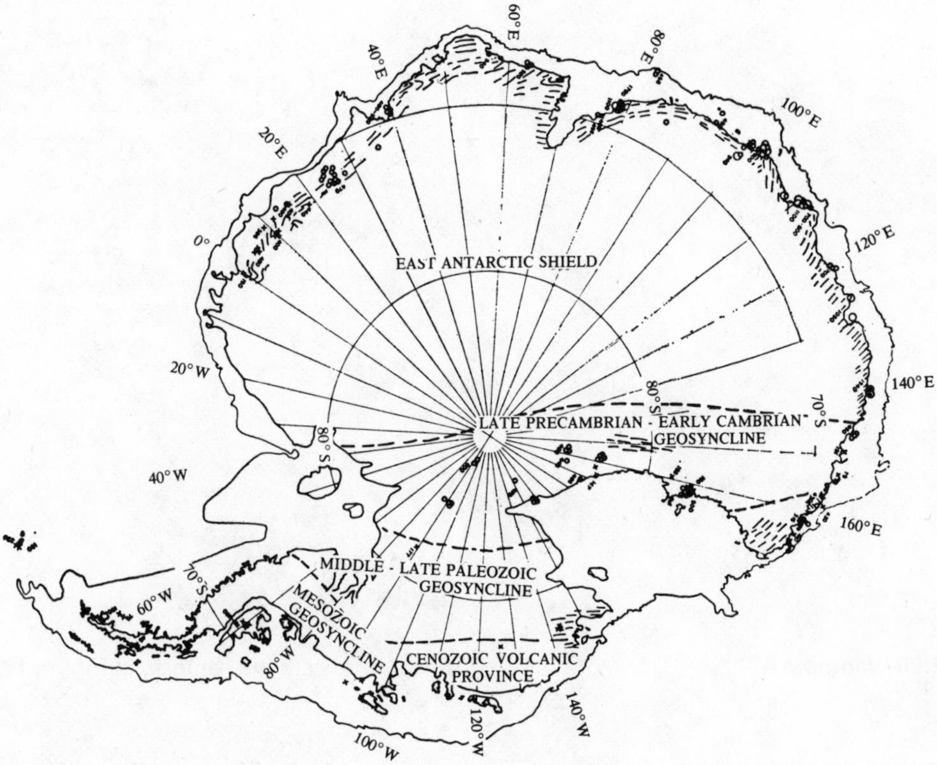

Figure 12.22. *Isotopic Age Dates for Antarctica.* Symbols and source are the same as in Figure 12.3.

REVIEW

1. Explain the differences between the original definitions of **Archeozoic** and **Proterozoic** and their modern usages.
2. What are the major **shields** of the earth today, and what is their significance?
3. What are the major Precambrian rock types, and what may they signify?
4. What do geochemists believe is the major significance of the **banded iron formations?**
5. What is the evidence for Precambrian and Eocambrian glaciations?
6. How can the presence of Eocambrian **tillites** near the equator be explained?
7. How do geochemists account for the rarity or absence of carbonates and sulfates in the Precambrian?
8. What are the Precambrian provinces of North America, and when were they last deformed?
9. Briefly describe what is known about Precambrian plate tectonics.

13

The Great Enigma: Precambrian versus Cambrian Life

13

The Oldest Fossils

We noted earlier that the oldest known sedimentary rocks, the approximately 3.5-billion-year-old **Swaziland Group** of South Africa, contain tiny spherical bodies resembling blue-green algae (Phylum Cyanophyta). The somewhat younger 3.1-billion-year-old **Fig Tree Group** which overlies it contains microscopic bodies (Fig. 13.1) that look like the remains of both **bacteria** and **blue-green algae.** In the 2.7–3.0-billion-year-old **Soudan Iron Formation** of Minnesota, similar bodies occur. Biologically produced organic compounds also occur in these rock units, supporting the biologic origin of the spheres.

Thus, at the beginning of the rock record, procaryotic, single-celled monerans had already evolved. The monerans even today are more resistant to ultraviolet radiation than other organisms, which supports the idea that when life arose, these harmful rays reached the earth's surface.

These fossils all occur in bands of cryptocrystalline quartz, or chert, originally gelatinous silica, probably precipitated by the organisms to act as an ultraviolet shield. The bands of chert alternate with the silicic iron oxides of the iron formation which probably represent the disposal of harmful oxygen by primitive photosynthesizers. Iron probably also served as a nonbiologic catalyst in the intracellular reactions that took place in the earlier cells which had not yet evolved their own enzymes.

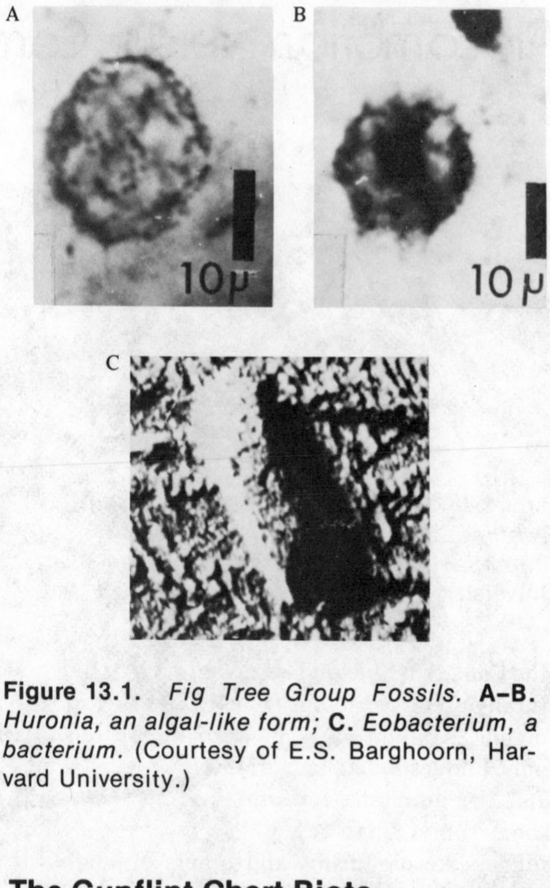

Figure 13.1. *Fig Tree Group Fossils.* **A–B.** *Huronia, an algal-like form;* **C.** *Eobacterium, a bacterium.* (Courtesy of E.S. Barghoorn, Harvard University.)

The Gunflint Chert Biota

The oldest unequivocal remains of a diversity of microorganisms occur in the **Gunflint Chert** of

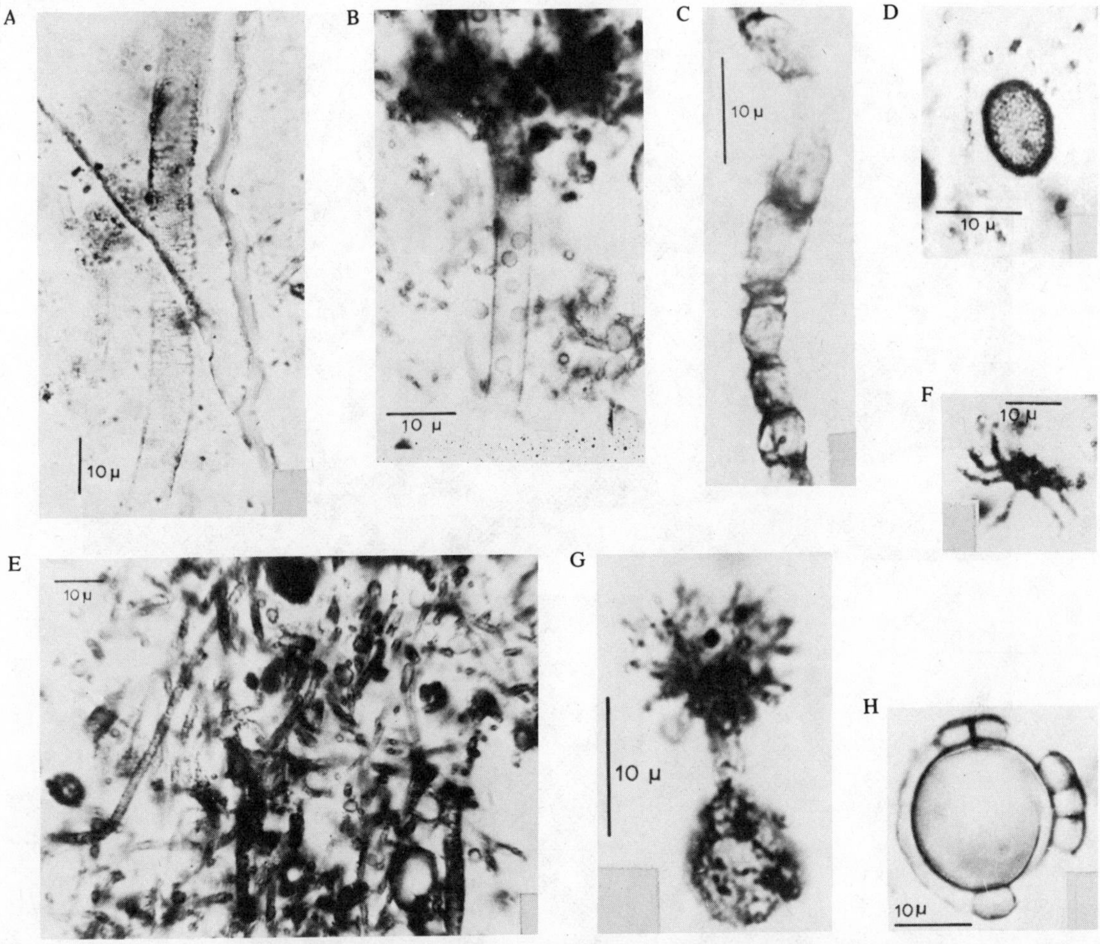

Figure 13.2. *Gunflint Chert Fossils.* **A–C.** *Blue-green algae: Animikia, Entosphaeroides, and Gunflintia;* **D.** *Huroniospora, an algal spore;* **E.** *Gunflintia and Hurionospora;* **F.** *Euastrion, a bacterium, and enigmatic forms;* **G.** *Kakabekia;* **H.** *Eosphaera.* (Courtesy of E.S. Barghoorn, Harvard University.)

the Canadian Shield. These abundant and structurally preserved fossils (Fig. 13.2) are approximately 2 billion years in age and include not only unquestioned bacteria, blue-green algae, and the ammonia-consuming *Kakabekia,* but some things that resemble green algae and fungus-like organisms and others of unknown origin. Some of these forms of uncertain relationships are probably algal cysts and are assigned to the unnatural group of **acritarchs,** pending determination of their affinities.

Stromatolites

At about this same time, 2 billion years ago, the oldest stromatolites appeared. **Stromatolites** (Fig. 13.3) are the concentrically layered mounds of calcium carbonate constructed by blue-green algal colonies as a by-product of their metabolism and trapped by their mucilage. The height of their colonies is indicative of the thickness of the intertidal zone in which this type of algae live (Fig. 13.4). The Precambrian colonies

Figure 13.3. *Precambrian Stromatolites.* **A.** *Lateral sections,* **B.** *Horizontal section.* (Courtesy Geological Survey of Canada, Ottawa.)

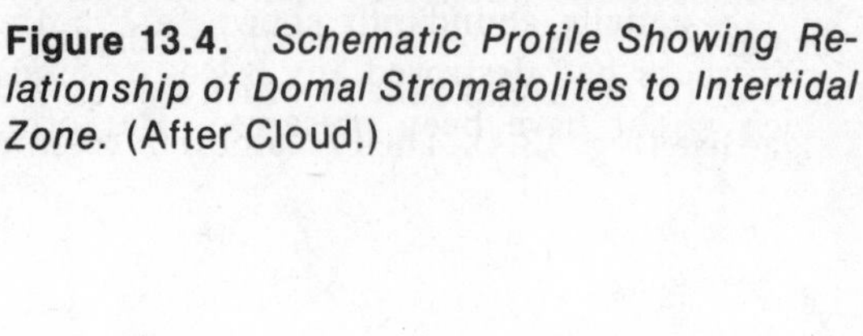

Figure 13.4. *Schematic Profile Showing Relationship of Domal Stromatolites to Intertidal Zone.* (After Cloud.)

are much higher than later ones, which supports the idea that the moon was much closer to the earth then and produced a larger tidal range. The calcareous layers of stromatolites serve as an effective ultraviolet shield today, enabling the cyanophytes to live in very shallow marine waters. (The living forms also occur in fresh waters and on land.)

Up until this time, organisms were undoubtedly virtually all pelagic, having originated in the organic soup of the oceans and occurring in the upper illuminated, photic zone where the energy to generate the first life and later support photosynthesis was available. Only where this narrow band of life in the oceans intersected the bottom at their margins in the sublittoral zone, could benthic organisms develop. Some workers, such as *Weyl,* have felt that the location of this narrow band was the **thermocline,** the zone of rapid temperature change that occurs in the upper levels of the ocean. Water in this zone may have had the proper density to support these floating organisms and been deep enough to be protected from ultraviolet radiation.

At any rate, by 2 billion years ago, oxygen from the layer containing life must have been abundant enough to diffuse widely throughout the oceans, for stromatolites and limestones gradually became abundant. They occur in profusion in Africa, Australia, Canada, China, Scandinavia, the Soviet Union, and the United States. Oxygen was also entering the atmosphere at this time as the banded iron formations ceased and terrestrial redbeds began to form. As free oxygen (O_2) invaded the atmosphere, ultraviolet radiation converted it to ozone (O_3) which began to shield the earth's surface from the harmful ultraviolet rays.

The Bitter Springs Biota

These conditions made possible the evolution of the more efficient and complex eucaryotic cell of protistans. In the 1-billion-year-old **Bitter Springs Formation** of Australia we find not only the usual bacteria, blue-green algae, and organisms of uncertain affinities, but structurally preserved **green algae** (Phylum Chlorophyta), displaying nuclei and even nuclear division (Fig. 13.5). From this point onward in the Precambrian, numerous microfossil localities with similar organisms and biologically produced organic compounds occur. They are known in China, France, Michigan, Montana, Poland, Scotland, Sweden, the Soviet Union, and elsewhere.

Summary of Precambrian Fossils

The Precambrian interval is paleontologically characterized, then, by pelagic microfossils of monerans first and by monerans and one-celled protistans later. Benthic stromatolites produced by blue-green algae also characterize its latter half. Search as they might, geologists have never been able to uncover any unquestioned megafossils of the larger, multicellular animals or metazoans in Precambrian rocks. Various markings and organic residues interpreted as jellyfish, brachiopods, worms, and arthropods were reported from time to time in Precambrian rocks, but all have been rejected because they are of nonbiologic origin, are not Precambrian in age, or were misidentified.

Cambrian Fossils

The Cambrian System stands in marked contrast to this dearth of fossils. At or near the base of the Cambrian, sponges, archaeocyathans, brachiopods, trilobites, and primitive groups of mollusks and echinoderms occur (Fig. 13.6). Worm tubes, perhaps of annelids, also are common. Before the Cambrian was over, the coelenterates (hydroids, conularoids), snails, clams, cephalopods, chelicerate and mandibulate arthropods, crinoids, graptolites, conodonts, and some less important groups appeared. The reason for this remarkable change from pelagic microfossils and benthic stromatolites to a host of shell-bearing benthic metazoans is one of the great mysteries of geology.

Causes of the Difference

Many reasons have been advanced through the years to explain this phenomenon. Some workers suggested that the metamorphism and igneous activity abundantly shown in Precambrian rocks had destroyed any traces of fossils which might have been preserved. But other

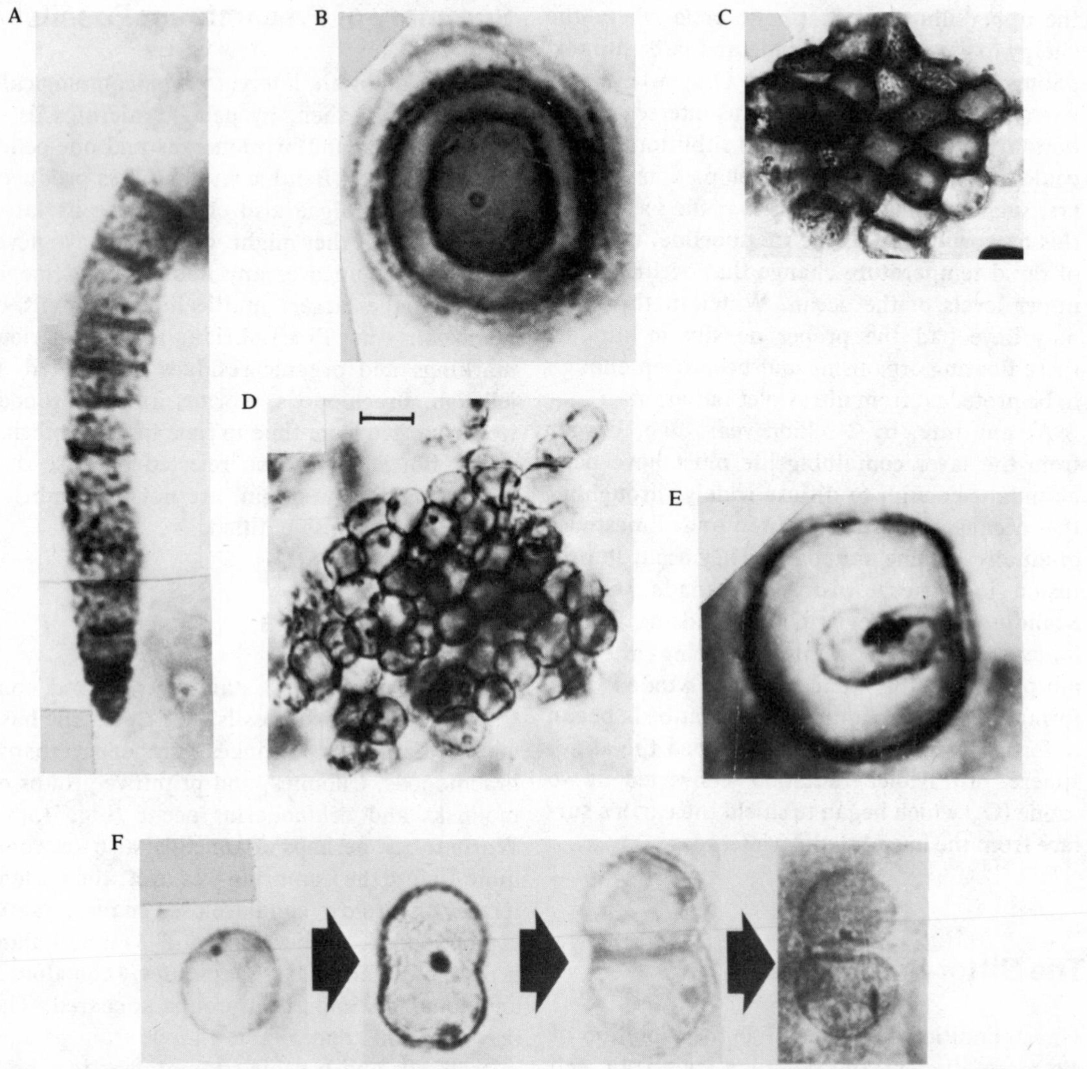

Figure 13.5. *Bitter Springs Chert Fossils.* **A, C, D.** *Blue-green algae: Cephalophytarion, Myxococcoides, and Palaeonacysts;* **B.** *?Dinoflagellate Gloeodiniopsis,* and **E, F.** *Green algae Caryosphaeroides, Glenobotrydion.* (Courtesy of J.W. Schopf, U.C.L.A.)

geologists pointed to the virtually unaltered sedimentary rocks of Montana's Belt Supergroup (Fig. 12.15) and Arizona's Grand Canyon Group (Fig. 12.1). Both contain stromatolites and no other megafossils. Other workers suggested that there was a lack of calcium carbonate in Precambrian seas or that the oceans were acidic. The presence of stromatolites and abundant Grenville carbonates refutes that notion. Some suggested life originated on land and that most Precambrian rocks are marine and thus do not record the event. As noted in Chapter 11, it appears that the physical-chemical conditions for the evolution of life as we know it could only have existed in the sea, however. In addition, Precambrian terrestrial redbeds are common in many areas. Other geologists suggested that Precambrian fossils had either been overlooked

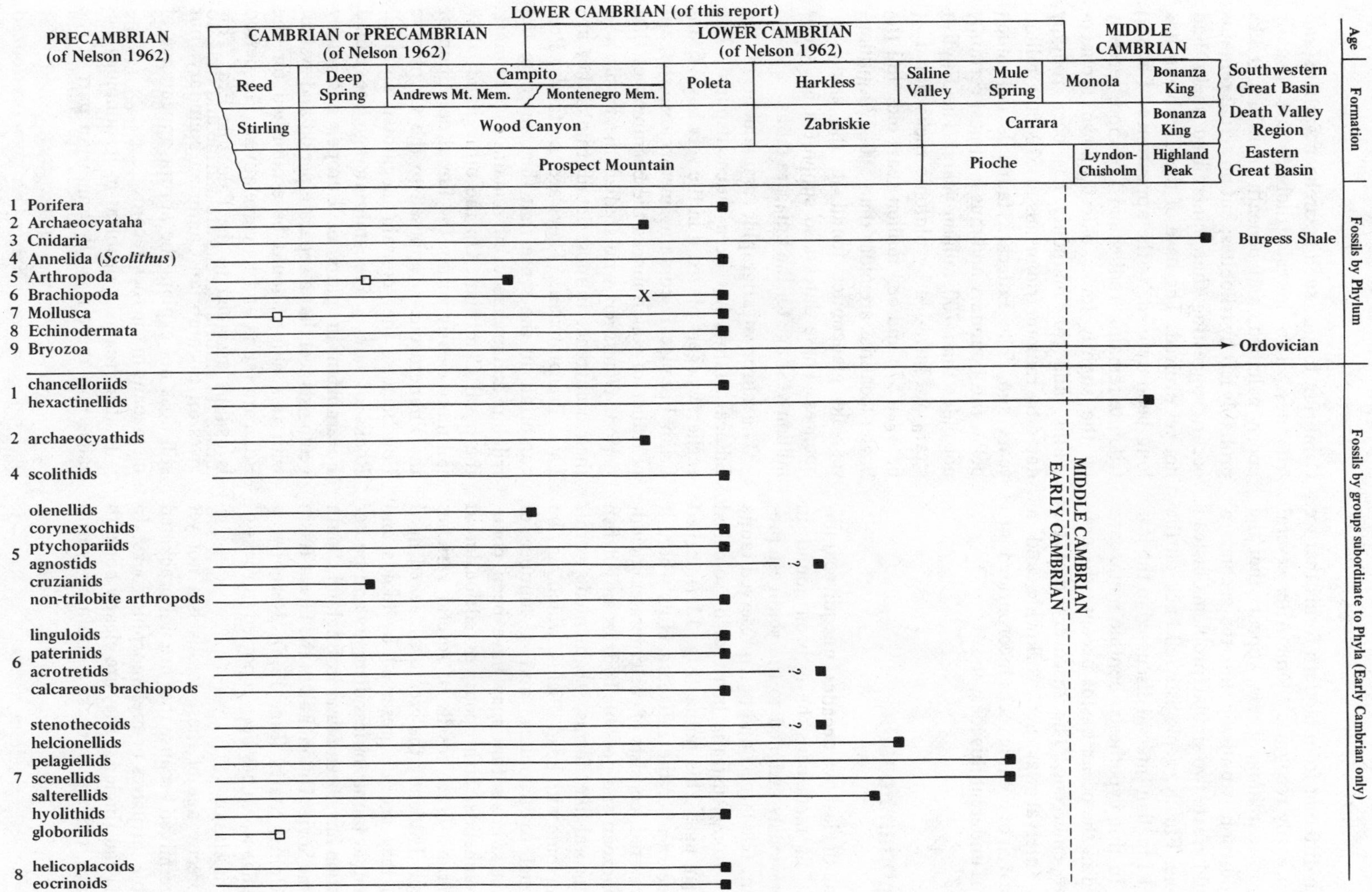

Figure 13.6. *Earliest Records of Animals in Western North America.* (From Preston E. Cloud, Jr., in Drake, *Evolution and Environment*, by permission of Yale University Press.)

or were so different from modern forms that we would not easily recognize them. After several centuries of searching, it now appears that the first supposition is unlikely and the second, if true, would create two similar problems instead of only one. Finally, geologists fell back on the idea of the Lipalian Interval discussed in the last chapter. In this hypothesis, benthic metazoans originated in the ocean basins because the cratons were emergent. The discovery that the Lipalian Interval was not worldwide and a careful search of oceanic rocks incorporated into mountains undercut this idea, too.

Eocambrian Fossils

Finally, in the last two decades, unequivocal impressions of metazoans have been found in strata previously assigned to the youngest Precambrian. The largest and best-preserved fauna is known from Australia in rocks of so-called Adelaidian age. This fauna (Fig. 13.7), called the **Ediacara** after the Ediacara Hills where it was discovered, consists of coelenterates including scyphozoan and possible hydrozoan jellyfish and octocoral-like forms, segmented worms including some with head shields, which may be transitional to trilobites, and a number of enigmatic animals that may have been conularoids, annelids, arthropods, or edrioasteroid echinoderms. Trace fossils of several varieties also occur. Evidently the creatures lacked hard parts as they are all preserved by molds and casts. Similar faunas of less diversity and poorer preservation have been found in England, South Africa, the Soviet Union, and in the Great Basin of the United States (Fig. 13.8). Recently, a possible flat-worm has been reported in rocks of this age in Alaska.

The precise age of these fossils has not yet been determined because they are not bracketed by datable intrusives or metamorphic rocks. In Australia and England, they are clearly beneath and, hence, older than rocks containing typical Cambrian fossils, such as archaeocyathans and trilobites. The lowest trilobites are usually poorly calcified, and beneath them are tracks probably left by trilobites; but no actual remains occur, suggesting that mineralized shells had not yet evolved. The base of the Cambrian has long been defined by the appearance of actual trilobite remains (redlichioids) or more recently by the slightly lower oldest known archaeocyaths, and this boundary has been roughly dated by radiometric means at 570–600 million years ago. The Ediacara fauna is not much below this boundary and thus is almost certainly not older than 700 million years. The English strata are intruded by plutonic rocks dated at between 574 and 684 million years old, and the Soviet localities are older than 590–670-million-year-old glauconites found in the same sequences. These dates also support a 600–700-million-year age for the Ediacara fauna.

The chronostratigraphic assignment of the Ediacara fauna has been much disputed. The earliest workers placed it in the very basal Cambrian because it contains metazoans, the appearance of which marks the beginning of that period. *Glaessner,* who studied the fossils and their occurrence in most detail, thinks they are Late Precambrian. They are certainly Precambrian in the sense that they antedate the trilobite and archaeocyath remains used to mark the basal Cambrian. On the other hand, they fall more clearly within the literal definition of the Phanerozoic as opposed to the Cryptozoic Eon because they contain megafossil animals. Rocks of latest Precambrian age are called **Eocambrian** in parts of Europe, and many geologists feel that a separate period and system with this name should be established for the Ediacara fauna and its correlatives. The period is usually assigned to the Paleozoic Era. This gives an Eocambrian-Cambrian boundary, but still does not explain the abrupt appearance of metazoans in the rock record.

It also creates a problem in terminology because now there is a Precambrian part of the

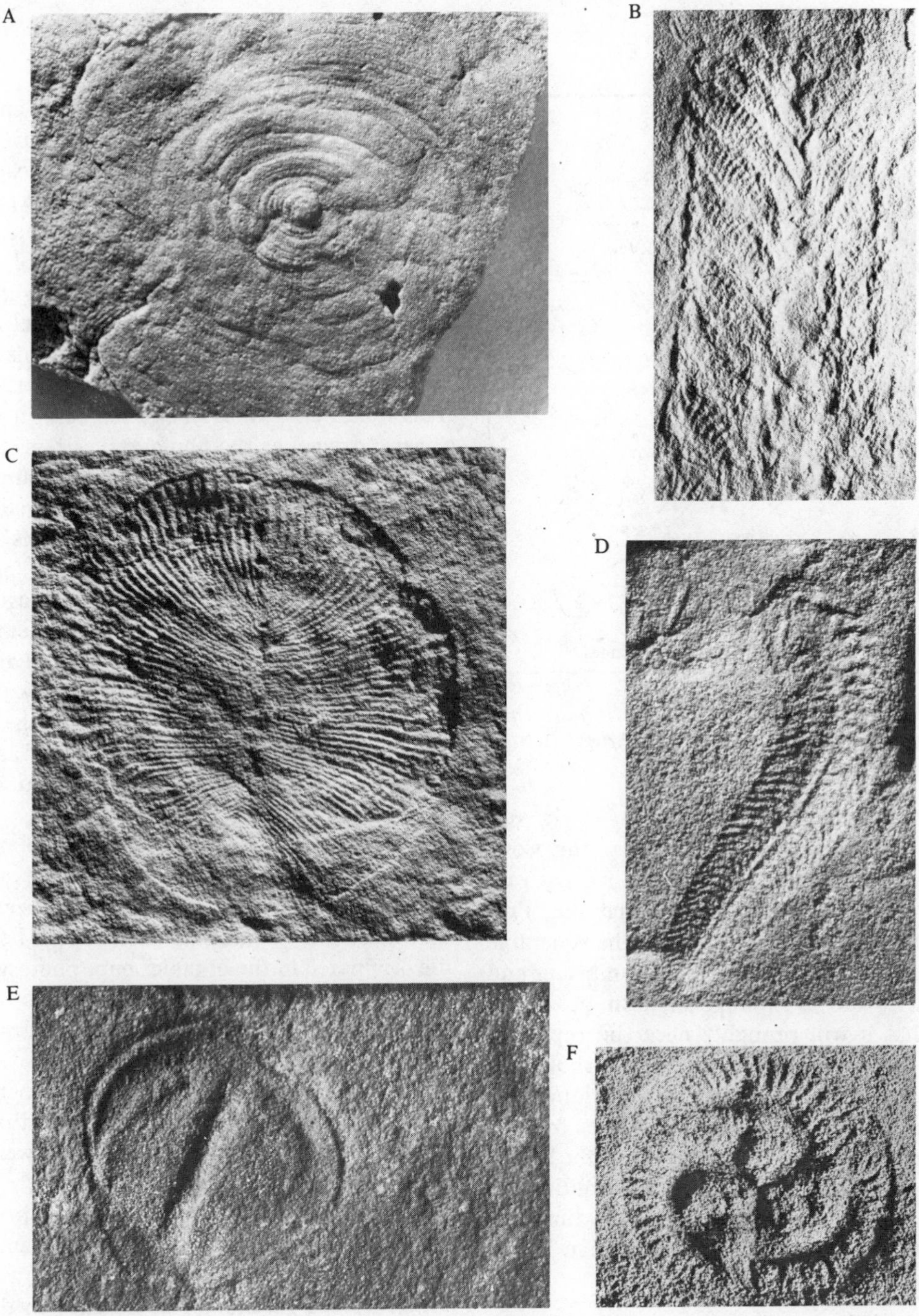

Figure 13.7. *Ediacara Eocambrian Fauna.* **A.** *Cyclomedusa,* a jellyfish; **B.** *Rangea,* an octocoral; **C.** *Dickinsonia,* a jellyfish or a worm; **D.** *Spriggina,* a segmented worm; **E.** *Parvancorina,* probably an arthropod; **F.** *Tribrachidium,* possibly an echinoderm. (Photographs courtesy of M.F. Glaessner.)

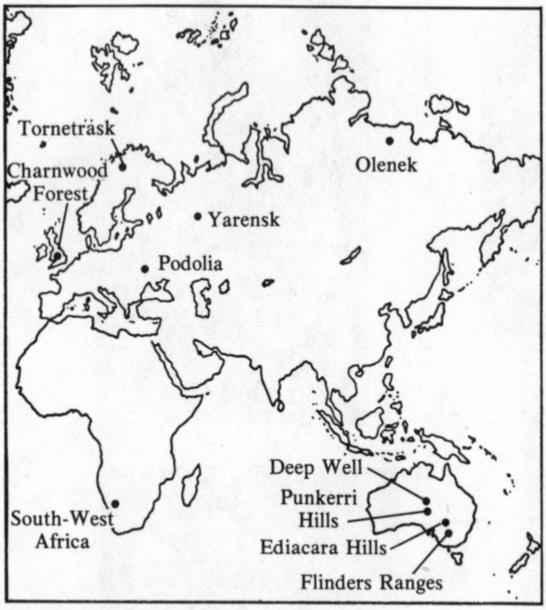

Figure 13.8. *Geographic Distribution of the Ediacara Fauna in the Old World.* (After Glaessner.)

Paleozoic Era. Strict interpretation of the derivation of a word means that now the word *Precambrian* is no longer available for the vast interval before the Eocambrian, and that Prepaleozoic must be substituted for the venerable term. But Prepaleozoic is still Precambrian, and the latter word is so ingrained in geological usage that it will probably never be replaced. Nor is there any strict reason why it should. If biologists changed the name of a phylum every time they added or subtracted a species from it, chaos would result. In the interest of the stability of nomenclature, this book will continue to use Precambrian for pre-Eocambrian time and rocks, while recognizing an Eocambrian Period and System assigned to the Paleozoic.

New Theories

This still does not resolve why the metazoans occur so late and relatively suddenly in geologic history. Their late occurrence has been correlated with the attainment of sufficient levels of atmospheric oxygen required for the metabolism of such large and complex organic entities. *Berkner* and *Marshall* and others have suggested that not until the amount of atmospheric oxygen reached one to three percent of its present level could metazoans evolve. Evolution could occur relatively rapidly once the level of atmospheric oxygen had built up to this point because when it did, the harmful ultraviolet radiation would be screened out and organisms could invade much more of the upper levels of the sea. This would cause a dramatic increase in atmospheric oxygen as the photosynthesizers rapidly filled the new space by increasing in numbers. The sudden increase in oxygen may well have made possible the evolution of metazoans. Likewise, the corresponding decrease in carbon dioxide, because of its uptake by photosynthesizers may have caused a decrease in temperature, leading to the ice age that occurred around the Precambrian-Eocambrian boundary. Thus, the physical and biological histories of the earth appear to be inextricably linked again as they are throughout geologic history. One summary of these theories is shown in Figure 13.9.

Weyl has recently revived another plausible explanation for the rapid appearance of the metazoans by giving it a new twist. He reasoned that most early organisms were pelagic because life originated in the organic soup, photosynthesizers had to remain near the surface to receive sunlight (but not at it because of ultraviolet radiation), and their density held them in the thermocline, or zone of rapid temperature change. There were no predators, with the possible exception of protozoans, so shells were not necessary for protection. Furthermore, a shell would cause an organism to sink, which would be lethal when the oxygen was still concentrated in the thermocline and above because it was not yet generated in enough abundance to diffuse throughout the oceans. Thus, most Precambrian life was shell-less and pelagic. Tiny floating microorganisms predominated except where the thermocline intersected the bottom in a relatively narrow band around the oceans where the early stromatolite-forming algae lived.

Only when oxygen had been produced in enough abundance and diffused downward

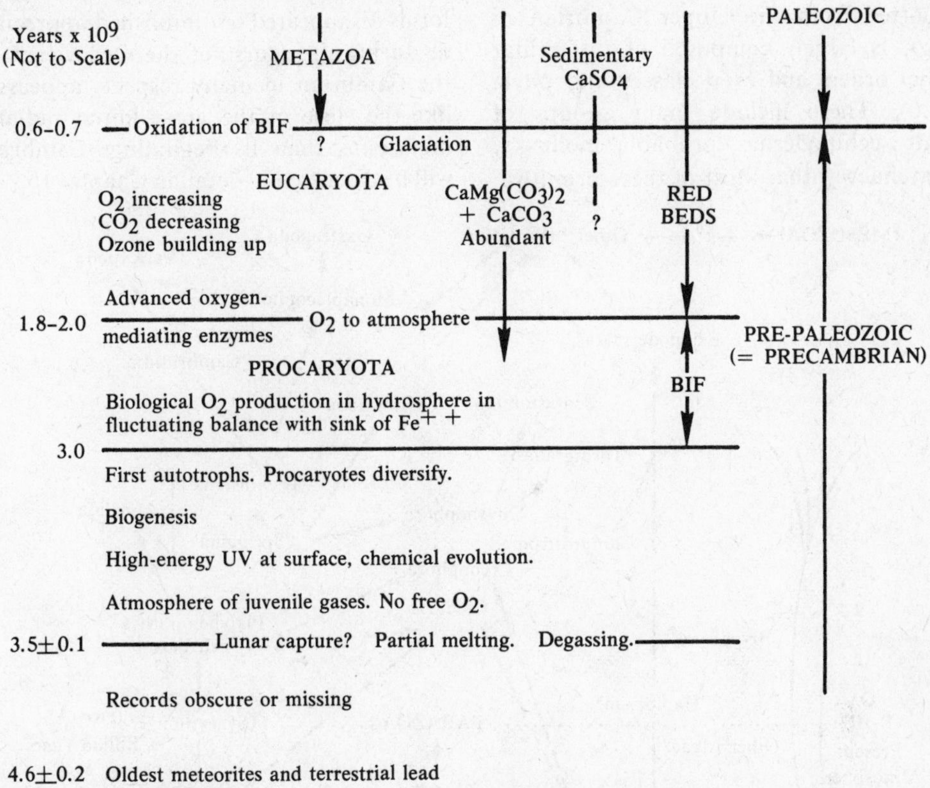

Figure 13.9. *Early Evolution of the Earth, Life, and Atmosphere.* (From Cloud, *Science*, vol. 160, May 17, 1968, Copyright 1968 by AAAS. Used by permission of author and AAAS.)

through the thermocline and throughout the water masses of the oceans could an extensive benthonic fauna develop. According to Weyl, this occurred at the beginning of the Eocambrian when benthos first appear. A shell was then a favorable adaptation to keep the animal on the bottom, and several lineages independently evolved the ability to mineralize shells and skeletons by or during the Cambrian. Large size was no longer the hindrance it is to floaters, so multicellular organisms could evolve.

Weyl's hypothesis is supported by the fact that there are no metazoan trails or tracks known before the Eocambrian. If metazoans had been around earlier without shells, they would certainly have left trails or tracks, but none are known. Yet they appear in some abundance in the Eocambrian, indicating that before that time, organisms were probably pelagic, hence, tiny and shell-less as postulated by Weyl. However, the presence of abundant and widespread blue-green algae, limestone, and redbeds in the latter half of the Precambrian rock record argues that oxygen was generated in some abundance considerably earlier than the Eocambrian. Perhaps, however, it diffused upward into the upper levels of the water and into the atmosphere before it could diffuse downward to any appreciable degree.

For whatever reason caused their apparently sudden appearance in the record, multicellular organisms spread widely in the Eocambrian and by and during the Cambrian had evolved into nearly all of today's major phyla and several probably totally extinct ones (Fig. 13.10). The Cambrian fauna, even though it contains all modern phyla with hard parts except possibly the vertebrates (although fish-like teeth have been reported from the Upper Cambrian of Wyoming), is largely composed of unfamiliar and extinct orders and even classes and phyla (Fig. 13.6). These include many groups of arthropods, echinoderms, probable mollusks, and the archaeocyatha. Most of these primitive forms disappeared or diminished appreciably by or during the course of the Ordovician. Thus, the Cambrian in many respects appears more like the close of the great initial radiation of metazoans than its beginning. Cambrian life will be discussed in detail in Chapter 15.

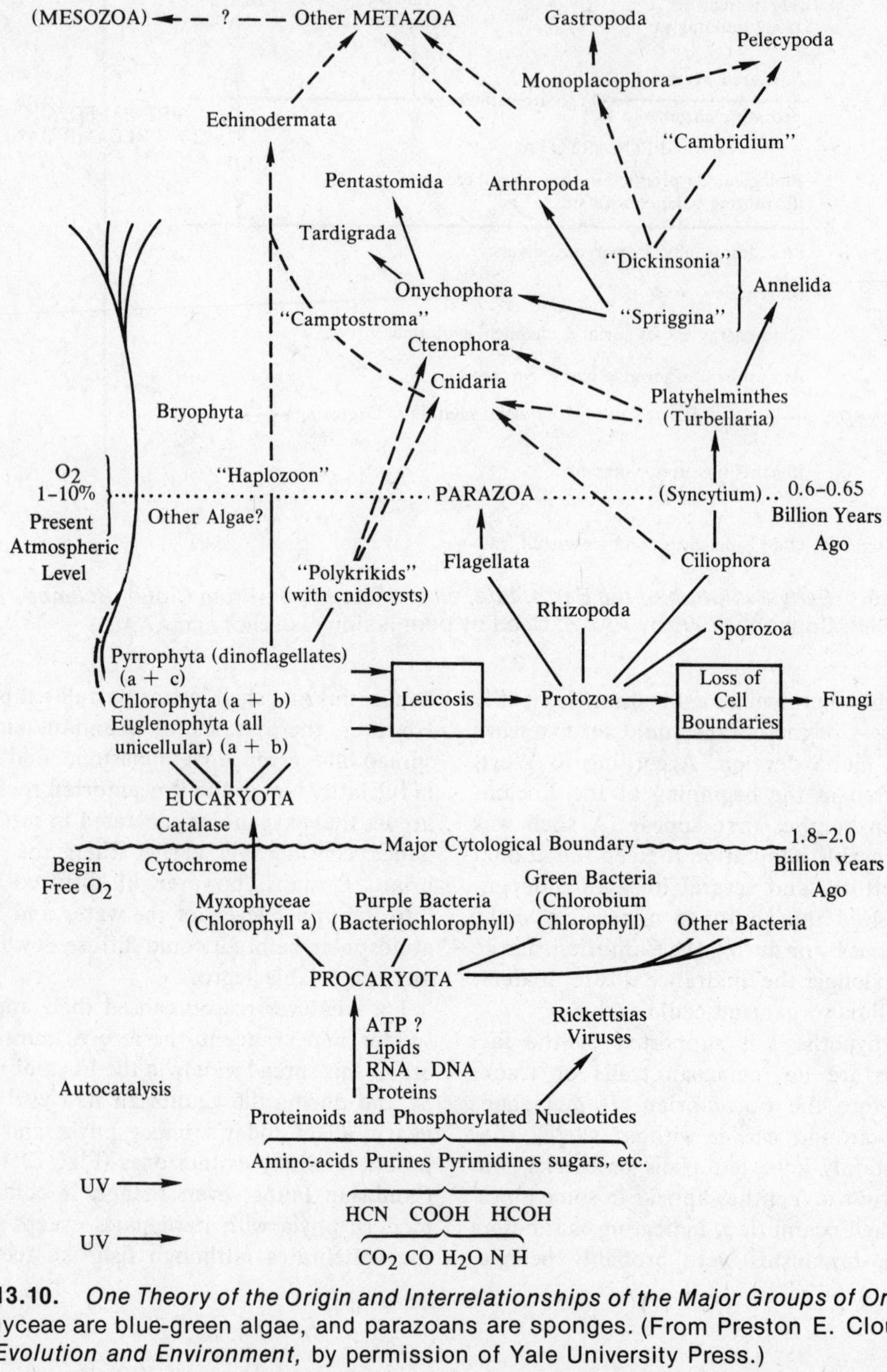

Figure 13.10. *One Theory of the Origin and Interrelationships of the Major Groups of Organisms.* Myxophyceae are blue-green algae, and parazoans are sponges. (From Preston E. Cloud, Jr., in Drake, *Evolution and Environment,* by permission of Yale University Press.)

REVIEW

1. What are the oldest known evidences of life on earth?
2. What is the significance of the **Gunflint Chert biota?**
3. What are **stromatolites,** and when do they appear in the geologic record?
4. What is the oldest unequivocal evidence of eucaryotic organisms?
5. Contrast the typical Precambrian biota with that of the Cambrian.
6. What is the significance of the **Ediacara fauna?**
7. What is the **Berkner-Marshall theory** of the origin of multicellular animals?
8. Describe the **Weyl theory** for the rapid appearance of the metazoans.

14

The Transition Mountains (Übergangsgebirge): Cambrian, Ordovician, and Silurian

14

The three periods of the Early Paleozoic that succeed the Eocambrian—the **Cambrian, Ordovician,** and **Silurian**—encompass approximately 200 million years of earth history from about 600 million to 400 million years ago. All three systems have their type areas in Wales, in or marginal to the Caledonian Mountain Belt (Fig. 2.19). The geologic history of individual regions in this interval is moderately well known; but geologists are just now starting to interpret this history in the larger framework of the new global tectonics, using paleomagnetism, computer fits of shorelines, and more conventional geologic tools, such as the matching of lithologic and biologic trends.

Continental Positions and Plate Tectonics

The probable positions of the continents in the Early Paleozoic are shown in Figures 14.1, 14.2, and 14.3. Plate boundaries are not indicated because knowledge of their locations is only fragmentary. Their possible positions are discussed at various places in the following text.

The evidence suggests that **Gondwanaland,** composed at least of present-day South America, Africa, India, Antarctica, and Australia, was in existence at this time and probably somewhat earlier as well. Late Precambrian-Early Paleozoic mountain belts match between Brazil and west Africa; between eastern South America, South Africa, west Antarctica, and east Australia; and between east Africa, India, east Antarctica, and west Australia (Fig. 14.4). A great mountain-building event, the **Pan-African Orogeny,** occurred between Middle Eocambrian and Middle Ordovician time. It produced these belts by deforming areas between all the pre-existing Gondwana continents, indicating their collision in this interval.

The polar wandering curves for the Early Paleozoic interval are superimposed if the present southern continents are hypothetically reunited into Gondwanaland (Fig. 14.5). Much of Gondwanaland was in the South, or Gondwana, polar region during the Cambrian-Silurian interval as indicated by the paleomagnetic data. The pole "migrated" southeasterly from northwest to south Africa during this time. (Actually, of course, Gondwanaland was moving to the northwest.) The present-day continent of Australia was at the northern edge of Gondwanaland and on the equator.

Deformed mountain belts completely encircle the Gondwana supercontinent when its constituents are reunited. It is uncertain whether the attendant trenches and subduction zones were all in existence in the Early Paleozoic, however. The **Andean-Cape-Borchgrevink-Tasman Trench** appears to probably have existed at this time (Figs. 10.5, 10.33).

The Early Paleozoic location of the various continental blocks which presently make up China is uncertain. Late Paleozoic faunal evidence indicates most may well have been part of

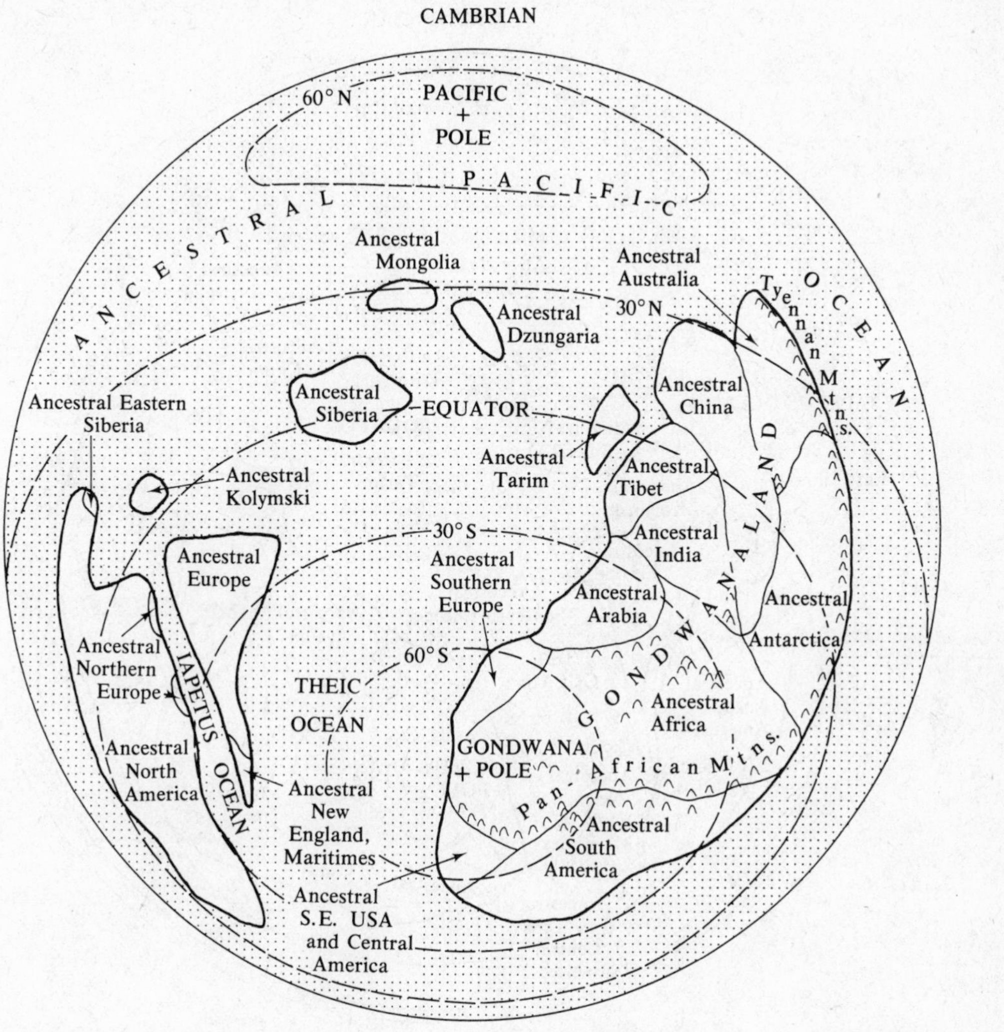

Figure 14.1. *Cambrian Paleogeography.* The positions of China, Tarim, Tibet, Dzungaria, and Mongolia are highly speculative. Plate boundaries and epeiric seas are not shown. (Material derived from Crawford, Dewey and Bird, Irving, Seyfert and Sirkin, and others.) 600 MILLION YEARS

Gondwanaland. They are so indicated on the paleogeographic maps, though it should be kept in mind that these locations are only very tentative.

The present Northern Hemisphere continental masses were many separate blocks in the Early Paleozoic. Paleomagnetic and paleoclimatic evidence suggests that North America was turned about 90° from its present position so that the equator passed through the continent in approximately the present north-south direction (Figs. 14.1–3). The continent gradually rotated counterclockwise through the Early Paleozoic.

The rock record indicates that a proto-Atlantic Ocean, the **Iapetus Ocean,** opened in the Eocambrian and Cambrian (Fig. 14.8), an event which may have triggered the extensive Cambrian epeiric seas of North America. In the

ORDOVICIAN

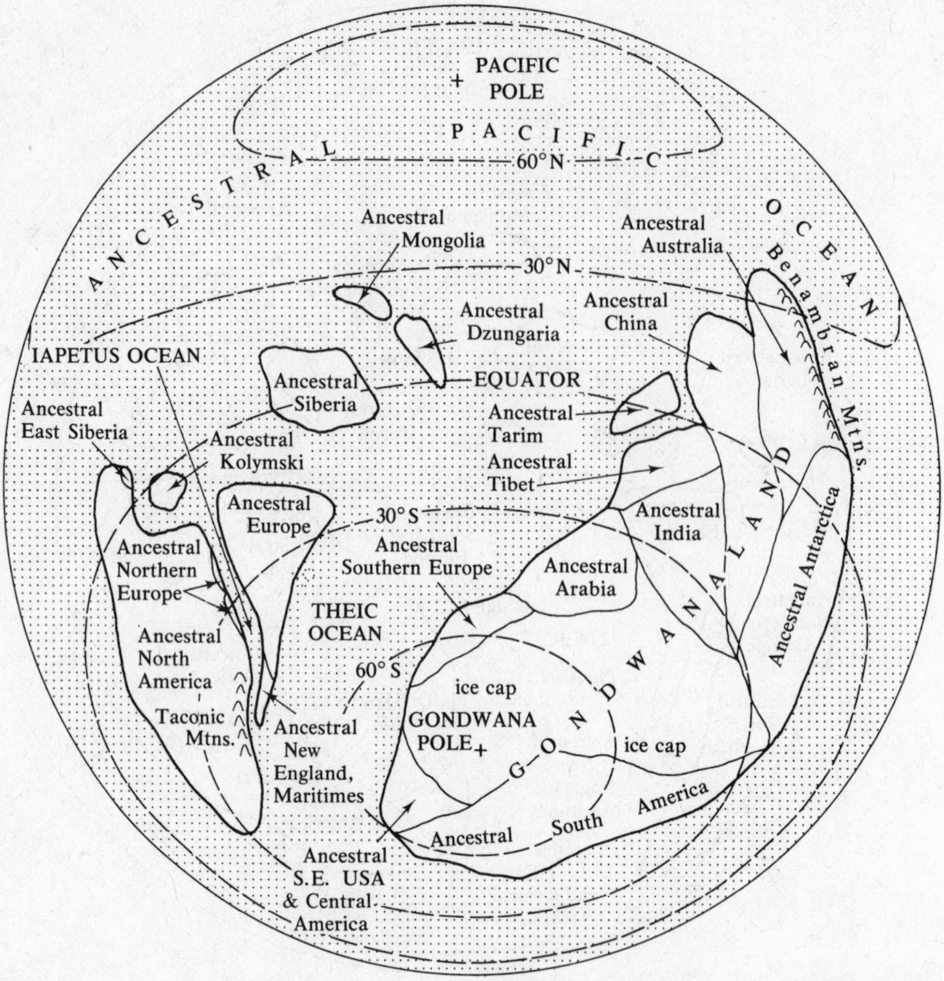

Figure 14.2. *Ordovician Paleogeography.* The positions of China, Tarim, Tibet, Dzungaria, and Mongolia are highly speculative. Plate boundaries and epeiric seas are not shown. (Material derived from Crawford, Dewey and Bird, Irving, Seyfert and Sirkin, and others.)

Ordovician and Silurian (and continuing through the Devonian), this ocean basin closed as the **European Craton** approached North America and Laurasia began to form. The zone occupied by this compressed ocean basin and crumpled continental margins formed the **Northern Appalachian-Caledonian Mountains,** a long Late Paleozoic belt that was split into its current eastern North American and northwest European parts by the Mesozoic rifting that produced the modern Atlantic Ocean. The Paleozoic zone of collision clearly did not parallel the modern shores because parts of both original continents are now found on both sides of the Atlantic.

Ancestral North America contained portions of Spitsbergen, Norway, Scotland, Ireland, and easternmost Siberia of the present day. Most of

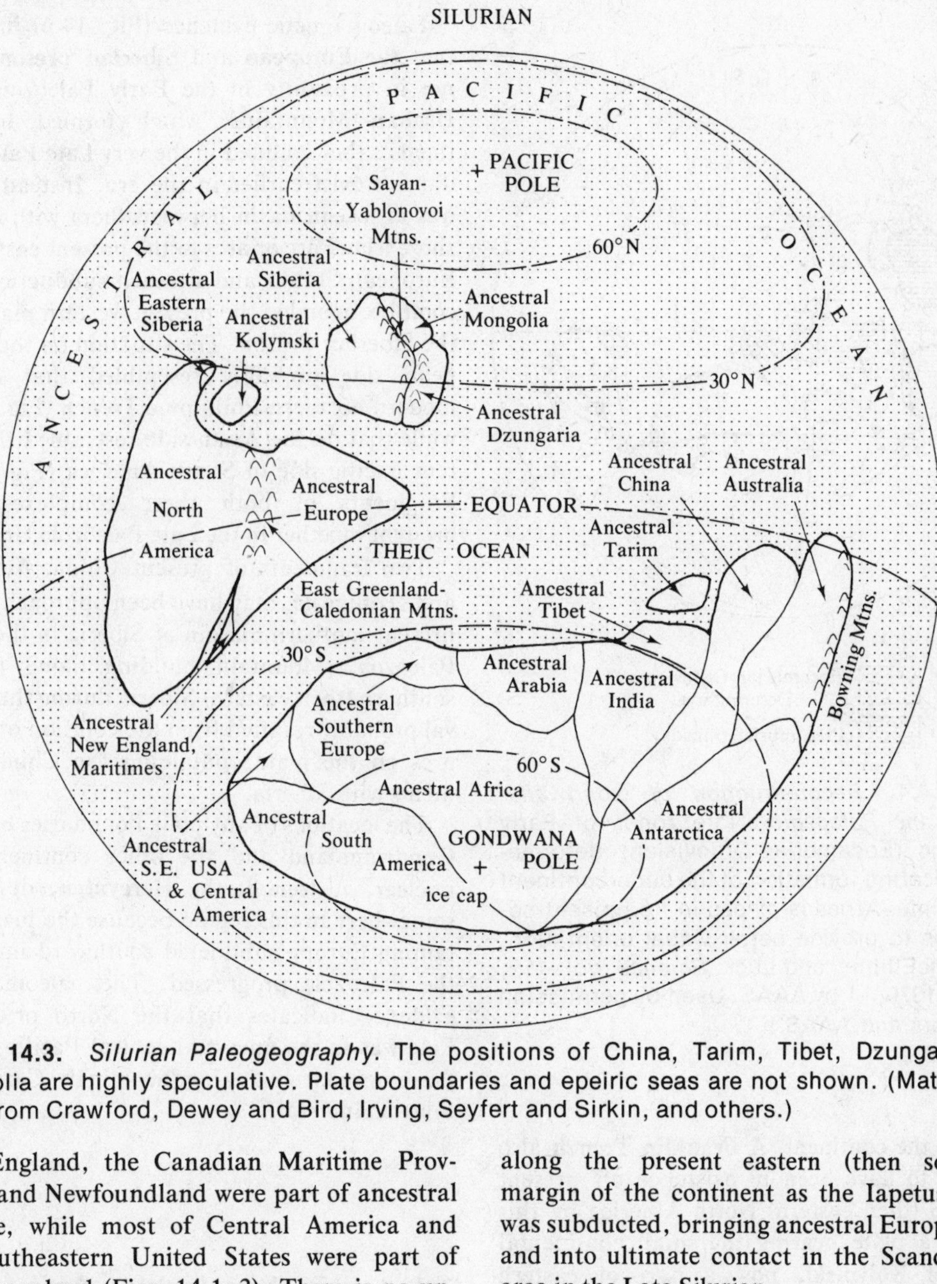

Figure 14.3. *Silurian Paleogeography.* The positions of China, Tarim, Tibet, Dzungaria, and Mongolia are highly speculative. Plate boundaries and epeiric seas are not shown. (Material derived from Crawford, Dewey and Bird, Irving, Seyfert and Sirkin, and others.)

New England, the Canadian Maritime Provinces, and Newfoundland were part of ancestral Europe, while most of Central America and the southeastern United States were part of Gondwanaland (Figs. 14.1–3). There is no unequivocal evidence that any margin of ancestral North America was at a plate boundary in the Cambrian. In the Ordovician, the Northern Appalachian-East Greenland Trench developed along the present eastern (then southern) margin of the continent as the Iapetus Ocean was subducted, bringing ancestral Europe closer and into ultimate contact in the Scandinavian area in the Late Silurian.

By the Ordovician, the **Cordilleran Trench** had developed off present western (then northern) North America, and part of the ancestral Pacific Ocean basin was being subducted

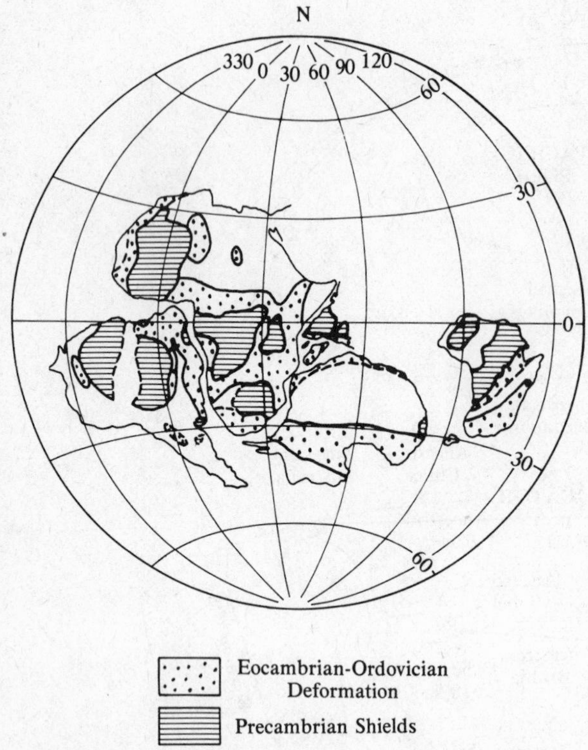

Figure 14.4. *Reconstruction of Gondwanaland in the Paleozoic.* Note zones of Early Paleozoic (Eocambrian-Ordovician) deformation indicating formation of the supercontinent at that time. Africa is placed in its present coordinates to provide better visual orientation. (From McElhinny and Luck, *Science*, vol. 152, May 15, 1970, © by AAAS. Used by permission of authors and AAAS.)

beneath the continent. A **Franklin Trench** also appears to have been in existence off present northern (then eastern) North America by this time as a plate bearing the small continental block of **Kolymski,** now a part of eastern Siberia, approached the ancestral continent. This plate probably was laterally bounded by transform faults in the approximate positions shown in Fig. 14.14. The present southern (then western) boundary of North America may have also had a trench along it or approaching it, but the present evidence is equivocal.

Paleomagnetic evidence (Fig. 14.6) indicates that the **European** and **Siberian cratons** were not in proximity in the Early Paleozoic. The Uralian Mountains, which formed between them as they collided in the very Late Paleozoic, did not exist earlier in the era. Instead, there was an ocean basin between them with an offshore island arc or arcs to the present east of the European Craton, and a trench-subduction zone complex bounded the present western margin of the Siberian Craton. The situation on the European side probably resembled that of the modern northern Philippine Trench (Fig. 14.7), while that on the Asian side resembled the current Pacific side of South America (Fig. 8.15). Fragments of both these complexes were brought together in the Late Paleozoic Ural belt.

Two fragments of present China, **Mongolia** and **Dzungaria,** may have been approaching the present southern margin of Siberia in the Early Paleozoic. Mountain building along present southern (then eastern) Siberia during this interval probably resulted from the collision of island arcs on the plate containing the China fragments with Siberia.

The locations of any plate boundaries between Gondwanaland and the other continents are unclear. A **trench,** the **Hercynian,** developed somewhere in this region because the plate containing Europe subducted southward into it as the Paleozoic progressed. The paleomagnetic evidence indicates that the North or Pacific Pole was in the ancestral central Pacific Ocean far removed from any landmass in the Cambrian-Silurian interval.

Caledonian-Appalachian Area

Several major physical events occurred in the Cambrian, Ordovician, and Silurian history of the earth. As noted earlier, the Cambrian was a time during which a proto-Atlantic or Iapetus Ocean opened (Fig. 14.8), an occurrence which

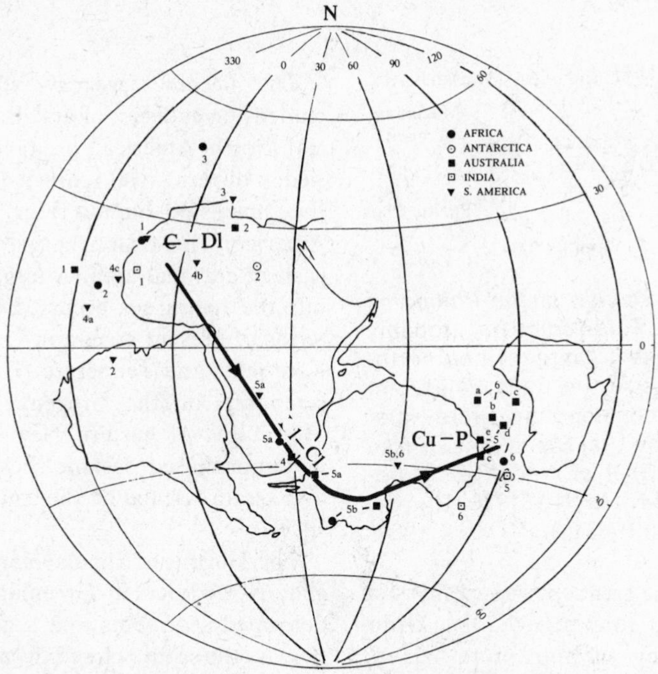

Figure 14.5. *Reconstruction of Gondwanaland by Superposition of Polar Wandering Curves.* Poles plotted for Cambrian-Lower Devonian, Lower Devonian-Lower Carboniferous, and Upper-Carboniferous-Permian. Africa is placed in its present coordinates for visual orientation only. Polar positions for various periods indicated. (From McElhinny and Luck, *Science*, vol. 152, May 15, 1970, © by AAAS. Used by permission of authors and AAAS.)

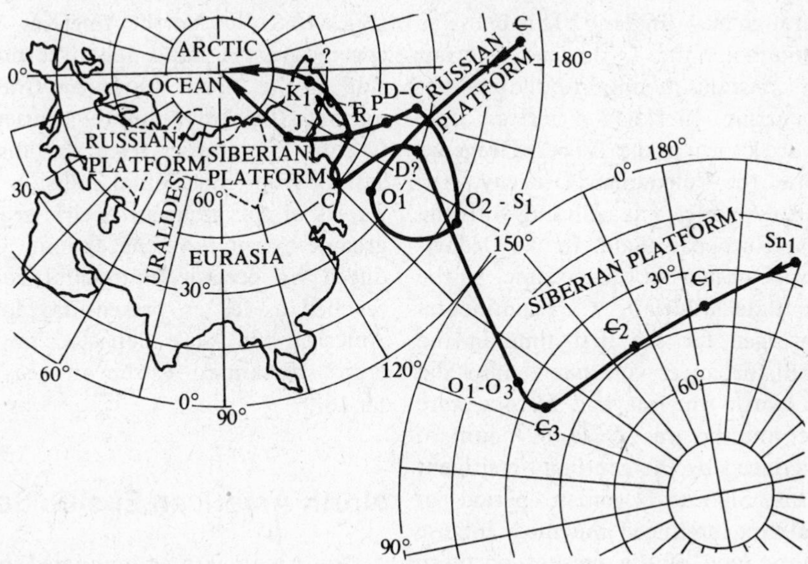

Figure 14.6. *The Disparity of Polar Wandering Curves for the European and Siberian Cratons in the Paleozoic.* (From Hamilton, *G.S.A. Bulletin*, vol. 81, Sept. 1970. By permission of the author and G.S.A.)

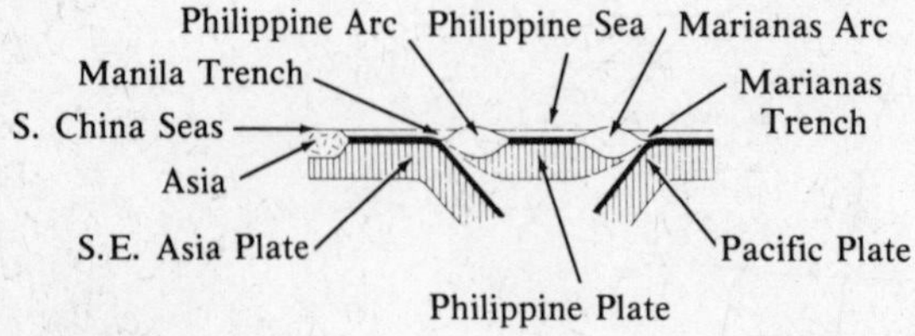

Figure 14.7. *Cross Section of the Philippine Plate Region Today.* This region is probably analogous to the present eastern (then northern) margin of Europe in the Early Paleozoic. The island arc-trench deposits were later brought together in the Ural Mountains. (From Dewey and Bird, *Journal of Geophysical Research*, vol. 75, no. 14, May 10, 1970. By permission of authors and AGU.)

may have triggered the great epeiric sea invasion of North America in that period. The Ordovician-Silurian closing of the proto-Atlantic produced the Caledonian-Northern Appalachian Mountain Chain (Fig. 14.9), which stretched from present-day northern Greenland and northern Norway, down along present east Greenland and the length of Norway, through Great Britain and Ireland, and down the present eastern side of Northern America from Newfoundland through New England. This belt was unstable throughout the Ordovician-Silurian interval with constant mountain-building disturbances occurring. In North America, these disturbances are known as the **Taconic Orogeny;** and in Europe, the **Caledonian Orogeny.** Apparently, North America was colliding with the margins of the European Plate. In the Ordovician, the European margin was oceanic. In the Middle and Late Silurian, the continental margins converged for the first time in the north. These disturbances continued into the Devonian as North America and Europe were fused together and the great Acadian mountain chain was produced by the continent-continent collision. This Silurian-Devonian period of mountain building produced the most intense and last deformation of the present northern Appalachians and the Caledonian Mountains as Europe and North America came into contact.

The Taconic Orogeny, which probably included the collision of an island arc with ancestral North America, produced a great clastic wedge of terrestrial redbeds (or molasse) called the **Queenston-Juniata** (Figs. 14.9–10) in North America. This drove the sea out of the present eastern cratonal area. A large influx of clastics into the epeiric sea occurred even as far west in terms of present geography as Michigan, Ohio, Kentucky, and Tennessee. The largest surviving remnants of the orogeny are the Taconic Mountains of eastern New York, a block of metamorphosed oceanic rocks thrust westward over strata formed on the continental shelf (Fig. 14.9).

The Northern Appalachian-Caledonian Orogeny is recorded in Greenland and Europe by deformed sediments and a great flood of clastics. As the suture closed toward the south in the Devonian, the rising **Acadian Mountains** continued to build this great clastic wedge of terrestrial redbeds in all directions. This wedge has been termed the **Old Red Sandstone "Continent"** (Fig. 14.11) in present northern Europe (British Isles, Scandinavia, northern Germany, Poland, and Russia) and the **Catskill "Delta"** (Fig. 16.9) in North America. It eventually covered a large band down the present eastern side of the North American Craton and the northern two-thirds of the present European Craton. The clastics from this disturbance resulting from continental collision muddied the waters of the adjacent epeiric seas to a much greater extent than the Taconic Orogeny produced by oceanic-continental collision. They reached as far as present-day Iowa in North America and were deposted in virtually the entire remaining European seas (see Chapter 16).

North American Epeiric Seas

North America was inundated by two major epeiric seas in the Lower Paleozoic (Figs. 14.12–14). The **Sauk Sea** began to spread westward

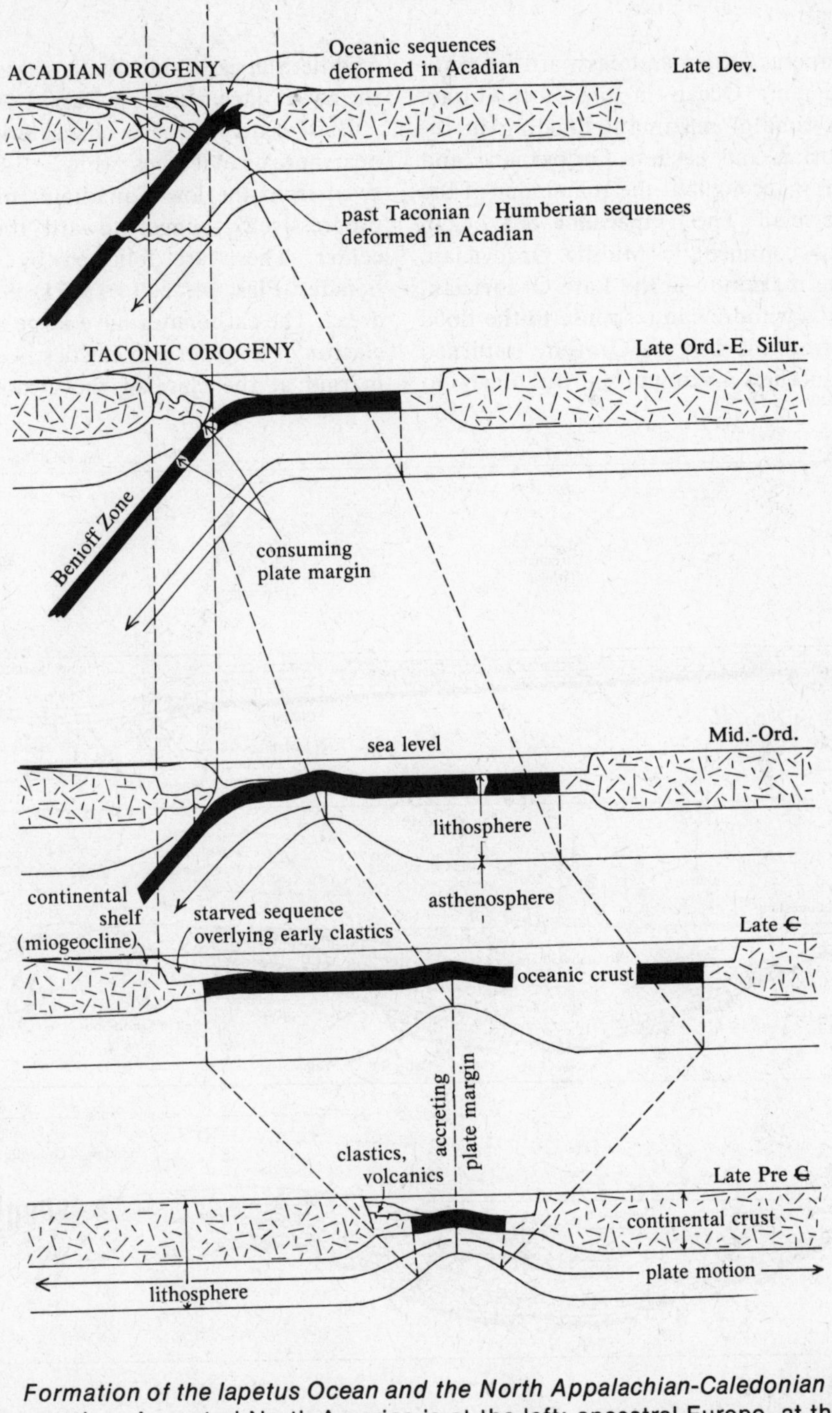

Figure 14.8. *Formation of the Iapetus Ocean and the North Appalachian-Caledonian Mountains.* Restored cross section. Ancestral North America is at the left; ancestral Europe, at the right; and the Iapetus (proto-Atlantic) Ocean, in between. Dashed lines connect the same points on all sections. (From Bird and Dewey, *G.S.A. Bulletin,* vol. 81, April 1970. By permission of the authors and G.S.A.)

from the Iapetus Ocean and eastward from the ancestral Pacific Ocean in the Eocambrian, reached its time of maximum flooding in the Late Cambrian and earliest Ordovician, and then withdrew throughout the remainder of the Early Ordovician. The **Tippecanoe Sea** began to flood the continent in Middle Ordovician, reached one maximum in the Late Ordovician, then partially withdrew in response to the flood of clastics from the Taconic Orogeny, returned to reach another maximum in the Early to Middle Silurian, and finally regressed through the Late Silurian into the Early Devonian.

Both sequences are characterized by basal nearshore sandstones (Figs. 14.15–18), derived from the low Canadian Shield, that get progressively younger toward the continental center. These are followed by offshore carbonates (Figs. 14.15, 14.19–21) that cover huge areas. The carbonates have large admixtures of clastics or even turn to clastics near the eastern margin at the close of the Ordovician as the

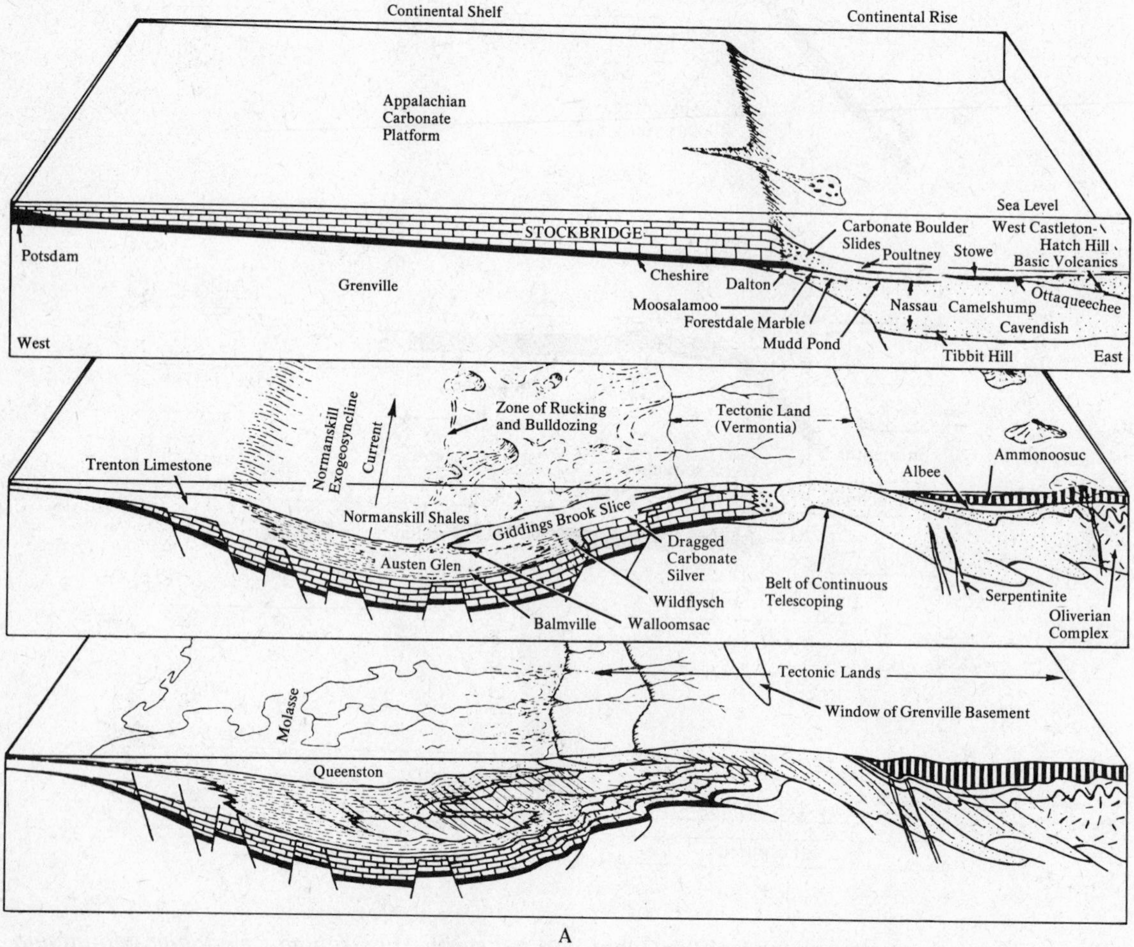

A

Figure 14.9. *Evolution of the Northern Appalachian Region in the Early Paleozoic.* **A.** *New England,* **B.** *Newfoundland.* In each sequence, the earliest stage is at the top and the latest, at the bottom. Allochthons, slivers, and slices are overthrust sheets. Geographic names refer to lithologic units. (From Bird and Dewey, *G.S.A. Bulletin,* vol. 81, April 1970. By permission of the authors and G.S.A.)

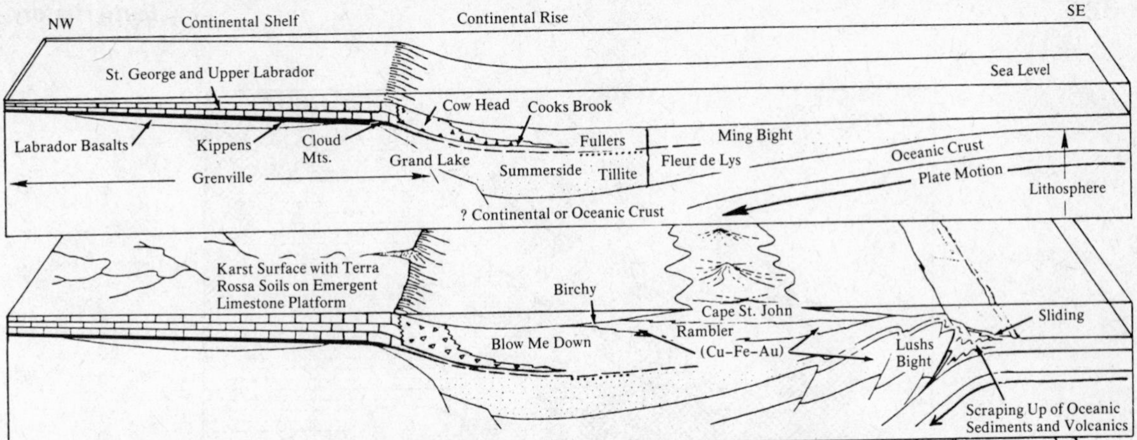

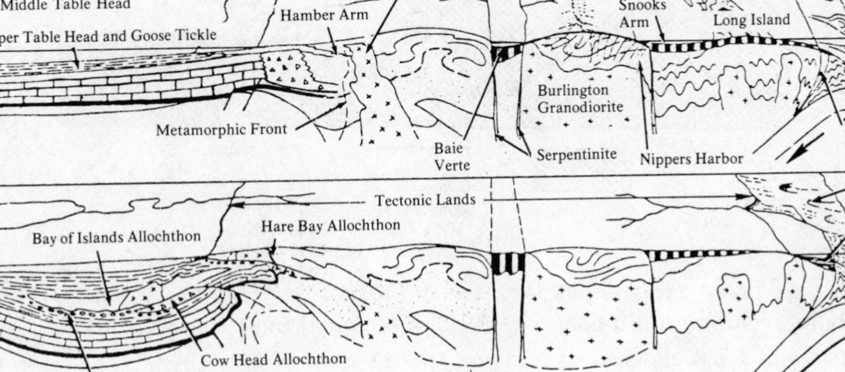

Figure 14.10. *Restored Cross Section through Queenston-Juniata Clastic Wedge.* Vertical scale greatly exaggerated. (From Kay and Colbert, *Stratigraphy and Life History,* © 1965 by John Wiley & Sons, Inc.)

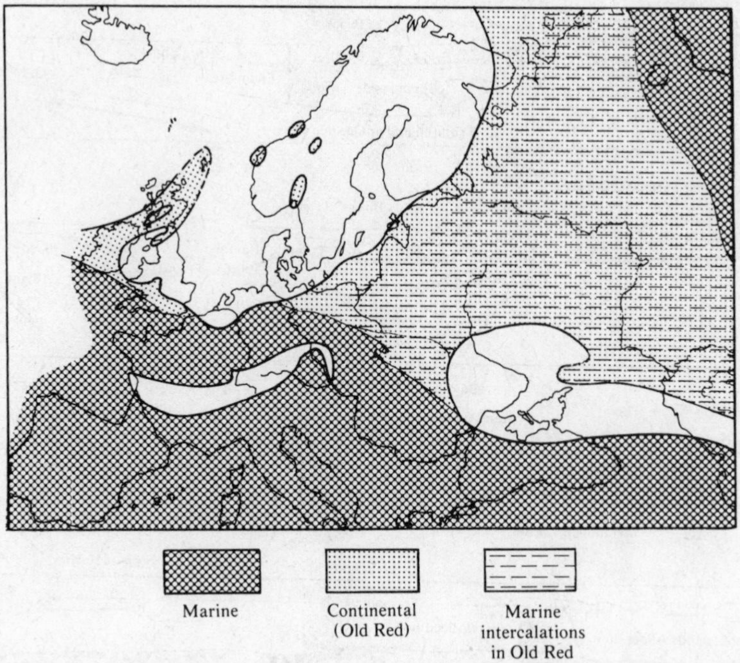

Figure 14.11. *Devonian Sedimentary Deposits of Present-Day Europe.* (Adapted from R. Brinkmann, *Geologic Evolution of Europe,* © 1960 by Ferdinand Enke Verlag, Stuttgart, W. Germany. English translation, J.E. Sanders.)

Taconic Mountains formed. The great Silurian peak of the Tippecanoe Sea is characterized by numerous large reefs, particularly around the Michigan Basin. This agrees with the paleomagnetic data which place North America in tropical regions along the equator. The regressive phases of the Tippecanoe Sea deposited extensive halite beds (Fig. 14.22) in the present Great Lakes area.

The North American Craton did not subside uniformly to receive sediments during the Lower Paleozoic. Some areas, such as the **Michigan Basin, Illinois Basin, Arbuckle Basin,** and **Williston Basin** (Fig. 14.23), subsided much more rapidly than the remainder and received greater thicknesses of sediment. This does not mean that the water was deeper at any time than in the adjacent areas; in fact, the facies indicate it was not, only that the basin subsided more rapidly and the sedimentary piles are thicker. Other regions, such as the **Cincinnati-Nashville Arch, Ozark Dome, Transcontinental Arch, Montana Dome,** and the **Peace River-Tathlina Uplift** (Fig. 14.23), did not subside as rapidly as the rest of the craton and have thinner sequences than the remainder. In some cases, such as the Ozark area, the sea did not completely cover a high area of the shield, and islands were left in the epeiric seas.

Events on Other Continents

The **European Craton,** embracing the areas between the present Caledonian, Uralian, and Hercynian mountains, was flooded by epeiric seas during most of the Early Paleozoic. The Lower Paleozoic rocks are now deeply buried

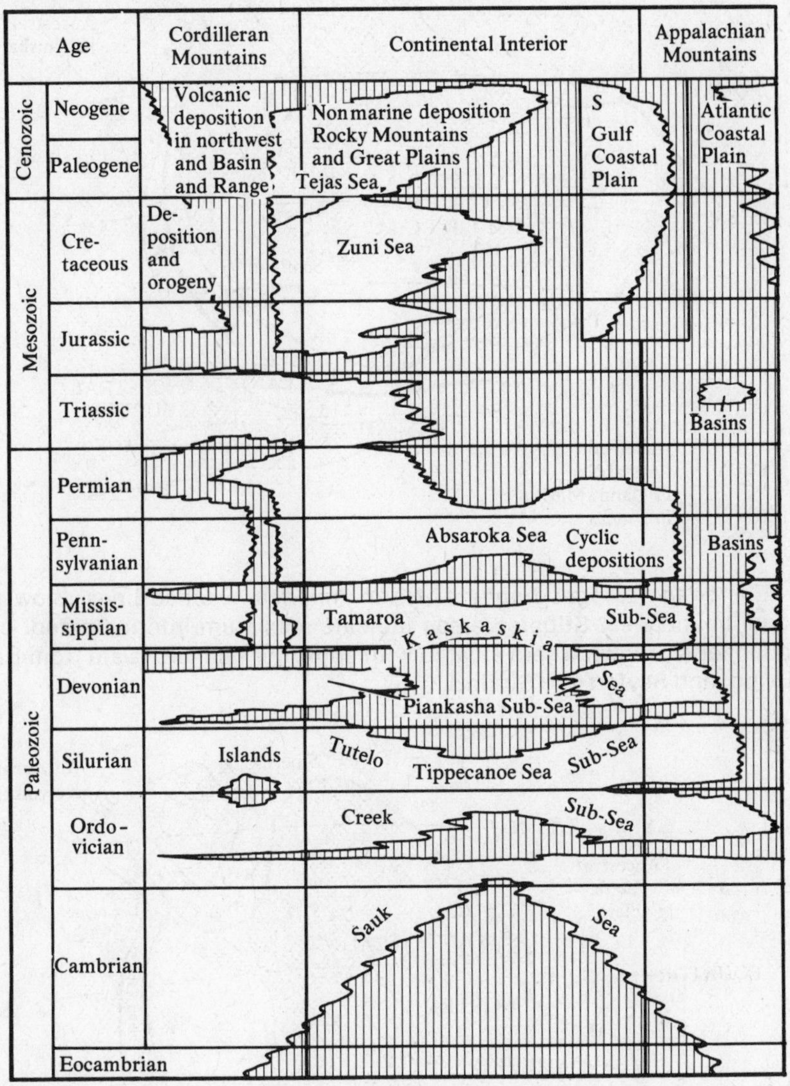

Figure 14.12. *Major Epeiric Seas of North America.* White areas show extent of marine invasions; lined areas represent nonmarine deposition or erosion. (After Foster.)

across most of the platform in European Russia but crop out in the Baltic Sea region. Many reefs are known, particularly in the Silurian when Europe was astride the equator. The **Siberian Craton** was flooded by epeiric seas throughout the Lower Paleozoic, but not thereafter. Evaporites and reefs occur, testifying to its position in tropical latitudes. The **Gondwanaland Craton** was not extensively flooded in the Cambrian-Silurian interval. Flooding occurred chiefly along the present northern margin, including the Amazon River area of present South America (Fig. 14.24), northern Africa, and northern and western Australia (Fig. 14.25).

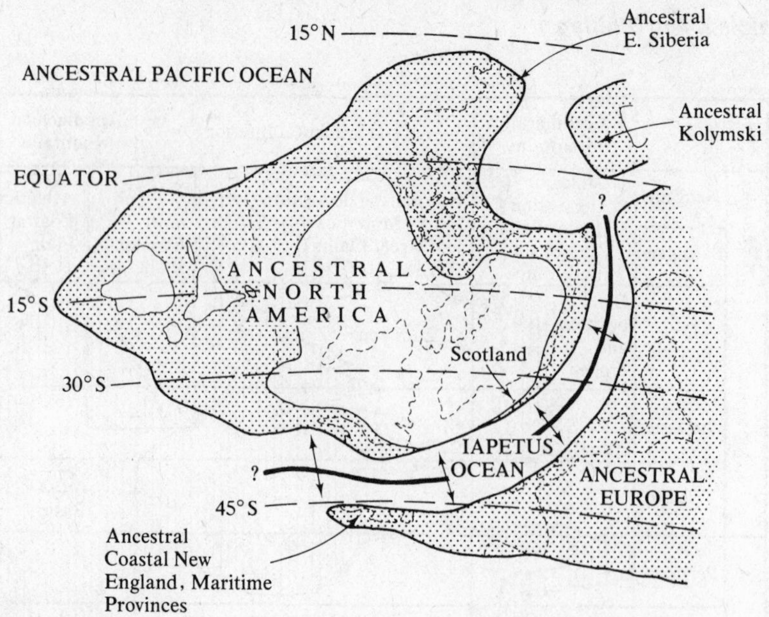

Figure 14.13. *Cambrian Paleogeography of North America.* Dashed lines show present shorelines and national boundaries. Stippled areas indicate maximum inferred extent of epeiric seas. Inferred plate boundaries shown where located with some confidence. (Data from Bird and Dewey, Herron et al., Irving, and Seyfert and Sirkin.)

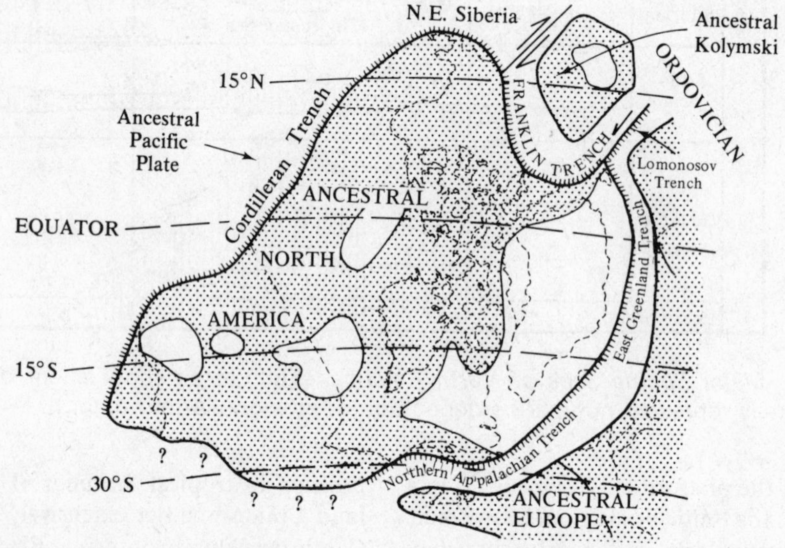

Figure 14.14. *Ordovician Paleogeography of North America.* Dashed lines show present shorelines and national boundaries. Stippled areas indicate maximum extent of epeiric seas. Inferred plate boundaries shown where located with some confidence. (Data from Bird and Dewey, Herron et al., Irving, Seyfert and Sirkin.)

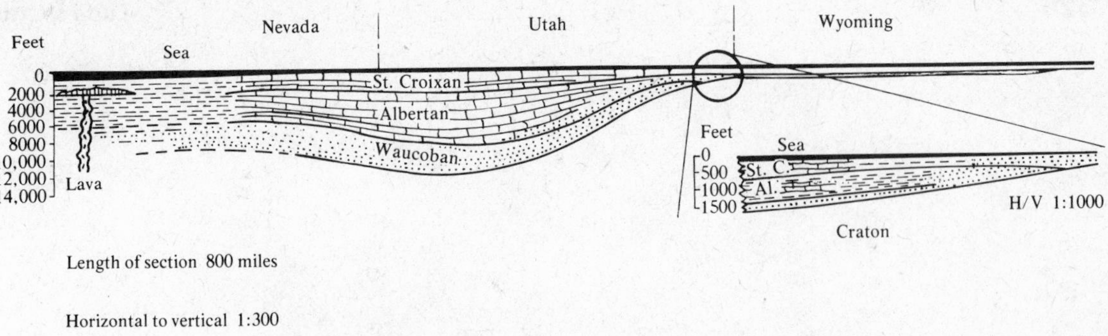

Figure 14.15. *Sauk Sea Transgressive Deposits of Cambrian Age.* Restored cross section across western North America. (From Kay and Colbert, *Stratigraphy and Life History,* © 1965 by John Wiley & Sons, Inc.)

Figure 14.16. *Glenwood Canyon, Colorado. Upper Cambrian Sandstones.* (Courtesy Colorado Dept. of Public Relations.)

Figure 14.17. *Wisconsin Dells. Upper Cambrian Sandstones.* (Courtesy of Wisconsin Natural Resources Dept.)

Evaporites and reefs occur in China, Australia, and Antarctica, supporting their position in the tropics.

With the South or Gondwana Pole situated in a large continental mass such as Gondwanaland, it is not surprising that continental glaciations occurred on the supercontinent during much of the Paleozoic. In the Early Paleozoic record, extensive evidence of continental glaciers occurs in the Ordovician of present North Africa, the region in which the pole was situated at that time. Features produced by both glacial erosion and deposition (Fig. 14.26) abound over a broad area from Morocco, through Mauri-

Figure 14.18. *Ausable Chasm, N.Y. Upper Cambrian Sandstones.* (Courtesy of William M. Mintz.)

tania, Algeria, Niger, to Libya and Chad. Some evidence of glaciation has been uncovered in present-day Brazil and South Africa as well.

Against and within this background, the life of the Early Paleozoic evolved. Fossils of the larger marine organisms became abundant for the first time as hard parts evolved. In the Lower Paleozoic epeiric sea deposits of North America, particularly the widespread limestones, the fossils occur in such abundance that they compose the bulk of many rock units. The Early Paleozoic seas were widespread yet not widely separated, for most of the continents were approaching. Let us now turn to the fossil record of the Cambrian, Ordovician, and Silurian periods and examine the effects of this situation.

A

B

Figure 14.19. *Lower Paleozoic Rocks in the Canadian Rockies.* **A.** *Mt. Stephen, Yoho National Park, British Columbia;* **B.** *Mt. Eisenhower, Banff National Park, Alberta.* Strata are chiefly Cambrian dolomites and limestones formed on the continental shelf. (Courtesy British Columbia Dept. of Travel Industry and Alberta Dept. of Industry and Tourism.)

A

B

Figure 14.20. *Lower Paleozoic Rocks in the Canadian Rockies.* **A.** *Moraine Lake, Valley of the Ten Peaks.* Strata are Lower Cambrian sandstones. **B.** *Lake Louise, Mt. Victoria, Banff National Park, Alberta.* Strata are chiefly Cambrian dolomites and limestones formed on the continental shelf. (Courtesy Alberta Dept. of Industry and Tourism.)

Figure 14.21. *Silurian Carbonates.* **A–B.** *Niagara Falls,* **C.** *Flowerpot Island, Georgian Bay Islands National Park, Ontario.* (Parts A, B courtesy N.Y. State Power Authority. Part C courtesy Ontario Dept. of Tourism and Information.)

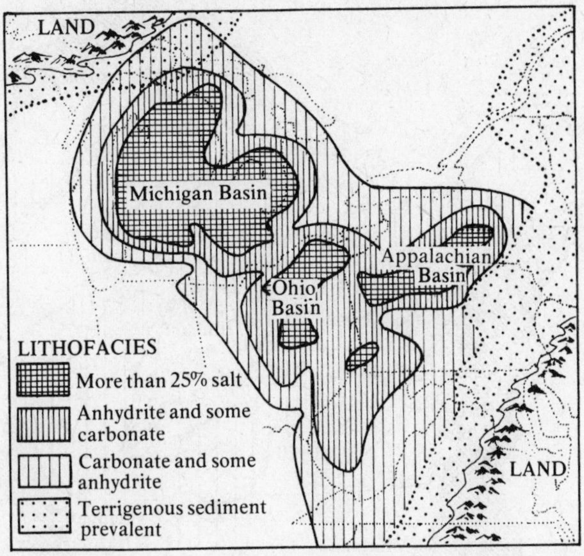

Figure 14.22. *Late Silurian Evaporite Basins of Great Lakes Region.* (After Alling and Briggs, 1961, Bulletin American Association of Petroleum Geologists, v. 45.)

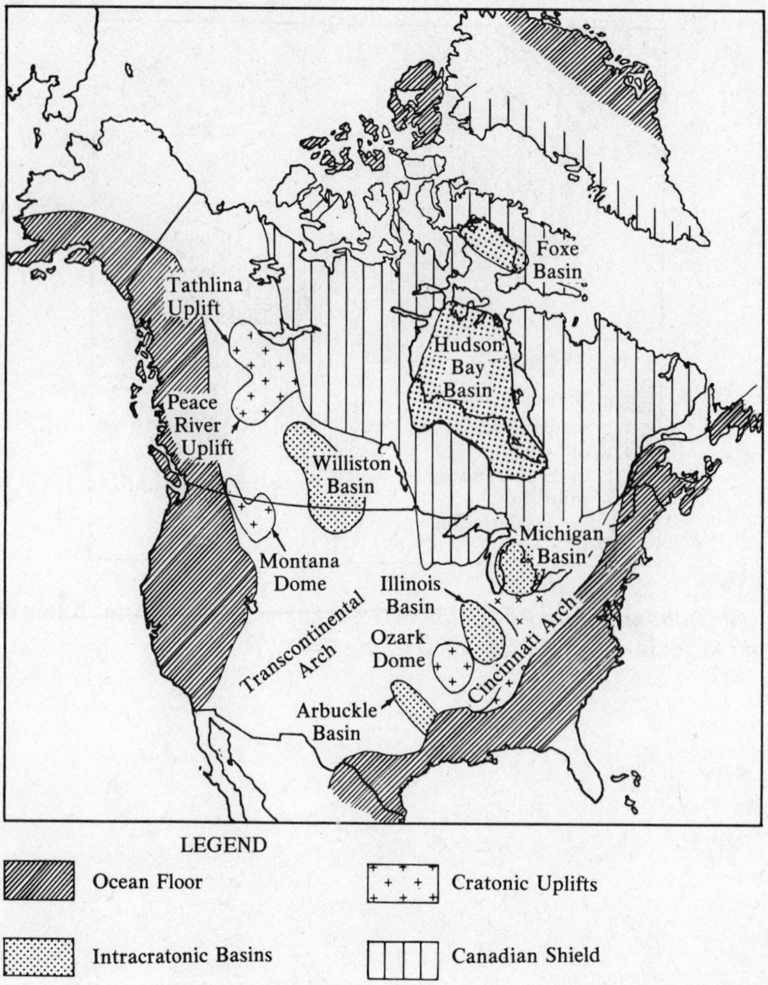

Figure 14.23. *Domes and Basins of Present-Day North America Which Existed in the Early Paleozoic.*

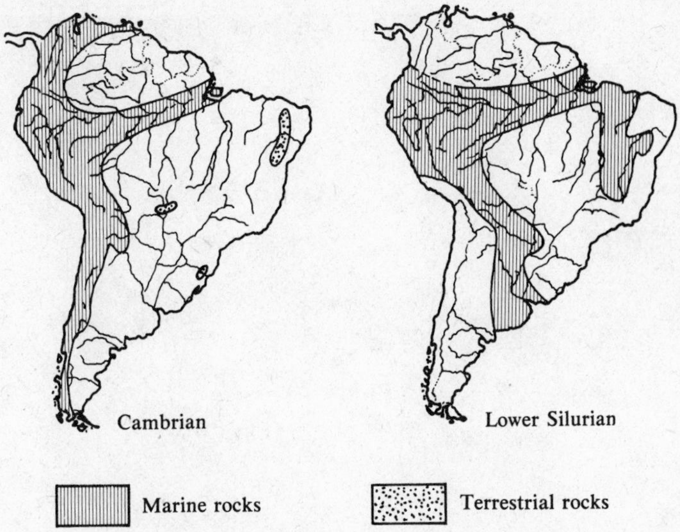

Figure 14.24. *Cambrian and Early Silurian Sedimentary Deposits of Present-Day South America.* (Adapted from Harrington, 1962.)

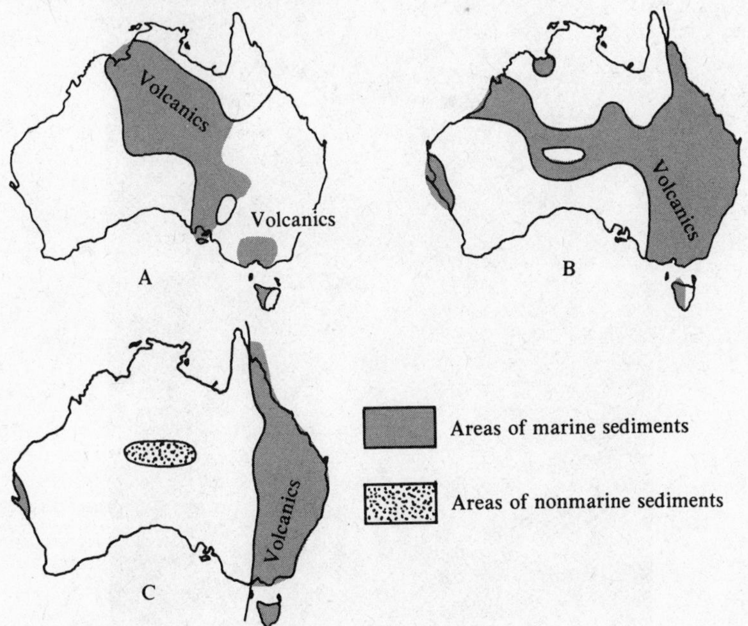

Figure 14.25. *Cambrian, Ordovician, and Silurian Sedimentary Deposits of Present-Day Australia.* (Adapted from Brown et al., 1968.)

A

B

Figure 14.26. *Saharan Glacial Features.* **A.** *Horseshoe pressure scour marks produced by high-velocity subglacial meltwater.* Open ends point in direction of water movement. **B.** *Glacial scour marks and grooves rippled by subglacial melting of permafrost.* Eastern Hoggar, S.E. Algeria. Upper Ordovician. (Courtesy of Rhodes W. Fairbridge, Columbia University.)

REVIEW

1. What is the significance of the **Pan-African Orogeny?**
2. Where were the Gondwana and Pacific poles in relation to the continents in the Early Paleozoic?
3. Why is it more difficult to determine plate positions and margins in the Early Paleozoic than in later geologic time? What plate boundaries can we recognize with some certainty?
4. What do the paleomagnetic and paleoclimatic evidence indicate about the geographic position of North America in the Early Paleozoic?
5. What lines of evidence indicate that Europe and Siberia were separate continents in the Early Paleozoic?
6. What produced the **Northern Appalachian-Caledonian Mountains?**
7. What was the **Iapetus Ocean,** and when did it exist?
8. What were the major epeiric seas of North America in the Early Paleozoic, and when did they reach maximum flood stage?
9. What were the major tectonic elements of the North American Craton in the Paleozoic?
10. What cratons other than the North American underwent extensive flooding in the Early Paleozoic?
11. What evidence is there of continental glaciation in the Early Paleozoic, and where does it occur?

15

Abundance in the Sea: Early Paleozoic Life

15

Life in the Early Paleozoic periods was apparently limited to the sea until the very end of the Silurian when the first land plants and animals appeared. Early Paleozoic marine life was also dominated by the invertebrates, judging from the few scraps of fish preserved before the Late Silurian. The Early Paleozoic is thus best described as the age of marine invertebrates (Figs. 15.1–4).

Major Marine Invertebrates

Dominating the invertebrate animal fossils in abundance and diversity are the **corals, bryozoans, brachiopods, trilobites,** and **graptolites.** Brachiopod and trilobite fossils appear right at the base of the Cambrian; in fact, the first appearance of redlichioid trilobites is used to mark that boundary in most areas. Brachiopods con-

Figure 15.1. *Diorama of Cambrian Marine Life.* Trilobites, jellyfish, and worms are on the bottom in the foreground. Branching tubular organisms are sponges. Swimming forms are various arthropods, some ancestral to merostomes and others to crustaceans. (Courtesy Field Museum of Natural History.)

tinued to flourish throughout the Paleozoic, but trilobites rose to a Late Cambrian-Early Ordovician peak and then declined. Bryozoans and corals did not appear until the Late Cambrian, but diversified rapidly through the Middle Paleozoic. Graptolites appeared late in the Cambrian and diversified greatly in the Ordovician and Silurian.

Corals

The major coral groups of the Paleozoic, the **tabulate** and **tetracorals,** appeared in the Late Cambrian and the Middle Ordovician respectively, but the tabulates (Figs. 15.2, 15.3, and 15.5) dominated the Early Paleozoic. By the late Ordovician, tabulates were constructing coral reefs, and in the Silurian, their zenith, they were the major framework builders of the abundant reefs of the period (Fig. 14.22). Such types as the honeycomb coral and the chain coral, as well as numerous others, occur in great abundance in Silurian rocks (Figs. 15.3, 15.5). The tetracorals (Figs. 15.2, 15.6) of the Early Paleozoic were chiefly small, horn or cup corals.

Bryozoans

The **bryozoans** appeared in the Late Cambrian, but did not become common until the Ordovician. The major Early Paleozoic groups were the twig-like or massive **trepostomes** and **cystoporoids,** the stony bryozoa (Fig. 15.7). In the Upper Ordovician, their remains are so abundant (Fig. 15.2) that they actually compose the bulk of many rock bodies. These two orders were reduced, but still common, in the Silurian.

Brachiopods

The brachiopods underwent a tremendous adaptive radiation in the Early Paleozoic. The **inarticulates** (Fig. 15.8) dominated the Cambrian brachiopod fauna, and, although the

Figure 15.2. *Diorama of Ordovician Marine Life.* Elongate swimmers and large benthonic form with tentacles are nautiloid cephalopods. Finely branched structures are bryozoans. Domical forms are tabulate corals, while solitary tentacle-bearing benthonic animals are tetracorals. A trilobite and a snail are crawling on the bottom, and a clump of brachiopods lies above the large nautiloid's head. (Courtesy Field Museum of Natural History.)

Figure 15.3. *Diorama of Silurian Marine Life.* Various tabulate corals cover the bottom at the left and center. Tentacle-bearing nautiloids are at upper left, right center, and lower right. Two trilobites and a variety of brachiopods also occur at the right. Stalked forms are crinozoan echinoderms. (Courtesy Field Museum of Natural History.)

Figure 15.4. *Diorama of Silurian Brackish-Water Life.* Large swimming forms are eurypterans; benthonic forms are snails. (Courtesy Field Museum of Natural History.)

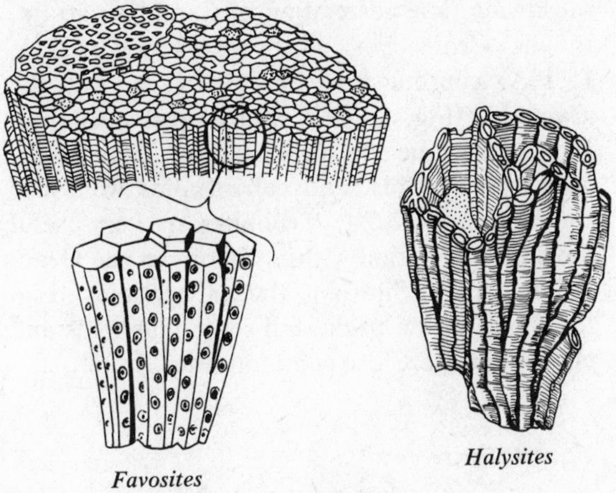

Figure 15.5. *Lower Paleozoic Tabulate Corals.* Except for enlargement of *Favosites*, approximately natural size. (After Kay and Colbert.)

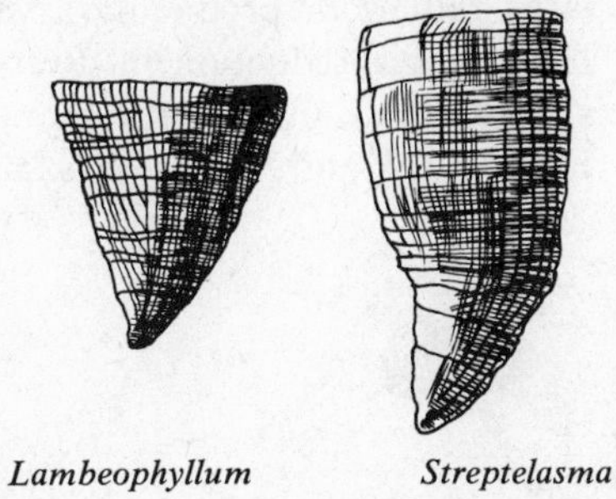

Figure 15.6. *Lower Paleozoic Tetracorals.* Natural size.

group was even more diverse in the Ordovician, inarticulates were no longer the major brachiopod group. Attached inarticulates (Fig. 15.8) were common in the Ordovician.

The **articulates** appeared in the Cambrian also and chiefly consisted of **orthoids** (Fig. 15.9). In the Ordovician, the articulate brachiopods diversified greatly. The orthoids (Fig. 15.10) and **strophomenoids** (Fig. 15.11) are the most abundant and widespread, but pentameroids, rhynchonelloids, and some atrypoids are also common fossils (Fig. 15.12). The Silurian brachiopod fauna was dominated by the **pentameroids** (Fig. 15.13). Pentameroids are useful fossils to geologists for time-correlating rocks

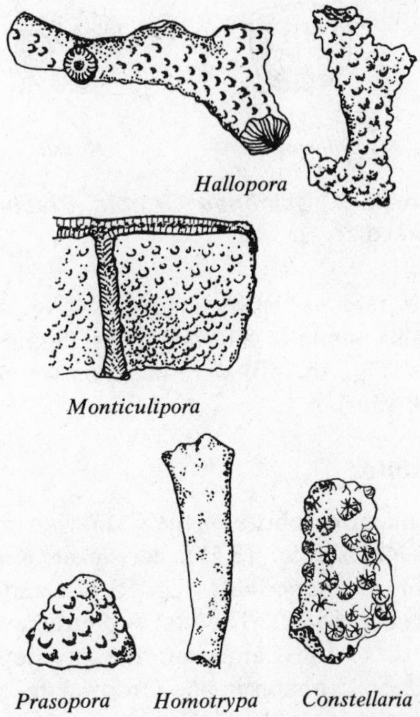

Figure 15.7. *Lower Paleozoic Bryozoans.* All but *Constellaria,* a probable cystoporoid, are trepostomes. Natural size. (After Cummings, Ulrich.)

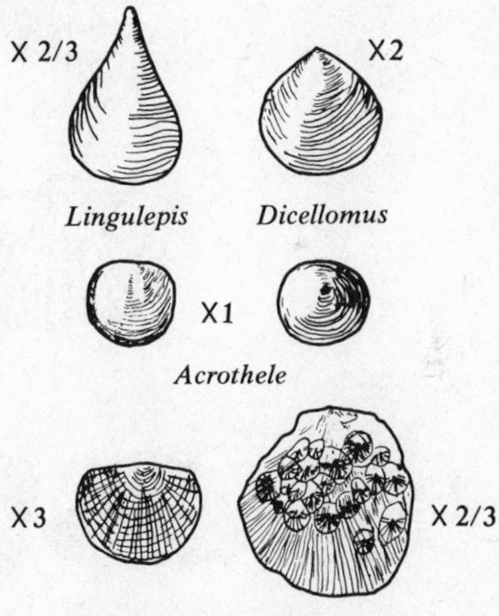

Figure 15.8. *Lower Paleozoic Inarticulate Brachiopods.* Magnifications are indicated. (Adapted from *Invertebrate Fossils* by Moore-Lalicker-Fischer, Copyright 1952 by McGraw-Hill, Inc., used by permission of McGraw-Hill Book Company.)

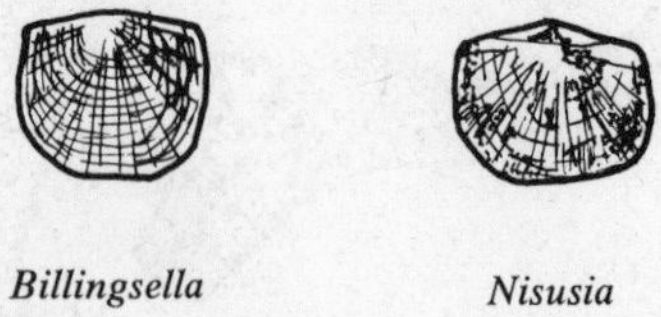

Figure 15.9. *Cambrian Orthoid Brachiopods.* Natural size.

within regions. Rhynchonelloids (Fig. 15.14A) are also common Silurian rocks. Some athyridoids (Fig. 15.14B) were occasionally common in the Silurian.

Trilobites

The major trilobites of the Cambrian were the **redlichioids** (Fig. 15.15), **corynexochoids** (Fig. 15.16), **ptychoparioids** (Fig. 15.17), and **agnostoids** (Fig. 15.18). The first two orders were restricted to the Cambrian; the ptychoparioids were also common in the Ordovician; and the agnostoids were chiefly Cambrian in age. Trilobites are the major fossils for both regional and worldwide time correlation of Cambrian strata. In the Ordovician, the ptychoparioids (Fig. 15.19A) continued to be important, but the **asaphoids** (Fig. 15.19B) and **trinucleoids** (Fig. 15.19C) became the major groups. Some **illaenoids, cheiruroids,** and **calymenoids** were also common (Fig. 15.20). Trilobites are very useful for time correlation within regions in the Ordovician. In the Silurian, the calymenoids (Fig. 15.21) reached a peak, and some illaenoids and **phacopoids** were also common (Fig. 15.22).

Graptolites

The graptolites appeared at the very end of the Cambrian but were only represented by rare attached dendroid types. In the Early Ordovician, however, the pelagic **graptoloid** graptolites (Figs. 15.23 and 15.25) appeared and underwent a large radiation, producing a variety of forms. There was a reduction in diversity and numbers in the Silurian, but the monograptid group (Figs. 15.24 and 15.25) was abundant

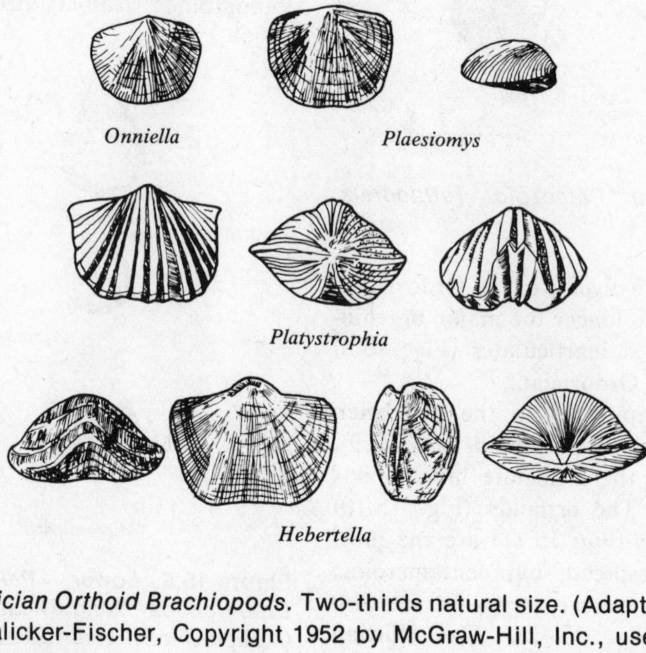

Figure 15.10. *Ordovician Orthoid Brachiopods.* Two-thirds natural size. (Adapted from *Invertebrate Fossils* by Moore-Lalicker-Fischer, Copyright 1952 by McGraw-Hill, Inc., used by permission of McGraw-Hill Book Company.)

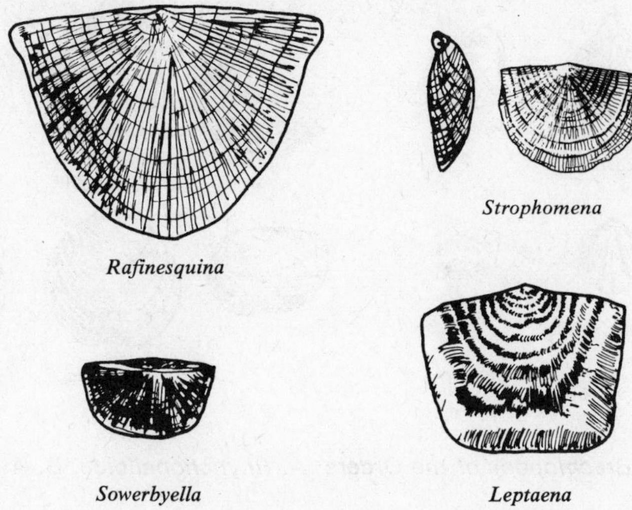

Figure 15.11. *Ordovician Strophomenoid Brachiopods.* Natural size.

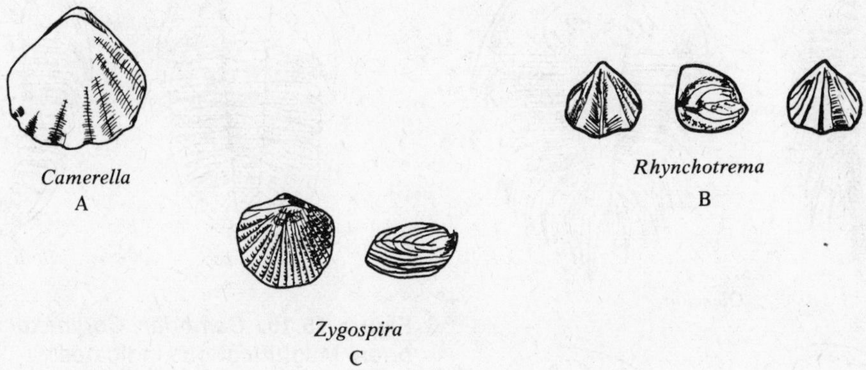

Figure 15.12. *Ordovician Brachiopods of the Orders* **A.** *Pentameroida,* **B.** *Rhynchonelloida,* **C.** *Atrypoida.* Parts A-B natural size. Part C X 2.

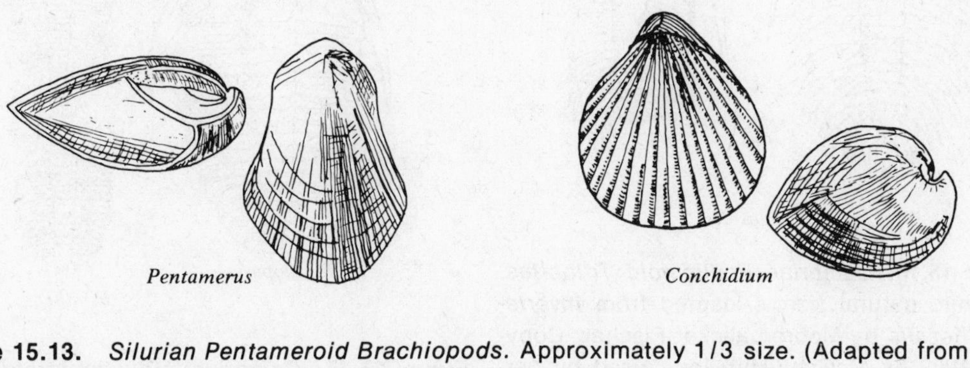

Figure 15.13. *Silurian Pentameroid Brachiopods.* Approximately 1/3 size. (Adapted from *Invertebrate Fossils* by Moore-Lalicker-Fischer, Copyright 1952 by McGraw-Hill, Inc., used by permission of McGraw-Hill Book Company.)

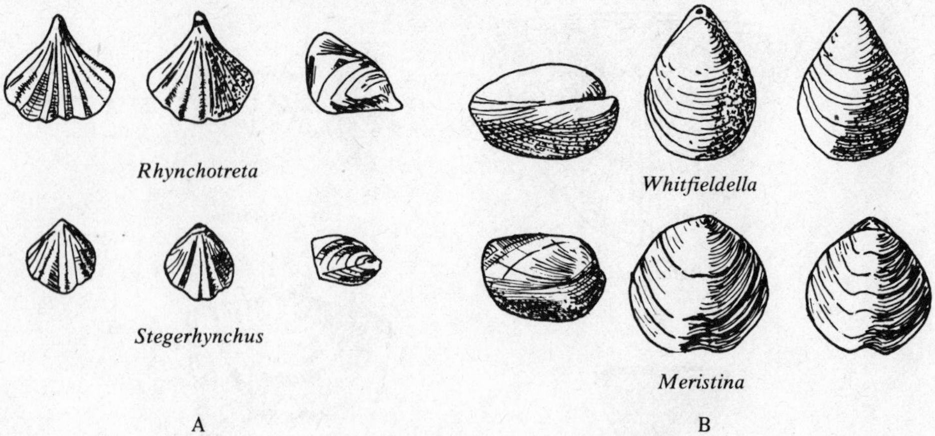

Figure 15.14. *Silurian Brachiopods of the Orders* **A.** *Rhynchonelloida,* **B.** *Athyridoida.* Natural size.

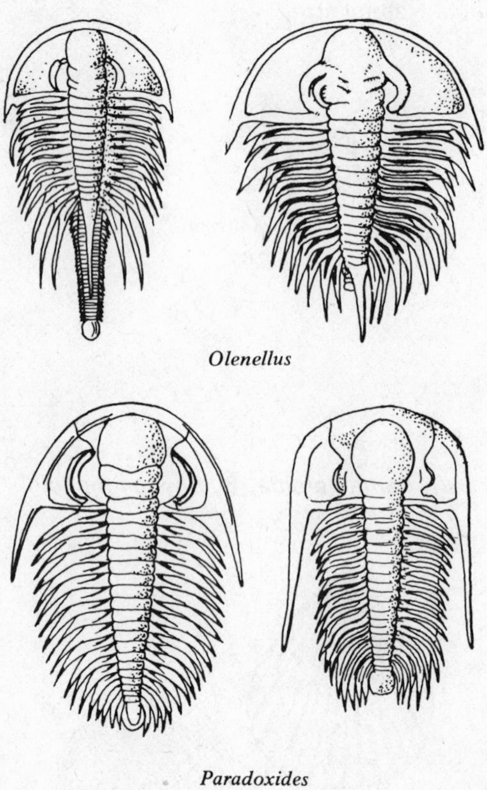

Figure 15.15. *Cambrian Redlichioid Trilobites.* One-third natural size. (Adapted from *Invertebrate Fossils* by Moore-Lalicker-Fischer, Copyright 1952 by McGraw-Hill, Inc., used by permission of McGraw-Hill Book Company.)

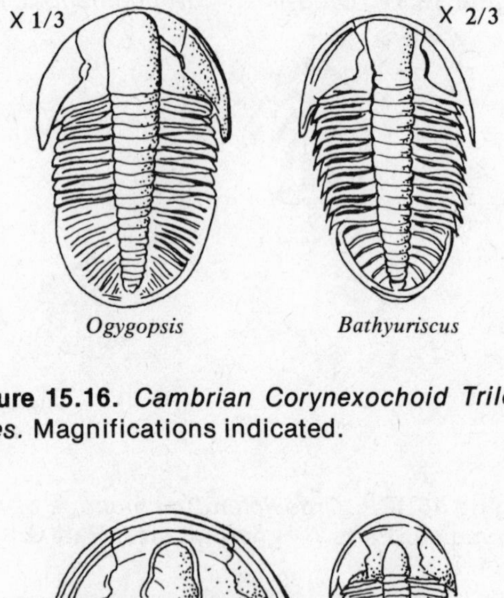

Figure 15.16. *Cambrian Corynexochoid Trilobites.* Magnifications indicated.

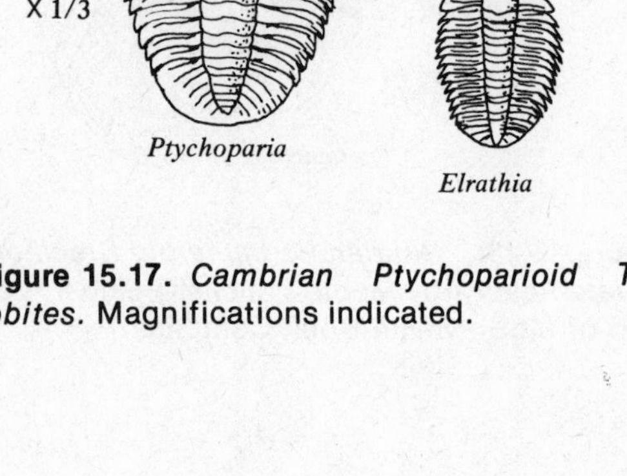

Figure 15.17. *Cambrian Ptychoparioid Trilobites.* Magnifications indicated.

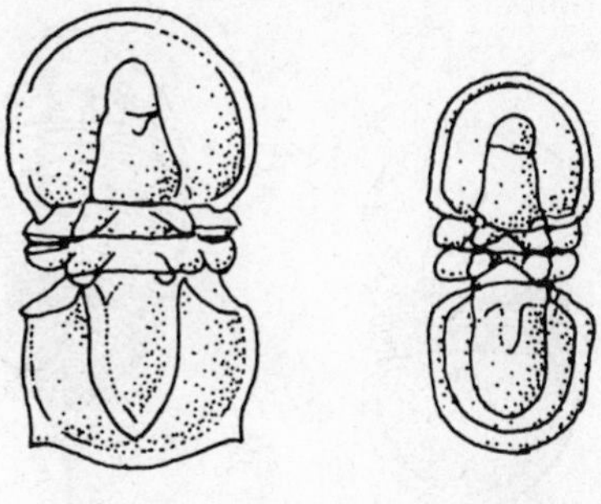

Figure 15.18. *Cambrian Agnostoid Trilobites.* X 4. (After Matthew, White.)

throughout the period, dying out in the Early Devonian.

Graptoloid graptolites are the major fossils used in time correlation of Ordovician, Silurian, and lowest Devonian strata. As planktonic forms, species and genera of this group have wide geographic ranges, which is vital for making long-distance time correlations. Unfortunately, they are chiefly found in black shales formed on toxic bottoms free of scavengers and decay bacteria. Their organic hard parts did not survive as readily on normal marine bottoms which supported a wealth of living things whose activities destroyed the scleroprotein stipes. Nevertheless, graptolites do occur in sufficient numbers even in these strata to make them the most important group of fossils for long-distance time correlation in the Lower Paleozoic.

Other Significant Marine Groups

Several other groups of fossil organisms are significant for various shorter intervals within the Cambrian-through-Silurian interval. They include the receptaculitids, archaeocyathans, stromatoporoids, nautiloids, eurypterans, ostracods, and conodonts.

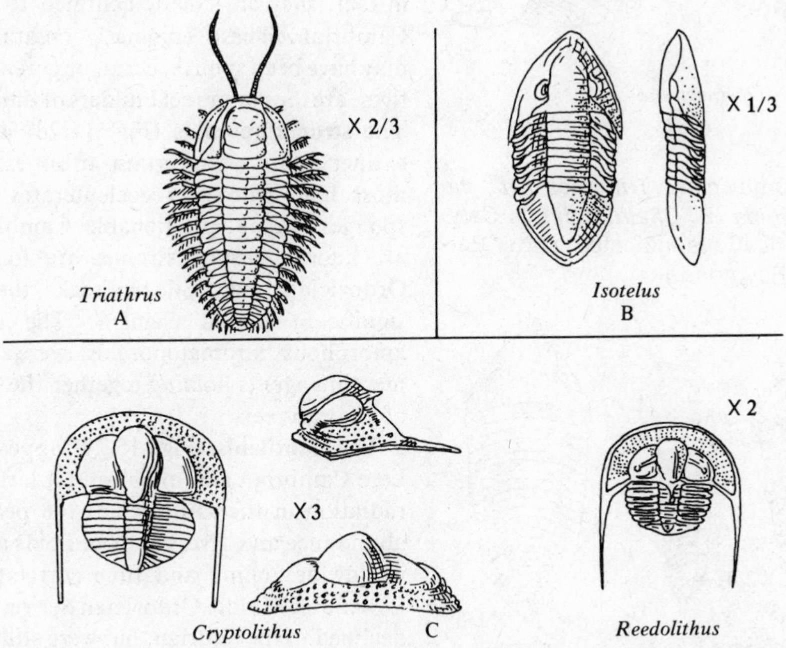

Figure 15.19. *Ordovician Trilobites of the Orders* **A.** *Ptychoparioida,* **B.** *Asaphoida,* **C.** *Trinucleoida.* Magnifications indicated. (After Beecher, Raymond.)

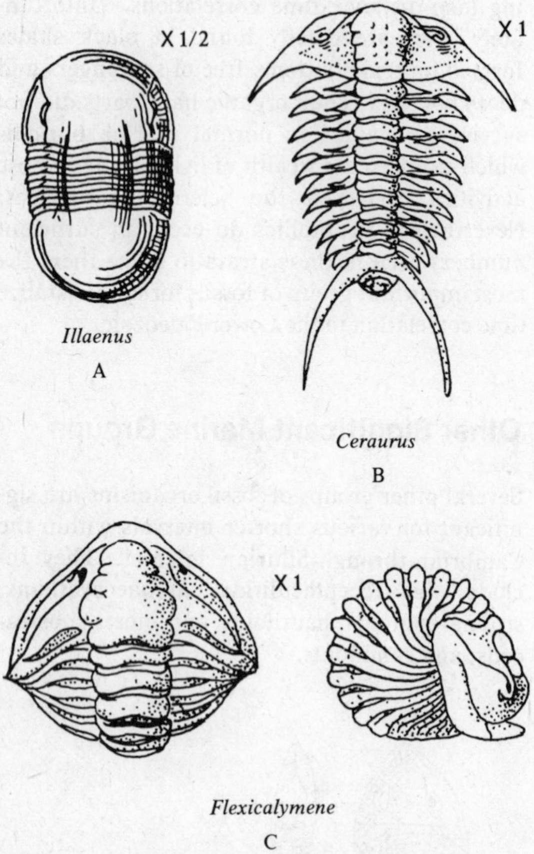

Figure 15.20. *Ordovician Trilobites of the Orders:* **A.** *Illaenoida,* **B.** *Cheiruroida,* **C.** *Calymenoida.* Magnifications indicated. (After Barton, Meek, and Raymond.)

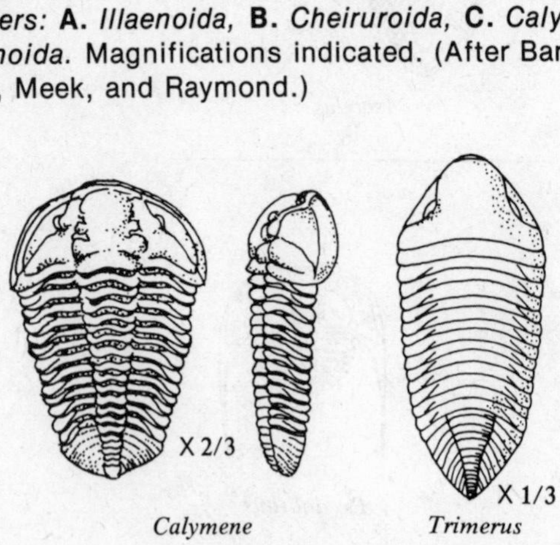

Figure 15.21. *Silurian Calymenoid Trilobites.* (After Clarke, Hall, and Harrington et al.)

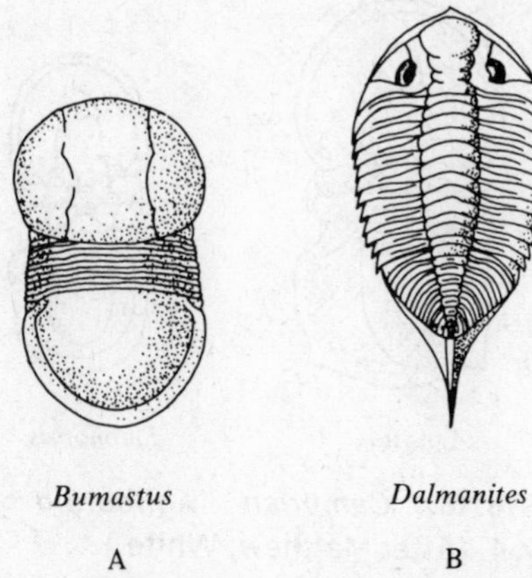

Figure 15.22. *Silurian Trilobites of the Orders:* **A.** *Illaenoida,* **B.** *Phacopoida.* Approximately one-third natural size. (After Hall.)

Receptaculitids (Fig. 15.26) were apparently large dasycladaceous green algae abundant in the Ordovician. The extinct **archaeocyathans** (Fig. 15.27) are limited to the Cambrian System; in fact, they are chiefly confined to the Lower Cambrian. These enigmatic creatures, which may have been sponge, coral, or green algae relatives, are the major reef builders of the Cambrian. The **stromatoporoids** (Fig. 15.28) are another extinct group of uncertain affinities, but seem most like hydrozoan coelenterates or sclerosponges. Some questionable Cambrian forms are known; certain stromes are found in the Ordovician; and in Silurian rocks, the group is a significant faunal element. The encrusting, amorphous stromatoporoids were major cementing agents holding together the framework of Silurian reefs.

The **nautiloids** (Fig. 15.29) appeared in the Late Cambrian and underwent a large adaptive radiation in the Ordovician, the peak of their abundance and diversity. Nautiloids are used for worldwide zoning and time correlation in the Lower and Middle Ordovician Series. Nautiloids declined in the Silurian, but were still a common faunal element.

The **eurypterans,** or sea scorpions (Fig. 15.30), were chiefly denizens of brackish waters,

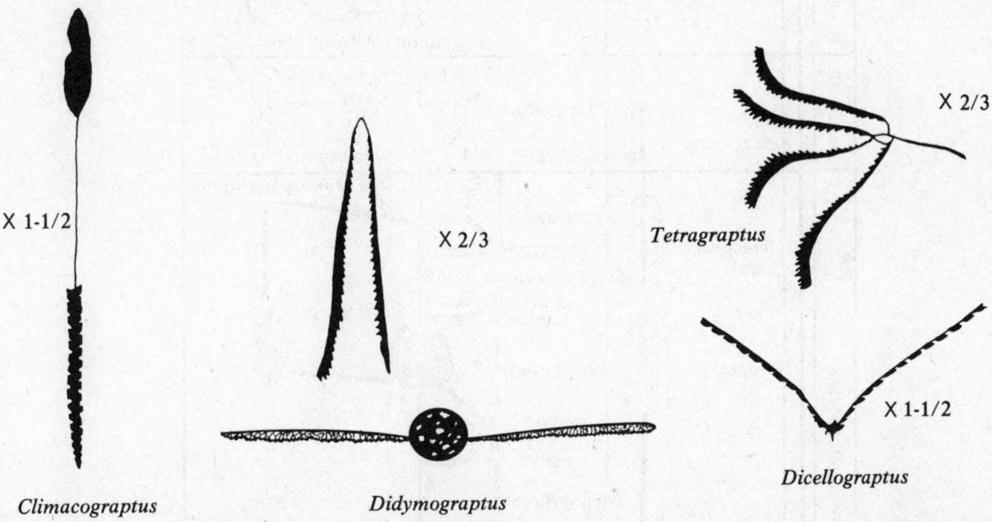

Figure 15.23. *Ordovician Graptoloid Graptolites.* Magnifications indicated. (Adapted from *Invertebrate Fossils* by Moore-Lalicker-Fischer, Copyright 1952 by McGraw-Hill, Inc., used by permission of McGraw-Hill Book Company.)

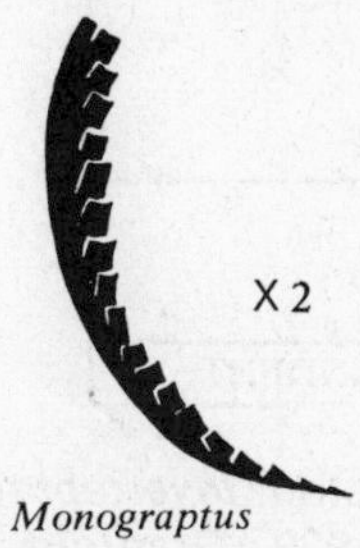

Figure 15.24. *Silurian Graptoloid Graptolite.* (Adapted from *Invertebrate Fossils* by Moore-Lalicker-Fischer, Copyright 1952 by McGraw-Hill, Inc., used by permission of McGraw-Hill Book Company.)

an environment not abundantly represented in the fossil record. They range from Ordovician-Permian, but it is only in the widespread brackish water deposits of the Upper Silurian that they are abundantly represented in the fossil record (Fig. 15.4). Most were less than a foot in length, but some may have reached nine feet with their appendages extended in front of them. The tiny crustaceans known as **ostracods** are rare in the Cambrian, but their bean-shaped microscopic shells are common in Ordovician and Silurian strata (Fig. 15.31).

The **conodonts** (Fig. 15.32) were common from mid-Cambrian times onward, but their peak was in the Upper Paleozoic.

Other lesser groups of fossils reach a peak in the Lower Paleozoic. These are the enigmatic hyolithans and tentaculites, and the cystoid and edrioasteroid echinoderms. The mollusk-like **hyolithans** (Fig. 15.33) reached a Cambrian peak, while the **tentaculites** (Fig. 15.33), which may also be mollusks, were locally common from the Ordovician through the Devonian. **Cystoids,** never abundant, were most common in the Ordovician and Silurian (Fig. 15.34). The long-ranging (Cambrian-Carboniferous) **edrioasteroids** (Fig. 15.35) reach an Upper Ordovician peak where they are commonly found cemented to brachiopods.

Lesser Groups

The groups just discussed are the major fossil invertebrate elements of the Lower Paleozoic.

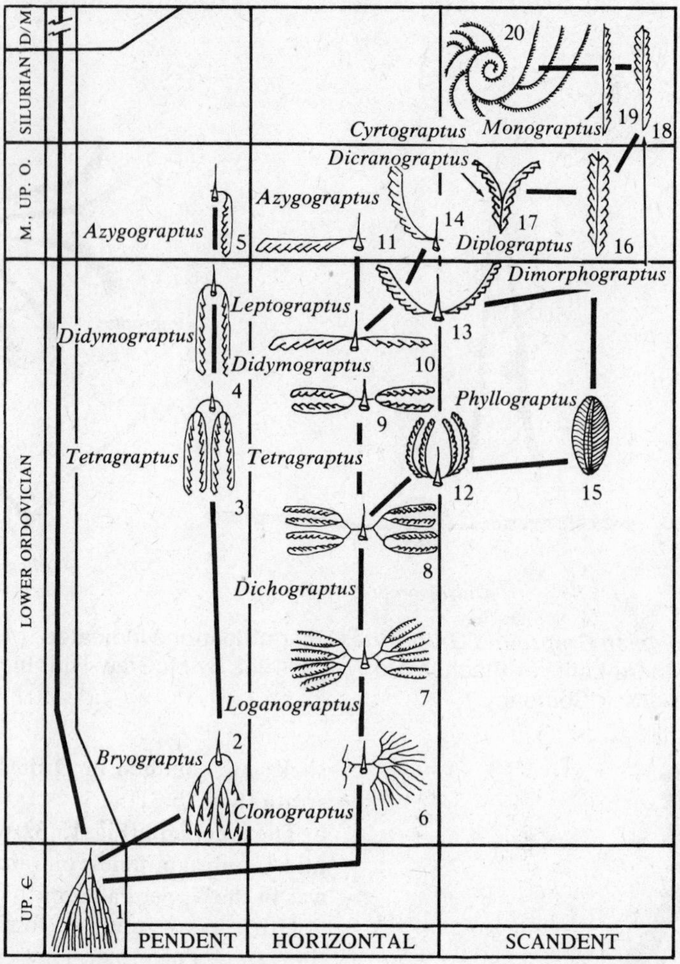

Figure 15.25. *Evolution of the Graptolites.* (From William H. Easton, *Invertebrate Paleontology,* Fig. 4.7 "Evolutionary Patterns of Graptozoans," After Elles, 1922, *Graptolite Faunas of the British Isles,* Proc. Geologists Assoc., and Swinnerton, 1938, *Development & Evolution,* British Assoc. Adv. Sci., Harper & Row, 1960.)

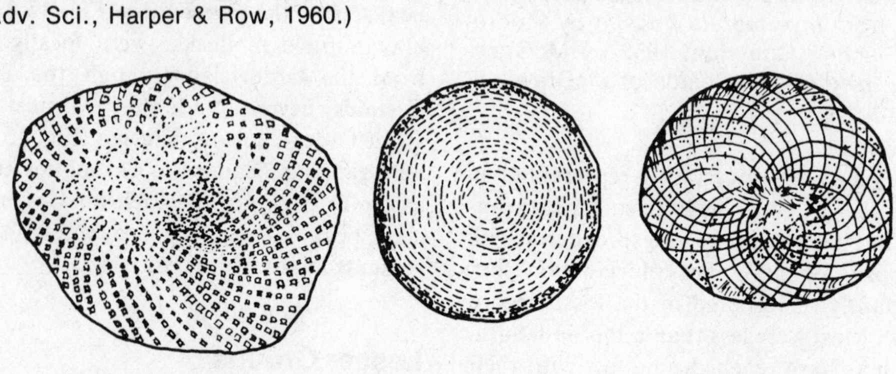

Figure 15.26. *Ordovician Receptaculitids.* Natural size. (After Schuchert, Winchell.)

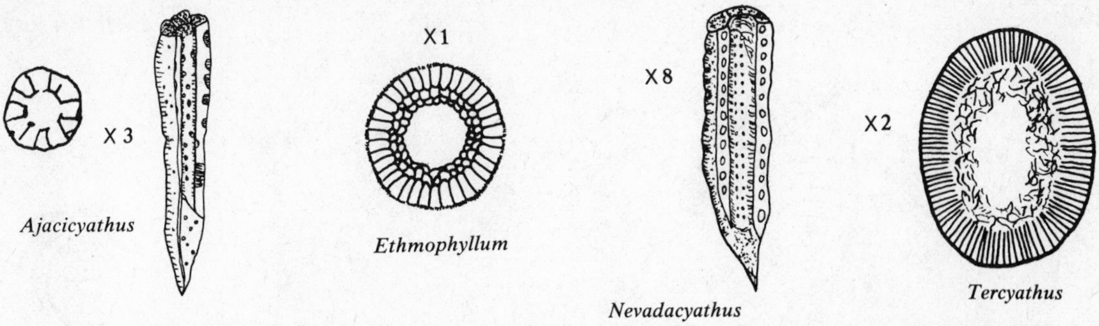

Figure 15.27. *Cambrian Archaeocyathans.* Magnifications indicated. (After Okulitch.)

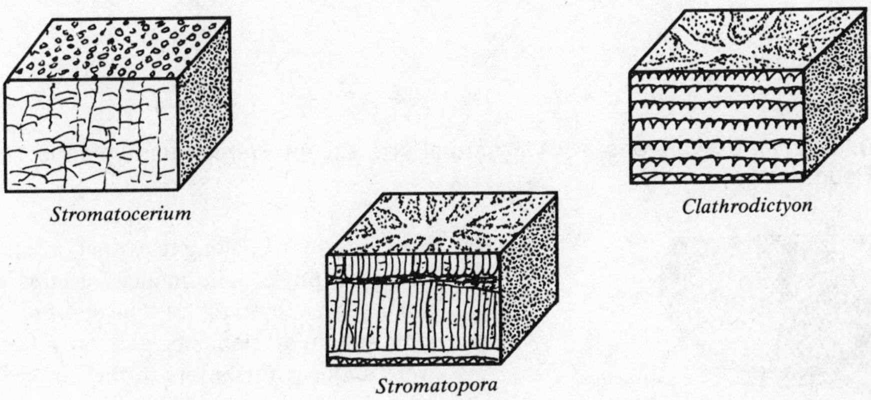

Figure 15.28. *Silurian Stromatoporoids.* Approximately X 8. (After Shimer and Shrock.)

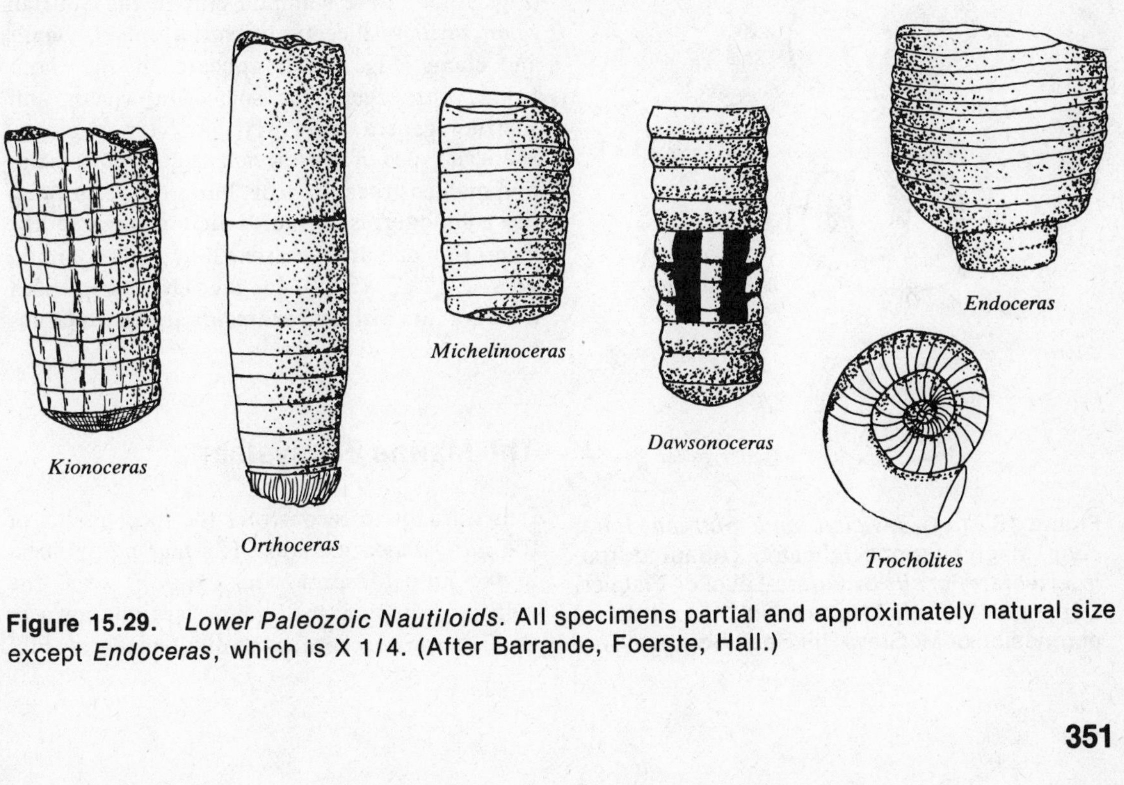

Figure 15.29. *Lower Paleozoic Nautiloids.* All specimens partial and approximately natural size except *Endoceras*, which is X 1/4. (After Barrande, Foerste, Hall.)

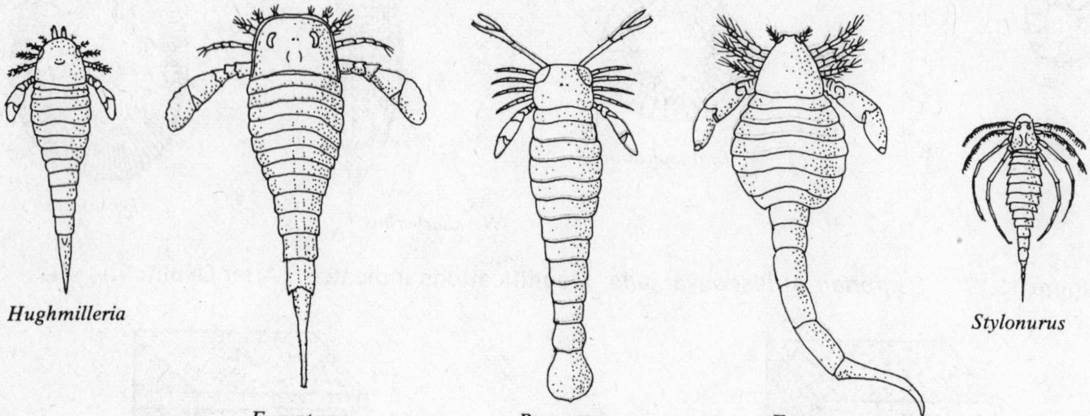

Figure 15.30. *Silurian Eurypterans.* All 1/5 natural size except *Pterygotus*, which is 1/40. (After Clarke and Reudemann.)

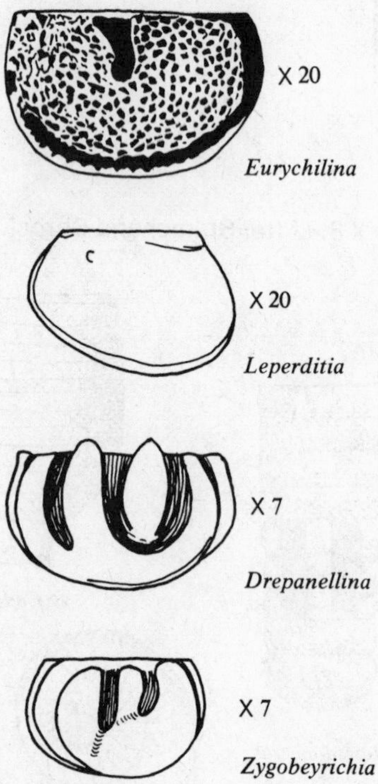

Figure 15.31. *Ordovician and Silurian Istracods.* Magnifications indicated. (Adapted from *Invertebrate Fossils* by Moore-Lalicker-Fischer, Copyright 1952 by McGraw-Hill, Inc. used by permission of McGraw-Hill Book Company.)

Note how many belong to extinct orders, classes, and even phyla. The modern groups appeared during this interval, but most were still subsidiary faunal elements and only occasionally were major contributors to the Lower Paleozoic record. The forams and radiolarians were uncommon throughout this interval. The **sponges** (Fig. 15.36) were common only in the Silurian when various siliceous types multiplied. **Snails** and **clams** (Fig. 15.37) appeared in the Cambrian, but except for some Ordovician and Silurian genera, they were not very common. Modern types of crustaceans such as barnacles and malacostracans occur, but are rare. Among the echinoderms, **crinoids** first appear in the Cambrian and are represented by some common genera (Fig. 15.38) in the Ordovician and Silurian. Starfish and echinoids appear in Ordovician rocks, but are rare.

The Marine Ecosystem

It is difficult to reconstruct the food chains of the Early Paleozoic seas. The major producers in the Late Mesozoic and Cenozoic seas, the pelagic phyloplankton groups, are unknown in the Paleozoic except for the dinoflagellates (Fig.

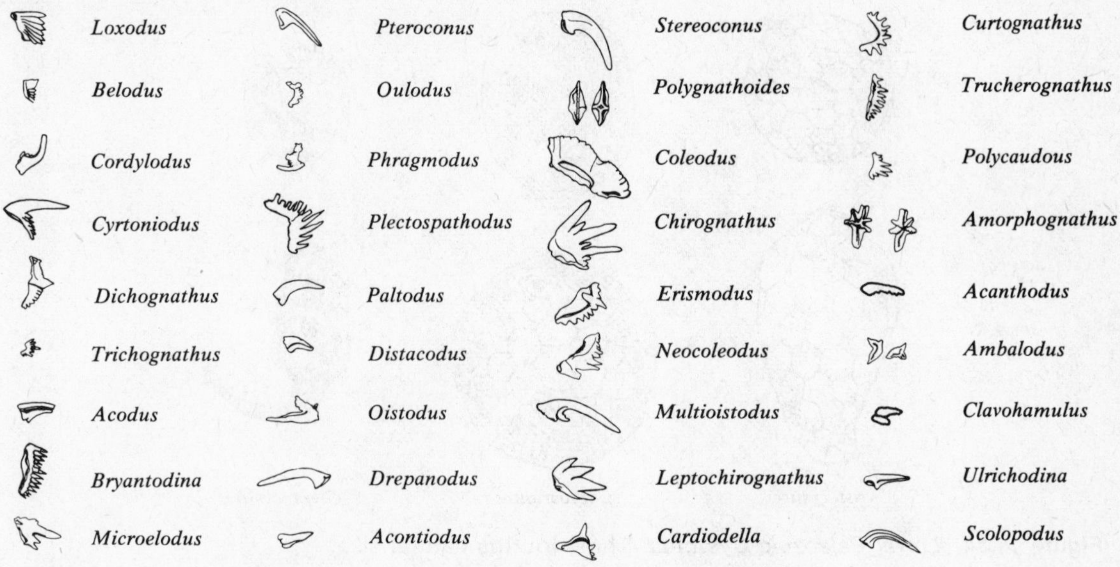

Figure 15.32. *Lower Paleozoic Conodonts.* X 10. (After Ellison.)

15.39). They may have been present, but lacked shells or had shells of easily destroyed organic materials. An enigmatic group of cysts called **acritarchs** (Fig. 15.40) are common in Lower and Middle Paleozoic strata. They may belong to an extinct group of phytoplankton that supported many Early Paleozoic food chains. The major groups of attached benthic algae, the greens, reds, and even the entirely soft-bodied browns, do occur; but it seems unlikely that they could support the amazing wealth of benthic animals of the Early Paleozoic seas.

Virtually all known Early Paleozoic animals were herbivores, scavengers, and mass detritus-plankton collectors. The abundance of benthos with hard shells in the fossil record is hard to reconcile with this fact. Perhaps the skeletons were of major value in protecting their shallow-water owners from the powerful rays of the sun. These may not yet have been adequately screened out by the still relatively low-oxygen atmosphere. The heavy shells may also have offered protection against the powerful waves which could develop in shallow epeiric seas.

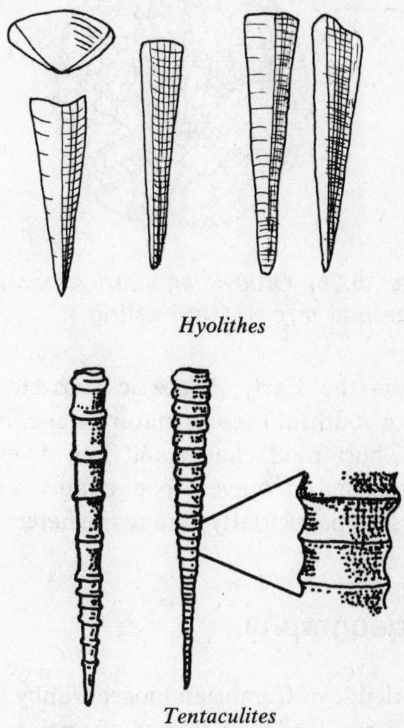

Figure 15.33. *Lower Paleozoic Hyolithans and Tentaculites.* Hyolithans are natural size; tentaculites are X 3.

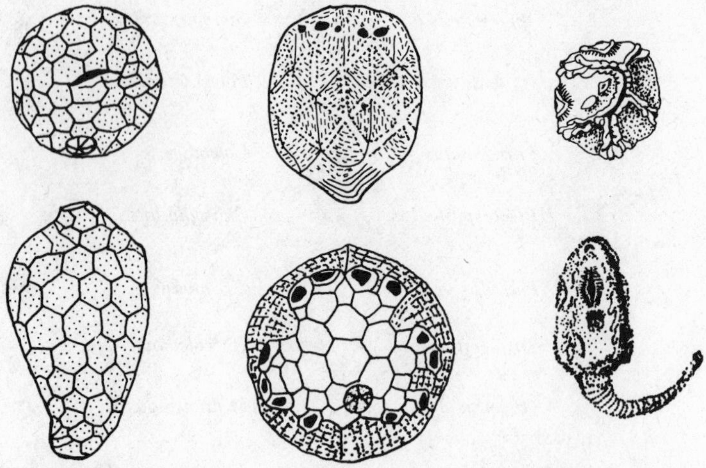

Aristocystites *Caryocrinites* *Glyptocystites*

Figure 15.34. *Lower Paleozoic Cystoids.* Three-fourths natural size.

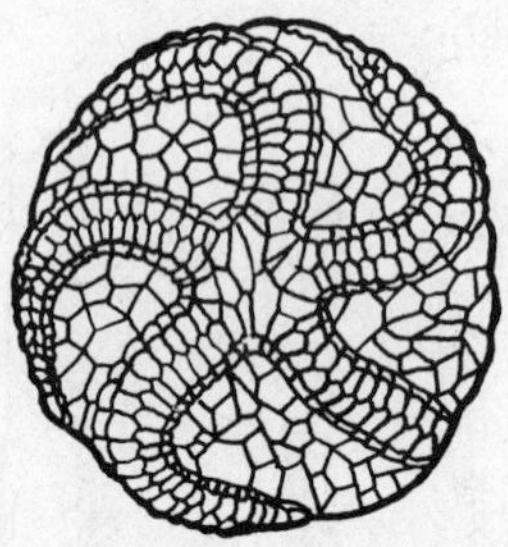

Figure 15.35. *Ordovician Edrioasteroid.* One-half natural size. (After Kesling.)

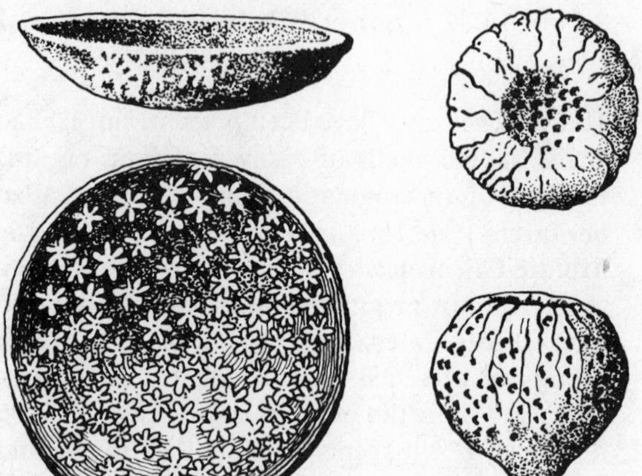

Astraeospongia *Astylospongia*

Figure 15.36. *Silurian Siliceous Sponges.* Natural size. (Adapted from *Invertebrate Fossils* by Moore-Lalicker-Fischer, Copyright 1952 by McGraw-Hill, Inc., used by permission of McGraw-Hill Book Company.)

Thus the Early Paleozoic ecosystem represents a youthful to early mature stage, beginning with short food chains and low diversity and progressing to longer food chains with more diversity, particularly among the heterotrophs.

Biogeography

Knowledge of Cambrian biogeography is still in its infancy. Trilobite faunas of ancestral North America and the South American part of Gondwanaland are very similar. Note on Figure 14.1 that both these areas lie along the line of 30° south latitude about where the present southern oceanic circulation flows from west to east. Thus, it is not surprising these areas resemble one another in their biota. Siberia, China, and Antarctica also have similar trilobite faunas,

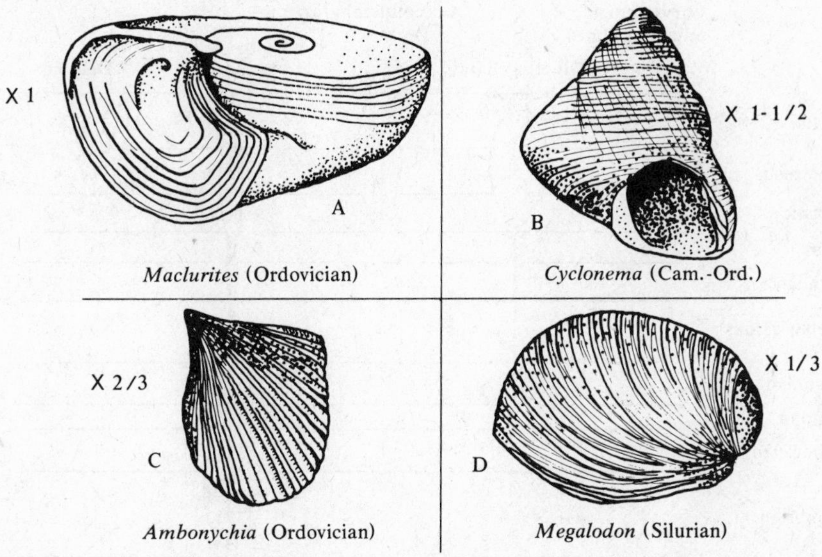

Figure 15.37. *Lower Paleozoic Mollusks.* **A-B.** *Gastropods,* **C-D.** *Bivalves.* Magnifications indicated.

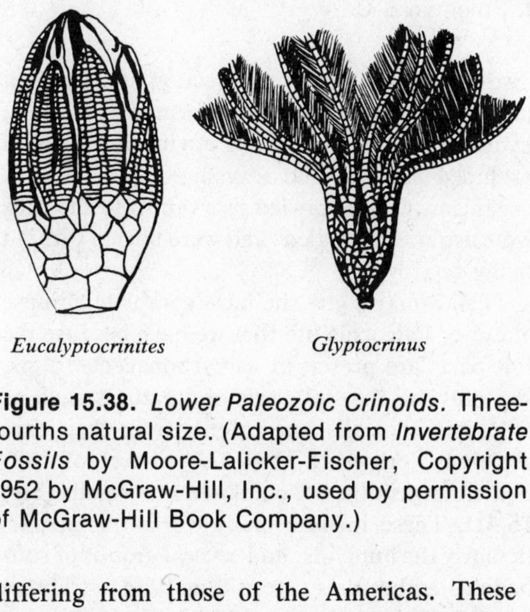

Figure 15.38. *Lower Paleozoic Crinoids.* Three-fourths natural size. (Adapted from *Invertebrate Fossils* by Moore-Lalicker-Fischer, Copyright 1952 by McGraw-Hill, Inc., used by permission of McGraw-Hill Book Company.)

differing from those of the Americas. These areas were situated just north of the Cambrian equator. At present in these latitudes, major currents flow from east to west and could transport organisms or their larvae between these areas. The difference in latitude could also explain the faunal differences between the Antarctic and South American portions of Gondwanaland.

The most thorough biogeographic study of the Early Paleozoic has been done for the Ordovician. The results agree well with the inferences about plate positions derived from physical criteria. Present-day North America, Siberia, China, and Australia have similar faunas, while present-day Europe and South America have similar faunas different from the other continents. As shown in Figure 14.2, the former three continents were in the tropics and equatorial currents could lead to faunal interchange. Present-day South America and Europe were in more temperate latitudes and not greatly distant. Related trilobites occur over much of Gondwanaland, including China, indicating the proximity of these continents at that time.

In the Late Ordovician and Silurian, there was a high degree of cosmopolitanism of faunas, probably reflecting the closing of the proto-Atlantic and the general proximity of all continents.

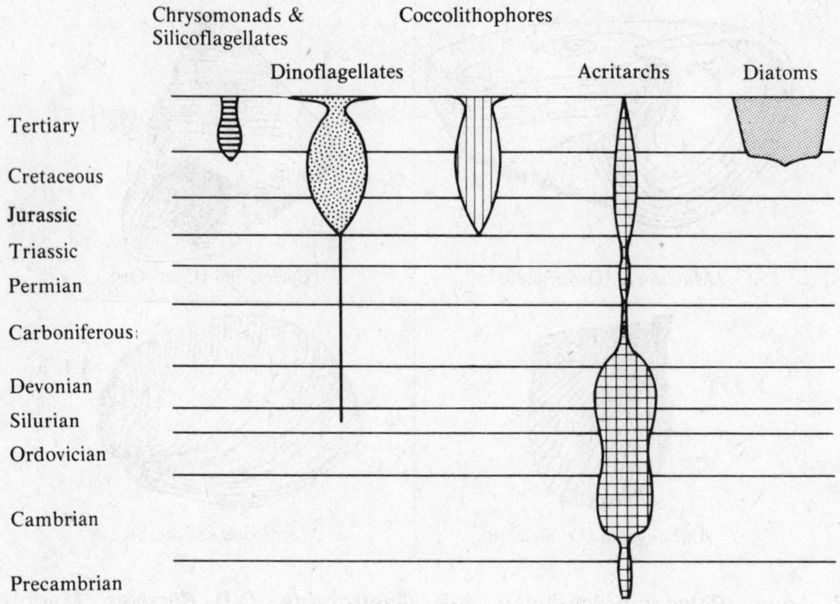

Figure 15.39. *Geologic Ranges and Diversity of Marine Phytoplankton Groups.* (From Raup and Stanley, *Principles of Paleontology*, © 1971, W.H. Freeman & Co.)

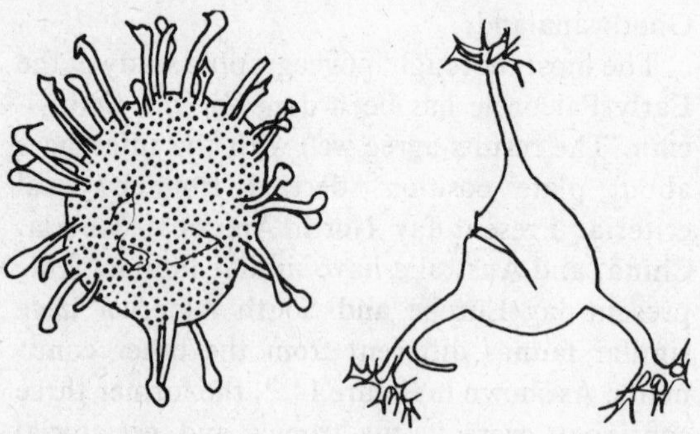

Figure 15.40. *Lower Paleozoic Acritarchs.* Highly magnified microfossils.

The Burgess Shale

No discussion of Early Paleozoic marine invertebrate life would be complete without mention of the famous Burgess Shale localities in the Middle Cambrian of the Canadian Rockies (Fig. 15.41). Known only from a few scattered outcrops high on the western ranges of the Rocky Mountains in Yoho National Park, British Columbia, this black shale unit is noted for its distillation fossils of a cross section of Cambrian life. The deposit apparently formed in a toxic, hydrogen sulfide-generating environment which excluded bacteria and scavengers. As a result, organisms which crawled or swam into the area were asphyxiated, died, and were buried without being destroyed.

Their remains give the most complete glimpse of Lower Paleozoic life that we have because the soft parts are preserved as carbonaceous films. Not only are the soft parts of creatures such as trilobites, typically known only from their exoskeletons, preserved, but also a host of soft-bodied forms otherwise unknown as fossils (Fig. 15.41). These include many worm phyla, particularly the annelids, and several groups of soft-bodied arthropods transitional in structure between the trilobites and the modern arthropod subphyla.

The Oldest Vertebrates

The **vertebrates** appeared in the Early Paleozoic, but were not common until the Late

Silurian. Some apparent tooth and jaw fragments from the Cambrian of Wyoming may be the oldest vertebrates. Other jaws are known in the very Early Ordovician of Missouri and Colorado, and then fragments of bony **ostracoderm** head shields or spines occur in several localities of Lower to Middle Ordovician age. These disarticulated remains occur in nearshore marine deposits, which has led to a spirited debate among vertebrate paleontologists about whether fish (and, hence, vertebrates) originated in the sea or in fresh water and were washed into the sea. Silurian fish were apparently rare until the end of the period when the rising Northern Appalachian-Caledonian Mountains produced extensive continental deposits. In lake and stream deposits interbedded in these terrestrial redbeds, both complete and partial fish remains are common for the first time. They are chiefly ostracoderms, but some fragments of placoderms and acanthodians also occur.

Invasion of the Land

The invasion of the land also occurred late in the Early Paleozoic. Although occasional carbonaceous smudges, spores, and supposed wood fragments have been reported from Cambrian, Ordovician, and Silurian rocks, it is not until the very close of the Silurian that unequivocal plant remains occur. These uncommon fossils are members of the simple tracheophyte group, the **psilophytes** (Fig. 15.42).

The animals also may have invaded the land at this time. **Scorpions** are known from Upper Silurian rocks, but it is uncertain if these earliest representatives of the group were fresh water or terrestrial. Recent studies suggest that these ancestral types were probably aquatic, unlike their descendants. Also found in Upper Silurian rocks is a member of the archipolypods, a group of arthropods related to the **millipedes.** The millipedes are entirely terrestrial today, but again it is not certain if these early fossils were aquatic or terrestrial because the preservation does not permit identification of breathing structures. Some Early Paleozoic malacostracan crustaceans may have been fresh water or even partially terrestrial.

A

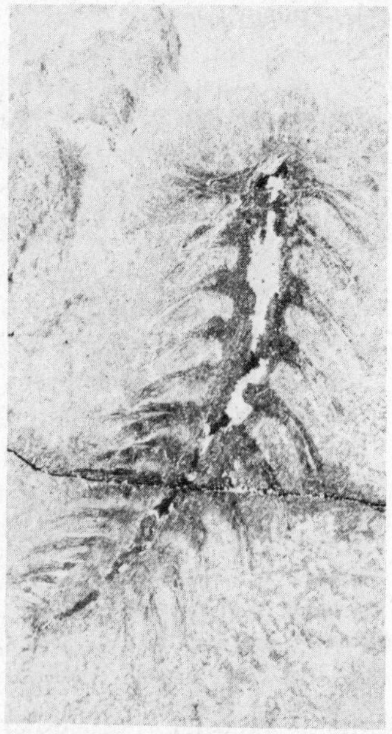

B

C

D

Figure 15.41. *Burgess Shale Locality and Fossils.* **A.** *The locality high in the Canadian Rockies of British Columbia;* **B.** *Canadia*, an annelid worm; **C.** *Wiwaxia*, an annelid worm; **D.** *Marrella*, a marrellomorph trilobitazoan; **E.** *Sidneyia*, a chelicerate-like trilobitozoan; **F.** *Emeraldella*, a chelicerate-like trilobitazoan; **G.** *Naraoia*, a chelicerate-like trilobitazoan; **H.** *Waptia*, a crustacean-like trilobitazoan; **I.** *Olenoides*, a corynexochoid trilobite; **J.** *Hymenocaris*, a primitive malacostracan crustacean. (Courtesy of U.S. National Museum.)

E

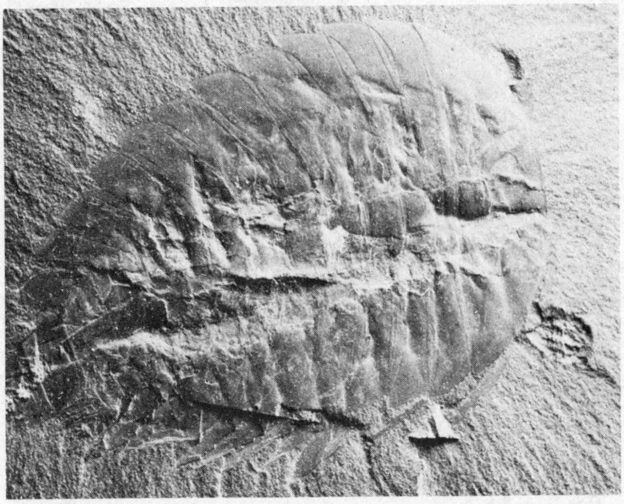

F

G

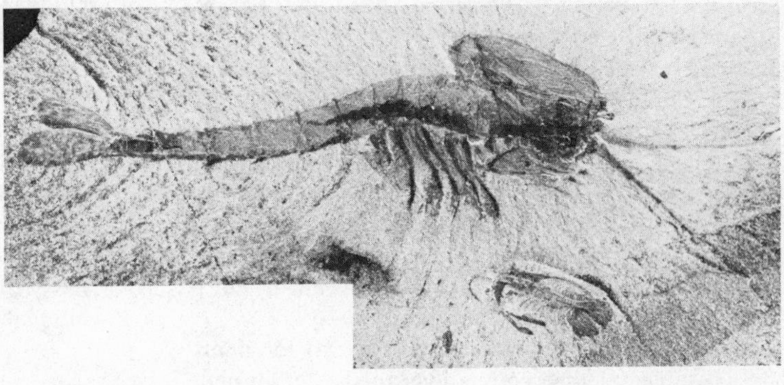

H

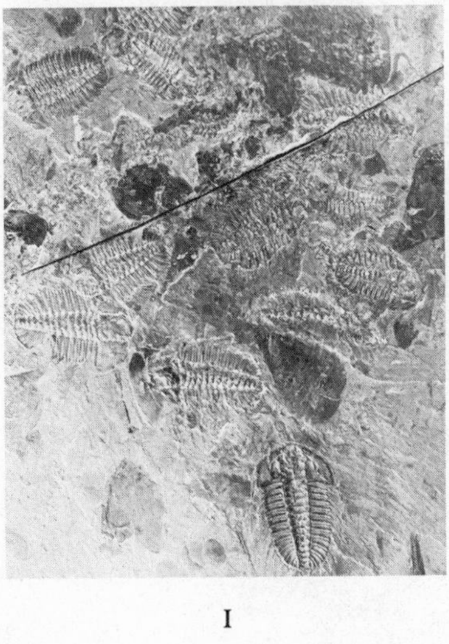

I

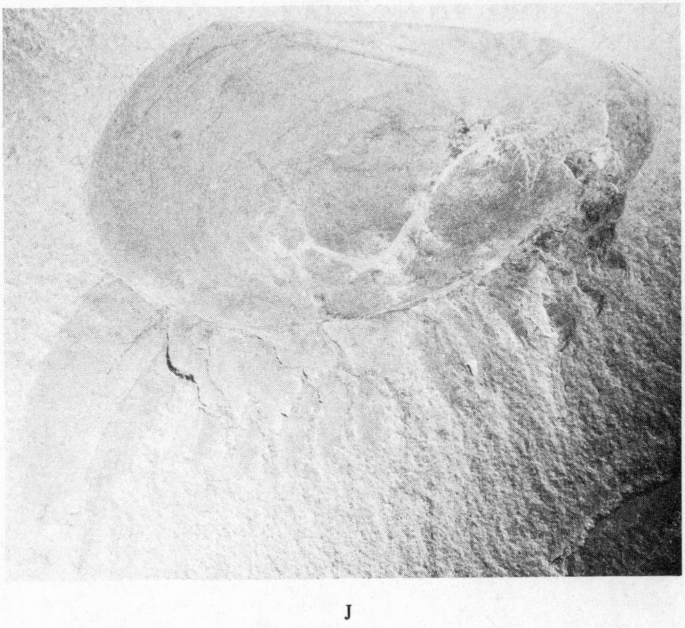

J

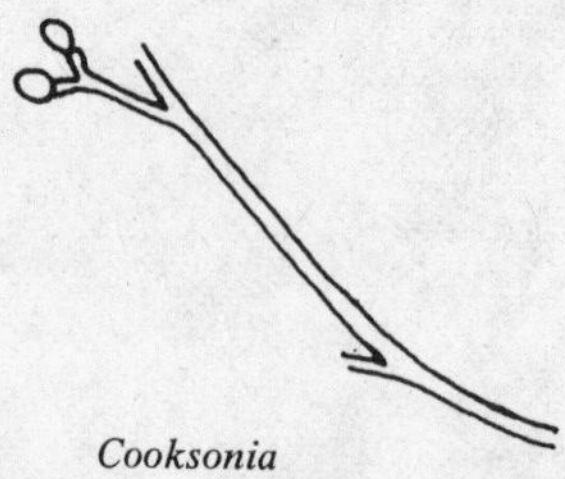

Figure 15.42. *Silurian Psilophyte.* Natural size. (After Banks.)

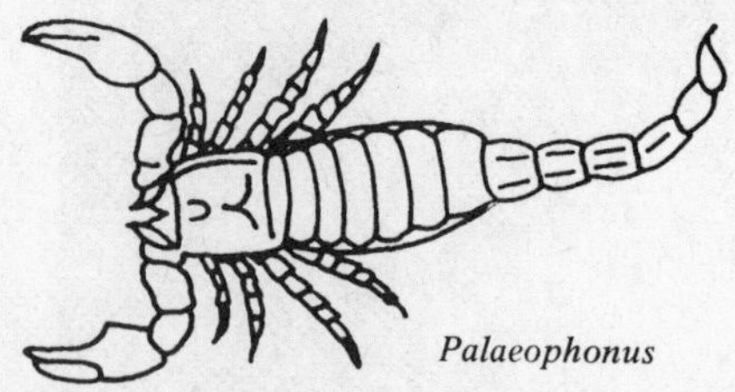

Figure 15.43. *Silurian Scorpion.* X 1/10. (After Pocock.)

REVIEW

1. Why is the Early Paleozoic called the **Age of Marine Invertebrates**?
2. What were the major marine invertebrate groups of the Early Paleozoic?
3. Which type of **corals** dominated Silurian reefs?
4. Which types of **bryozoans** reached an Ordovician peak?
5. What were the four major **brachiopod** groups of the Early Paleozoic?
6. How could you distinguish a Cambrian **trilobite** fauna from an Ordovician one?
7. Which group of fossils is most useful for intercontinental time correlations in the Ordovician and Silurian? Why?
8. In which Early Paleozoic period did each of the following groups reach a peak: receptaculitids, archaeocyaths, stromatoporoids, nautiloids, eurypterans?
9. Which significant modern groups of marine invertebrates were relatively unimportant in the Early Paleozoic?
10. Why is it difficult to reconstruct Early Paleozoic marine food chains?
11. What is the significance of the **Burgess Shale fauna?**
12. What evidence is there of vertebrates and land life in the Early Paleozoic?
13. How well does Early Paleozoic biogeography fit with plate tectonics models?

16

Limestone, Coal, and Redbeds: Devonian, Carboniferous, and Permian

16

The three periods of the Late Paleozoic, the Devonian, Carboniferous, and Permian, encompass approximately 175 million years of earth history from about 400 to 225 million years ago. The **Devonian** has its type area in the ancient Hercynian mountain belt of southwest England; the **Carboniferous,** in central Great Britain on a shelf adjacent to the Caledonian Mountains; and the **Permian,** in Russia on the European Craton just west of an ancient ocean basin where the Ural Mountains now stand (Fig. 2.19).

In North America, the Carboniferous is typically subdivided into two periods—the **Mississippian,** named from the Upper Mississippi Valley of Illinois and Iowa, and the **Pennsylvanian,** named from the state of Pennsylvania. Both type areas are on the craton, but the latter is near the edge of the Appalachian Mountains. Geologists have a considerably better picture of the global tectonics of the Late Paleozoic than they do of the Early Paleozoic interval.

Gondwanaland

Gondwanaland was still a single entity as indicated by the paleomagnetic, geological, and paleontologic data (Figs. 16.1–3). However, it had shifted position because the location of the Late Paleozoic Gondwana Pole (Fig. 14.5) changed from present South Africa in the Devonian, across present Antarctica toward present southern Australia in the Permian. The situation of the pole in Gondwanaland is reflected by the widespread Carboniferous and Permian glacial deposits and erosional features (Fig. 16.4) of eastern South America, southern Africa, Antarctica, India, and Australia.

The succeeding terrestrial redbed sequences with coal (Fig. 16.5) are strikingly similar on all the present southern continents that once composed Gondwanaland. The best known is the Karroo Group of South Africa with its large mammal-like reptile fauna. Nearshore marine rocks of the same lithofacies and with the same fossil assemblages occur in the Permocarboniferous sequence near Buenos Aires, Argentina, in Southwest Africa, the Salt Range of Pakistan, and southeastern Australia. The marine *Eurydesma* fauna (Fig. 16.6) contained in these rocks is only one of the impressive paleontologic links between the now widely separated southern continents. We have already mentioned the small fresh-water reptile *Mesosaurus* (Fig. 17.59) found in the Permocarboniferous of South America and South Africa. Most striking of all in stimulating early thought about continental drift, however, is the *Glossopteris* flora (Fig. 16.7) of Late Carboniferous-to-Early Triassic age. This apparent seed fern and its attendant plants are the major contributors to the important coal beds of this interval on today's southern continents.

A Pacific plate(s) must have been moving against Gondwanaland along the present eastern side of Australia. This region of the Tasman Trench underwent continuous disturbances which culminated in the mid-Carboniferous **Kanimblan Orogeny.** (After the Paleo-

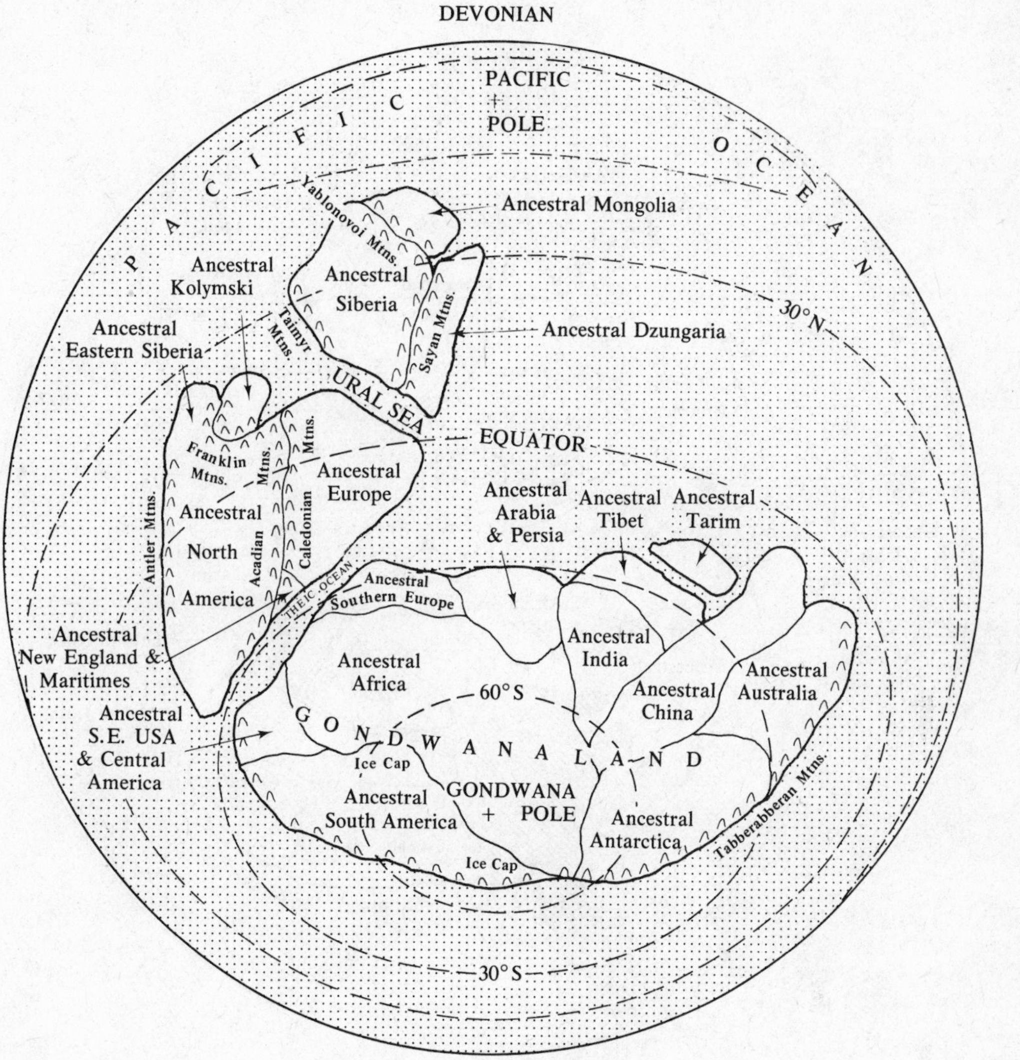

Figure 16.1. *Devonian Paleogeography.* The positions of the various continental fragments of China are highly speculative. Plate boundaries and epeiric seas are not shown. (Material derived from Badham and Hall; Crawford; Dewey and Bird; Irving; Seyfert and Sirkin; and others.)

zoic, there ceased to be a trench along the present eastern side of Australia.) The effects of this orogeny remain today in the Great Dividing Ranges (Fig. 16.8) of eastern Australia. The Antarctic continuation of this trench, the Borchgrevink, was also affected. In the Permian, terrestrial beds with interbedded coal and tillites succeeded this orogenic disturbance. In the present east Pacific area, a Pacific plate(s) must have been impinging against South America as the Andean Trench underwent deformations throughout much of the Late Paleozoic. The Andean belt of the Late Paleozic bends eastward through present northern Argentina (Sierra de la Ventana) and the Cape Mountains of South Africa, also deformed in the Permo-Triassic interval, and continues into the present Antarctica and eastern Australian belt discussed

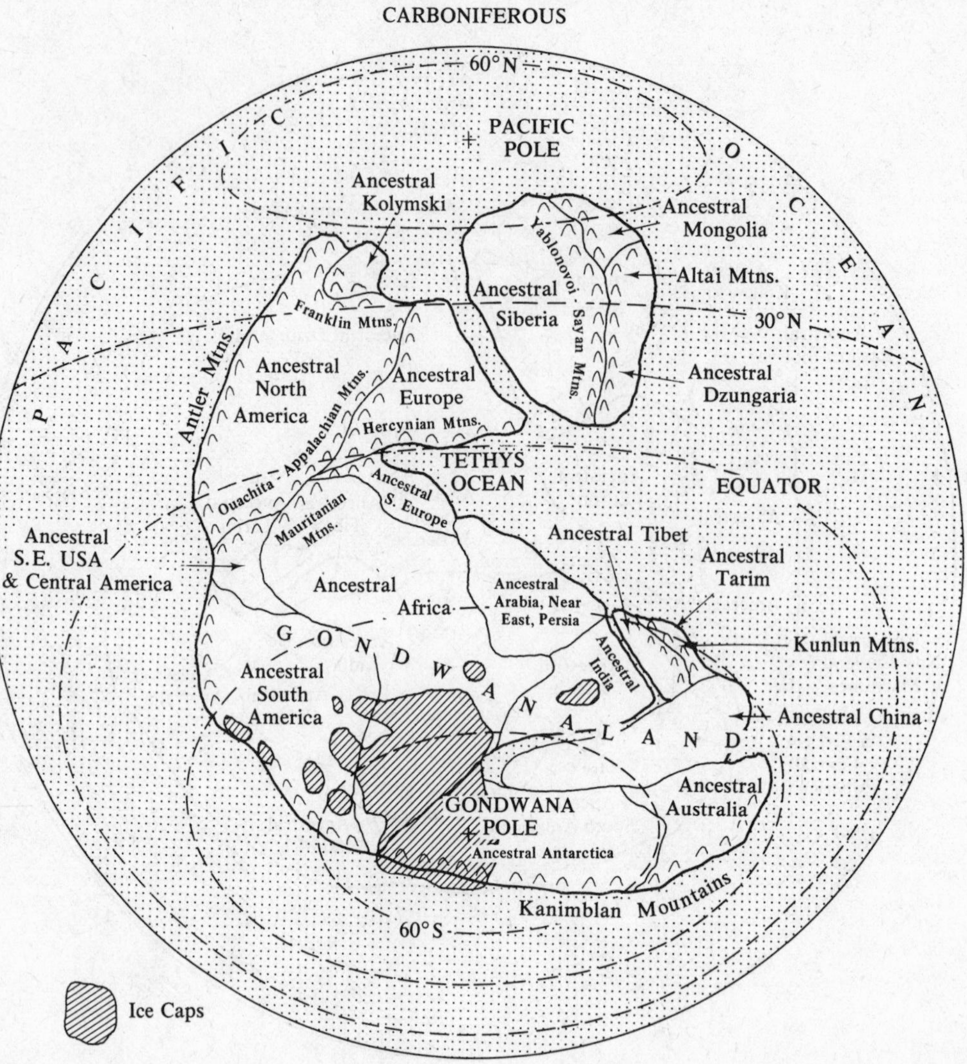

Figure 16.2. *Carboniferous Paleogeography.* The positions of the various continental fragments of China are highly speculative. Plate boundaries and epeiric seas not shown. (Material derived from Badham and Hall; Crawford; Crowell and Frakes; Irving; Seyfert and Sirkin; and others.)

earlier. Geologists call this great mountain chain (Fig. 10.5) the **Samfrau** after South America, Africa, and Australia. The plate-margin trench remained along the South American and Antarctic coasts of Gondwanaland after the Paleozoic.

The present northwest part of Gondwanaland, what is now Africa and the southeastern United States-Central American region, approached present-day North America in the Devonian and collided with it in the Carboniferous and Permian, producing the **Ouachita-Appalachian-Mauritanian Mountains.** Farther east, northern Gondwanaland (what is now southern Europe) collided with the European Craton, producing the **Hercynian Mountains,** an eastward continuation of the same chain. The **Tarim Craton** collided with Tibet, producing the

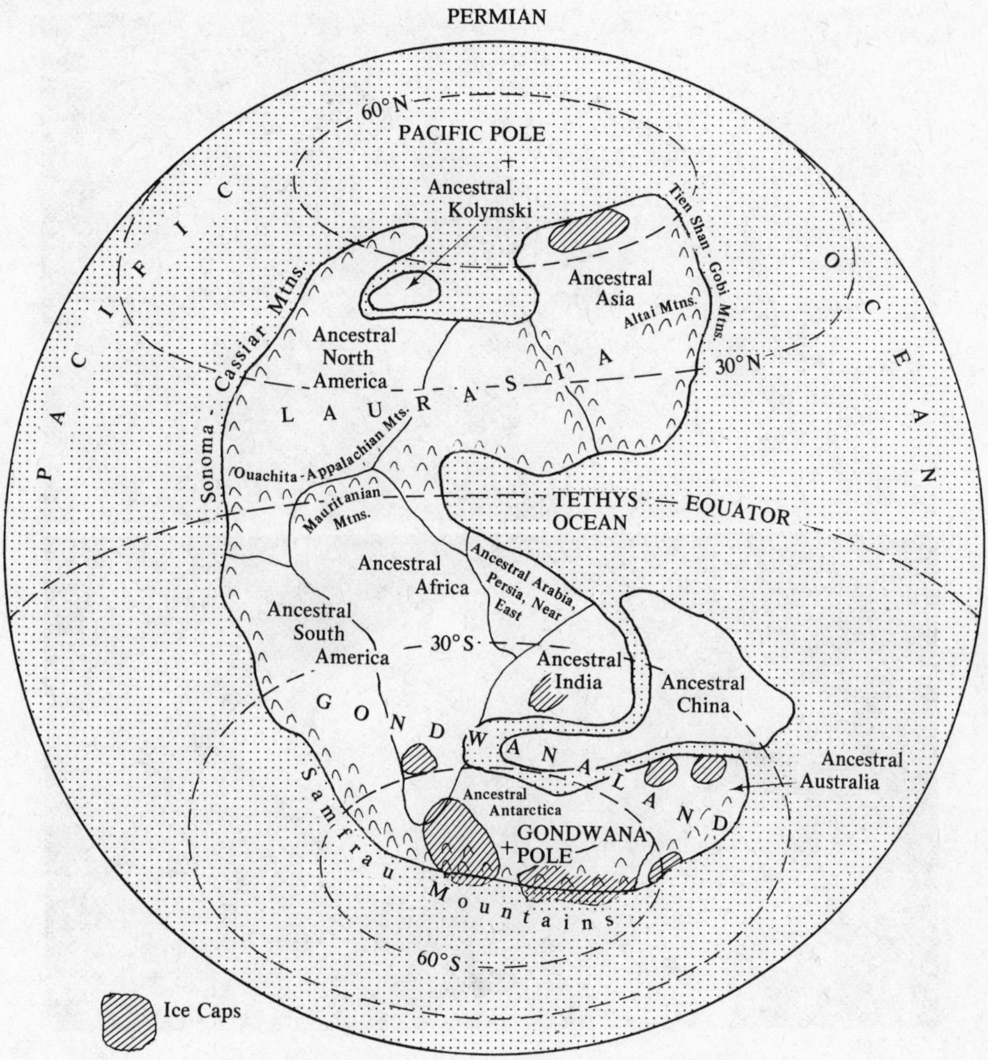

Figure 16.3. *Permian Paleogeography.* The positions of the various continental fragments of China are highly speculative. Plate boundaries and epeiric seas not shown. (Material derived from Badham and Hall; Crawford; Crowell and Frakes; Seyfert and Sirkin; and others.)

first-generation **Kunlun Mountains.** By the Permian, China had begun to separate from Gondwanaland and move northward.

Laurasia

Laurasia also took form as a single supercontinent in the Late Paleozoic. The collision of present eastern North America and northwest Europe which had begun in the Late Silurian continued into the Devonian, producing the **Acadian Mountains** (Fig. 16.1) as they are called in North America or the **Old Red Sandstone "Continent"** (Fig. 14.11) as it is known in Europe. Acadia refers to a region in the Canadian Maritime provinces where the effects are well displayed. From this region to the present

Figure 16.4. *Gondwana Glacial Deposits and Erosional Features.* **A.** *Polished pavement on quartzite, friction cracks, South Australia.* **B.** *Striated and faceted stone in tillite, Tasmania.* (Courtesy of Warren Hamilton, U.S.G.S. from Hamilton and Krinsley, *G.S.A. Bulletin*, vol. 78, June 1967. By permission of authors and G.S.A.)

south, the **Catskill Clastic Wedge** (Fig. 16.9), which extends from southeastern New York to Virginia, is the main expression of the Acadian Orogeny in the United States. Deformation continued later into the Paleozoic as North America collided with the present northwestern edge of Africa (Figs. 16.2–3). Thick sequences of clastics formed in present-day eastern North America in the Mississippian. The **Pocono Clastic Wedge** of Pennsylvania is one of the best known. Several fault-block basins developed in the Canadian Maritime Provinces (Fig. 16.10).

The final phases of orogenic deformation in the Appalachian mountain belt occurred toward the close of the Paleozoic Era. In North America, this major deformation is called the **Appalachian** or **Alleghenian Orogeny** and extends from the Pennsylvanian, through the Permian, and into the Triassic. Its major expressions today are the folds, thrust faults, and intrusions preserved in our present Appalachian Mountains (Fig. 16.11) and the **Dunkard Clastic Wedge** which extends westward from them. The causes of the Appalachian-Alleghenian Orogeny are unclear. Some authorities consider it to be the final result of the collision of North America and the African edge of Gondwanaland, while others believe it represents the initial stages in the formation of the rift that eventually formed the present Atlantic Ocean basin. The fact that the Late Paleozoic structural trends parallel the present coastlines and cut across older Appalachian structures tends to support the latter contention. Gentle deformations in the North American and European cratons in the Pennsylvanian and Permian align parallel to both the present Atlantic rift and the Labrador Sea rift between Greenland and North America which is no longer active.

In Europe, the collision of Gondwanaland (present southern Europe and northwest Africa) with the European Craton produced extensive mountain building (Figs. 16.2–3). This episode is named the **Hercynian** (after the Harz Mountains, Germany) or the **Variscan** (after a region in Saxony, Germany) **Orogeny.** Culmination was in the Late Carboniferous and Early Permian. Its present remnants (Fig. 16.12) occur in isolated low mountain massifs in southwestern England, northern France and Germany, Bohemia, and the Iberian Peninsula (Spain and Portugal). The disturbance is also recorded in the Alps of southern Europe and the Atlas Mountains of northwest Africa, but the effects are largely obscured by later orogeny. In the early phases of this orogeny, the plate containing northern Europe and the **Theic Ocean** was being subducted beneath northern Gondwanaland (southern Europe). Later, a rift may have opened within southern Europe, and the northern section was split off, moved north, and collided with northern Europe, producing the main Hercynian Orogeny.

The collision of the present continental masses of Europe and Siberia also occurred in the Late Paleozoic and produced the **Uralian Orogeny** (Fig. 16.3). In the Early Devonian, an island arc collided with the present eastern edge of the European Craton. This collision brought a subduction zone to the edge of the craton, but this zone then reversed its Early Paleozoic dip away from the continent to a direction toward and beneath the continent. (From Middle Devonian through Early Permian, the situation probably resembled that in eastern Asia today, where the Ryukyu arc is being driven against the craton by the Pacific.) On the Siberian side, a subduction zone dipping beneath the continent had probably been present since the Eocambrian. Both subduction zones collided as Europe and Siberia joined in the Permian.

Southern Laurasia was deformed along its entire southern and eastern margins in the Late Paleozoic (Fig. 16.3). As stated earlier, the westernmost part of this belt in present-day North America and Europe resulted from the collision of continental plate edges (Figs. 16.2–3) as the ancient Theic Ocean was consumed. To the present east, through eastern Europe and across southern Asia, however, the colliding margins were continental on the Laurasian side, but oceanic on the Gondwana side. The original **Tien Shan** and **Gobi mountain ranges** were produced by this disturbance. Along the present eastern side of Siberia, an oceanic plate was impinging against the craton along a subduction zone. This area was in high latitudes, and there is evidence of Permian glaciation (Fig. 16.3).

The Devonian-Mississippian **Innuitian Orogeny** deformed the present northern margin of North America, producing the **Franklin** or **Innuitian Mountains** (Fig. 16.13). This compression resulted from the collision of **Kolymski,** a small piece of present eastern Siberia, with the

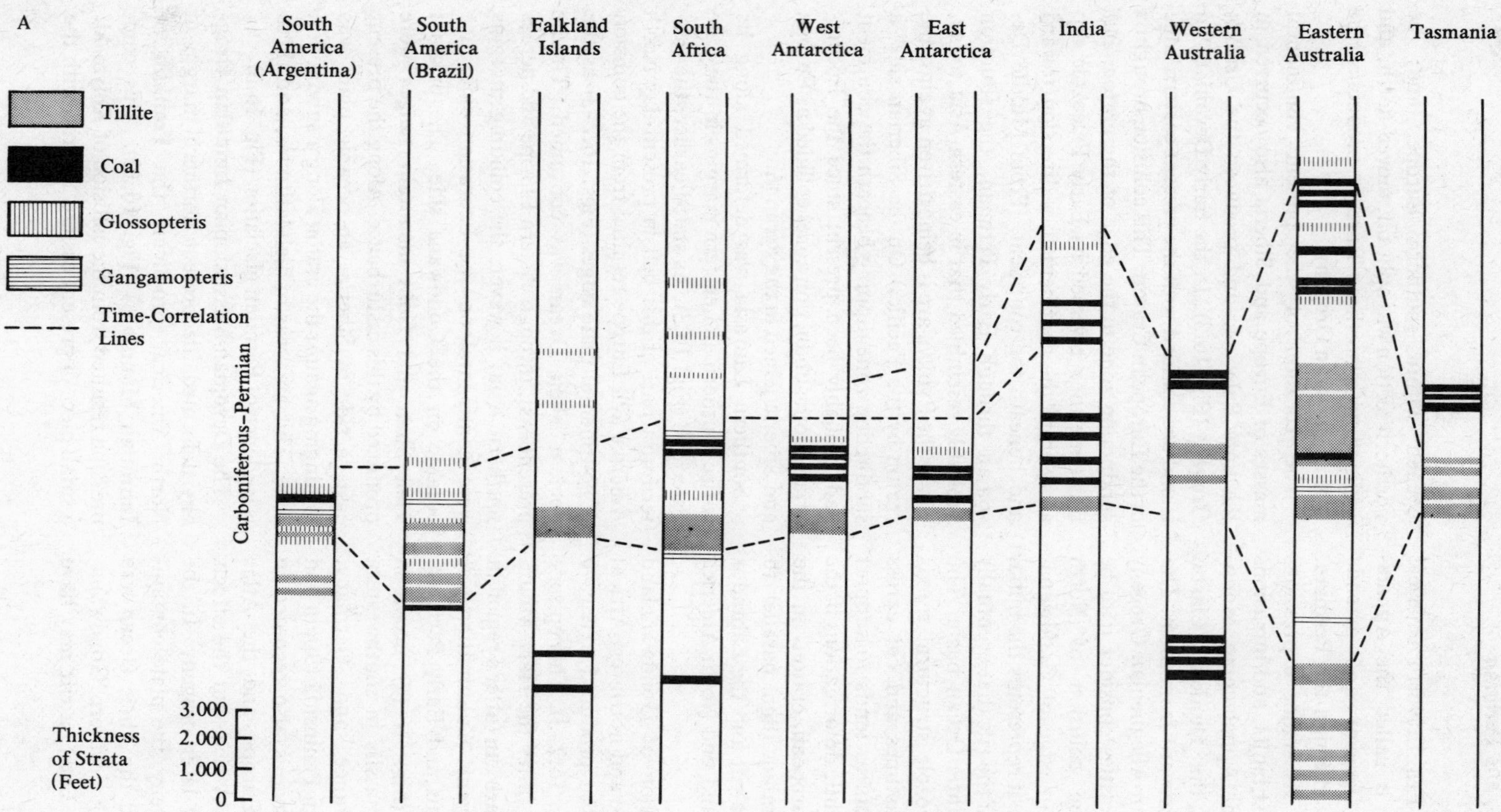

Figure 16.5. *The Gondwana Succession on the Southern Continents.* **A.** *Generalized cross sections of all Gondwana areas.* (After Hurley.)

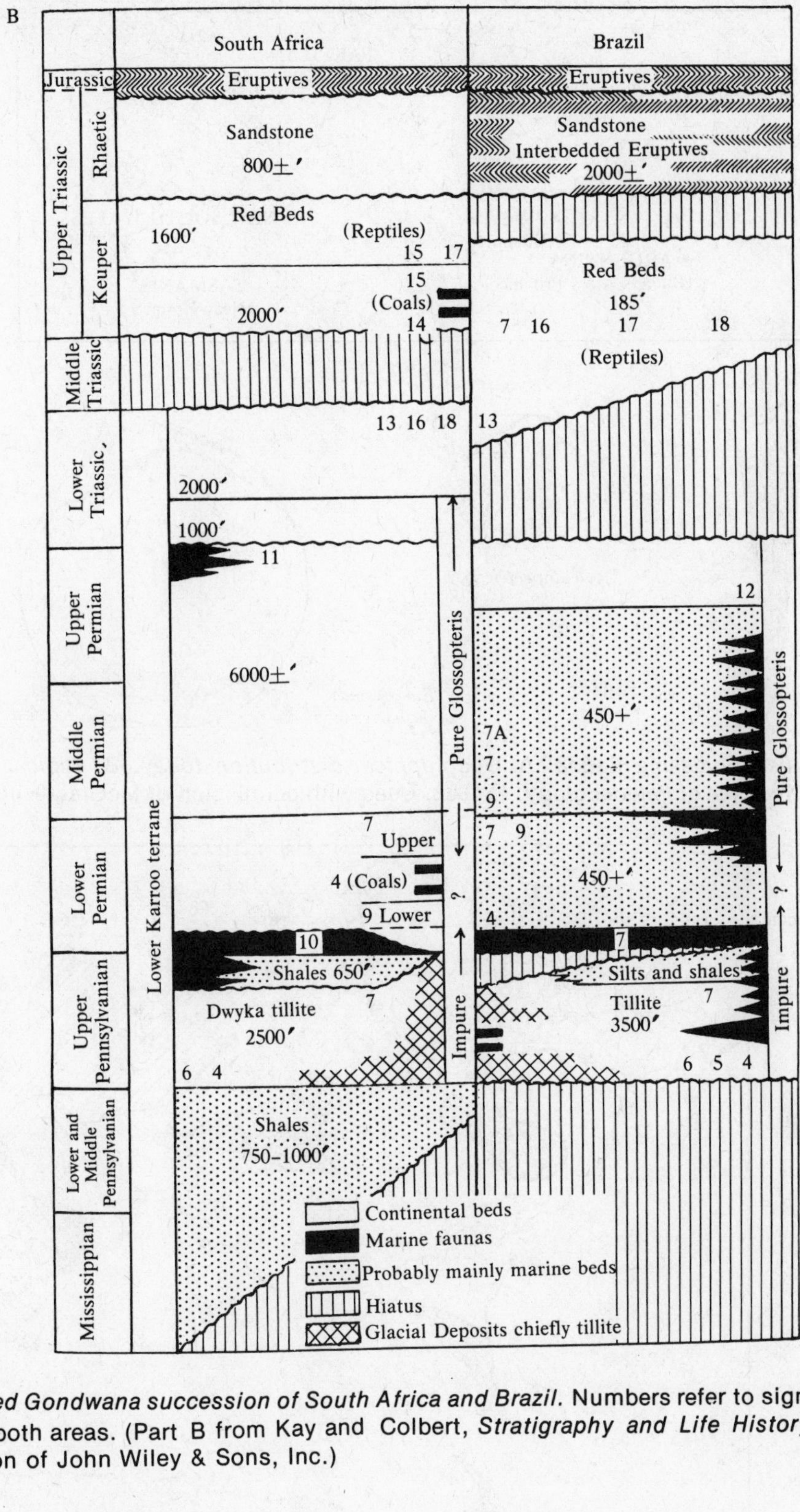

B. *Detailed Gondwana succession of South Africa and Brazil.* Numbers refer to significant fossils found in both areas. (Part B from Kay and Colbert, *Stratigraphy and Life History*, © 1965, by permission of John Wiley & Sons, Inc.)

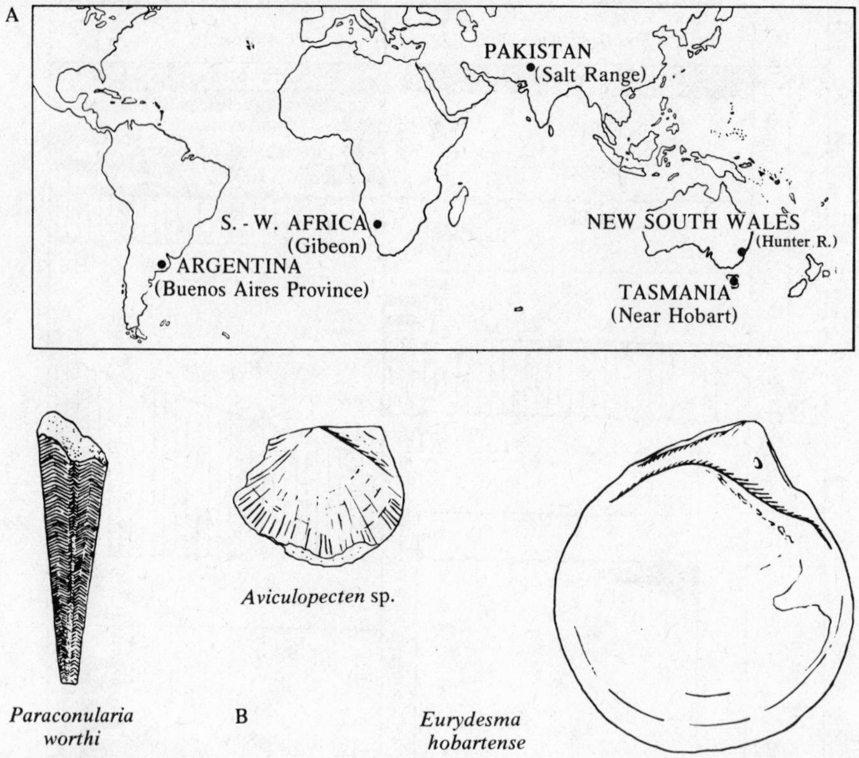

Figure 16.6. *The Eurydesma Fauna.* **A.** *Geographical distribution today,* **B.** *Typical members.* (From Ager, *Principles of Paleoecology*, © 1963. Used with permission of McGraw-Hill Book Co.)

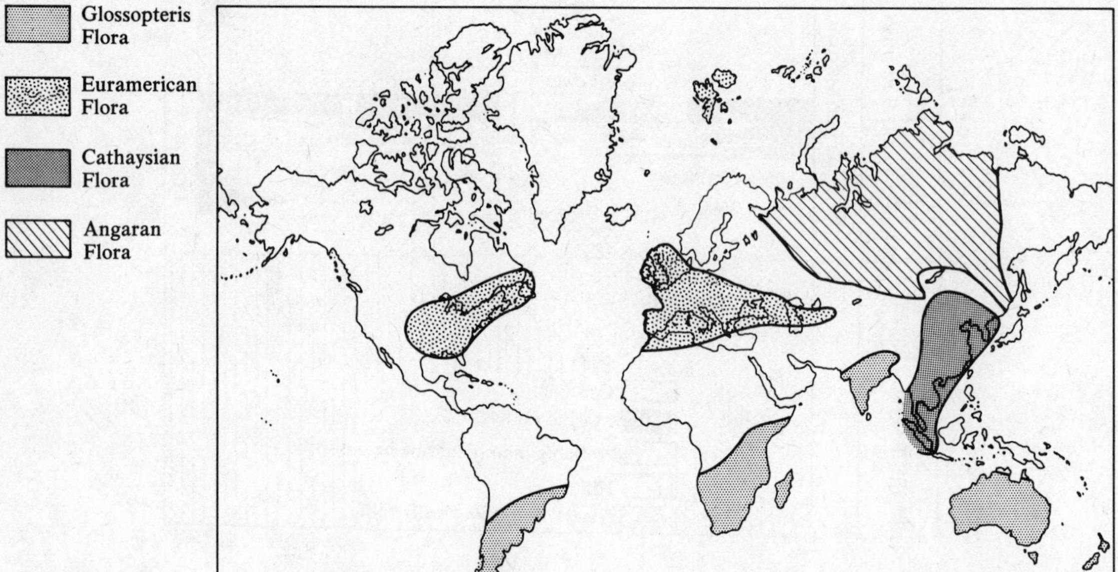

Figure 16.7. *Present-Day Distribution of Upper Paleozoic Floras.* (After Halle, Just.)

A

B

Figure 16.8. *Great Dividing Ranges of Australia.* **A.** *Mt. Kosciusko, highest mountain in Australia,* **B.** *The Three Sisters in the Blue Mountains.* (Courtesy Australia Tourist Commission.)

present northern margin of North America. Postorogenic block faulting and basin filling succeeded the mountain building. This sequence probably signals the opening of a new ocean basin as Kolymski split away from ancestral North America and moved toward Siberia with which it collided in the Mesozoic.

A subduction zone continued to exist along the present western margin of North America, producing Late Paleozoic deformations called the **Antler Orogeny** (chiefly Mississippian) and the **Sonoma** or **Cassiar Orogeny** (chiefly Permian). The continent was colliding with a Pacific Ocean plate(s) to the present southwest (Fig. 16.14).

The present southern edge of the North American Craton was buckled, chiefly in the Pennsylvanian, into the so-called **Ancestral Rockies** and

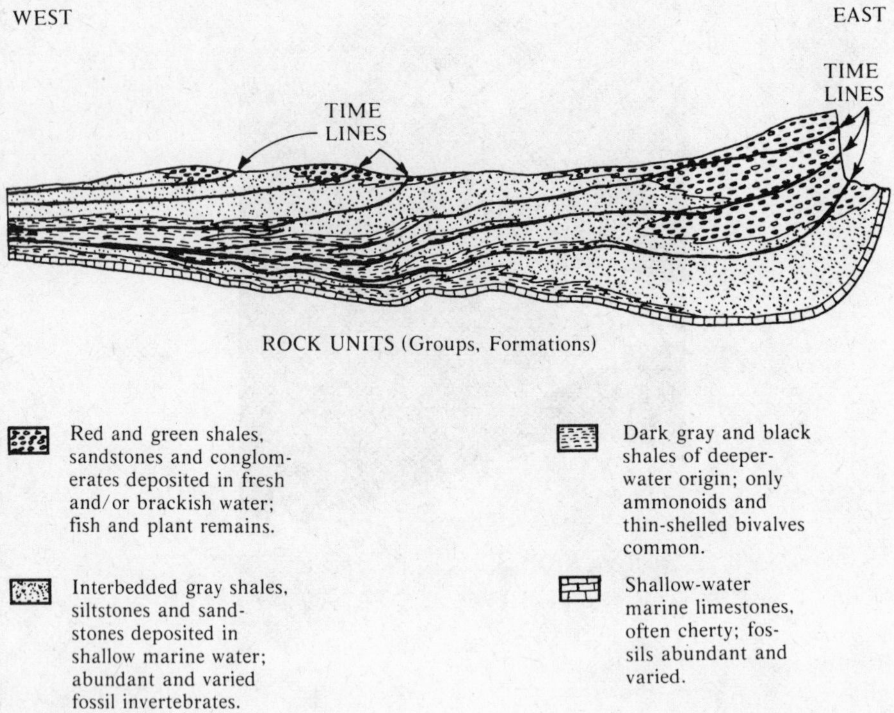

Figure 16.9. *Cross Section of Devonian System of Present-Day Southern New York Showing Catskill Clastic Wedge.*

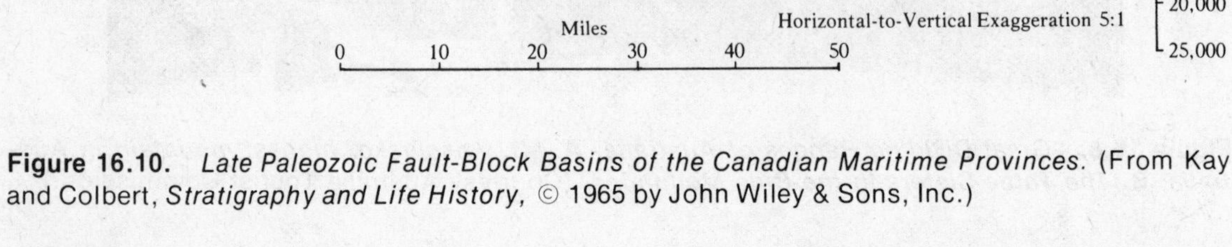

Figure 16.10. *Late Paleozoic Fault-Block Basins of the Canadian Maritime Provinces.* (From Kay and Colbert, *Stratigraphy and Life History,* © 1965 by John Wiley & Sons, Inc.)

Oklahoma Mountains (Fig. 16.15). Great thicknesses of marine and terrestrial clastics accumulated in the basins between the various ranges produced by this deformation. Later deformed by the disturbances that produced the present western Cordillera, some of the arkoses now stand with steep eastward slopes along the flanks of the Colorado Rockies, forming the picturesque Flatirons, Red Rocks, and Garden of the Gods (Fig. 16.16). Some of the marine strata have been subsequently elevated to form high mountains and plateaus (Fig. 16.17).

Completing our circle of Laurasia was a westward continuation of the southern Appalachian

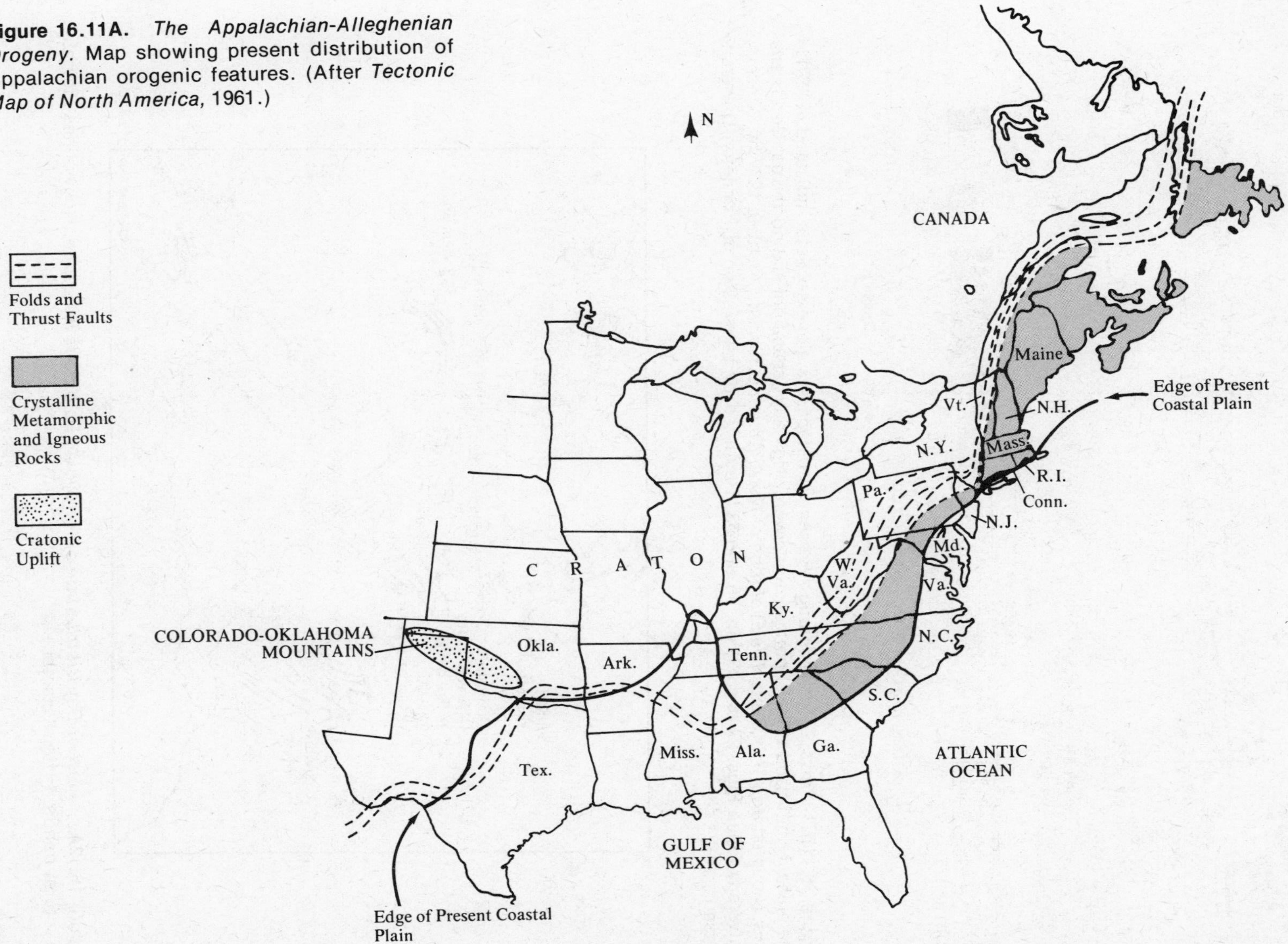

Figure 16.11A. *The Appalachian-Alleghenian Orogeny.* Map showing present distribution of Appalachian orogenic features. (After *Tectonic Map of North America,* 1961.)

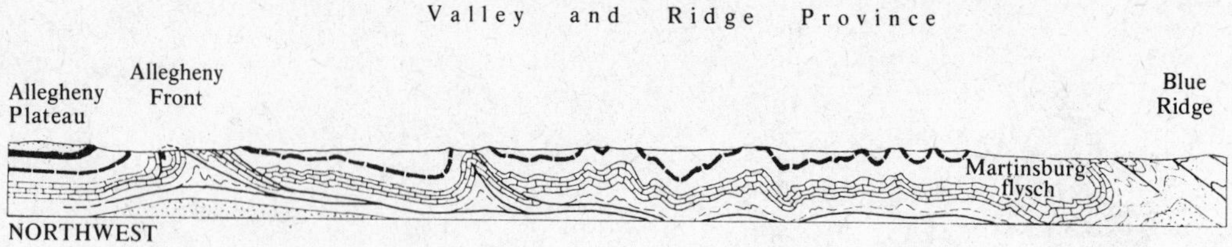

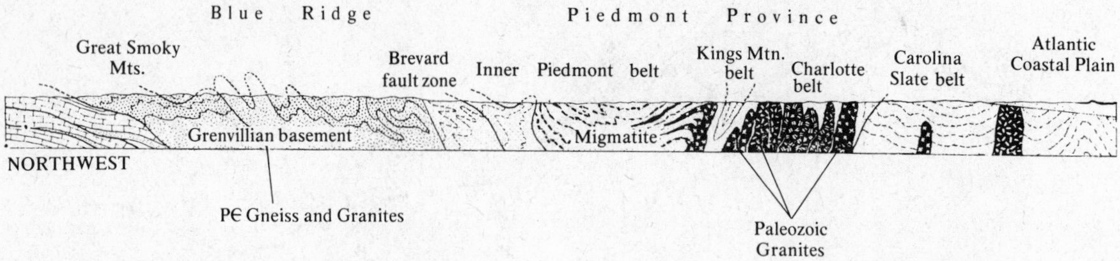

Figure 16.11B. *Cross Sections through Present Appalachians* along sections indicated in 16.11A The present right-hand sides of both sections were originally ocean floor and northwest Gondwanaland. (Adapted from P.B. King, *The Evolution of North America*, rev. ed., © 1959, 1977 by Princeton University Press Figs. 29 and 33, pp. 48 and 53. Reprinted by permission of Princeton University Press.)

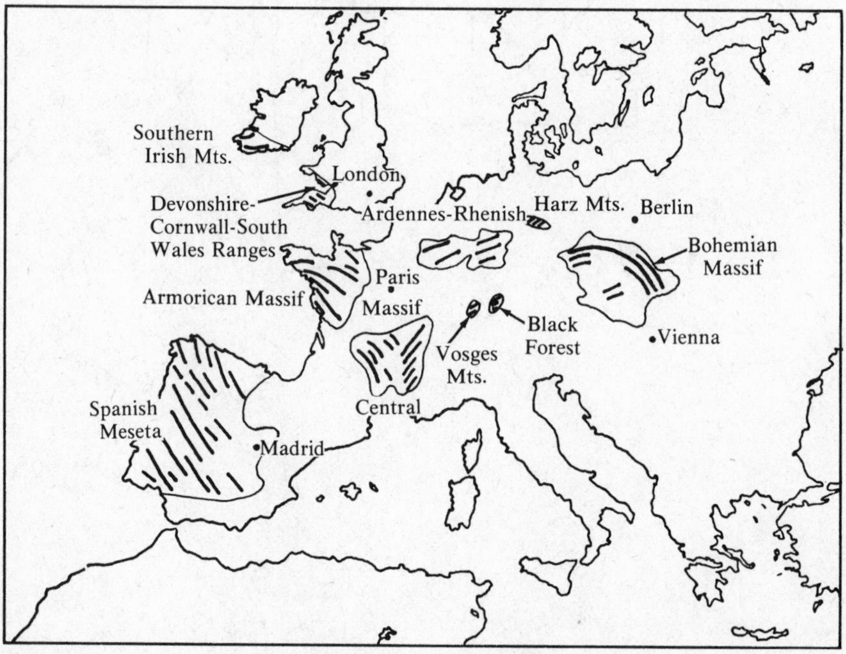

Figure 16.12A. *Present-Day Distribution of Hercynian Fold Belts in Europe.* Heavy lines indicate structural trends. (After Kummel.)

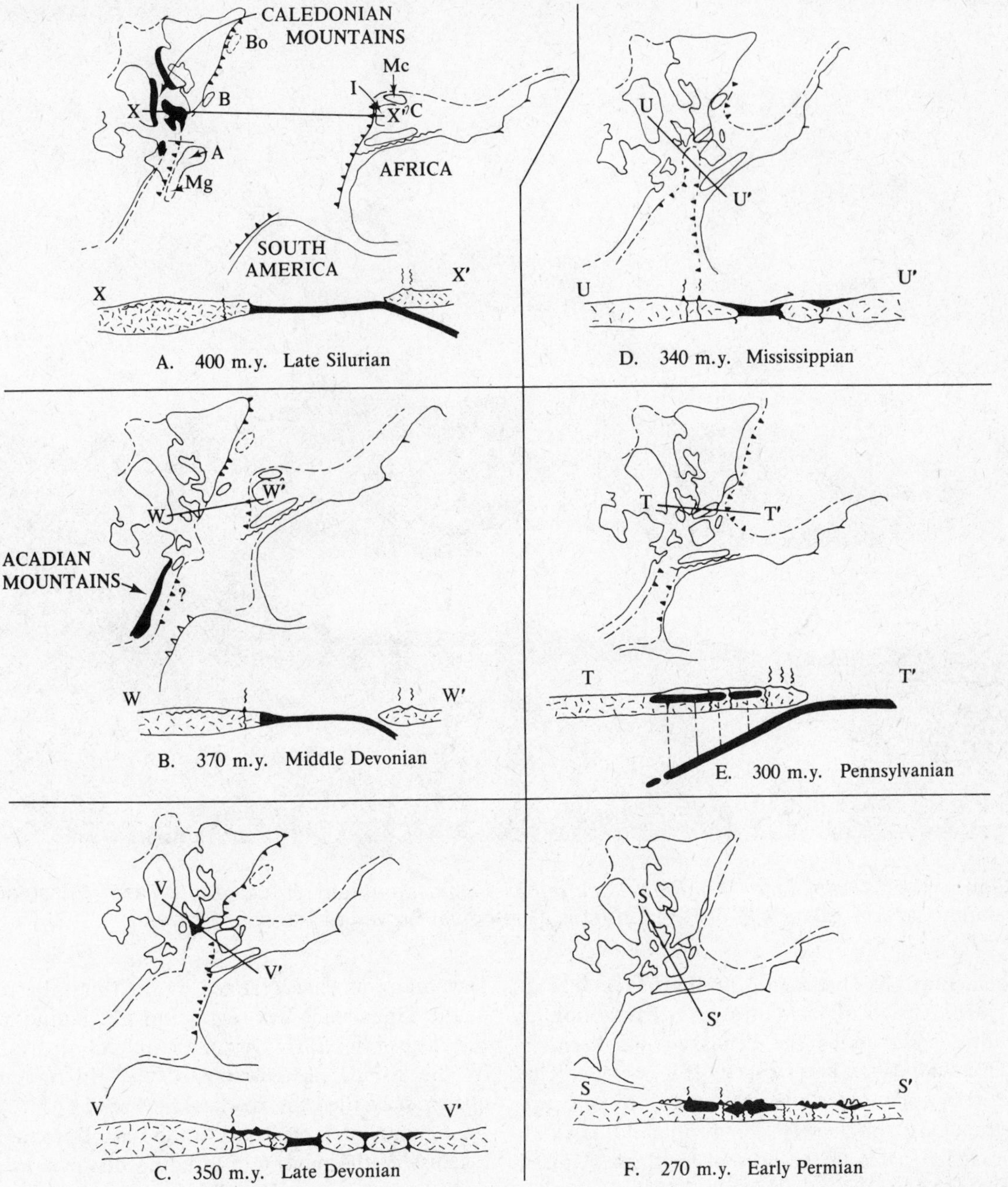

Figure 16.12B. *Interpretation of Plate Tectonics of the Hercynian-Appalachian Region.* A, Avalon; B, Brittany; Bo, Bohemia; C, Corsica-Sardinia; I, Iberia; M, Morocco; MC, Massif Central and Montagne Noire; Mg, Meguma. Barbed line marks subduction zone; dash-dot line marks edge of continental plate. (After Badham and Olds, *Geology*, vol. 3, July 1975.)

Figure 16.13. *Franklin or Innuitian Mountains.* Castle Mountain at Yelverton Pass, Ellesmere Island, Canada. (Courtesy of R.L. Christie, Geological Survey of Canada, Ottawa.)

belt called the **Ouachita-Marathon Trench** (Fig. 16.18), which underwent a great Mississippian filling followed by the Pennsylvanian-Permian **Ouachita-Marathon Orogeny** (Fig. 16.19). This deformation probably was caused by the approaching edge of Gondwanaland, which included parts of the present southeast United States and South America (Figs. 16.1–3).

The collision of Gondwanaland with Laurasia in the present Marathon-Ouachita-Appalachian-Hercynian region produced a true Pangaea in the Triassic (Fig. 16.3).

North American Epeiric Seas

The North American Craton portion of Laurasia was widely flooded in the Devonian. The Early Devonian was characterized by the final phases of the Tippecanoe Sea regression. Beginning at the close of the Early Devonian and culminating in the Middle Mississippian was the major epeiric sea called the **Kaskaskia** (Fig. 14.12). A minor regression occurred in the Late Devonian because of the mountain-building disturbances to the present east. The Devonian epeiric seas were characterized by widespread limestones containing abundant coral reefs (Fig. 16.20). The deposits along the eastern side of the sea consist chiefly of shales, reflecting the rising Acadian Mountains. This relatively flat-lying and exceedingly fossiliferous sequence of this system, particularly in western New York State (Fig. 16.21), is probably the finest Devonian section in the world.

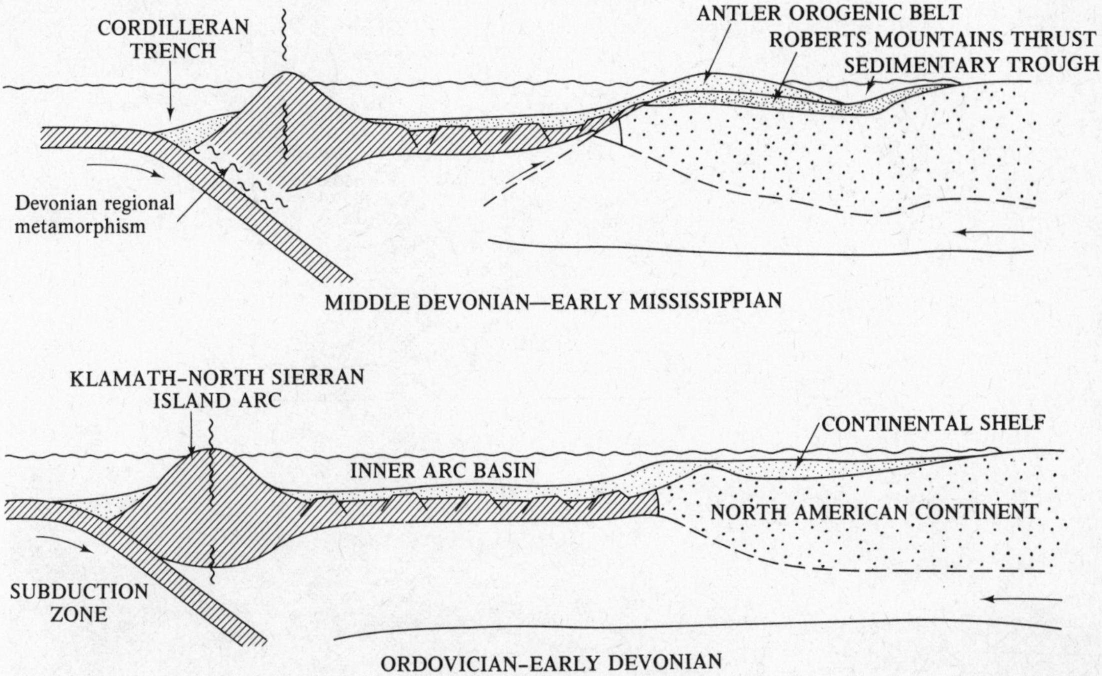

Figure 16.14. *Cross Sections of Cordilleran Trench and Western North America in the Paleozoic.* Pressure of westward-moving continent drives it under inner arc basin, producing thrust faulting and mountain building. (After Burchfiel, Davis, Poole.)

The re-advance of the Kaskaskia Sea is shown by a basal black shale (Fig. 3.3B), an atypical basal deposit, indicating that the sea progressively engulfed its swampy shoreline as it slowly covered the craton again. Geologists who have made the mistake of substituting rock correlation for time correlations have been puzzled by what conditions could cause a widespread black-mud-producing marine environment. The answer, of course, is that the environment was not widespread at any one time, but varied in age depending on the shifting position of the shore zone. The Early and Middle Mississippian deposits represent the culmination of the Kaskaskia Sea with widespread fossiliferous limestone (Fig. 16.22). The Kaskaskia Sea regression occurred in the Late Mississippian.

It was followed by the **Absaroka Sea** (Fig. 14.12), the last major epeiric sea invasion of North America in the Paleozoic. The Absaroka Sea reached its maximum extent near the close of the Pennsylvanian and then withdrew throughout the Permian and Triassic.

A peculiar characteristic of the transgressive phase of the Absaroka Sea is the presence of **cyclothems.** Cyclothems (Fig. 16.23) are cyclically repeated minor transgressive and regressive deposits that go upward from basal deltaic sandstones through nonmarine shale and freshwater limestone to underclay and coal to nearshore marine shale and offshore marine limestone, followed by nearshore limestone and shale, and finally nonmarine shale. To the east, as in Pennsylvania, the terrestrial parts of the sequence are more abundant; to the southwest, as in Kansas, the marine phases dominate; while in the mid-continent, such as Illinois, the complete sequence is best developed.

These sequences are repeated dozens of times, indicating a constantly fluctuating sea level. Apparently, cyclothems were caused by the coincidence of a broad, almost flat lowland over the

A

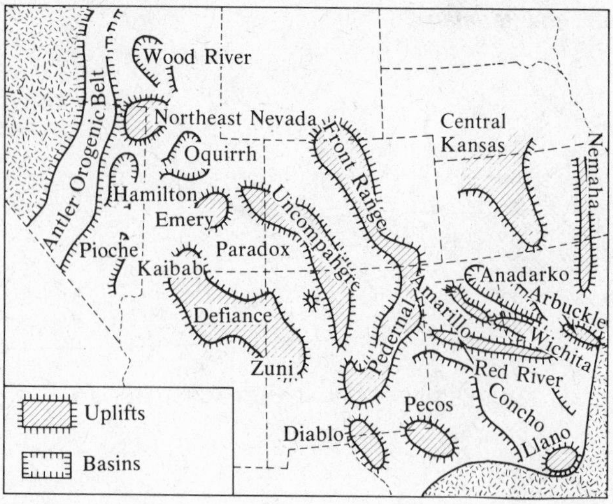

B

Figure 16.15. *Ancestral Rockies—Oklahoma Mountains of Craton.* **A.** *Present-day distribution of the mountains and their associated basins.* (After Clark and Stearn.) **B.** *Wichita Mountains near Lawton, Oklahoma.* (Courtesy Oklahoma Industrial Development and Park Dept.)

Figure 16.16. *The Garden of the Gods, Colorado.* Top photograph also shows Pikes Peak, a block of Precambrian granite elevated during the Mesozoic-Cenozoic Cordilleran Orogeny. (Courtesy of William M. Mintz.)

A

B

Figure 16.17. *Pennsylvanian Strata in the West.* **A.** *Ouray, San Juan Mountains, Colorado.* (Courtesy Colorado Dept. of Public Relations.) **B.** *Goosenecks of the San Juan River, Utah.* (Courtesy of William M. Mintz.)

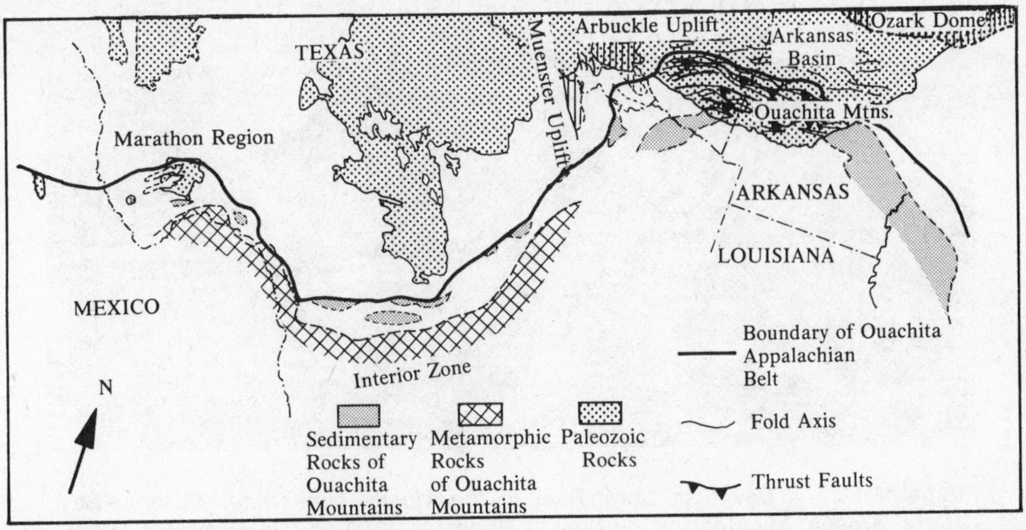

Figure 16.18. *Present-Day Distribution of Ouachita-Marathon Orogenic Features.* (After Flawn, Clark and Stearn.)

Figure 16.19. *The Ouachita Mountains,* Eastern Oklahoma. (Courtesy of Mike Shelton, State of Oklahoma.)

Figure 16.20. *A Devonian Coral Reef in the Alpena Limestone,* Thunder Bay Quarry, Alpena, Michigan. (Courtesy of Robert V. Kesling, University of Michigan.)

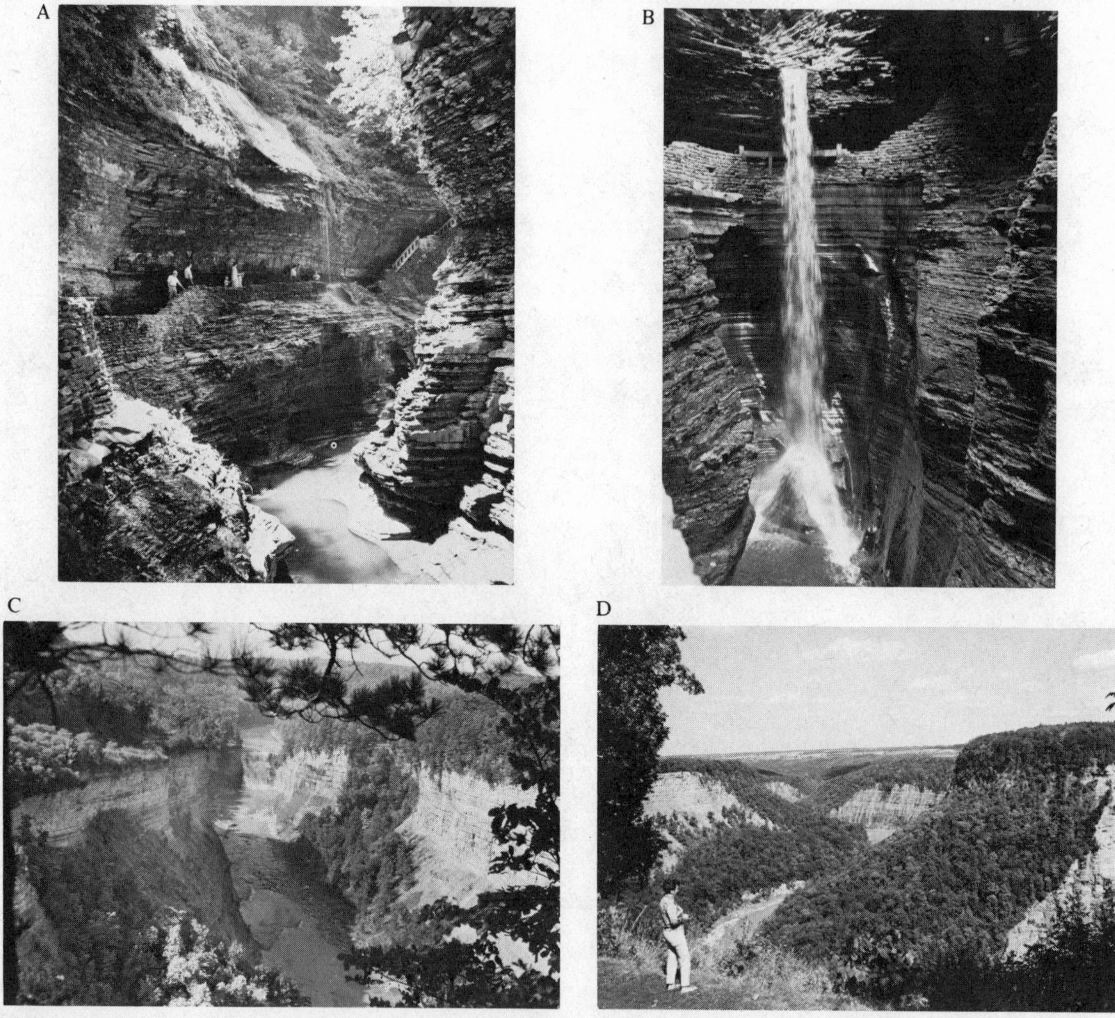

Figure 16.21. *Kaskaskia Sea (Devonian) Rocks of New York.* **A–B.** *Watkins Glen State Park.* Fine-grained sandstones and siltstones of Late Devonian age. **C–D.** *Letchworth State Park.* Upper Devonian sandstones and shales in the Genesee River Gorge. (Courtesy State of New York.)

Figure 16.22. *Mississippian Limestones.* **A.** *In Mississippi River Bluffs near Alton, Illinois*; **B.** *In Grand Canyon* (Redwall Limestone forms prominent cliff in mid-canyon); **C.** *In Cascade Mountain, Banff National Park, Alberta, Canada.* (Part A courtesy of Illinois State Geological Survey; Part B courtesy of William M. Mintz; Part C courtesy of Alberta Dept. of Industry and Tourism.)

North American interior with glacial-interglacial sequences of Gondwanaland which alternately withdrew from and added to the oceans large amounts of water. A rise or fall of the sea of a few tens of meters could thus flood vast areas. Note that the coal formed just before the onset of marine conditions, indicating that the coal swamp forests occurred in broad bands along or near the seashore. Thicker sections accumulated in the more rapidly subsiding Illinois and Forest City basins. The major coal beds of eastern North America today are of Pennsylvanian age (Fig. 16.24).

The regressive phase of the Absaroka Sea occupied the entire Permian. The sea gradually retreated to the southwest, so that by Late Permian times it remained only in southwest Texas and adjacent New Mexico. In the late Middle Permian, great reefs grew in this region which have produced a wealth of fossils (Fig. 7.9) and one of the clearest demonstrations of intertonguing facies exposed anywhere (Fig. 16.25). At the same time in the present Rocky Mountains area of Idaho and adjacent states, another remnant of the regressing sea deposited the phosphate-rich Phosphoria Formation. By the end of the Permian, virtually the entire North American Craton was surrounded by mountains pouring clastics onto it. It was covered in many areas, particularly the present west, by great thicknesses of terrestrial redbeds that continued to form through the Triassic (Fig. 16.26).

In the Late Paleozoic, the equator ran roughly east-west through North America in the northern United States and southern Canada, thus

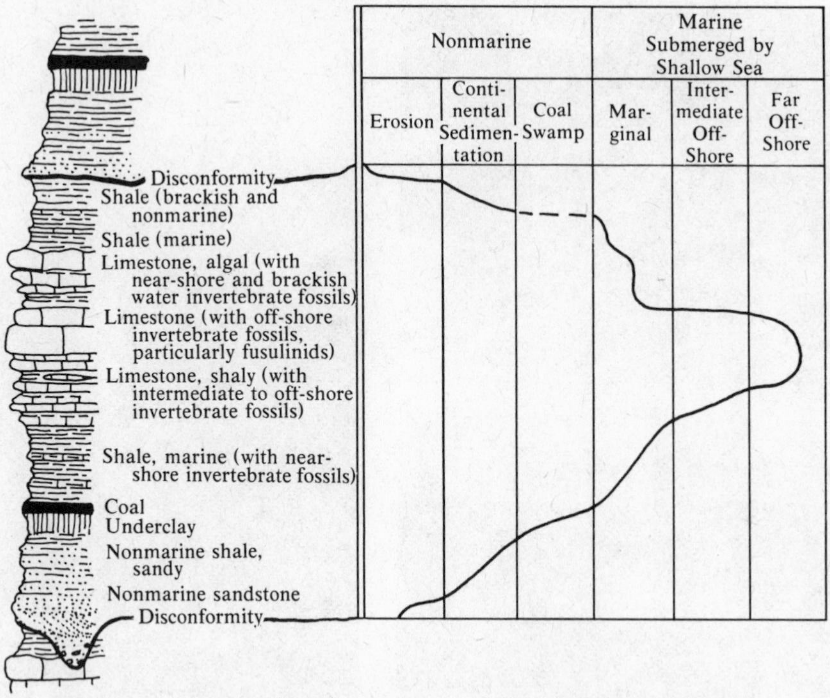

Figure 16.23. *Cross Section of a Complete Cyclothem.*

Figure 16.24. *Pennsylvanian Coal Fields of the Eastern United States.*

making the coal forests tropical and the terrestrial redbed areas comparable to today's low-latitude deserts (Figs. 16.1–3).

European Craton

The European Craton was flooded by terrestrial clastics from the Caledonian Mountains in the Devonian. Only by the Late Devonian did epeiric seas again return, covering much of Russia (Fig. 16.27). In the Rhine-Meuse area of Germany, a marginal shelf sea in the Devonian, is an exceptionally complete and fossiliferous section of the system that is better than the more highly deformed type area in England. The Late Devonian epeiric sea expanded in the Early Carboniferous, even flooding the area of the former Caledonian Mountains (Fig. 16.28),

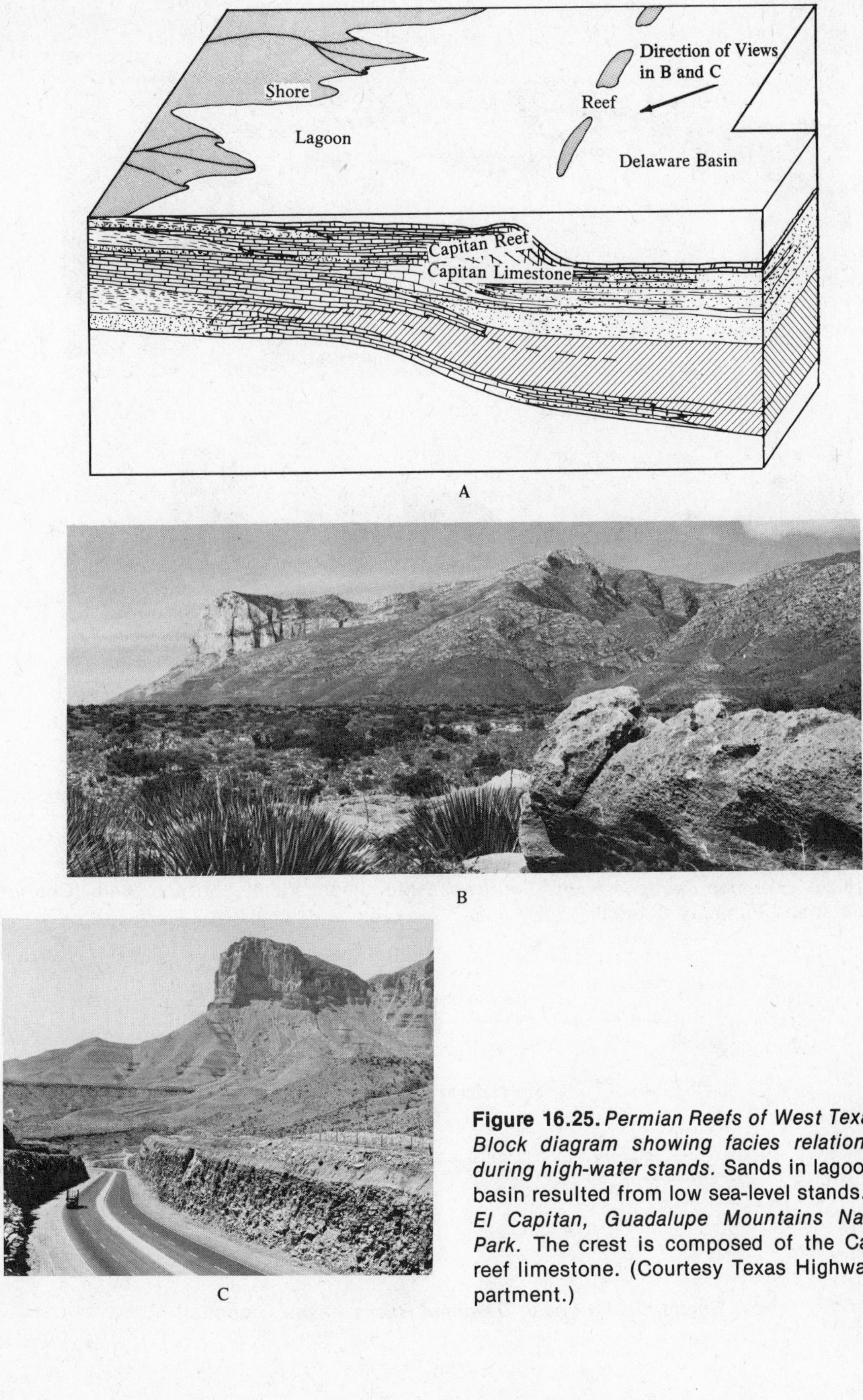

Figure 16.25. *Permian Reefs of West Texas.* **A.** *Block diagram showing facies relationships during high-water stands.* Sands in lagoon and basin resulted from low sea-level stands. **B-C.** *El Capitan, Guadalupe Mountains National Park.* The crest is composed of the Capitan reef limestone. (Courtesy Texas Highway Department.)

Figure 16.26. *Permian Redbeds in the Southwest.* Monument Valley, Arizona-Utah. (Courtesy Utah Tourist and Publicity Council.)

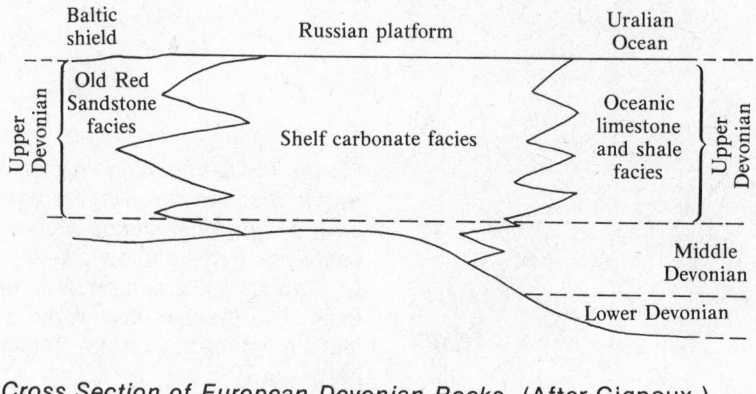

Figure 16.27. *Cross Section of European Devonian Rocks.* (After Gignoux.)

Figure 16.28. *Lower Carboniferous Limestone in Great Britain* near Llangollen, Wales. (Crown Copyright. Geological Survey photograph. Reproduced by permission of the Controller, Her Britannic Majesty's Stationery Office, London.)

which was then greatly reduced and, thereafter, was largely a stable region.

In the Late Carboniferous, the culmination of the **Hercynian Orogeny** drove the epeiric sea from the European Craton for the remainder of the Paleozoic. The terrestrial clastics (Fig. 16.29) that accumulated around these mountains and on the shelf to the north include the great coal beds of present-day Europe. Coal swamp forests flourished there, as they did farther to the present southwest in what is now North America. In the Permian, most of western Europe was the site of terrestrial redbed deposition broken only by the Late Permian hypersaline **Zechstein Sea,** which flooded most of its northern margin. The Russian part of the craton remained flooded by an epeiric sea until the very close of the period when the influx of clastics from the rising **Ural Mountains** drove it out (Fig. 16.30).

Beginning in the Late Carboniferous and Permian, the plate (later to become plates) which contained Gondwanaland began to rotate counterclockwise toward present southern Eurasia. As it did, it began to close off the western end of the broad seaway which extended into western Pangaea (Figs. 16.3, 18.1–3, 20.1). This great east-west equatorial seaway is known as the **Tethyan Ocean.** As the Gondwana plates impinged upon southern Eurasia, they produced

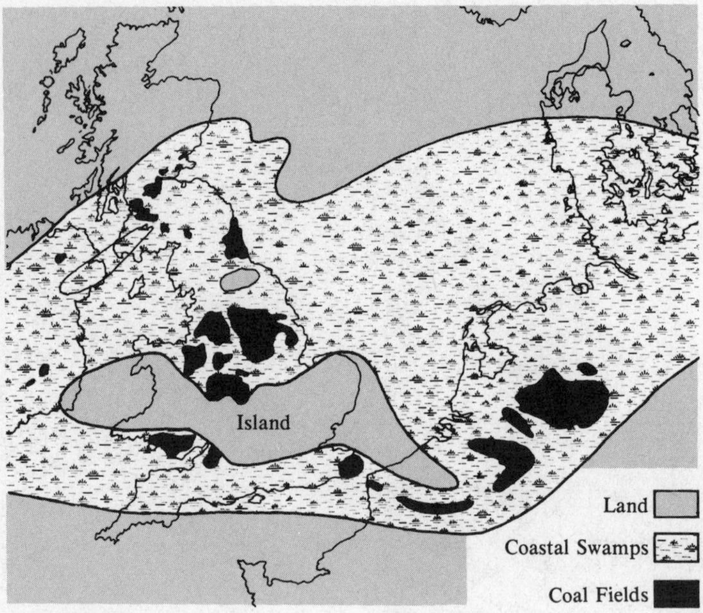

Figure 16.29. *Paleogeography of Present-Day Northwest Europe in the Late Carboniferous.* (Adapted from Wills, 1951.)

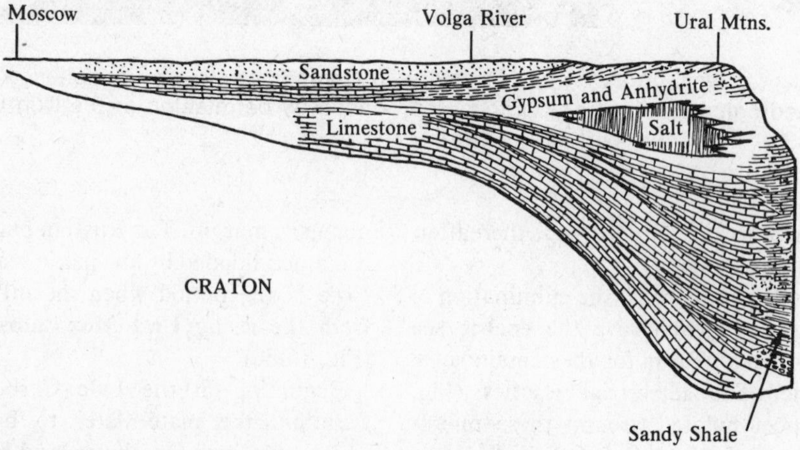

Figure 16.30. *Cross Section of Russian Permian Rocks.* Vertical dimension highly exaggerated. (After Kummel and Dunbar.)

deformation throughout the Mesozoic, reaching a culmination of orogeny in the Cenozoic as India, the Near East, and probably some of the Chinese continental fragments collided with Eurasia. In the Late Paleozoic, the equator ran diagonally from the present northwest to southeast across present-day extreme western Europe (Fig. 16.3), thus placing Tethys in the tropics.

Siberian Craton

The Siberian Craton was largely emergent throughout the entire Late Paleozoic interval. Surrounded on all sides by mountain belts undergoing deformation, its deposits are of continental origin. This part of Laurasia was in middle latitudes in the Late Paleozoic and at approximately right angles to its present position. Present eastern Siberia attained polar latitudes in the Permian, and continent glaciers formed (Fig. 16.3). The collision of the present eastern edge of Asia with the Pacific Plate was probably responsible for much of the Late Paleozoic instability in that zone. The Dzungarian and Mongolian Cratons apparently collided with the Siberian one in the beginning of the Late Paleozoic, producing the **Sayan-Yablonovoi Mountains.** The two cratons later collided with each other, producing an early version of the **Altai Mountains.** The main Chinese Craton, probably part of Gondwanaland just breaking away to the north in the Late Paleozoic, was flooded by epeiric seas during much of this interval. At the close, however, it was covered by continental deposits rich in coal.

The Late Paleozoic was a time of extreme continental unification. The conditions produced by this unique geography had significant effects on the life of the Late Paleozoic, as we shall see in the next chapter.

REVIEW

1. Where were the Gondwana and Pacific Poles situated in the Late Paleozoic according to paleomagnetic data, and what geologic evidence supports these positions?
2. Where were the **Samfrau Mountains**; how were they produced and when?
3. What is the significance of the **Acadian Orogeny**, and when did it occur?
4. How are today's **Catskill Mountains** related to the Late Paleozoic **Acadian Mountains**? What was the **Old Red Sandstone "Continent,"** and what was its relationship to these mountains?
5. What are two possible explanations for the **Appalachian-Alleghenian Orogeny**, and when did it occur?
6. What is the significance of the **Hercynian** and **Uralian orogenies** in the evolution of Eurasia?
7. What major orogenies affected the Cordilleran region of North America in the Late Paleozoic, and what was their probable origin?
8. What is the significance of the **Ancestral Rockies-Oklahoma Mountains uplifts**, and when did they occur?
9. What do the **Ouachita-Marathon Orogenies** indicate about plate positions in the Late Paleozoic? How are they related to the Appalachian and Hercynian orogenies?
10. When did the **Kaskaskia** and **Absaroka epeiric seas** advance, peak, and retreat?
11. What are **cyclothems**, and what factors may have caused their formation?
12. Where and when did major redbeds form in the Late Paleozoic of Gondwanaland and Laurasia?
13. When was the European Craton flooded by seas in the Late Paleozoic?
14. What is known about the Late Paleozoic positions of the various continental fragments composing modern Asia?
15. What were the **Theic** and **Tethyan oceans?**
16. What plate boundaries can be recognized for the Upper Paleozoic, and what evidence indicates their existence?

17

Invasion of the Land: Life of the Late Paleozoic

17

Beginning with Devonian, there was widespread invasion and occupation of the earth's land masses by organisms. Among the invertebrates, mollusks and particularly arthropods successfully invaded the land, while the vertebrates succeeded with the amphibians, transitional forms between fish and typical tetrapods. Before the end of the Paleozoic, reptiles dominated the animal world on land. The plants covered the landscape in the Late Paleozoic and diversified into all the present phyla with the probable exception of the flowering plants.

Marine Life

This is not to say that life in the sea (Figs. 17.1–4) did not continue to flourish in the Late Paleozoic. Indeed, it continued to diversify and multiply, but many of the orders and some of the classes were different from those of the Early Paleozoic. The major Late Paleozoic invertebrate groups were the **foraminifera, corals, bryozoans, brachiopods, ammonoids, ostracods, crinozoan echinoderms,** and **conodonts.** There was a high degree of cosmopolitanism in Late Paleozoic seas because of the close concentration of all the continental masses.

Foraminifera

The **foraminifera** became abundant for the first time in the Mississippian. The common types of that period were planispirally coiled and had granular calcareous walls (Fig. 17.5A). From these evolved the elongate, spindle-shaped **fusulines** (Fig. 17.5B–C) of the Pennsylvanian and Permian. These abundant, rapidly evolving foraminifera are major correlation fossils for rocks of those systems. No fusuline forams are known to survive the Permian.

Corals

The **coral** fauna of the Late Paleozoic was characterized by the abundance of **tetracorals,** an important group for regional correlation in the Devonian and Mississippian. The tabulate corals that dominated the Ordovician and Silurian were abundant only in the Early Devonian regressive places of the Tippecanoe Sea. Thereafter, the tetracorals underwent a great expansion, reaching a Devonian peak (Figs. 17.6–7) and building reefs such as the one that today forms the Falls of the Ohio at Louisville, Kentucky. The tetracorals were somewhat diminished, but still common, in the Carboniferous (Fig. 17.8). The tabulate coral *Chaetetes* (Fig. 17.9), with its very slender corallites, is locally common in Carboniferous strata. Tetracorals diminished greatly in the Permian as widespread continental conditions caused the retraction of the epeiric seas in which they thrived and continental unification reduced the number of faunal provinces. A few Permian

Figure 17.1. *Diorama of Devonian Marine Life.* The majority of the tentacle-bearing sessile forms are tetracorals. Nautiloids, trilobites, and brachiopods lie on the bottom. On the corals to the left are found lacy bryozoans, snails, trilobites, and brachiopods. Some tabulate corals and brachiopods occur on the right. (Courtesy Field Museum of Natural History.)

Figure 17.2. *Diorama of Mississippian Marine Life.* Crinoids predominate in this reconstruction. (Courtesy Field Museum of Natural History.)

Figure 17.3. *Diorama of Pennsylvanian Marine Life.* The upright vase-like forms with snails crawling on them are sponges. From left to right in the foreground are nautiloids, snails and brachiopods, a sea urchin, more gastropods, and an ammonoid. Clams, another ammonoid, and tetracroals appear in the distance. (Courtesy U.S. National Museum.)

Figure 17.4. *Diorama of Permian Marine Life.* The tall tubular bodies at the left and the bead-like structures to the right are sponges. A great variety of brachiopods appear at the center with some coiled cephalopods. A few snails and tetracorals are also shown. (Courtesy Field Museum of Natural History.)

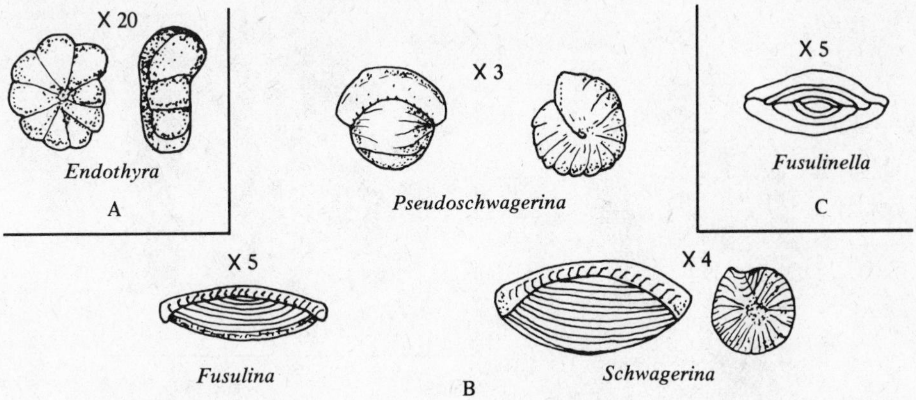

Figure 17.5. *Late Paleozoic Foraminifera.* **A.** *Mississippian Genus Endothyra.* **B.** *Pennsylvanian-Permian Fusulinaceans.* **C.** *Cross section of Fusulinacean.* Magnifications indicated. (After Dunbar, Henbest, Skinner, and Wilde.)

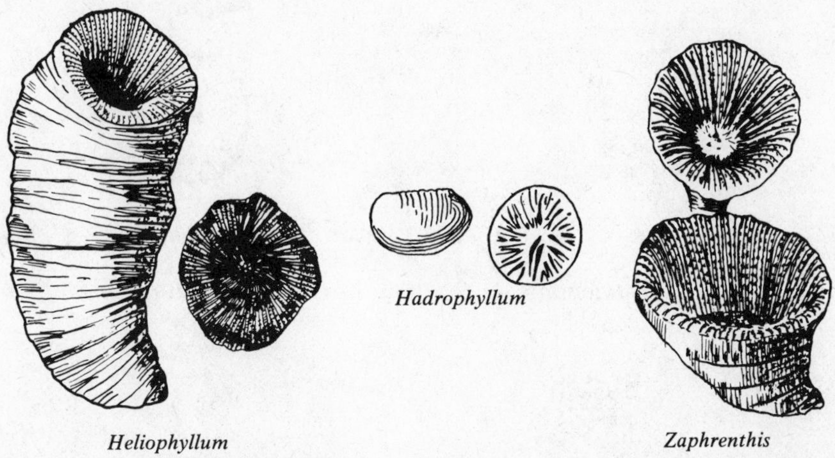

Figure 17.6. *Devonian Solitary Tetracorals.* Lateral and top views. Two-thirds natural size. (Adapted from *Invertebrate Fossils* by Moore-Lalicker-Fischer, Copyright 1952 by McGraw-Hill, Inc., used by permission of McGraw-Hill Book Company.)

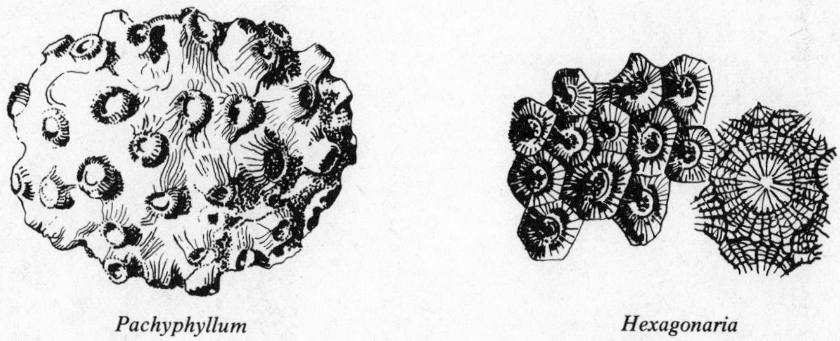

Figure 17.7. *Devonian Colonial Tetracorals.* Two-thirds natural size except enlargement at right. (Adapted from *Invertebrate Fossils* by Moore-Lalicker-Fischer, Copyright 1952 by McGraw-Hill, Inc., used by permission of McGraw-Hill Book Company.)

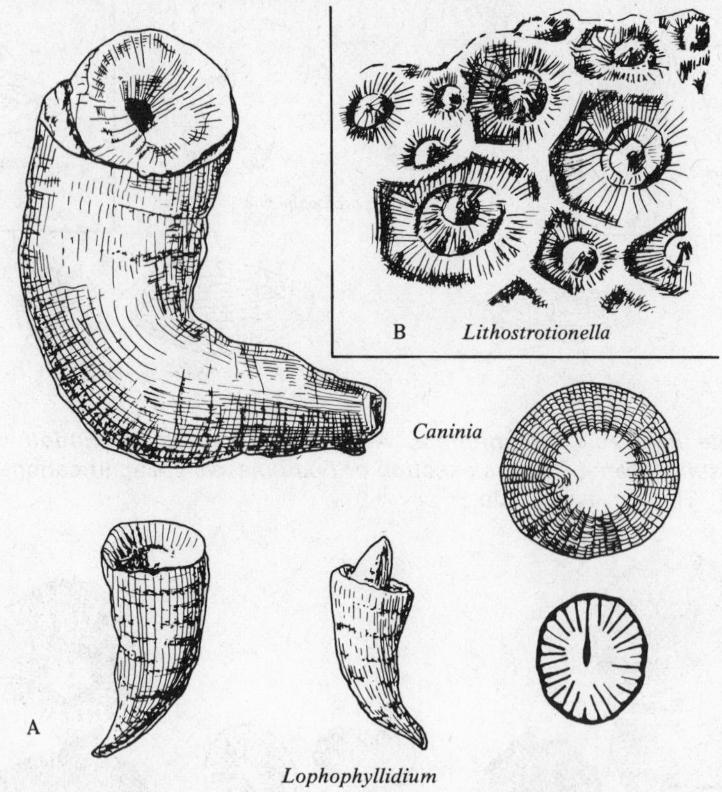

Figure 17.8. *Carboniferous Tetracorals.* **A.** *Solitary,* **B.** *Colonial.* Natural size.

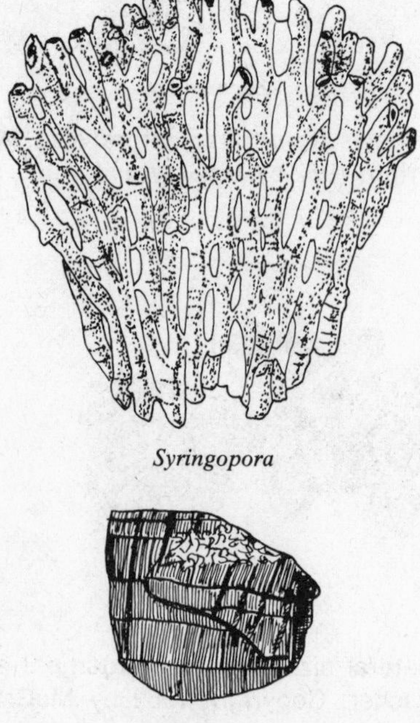

Figure 17.9. *Late Paleozoic Tabulate Corals.* Natural size. (After Easton, Kay, and Colbert.)

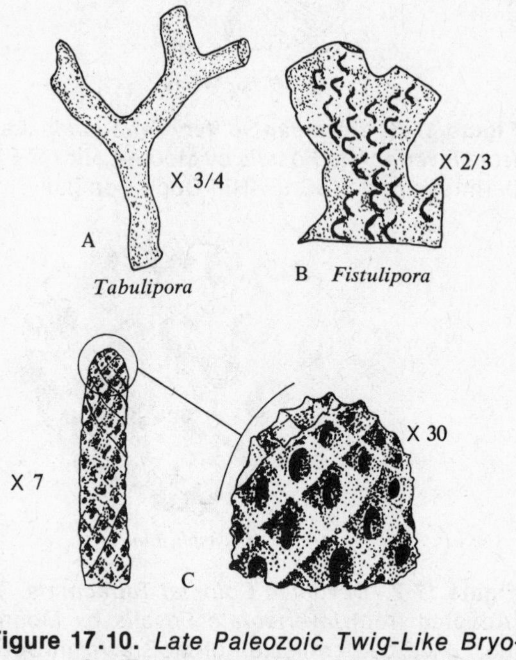

Figure 17.10. *Late Paleozoic Twig-Like Bryozoans.* **A.** *Trepostome,* **B.** *Cystoporoid,* **C.** *Cryptostome.* Magnifications indicated. (After Levin, Ulrich.)

tetracorals show the insertion of new septa into all six spaces between the original septa and may be transitional to hexacorals. Until recently, it was thought that the tetracorals failed to survive the Paleozoic, but recent finds in the Soviet Union and the Indian subcontinent indicate they survived into rocks of earliest Triassic age. The tabulate corals may also have failed to survive the Paleozoic, although some possible Mesozoic finds have been reported.

Bryozoans

The **bryozoans** are another significant faunal element which displayed a change in the dominance of orders from Early to Late Paleozoic. The stony bryozoa, the trepostomes and cystoporoids, so abundant in the Ordovician and to some extent, the Silurian, decline to insignificance in Upper Paleozoic rocks. There were only a few common Late Paleozoic genera (Fig. 17.10A–B). The **cryptostome** bryozoans, particularly the **lacy** fenestellids, diversified greatly and dominate Upper Paleozoic bryozoan assemblages. The lacy forms (Fig. 17.11) reached a Devonian-Carboniferous peak and include the chiefly Mississippian *Archimedes* which had a screw-shaped axis supporting its lacy fronds. Delicate, twig-like cryptostomes (Fig. 17.10C) also were common. The cryptostomes declined greatly in the Permian, and only a few survived into the earliest Triassic beds. The trepostomes and cystoporoids also appear not to have survived the Paleozoic, but some later bryozoans may be their descendants.

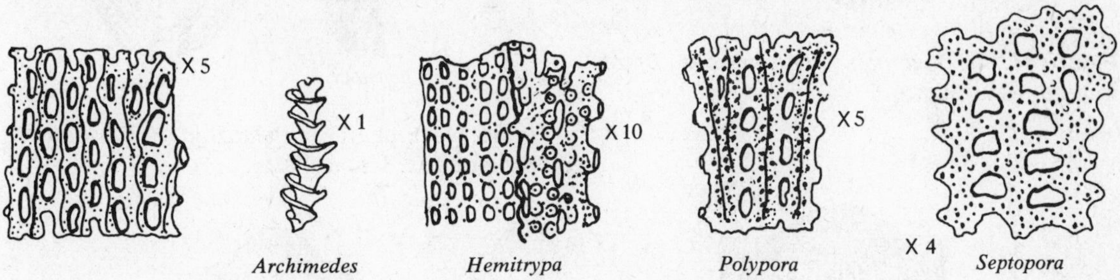

Figure 17.11. *Late Paleozoic Lacy Cryptostome Bryozoans.* Magnifications indicated. (After Ulrich.)

Brachiopods

Brachiopods flourished in the Late Paleozoic, but once again we note a shift in the abundance of orders. Of the Early Paleozoic brachiopod groups, the inarticulates declined; the orthoids (Fig. 17.12) diminished but still included some common genera; the strophomenoids (Fig. 17.13) did well only through the Devonian and became extinct in the Carboniferous; the pentameroids (Fig. 17.14) became extinct in the Devonian with only a few forms common in that period; and the **rhynchonelloids** (Fig. 17.15) continued to flourish.

The Late Paleozoic saw the dominance of the **chonetoids, productoids, davidsonioids, atrypoids, athyridoids,** and the **spiriferoids.** The chonetoids (Fig. 17.16) were particularly abun-

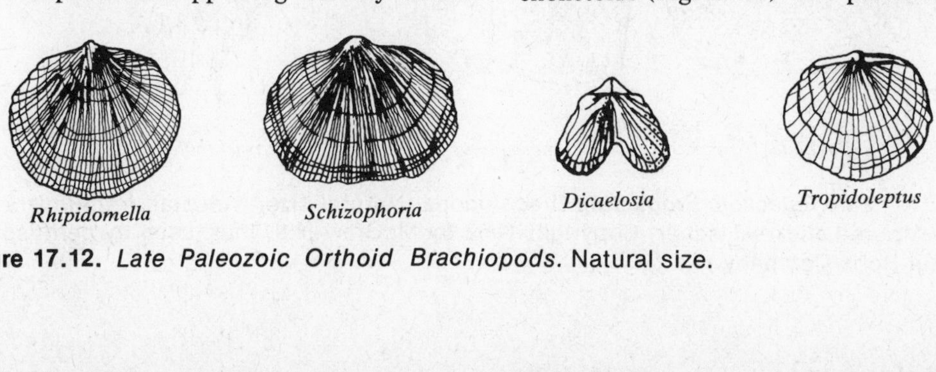

Figure 17.12. *Late Paleozoic Orthoid Brachiopods.* Natural size.

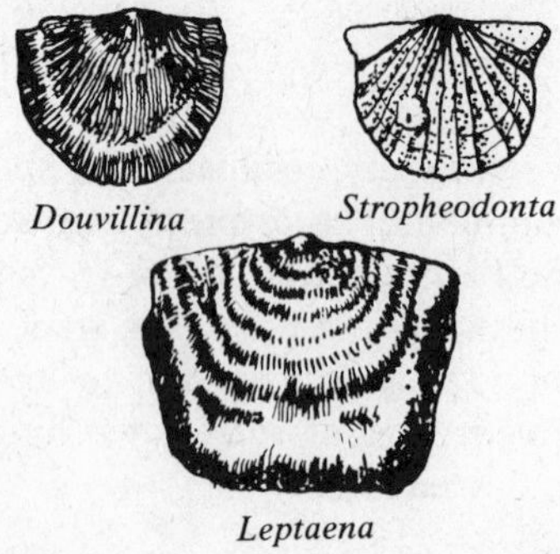

Figure 17.13. *Late Paleozoic Strophomenoid Brachiopods.* Natural size.

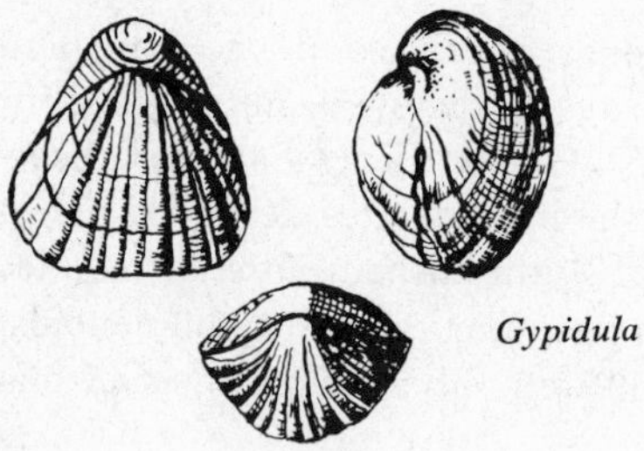

Figure 17.14. *Devonian Pentameroid Brachiopods.* Natural size.

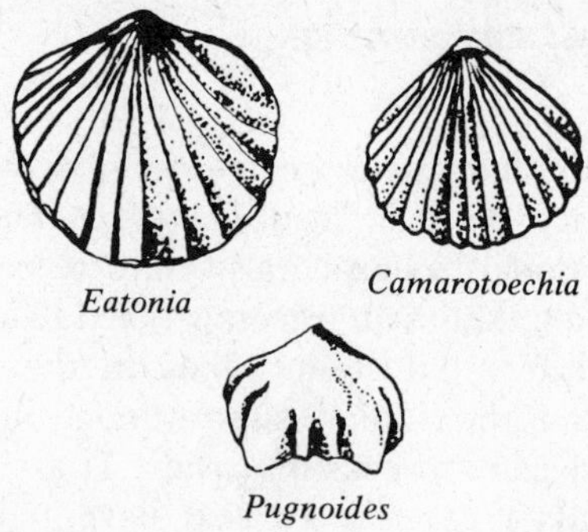

Figure 17.15. *Late Paleozoic Rhynchonelloid Brachiopods.* Natural size.

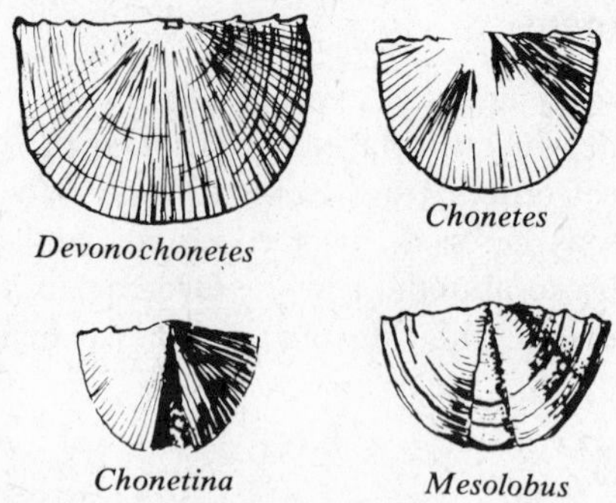

Figure 17.16. *Late Paleozoic Chonetoid Brachiopods.* Natural size.

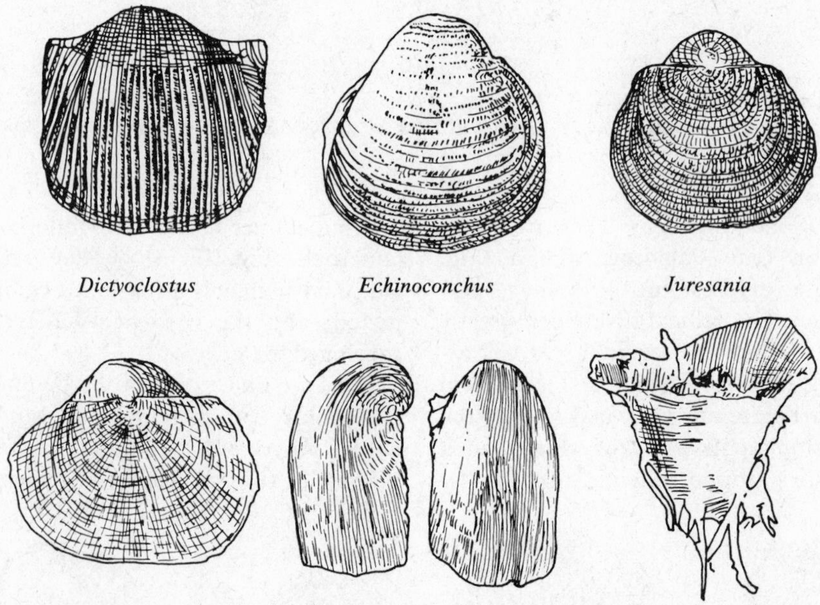

Figure 17.17. *Late Paleozoic Productoid Brachiopods.* Natural size. (Adapted from *Invertebrate Fossils* by Moore-Lalicker-Fischer, Copyright 1952 by McGraw-Hill, Inc., used by permission of McGraw-Hill Book Company.)

dant in the Devonian and Carboniferous. The spiny productoids (Fig. 17.17) appeared in the Devonian and became the dominant Carboniferous and Permian brachiopod group. In the Permian, genera such as *Prorichthofenia* (Fig. 17.17) converged on corals as they inhabited the same reef environment. Another bizarre brachiopod order, the **oldhaminoids** with their peculiar slotted shells (Figs. 17.4, 17.18), also inhabited these reefs. The davidsonioids (Fig. 17.19) were common in the Permocarboniferous. Atrypoids (Fig. 17.20) reached a Devonian climax and then became extinct. Athyridoids (Fig. 17.21) were common throughout the Late Paleozoic. The spiriferoids (Fig. 17.22) reached a Devonian peak, but continued to be important through the remainder of the era. The dominant modern order of articulate brachiopods, the terebratuloids (Fig. 17.23), appeared in the Devonian, but only a few genera were common. Brachiopods are important for making regional correlations throughout the Upper Paleozoic.

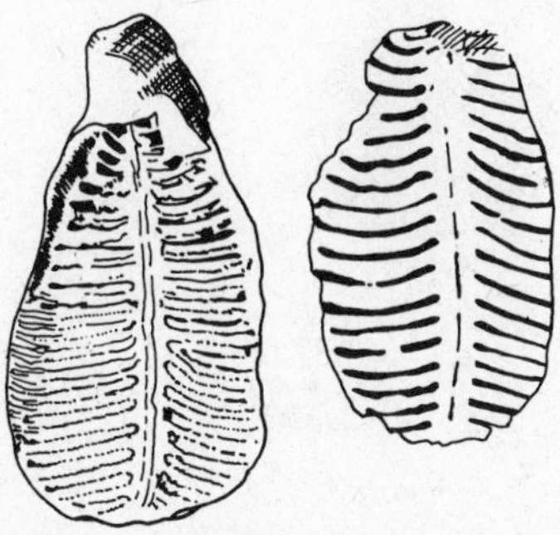

Leptodus

Figure 17.18. *Peculiar Slotted Oldhaminoid Brachiopod.* X 1/2. (Adapted from *Invertebrate Fossils* by Moore-Lalicker-Fischer, Copyright 1952 by McGraw-Hill, Inc. used by permission of McGraw-Hill Book Company.)

Most of the abundant Late Paleozoic brachiopod groups did not fare well thereafter. As noted, atrypoids did not survive the Devonian. The orthoids, declining for a long time, failed to survive the Permian; while the also declining strophomenoids ended in the Carboniferous. Except for some questionable Mesozoic descendants, the chonetoids did not live beyond the Permian. Productoids survived into the earliest Triassic, and a few peculiar Lower Mesozoic brachipods may have been their descendants. Davidsonioids and athyridoids may have had some Triassic survivors, and the spiriferoids even lasted into the Jurassic; but Mesozoic representatives of all three groups are rare and atypical. Only the rhynchonelloids and terebratuloids, the two surviving articulate groups, continued to diversify in the Mesozoic.

Mollusks

The **mollusks** became more important in the Late Paleozoic, chiefly because of the appearance and radiation of the **ammonoids** (Figs. 17.24–25). Ammonoids are relatively common

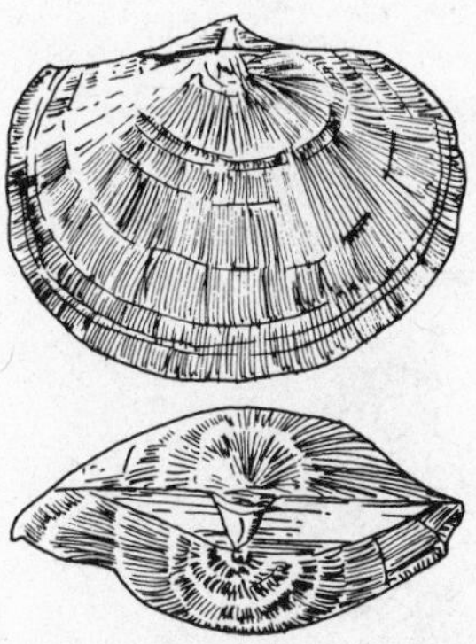

Derbyia

Meekella

Figure 17.19. *Late Paleozoic Davidsonioid Brachiopods.* Natural size. (Adapted from *Invertebrate Fossils* by Moore-Lalicker-Fischer, Copyright 1952 by McGraw-Hill, Inc., used by permission of McGraw-Hill Book Company.)

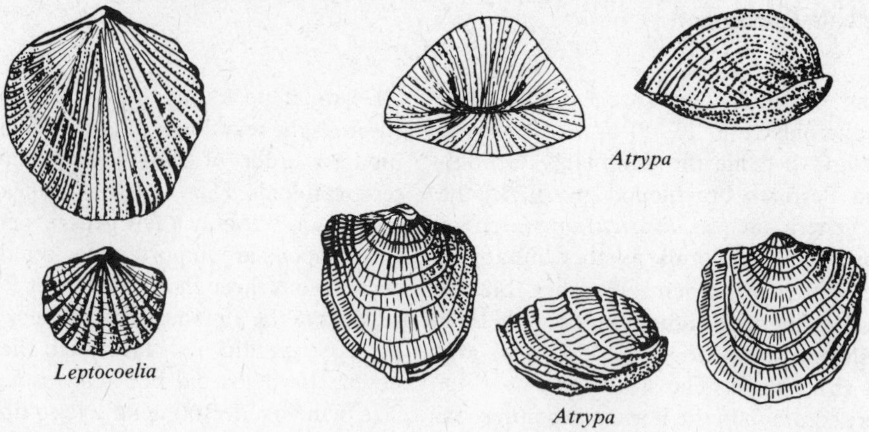

Figure 17.20. *Devonian Atrypoid Brachiopods.* Natural size. (Adapted from *Invertebrate Fossils* by Moore-Lalicker-Fischer, Copyright 1952 by McGraw-Hill, Inc., used by permission of McGraw-Hill Book Company.)

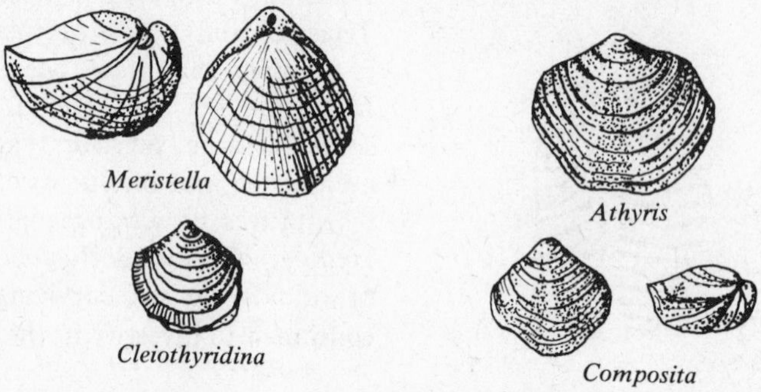

Figure 17.21. *Late Paleozoic Athyridoid Brachiopods.* Two-thirds natural size.

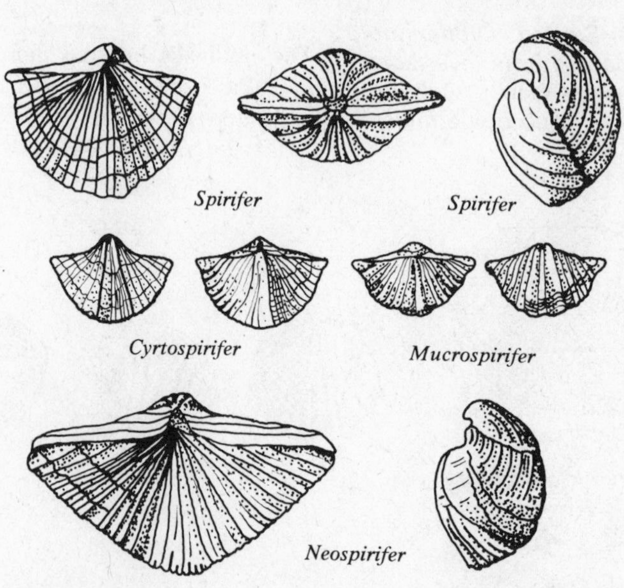

Figure 17.22. *Late Paleozoic Spiriferoid Brachiopods.* Two-thirds natural size.

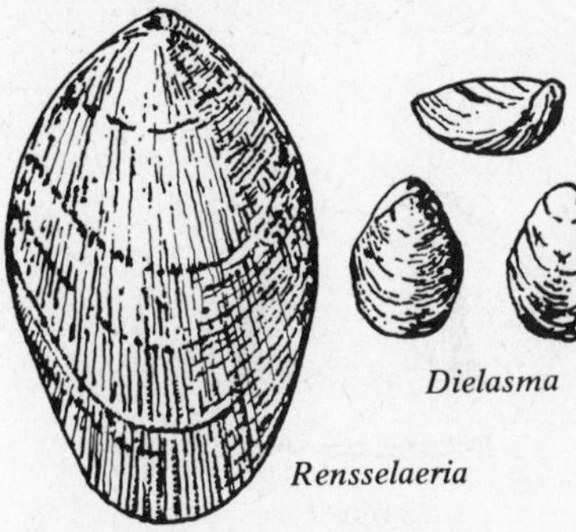

Figure 17.23. *Late Paleozoic Terebratuloid Brachiopods.* Natural size.

fossils from the Upper Devonian onward, though a diminution appears to occur near the end of the era. Most Paleozoic ammonoids are **goniatites** (Fig. 17.24). Ammonoids with ceratite sutures appear in the Mississippian and with ammonite sutures, in the Permian (Fig. 17.25). Because their pelagic life habits enabled them to spread widely, goniatites are major fossils for making long-distance correlations of Upper Paleozoic rocks. Goniatites nearly all disappeared before the close of the Permian. Only a few survived into the earliest Triassic.

Ostracods

The tiny, bivalved **ostracod** crustaceans (Fig. 17.26) increase in abundance in the Upper Paleozoic strata. Highly ornamented genera are common as well as the smoother forms. In common with most marine groups, ostracods declined in the Late Permian when the seas were contracted by the extreme emergence of continents and continental unification reduced the number of faunal provinces.

Echinoderms

The stalked or **crinozoan echinoderms** reached a peak in the Late Paleozoic seas. Most abundant were the **crinoids.** Crinoids were common in the Devonian seas (Fig. 17.27), but the great peak occurred in the widespread limestone-depositing epeiric seas of the Mississippian (Figs. 17.2, 17.28). The Pennsylvanian saw a decline in crinoids, and only a few members of

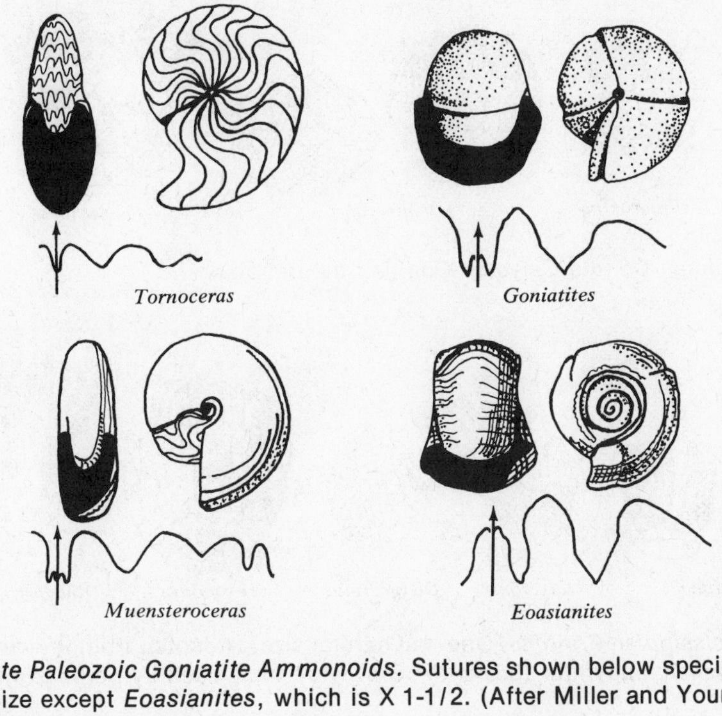

Figure 17.24. *Late Paleozoic Goniatite Ammonoids.* Sutures shown below specimens. Specimens are 3/4 natural size except *Eoasianites*, which is X 1-1/2. (After Miller and Youngquist.)

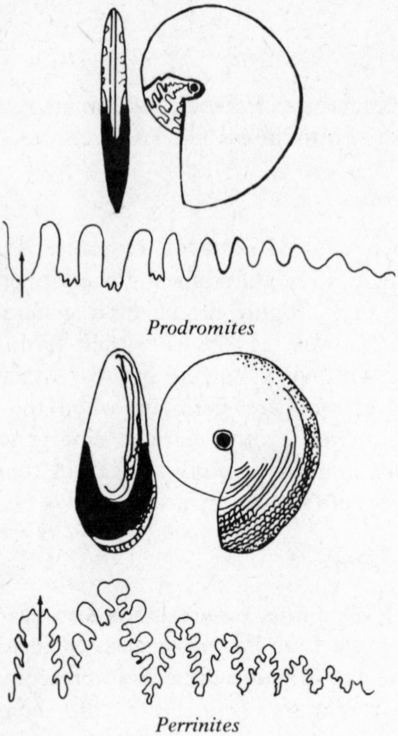

Figure 17.25. *Late Paleozoic Ammonoids with Ceratite and Ammonite Sutures.* One-half natural size except sutures.

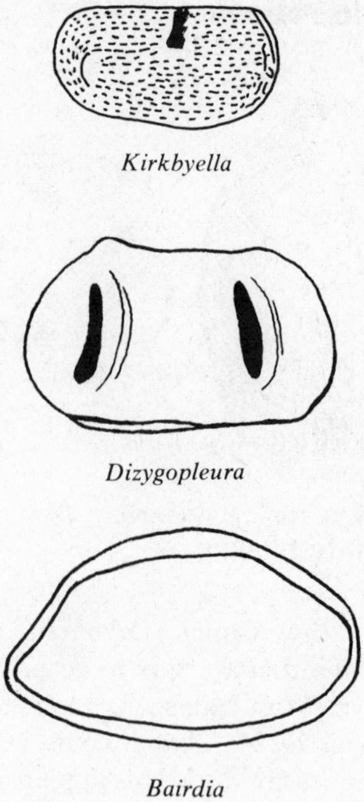

Figure 17.26. *Late Paleozoic Ostracods.* X 30.

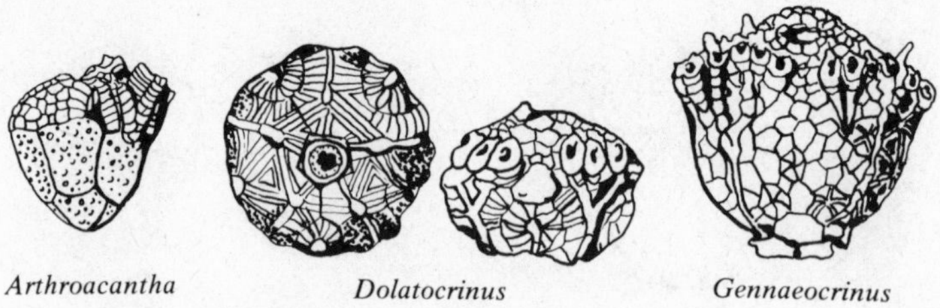

Figure 17.27. *Devonian Crinoid Calyces.* One-half natural size.

Figure 17.28. *Mississippian Crinoids.* One-half natural size. (Adapted from *Invertebrate Fossils* by Moore-Lalicker-Fischer, Copyright 1952 by McGraw-Hill, Inc., used by permission of McGraw-Hill Book Company.)

one of the Paleozoic subclasses survived the Permian. The **blastoids** (Fig. 17.29) also reached a Mississippian peak. Blastoids declined greatly in the Pennsylvanian and Permian and failed to survive the latter period.

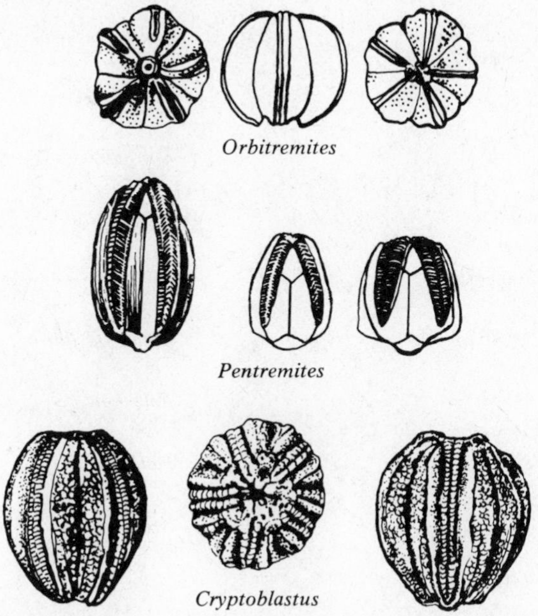

Figure 17.29. *Late Paleozoic Blastoids.* Natural size.

Conodonts

Conodonts (Fig. 17.30), the tiny tooth- or plate-like structures belonging to a pelagic, worm-shaped animal, are among the most common and geologically useful fossils of the Upper Paleozoic. They are particularly abundant in Devonian and Carboniferous rocks where they have been used to zone the strata. Conodonts were of lesser importance in Permian and Triassic times when the shallow seas and provinciality were very restricted.

Other Groups

Other marine invertebrate groups were occasionally common in the Late Paleozoic. **Glass sponges** (Fig. 17.31), siliceous forms with an interlocking gridwork of spicules, were very common in shallow Devonian seas, such as occurred in western New York. A peculiar group of calcareous sponges whose members resemble beads on a string (Fig. 17.32) were common in the Pennsylvanian of many areas. The **stromatoporoids** (Fig. 17.33) were equally as abundant in the Devonian as in the Silurian, but thereafter are represented only by atypical forms. **Conularoids** (Fig. 17.34A), possible scyphozoan coelenterates, were at their peak in the Late Paleozoic, but declined to extinction shortly thereafter. **Tentaculites** (Fig. 17.34B) were still common in the Devonian. **Gastropods** (and their monoplacophoran ancestors) and **pelecypods** continued to increase and produce many common genera (Figs. 17.35–36). The **trilobites** (Fig. 17.37) were common for the last time in the Devonian with the **phacopoids** the major group. Local correlation is often done with phacopoid trilobites. Thereafter, trilobites were represented only by occasional **illaenoids,** and they too finally passed from the scene in the Late Permian. Trilobite decline and extinction was a long process and was undoubtedly related to replacement by superior competitors, such as mollusks, other arthropods, echinoderms, and fish. Finally, certain large globular sea urchins called **melonechinoids** (Fig. 17.38) are common in some Mississippian strata.

Marine Biogeography

In the Devonian, the brachiopod faunas of Europe, Siberia, Australia, and the western United States were similar. As shown in Figure 16.1, these regions were in the tropics and, with the exception of Australia, very near. Equatorial currents could explain the Australian faunal similarities. Eastern and southern North America formed another, probably more temperate, brachiopod faunal province; while the Gondwanaland continents of South America, South Africa, and Antarctica formed a third, probably cold-water province.

In the Carboniferous, the southern part of North America and northern South America

Figure 17.30. *Late Paleozoic Conodonts.* X 10. (After Ellison.)

had similar foram faunas. As seen in Figure 16.2, these areas were adjacent in the tropics. Another Carboniferous foram province included Europe, Siberia, and Alaska, which were near each other in the north warm-temperate zone.

Carboniferous reef corals occurred in the Ural Sea and Siberia, which were at that time near the northern edge of the tropics in latitudes where reef corals live today. Permian fusuline forams formed a boreal cold-water province, a temperate western North American province, and a tropical Tethyan province which agree with plate tectonic positions.

Extinctions

From the preceding discussion of Late Paleozoic marine invertebrates, it becomes apparent why early geologists were struck by the number of disappearances of Paleozoic groups at or near the close of the Permian. Extinctions occurred in many important lineages of forams, corals, bryozoans, brachiopods, arthropods, and stalked echinoderms. Triassic faunas contained representatives of other lineages in these phyla and the first great abundance of the mollusks.

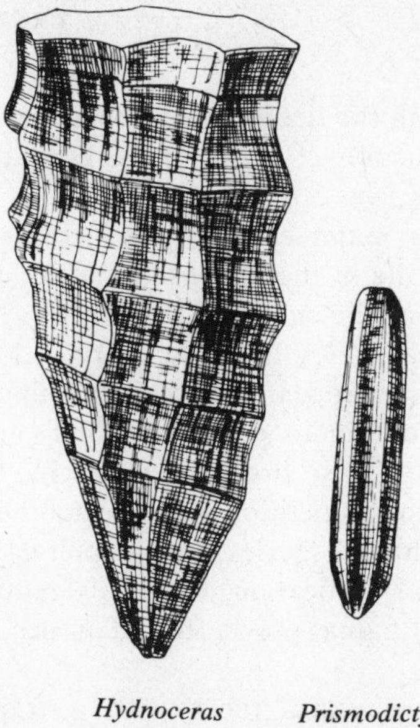

Figure 17.31. *Devonian Glass Sponges.* X 1/3. (Adapted from *Invertebrate Fossils* by Moore-Lalicker-Fischer, Copyright 1952 by McGraw-Hill, Inc., used by permission of McGraw-Hill Book Company.)

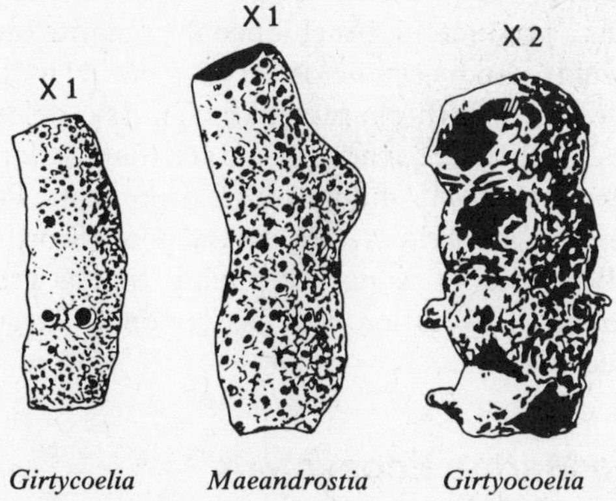

Figure 17.32. *Carboniferous Calcareous Sponges.* Magnifications indicated. (Adapted from *Invertebrate Fossils* by Moore-Lalicker-Fischer, Copyright 1952 by McGraw-Hill, Inc., used by permission of McGraw-Hill Book Company.)

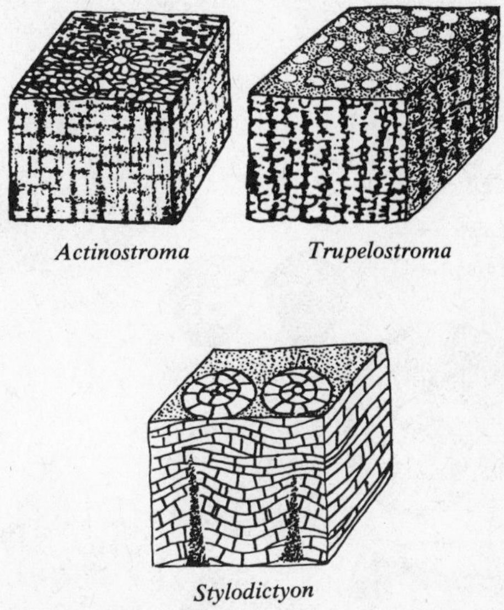

Figure 17.33. *Devonian Stromatoporoids.* X 8. (After Shimer and Shrock.)

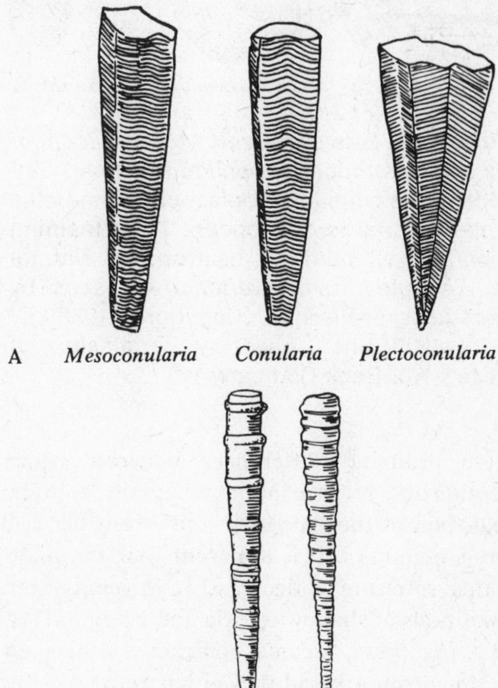

Figure 17.34. *Enigmatic Conical Fossils of the Late Paleozoic.* **A.** *Conularoids,* **B.** *Tentaculites.* Part A natural size and Part B X 4.

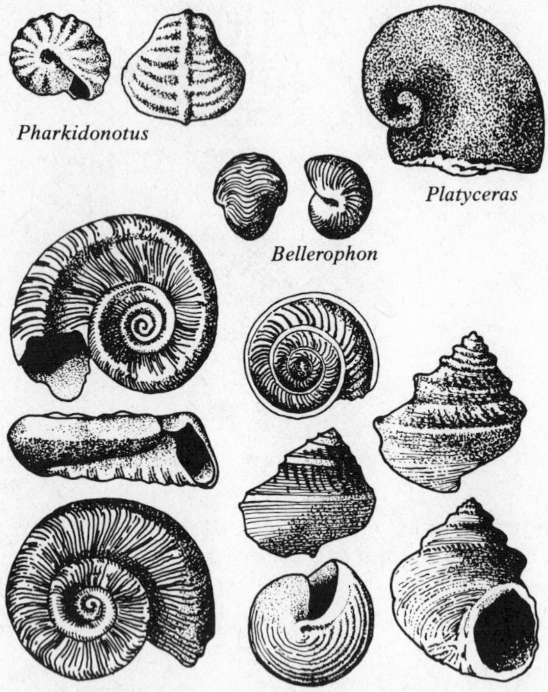

Figure 17.35. *Late Paleozoic Monoplacophorans and Gastropods. Pharkidonotus* and *Bellerophon* are coiled monoplacophorans of the group ancestral to gastropods. The remaining genera are all primitive gastropods. Natural size. (Adapted from *Invertebrate Fossils* by Moore-Lalicker-Fischer, Copyright 1952 by McGraw-Hill, Inc., used by permission of McGraw-Hill Book Company.)

These profound differences between typical Permian and Triassic faunas originally led to the separation of the Paleozoic and Mesozoic eras at this boundary. It is apparent that the major groups suffering a decrease in diversity were inhabitants of shallow epeiric and marginal seaways. As these became contracted, their environment shrank and they either retreated into the deeper waters of the ocean basins or became extinct. Many became extinct, such as the fusulines, trilobites, and blastoids. Others must have survived because they have Triassic descendants, either as struggling survivors or new groups that evolved from them. We have little direct record of events in the deeper ocean basins because of the destruction of their floors by plate movements.

Another major cause of the decrease in diversity of life in the Permian-Triassic boundary interval was the existence of Pangaea. Not only did the seas which formerly intervened between continents disappear, but the remaining shelves were in continuous contact around the supercontinent. This brought formerly separate faunas into competition with one another with a resultant loss of species (see Chapter 21). Thus, continental unification drastically reduced the number of faunal provinces and, hence, organic diversity.

In recent years, rocks with a transitional fauna between the typical Permian and Triassic have been found, chiefly in mountain belts where marginal marine and ocean basin deposits are preserved. The record shows, as we might expect from our knowledge of modern life, that there was no sudden extinction or catastrophe, just a gradual disappearance of Permian forms and replacement by Triassic forms. Groups formerly thought to have become extinct in the Permian, such as tetracorals, cryptostome bryozoans, productoid brachiopods, goniatite ammonoids, and certain ostracods, occur in the same strata with characteristic Traissic pelecypods. When we keep in mind that a large interval of 25-35 million years from Late Permian to Late Triassic has a very poor record of shallow seas, it is not surprising that a great amount of evolution and extinction is largely unrecorded.

The Marine Ecosystem

The lack of the present phytoplankton producers, except for occasional dinoflagellates, in the Late Paleozoic seas poses the same problem about the ecosystem as it did in the Early Paleozoic. Dinoflagellates occur, but are uncommon fossils. Perhaps they were abundant, but most lacked preservable hard parts. The enigmatic acritarchs were abundant in the Devonian, but declined greatly thereafter (Fig. 15.39).

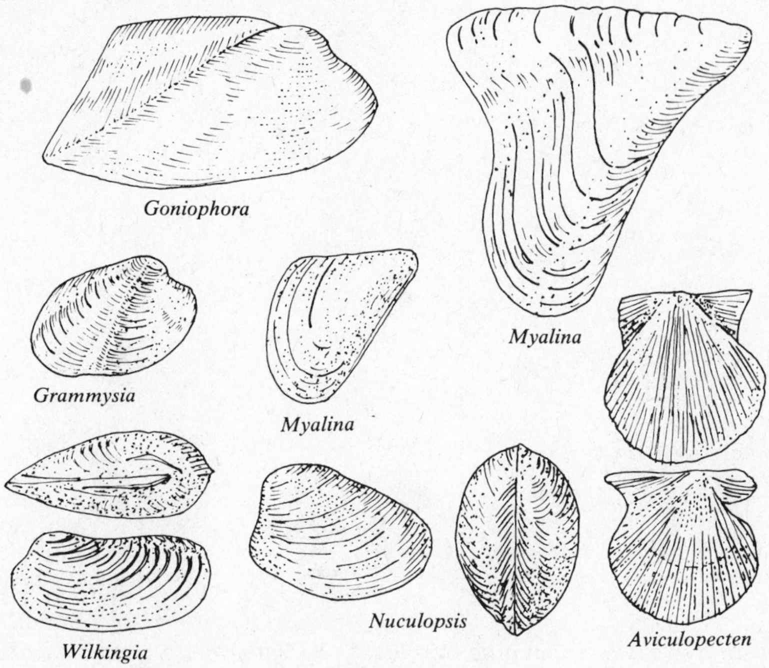

Figure 17.36. *Late Paleozoic Bivalves.* Natural size.

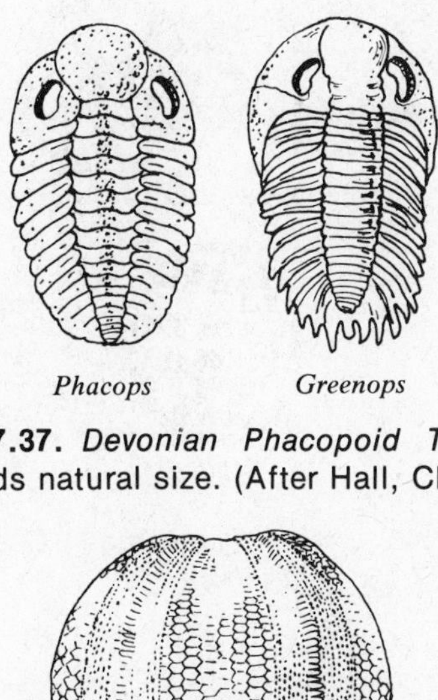

Figure 17.37. *Devonian Phacopoid Trilobites.* Two-thirds natural size. (After Hall, Clark.)

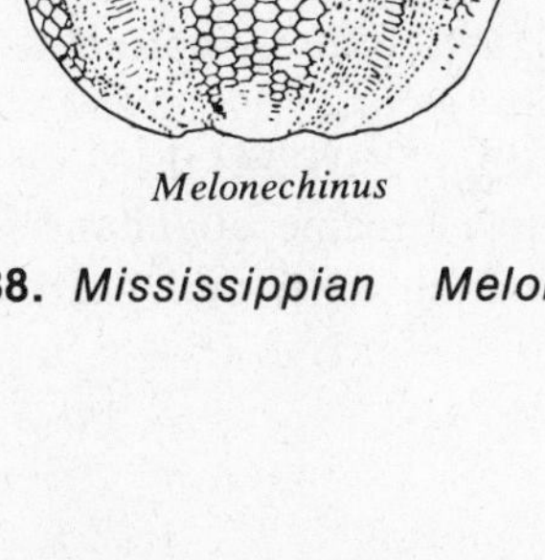

Figure 17.38. *Mississippian Melonechinoid.* X 1/2.

In contrast to the Early Paleozoic, the Late Paleozoic ecosystem represents a highly mature stage with great diversity and long food chains. This situation, characterized by complex interactions among many small, narrowly adapted populations, is favored by stable conditions and, hence, is sensitive to physical and biologic changes. The continental unification, more rigorous climates, and disappearing epeiric seas at the close of the Permian could easily have toppled such an ecosystem, producing the great wave of extinctions marking the end of the Paleozoic.

Fresh-Water Invertebrates

Fresh-water invertebrate life expanded in the Upper Paleozoic also as the **pelecypods** invaded nonmarine waters in the Devonian and the **gastropods** in the Carboniferous (Fig. 17.39). Some of the gastropods may well have been terrestrial. The **branchiopod** crustaceans (Fig. 17.40), a predominantly fresh-water group, appeared in

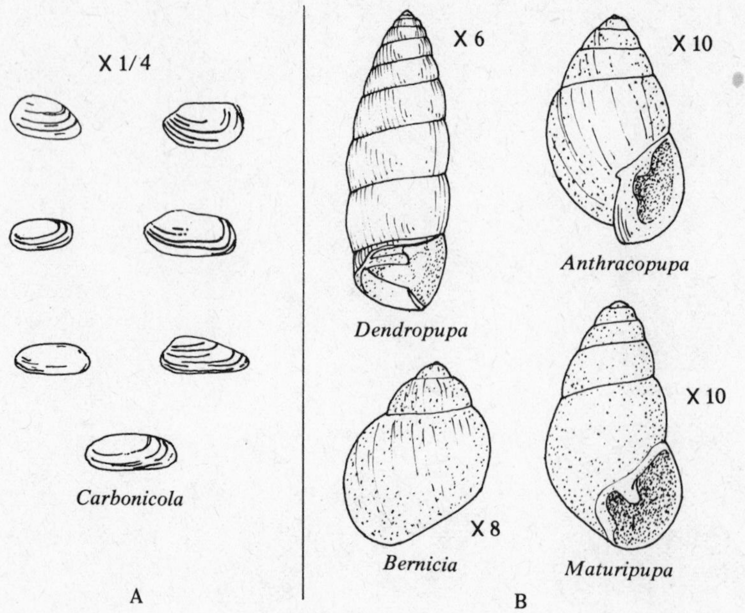

Figure 17.39. *Late Paleozoic Nonmarine Mollusks.* **A.** *Variation in a species of Carboniferous bivalve.* **B.** *Carboniferous gastropods.* Magnifications indicated. (After Eagar, Knight et al.)

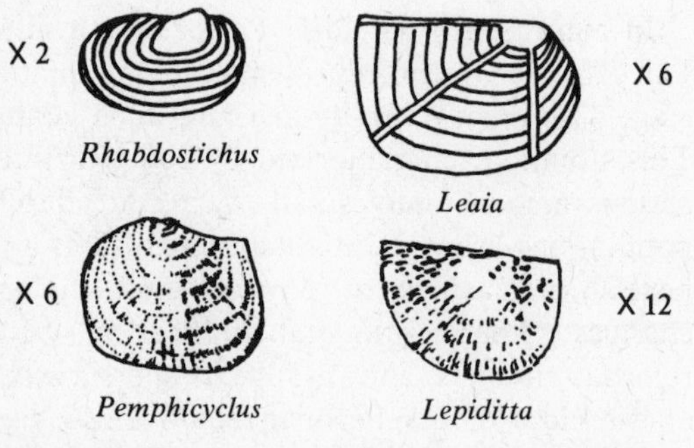

Figure 17.40. *Late Paleozoic Branchiopod Crustaceans.* Magnifications indicated. (After Kobayashi, Jones, Raymond.)

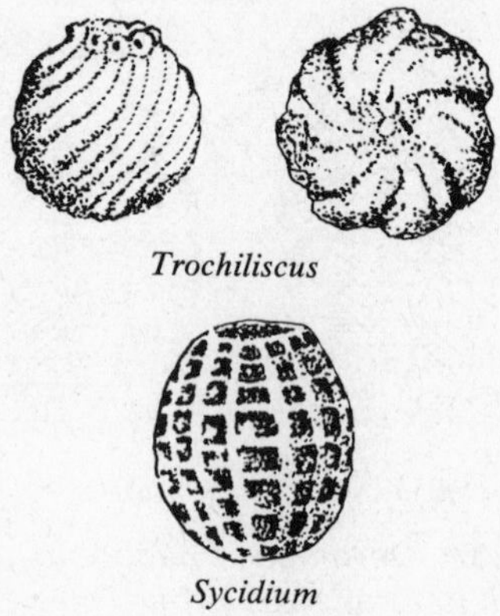

Figure 17.41. *Late Paleozoic Charophyte Oogonia.* X 15. (After Peck.)

the Devonian. **Ostracods** invaded fresh waters in the Pennsylvanian (some questionable Devonian forms have been reported). Some of the Paleozoic malacostracans may also have been fresh water. The protistan **charophytes,** which first appeared in the Late Silurian, were occasionally common in the Devonian (Fig. 17.41).

Fish

The aquatic vertebrates, represented chiefly by the **fish,** were important denizens of Late Paleozoic waters (Fig. 7.108). At the close of the Early Paleozoic, fish became abundant for the first time in the fresh-water and nearshore marine

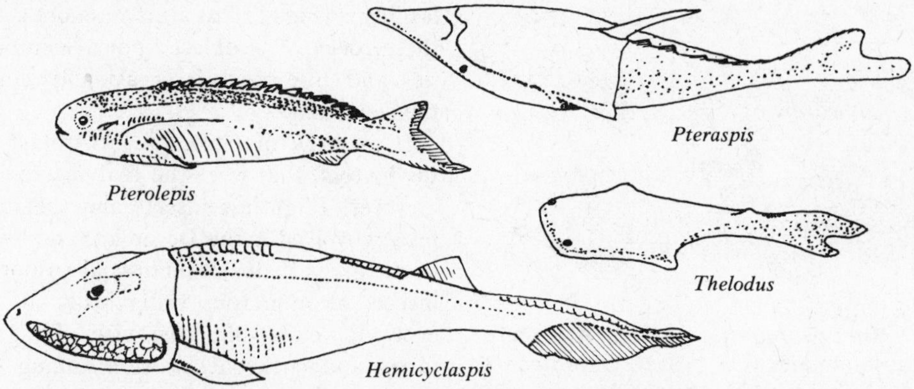

Figure 17.42. *Devonian Ostracoderms.* X 1/2 (After Colbert.)

areas of the Old Red Sandstone "Continent." This expansion continued through the Devonian, often called the "Age of Fishes" because every fish class was present and common during the period. The chiefly fresh-water **ostracoderms** (Fig. 17.42) occurred, but were unimportant after the Early Devonian. **Placoderms** (Fig. 17.43) reached their peak of abundance and diversity in the Devonian and left only a few last Carboniferous survivors. Placoderms include a host of types, such as the largely marine **arthrodires,** shark-like **petalichthyoids,** skate or ray-like **rhenanoids,** and the fresh-water **antiarchs** which resembled ostracoderms. The **sharks** (Fig. 17.44), which arose in the Devonian and were very common in the Carboniferous, are known chiefly from teeth and spines because of their cartilage skeletons. Except for one group, they were all marine.

The largely fresh-water **acanthodeans** (Fig. 17.45) culminated in the Devonian and rapidly declined thereafter. They were the probable ancestors of the **bony fish** (Fig. 17.46), a group

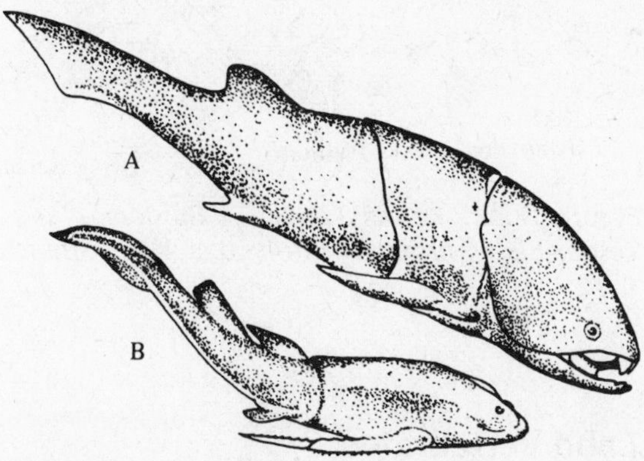

Figure 17.43. *Devonian Placoderms.* **A.** *Coccosteus,* an arthrodire, **B.** *Bothriolepis,* an antiarch. X 1/4. (After Colbert.)

Figure 17.44. *A Primitive Devonian Shark.* X 1/12. (After Colbert.)

Figure 17.45. *A Devonian Acanthodean.* Two-thirds natural size. (After Traquair and Watson.)

Figure 17.46. *Devonian Bony Fish.* **A.** *Cheirolepis,* a ray-finned form, X 1/3; **B.** *Osteolepsis,* a lobe-finned form, X 1/5. (After Colbert.)

which first appeared in Devonian fresh waters. Both ray-fins (actinopterygians) and lobe-fins (sarcopterygians) occurred. The ray-fins diversified greatly thereafter, chiefly by invading the sea, but the lobe-fins were unimportant after the Devonian and barely survived the Paleozoic (Fig. 7.108).

Land Life

Life on land is characterized by an explosive expansion in Late Paleozoic times. The **arthropods** were the chief invertebrate group to exploit this environment. The **arachnean** chelicerates evolved from the aquatic eurypterans via the scorpions. By the Carboniferous Period, unequivocal terrestrial **scorpions** occur in the coal swamp deposits and associated terrestrial or near-shore beds (Fig. 17.47). **Spiders** and similar forms (Fig. 17.48) are even more common in these rocks. The oldest known member of the tick and mite group is occasionally common in the Devonian.

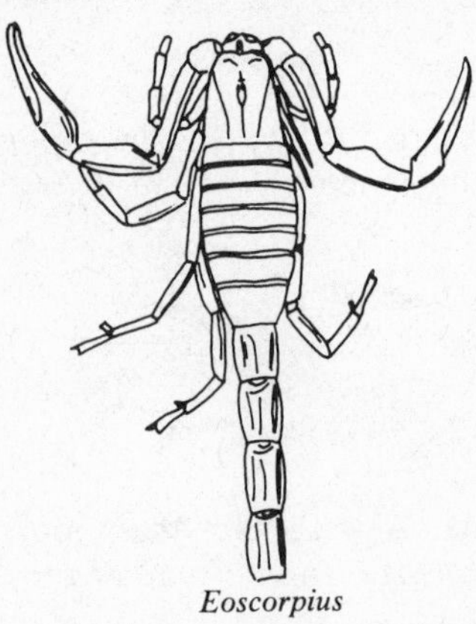

Figure 17.47. *A Pennsylvanian Scorpion.* X 1.

It was the mandibulate arthropods, chiefly the **insects,** that were the real success on land, however. Flightless insects and the first flying insects evolved in the Devonian; and by the time of the great Pennsylvanian coal swamps, flying insects of numerous orders (Fig. 17.49) had evolved, many of large size. Dragonfly-like forms and cockroaches were among the most spectacular of these. Some of the former group had two-foot wing spans. Flying insects were also abundant in some of the Permian terrestrial and fresh-water beds, such as those in central Kansas. The **myriapods,** another terrestrial mandibulate group, are represented by archipolypods and millipedes (Fig. 17.50) in the Pennsylvanian coal swamp deposits.

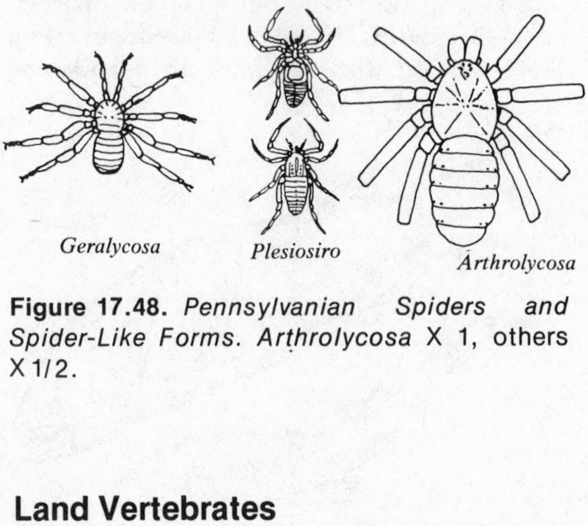

Figure 17.48. *Pennsylvanian Spiders and Spider-Like Forms. Arthrolycosa* X 1, others X 1/2.

Land Vertebrates

The **vertebrates** invaded the land for the first time in the Late Devonian. Many Devonian **fish** display adaptations fitting them for the land. Most fresh-water forms had lungs; and several, particularly the **rhipidistian sarcopterygians** (Fig. 17.51), had bones which attached to supporting shoulder and hip girdles and supported their paired fins (Fig. 7.111). The rhipidistians also had internal nostrils connecting their nasal passages with the passageway to the lungs. These features were probably very useful to the fish in crawling from pool to pool in the dry

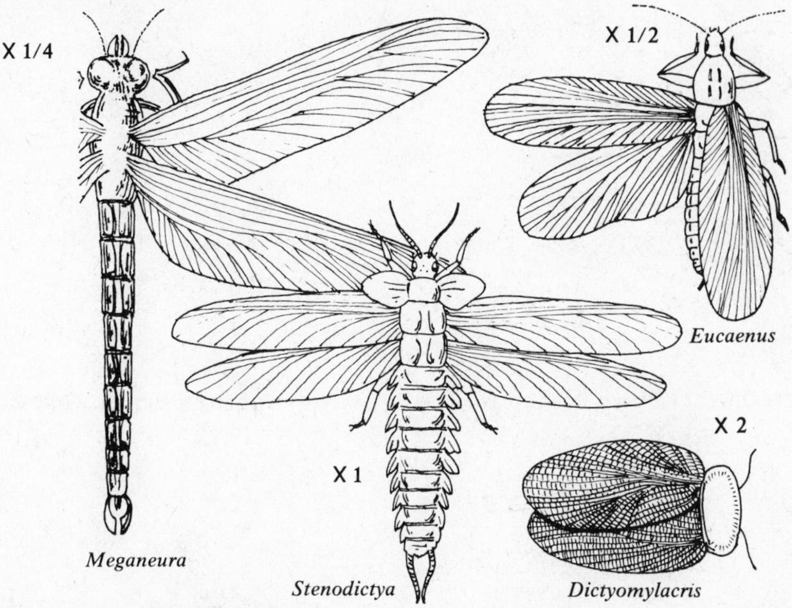

Figure 17.49. *Pennsylvanian and Permian Insects.* The two forms at the left resemble dragonflies but belong to extinct groups; the two forms on the right are cockroaches. Magnifications indicated. (Adapted from *Invertebrate Fossils* by Moore-Lalicker-Fischer, Copyright 1952 by McGraw-Hill, Inc., used by permission of McGraw-Hill Book Company.)

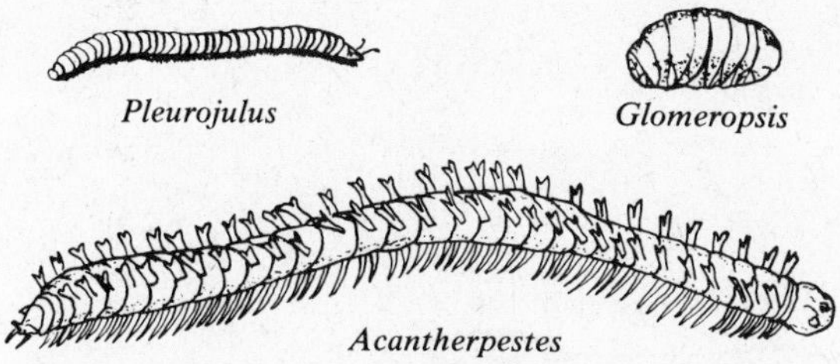

Figure 17.50. *Late Carboniferous Myriapods.* X 2/3. (After Fritsch.)

season, either in search of food or to escape a drying pond. These fish, such as *Eusthenopteron,* are best known from the Old Red Sandstone "Continent," which was probably a low-latitude, savanna-like area with a pronounced dry season as indicated by the paleomagnetic and lithologic data. Selection favored the gradual strengthening of the limbs and girdles and the improvement of the lungs until ultimately the fish-amphibian boundary was bridged.

The oldest known **amphibian**, *Ichthyostega* (Fig. 17.52), from the Late Devonian of the east Greenland part of the Old Red Sandstone region, still has a fish-like tail and weak limbs. The bones of the skull and the teeth of the earliest amphibians are very similar to those of the most amphibian-like fish. In the Pennsylvanian coal swamps of Laurasia, the amphibians underwent a major radiation, including several types of large, bulky, flat-headed **labyrinthodonts** (Fig. 17.53) and small, salamander-like **lepospondyls** (Fig. 17.57). Most of these forms still spent most or all of their lives in the water. Some larval forms with fish-like gills are known. Both

Figure 17.51. *Rhipidistian Lobe-Finned Fish,* precursor of the Tetrapods.

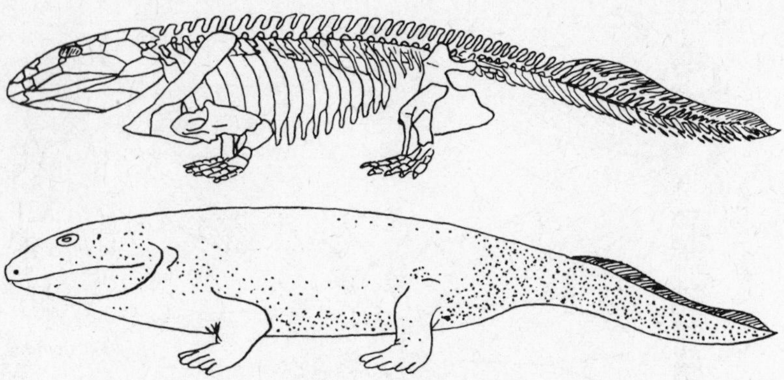

Figure 17.52. *The Oldest Known Amphibian, Ichthyostega,* Upper Devonian, Greenland. X 1/10 (After Jarvik.)

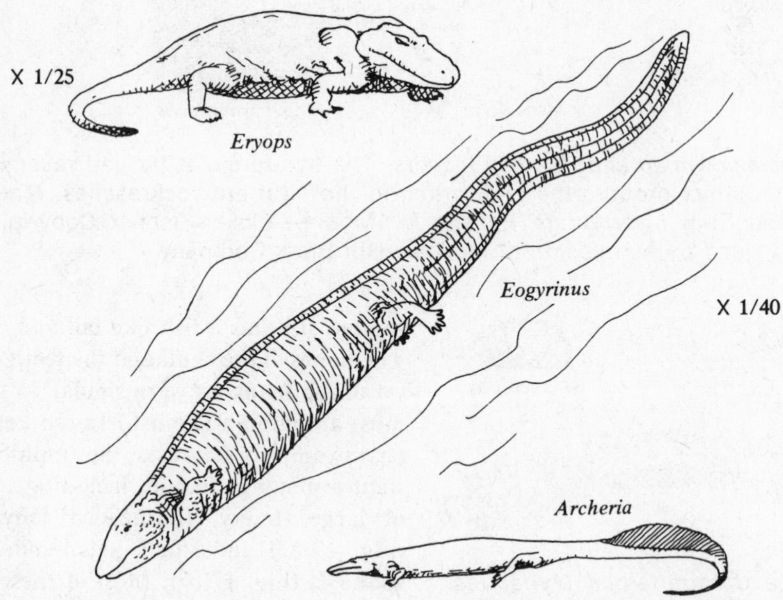

Figure 17.53. *Late Paleozoic Labyrinthodont Amphibians.* Magnifications indicated. (After Fenton, Romer.)

types of Paleozoic amphibians continued into the Permian with some reduction as the coal-producing swamps gradually disappeared. The labyrinthodonts succeeded in invading Gondwanaland in the Permian as Pangaea formed. The lepospondyls did not survive the Permian.

Early in the Pennsylvanian, the **reptiles** arose from the labyrinthodont amphibians and began to diversify. Only a few imperfect specimens of Pennsylvanian reptiles are known, chiefly of the stem-reptile or **cotylosaur** group (Fig. 17.54), though a few **mammal-like** reptiles are also known. The latter group includes the first herbivorous reptiles. The cotylosaurs (Fig. 17.55) did not differ greatly from their labyrinthodont ancestors.

In the Permian, the reptile record greatly improves with the widespread occurrence of terrestrial redbeds. In these rocks, survivors of the transitional amphibian-reptile forms occur (Fig. 17.56), forming such perfect connecting links that it is impossible to decide in which class to

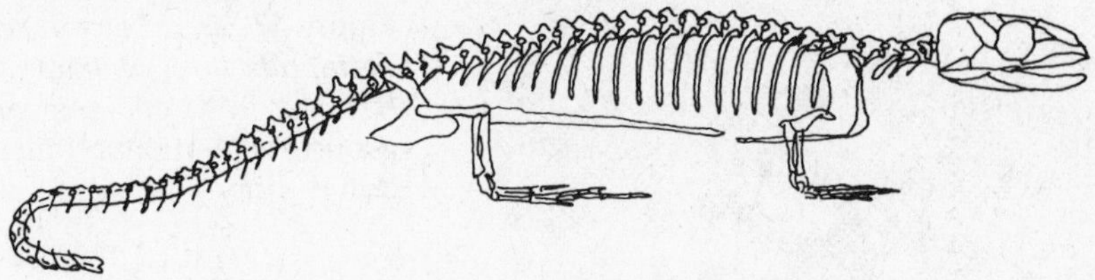

Figure 17.54. *Hylonomus, a Pennsylvanian Cotylosaur.* One of the oldest known reptiles. (After Carroll.)

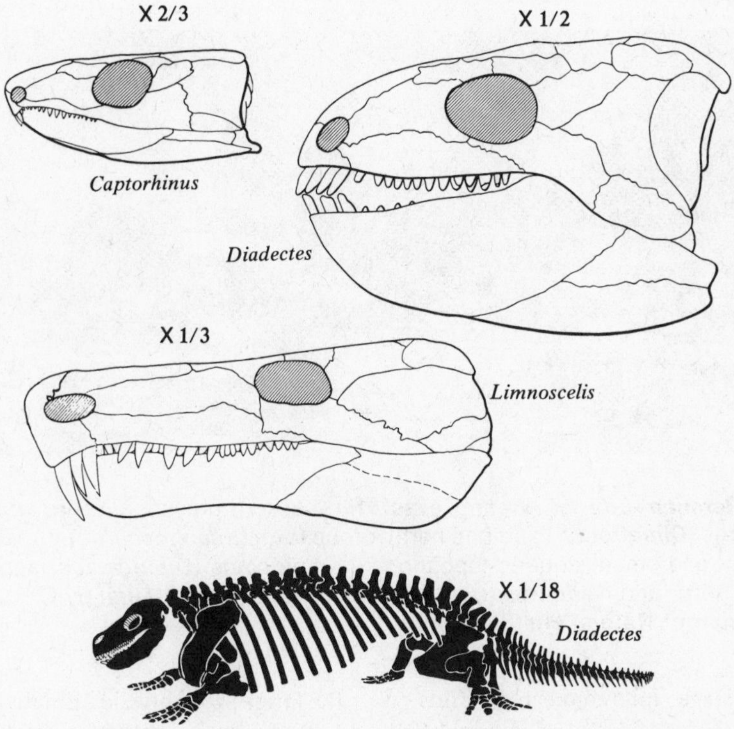

Figure 17.55. *Late Paleozoic Cotylosaur Reptiles. Diadectes* is close to the amphibian-reptile boundary. Magnifications indicated. (From Edwin H. Colbert, *Evolution of the Vertebrates,* Copyright ©1969, by permission of John Wiley & Sons, Inc.)

place them. Fossil reptilian amniotic eggs are first known in the Permian as well. Cotylosaurs (Fig. 17.55) were still common in the Permian and spread from Laurasia to the more northern tropical parts of Gondwanaland.

It is the mammal-like reptiles, however, that dominate Permian land faunas. The Permian forms of **pelycosaurs** include the spectacular fin-backed genera *Dimetrodon* and *Edaphosaurus* (Fig. 17.57). Originating in the North American part of Laurasia, they gradually spread to present-day Europe and to Gondwanaland by the later Permian. All these large reptiles were restricted to warmer climatic zones in the Early Permian. The later Permian and Triassic forms belong to the more advanced group, the **therapsids** (Fig. 17.58). Large faunas are known from Russia and particularly South Africa. Although most of the genera and species differed between the Late Permian and Early Triassic, the major groups were the same across the era boundary. Terrestrial life migrated widely at this time be-

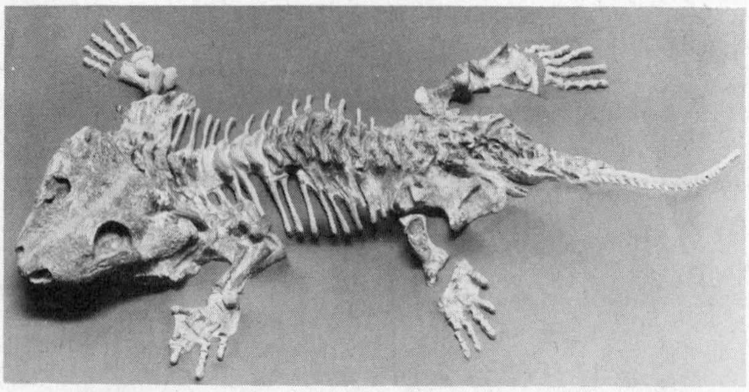

Figure 17.56. *Seymouria, a Permian Tetrapod Intermediate in Structure Between Amphibians and Reptiles.* (Courtesy National Museum of Natural History.)

Figure 17.57. *Permian Life of West Texas.* Fin-backed pelycosaurs (mammal-like reptiles) include carnivores (*Dimetrodon*) and one herbivorous form (*Edaphosaurus*). Small pelycosaurs are seen on the left; and small, aquatic lepospondyl amphibians (*Diplocaulus*) appear on the right. Conifers, ferns (left), and calamites (right) dominate vegetation. (Mural by C.R. Knight, courtesy American Museum of Natural History.)

cause of the large interconnected lands of Pangaea. The therapsids were undoubtedly warm-blooded, and probably hairy, by the Late Permian because South Africa was in the polar regions. Furthermore, their bone structure is characteristic of warm-blooded animals because it lacks growth rings and contains many canals. The ratio of prey to predator weight in the Late Permian faunas also falls in the range characteristic of warm-blooded animals.

A few other Permian reptilian types (Fig. 7.115) are known. These are the protorosaurs, perhaps ancestral to the sauropterygians, and the eosuchians, perhaps ancestral to the lizards. Both had lizard-like proportions and were relatively small. The former arose in Laurasia, and the latter probably did. Finally, no discussion of Late Paleozoic reptiles would be complete without mention of the enigmatic **mesosaurs** (Fig. 17.59) of the polar regions of South America and South Africa. The present disjunct distribution of this Late Carboniferous or Early Permian form was one of the lines of evidence that originally led to the formulation of ideas on continental drift.

Land Floras

The first great land floras developed in the Late Paleozoic. The simple **psilophytes** (Fig. 17.60),

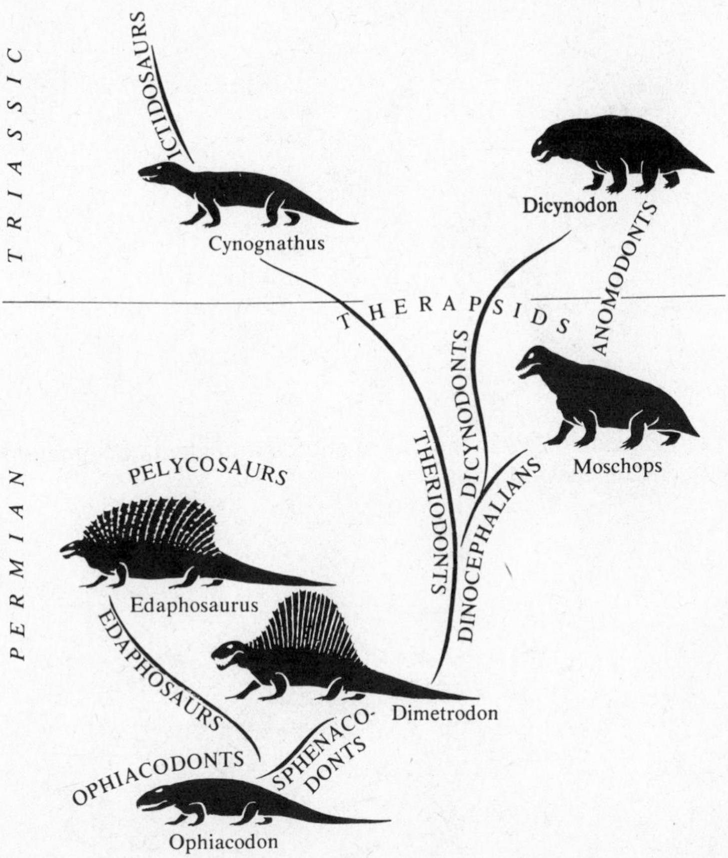

Figure 17.58. *Mammal-like Reptiles.* (From Edwin H. Colbert, *Evolution of the Vertebrates,* © 1969, by permission of John Wiley & Sons, Inc.)

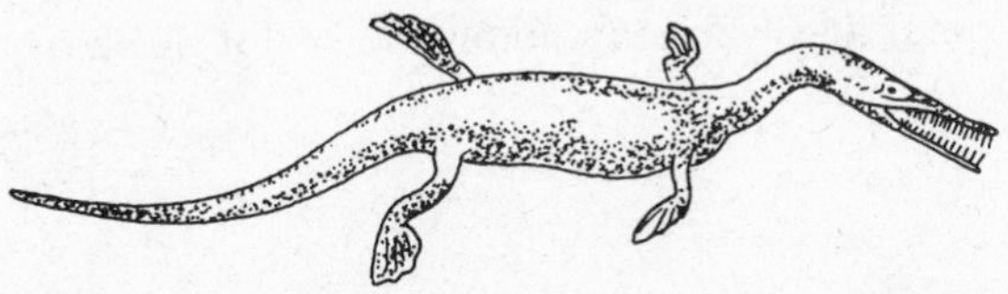

Figure 17.59. *Mesosaurus.* X 1/5. A small, aquatic reptile of Late Paleozoic age. Mesosaurus lived in cold climates and probably burrowed into the mud of lake bottoms to hibernate through the long winters of southern Gondwanaland. (After Colbert.)

which first appeared near the close of the Silurian, diversified in the Devonian, their peak of abundance. At least three major lineages existed, represented by *Rhynia, Zosterophyllum,* and *Trimerophyton* (Fig. 7.120). Before the close of the Devonian, however, lycopods, articulates, and ferns each evolved separately from the psilophytes and became common (Fig. 17.61). Although most were low-growing herbaceous forms in the Devonian, some of them, such as the lycopod *Archaeosigillaria* and the fern *Eospermatopteris,* reached tree size.

In the great Pennsylvanian coal swamps of Laurasia (Fig. 17.62), the **lycopods, articulates,** and **cordaitean conifers** all reached the size of large trees. The **seed ferns,** which also reached a peak in the Pennsylvanian, were common low trees. The undergrowth consisted of herbaceous lycopods, articulates, and ferns. The lack of growth lines in the coal swamp plants and the paleomagnetically determined continental posi-

Figure 17.60. *Early Devonian Flora.* The low, dichotomous plants dominating the vegetation are chiefly psilophytes. (After Seward.)

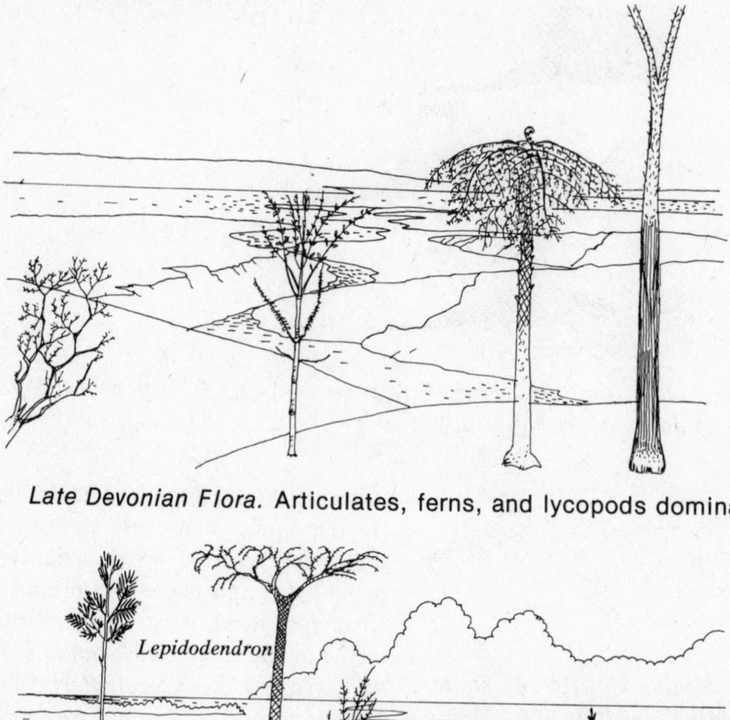

Figure 17.61. *Late Devonian Flora.* Articulates, ferns, and lycopods dominate. (After Seward.)

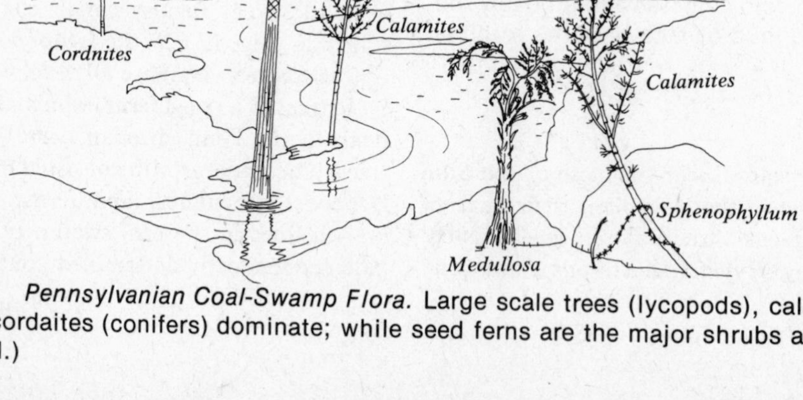

Figure 17.62. *Pennsylvanian Coal-Swamp Flora.* Large scale trees (lycopods), calamites (articulates), and cordaites (conifers) dominate; while seed ferns are the major shrubs and low trees. (After Seward.)

tions for the Pennsylvanian leave little doubt that this vegetation lived in moist tropical latitudes. Distinctive temperate floras flourished in present western North America, Siberia, and China (Fig. 16.7). The northern Chinese (Cathaysian) flora was probably subtropical as there are no growth rings in the wood.

In the Permian, as the coal swamps gradually disappeared, the great tropical Laurasian plant groups of the Paleozoic declined. Conifers became the dominant form of plant life in the Permian (Fig. 17.63). The differentiation of floral provinces was much more pronounced in the Permian.

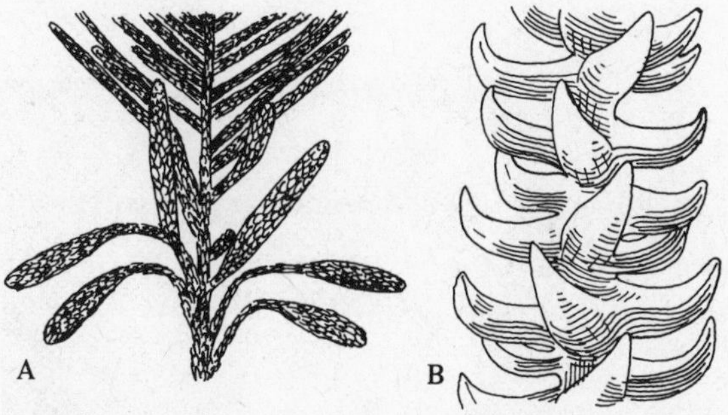

Figure 17.63. *Late Paleozoic Conifer. Ernestiodentron.* **A.** *Leafy branch with cones.* **B.** *Leaves, X3.* (After Florin.)

The Gondwanaland flora of the Late Paleozoic is dominated by the seed fern *Glossopteris* and its allies (Fig. 17.64). Apparently low shrubs, these plants with prominent growth rings thrived in the cool climates near the Late Paleozoic glaciers and accumulated to sufficient thicknesses in the interglacials to form the major coal beds of the present Southern Hemisphere continents.

As with land animals, there is no profound Permotriassic floral break, only the continuing rise of the conifers and also the cycadophytes and gingkos which had appeared in the Late Paleozoic but were uncommon. (Cycadophytes apparently originated in Gondwanaland.) The large scale trees, calamites, and cordaites, were essentially gone well before the Mesozoic began. In Gondwanaland, the distinctive flora continued into the Triassic, but as with mammal-like reptiles, the species content was different.

Mazon Creek Fossils

A significant fossil locality of the Late Paleozoic is the Mazon Creek area of Illinois. In Pennsylvanian concretions (Fig. 17.65) are preserved impressions of a sample of nearly the entire biota of near-shore marine and terrestrial communities of the area. The marine fossils include numerous soft-bodied worm and arthropod groups, a tadpole larva of the tunicates (possible vertebrate relatives), and the oldest fossil lamprey (the soft-bodied, parasitic survivor of the Agnathea). Also included are enigmatic creatures such as the Tully Monster, *Tullimonstrum.* The terrestrial biota includes insects, arachneans, small vertebrates, and plant leaves. With discoveries such as these, the paleontologist can reconstruct evolutionary histories and ecosystems which could hardly be guessed at otherwise.

Figure 17.64. *Glossopteris Flora. Glossopteris* and the larger *Gangamopteris* dominate. This flora inhabited cool climates in southern Gondwanaland. (After Seward.)

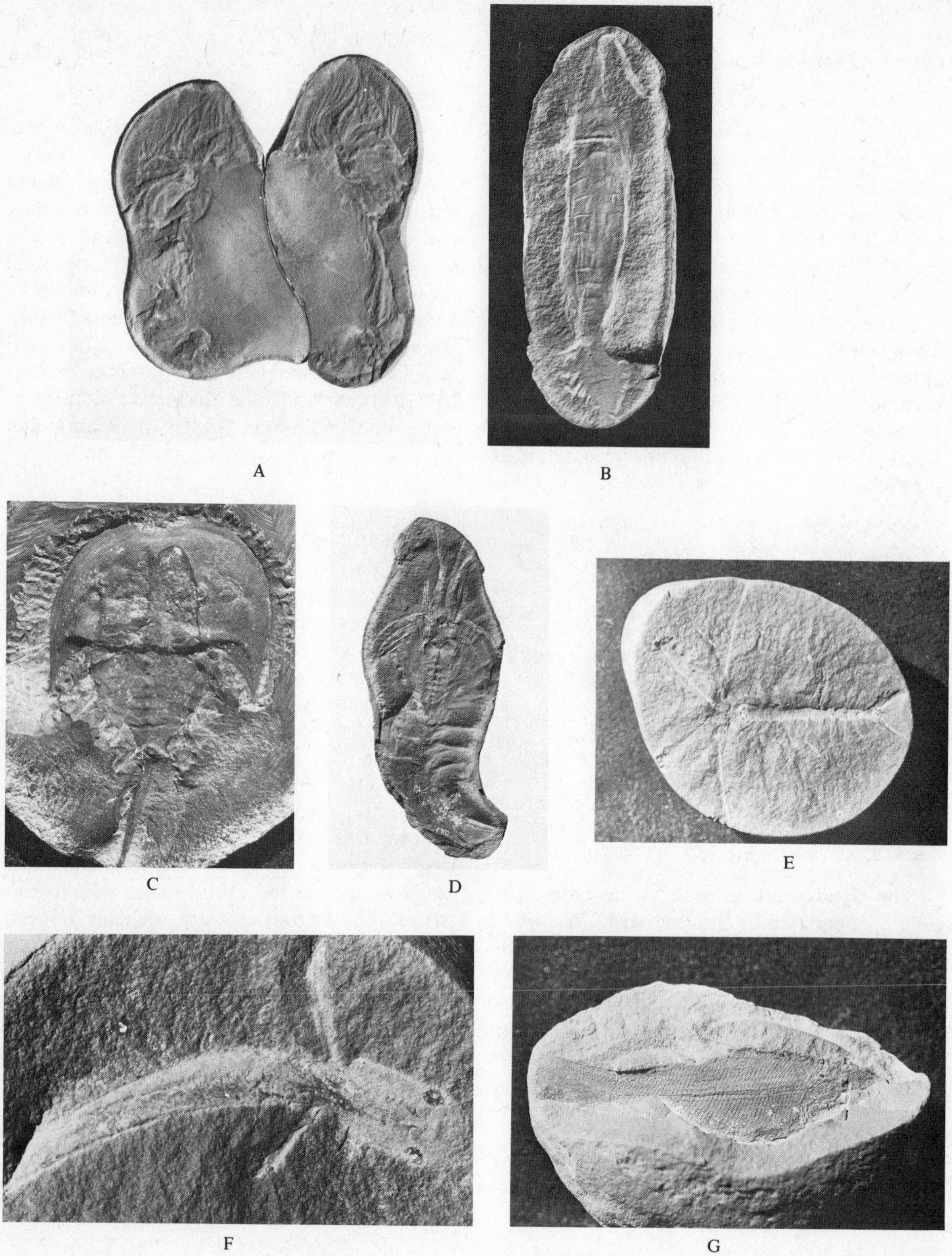

Figure 17.65. *The Mazon Creek Fauna.* **A.** *Medusoid coelenterate Antracomedusa*; **B.** *Tully Monster, Tullimonstrum, a worm-like form of unknown affinities*; **C.** *Xiphosuroid arthropod Palaeolimulus*; **D.** *Malacostracan (eocarid) crustacean Belotelson*; **E.** *Malacostracan (syncarid) crustacean Palaeocaris*; **F.** *Larval fish*; **G.** *Paleoniscoid ray-finned bony fish.* (Courtesy of Ralph G. Johnson, University of Chicago.)

REVIEW

1. When did **foraminifera** become important in the geologic record, and what was the major Late Paleozoic group?
2. What **coelenterates** were the major reef builders of the Late Paleozoic?
3. Which group of **bryozoans** dominated Late Paleozoic faunas?
4. What were the six major **brachiopod** groups of the Late Paleozoic? What happened to each of them by the close of the era?
5. What significant group of **mollusks** arose and became common in the Late Paleozoic?
6. What two major types of **crinozoan echinoderms** reached a peak in the Mississippian?
7. Why are **conodonts** geologically important in the Late Paleozoic?
8. Which major groups of Paleozoic organisms died out at or near the close of the era?
9. Give at least one plausible explanation for the large number of extinctions that occurred near the Paleozoic-Mesozoic boundary.
10. What major fresh-water groups appeared in the Late Paleozoic?
11. Why is the Devonian known as the **Age of Fishes**?
12. Which groups of organisms invaded the land in the Late Paleozoic?
13. Discuss the Late Paleozoic evolution of the amphibians.
14. What were the major Late Paleozoic reptile groups?
15. What evidence is there that Late Permian therapsid mammal-like reptiles were warm-blooded?
16. What were the dominant plant types of the Carboniferous coal swamps?
17. What differences are there between the evolution of land and marine life at the Paleozoic-Mesozoic boundary?

18

The Fountainhead of Stratigraphy: The Mesozoic Erathem

18

The three periods of the Mesozoic Era, the Triassic, Jurassic, and Cretaceous, encompass approximately 160 million years of earth history from about 225 to 65 million years ago. The type area of the **Triassic** is in the clastic wedge, terrestrial redbed sequence derived from the Hercynian Mountains in central Germany; the type area of the **Jurassic** is in the Jura Mountains of northwest Switzerland, a former marine shelf area along the southern margin of Europe; and the type **Cretaceous** is on a former shelf area in the present Paris Basin of northern France (Fig. 2.19). At its beginning, the Mesozoic was characterized by the extreme concentration of the continental masses into Laurasia and Gondwanaland, themselves interconnected to form Pangaea; at its close, it was characterized by extreme fragmentation as a new pattern of plates and their movements developed (Figs. 18.1–4).

Triassic Events

The pattern of magnetic bands in the present North Atlantic Ocean basin suggests that it may have begun to form off the present eastern United States as early as the Triassic, though no oceanic rocks of that age have been found. The oldest known sediments, from the basin margins, date from the Jurassic, but there may be older material beneath them. Triassic basalts are common in present eastern North America, indicating the rift was in existence in that period; and the Late Paleozoic Appalachian disturbances may indicate an even earlier initiation of rifting. The paleomagnetic data suggest a Late Jurassic opening for South Atlantic, and the oldest rocks thus far found are Middle Cretaceous in age. (The extreme southern Atlantic may have opened even earlier.) Thus, before the end of the Mesozoic, North America and South America were separating from Gondwanaland (Figs. 18.3, 18.4C).

Gondwanaland suffered further fragmentation as new rifts developed and another new ocean basin, the Indian, began to form. Present-day Africa split off from the remainder of Gondwanaland, composed of present-day India, Australia, and Antarctica, beginning in the Late Triassic (Figs. 18.1, 18.4A). An extension of this rift continued up the present east side of India, separating it from the other former Gondwanaland continents. In the Mesozoic, the former western Gondwanaland continents—South America, Africa, India—and probably China as well moved into equatorial regions, while the Antarctic-Australian piece remained in high latitudes. The more oceanic conditions in the South polar area apparently moderated conditions as no continental glaciations are known.

As the North Atlantic rift began to form, the area of the Old Appalachian-West African (Mauritanian) Mountains became a zone of tension with its attendant block faulting, intrusion, and terrestrial deposition characterizing the Triassic as the Atlantic rift opened. In North America, the surviving terrestrial redbeds and associated igneous rocks, the Newark Group (Figs. 18.5–6), are preserved in downfaulted

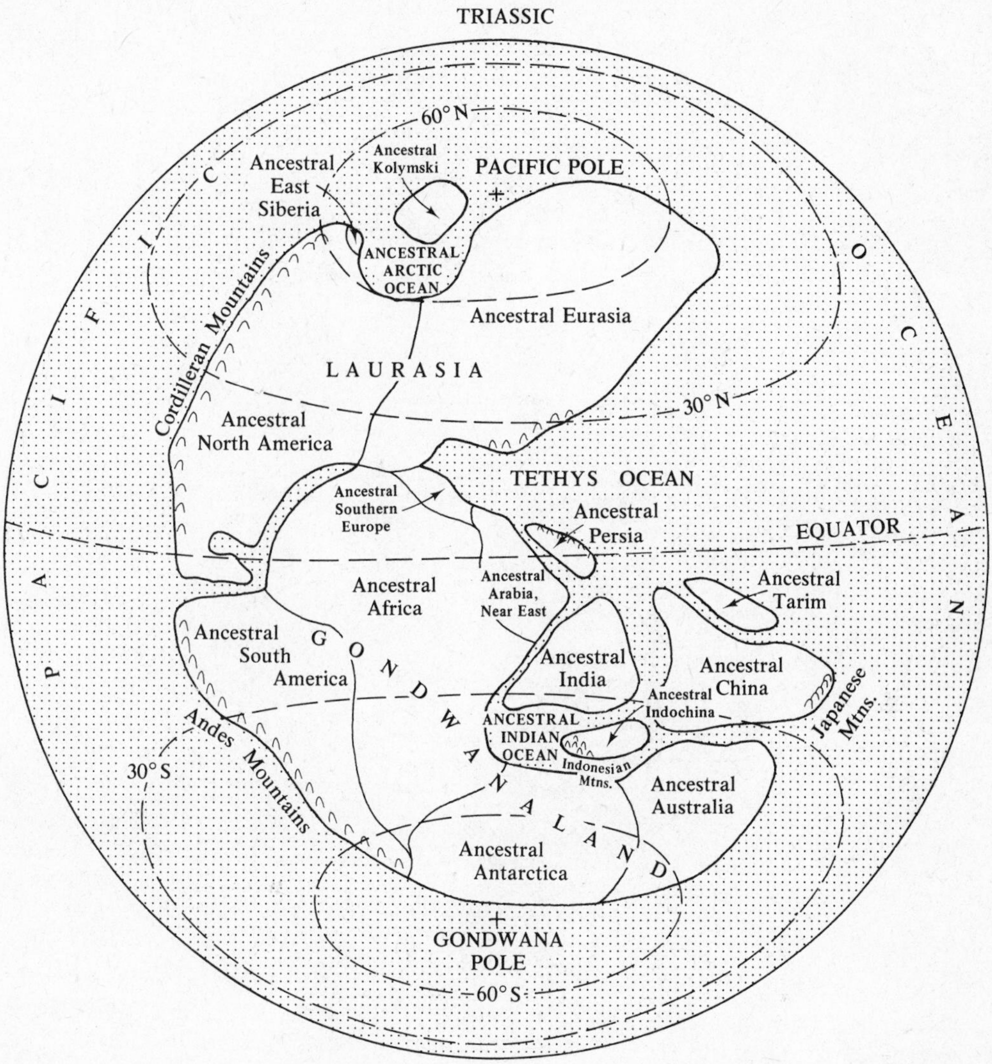

Figure 18.1. *Triassic Paleogeography.* The position of main mass of China is speculative. Plate boundaries and epeiric seas are not shown. (Material from Bird, Dewey, Dietz, Holden, Irving, McKenzie, Sclater, Seyfert, Sirkin, and others.)

basins from the Canadian Maritime Provinces through the Carolinas. (They are known as far south as Florida in drill holes.) The Triassic equator passed just south of the present southern United States in Mexico (Fig. 18.1), and the Pacific Pole was situated in the area presently east of Siberia.

The extreme continental conditions that characterized the later Permian continued through much of the Triassic as well. The cratons were all emergent except for some marginal seas. There were no epeiric seas that covered North America in this period, only the final marginal seaways in the Idaho-Wyoming-Utah areas of the regressing Absaroka Sea. By the Late Triassic, the terrestrial redbeds covered all the Southern Rocky Mountains-Colorado Plateau area except for the remnant peaks of the Ancestral Rockies.

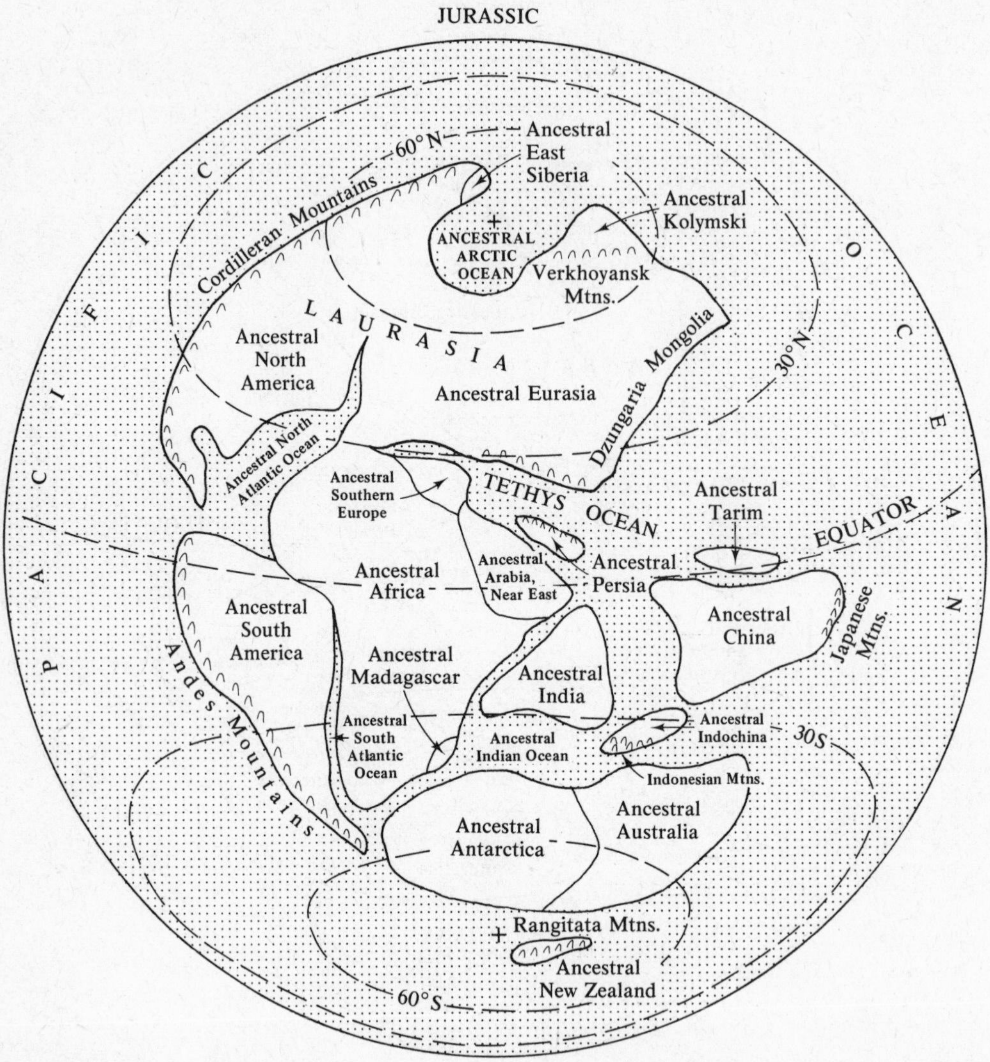

Figure 18.2. *Jurassic Paleogeography.* The positions of the Chinese plates are speculative. Plate boundaries and epeiric seas are not shown. (Material from Bird, Dewey, Dietz, Holden, Irving, McKenzie, Sclater, Seyfert, Sirkin, and others.)

This sequence, which has been carved into much of the present spectacular scenery of the Colorado Plateau (Fig. 18.7), passes without interruption or evidence of a physical boundary right into the Jurassic.

All over northern and western Europe, a terrestrial redbed or Germanic facies (Fig. 18.8) characterizes this system. The Permian-Triassic boundary is difficult to establish in the nearly continuous and unfossiliferous sequence of this New Red Sandstone "Continent." A minor marine invasion of central Europe, the **Muschelkalk Sea,** occurred in the Middle Triassic, and at the close of the period another incursion began. (These seaways reflect the beginning phases of the opening of the Atlantic Ocean.) The Asian continental block was also emergent in the Triassic. In the Gondwanaland continents,

CRETACEOUS

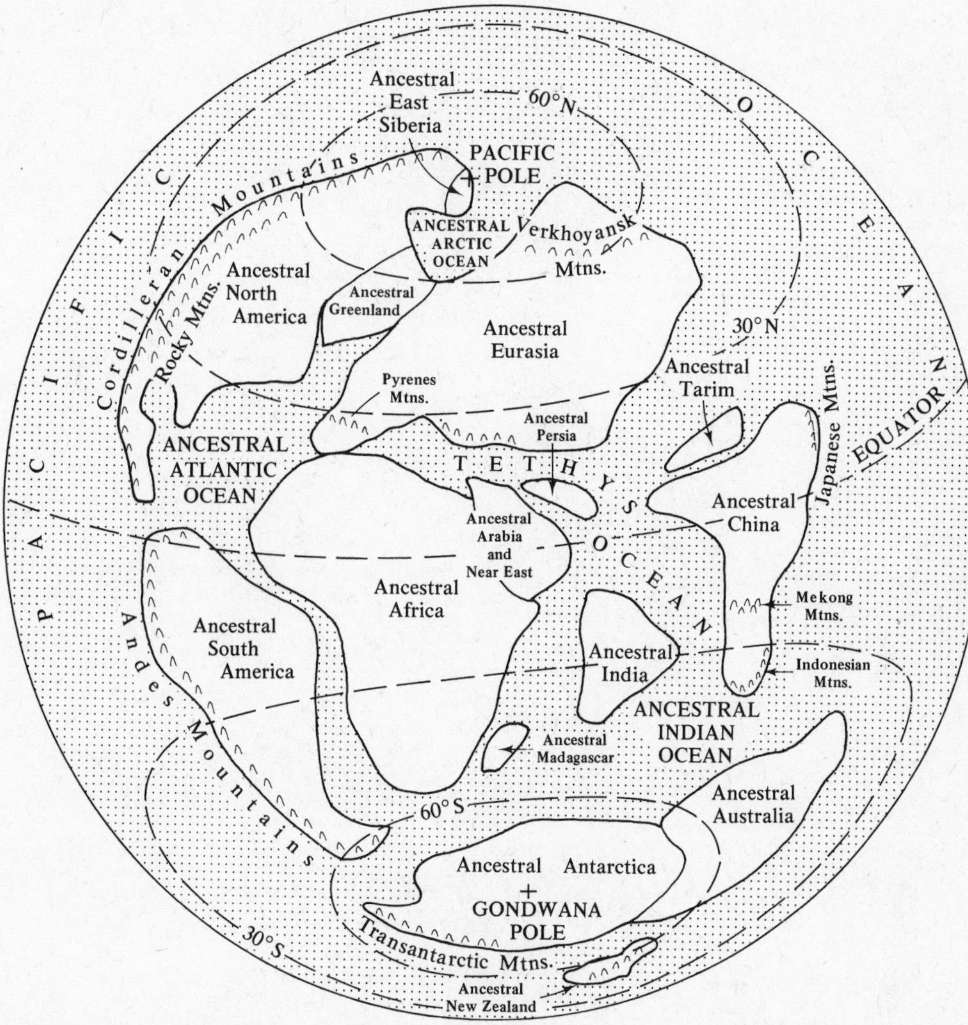

Figure 18.3. *Cretaceous Paleogeography.* The position of the main mass of China is uncertain, although Mongolia's position as part of Asia is well established. Epeiric seas and plate boundaries are not shown. (Material from Bird, Dewey, Dietz, Holden, Irving, McKenzie, Sclater, Seyfert, Sirkin, and others.)

the Karroo-type terrestrial deposition continued from the Permian without break. Marginal Triassic seas in western Australia reflect the opening of the Indian Ocean at that date.

Mesozoic mountain building was primarily the result of oceanic plates and island arcs colliding with continents, rather than the continent-continent collisions of the Paleozoic which produced Pangaea. As noted, the Mesozoic was primarily a time of continental breakup, so this is not surprising. The Cordilleran and Andean unstable belts continued from the Paleozoic and resulted from the collision of east Pacific plates with the continental margins of the new American plates (Fig. 18.1) along a trench. Consumption of Pacific crust and flysch, volcanism, and

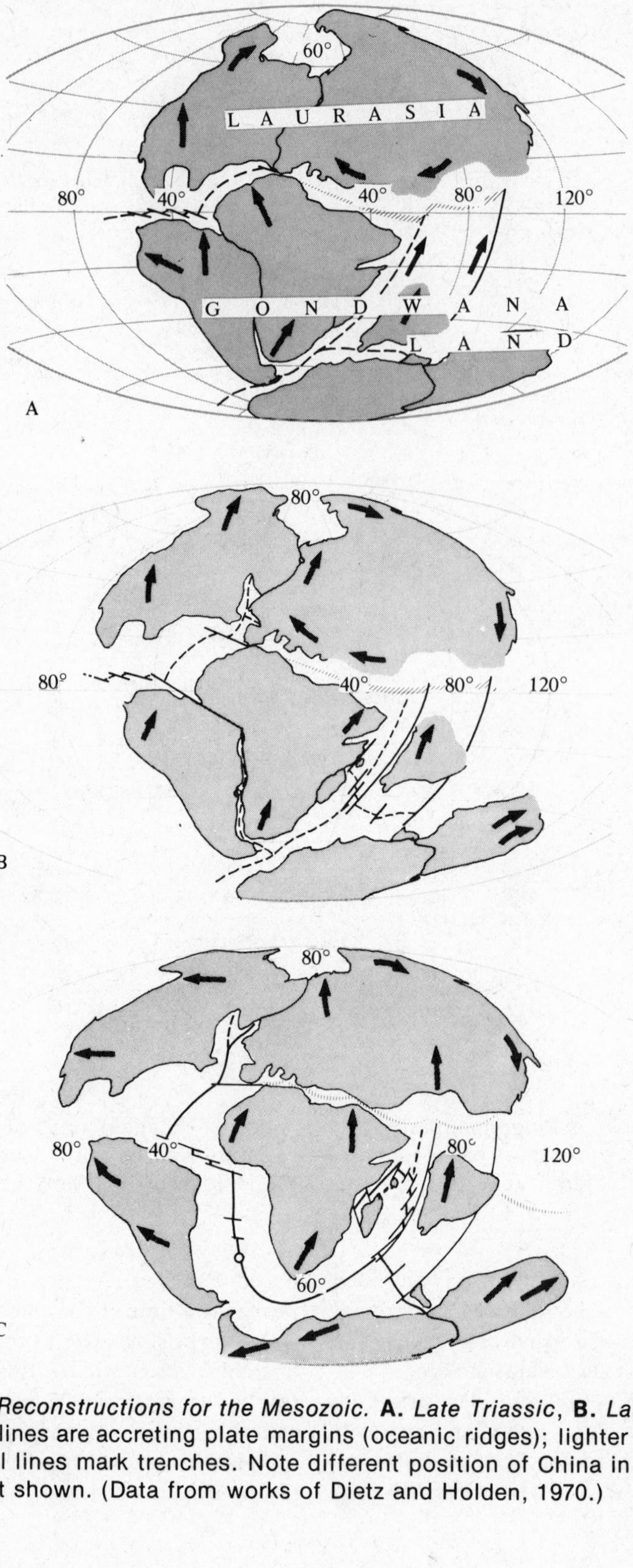

Figure 18.4. *Plate Reconstructions for the Mesozoic.* **A.** *Late Triassic*, **B.** *Late Jurassic*, **C.** *Late Cretaceous.* Heavy lines are accreting plate margins (oceanic ridges); lighter lines are transform faults; short parallel lines mark trenches. Note different position of China in this interpretation. Epeiric seas are not shown. (Data from works of Dietz and Holden, 1970.)

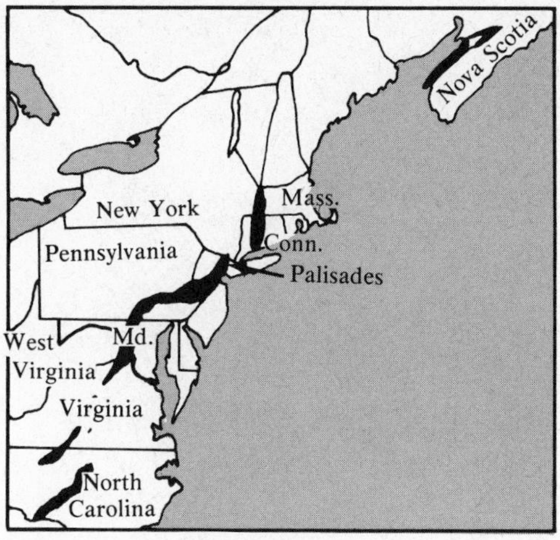

A

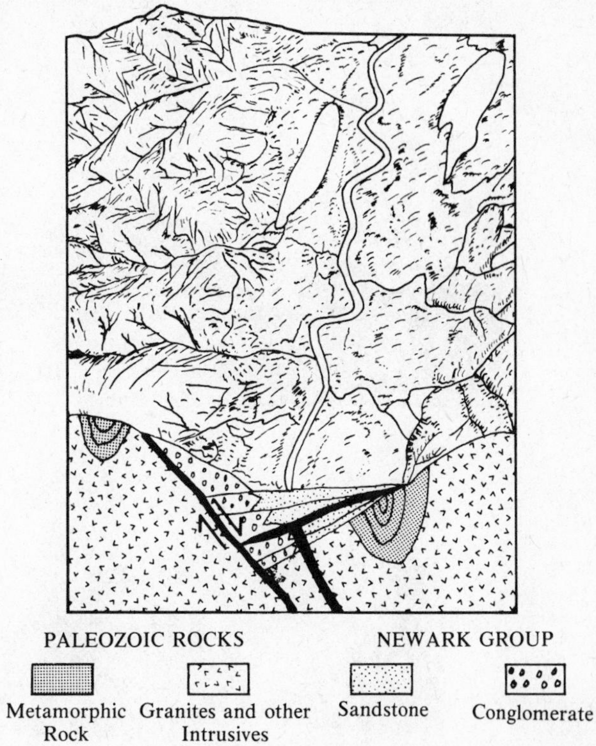

B

Figure 18.5. *Triassic Fault-Block Basins of Eastern North America.* **A.** *Map of surface distribution,* **B.** *Block diagram of typical Triassic fault-block basin.* Compare with Fig. 10.20. (After Clark and Stearn.)

Figure 18.6. *Palisades of the Hudson River,* Palisades Interstate Park, New Jersey. A sill intruded in Newark Group rocks. (Courtesy State of New Jersey.)

A

B

C

Figure 18.7. *Triassic-Jurassic Redbeds and other Continental Strata in Colorado Plateau Country.* **A.** *Echo Cliffs, northern Arizona;* **B.** *Deadhorse Point, Canyonlands area, eastern Utah;* **C.** *Checkerboard Mesa, Zion National Park, southwest Utah;* **D.** *Colorado National Monument, western Colorado;* **E.** *Arches National Park, eastern Utah.* (Courtesy of William M. Mintz.)

D

E

orogeny continued throughout the Mesozoic. In the present North Pacific area in the Late Triassic, the small east Siberian plate including Kolymski was moving toward Asia. The Tethyan Ocean area (Fig. 18.9) was a complex of small plates, ridges, trenches, and transform faults. Excellent sequences for all the Mesozoic systems are developed in this area of the present Mediterranean. The abundantly fossiliferous Triassic section, best known in the Alps, is the world standard for the system which is chiefly represented by terrestrial rocks in its type area. Tethys was in the Triassic tropics and contained abundant reefs of hexacorals and dasycladaceous

Figure 18.8. *New Red Sandstone of Europe.* Permotriassic redbeds in Scotland. (Crown Copyright. Geological Survey photograph. Reproduced by Permission of the Controller, Her Britannic Majesty's Stationery Office, London.)

green algae, which today form the great stone blocks of the Dolomite and other limestone Alps (Fig. 18.10).

Jurassic Events

In the Jurassic, as the Atlantic and Indian ocean basins developed further (Figs. 18.2, 18.4B) and the Late Paleozoic mountains were largely reduced, there was a return to more widespread epicontinental seas. In North America, the **Zuni Sea** (Fig. 14.12) began transgressing in the Late Triassic-Early Jurassic interval, from both the Pacific Ocean and the newly formed Gulf of Mexico part of the Atlantic on the southeast. As the Atlantic rift opened (Fig. 18.11), the crust in the rift area thinned and the trailing edge of the continent foundered. Thus, much of the site of the old Ouachita-Appalachian Mountains became flooded by seas in the Mesozoic. In the Jurassic, eastern Mexico, southern Texas and Louisiana, and perhaps additional areas to the east were flooded; and the sea probably connected with the one coming from the Cordilleran region.

The northern incursion of the Zuni Sea flooded much of the present Great Plains and southern Rocky Mountain area of the craton in the Late Jurassic. The deposits of this marine invasion, often called the Sundance Sea, are chiefly clastics, reflecting the instability in the Cordilleran region to the west.

A mantle plume or hot spot which may have triggered the opening of the North Atlantic, produced the White Mountain intrusions of New England, principally of Jurassic age. The nematath of this hot spot forms the New England Seamounts extending from the White Mountains to the Azores Islands.

Near the close of the Jurassic, compression between an east Pacific plate (some authorities believe it contained a microcontinent) and the northwestward moving margin of North America produced major deformation of the western continental margin, the **Nevadan Orogeny** (Fig. 18.12). A flood of clastics, the **Morrison Clastic Wedge,** temporarily drove the Zuni Sea off much of the craton. Great granite intrusions of Jurassic-Cretaceous age began to form along the western margin of North America from Alaska through Baja California (Fig. 18.13). These include the huge Sierra Nevada batholith of California whose name is given to the orogeny. Near the Jurassic-Cretaceous boundary, 135 million years ago, spreading rates slowed in response to the new plate organization engendered by the separation of Africa from South America. This produced another phase of intrusion and thrust faulting. In Nevada and Utah, this deformation is known as the **Sevier Orogeny;** in Canada, the **Columbian Orogeny.** As the east Pacific Plate continued to be thrust under the western, advancing continental edge of the North American Plate, a mélange of oceanic and continental rocks called the Franciscan Formation developed in Cali-

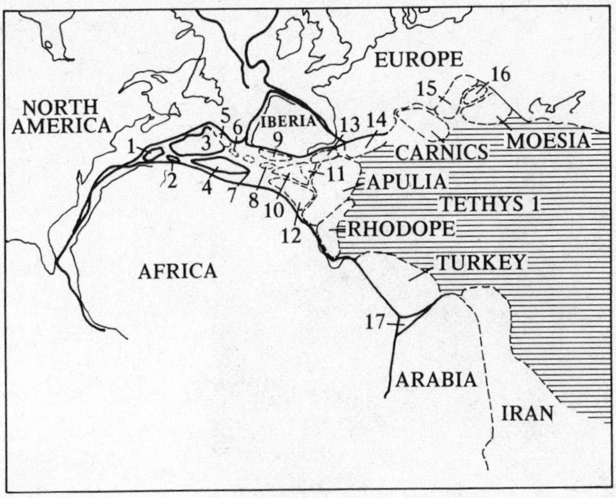

Legend

1. Lanzarote-(Fuerteventura), 2. Haha Block, 3. Moroccan Meseta, 4. Oran Meseta, 5. Rif Block, 6. Grand Kabylie Block, 7. Petit Kabylic Block, 8. Calabrian Block, 9. Betic Block, 10 and 11. Iblean Block, 12. Chott Sahel, 13. Balearic Block, 14. Corsica-Sardinia Block, 15. Tatra Block, 16. Tirgu Mures Block, 17. Sinaii Block.

A

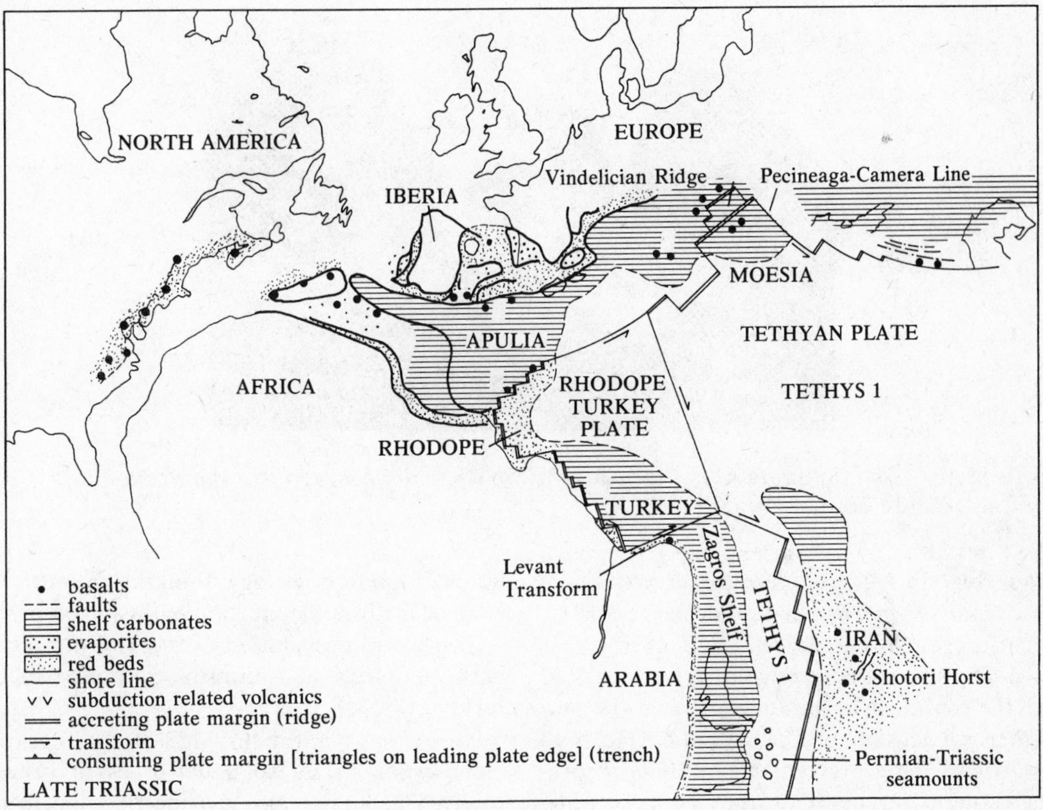

B

Figure 18.9. *Triassic Paleogeography of Tethyean Region.* **A.** *Position of small "microplates" in Middle Triassic.* Present positions given in Fig. 20.32. **B.** *Late Triassic plate boundaries and motions.* (After Dewey et al., *G.S.A. Bulletin*, vol. 84, October 1973.)

A

B

Figure 18.10. *The Dolomite Alps, Italy.* **A.** *The Tre Cime di Lavaredo,* **B.** *The Sassolungo.* (Courtesy Italian State Tourist Office.)

fornia (Fig. 18.14). Compare these events with those discussed in Chapter 8, and Figure 8.17.

In Europe, an epeiric sea (Fig. 18.15) that began forming at the end of the Triassic spread over the continent in the Jurassic from the new northerly extension of the Atlantic Ocean to Tethys except for Hercynian remnants in present Eastern Europe. The study of these widespread, abundantly fossiliferous Jurassic rocks (and the Cretaceous strata of the area to some extent also) by early geologists led to the formation of many fundamental stratigraphic principles, such as faunal succession, facies, and stage and zone terminology and correlation. Thus, the Mesozoic Erathem and the Jurassic System in particular are really the wellsprings of much of modern geologic thinking. Toward the close of the Jurassic, partial regression occurred.

The complex evolution of the Tethyan region between Africa and Eurasia continued in the Jurassic (Fig. 18.16).

The African Craton (Fig. 18.15), the interior of Gondwanaland for so much of earth history, received its first major marine flooding in the Jurassic as the Atlantic and Indian oceans opened. The present north coast, particularly Algeria, and the east coast were flooded. Marine embayments developed in western Australia as the Indian Ocean developed further in the Jurassic. As the rifts opened in Gondwanaland, great amounts of basaltic lava poured from fissures. These Jurassic-Cretaceous basaltic rocks

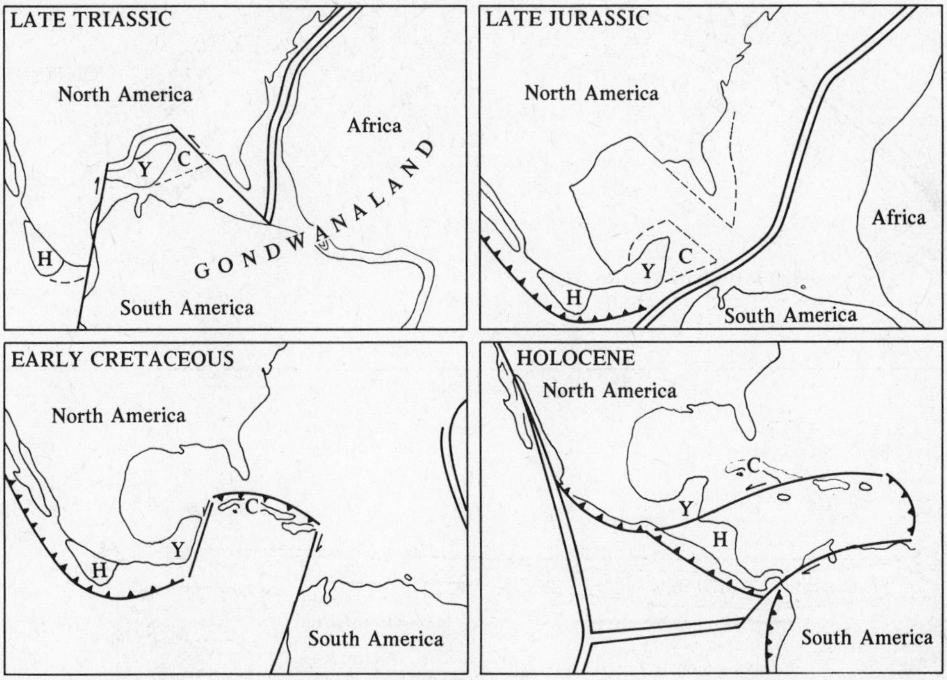

Figure 18.11. *Tectonic History of Gulf of Mexico–Caribbean Area.* Double line, line spreading; toothed line, subduction zone; heavy single line, transform fault; *H*, Honduras; *Y*, Yucatan; *C*, Cuba. Present orientation and shoreline used to facilitate geographic reference. (After Moore and del Castillo, *G.S.A. Bulletin*, vol. 85, no. 4, April 1974.)

accumulated to great thicknesses on all the present southern continents. In some areas such as South Africa, these flat-lying rocks have subsequently been elevated to great heights and deeply eroded, producing spectacular mountains such as the **Drakensberg** (Fig. 18.17). The formation of the Walvis and Chagos hot spots caused these flood basalts, and their locations have left nemataths that allow accurate positioning of the continents by longitude from the Late Jurassic onward. The Gondwana Pole was near the margin of the Antarctic-Australia continent in the Jurassic (Fig. 18.2).

The Jurassic also saw the collision of the small east Siberia plate of Kolymski with present east Asian Eurasia, the **Kimmerian Orogeny,** producing the **Verkhoyansk Mountains.** Batholiths generated by the Kimmerian orogenic welt abound in eastern Siberia (Fig. 18.13). Batholiths also occur in eastern and southeastern Asia, reflecting plate collisions in those regions. India and probably a number of the Chinese plates were moving north across Tethys toward the rest of Eurasia (Fig. 18.2).

Although the Pacific Ocean clearly must have been in existence at least since the Precambrian to take up all the crustal space left by the bunching together of the continents in the Paleozoic, none of its crust of that age is currently known to have survived to the present in the ocean basin. It has been deformed and/or destroyed as its margins have descended into the various trenches along the present-day west coasts of the Americas and Antarctica and the present-day east coasts of Asia and Australia. The oldest rocks yet known from the Pacific Ocean basin are uppermost Jurassic carbonate oozes from the western margin of the present Pacific Plate along the Mariana and Japan trenches. Older rocks may occur beneath these sediments.

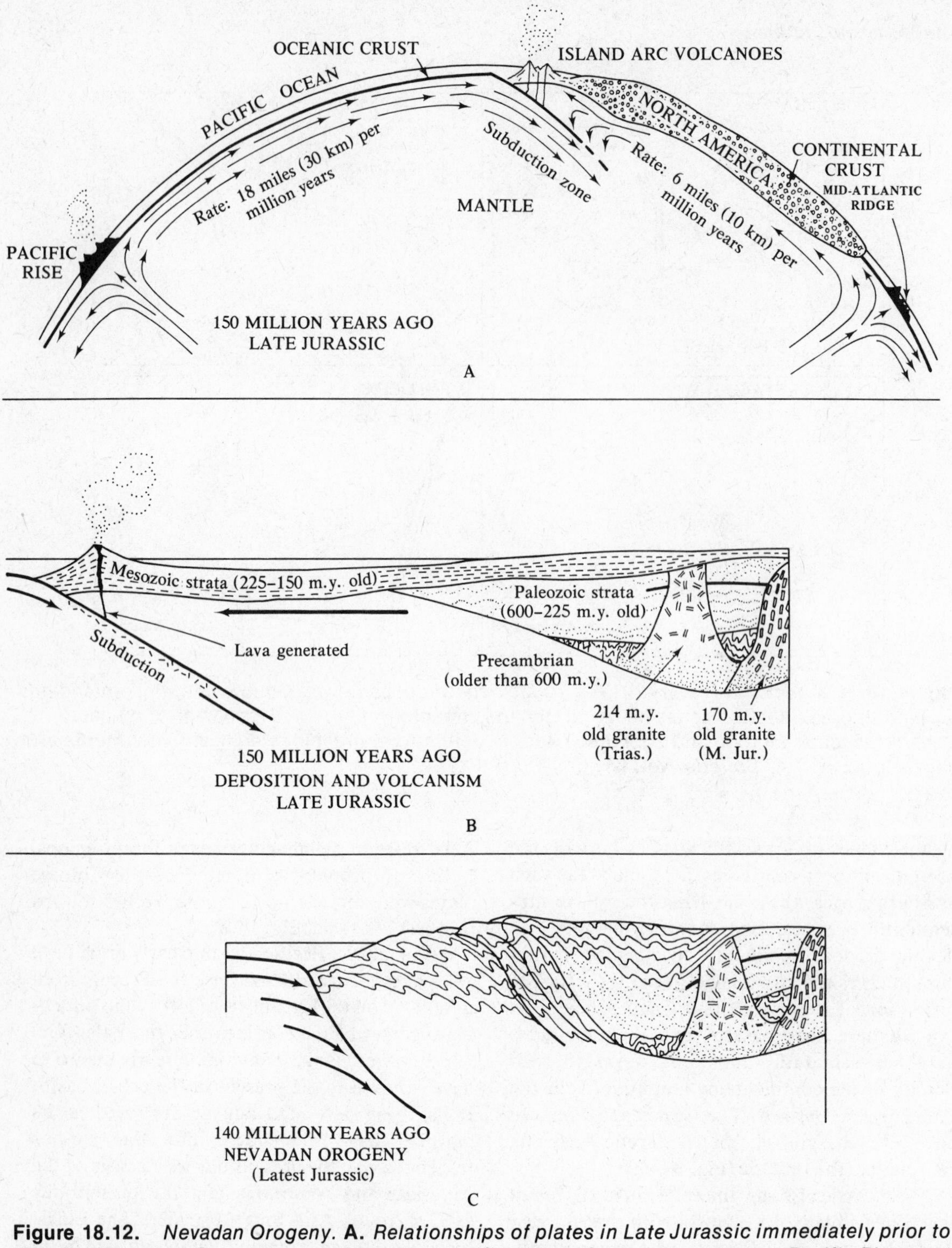

Figure 18.12. *Nevadan Orogeny.* **A.** *Relationships of plates in Late Jurassic immediately prior to Nevadan Orogeny.* Remaining sections are magnifications of area where eastern Pacific Plate and North American Plate met. **B.** *Immediately preceding Nevadan Orogeny in Late Jurassic.* **C.** *Nevadan Orogeny at close of Jurassic.* The trench and island have collided with the continent. **D.** *Generation of Sierran granites as orogenic welt forms in Cretaceous.* **E.** *Present situation of Sierra Nevada in California.* (After Bateman, *California Geology,* Jan. 1974.)

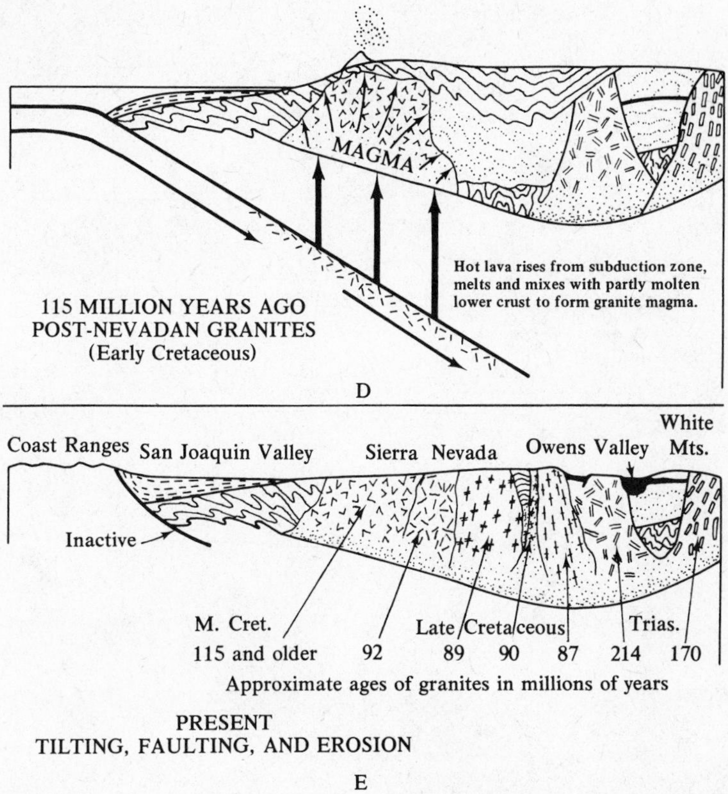

Cretaceous Events

The Cretaceous was a long (60–70 million years) and eventful interval in earth history. It began (Fig. 18.2) with only the incipient Atlantic and Indian oceans and, hence, the continents still largely interconnected, and ended with a time of extreme fragmentation and isolation of land masses (Fig. 18.3).

North America

In North America, which began to separate from the Eurasian part of old Laurasia (Fig. 18.4C), the Cretaceous was a time of great flooding by the Zuni epeiric sea. After a slight withdrawal in the latest Jurassic and Early Cretaceous, the sea began advancing again from both the Cordilleran region and the Gulf of Mexico. The two branches eventually joined in the Late Cretaceous, flooding the entire present Great Plains and Gulf Coastal Plain areas. Marginal flooding continued around the eastern edge of the continent to present Long Island. The part of the sea toward the Gulf of Mexico produced limy deposits including chalk (Fig. 18.18), but clastics became prevalent in the Great Plains area adjacent to the Cordilleran Mountains.

These reflect the beginning stages of the great **Laramide** or **Cordilleran Orogeny** that deformed the entire west coast in the Late Cretaceous and the Paleogene. Collision of an east Pacific plate continued to produce underthrusting of the advancing western margin of North America, mirror image thrust faults in the continental margins from the Canadian Rockies through western Utah and eastern Nevada (Fig. 18.19), extensive igneous intrusion forming a rising orogenic welt along the present west coast (Fig. 18.12D), and buckling of the craton to produce the southern Rocky Mountains stretching from southern Wyoming through New Mexico (Fig. 18.20). The Laramide Orogeny reflects a shift in the motion of North America from northwest to southwest as the North Atlantic basin opened. A great clastic wedge,

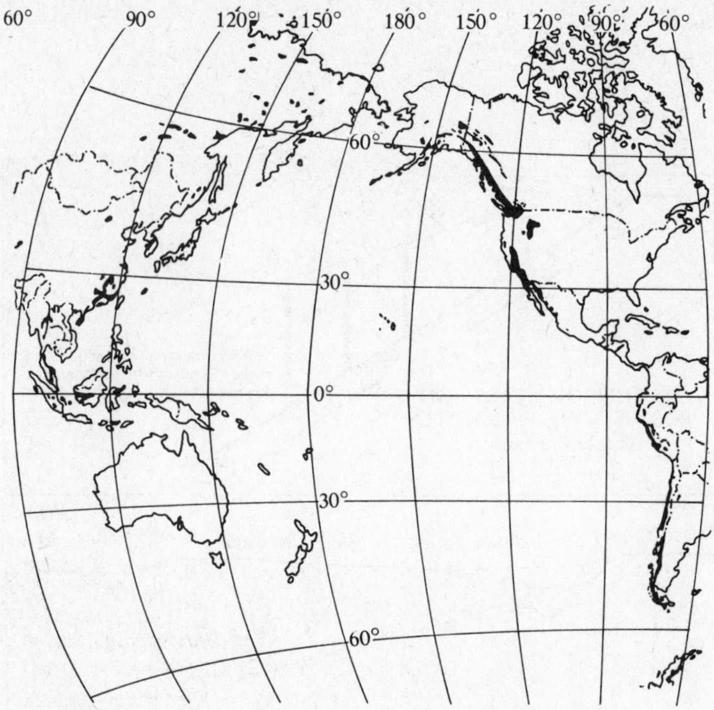

A

B

Figure 18.13. *Mesozoic Intrusions of the Circumpacific Area.* **A.** *Map of distribution.* The North American ones are chiefly Jurassic-Cretaceous age. **B.** *The Sierra Nevada Batholith, Yosemite Valley, Yosemite National Park, California.* (Part A from Bateman and Eaton, *Science*, vol. 158, Dec. 15, 1967, © by AAAS. Used by permission of the authors and AAAS. Part B courtesy of William M. Mintz.)

A

B

C

D

E

Figure 18.14. *Franciscan Rocks, California.* **A.** *Weathered serpentine, Diablo Range* (courtesy of G. Oakeshott); **B.** *Pillow basalts, Point Bonita* (courtesy of C.W. Jennings); **C.** *Chert, Wolfback Ridge* (courtesy of Mary Hill); **D-E.** *Blueschists* (courtesy of Edgar H. Bailey). (From *California's Changing Landscapes* by Gordon Oakeshott, © 1971 by McGraw-Hill, Inc.)

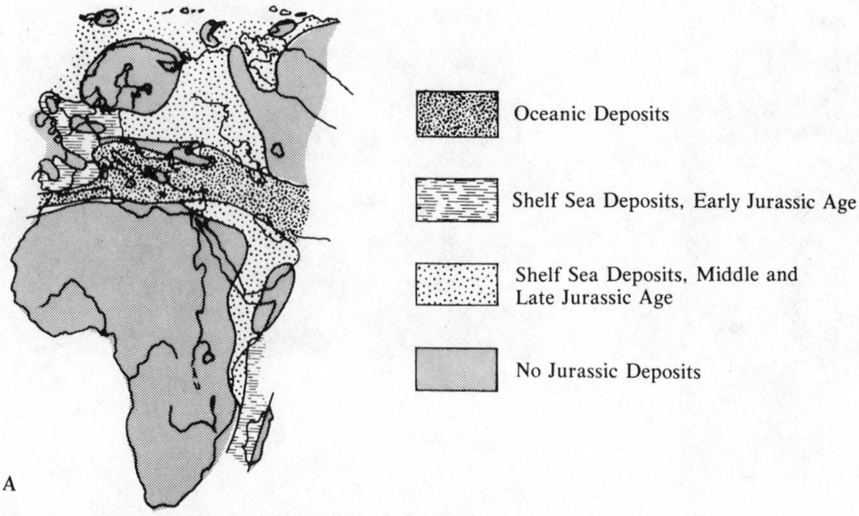

A

B

Figure 18.15. *Jurassic Deposits of Present-Day Europe and Africa.* **A.** *Map showing distribution,* **B.** *Limestone and shale strata of Early Jurassic age near Lyme Regis, England.* (Part A after Kummel. Part B Crown Copyright. Geological Survey photograph. Reproduced by permission of the Controller, Her Britannic Majesty's Stationery Office, London.)

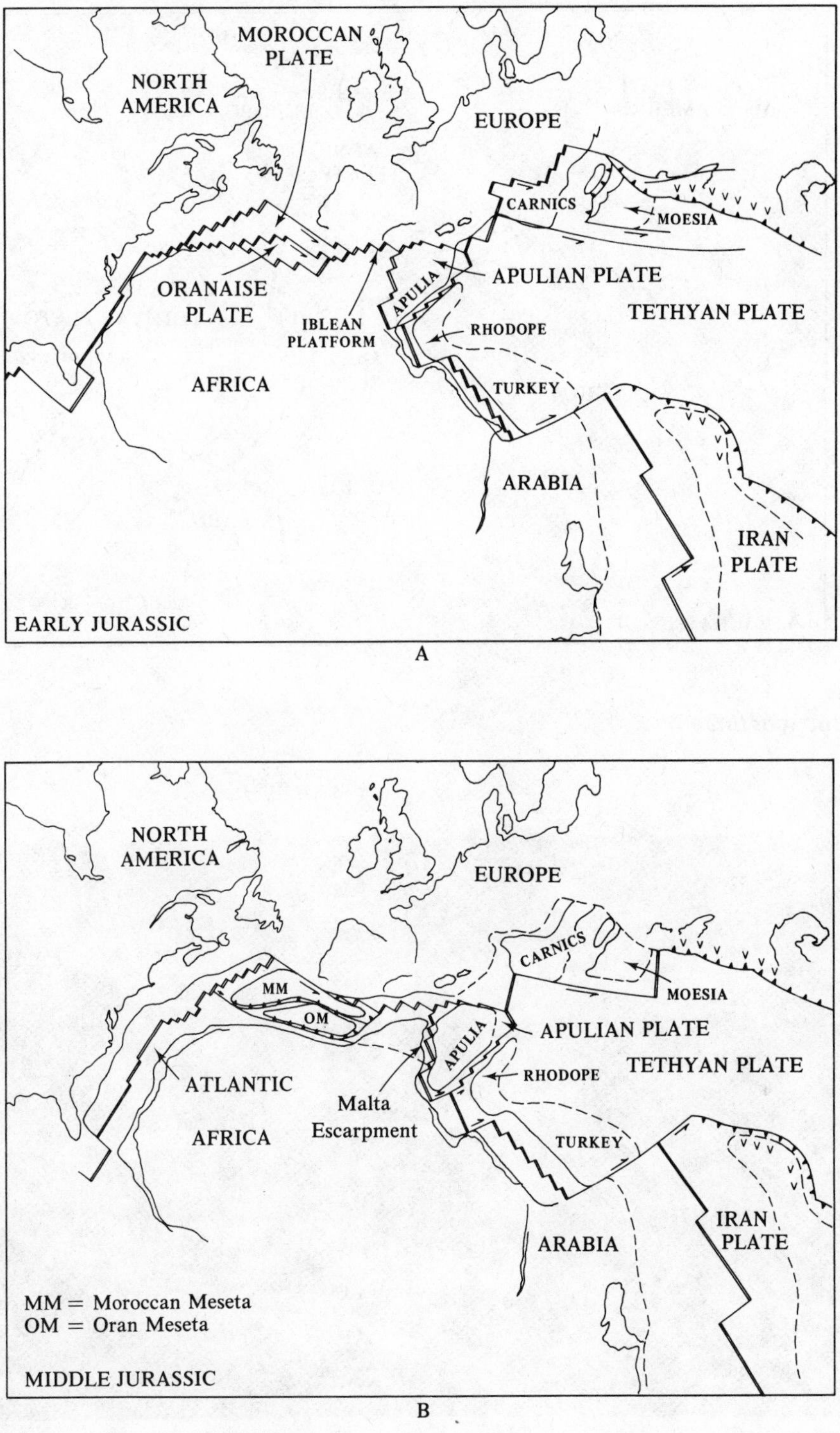

Figure 18.16. *Tethyean Paleogeography in the Jurassic.* **A.** *Early,* **B.** *Middle,* and **C.** *Late Jurassic.* Symbols as in Fig. 18.9. (After Dewey et al., *G.S.A. Bulletin*, vol. 84, October 1973.)

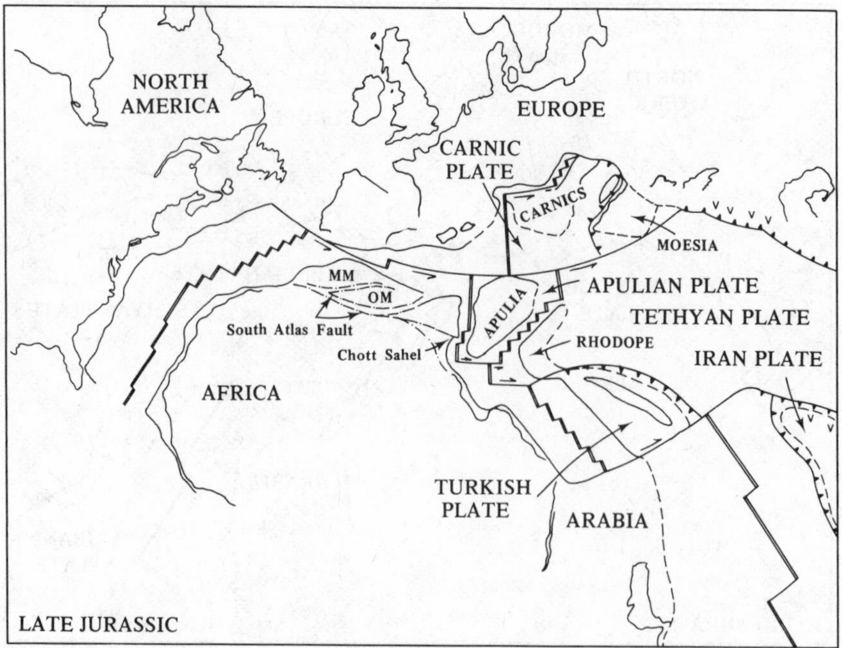

C

Figure 18.16. (*continued*)

Figure 18.17. *Drakensberg Escarpment, South Africa.* (Courtesy South African Tourist Corporation.)

A

B

Figure 18.18. *Cretaceous Chalk Deposits of Kansas.* **A.** *Castle Rock,* **B.** *The Sphinx.* (Courtesy Kansas Dept. of Economic Development.)

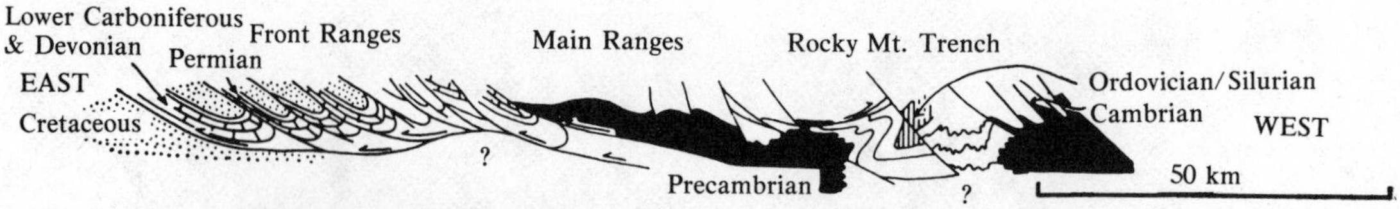

A

B

C

D

Figure 18.19. *The Cordilleran Orogeny.* **A.** *Cross section of deformed belt today;* **B-C.** *Folded and thrust-faulted ranges of Canadian Rockies, Banff National Park;* **D.** *Yosemite Falls, Yosemite National Park, California.* Valley walls are composed of Late Mesozoic granite intrusives. (Part A from Dewey and Bird, *Journal of Geophysical Research*, vol. 75, no. 14, May 10, 1970. By permission of authors and AGU. Parts B-C courtesy of Alberta Dept. of Industry and Tourism. Part D courtesy of William M. Mintz.)

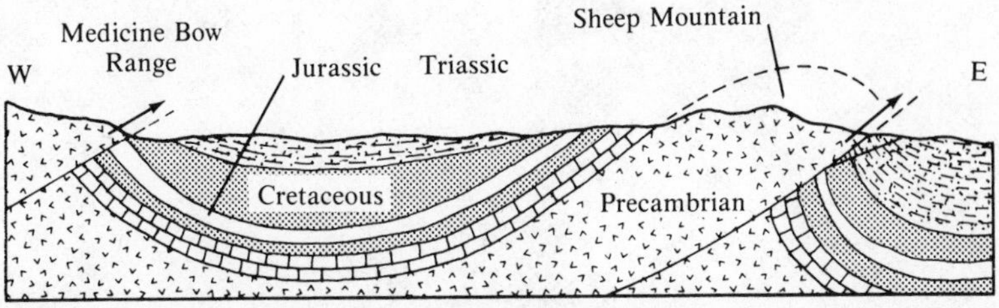

Figure 18.20. *Cross Section of the southern Rockies,* a deformed part of the craton. These open folds occur in southern Wyoming. (After Eardley.)

often called the **Mesa Verde** (Figs. 18.21–22) built eastward, driving out the Zuni Sea by the close of the Cretaceous.

By the end of the Cretaceous, North America was still connected with present Europe via Greenland (Fig. 18.3). North America and South America were separated by the Gulf of Mexico-Caribbean oceanic area, probably with a narrow isthmus or island chain in between (Fig. 18.23). North America was in the middle latitudes in the Cretaceous, but tipped from its present position so that the pole was just to the present west of Alaska and the latitude lines ran diagonally across the continent from what is currently southwest to northeast.

Eurasia

In present Europe, the epeiric seas also returned after an Early Cretaceous regressive phase. Spreading northward from Tethys, they engulfed nearly all of Europe except the Baltic shield by the Late Cretaceous (Fig. 18.24). Limy deposits were widespread, particularly in the area of northern France and southern England where the famous Cretaceous chalk formed (Fig. 18.25). The seas regressed at the close of the Cretaceous. Complex subduction and transform faulting gradually closed Tethys during the Cretaceous (Fig. 18.26).

In present-day Asia, the craton was flooded for the first time since the Early Paleozoic. Seas spread south from the Arctic Ocean along the present Ob River Valley to join the Tethyan Ocean (Fig. 18.27). This seaway is known as the Turgai Strait. Much of the southern margin of Eurasia was the site of a Tethyan trench which continued to receive sediments and volcanics and undergo intermittent disturbances in the Cretaceous as plates containing Africa, India, and China moved northward (Fig. 18.3). The seaway was narrower in the present west (Fig. 18.26) where Europe and Africa had long been

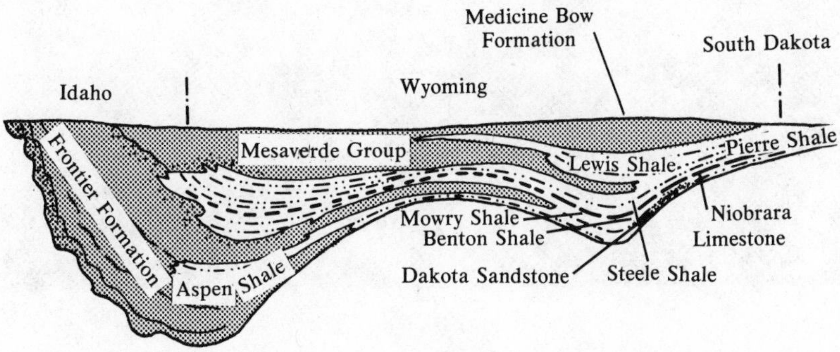

Figure 18.21. *Cross Section of the Mesa Verde Clastic Wedge.* Dashed lines are time lines. Stippled pattern indicates terrestrial clastics; white units formed in the marine environment. (After King.)

A

B

Figure 18.22. *Mesa Verde Group Rocks.* Sandstones form prominent cliffs; shales form slopes. **A.** *Point Lookout, Mesa Verde National Park, Colorado.* **B.** *Book Cliffs, eastern Utah.* (Courtesy of William M. Mintz.)

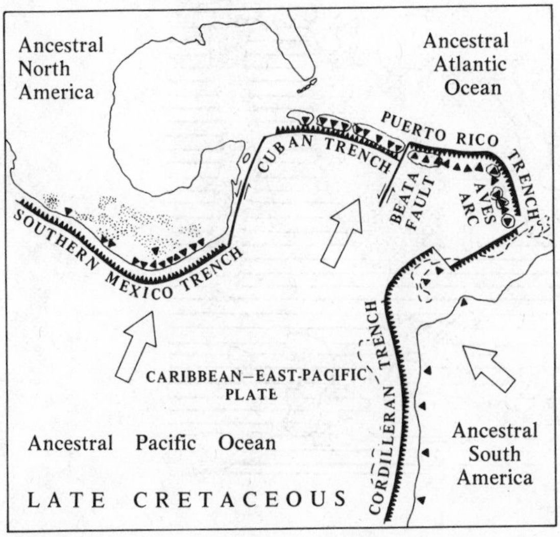

Figure 18.23. *Caribbean Region in Late Cretaceous.* Triangles represent volcanoes. Large arrows indicate direction of plate movement. Sawteeth along trench point in direction of underthrusting. (After Malfait and Dinkelmann, *G.S.A. Bulletin*, vol. 83, February 1972.)

in proximity, but widened to the present east where the continental edges of what are now Arabia and India had not yet encountered Asia. Tethys was in tropical latitudes just north of the equator in the Cretaceous.

Gondwanaland

The fragmentation of Gondwanaland continued (Figs. 18.3, 18.4C). South America continued to move away from Africa toward the present southwest as the South Atlantic basin expanded. Cretaceous marginal seas flooded portions of the eastern edge of South America. In the Late Cretaceous, two impinging east Pacific plates, the Farallon and Phoenix, strongly deformed the Andean region (Fig. 18.28). This belt of deformation continued into the Palmer Peninsula of Antarctica, which was still in close proximity.

Africa, a separate continent at this time also, was extensively submerged in the Cretaceous. A large epeiric sea covered most of the western bulge of Africa from the Tethyan Ocean southward to the Gulf of Guinea and also extended eastward to Arabia (Fig. 18.29). Marginal embayments from the Indian Ocean flooded east Africa. India was an isolated continent and marginally flooded (Fig. 18.27). Western Australia was flooded by an epeiric sea from the Indian Ocean in the Middle Cretaceous (Fig. 18.30). A Pacific plate was colliding with the northern edge of the plate including Australia, for the New Guinea region, a subduction zone, was deformed in the Cretaceous. The tiny continental fragment of New Zealand also was deformed as it separated from Australia by spreading in the Tasman Sea. Australia and Antarctica were still one continent in the Cretaceous, and the Gondwana Pole was situated in present Antarctica (Fig. 18.3).

The Cretaceous Earth

In the Late Cretaceous, we see the reverse conditions of those that closed the Paleozoic Era. The Late Cretaceous was a time of widespread flooding of continental masses and of numerous ocean basins separating the continents and, in turn, separated by them. The major land areas of North America, South America, Europe, Asia, Africa, India, and Australia-Antarctica were all isolated at this time. The oceans were interconnected in the equatorial regions by Tethys and the Central American seaway, a condition which provided a circumglobal tropical seaway. The geography of the Mesozoic, with its fragmentation as opposed to the consolidation of the Paleozoic, certainly must have had significant effects on the life of the era. We shall now turn to the fossil record of the Mesozoic.

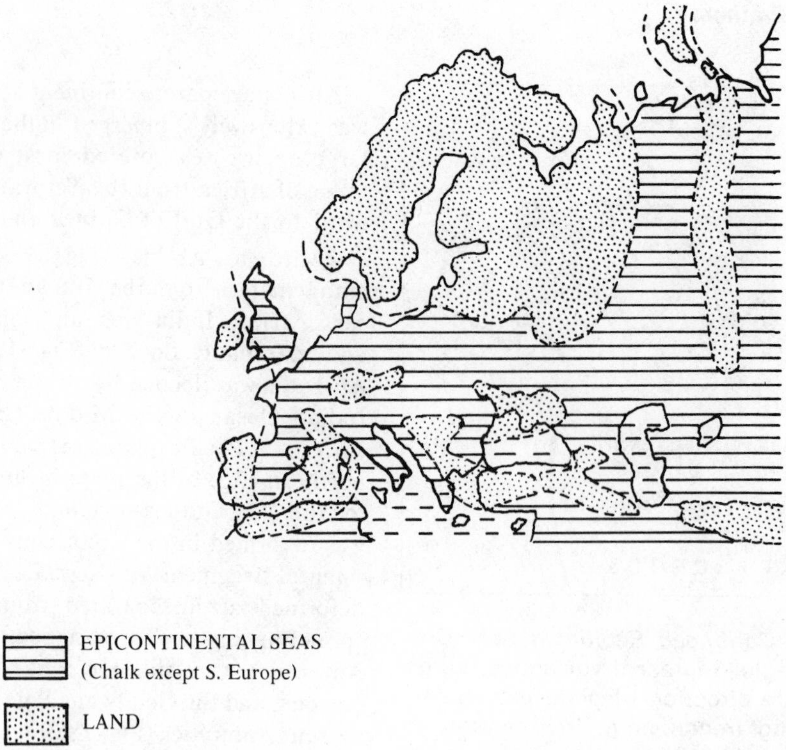

Figure 18.24. *Cretaceous Chalk Sea Deposits of Present-Day Europe.* (Adapted from J.F. Umbgrove, *The Pulse of the Earth*, © 1947 by Martinus Nyhoff, Leiden, The Netherlands.)

Figure 18.25. *The White Cliffs of Dover, England.* (Crown Copyright. Geological Survey photograph. Reproduced by Permission of the Controller, Her Britannic Majesty's Stationery Office, London.)

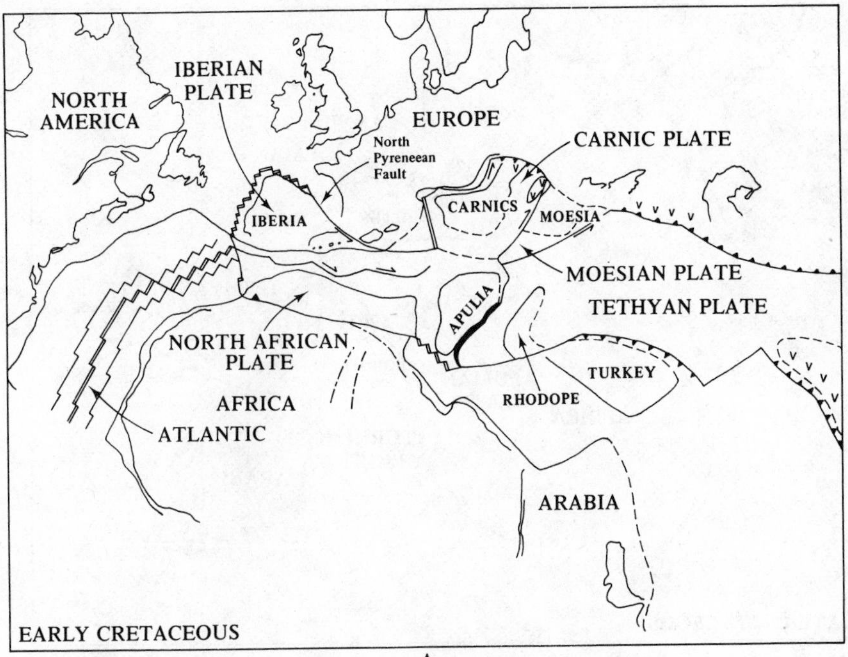

A

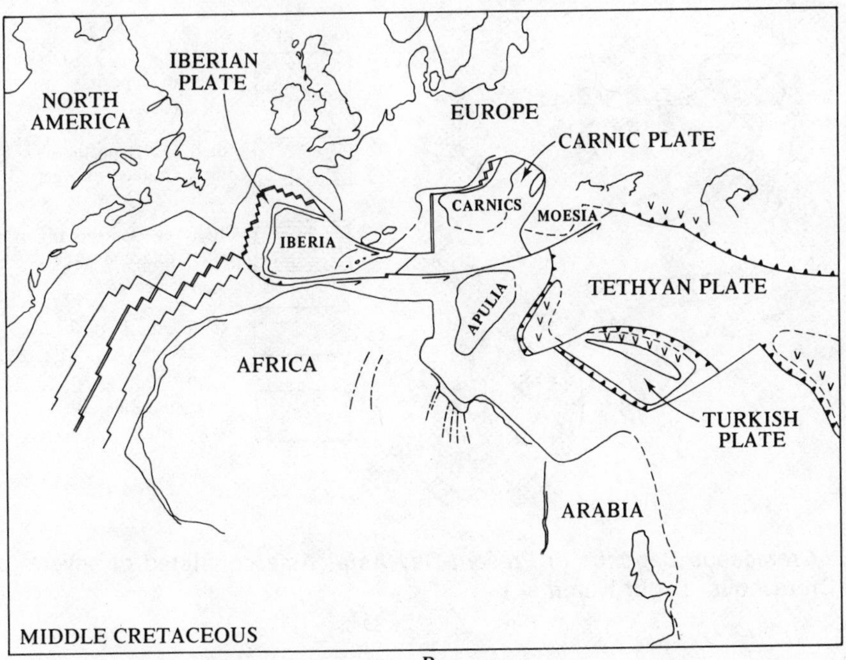

B

Figure 18.26. *Cretaceous Paleogeography of the Tethyean Region.* **A.** *Early,* **B.** *Middle,* and **C.** *Late Cretaceous.* Symbols as in Fig. 18.9. (After Dewey et al., *G.S.A. Bulletin*, vol. 84, October 1973.)

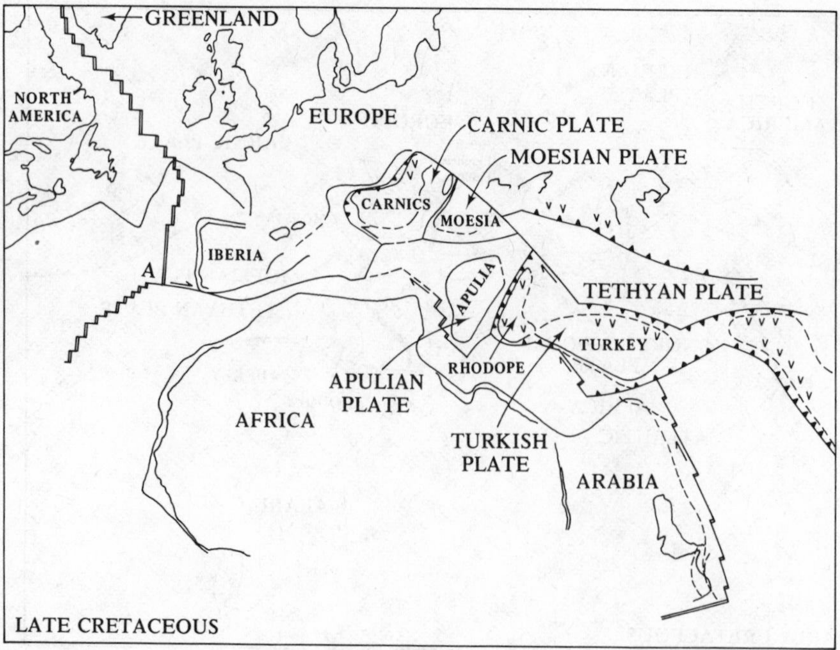

C

Figure 18.26. *(continued)*

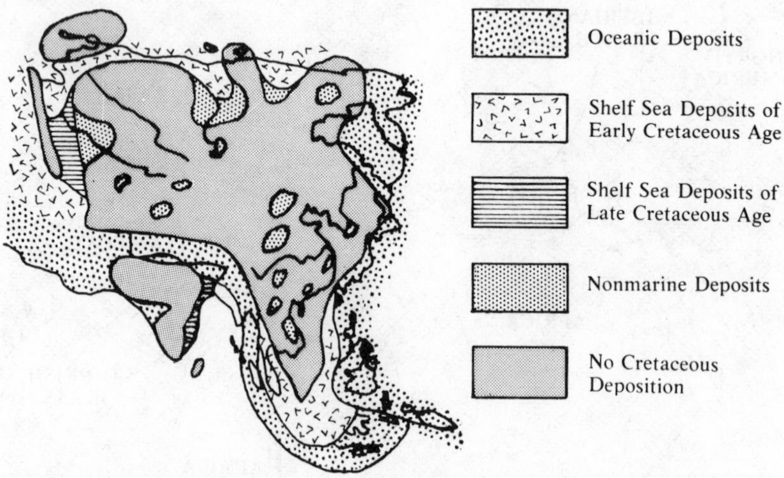

Figure 18.27. *Cretaceous Deposits of Present-Day Asia.* Asia consisted of several continental blocks in the Cretaceous. (After Kummel.)

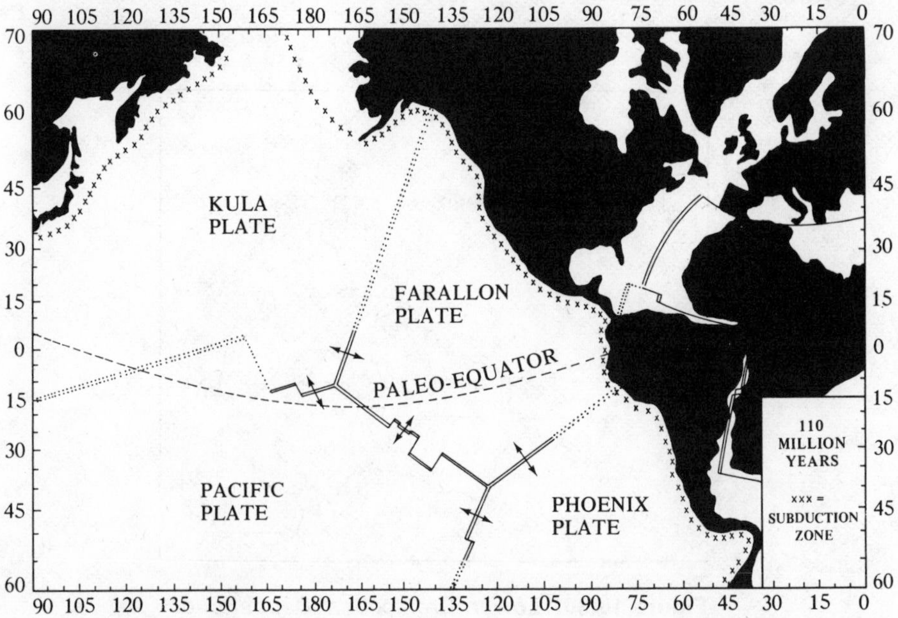

Figure 18.28. *Pacific Ocean in Mid-Cretaceous.* Plate arrangement in the Middle Cretaceous, 110 million years ago. (From Larson and Pitman, *G.S.A. Bulletin*, vol. 83, Dec. 1972.)

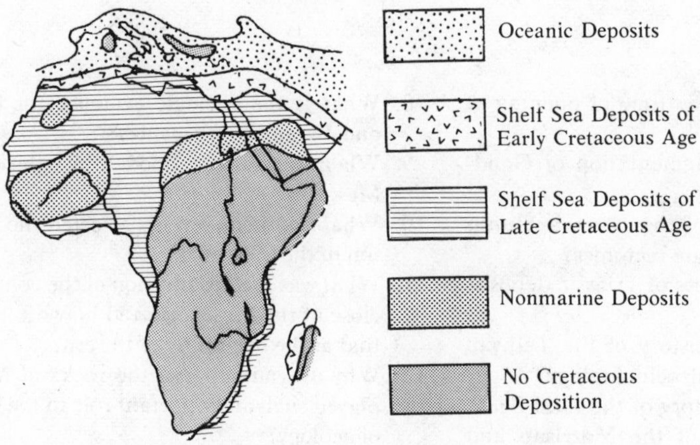

Figure 18.29. *Cretaceous Deposits of Present-Day Africa.* (After Kummel.)

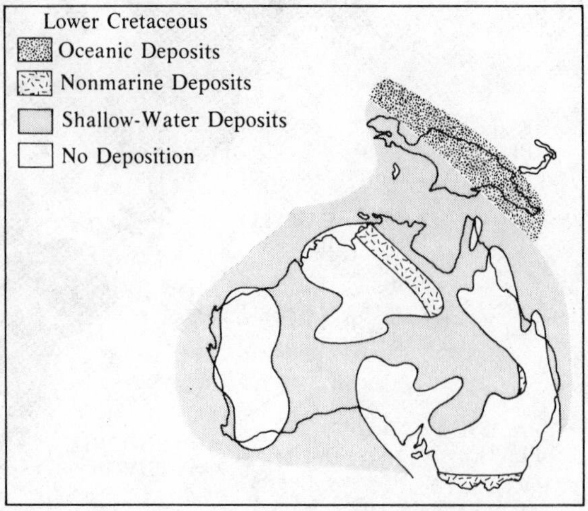

Figure 18.30. *Lower Cretaceous Deposits of Present-Day Australia.* (Adapted from Brown et al., 1968.)

REVIEW

1. What is the evidence for the time of opening of the Atlantic Ocean?
2. How did the Mesozoic fragmentation of Gondwanaland occur?
3. What is the significance of the **Newark Group fault basins** in terms of plate tectonics?
4. What are the dominant types of Triassic deposits on the Laurasian cratons?
5. What was the Mesozoic history of the Tethyan Ocean, and what was its paleoclimatology?
6. What was the geologic history of the **Zuni Sea**?
7. What is the significance of the **Morrison** and **Mesa Verde clastic wedges**?
8. What is the geologic evidence for the **Nevadan** and **Laramide orogenies**?
9. When was Europe flooded by epeiric seas in the Mesozoic?
10. What significant events affected the African Craton in the Mesozoic?
11. What was the distribution of the continents at the close of the Mesozoic, and how did it differ from that at the beginning of the era?
12. Why do you suppose the rocks of Mesozoic age played such an important role in the development of geology?

19

The Age of Reptiles and Ammonoids: Mesozoic Life

19

The drama of Mesozoic life is characterized by the dominance of reptiles among land animals and of mollusks, particularly the cephalopods, among the larger marine invertebrates. The fossil record of this era closes with another apparent "sudden disappearance" or "mass extinction" whose explanation is more difficult to ascertain than the Permian one.

Marine Life

Life in the sea flourished again in the Mesozoic Era (Figs. 19.1–2) after the temporary contraction of epeiric seas induced by the extreme continental conditions of the Middle Permian through Middle Triassic. As the Atlantic and Indian oceans formed and epeiric seas spread from them over the now-lowered continents, marine life returned with a flourish. Its appearance in most areas is quite dramatic, however, because it is so different from that of the preceding Paleozoic. This is not surprising in view of the long missing interval and the fact that a major life habitat, the epeiric sea, had been virtually eliminated in the Permotriassic boundary interval. Many lineages of organisms were terminated by the loss of this habitat; so that when

Figure 19.1. *Diorama of Jurassic Marine Life.* Pelagic belemnoids swim over a bed of oyster pelecypods. (Courtesy Field Museum of Natural History.)

Figure 19.2. *Diorama of Cretaceous Marine Life.* The straight and coiled cephalopods are ammonites. Several small snails and clams occur on the bottom. (Courtesy Field Museum of Natural History.)

the environment returned, it was open to a new adaptive radiation and colonization. In some mountain ranges, where pieces of ocean basins, such as the Tethyan, are preserved, marine sediments are continuous across the boundary zone and the faunal changeover is more gradual. In the Jurassic we get our first good look at life in the ocean basins, for the oldest known oceanic rocks not destroyed or deformed by subduction date from that period.

Protistans

The major marine protistan groups of the Mesozoic were the **dasycladaceous green algae, coccolithophores, diatoms, dinoflagellates, foraminifera, radiolaria,** and **tintinnoids.** Most coccoliths, diatoms, dinoflagellates, radiolaria, and the planktonic forams appear late in the era once the oceanic basin record is preserved without major alteration.

The benthonic dasycladaceous **green algae** (Fig. 19.3) were abundant in the warm, tropical waters of Tethys. They were major contributors to the reefs which today stand as the isolated blocks of the Dolomite Alps after the less resistant intervening facies had been stripped away. The group reached a peak in the Triassic-Jurassic interval.

Coccolithophores (Fig. 19.4) first appeared in the Jurassic and expanded greatly in the Cretaceous. The calcareous platelets of these microscopic pelagic protistans collected in tremendous numbers on the ocean floors, giving rise to the great Cretaceous chalk beds. They are also used in correlating Mesozoic ocean basin sediments. Coccolithophores have continued to be abundant down to the present and are major producers supporting life in modern oceans.

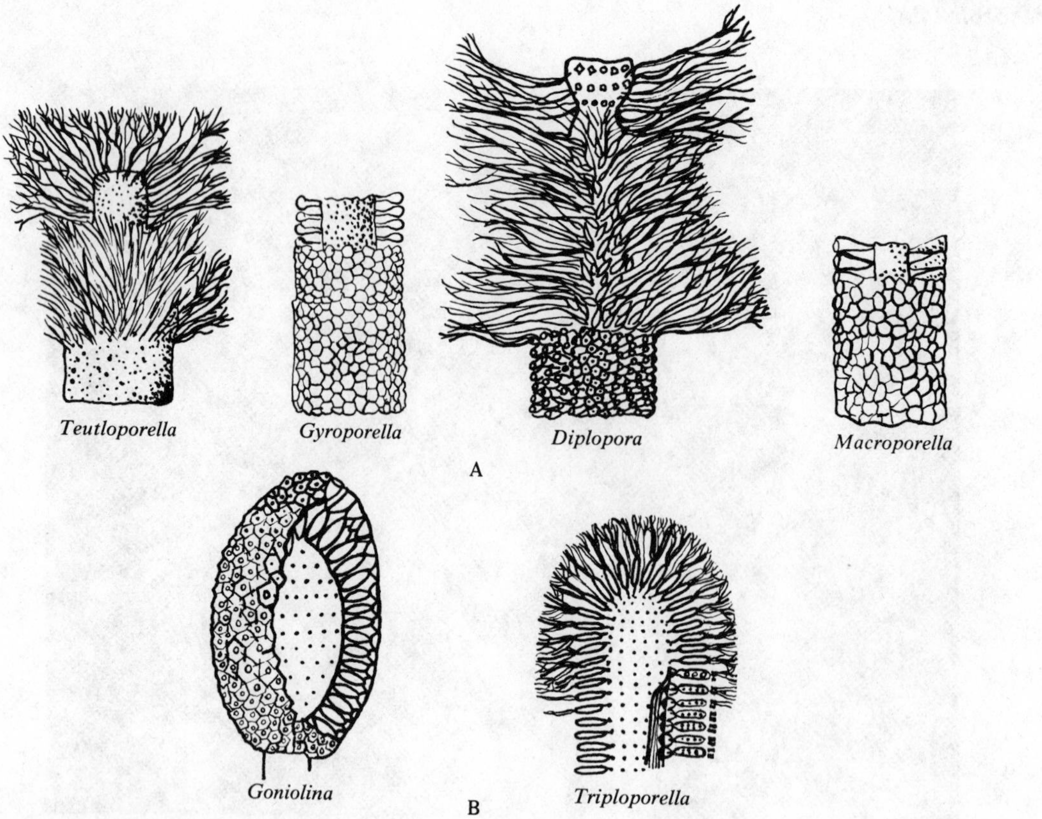

Figure 19.3. *Mesozoic Dasycladaceous Green Algae.* **A.** *Triassic,* **B.** *Jurassic.* Approximately natural size. Side branches restored on some. (After Pia.)

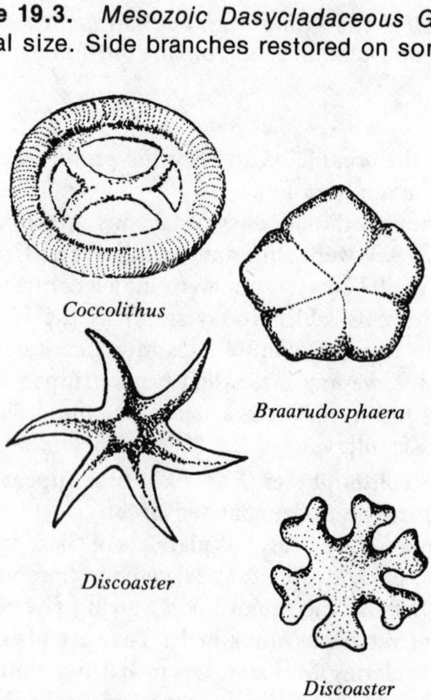

Figure 19.4. *Late Mesozoic Coccoliths.* X 1200. (After Bramlette and Riedel.)

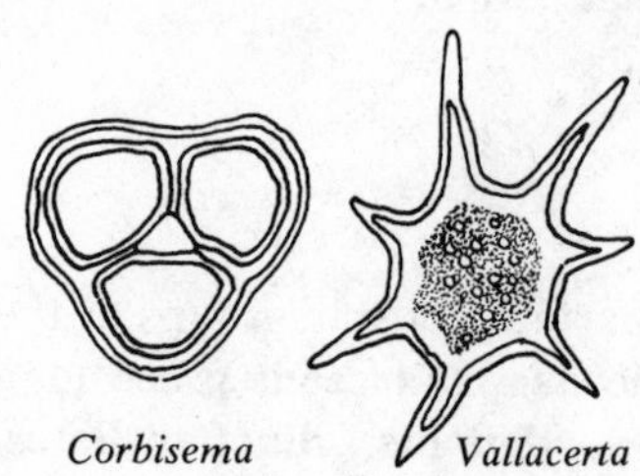

Figure 19.5. *Cretaceous Silicoflagellates.* X 1200. (After Piveteau.)

Their absence before the Jurassic is hard to explain unless they lacked the ability to secrete hard parts before this, or their niche was occupied by some shell-less microorganism such as the acritarchs. If either of the two hypothesis is true, the prodigious amounts of CO_2 coccolithophores extracted from the sea once they began to secrete coccoliths surely must have had repercussions in the earth's ecosystem. If they really did not exist or had no ecologic predecessors before the Late Mesozoic, their appearance and spread, particularly with the widespread Cretaceous seas, may have triggered a significant increase in the earth's oxygen supply at that time. Considering that diatoms, the second greatest group of modern photosynthesizers, do not appear in the fossil record until the Cretaceous, this increase may have been so significant that it somehow may have controlled the extinctions at the end of the era. The **silicoflagellates** (Fig. 19.5), also important producers today and near relatives of the coccolithophores, appeared in the Cretaceous.

The siliceous **diatoms** (Fig. 19.6) are not known until the Cretaceous, but were abundant enough by the end of that period to form siliceous shales such as the Moreno of California. The group is chiefly planktonic, and the Mesozoic forms were all marine.

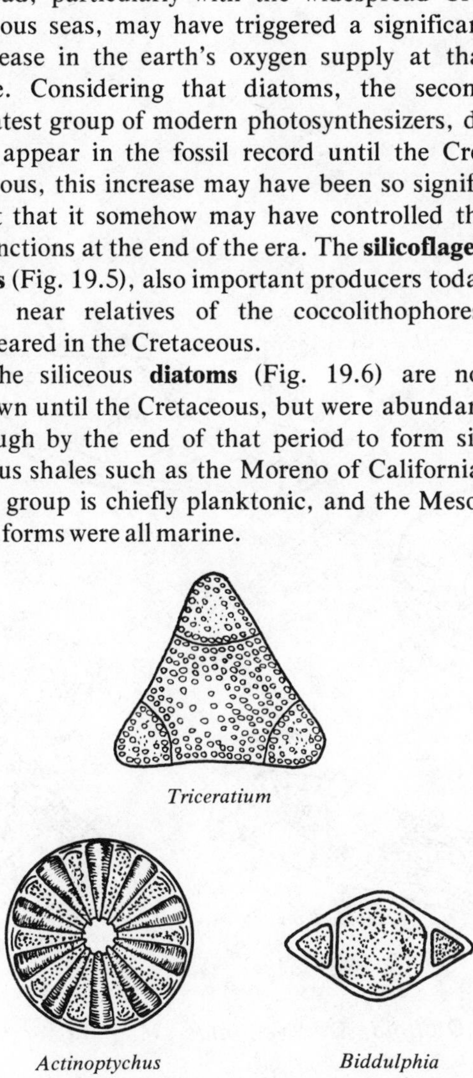

Figure 19.6. *Cretaceous Diatoms.* X 500. (After Long, Fuge, Smith.)

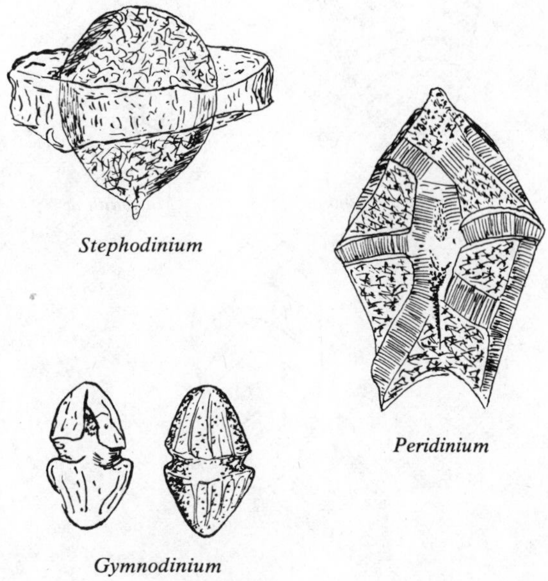

Figure 19.7. *Mesozoic Dinoflagellates.* X 500. (After Piveteau, Termier.)

Dinoflagellate cysts (Fig. 19.7) become common for the first time in the Upper Mesozoic. This pelagic group is known from rare finds back into the Silurian and is the only known group of major producers in the oceans today that appears in the fossil record before the Jurassic.

Foraminifera (Figs. 19.8–10) underwent a great radiation in the Jurassic and Cretaceous that continues down to the present. The small, benthonic types (Fig. 19.8) diversified greatly and gave rise to the pelagic Superfamily Globigerinacea in the Jurassic. This group (Fig. 19.9) multiplied so profusely in the Cretaceous that the tests of individuals were major contributors to marine sediments. These include not only the ocean basin oozes, but also the great shelf chalks that accumulated in the widespread epeiric seas of the Cretaceous. From the Late Cretaceous onward, pelagic forams are important in long-range correlations. The large, disc-shaped orbitoids (Fig. 19.10A) and the large, spindle-shaped alveolinids (Fig. 19.10B) appeared and became common in the shelf seas of the Late Cretaceous. They are probably the ecologic successors of the fusulines which disappeared with the last Paleozoic epeiric sea.

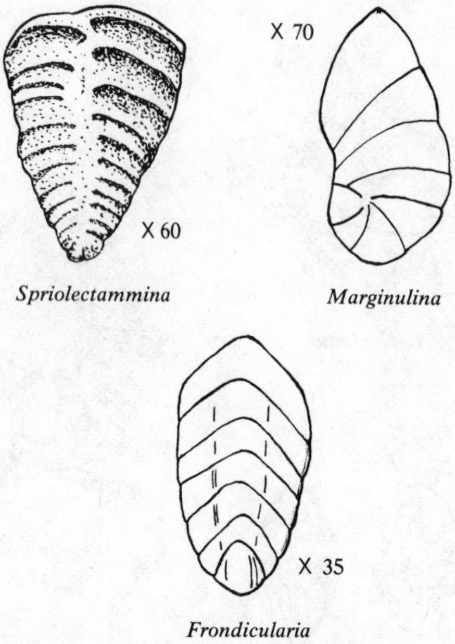

Figure 19.8. *Mesozoic Small Benthonic Foraminifera.* Magnifications indicated. (After Cushman.)

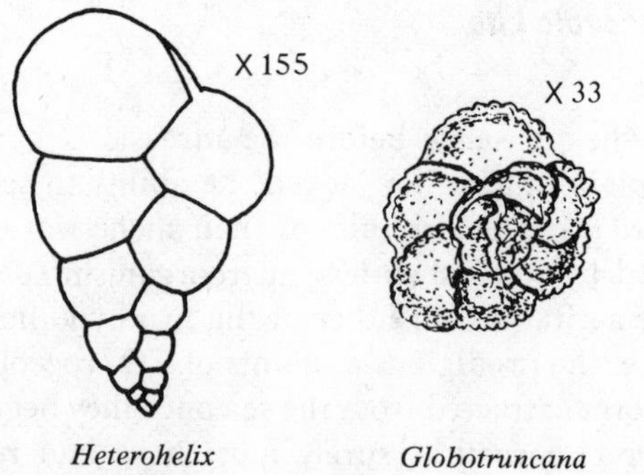

Figure 19.9. *Mesozoic Planktonic Foraminifera.* Magnifications indicated.

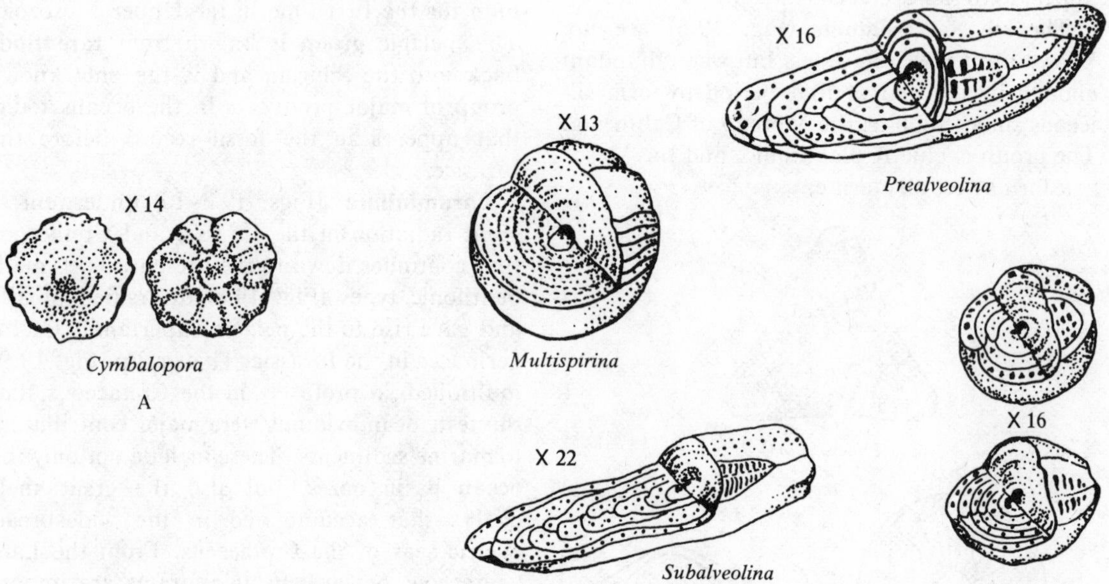

Figure 19.10. *Cretaceous Large Foraminifera.* **A.** *Orbitoid,* **B.** *Alveolinids.* Magnifications indicated. (After Cushman, Sigal.)

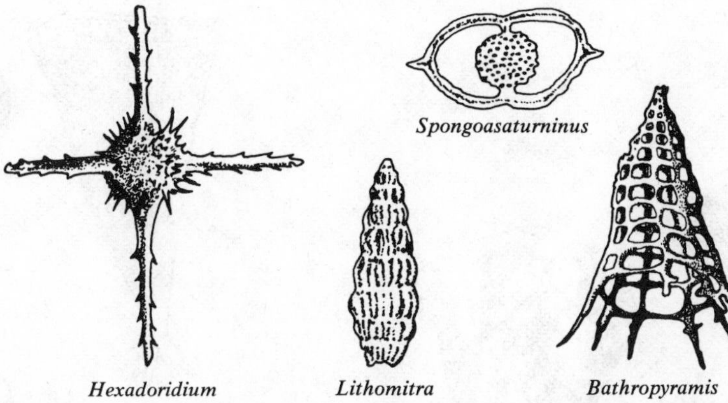

Figure 19.11. *Mesozoic Radiolaria.* X 75. (after Campbell, Clark, and Riedel.)

The **radiolaria** (Fig. 19.11) were abundant for the first time in the Jurassic and Cretaceous. Chiefly open-ocean, pelagic forms, they are not common in older rocks except in areas where oceanic crust has been incorporated in mountain ranges by underthrusting. In the Jurassic and Cretaceous, there are large-scale examples of this occurrence as well as an abundance of former oceanic siliceous oozes where they are used for time correlation.

The pelagic ciliate group of **tintinnoids** (Fig. 19.12) reaches a Jurassic-Cretaceous peak in the fossil record.

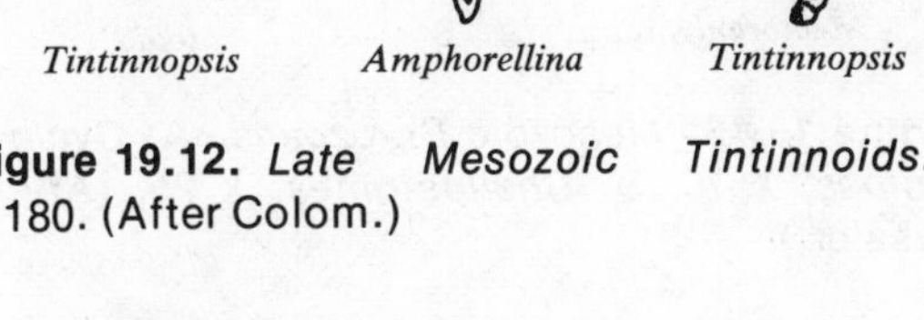

Figure 19.12. *Late Mesozoic Tintinnoids.* X 180. (After Colom.)

Invertebrate Animals

The major marine invertebrate groups of the Mesozoic were the **sponges, hexacorals, bryozoans, brachiopods, gastropods, pelecypods, ammonoids, belemnoids, ostracods,** and **echinoids.** Almost all Mesozoic time correlations are made with the abundant, nektonic, and rapidly evolving ammonoids, the single most important group of Mesozoic fossils to geologists.

The **sponges** (Fig. 19.13) were common in the tropical seas of Tethys in the Jurassic and particularly the Cretaceous. Both calcareous and siliceous types occurred, with the latter predominating.

Hexacorals (Fig. 19.14) first appeared in the Middle Triassic from unknown ancestors in tropical Tethys. By the Late Triassic, they extended worldwide and were building reefs. From the Middle Jurassic onward, hexacorals are important faunal elements, particularly in reefs, with their greatest abundance and diversity in Tethys, that great east-west seaway near the equator. The Atlantic was not very wide at that time, and the fauna extended to southern North America.

Bryozoans, like most shallow-water marine invertebrates, were rare in the Triassic, so we lack the necessary evolutionary links between the last Paleozoic types and the new Mesozoic lineages. The surviving cyclostomes (Fig. 19.15A) underwent a Jurassic and Cretaceous expansion and, until the Late Cretaceous, dominated bryozoan faunas. The cheilostomes

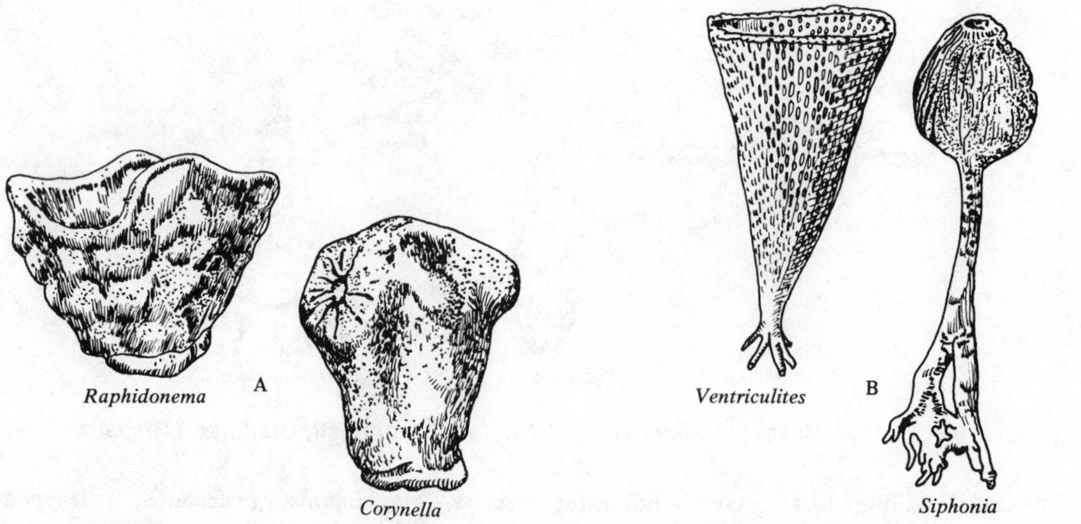

Figure 19.13. *Late Mesozoic Sponges.* **A.** *Calcareous forms, X 1;* **B.** *Siliceous forms, X 1/2.* (After Okulitch.)

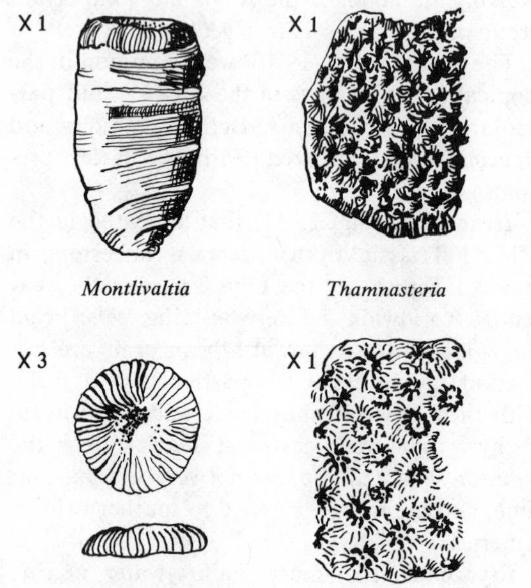

Figure 19.14. *Mesozoic Hexacorals.* Magnifications indicated. Two solitary and two colonial genera shown.

(Fig. 19.15B) appeared in the Late Jurassic and by the middle of the Cretaceous had become the major group of bryozoans, a position they maintain today.

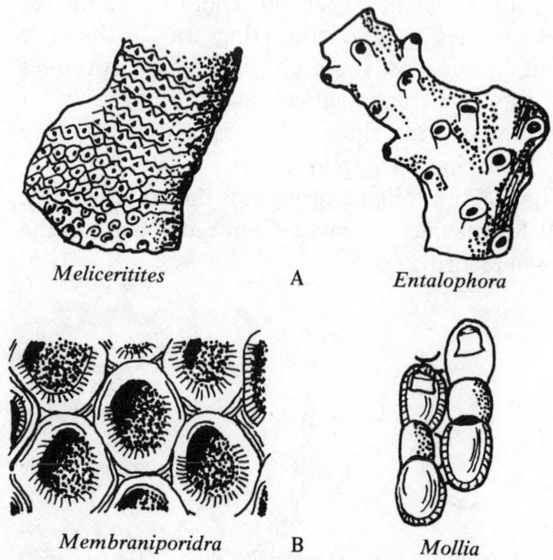

Figure 19.15. *Mesozoic Bryozoans.* **A.** *Cyclostomes, X 10;* **B.** *Cheilostomes, X 25.* (After Bassler.)

Brachiopods had dwindled considerably since the Paleozoic, and most of the dominant orders of that era were either gone or nearly so. The **terebratuloids** and **rhynchonelloids** (Fig. 19.16), the two living articulate orders, continued to diversify, however. Both groups reached a Jurassic peak and declined thereafter.

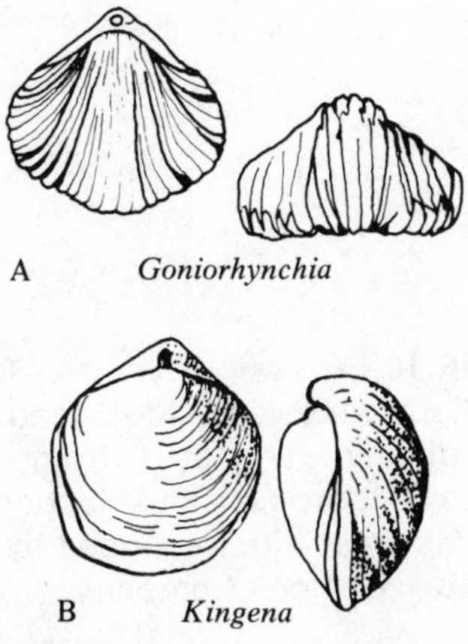

Figure 19.16. *Mesozoic Articulate Brachiopods.* **A.** *Rhynchonelloid,* **B.** *Terebratuloid.* X1. (After Buckman.)

The **mollusks** were the major phylum of large invertebrates in the Mesozoic. **Gastropods** (Fig. 19.17) increased steadily throughout the era, being most abundant in the Cretaceous, the period when the modern carnivorous types appeared. Especially noteworthy is a group of huge, high-spired forms with internal plications called **nerineids** found only in Jurassic and Cretaceous strata, chiefly the latter. The *Turritella* group of high-spired snails that was abundant in the Paleogene became common in the Cretaceous along with many other snails.

The **pelecypods** were even more abundant in the Mesozoic, though they too increased in importance as the era went on. In the Triassic, a group of scallops, or pectens (Fig. 19.18), is widespread and used for correlation within regions. They may have attached themselves to floating organisms such as seaweed or logs to achieve their wide distribution. In the Jurassic and Cretaceous, many clams, including the scallops and oysters, were extremely abundant (Fig. 19.19). Most of these genera were greatly reduced or disappeared by the end of the Cretaceous, but the clam group as a whole, particularly the more advanced burrowing clams, continued to expand down to the present. A very significant group of reef-dwelling clams, the

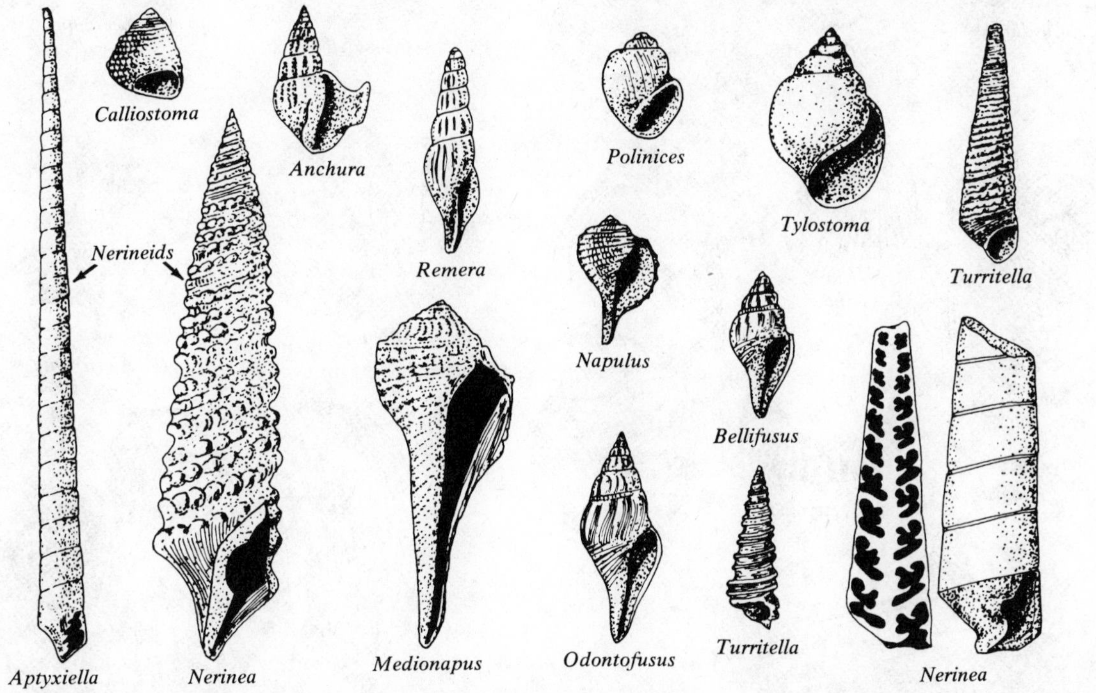

Figure 19.17. *Mesozoic Gastropods.* X 1/3. (Adapted from *Invertebrate Fossils* by Moore-Lalicker-Fischer, Copyright 1952 by McGraw-Hill, Inc., used by permission of McGraw-Hill Book Company.)

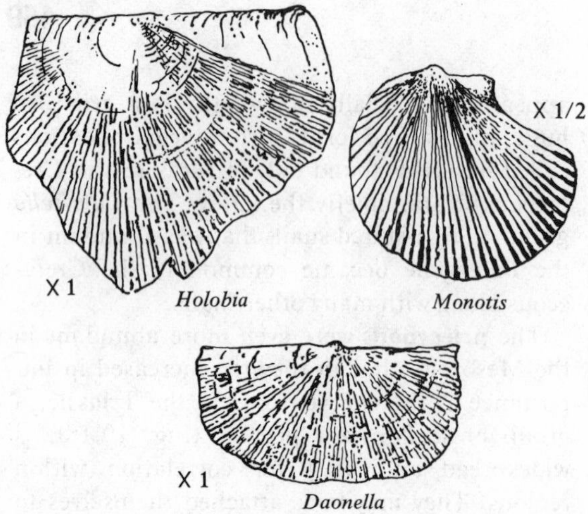

Figure 19.18. *Triassic Bivalves.* These primitive scallops were very widespread. Magnifications indicated. (Adapted from *Invertebrate Fossils* by Moore-Lalicker-Fischer, Copyright 1952 by McGraw-Hill, Inc., used by permission of McGraw-Hill Book Company.

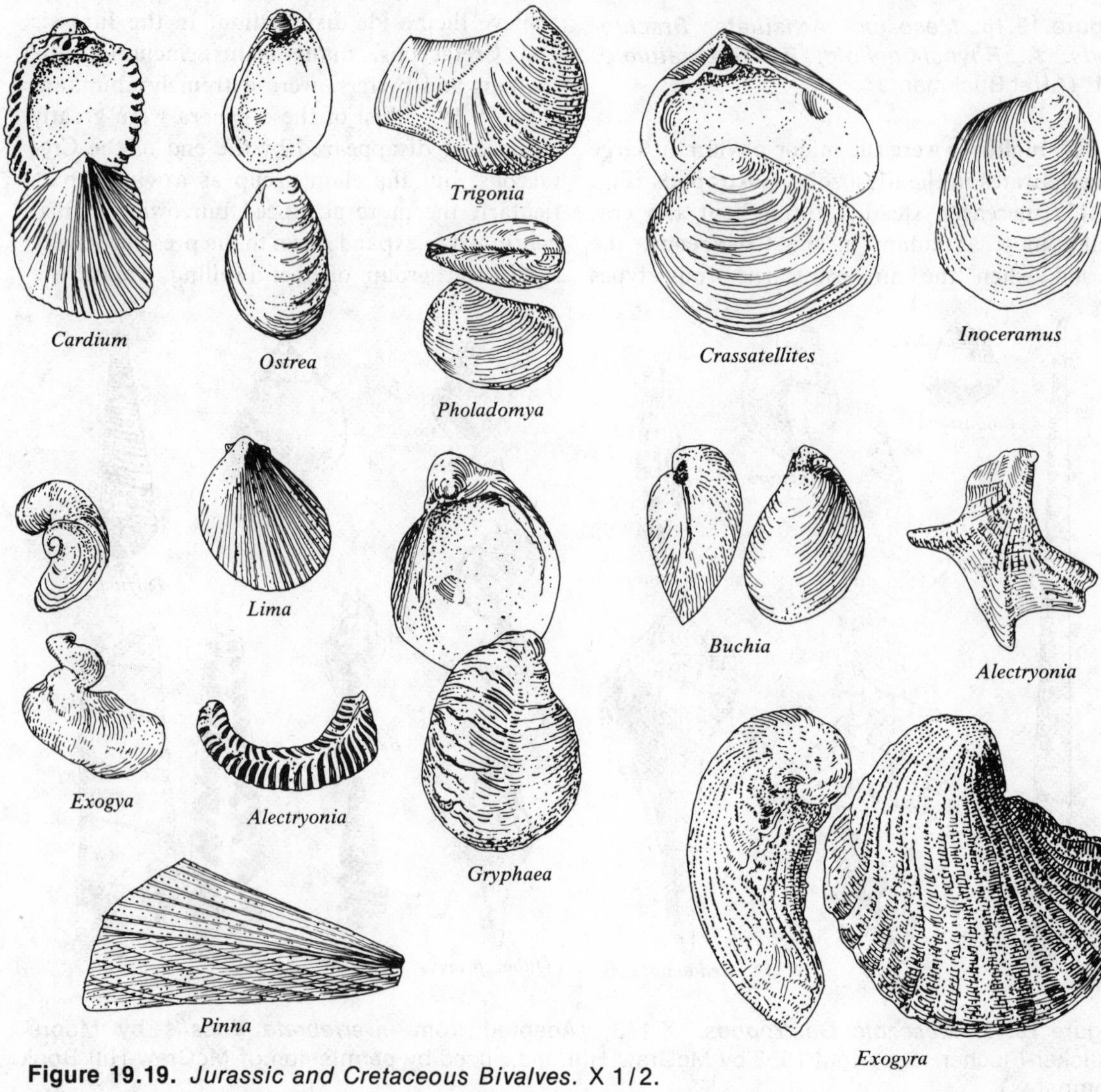

Figure 19.19. *Jurassic and Cretaceous Bivalves.* X 1/2.

rudists or **hippuritoids** is restricted to Upper Jurassic and Cretaceous rocks. These large bivalves converged on corals by developing a conical right or lower valve and a lid-like left or upper valve (Fig. 19.20). Gregarious tropical forms, they lived in the same environment as did reef corals which accounts for the similar shape. None survived the Cretaceous and there is no apparent reason for their extinction.

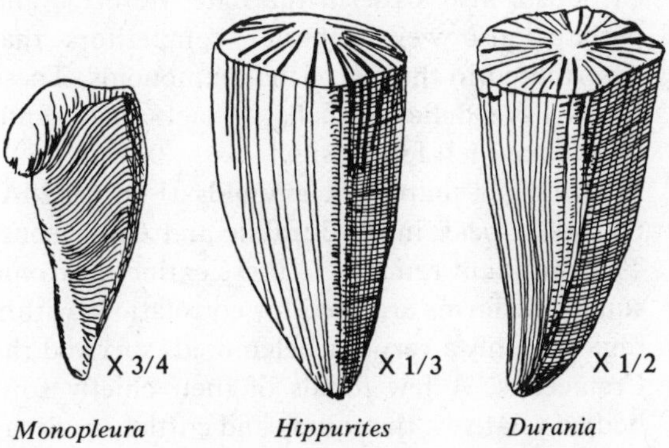

Figure 19.20. *Rudist Bivalves.* Magnifications indicated.

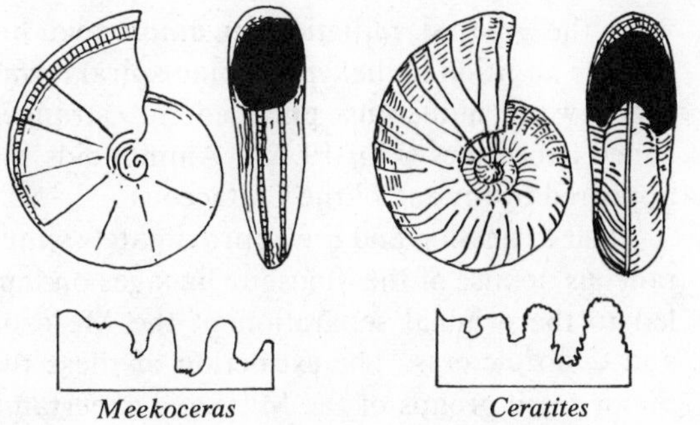

Figure 19.21. *Triassic Ceratite Ammonoids.* Specimens are 2/3 natural size. Sutures shown below specimens.

The **cephalopods** are the most geologically important group of mollusks because their chiefly nektonic mode of life ensures the wide geographic distribution so vital for long-distance time correlations. Far and away the most abundant Mesozoic cephalopods were the forms with complex sutures, the **ammonoids.** In the Triassic, **ceratites** (Fig. 19.21) dominated, but they did not survive the period. In the Jurassic and Cretaceous, the **ammonites** (Fig. 19.22) under-

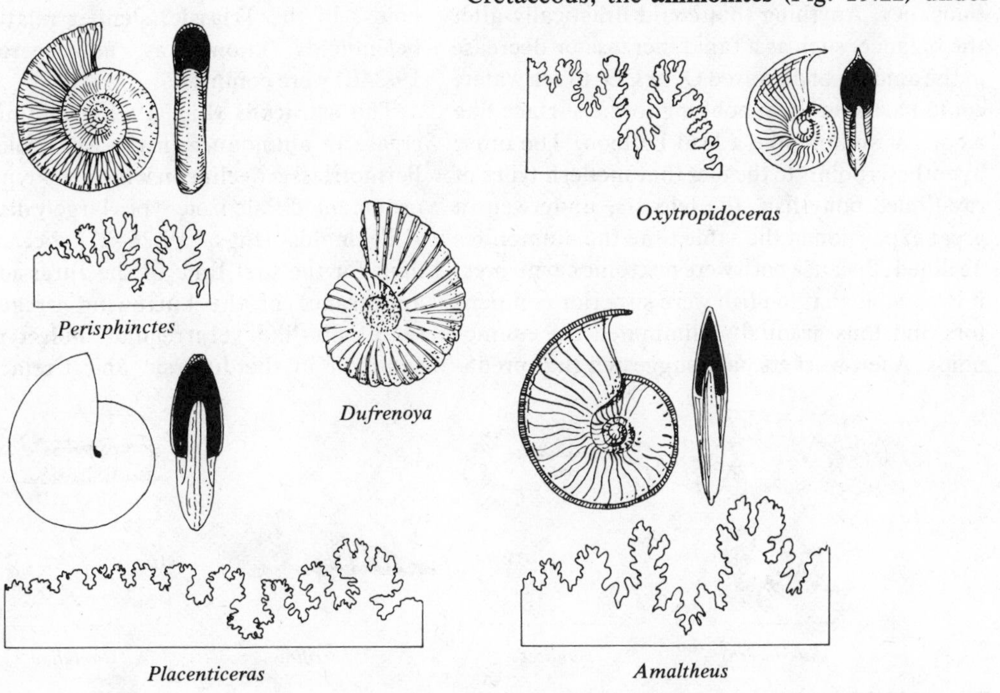

Figure 19.22. *Jurassic-Cretaceous Ammonitic Ammonoids.* Specimens are approximately 1/3 natural size. Sutures shown below specimens except *Dufrenoya.*

went the greatest radiation in ammonoid history. In addition to the typical planispiral forms, there were numerous uncoiled or strangely coiled lytocerines (Fig. 19.23). Ammonoids disappeared by the end of the Cretaceous.

Their extinction and the approximately simultaneous demise of the dinosaur lineages on land led to the original separation of the Mesozoic and Cenozoic eras. The extinction of these two major fossil groups of the Mesozoic is certainly real and puzzling. Although both groups declined toward the close of the Mesozoic, some members survived in fair abundance until the very end. Late Cretaceous seaways were widespread and ammonites were largely pelagic anyway, so we cannot invoke restriction of the seas as a cause for their extinction.

Two hypotheses for their extinction appear reasonable. One is the possible change in the proportions of gaseous constituents of the earth's atmosphere and oceans mentioned earlier. The pelagic ammonoids, like the living nautilus, depended on a delicate adjustment of gases and fluids in their chambers for the proper buoyancy. Anything that could drastically alter the balance, such as a rapid increase or decrease in the amount of dissolved O_2 or CO_2 in sea water, could have sent them bobbing to the surface like a cork or sinking like a lead balloon. The other hypothesis points to the fact that modern types of ray-finned bony fish, the teleosts, underwent a great expansion at the same time the ammonites declined. Because both were nektonic carnivores, it is possible that the fish were superior competitors and thus gradually eliminated the ammonoids. A few workers have suggested that predation by fish eliminated the ammonoids. Certain fish did eat ammonoids, but it seems unlikely that any successful group could consume another group to extinction as studies of the checks and balances of modern ecosystems show. Besides, the fish continued expanding into the Cenozoic, a feat they hardly could have accomplished if they had eliminated their own food supply. Modern cephalopods, such as the squid, octopus, and cuttlefish, also arose in the Late Mesozoic and perhaps also were superior competitors that contributed to the fall of the ammonoids. These forms have delicate shells or lack shells and, hence, have left few fossils.

The cigar-shaped **belemnoids** (Fig. 19.24A) reached a peak in the Jurassic and Cretaceous. The abundant remains of these extinct nektonic squid-like forms are used for correlations within regions. Only a very few belemnoids survived the Cretaceous. A few fossils of their chiefly soft-bodied relatives, the squids and cuttlefish, occur in the Upper Mesozoic, indicating that these lighter, perhaps more agile, forms were superior competitors and eliminated the belemnoids. In the Triassic, slender relatives of the belemnoids known as **aulacoceroids** (Fig. 19.24B) were common.

The **ostracods** (Fig. 19.25) continued to increase in abundance in the Mesozoic after the Permotriassic decline in which the typical highly ornamented Paleozoic types largely disappeared.

Echinoids (Figs. 19.26–27) became abundant for the first time in the Jurassic, with the appearance of the burrowing irregulars. The sea-urchin-like, yet irregular, **holectypoids** were common in the Jurassic and Cretaceous, but

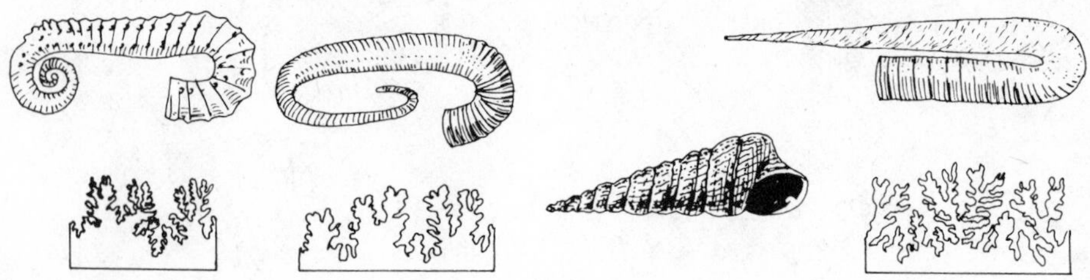

Figure 19.23. *Lytocerine Ammonoids.* Specimens are 1/4 natural size. Sutures shown below specimens. See also *Baculites* in Fig. 19.2.

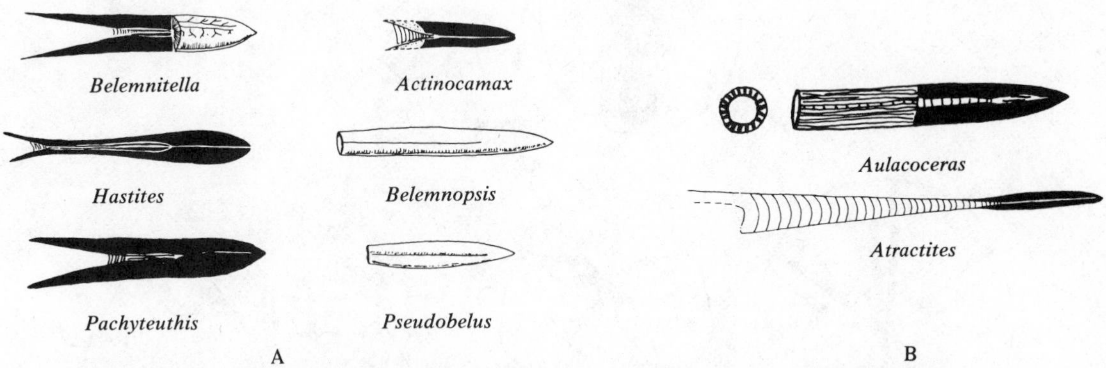

Figure 19.24. *Mesozoic Coleoid Cephalopods.* **A.** *Belemnoids,* **B.** *Aulacoceroids.* Black areas are sections. X 1/3. (Adapted from *Invertebrate Fossils* by Moore-Lalicker-Fischer. Copyright 1952 by McGraw-Hill, Inc., used by permission of McGraw-Hill Book Company.)

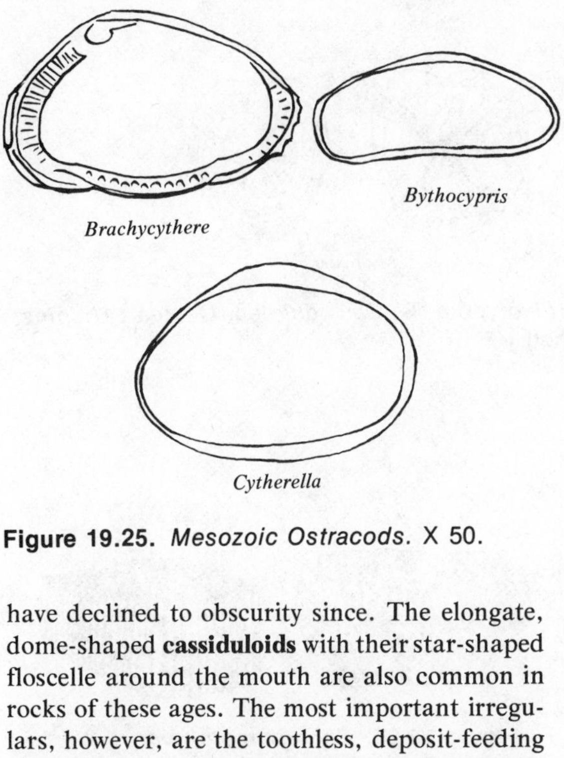

Figure 19.25. *Mesozoic Ostracods.* X 50.

have declined to obscurity since. The elongate, dome-shaped **cassiduloids** with their star-shaped floscelle around the mouth are also common in rocks of these ages. The most important irregulars, however, are the toothless, deposit-feeding **heart urchins.** Several major types evolved in the Jurassic, and the modern orders became dominant in the Cretaceous. Some are particularly well-represented in chalk facies. The regular, sea-urchin types of echinoids were occasionally common in the Late Mesozoic.

Other marine invertebrate groups are common for shorter intervals in the Mesozoic Era. The **spongiomorphoids** (Fig. 19.28), an extinct group related to the stromatoporoids and thus probably hydrozoan coelenterates or sclerosponges, were common reef builders in the Traissic of Tethys. **Malacostracan crustacean** fragments are abundant, and complete specimens are occasionally encountered in Jurassic and Cretaceous rocks (Fig. 19.29). **Annelid** tubes (Fig. 19.30) are particularly common in the Cretaceous. The star-shaped columnals of some crinoids (Fig. 19.31) are occasionally common in Triassic and Jurassic strata, and the modern floating-swimming **crinoids** (Fig. 19.31) underwent a short, but widespread evolutionary burst in the Late Cretaceous seas that makes them useful for long-distance time correlation of a thin stratigraphic interval.

Marine Biogeography

Marine biogeography of the Mesozoic is characterized by a broad tropical east-west Tethyan faunal realm between the Laurasian and Gondwanaland continents, a north temperate faunal realm of the somewhat cooler epeiric seas of northern Laurasia, and a south temperate faunal realm of the southern Gondwana continents. In the Cretaceous, numerous provinces developed as the continents dispersed. The Tethyan fauna showed occasional provincialism along its east-

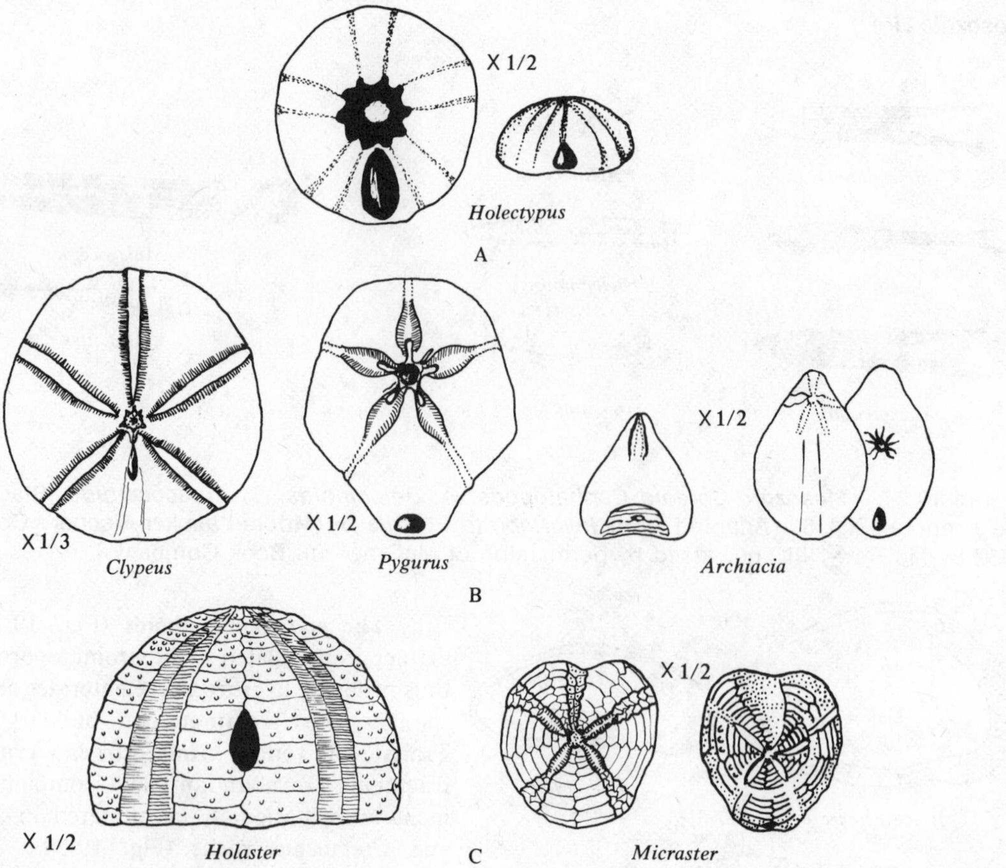

Figure 19.26. *Mesozoic Irregular Echinoids.* **A.** *Holectypoid,* **B.** *Cassiduloids,* **C.** *Heart Urchins.* Magnifications indicated. (After Clark and Twitchell.)

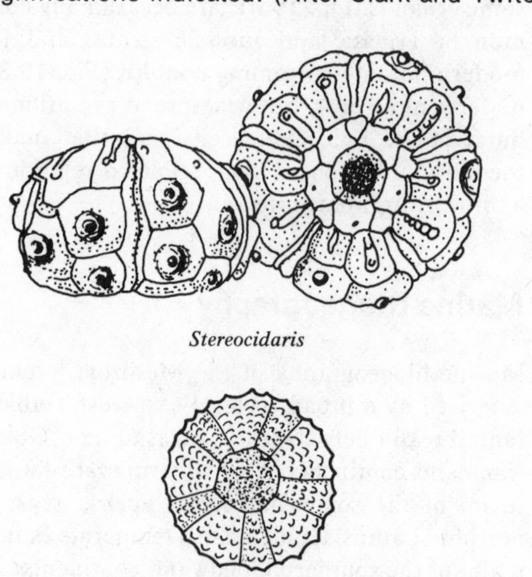

Figure 19.27. *Mesozoic Regular Echinoids.* X 2/3.

Figure 19.28. *Spongiomorphoids.* X 2.

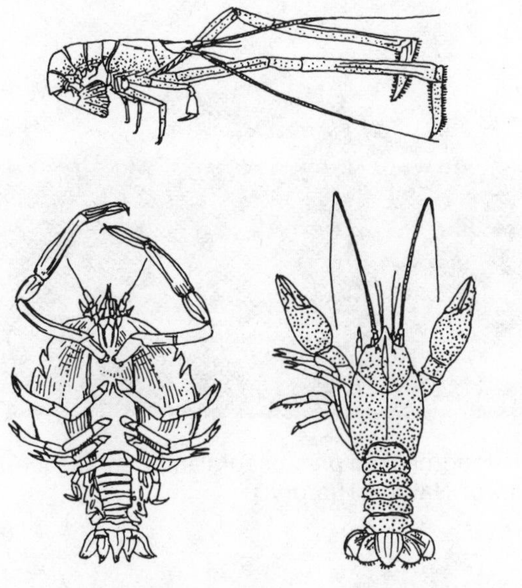

Figure 19.29. *Mesozoic Malacostracans.* X1/3. (After Oppel.)

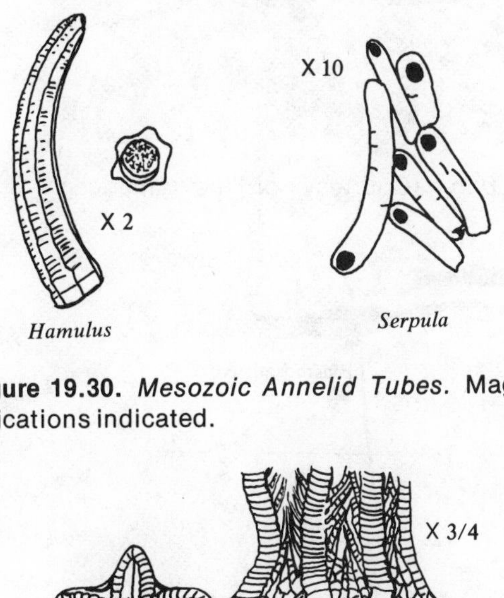

Figure 19.30. *Mesozoic Annelid Tubes.* Magnifications indicated.

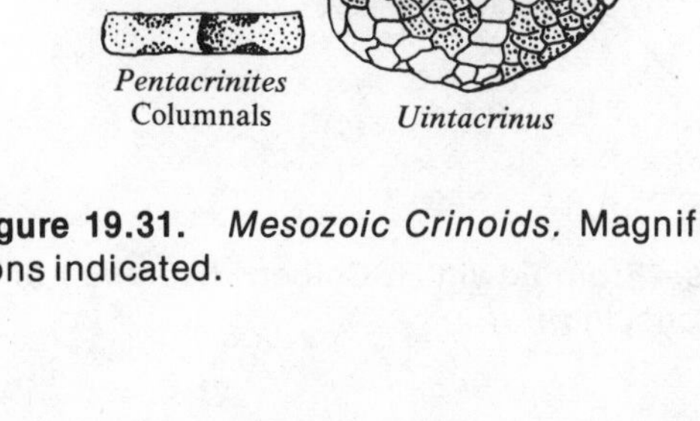

Figure 19.31. *Mesozoic Crinoids.* Magnifications indicated.

west length. For example, Late Jurassic Tethyan belemnites were arrayed into Mediterranean, Ethiopian, and Indo-Pacific provinces occupying respectively southern Europe-northern Africa, northeast Africa-southwest India, and northern India-Australia. The Caribbean region, a portion of western Tethys during much of the Mesozoic, became differentiated as a separate province for shallow-water benthos in the Late Cretaceous because of the widening Atlantic Ocean.

Marine Vertebrates

As noted earlier, the ray-finned bony fish underwent a great radiation in Mesozoic seas, particularly in the Cretaceous (Fig. 7.108).

The reptiles, so abundant on the land in the Mesozoic, invaded the seas and achieved great success there also. The fish or dolphin-like **ichthyosaurs** (Figs. 19.32–34) were common in the Triassic and reached a Jurassic peak. These viviparous forms have an unknown ancestry and died out by the Middle Cretaceous. The **sauropterygian** transition to the sea is better documented. The terrestrial, lizard-like protorosaurs gave rise to the amphibious Triassic **nothosaurs** (Fig. 19.34). From these evolved the Triassic seacow-like, but mollusk-crushing **placodonts** and the chiefly Jurassic-Cretaceous sea-serpent-like **plesiosaurs** (Figs. 19.32, 19.34). Both long- and short-necked varieties of these huge marine reptiles lived during their Late Cretaceous peak. All died out by the end of the Cretaceous. The **lizards** and **turtles** (Figs. 19.34–35) went to sea in the Cretaceous. The **mosasaurs,** or huge marine lizards, did not survive the end of the Cretaceous, but marine turtles of smaller size than their Cretaceous relatives persist to the present.

Marine Ecosystems

The rejuvenated marine ecosystem of the Mesozoic passed through the entire sequence from youthful to fully mature in a much shorter time

Figure 19.32. *Mesozoic Marine Reptiles.* Jurassic long-necked plesiosaurs and fish-like ichthyosaurs. (C.R. Knight mural, courtesy Field Museum of Natural History.)

Figure 19.33. *Fossil Ichthyosaur with Young.* (Courtesy of the American Museum of Natural History.)

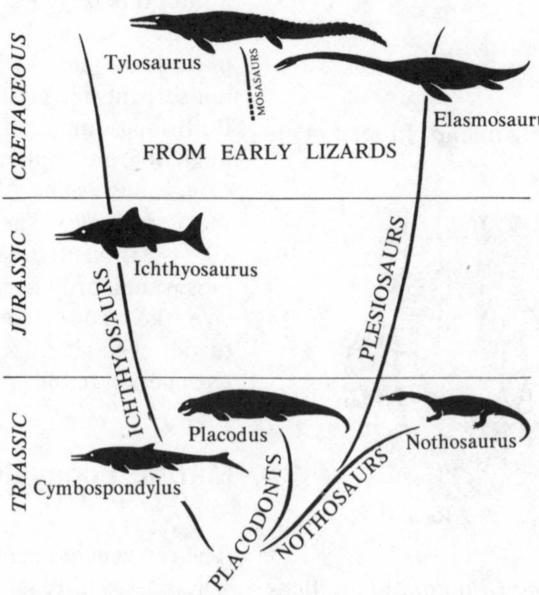

Figure 19.34. *Evolution of Mesozoic Marine Reptiles.* (From Edwin H. Colbert, *Evolution of the Vertebrates,* © 1969, by permission of John Wiley & Sons, Inc.)

Figure 19.35. *Mesozoic Marine Reptiles and Flying Reptile.* Marine turtle and lizard (mosasaur) and a flying reptile, or pterosaur. Cretaceous forms from North America. (C.R. Knight mural, courtesy American Museum of Natural History.)

span than the Paleozoic equivalent. The Triassic, Jurassic, and Early Cretaceous were the youthful stage of relatively low diversity and short food chains. By mid-Cretaceous times, a mature ecosystem had been attained, particularly with the advent of the modern photosynthetic phytoplanton. By the Late Cretaceous, the ecosystem was in the fully mature or climax stage with numerous small, environmentally sensitive populations having complex interactions. At the close of the Cretaceous, this narrowly adapted ecosystem was affected by extinctions. Some of the environmental causes of this disruption were previously noted.

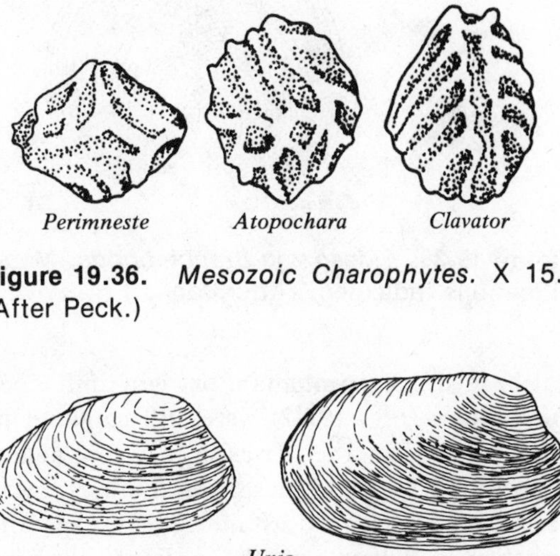

Figure 19.36. *Mesozoic Charophytes.* X 15. (After Peck.)

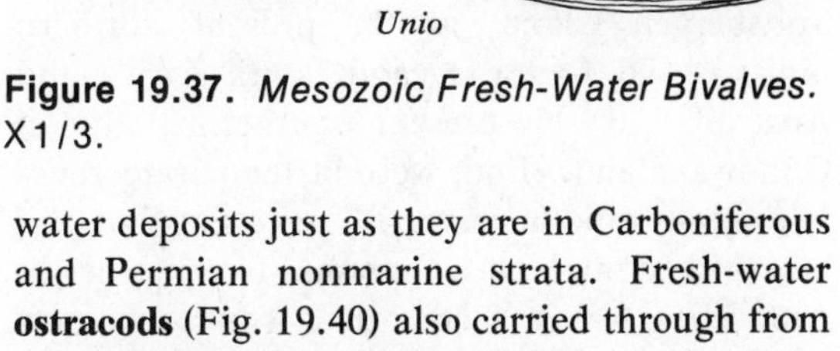

Figure 19.37. *Mesozoic Fresh-Water Bivalves.* X1/3.

Fresh-Water and Land Life

Protistans and invertebrates continued to diversify in fresh waters and on land during the Mesozoic. The **charophytes** become common in fresh-water continental deposits of the Jurassic and Cretaceous (Fig. 19.36). Fresh-water **clams** (Fig. 19.37) continue, and the oldest unequivocal fresh-water and terrestrial **snails** (Fig. 19.38) are found in Late Mesozoic strata; aquatic forms, in the Jurassic; and terrestrial, in the Cretaceous. The **branchiopod crustaceans** (Fig. 19.39) continue to be common in Triassic fresh-water deposits just as they are in Carboniferous and Permian nonmarine strata. Fresh-water **ostracods** (Fig. 19.40) also carried through from the Late Paleozoic and increased in numbers. The terrestrial arthropods continued their evolution, particularly the **insects** (Fig. 19.41). Most of the modern orders appeared in the Mesozoic. Among the lesser arthropod groups, the terrestrial centipedes appeared in the Cretaceous.

Among the vertebrates, ray-finned **fish** continued to be common in fresh waters. The large,

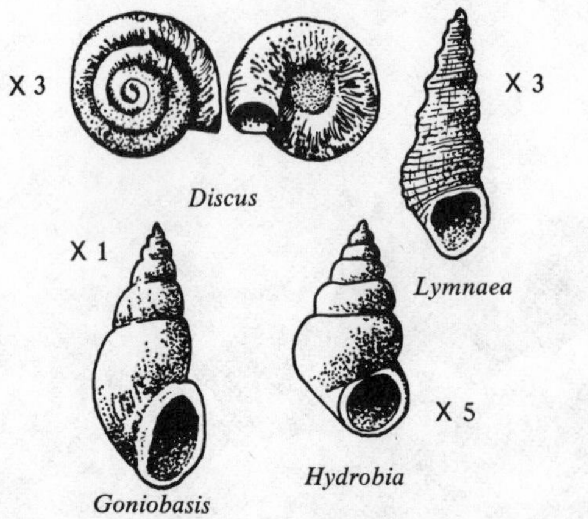

Figure 19.38. *Mesozoic Fresh-Water Gastropods.* Magnifications indicated.

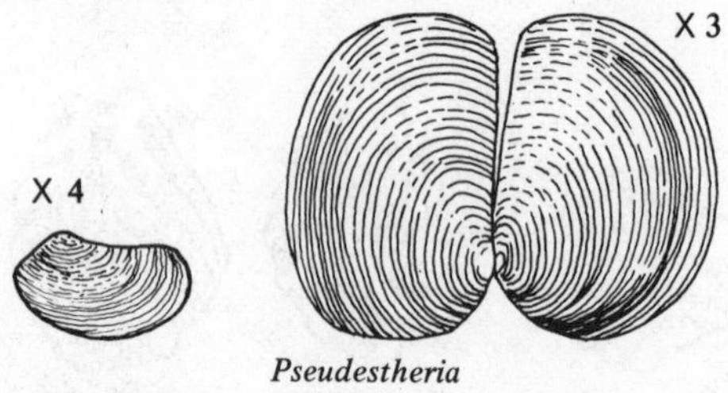

Figure 19.39. *Mesozoic Branchiopods.* Magnifications indicated. (After Jones.)

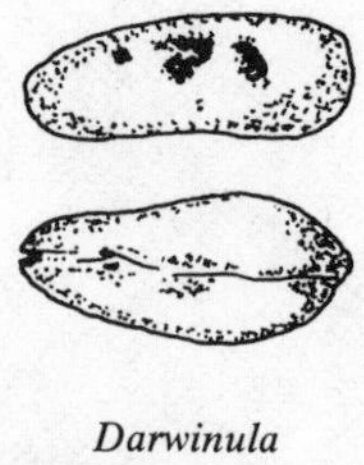

Figure 19.40. *Mesozoic Fresh-Water Ostracod.* X 30. (After Sars.)

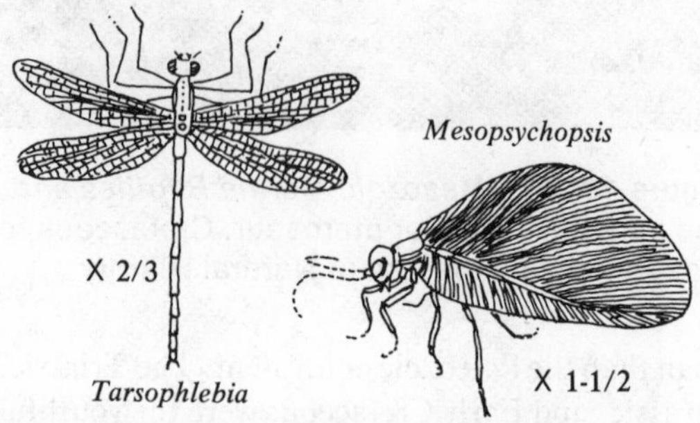

Figure 19.41. *Mesozoic Insects.* These forms are from the famous Solenhofen area. Magnifications indicated.

flat-headed stegocephalian or labyrinthodont **amphibians** (Fig. 19.42) were still common in many areas in the Triassic as continental deposition remained widespread. Similar genera occurred in such distant areas as the Spitsbergen Island, at the present northern limits of old Laurasia, and South Africa and Australia, at the present southern limits of Gondwanaland. (Both were in temperate zones in the Mesozoic, however.) The recent find of a labyrinthodont (and a mammal-like reptile) in Antarctica was a factor that won over most vertebrate paleontologists to the idea of moving continents. The ancestor of the frogs has been found in the Triassic of Gondwanaland on the present island of Madagascar, and ancestral salamanders are known from the Jurassic and Cretaceous of North America.

Reptiles

In the terrestrial environment, the **reptiles** dominated the faunas. The early Mesozoic reptile assemblage consisted mainly of the **mammal-like** or synapsian group (Figs. 17.58, 19.43) which continued from the Paleozoic. These reptiles, which were abundant until the Late Triassic, became more and more mammal-like as the period progressed, and near its close a carnivorous lineage crossed the boundary and became the first small mammals. Many lineages so closely approximate the mammalian condition that not all the so-called primitive "mammals" may actually be mammals, but just independent offshoots of the synapsian reptiles. It is probable that the early Mesozoic mammal-like reptiles had a furry covering and were able to regulate internally their own temperature. Their presence in subpolar regions, their highly

Figure 19.42. *A Triassic Labyrinthodont Amphibian from Europe.* (Augusta-Burian mural.)

Figure 19.43. *Mammal-Like Reptiles from South Africa.* Three carnivores surround a herbivore. This restoration is not totally accurate because these reptiles were probably covered with fur as a protection against the harsh winters of southern Gondwanaland. (C.R. Knight mural, courtesy Field Museum of Natural History.)

porous bones, and the low predator-prey weight ratio strongly support this inference.

In the Late Triassic, the mammal-like reptiles, which included both carnivores and herbivores and had successfully invaded all the present continental areas, were ecologically replaced by other orders of reptiles, chiefly the dinosaur groups (Fig. 19.44). The synapsian reptiles persisted into the Jurassic in declining numbers and then became extinct, unless some of the so-called primitive Mesozoic and Cenozoic mammal orders are not ancestral to other mammals. In that case, the group may actually survive down to the present in the form of the egg-laying monotreme "mammals," the platypus and spiny "anteaters." The true mammals, which descended from the synapsian reptiles, also may have contributed to the demise of the reptile group by ecologically replacing the smaller carnivorous, insectivorous, and herbivorus types.

The **dinosaurs** actually belong to two (or perhaps three) separate orders of reptiles that arose from a group of reptiles called the **thecodonts** (Fig. 19.45). Thecodonts were common in the Triassic and included several adaptive types including the phytosaurs, the ecologic but not evolutionary predecessors of the crocodiles (Fig. 19.45D). Thecodonts, a diverse lot, gave rise to many reptilian groups in the Mesozoic, including the two dinosaur orders and the crocodiles in the Triassic and the pterosaurs and the true birds in the Jurassic. The Triassic group of bipedal thecodonts called **pseudosuchians** Fig. 19.45A) probably gave rise to the dinosaur orders as both types are primitively bipedal and the ancestral condition shows in the large hind limbs and pelvis of the later, quadrupedal types (Fig. 19.46). The oldest members of both kinds of dinosaurs occur in Gondwanaland, indicating they originated there (Fig. 19.57), but dinosaurs spread to all continents before there was any appreciable fragmentation of the Early Mesozoic continental unity.

Recently, a convincing case has been made by *Bakker* and others that dinosaurs, contrary to their usual image, were active, warm-blooded animals probably covered with fur or feathers in many cases. (Pseudosuchian thecodonts and the advanced synapsian reptiles were probably warm-blooded also.) Large dinosaurs have been found in areas such as northern Canada that were north of the Arctic Circle in Cretaceous times. No large cold-blooded animal is capable of living that far north today because it could not cope with the long, sunless, cold winters. The density of porous passageways in dinosaur bones exceeds even that of the warm-blooded mammals, while cold-blooded organisms have relatively few canals and other passageways in their bones. If one computes the actual weight of predator and prey animals in a community and compares them, this biomass ratio is higher in cold-blooded communities than in warm-blooded ones. This is because more prey is needed to fuel the higher internal temperatures of warm-blooded animals; hence, fewer of them can exist relative to the same amount of prey. Dinosaur communities fall in the warm-blooded range.

Some dinosaur fossils have now been found which show the presence of a feathery or hair-like covering characteristic of warm-blooded animals. Indeed, the oldest bird, ***Archaeopteryx*** (discussed later) is essentially a feathered dinosaur with wings for gliding. Therefore, feathers were probably present in the dinosaurs ancestral to birds. (Thus in a sense, modern birds may be surviving dinosaurs; the large flightless forms, such as ostriches, may actually be living dinosaurs!) Dinosaurs have a locomotor anatomy and limb proportions indicating they were capable of rapid running movements such as those of mammals and unlike those of large cold-blooded animals. Finally, dinosaur brains, though not large in absolute terms, equal or exceed those of comparably-sized birds. In living organisms, the ability to maintain internally a high body temperature is a prerequisite for the development of large brains.

One group of dinosaurs, which may not be a natural unit, is called the **lizard,** or **reptile-pelvis,** type and is characterized by the typical triradiate reptilian pelvis (Fig. 19.46). Its early, Triassic members were the bipedal **prosauro-**

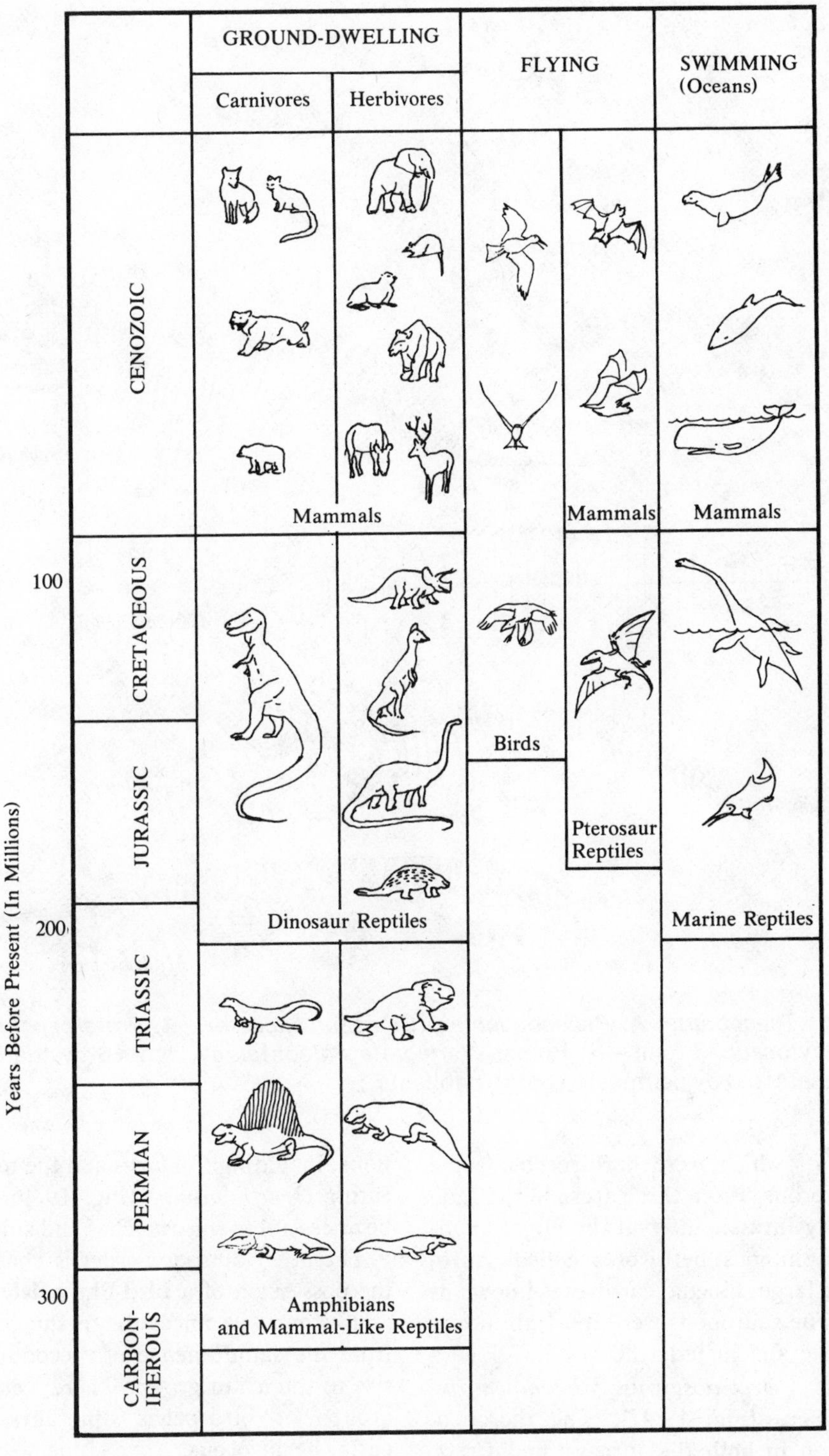

Figure 19.44. *Ecologic Replacements in the Mesozoic and at Its Close.* (From A. Lee McAlester, *The History of Life,* © 1968. Reprinted by permission of Prentice-Hall, Inc., Englewood-Cliffs, New Jersey.)

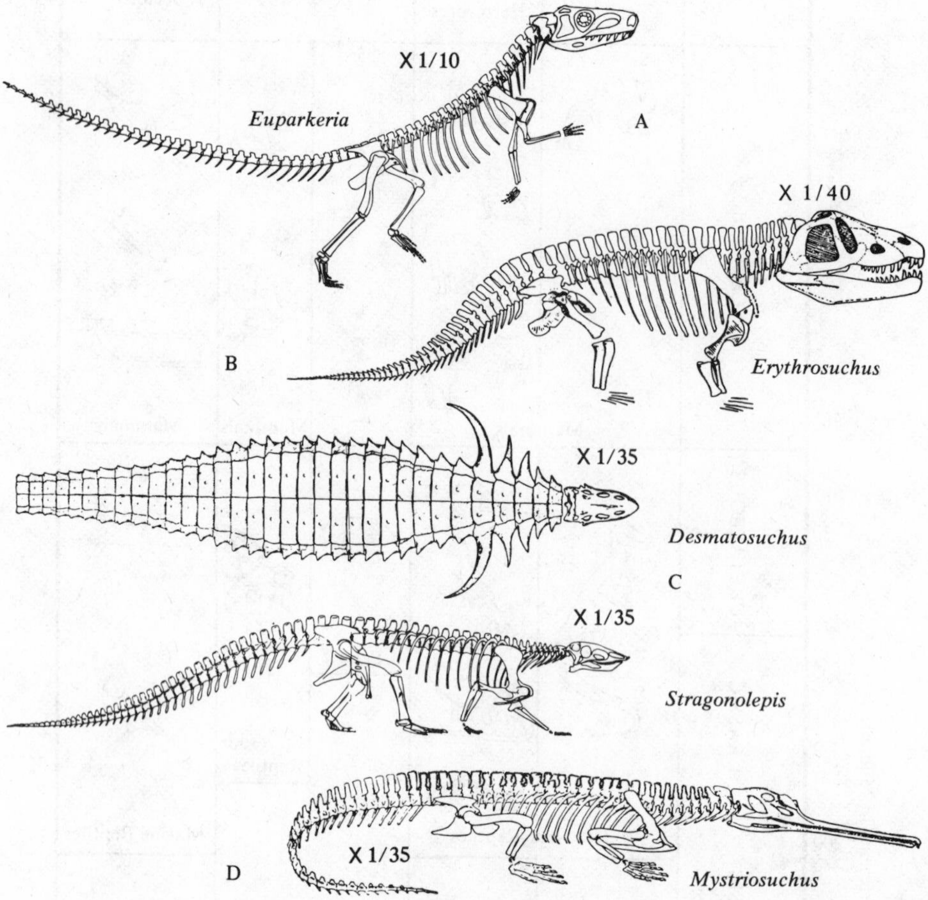

Figure 19.45. *Thecodonts.* **A.** *Pseudosuchian (Dinosaur ancestor),* **B.** *Proterosuchian,* **C.** *Aetosaurs,* **D.** *Phytosaur.* (From A.S. Romer, *Vertebrate Paleontology,* © 1966 by the University of Chicago Press. Used by permission of the publisher.)

pods, most of which were herbivorous, some being carnivorous. From these arose in the Late Triassic-Early Jurassic interval the huge, quadrupedal, amphibious herbivores called **sauropods** and the large, bipedal carnivores known as **theropods.** The sauropods were most abundant in the Jurassic and include the familiar genera *Apatosaurus, Brachiosaurus, Brontosaurus,* and *Diplodocus* (Fig. 19.47). The theropods were common in both the Jurassic and Cretaceous with *Allosaurus,* a common Jurassic form, and *Tyrannosaurus,* a significant Cretaceous genus (Fig. 19.48). There were also smaller and more slender theropods of ostrich-like proportions, including *Coelurus* and the toothless, egg-eating *Ornithomimus* (Fig. 19.46), which may be ancestral to the ostriches and similar birds.

The other dinosaur order is characterized by the possession of a **bird-like pelvis** (Fig. 19.46) and either was ancestral to the birds or arose from the same lineage of thecodonts that gave rise to the avian group. There were four major groups of bird-pelvis dinosaurs, all apparently herbivorous. The bipedal **ornithopods** culminated in the amphibious Cretaceous hadrosaurs or duck-billed dinosaurs (Fig. 19.49). The armored **stegosaurs** (Fig. 19.50) were common in the Jurassic; and the armored

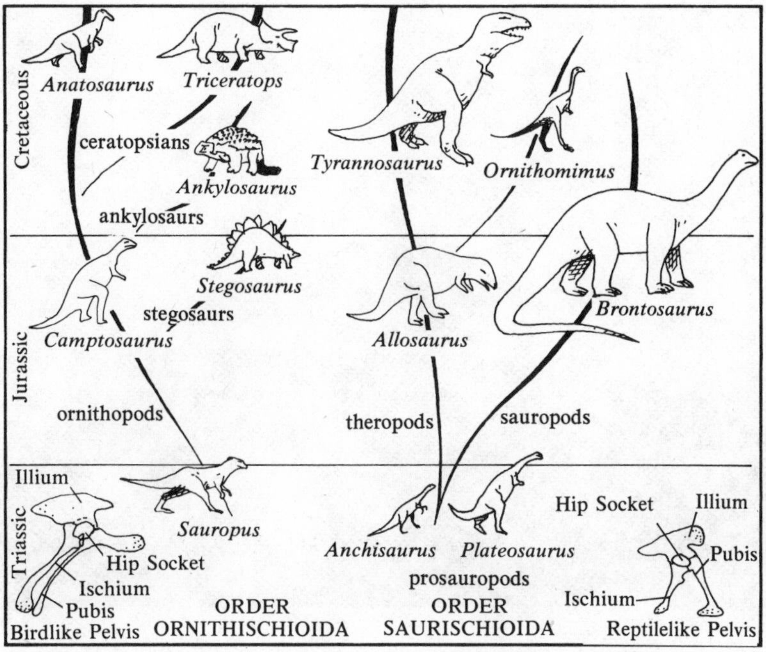

Figure 19.46. *Dinosaur Evolution.*

Figure 19.47. *Jurassic Sauropods. Brontosaurus* is a common North American genus. (C.R. Knight mural, courtesy American Museum of Natural History.)

Figure 19.48. *Cretaceous Ceratopsian and Theropods. Triceratops* and *Tyrannosaurus* prepare to battle in region of present Montana. (C.R. Knight mural, courtesy Field Museum of Natural History.)

Figure 19.49. *Cretaceous Ornithopods and Ankylosaur, Ostrich-Like Dinosaurs.* Various duck-billed dinosaurs are shown at the left and right sides. An ankylosaur is in the center foreground with the ostrich-like forms behind. Common North American forms are shown. (C.R. Knight mural, courtesy Field Museum of Natural History.)

Figure 19.50. *Jurassic Stegosaurs. Stegosaurus* is surrounded by various cycadophytes in this North American scene. (C.R. Knight mural, courtesy American Museum of Natural History.)

ankylosaurs (Fig. 19.49), in the Cretaceous. Both forms were quadrupeds. Also quadrupedal were the great horned dinosaurs or **ceratopsians** (Fig. 19.48) of the Late Cretaceous. *Triceratops* is a common North American genus. The powerful jaws of ceratopsians may have been adapted to a carnivorous diet.

All the dinosaur groups apparently became extinct by the end of the Cretaceous, with theropods and ceratopsians persisting the longest. (As noted earlier, birds may be surviving dinosaurs.) The Mesozoic-Cenozoic boundary is marked in terrestrial deposits by the remains of the highest dinosaurs. Their extinction still remains a mystery because there were no immediate, large mammalian successors, thus ruling out superior competitors as a possible cause of the extinction. In a few areas, such as the Bug Creek region of Montana, a transitional sequence across the boundary occurs. This sequence reveals a gradual reduction in dinosaur abundance and a concomitant increase in mammalian remains. There is also a progressive change in the plant communities in these strata. Such evidence suggests that the food chain that dinosaurs dominated may have been gradually replaced by a food chain based on different plants which interacted with the rapidly increasing modern types of insects and small seed- and insect-eating mammals. The diminution of the dinosaurs through the whole Late Cretaceous interval suggests that some sort of ecologic change such as this was occurring. On the other hand, it is hard to reconcile the complete disappearance of a diversified group such as the dinosaurs with the slow replacement of plant communities. This theory also fails to explain the simultaneous extinction of all the successful groups of large reptiles that invaded the seas in the Mesozoic. Only something that would cut across many environments and food chains could cause the Late Mesozoic extinction.

A change in the atmospheric O_2/CO_2 ratio could have triggered such a reaction and is also an attractive hypothesis because it could explain the simultaneous extinction of the ammonoids. As stated earlier, the appearance and great increase in abundance of the large groups of modern photosynthesizers or their acquisition of the ability to secrete shells in the Late Mesozoic could have caused large-scale changes in the ratio of the gases. The exact effect on the reptiles is uncertain, but it could have affected their rate of metabolism or it could have altered the earth's heat budget through changes in the greenhouse effect. A cooling trend at the end of the Mesozoic is indicated by several lines of geologic, paleontologic, and geochemical evidence. This cooling trend could have been initiated by a large decrease in the CO_2 content of the atmosphere which controls the amount of atmospheric heat retained.

Other, chiefly terrestrial, reptilian groups such as the turtles, lizards, snakes, rhynchosaurs, and crocodiles arose in the Mesozoic and were occasionally common but less spectacular than the preceding groups. It is notable, however, that these smaller reptiles are the ones that survive today (Fig. 7.115).

Aerial Vertebrates

The thecodonts or early dinosaurs gave rise to two groups of flying vertebrates. The flying reptiles, or **pterosaurs** (Figs. 19.35, 19.51–2), appeared in the Jurassic, their exact ancestry unknown, and culminated in the great, seagoing, toothless *Pteranodon* of the Cretaceous. Some individuals of this species had a wing span

Figure 19.51. *Jurassic Life in Solenhofen Area of Germany.* Pterosaurs, or flying reptiles, above; earliest birds in middle; small dinosaurs and arthropod below. Plants are chiefly cycadeoids. (C.R. Knight mural, courtesy Field Museum of Natural History.)

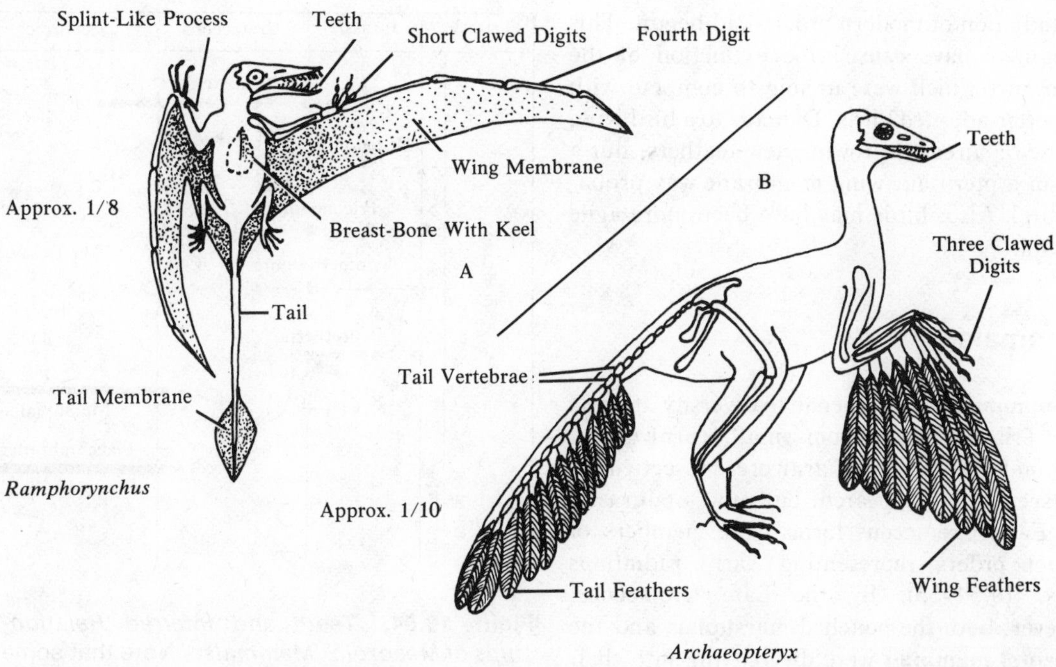

Figure 19.52. *Mesozoic Flying Vertebrates.* **A.** *Jurassic pterosaur,* **B.** *Jurassic bird.* (From R.M. Black, *The Elements of Paleontology*, © 1970, Cambridge University Press.)

in excess of 25 feet. Recently, some bones of a pterosaur which may have had a wing span twice this size were unearthed in Texas. A Jurassic pterosaur covered with hairy or hair-like feathers has recently been found in Russia. This find supports the idea that the flying reptiles must have been warm-blooded to sustain the high metabolism necessary for flight. The pterosaurs died out near the close of the Cretaceous.

In the Jurassic, another group of early dinosaurs or thecodonts gave rise to the **birds.** The birds underwent a great radiation that continues to the present day, thus entitling them to rank as a separate vertebrate class. The Jurassic "bird" *Archaeopteryx* (Figs. 19.51–2), is actually a connecting link between the two classes. With its teeth, front claws, long bony tail, and lack of hollow bones and a beak, its skeleton is that of a reptile and not a bird. But the impressions of large feathers covering the body, forelimbs, and tail indicate that this creature was well on its way to becoming a typical bird. Although many dinosaurs may also have had feathers, the long feathers of the forelimbs of *Archaeopteryx* form wing-like structures, indicating that the animal must have done some gliding. In the Cretaceous, a few toothed birds persisted (Fig. 19.53), but

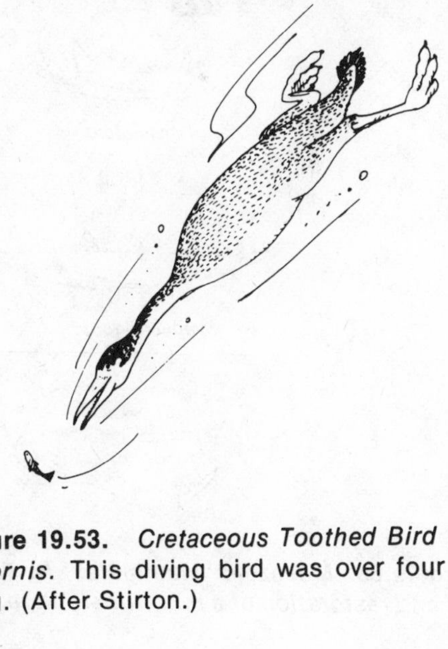

Figure 19.53. *Cretaceous Toothed Bird Hesperornis.* This diving bird was over four feet long. (After Stirton.)

the radiation of modern orders had begun. This event may have caused the extinction of the pterosaurs which were unable to compete with the better-adapted birds. Damage to a bird wing can be repaired by growing new feathers, but a tear in a pterosaur wing membrane was probably fatal. Also, birds may have been more agile on land.

Mammals

The **mammalian** class began to diversify after its Late Triassic origin from small, carnivorous, mammal-like reptiles. Carnivores, insectivores, and seed-eaters appeared, but most of Jurassic and Early Cretaceous forms were members of extinct orders representing early radiations (Figs. 19.54–55). By the Late Cretaceous, however, both the pouched **marsupials** and the **placental mammals** were diversifying into their modern orders, at least five of which appeared before the end of the period (Fig. 19.56). The condition of the earth at the time of the mammalian radiation was quite different from that of the Late Paleozoic and Early to Middle Mesozoic when the reptiles underwent their radiation.

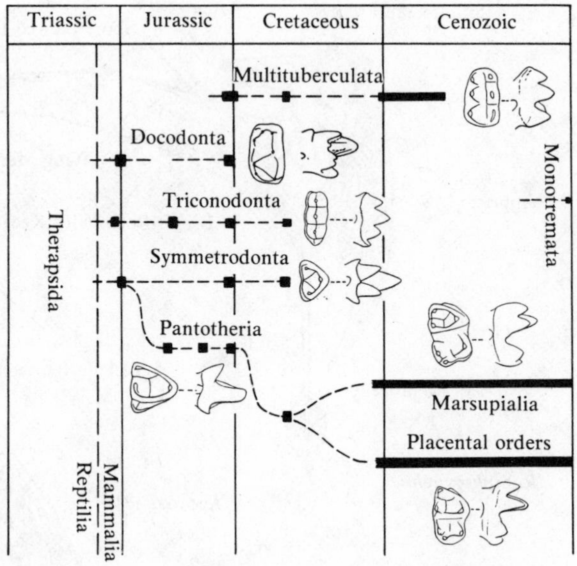

Figure 19.54. *Teeth and Inferred Relationships of Mesozoic "Mammals."* Note that some of the groups called mammals may actually be of separate origin from the true mammals. (From B. Kummel, *History of the Earth,* 2nd ed. W.H. Freeman & Co., © 1970.)

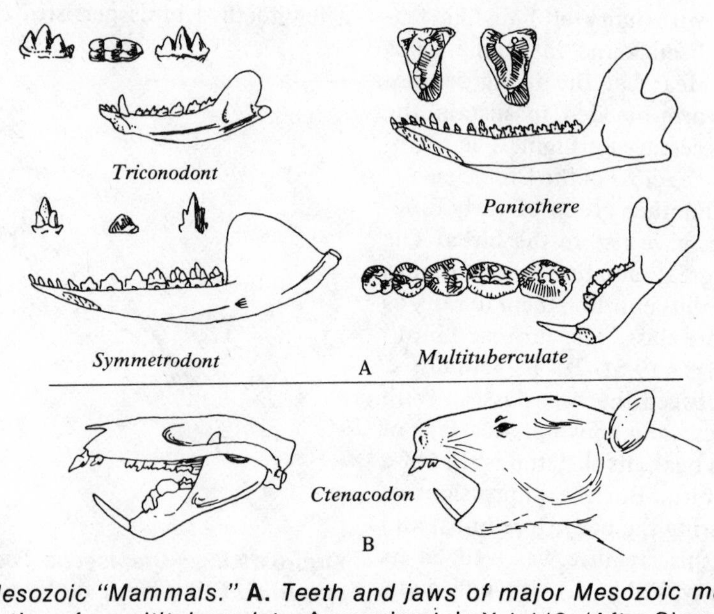

Figure 19.55. *Mesozoic "Mammals."* **A.** *Teeth and jaws of major Mesozoic mammal groups,* **B.** *Skull and restoration of a multituberculate.* Approximately X 1-1/2. (After Simpson.)

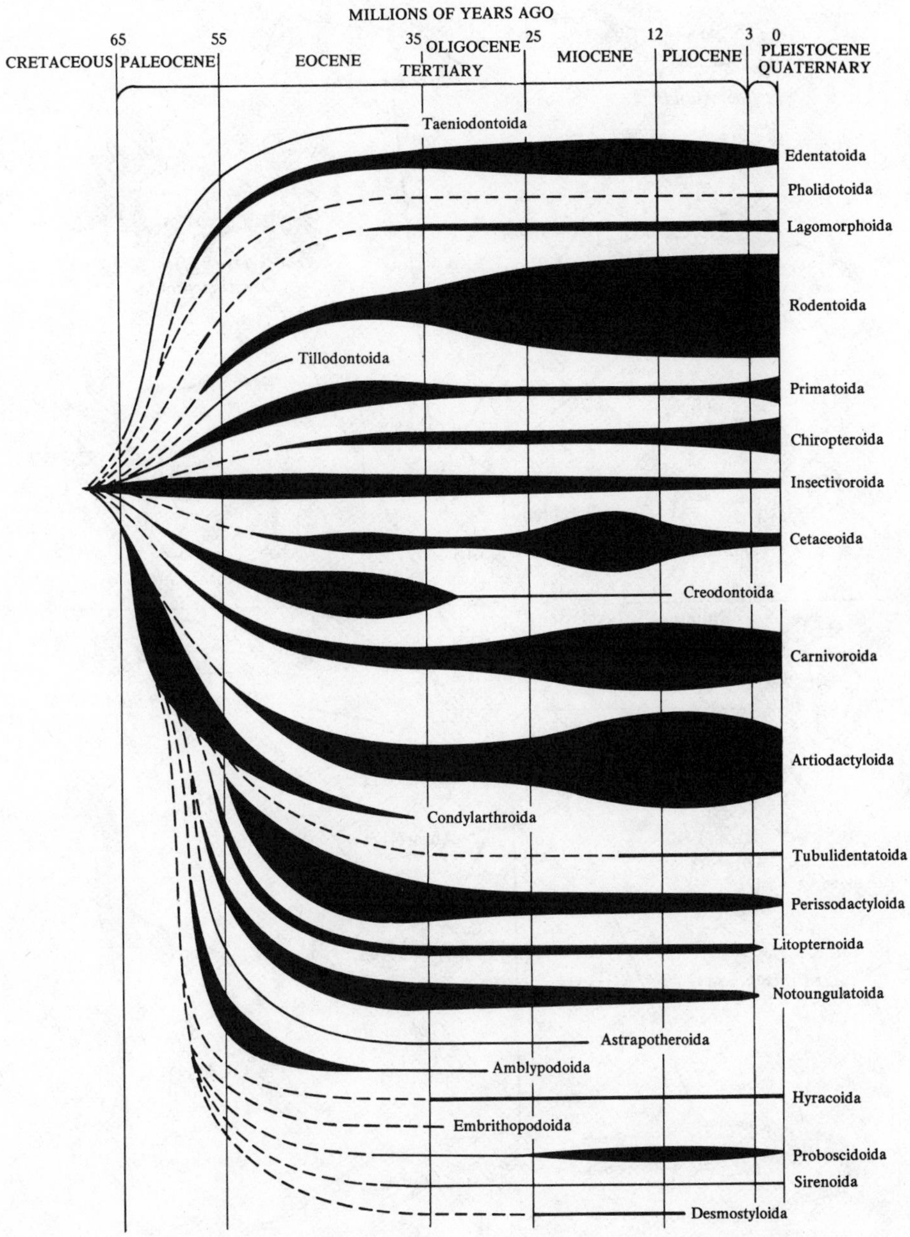

Figure 19.56. *The Mammalian Radiation.* Note the beginnings of most mammalian orders in the latest Mesozoic or earliest Cenozoic. (Adapted from Björn Kurten, "Continental Drift and Evolution," *Scientific American*, March 1969.)

In that interval, the continents were all together or in close proximity, allowing the rapid and easy intercommunication between distant areas that resulted in the great cosmopolitanism of Carboniferous through Early Cretaceous terrestrial life (Figs. 16.2–3, 18.1–2, 19.57). As we saw in the preceding chapter, however, the Late Cretaceous was a time of extreme fragmentation

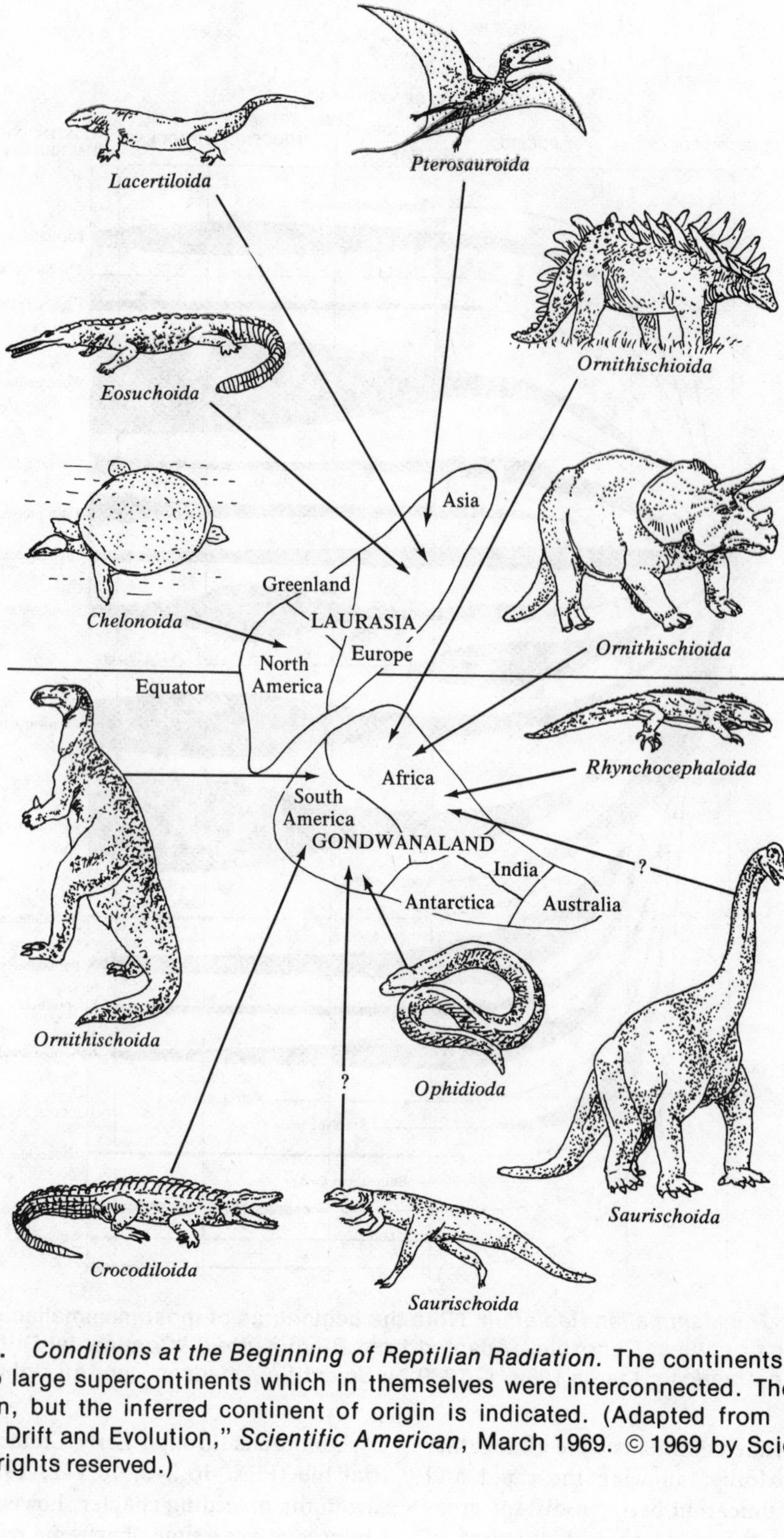

Figure 19.57. *Conditions at the Beginning of Reptilian Radiation.* The continents were concentrated in two large supercontinents which in themselves were interconnected. The faunas were cosmopolitan, but the inferred continent of origin is indicated. (Adapted from Björn Kurten, "Continental Drift and Evolution," *Scientific American*, March 1969.)

and isolation, a condition which persisted throughout the Paleogene and into the Neogene (Figs. 18.3, 20.2–3). The Early to Middle Cretaceous mammals were able to reach all the continents, but thereafter their evolution occurred in isolation, producing several distinct provincial faunas such as those of South America, Africa, Australia, and the three northern continents, North America, Europe, and Asia, which were in contact much of the time and, hence, had faunal interchange.

Plants

The plant life of the Mesozoic Era was characterized by two distinct floras—the **gymnosperm** flora, which continued from the Permian and dominated into the Early Cretaceous, and the **angiosperm** flora, which became dominant in the Middle Cretaceous and has persisted in that role to the present. The gymnosperms, or naked-seed plants, have no fruit covering their seeds. They include the **conifers, cycadophytes,** and **gingkos,** the major elements of the plant world through much of the Mesozoic (Figs. 19.58–59). The conifers had attained dominance of the plant world in the Permian and continued in this status through much of the Mesozoic. The large Triassic *Araucarioxylon* (Fig. 19.60), so abundant in the Petrified Forest National Park of Arizona, represents a group now restricted to the southern continents by continental drift.

Figure 19.58. *Mesozoic Flora.* In foreground, note dominance of cycadeoids. Conifers dominate in the distance and include, at center, araucarians. Jointed arthrophytes occur in the middle distance. (Augusta-Burian mural.)

Figure 19.59. *Mesozoic Flora of Late Triassic Age.* Note gingkos (lower left, upper right), conifers (right distance), cycadeoids (lower right), and ferns (lower center). (After Seward.)

Figure 19.60. *Petrified Araucarioxylon Logs.* Petrified Forest National Park, Arizona. These logs are of Triassic age. (Courtesy of William M. Mintz.)

This freed it from competition with the more advanced conifers, such as the pines and sequoias that evolved in the northern continents later in the Mesozoic. Cycadophytes, which arose in Gondwanaland in the Carboniferous, became very abundant and widespread in the Mesozoic. Both the surviving cycads and the totally Mesozoic cycadeoids were present. The gingko trees also arose in the Late Paleozoic, but reached a Mesozoic peak. The undergrowth of this Mesozoic flora consisted chiefly of ferns and some peculiar seed ferns. The latter group died out in the Cretaceous.

The origin of angiosperms is uncertain. Some Carboniferous and Jurassic pollen resembles that of angiosperms; a Triassic leaf is apparently that of a palm (Fig. 19.61); and palm wood is

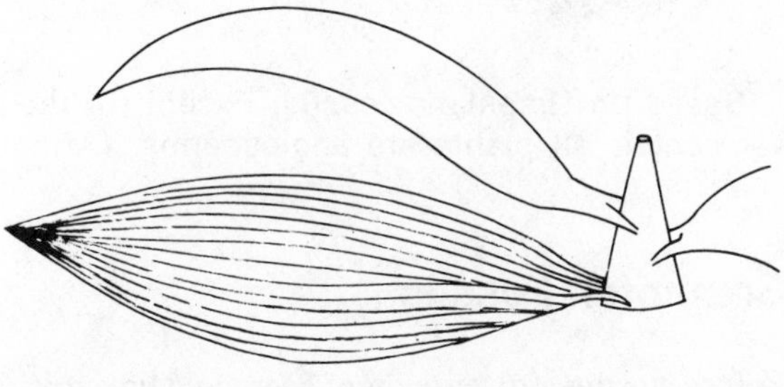

Figure 19.61. *Sanmiguelia, a Possible Triassic Palm.* (After Brown, Stirton.)

known from rocks of possible Jurassic age in Utah. Because flowers and fruits, two of the most diagnostic parts of angiosperms, are rarely preserved as fossils, it is virtually impossible to identify the ancestors of the group. In addition, it appears that angiosperms may have arisen in upland areas where deposition is not the rule and, hence, could only be represented in the preserved lowland record by pollen. Unquestioned angiosperm leaves, apparently belonging to modern genera, appeared in the Early Cretaceous (Fig. 19.62). By the Middle and Late Cretaceous, angiosperms dominated the earth's floras (Fig. 19.63), either replacing or forcing into more vigorous climatic regions members of the other plant divisions.

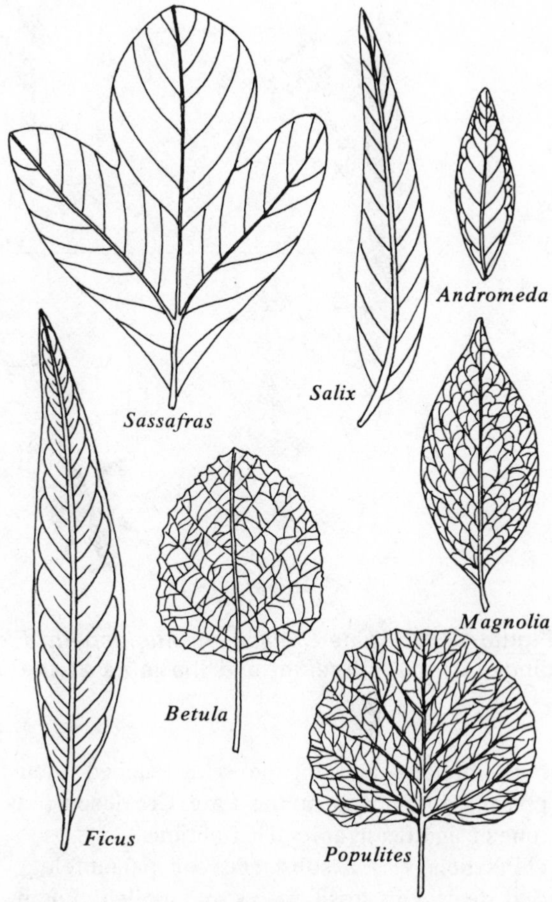

Figure 19.62. *Cretaceous Angiosperm Leaves.* (After Berry and Ward.)

The rise of the angiosperms was undoubtedly closely coupled with that of the pollenizing insects. Indeed, the angiosperm-insect relationship is probably the most important ecologic interaction in terrestrial ecosystems today. The Late Mesozoic floral change may have contributed to the extinction of the dinosaurs, but it apparently cannot explain the extinction of marine reptiles and ammonoids unless it affected atmospheric gas ratios. The floral changeover on land may have either triggered or been triggered by changes in the O_2/CO_2 ratio. It certainly appears to be more than a coincidence that the major producers of the earth today,

Figure 19.63. *Late Cretaceous Angiosperm Flora.* Based on Greenland fossils. Except for the gingko at the left center and the ferns at the lower center, all plants are angiosperms. (After Seward.)

both on land and in the sea, reached their present dominance in the Late Cretaceous, a time of significant animal extinctions.

Palynology is a subscience of paleontology that deals with fossil spores and pollen. Fossil pollen became common in the Late Jurassic and has continued to increase down to the present (Fig. 19.64), making palynology an important discipline.

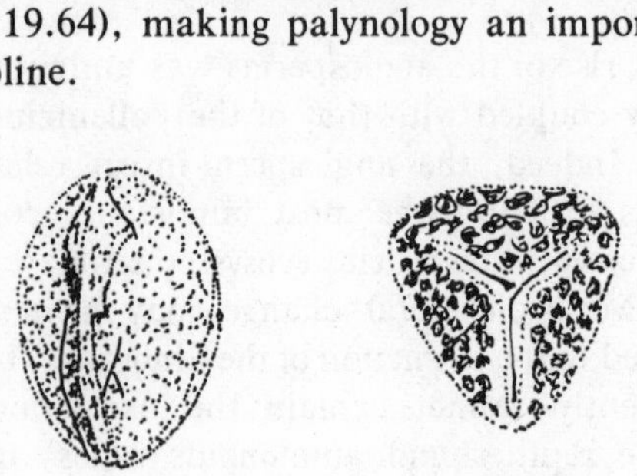

Figure 19.64. *Mesozoic Pollen Grains.* X 300. (After Kuyl, Muller, and Waterbolk.)

Solnhofen Fossils

Before closing a discussion of Mesozoic life, it is necessary to mention one of those rare fossil occurrences that gives us a glimpse of the soft-bodied life of the past so rarely preserved in the geologic record. Near Solnhofen, in southern Germany, are deposits of a very fine-grained lithographic limestone that originally formed as a chemical precipitate in quiet lagoons along the shores of the regressing Late Jurassic epeiric sea (Fig. 19.51). Into this quiet, lime-mud environment settled the remains of jellyfish, insects, pterosaurs, birds, and many other generally unpreserved forms. The body outlines showing jellyfish tentacles, insect wings, pterosaur wing membranes, and the feathers of the oldest birds are preserved as impressions (Fig. 19.65) in this remarkable deposit. A similar deposit which preserves hairy and feathery reptilian coverings has recently been found in the U.S.S.R.

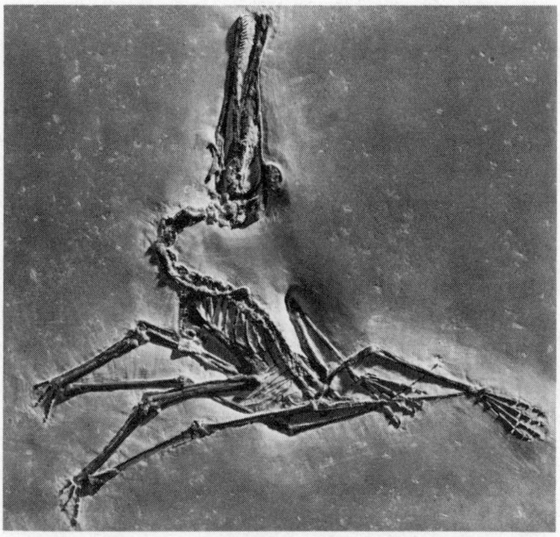

Figure 19.65. *Pterodactylus,* a fossil pterosaur from the Solnhofen area of Germany. The wing membranes were not preserved on this specimen, but note the otherwise excellent preservation of this articulated skeleton. (Courtesy of the American Museum of Natural History.)

REVIEW

1. What major modern groups of oceanic producers appeared in the Mesozoic Era?
2. Where and when were **dasycladaceous green algae** significant in the Mesozoic?
3. What significant ecologic breakthrough was made by the foraminifera in the Late Mesozoic? What were the significant large forams of the Mesozoic?
4. What was the major group of reef-building corals in the Mesozoic?
5. Which three classes of mollusks dominate Mesozoic faunas?
6. Why are the **ammonoids** so biostratigraphically significant?
7. What types of sutures do most Mesozoic ammonoids possess?
8. What is the major Mesozoic group of echinoderms?
9. Why is the Mesozoic Era called the **Age of Reptiles?**
10. What significant extinctions occurred near the end of the Mesozoic, and what are some possible causes?
11. What was the evolutionary pattern of the major dinosaur groups?
12. What were the major plant groups of the Mesozoic, and what major changeover occurred in the Cretaceous?
13. What two major vertebrate classes originated in the Mesozoic?
14. What evidence is there that dinosaurs, flying reptiles, and advanced mammal-like reptiles were warm-blooded?
15. How did plate tectonics affect the evolution of Mesozoic life?

20

Just Yesterday: The Cenozoic Era

20

The Cenozoic Era encompasses the last 65 million years of earth history, the time when the present physical features, plate distribution, and organisms of the planet developed. The Cenozoic Era is usually subdivided into seven epochs: from oldest to youngest, **Paleocene** (11.5 million years), **Eocene** (15 million years), **Oligocene** (15 million years), **Miocene** (18.5 million years), **Pliocene** (3 million years), **Pleistocene** (2 million years), and **Holocene** (10,000 years). The type areas of the first four are situated on the sites of marginal shelf seas in northern Europe. The Paleocene type area is in northern France; the Eocene, in the Paris and London basins; the Oligocene, in the Hanover Basin of northern Germany; and the Miocene, in western France. The Pliocene and Pleistocene type areas are situated in the former Tethyan region in Italy, the former in the north and the latter in the south. The Holocene or Recent has no type area; it encompasses the whole present surface of the earth.

The distribution of these epochs into periods, however, is not universal (Fig. 20.1). Most American and many European geologists use the older terms **Tertiary** and **Quaternary** with the boundary located between the Pliocene and Pleistocene. The type area of the Tertiary is that of the Pliocene in northern Italy, while the type Quaternary is in northern France. This separation is very unequal, for the Quaternary encompasses only about 2 million years, but it does emphasize the striking worldwide Late Cenozoic glaciations. Geologists using these terms emphasize that the whole era is so short, including less time than the Cretaceous Period, that other period subdivisions are not needed.

Periods	Epochs	Periods
Quaternary 2 Million Years Long	Holocene	Neogene 23.5 Million Years Long
	Pleistocene	
Tertiary 63 Million Years Long	Pliocene	
	Miocene	
	Oligocene	Paleogene 41.5 Million Years Long
	Eocene	
	Paleocene	

Figure 20.1. *Cenozoic Time Scales.* Spacing is not proportional to time.

Most European and some American geologists, on the other hand, prefer the **Paleogene** (or **Nummulitic**) and **Neogene** periods as subdivisions of the Cenozoic. In this case, the boundary falls between the Oligocene and Miocene and is near the climax of the orogeny in the Tethyan region. This subdivision gives periods of more equal length, 41.5 million and 23.5 million years respectively. There are no type areas of these two systems, only the type areas of the epochs assigned to them. Recently some geologists have developed a compromise that incorporates the best features of both. In this scheme, Paleogene, Neogene, and Quaternary periods are used, the first including the Paleocene, Eocene, and Oligocene; the second, the Miocene and Pliocene; and the third, the Pleistocene and Holocene. Thus, the long Tertiary, which was almost synonymous with Cenozoic, is subdivided; yet the uniqueness of the Late Cenozoic ice age is recognized. In our discussion, we will discuss the history as it unfolds epoch by epoch and leave the debate over periods to the specialists.

Plate Positions and Movements

In the Cenozoic Era, the plates gradually assumed their present form and positions (Figs. 20.2–3). The northern continents of North America and Eurasia clustered around the North or Pacific Pole which became situated in the Arctic Sea (or Ocean). More of western North America was rotated northward into cooler zones as the Cenozoic progressed (Figs. 20.2–3). Greenland, part of the American Plate, began to separate from Norway in the Paleocene, thus opening an oceanic connection between the Arctic Sea and the Atlantic Ocean. (For a short time in the early Cenozoic, the Labrador Sea was a site of spreading between Greenland and North America.) Eventually, as this rift widened and the rotation of North America largely closed off the Arctic from the Pacific, the Arctic Sea became a northward extension of the Atlantic (Fig. 20.3). Large-scale strike-slip faulting has also apparently affected the Arctic. North America probably collided with Asia somewhere in the East Siberian region, but geologists still disagree as to the exact boundary between the North American and Eurasian plates. The geology of this region is very difficult to decipher as it is the point of collision of at least three plates, the Eurasian, American, and Pacific, and possibly another, called the **Kula,** which has disappeared into the Aleutian Trench (Fig. 20.4).

In the Early Cenozoic, the western boundary of North America was a trench in which the small **Farallon Plate,** spreading eastward from the East Pacific Rise, was being consumed (Figs. 20.4–5). In the Late Oligocene, western North America began encountering the spreading ridge (Figs. 20.5–6), and the boundary between the Pacific and North American plates became a transform fault which eventually became the present San Andreas. This fault lengthened with time as the Farallon Plate disappeared. The formation of this boundary in western North America apparently converted the whole western Cordillera into a broad zone of transform faulting which triggered new unrest throughout the area. One of the most spectacular pieces of evidence of these latter features was the opening of the Gulf of California as Baja California split off North America. Rejuvenation of spreading in the Atlantic, after a mid-Tertiary lull, may also have contributed to the late Cenozoic tectonic activity of North America. Broken tips of plates subducted beneath western North America may also have been factors.

Gondwanaland continued its fragmentation in the Cenozoic (Figs. 20.2–3). South America and Africa further separated as the Atlantic Ocean basin widened. Africa itself started to be sundered by the development of the great rift valleys. Arabia has already split off, and an incipient ocean basin, the Red Sea, has formed. Eastern Africa is now beginning to separate from the remainder of the continent as the rift valleys, marking a new plate margin, develop there. Australia and Antarctica began their separation in the Eocene with Australia shifting northward into temperate and tropical climates, while Antarctica became centered on the Gondwana or South Pole. The opening of the direct Indian and Pacific Ocean link and the widening separation of Antarctica and South America in the Cenozoic completed the development of the present world-encircling ocean in high latitudes of the Southern Hemisphere. The collision of the African, Indian, and various smaller plates with Eurasia in the Cenozoic caused the great deformation of the Tethyan region known as the **Alpine Orogeny.** China also became welded to the rest of Asia as its various fragments came together against Mongolia. China is now being pushed eastward as India continues its northward press against Asia.

A high rate of ocean spreading in the southeastern Pacific not only continued the Andean

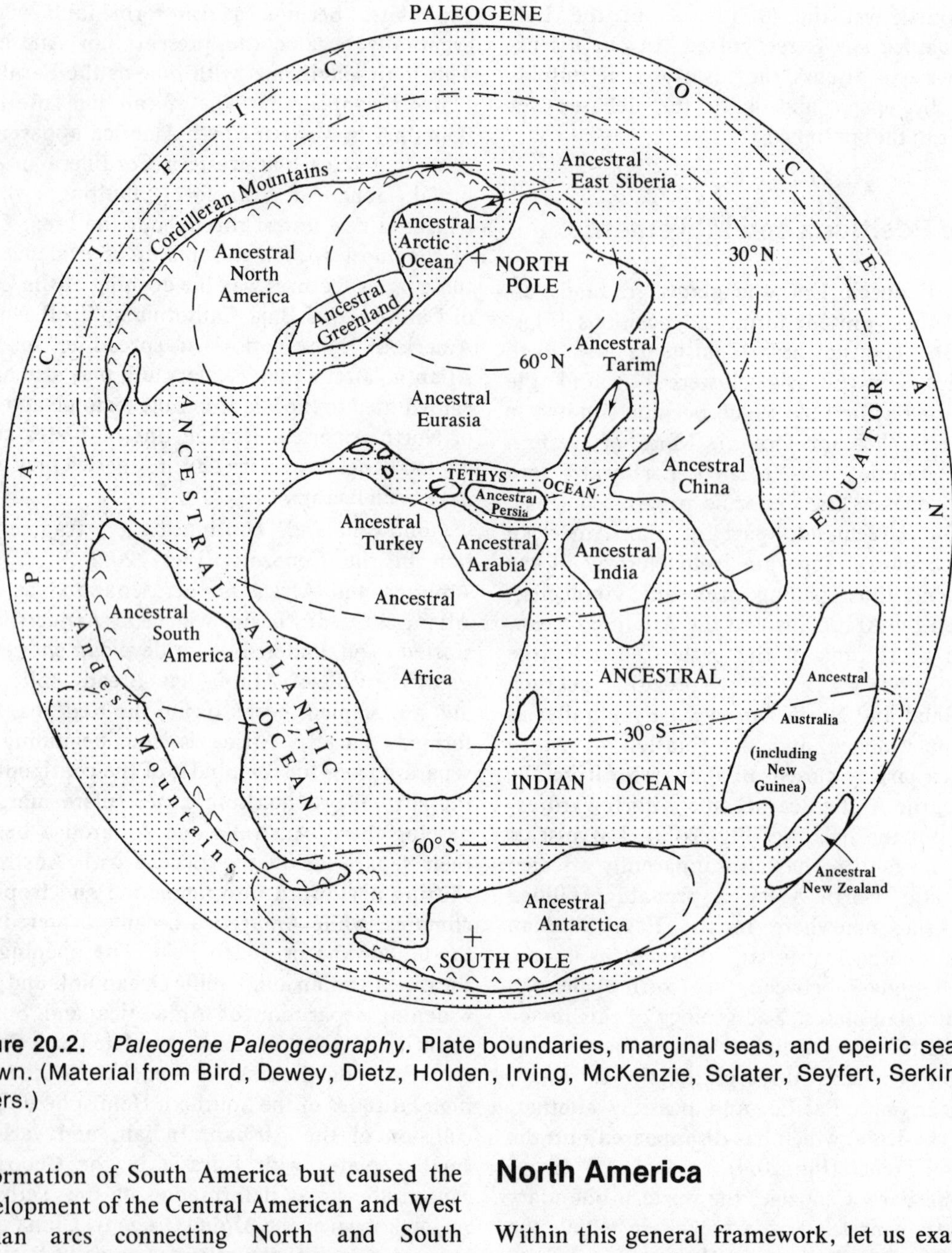

Figure 20.2. *Paleogene Paleogeography.* Plate boundaries, marginal seas, and epeiric seas not shown. (Material from Bird, Dewey, Dietz, Holden, Irving, McKenzie, Sclater, Seyfert, Serkin, and others.)

deformation of South America but caused the development of the Central American and West Indian arcs connecting North and South America (Fig. 20.3). The Pacific Plate continued to impinge against eastern Asia in a series of trenches, thus producing deformation in this area down to the present.

North America

Within this general framework, let us examine some of the details of Cenozoic physical history. In North America, the Laramide or Cordilleran Orogeny continued well into the Paleogene and produced the Rocky Mountains (Fig.

NEOGENE

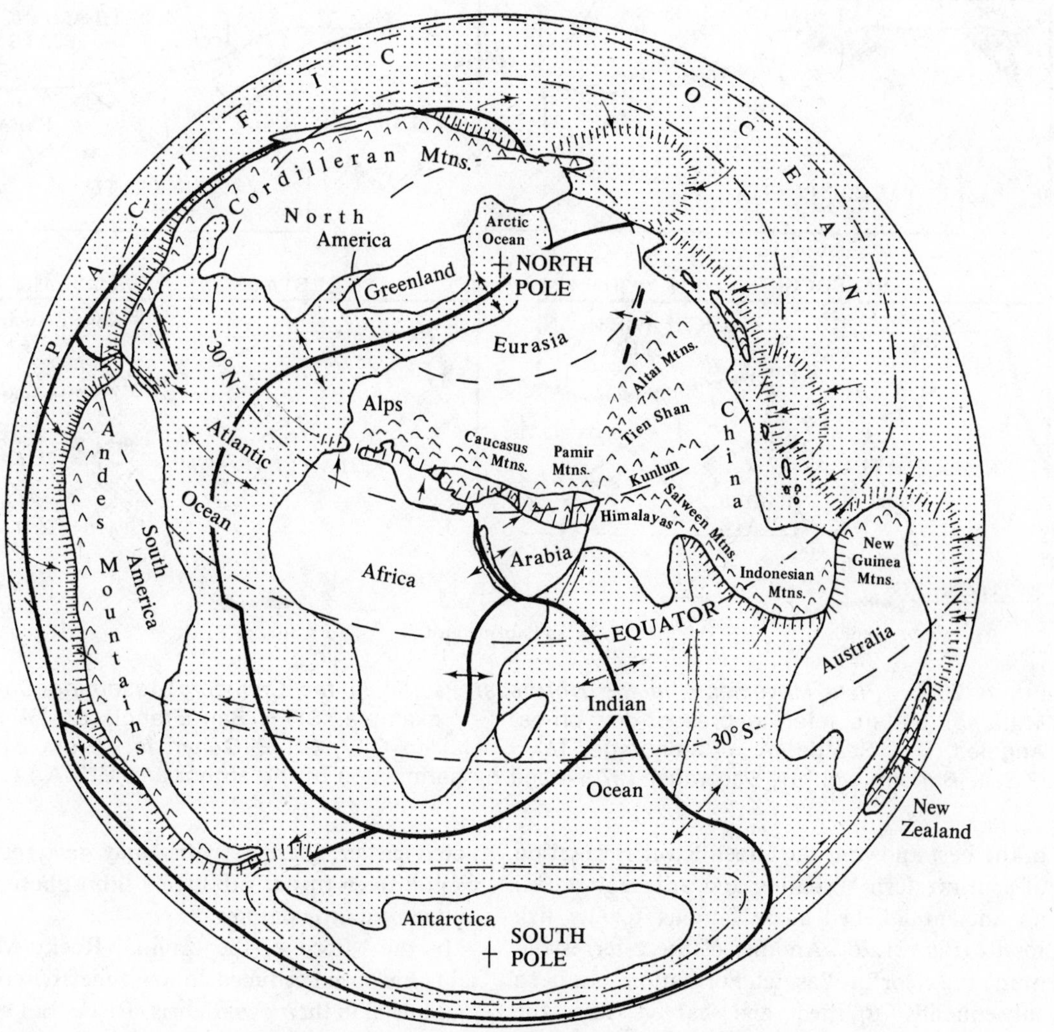

Figure 20.3. *Neogene Plate Configuration.* Heavy lines with diverging arrows are spreading ridges (dashed where uncertain); finer lines are transform faults with arrows indicated directions of movement; and closely spaced parallel lines mark trenches with arrows pointing in direction of subduction. Major Cenozoic mountain belts are indicated by mountain symbol.

20.7) and other ranges. The last epeiric sea to invade the craton, the **Tejas Sea** (Fig. 14.12), made only a brief Paleocene incursion into the Great Plains area, but was soon driven out by the **Fort Union Clastic Wedge** of terrestrial sediments coming from a strong pulse of the Cordilleran mountain building. The Tejas Seas gradually withdrew out through the Gulf Coastal region during the remainder of the Cenozoic.

Intermontaine basins within the rising Rocky Mountains (Fig. 20.7) contain numerous terrestrial deposits of Paleocene and Eocene age. One

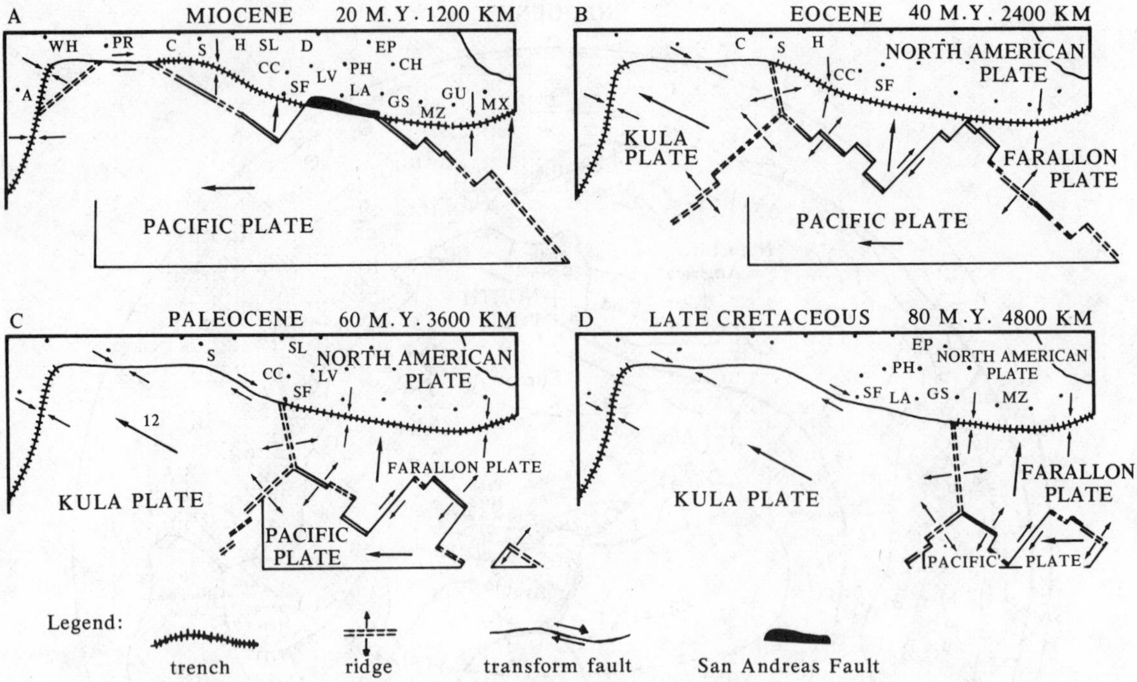

Figure 20.4. *Plate Movements in the Northeast Pacific in the Cenozoic Assuming Constant Motions.* Letters refer to present-day cities. For example, SF = San Francisco, LA = Los Angeles, S = Seattle, A = Anchorage, MX = Mexico City, GS = Guaymas. (From Atwater, *G.S.A. Bulletin*, vol. 81, December 1970. Used by permission of the author and G.S.A.)

of the best known is the Green River Formation of southwestern Wyoming and adjacent states, an ancient lakebed shale famous for its fish fossils (Fig. 21.26). Another of these terrestrial units, the colorful Wasatch Formation, has been subsequently uplifted and carved into the fantastic pillars of Bryce Canyon National Park (Fig. 20.8).

In the Oligocene, another phase of deformation in the mountains sent a sheet of clastics, the **White River Clastic Wedge**, over the Great Plains. This deposit, well known for its mammalian fossils, has been carved by erosion into the noted Badlands of South Dakota (Fig. 20.9).

Volcanic activity, including great outpourings of lava, occurred in the Oligocene of some areas of the Rockies. These have since been uplifted and carved into the jagged San Juan Mountains of Colorado and the Absaroka Mountains of Wyoming and Montana (Figs. 20.10–11). As noted earlier, these activities may have resulted from broken tips of subducted lithosphere slabs or from mantle plumes.

In the Miocene, the original Rocky Mountains had been reduced to low relief by erosion and burial in their own debris. Rivers ran across the countryside without regard for the underlying rocks. There was broad uplift and block faulting (Fig. 20.12) in the Late Pliocene and Pleistocene. During this rise of the present Rockies, the major rivers were able to maintain their courses across the exhumed mountain ranges, forming numerous spectacular canyons (Figs. 20.13–14).

Vulcanism was widespread in the Pacific Northwest throughout the Cenozoic. Submarine lava flows abounded in the Paleogene. There were great outpourings of basalt, forming the Columbia River Plateau (Fig. 20.15) and Snake River Plain chiefly in the Miocene. Many geolo-

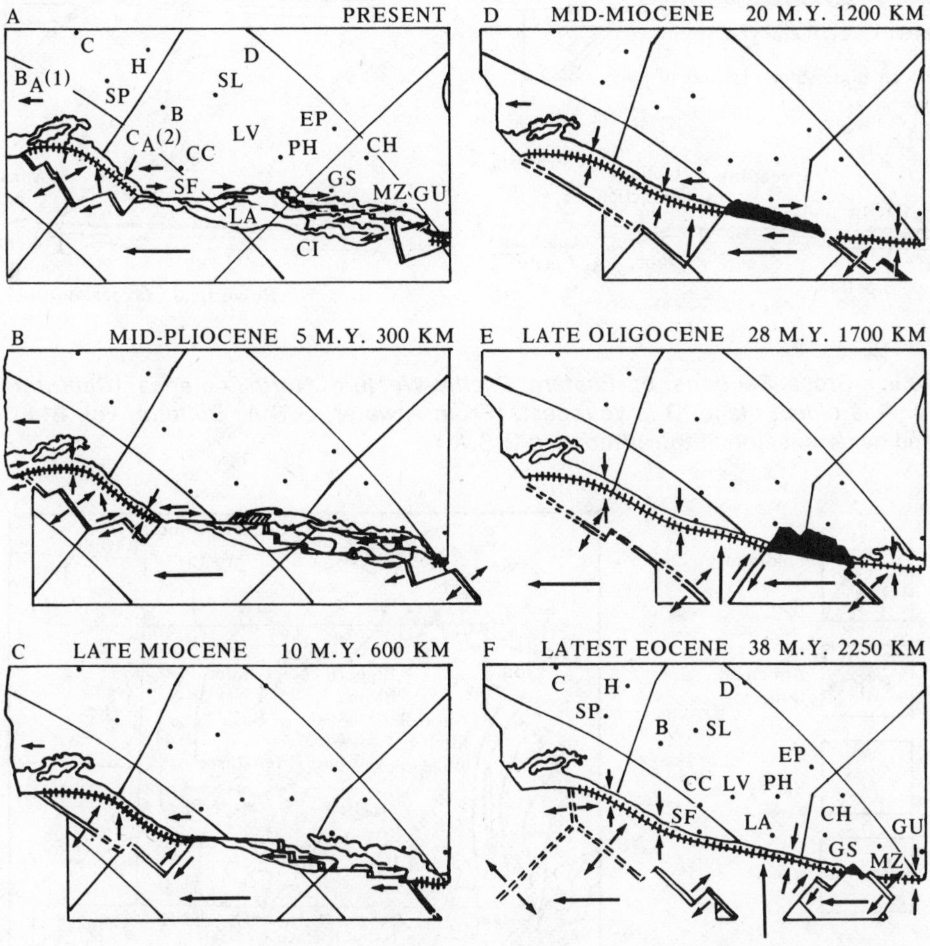

Figure 20.5. *Plate Movements off Western North America in the Cenozoic.* Constant motion of 6 cm/year is assumed. Black areas are regions of overlap, which probably can be accounted for by a slight change in plate motions. Letters indicate modern cities for reference. (From Atwater, *G.S.A. Bulletin*, vol. 81, December 1970. Used by permission of the author and G.S.A.)

gists attribute this eastward moving band of vulcanism to the passage of North America over a hot mantle plume which is now situated in Yellowstone National Park. In the Late Neogene, the high volcanoes of the Cascade Mountains developed (Fig. 20.16), probably from magmas generated by a subduction zone beneath the Pacific Northwest. There was also extensive vulcanism in the Colorado Plateau and Sierra Madre regions in the Late Cenozoic.

In California, the geology is extremely difficult to decipher because of the great dislocations caused by the San Andreas fault system, a great transform fault zone (Fig. 20.17). This fault system connects the offset portions of the East Pacific Rise in the Gulf of California and off Northern California (Fig. 20.18). The missing pieces of granite (and associated metamorphics) connecting the southern Sierra Nevada and northern Baja California batholiths were moved northwestward so that they now lie in the Coast Ranges near and to the south of San Francisco (Fig. 20.19). The entire west coast was the scene of marginal embayments and seas

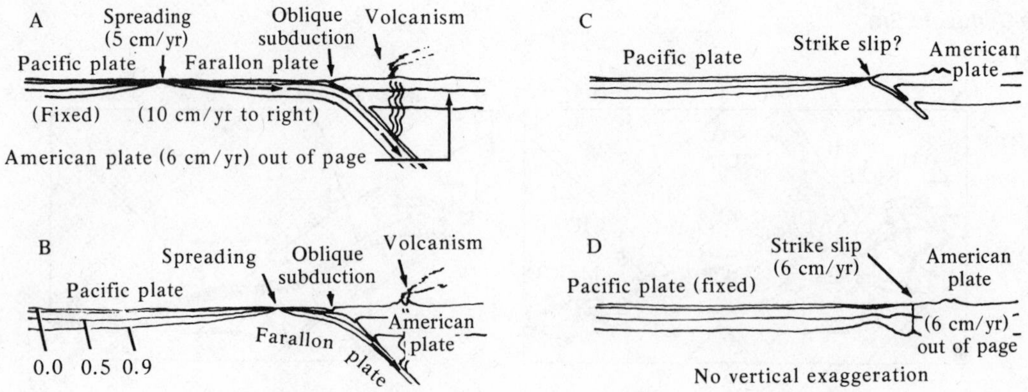

Figure 20.6. *Cross Sections of Eastern Pacific-Western North America* (*California*) *in the Cenozoic.* **A** *is oldest stage*; **D** *is youngest.* (From Atwater, *G.S.A. Bulletin*, vol. 81, December 1970. Used by permission of the author and G.S.A.)

Figure 20.7. *Early Cenozoic Mountains and Intermontane Basins of the Central and Southern Rocky Mountains Area.*

A

B

Figure 20.8. *Bryce Canyon, Bryce Canyon National Park, Utah.* **A.** *General view of amphitheater cut in Wasatch Formation,* **B.** *Close-up of a single pillar.* (Courtesy William M. Mintz.)

Figure 20.9. *The White River Badlands of South Dakota,* Badlands National Monument. (Courtesy of William M. Mintz.)

Figure 20.10. *Yellowstone National Park in the Absaroka Mountain Area of Wyoming.* Active evidences of internal heat are shown by **A** *the Norris Geyser Basin* and **B** *Old Faithful Geyser.* Volcanic deposits are displayed in **C** *the Grand Canyon of the Yellowstone.* (Part A courtesy of William M. Mintz. Parts B-C courtesy of National Park Service.)

through much of the Cenozoic, with major incursions occurring in the Eocene and Miocene. The shearing engendered by North America's colliding with the Pacific rift zone and thus forming a transform fault zone, however, produced folding, faulting, and uplift along the west coast in the Neogene (Fig. 20.19). This deformation gradually drove the seas out of all but southern California by the present. This disturbance is called the **Cascadian Orogeny** in the northwest and the **Pasadenan** or **Coast Range Orogeny** in California (Figs. 20.20–21). Block-faulting occurred on a grand scale from the Sierra Nevada to the Rockies in the Late Pliocene and Pleistocene. This elevated the great two-mile-high east scarp of the Sierra (Fig. 20.22) and the numerous high ranges of the Great Basin (Fig. 20.23). The cause of this extension in the Cordillera is uncertain. Current hypotheses include the shearing induced by transform faulting (Fig. 20.19), the presence of hot rising mantle plume under the western states, and the subsurface presence of the hot broken tip of the Farallon Plate formerly subducted beneath the western continental margin.

Eastern North America had been foundering since the Atlantic rift opened. The crust thins beneath a rift zone, and the trailing edges of

Figure 20.11. *The San Juan Mountains, Colorado.* Two scenes in the Mt. Sneffels area near Ridgeway. (Courtesy Colorado Dept. of Public Relations.)

plates subside. The Atlantic and Gulf Coastal Plains (Figs. 20.24, 20.27) of North America subsided throughout the Late Mesozoic and Cenozoic. Most of the Mesozoic and many of the Cenozoic sediments on these plains are of marine origin, but as the Rockies and even the Appalachians underwent Late Cenozoic rejuvenation, the flood of clastics was so great that all later ones are terrestrial in origin (Fig. 20.24). (Cenozoic rejuvenation of the Appalachians has probably been caused by isostatic adjustment.) Thus, the broad uplift that produced today's topographic Cordilleran and Appalachian mountains (Figs. 20.25–26) is of the same age, although the earlier times of rock accumulation and deformation differed. The isostatic rise of salt domes from Mesozoic evaporate deposits has also characterized the Gulf Coast in Cenozoic times (Fig. 20.27). These evaporites formed in the narrow, restricted basins of the incipient Mesozoic Atlantic Ocean.

North America and South America had been separated by the opening of the Atlantic rift in the Early Mesozoic. It not only opened the ocean basin, but also rotated the two continents away from each other in the Late Mesozoic and the Paleogene (Figs. 18.2–3, 20.2). In the Neogene, however, the competing compression of the westward-moving western Atlantic and eastward-moving east Pacific produced a great deal of deformation in the region between the two continents. Previous to this time, an island

Figure 20.12. *The Grand Tetons, Wyoming.* The steep eastern face of these mountains was produced by faulting. The rocks are Precambrian in age. (Courtesy of William M. Mintz.)

Figure 20.13. *Physiographic Diagram of a Part of the Western States.* Note canyons cut through the ranges of the Rockies, block-fault mountains, and volcanic mountains. (After Clark and Stearn.)

arc-trench system developed from Honduras through the West Indies into northern Venezuela (Fig. 18.23). The Atlantic part of the North American Plate still dives beneath the trench on the eastern side of the arc today (Fig. 20.28D). In the Cenozoic, an east Pacific plate spreading from the East Pacific Rise impinged on a continental side of the South American Plate in South America and triggered renewed mountain building in the Andes (Figs. 8.15, 20.29). In the narrow sea between the Americas, however, two oceanic plate edges collided. In

A

B

Figure 20.14. *Canyons Cut Through the Rocky Mountains of Colorado.* **A.** *Black Canyon of the Gunnison,* **B.** *Royal Gorge.* (Courtesy of William M. Mintz.)

Figure 20.15. *Columbia River Plateau Basalts at Dry Falls State Park, Washington.* (Courtesy of Washington State Dept. of Commerce and Economic Development.)

Figure 20.16. *High Cascade Volcanoes.* **A.** *Mt. Shasta, California;* **B.** *Mt. Mazama, Oregon, caldera now occupied by Crater Lake;* **C.** *Mt. Hood, Oregon;* **D.** *Mt. Rainier, Washington.* (Part A courtesy of William M. Mintz. Parts B-C courtesy of Oregon State Highway Travel Division. Part D courtesy Washington State Dept. of Commerce and Economic Development.)

A

B

Figure 20.17. *The San Andreas Fault, California.* **A.** *Northern California just north of San Francisco.* Rift zone from Tomales Bay (foreground) to Bolinas Lagoon in the distance. **B.** *Southern California. Elkhorn Hills and Carrizo Plain area.* (Courtesy of T.W. Dibblee, Jr., *California State Division of Mines Bulletin 190.*)

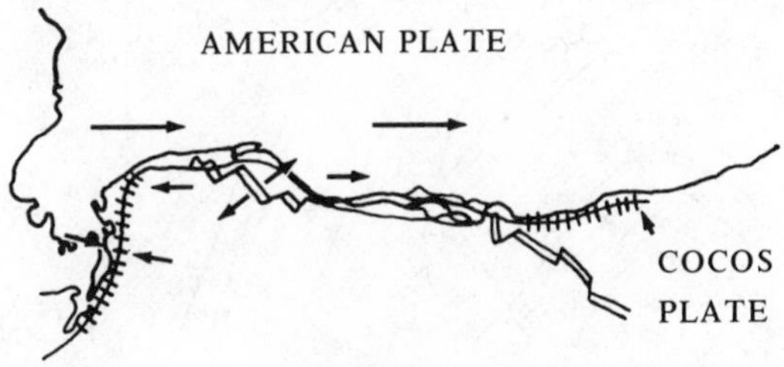

Figure 20.18. *Relationship of the San Andreas Fault to Other Plate Boundaries in the Northeast Pacific Area.* (From Atwater, *G.S.A. Bulletin,* vol. 81, December 1970. Used by permission of the author and G.S.A.)

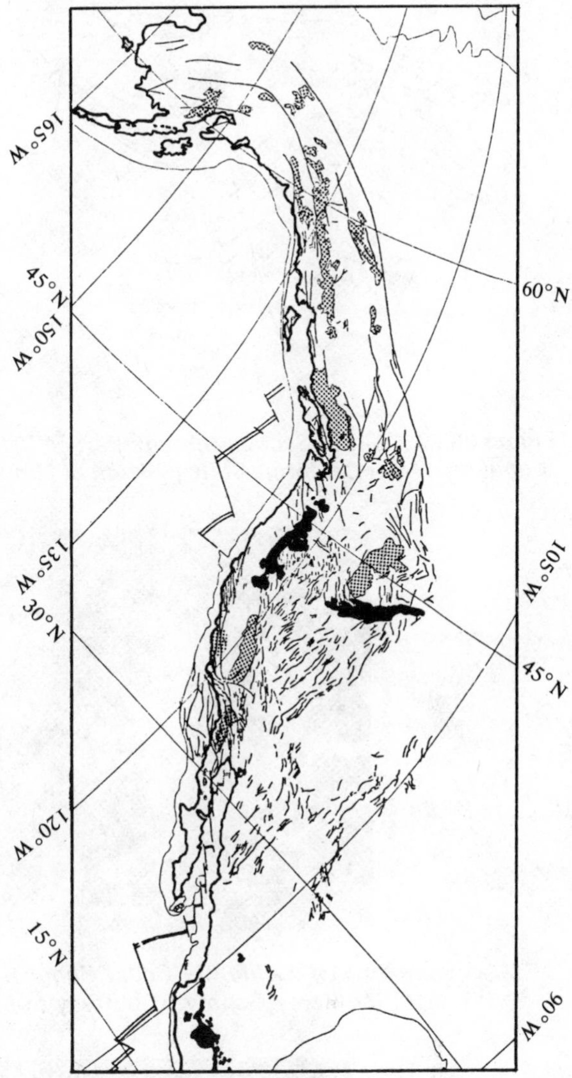

Figure 20.19. *Major Tectonic Features of Western North America Related to the Motion between the American and Pacific plates.* Fine stippling indicates igneous intrusions (batholiths); double lines are spreading ridges; single lines are faults; and black areas are lava fields. (From Atwater, *G.S.A. Bulletin,* vol. 81, December 1970. Used by permission of the author and G.S.A.)

this compression, the Caribbean became an isolated plate with the east Pacific plate diving beneath its western edge (Fig. 20.28). A volcanic arc above this subduction zone formed the Panama land bridge, linking the two continents for the first time since the Mesozoic. It also interrupted, along with the Tethyan mountain building of the Old World, the great east-west flow of water in the equatorial regions which had occurred in the Late Mesozoic and Early Cenozoic.

Europe

In Europe, the most important Cenozoic physical event was the **Alpine Orogeny** (Figs. 20.30–31). After the Late Cretaceous-Paleocene regression, the craton was marginally flooded in the Eocene in both the northwest and southeast. In the Oligocene, the major compression resulting from the rotation of Africa against southern Eurasia occurred. The great folding and thrust faulting of the Alps and other southern European ranges (Fig. 20.31) reached its peak at this time, though deformation had been continuous since the Mesozoic and continues today in many areas of the Mediterranean region (Fig. 20.32). There was also much flooding of the northern craton in the Oligocene. In the Neogene, the

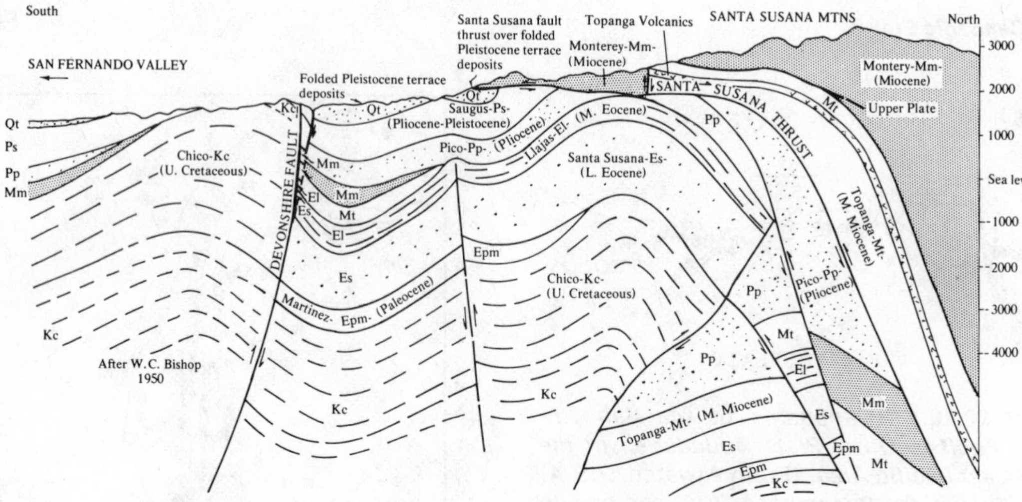

Figure 20.20. *Cross Section through Part of California Coast Ranges Showing Late Cenozoic Deformation.* (After *California State Division of Mines Bulletin 170*, 1954).

Figure 20.21. *California Coast Ranges.* Peachtree Valley in southern Gabilan Range, Monterey County. (Courtesy of Mary Hill.)

Figure 20.22. *Sierra Nevada Eastern Scarp.* Mt. Williamson and Mt. Barnard. (Courtesy of William M. Mintz.)

Figure 20.23. *The Basin-Ranges of Nevada.* Wheeler Peak in the Snake Range. Upfaulted Lower Paleozoic sedimentary rocks. (Courtesy Nevada Dept. of Economic Development.)

A

B

Figure 20.25. *The Rocky Mountains.* **A.** *Hallett Peak from Bear Lake, Rocky Mountain National Park, Colorado.* Rocks are Precambrian gneisses and granites of the craton. **B.** *Garden Wall, Glacier National Park, Montana.* Rocks are Precambrian Belt Supergroup sedimentary strata originally deposited in a shelf sea. (Courtesy of William M. Mintz.)

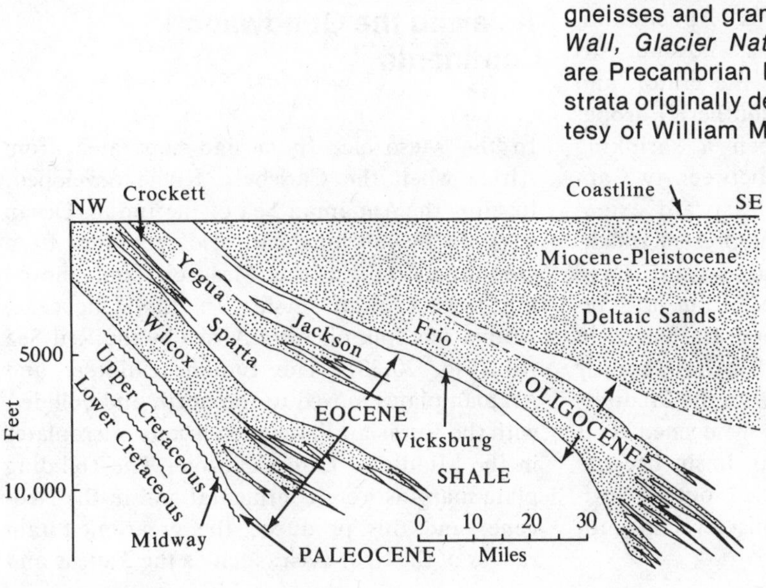

Figure 20.24. *Cross Section of Gulf Coastal Plain Showing Regressive Sequence.* (After Colle et al.)

Figure 20.26. *The Appalachian Mountains.* **A.** *The Adirondacks of New York,* **B-D.** *The Great Smoky Mountains of North Carolina and Tennessee.* (Part A courtesy of William M. Mintz. Parts B-D courtesy of National Park Service.)

flood of clastics from the Alps, Appennines, Dinaric Alps, Balkan Mountains, and Caucasus gradually drove the seas from the craton and welded many microplates with southern Europe.

The Mediterranean has been a shrinking ocean basin caught in a vise between two approaching continents. A long eastward extension of Tethys encompassing the present Black, Caspian, and Aral seas (and several large Balkan lakes) became isolated from the remainder and broke up into separate basins in the Neogene. Only the Black Sea has recently regained a tenuous connection with the Mediterranean. At the close of the Miocene, it is possible that the Mediterranean basin became completely landlocked and dried out, as evidenced by evaporite salt at this level in submarine cores.

Asia and the Gondwana Continents

In the Mesozoic, India had separated from Africa when the Carlsberg Ridge developed, forming the Andaman Sea of the Indian Ocean Figs. 18.1–3). China had also separated from Gondwanaland in the Mesozoic and moved northward toward Asia. In the Oligocene, Arabia was split off from Africa by the Red Sea rift (Fig. 20.33). The Indian, Chinese, and Arabian plates moved to the north and collided with the Eurasian Plate and various microplates in the Middle to Late Cenozoic. The colliding plate margins were continental and in the Neogene, and this produced the great mountain ranges of the Near East (such as the Taurus and

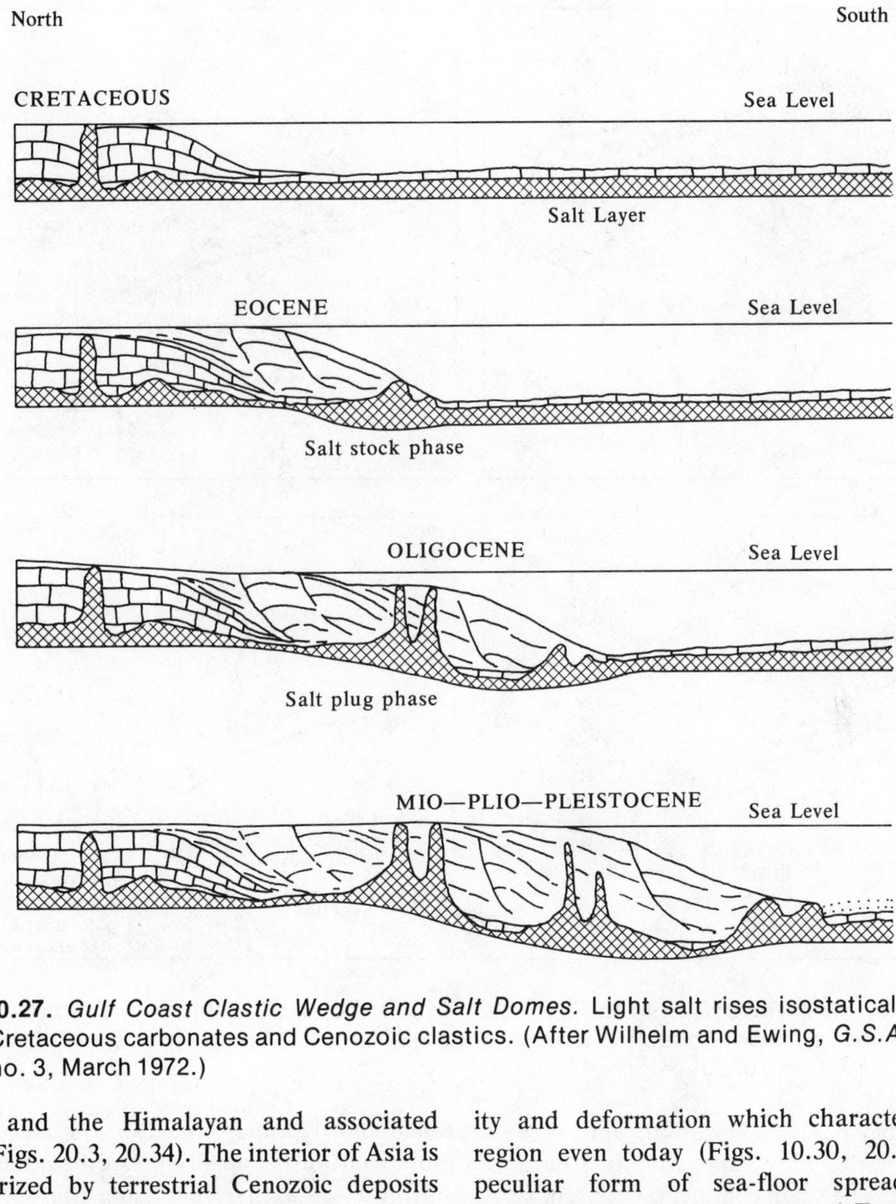

Figure 20.27. *Gulf Coast Clastic Wedge and Salt Domes.* Light salt rises isostatically through heavier Cretaceous carbonates and Cenozoic clastics. (After Wilhelm and Ewing, *G.S.A. Bulletin*, vol. 83, no. 3, March 1972.)

Zagros) and the Himalayan and associated ranges (Figs. 20.3, 20.34). The interior of Asia is characterized by terrestrial Cenozoic deposits shed by the high mountains elevated by continental collisions. Large strike-slip faults in central Asia seem to indicate that western China is being squeezed out of the way to the east as India pushes northward into Eurasia. Eastern Asia is bounded by trenches where the oceanic Pacific Plate and associated minor plates plunge beneath the continental Eurasian Plate (Fig. 8.15). The compression in this contact zone of island arcs has produced the continual instability and deformation which characterizes the region even today (Figs. 10.30, 20.3). Some peculiar form of sea-floor spreading has occurred between these arcs and Eurasia, thus opening back-arc oceans. This spreading may result from the rise of heat from subduction slabs or tension caused by the subducting slab pulling on the outer edge of the marginal sea.

Australia and Antarctica became separate continents in the Cenozoic (Fig. 20.3). Both were largely emergent and characterized by terrestrial and fresh-water deposits, with volcanics common in some areas. There were marginal

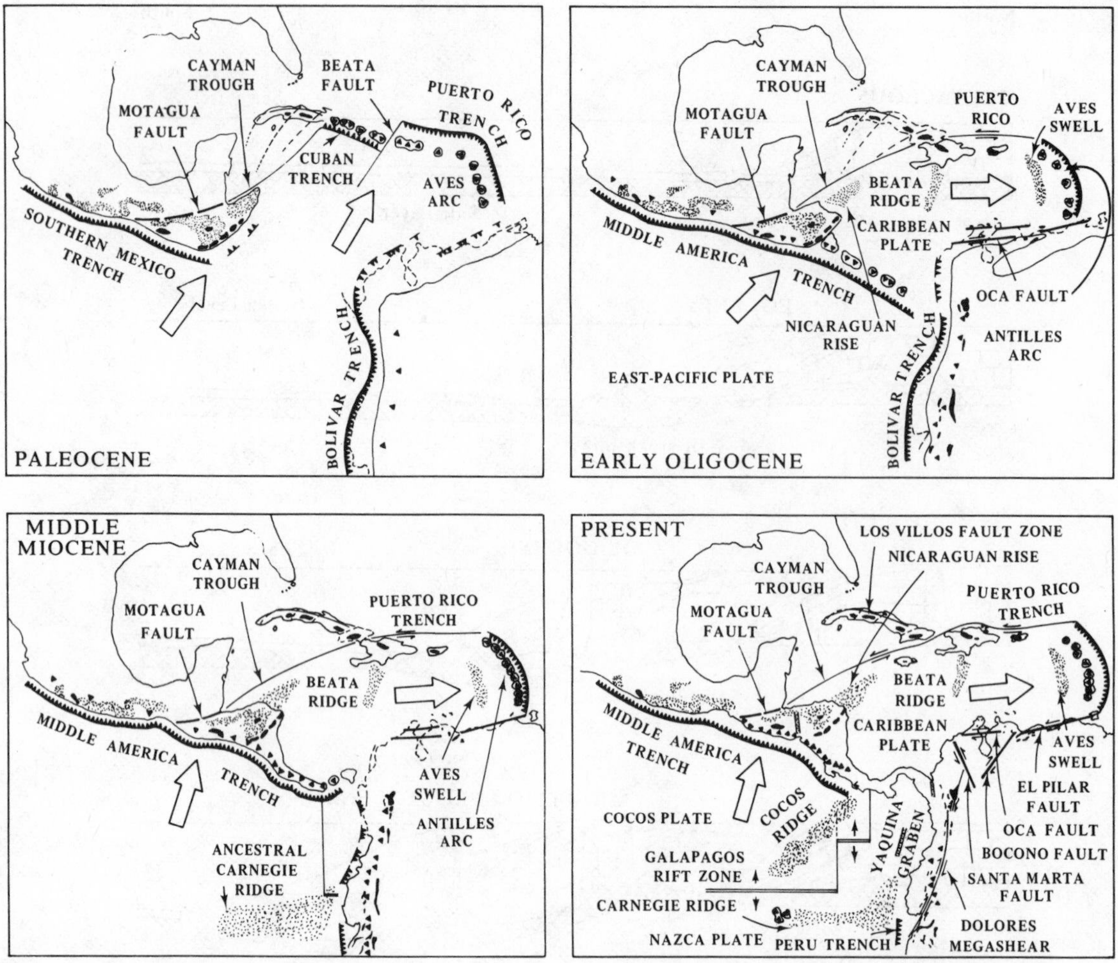

Figure 20.28. *Plate Tectonics of the Caribbean Region.* Sawteeth point in direction of subduction; large arrows indicate direction of plate movement. (After Malfait and Dinkelman, *G.S.A. Bulletin,* vol. 83, Feb. 1972.)

marine embayments in the south and west of Australia in the Early and Middle Cenozoic (Fig. 20.35). The small continental block of New Zealand became situated at the colliding edges of the Australian and Pacific plates in the Cenozoic, and the ensuing deformation produced the Southern Alps of that country (Fig. 20.36).

The Ocean Basins

The ocean basins continued to receive a sedimentary record during the Cenozoic Era. The calcareous deposits formed on the ridge flanks moved outward and downward, gradually dissolving away as they went into the deep-water oceanic waters. The older Cenozoic strata, situated near the margins of the present basins, are less calcareous and contain more residual clays. The large Pacific Plate, entirely oceanic, is notable because its western edge is bounded by several smaller plates, occupied by the Philippine Sea and other ocean basins, which lie between volcanic island arcs and Asia. These basins appear to have undergone a unique variety of sea-floor spreading as noted earlier.

Figure 20.29. *The Andes Mountains of South America.* The Cordillera de Huayhuash, Peru. Strata are of Mesozoic and Cenozoic age. (Courtesy of Peter J. Coney, from *G.S.A. Bulletin*, vol. 82, July 1971.)

The Pleistocene Ice Age

An event of profound physical and biological importance for the entire earth is the glacial ice age that occurred in the most recent part of the Cenozoic. Not until the 1820s and 1830s did geologists first begin to perceive the connection between various geological features of areas distant from present glaciers and the action of glaciers. The Swiss, naturally enough, noted that the erosional and depositional features produced by present Alpine glaciers extend far beyond the present extent of ice, even onto the plains. Before that time, European geologists attributed the poorly sorted clastic debris called "drift" found over much of the surface of northern Europe to Noah's Flood. A noted skeptic of the glacial theory, naturalist ***Louis Agassiz,*** went to the Alps to prove *Venetz-Sitten, de Charpentier,* and the other Swiss wrong, but came away convinced of the reality of an ice age in the not distant past. Agassiz discovered even more abundant evidence of glaciation in the British Isles, Scandinavia, and the north European Plain (Fig. 20.37). In 1846, he immigrated to America and found that an even greater area, virtually all of the continent north of the Ohio and Missouri rivers, had been glaciated (Fig. 20.37).

Also in 1846, *Edward Forbes* suggested that Lyell's Pleistocene "Period," originally based on the percentage of species of living mollusks it contained, was synonymous with this recent glacial age, the Quaternary. Most geologists, unfortunately, ultimately accepted this argument, including Lyell himself in 1873. As a result, they have been perpetually plagued by the problem of assigning boundaries to an epoch or period originally defined in two entirely different ways. The type Pleistocene of Lyell is in the marine rocks of southern Italy, while the type Quaternary is the terrestrial glacial debris of northern France. Recently obtained radiometric dates indicate both the type Pleistocene and the onset of widespread continental glacial conditions in middle latitudes began between two and three million years ago. Continental glaciation in high latitudes, such as Antarctica, began much earlier in the Cenozoic, judging from ocean floor sediments.

Geologists now know that Quaternary continental ice caps (Fig. 20.37) covered not only North America (including Greenland) north of the Ohio and Missouri rivers, northern Europe, and the Alps, but also western and southern South America, all of Antarctica, northern Siberia, the central Asian mountains and plateaus, Tasmania, and New Zealand. In

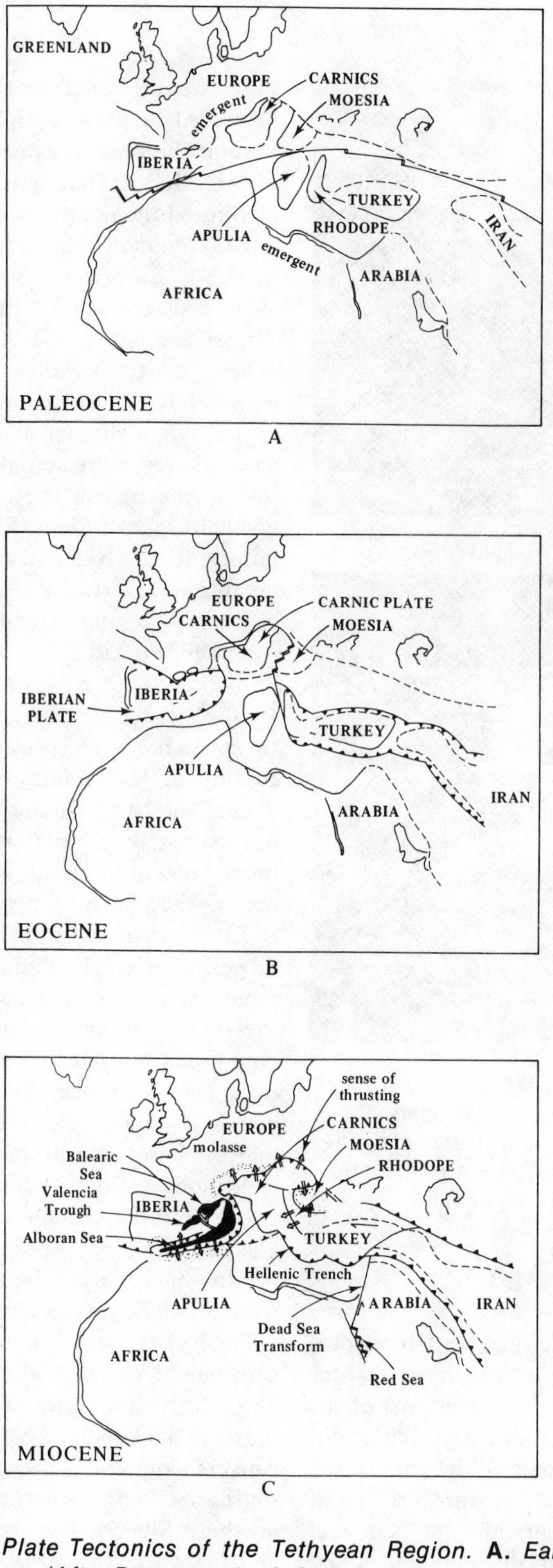

Figure 20.30. *Cenozoic Plate Tectonics of the Tethyean Region.* **A.** *Early Paleocene,* **B.** *Early Eocene,* **C.** *Middle Miocene.* (After Dewey et al., *G.S.A. Bulletin,* vol. 84, October 1973.)

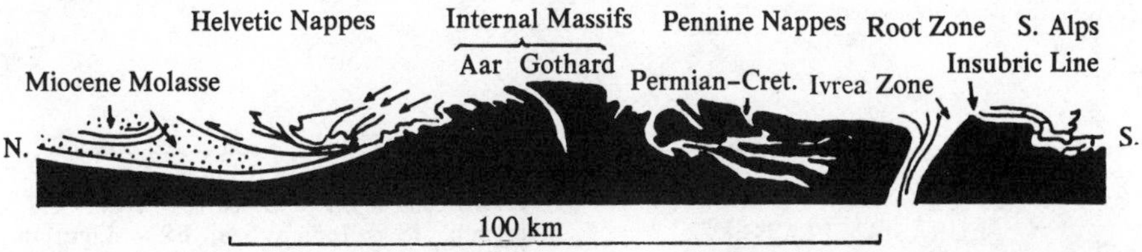

A

B

C

D

E

Figure 20.31. *The Alpine Orogeny.* **A.** *Diagrammatic view of major tectonic elements*; particularly the nappes or large overthrust sheets; **B.** *Pennine Alps, Weisshorn in foreground, Matterhorn in distance*; **C.** *Matterhorn from Zermatt*; **D.** *Pilatus from Rigi, Lake Lucerne*; **E.** *Jungfraujoch, Bernese Alps.* (Part A from Dewey and Bird, *Journal of Geophysical Research*, vol. 75, no. 14, May 10, 1970. By permission of authors and AGU. Parts B-E courtesy Swiss National Tourist Office.)

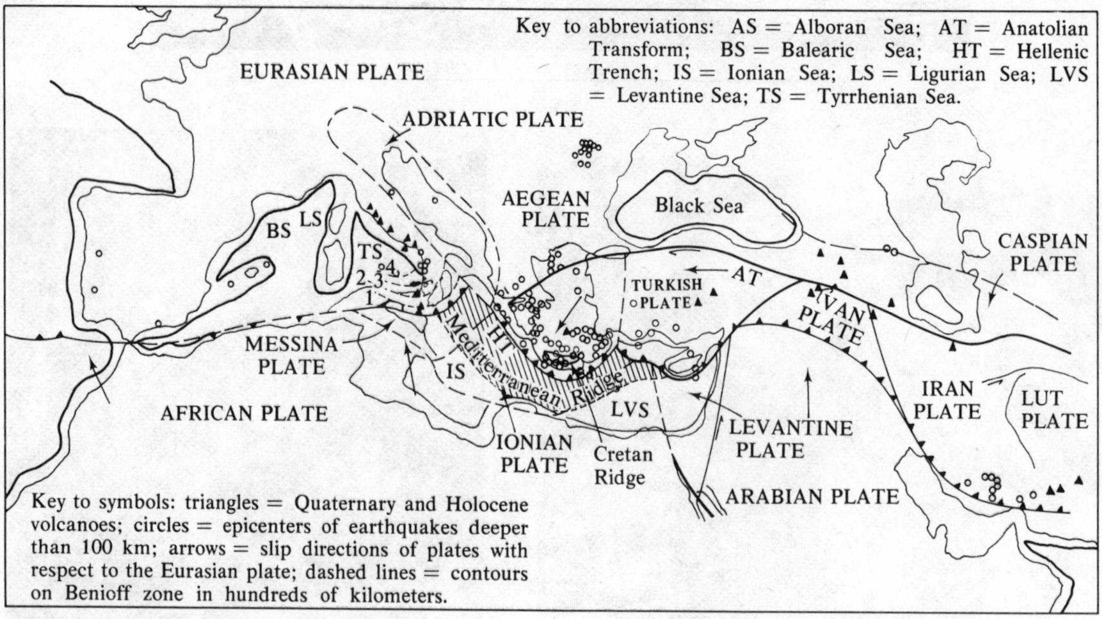

Figure 20.32. *Major Tectonics Features of the Present Tethyean Region.* Key to symbols: triangles = Quaternary and Holocene volcanoes; circles = epicenters of earthquakes deeper than 100 km; arrows = slip directions of plates with respect to the Eurasian Plate; dashed lines = contours on Benioff zone in hundreds of kilometers. Key to abbreviations: AS = Alboran Sea; AT = Anatolian Transform; BS = Balearic Sea; HT = Hellenic Trench; IS = Ionian Sea; LS = Ligurian Sea; LVS = Levantine Sea; TS = Tyrrhenian Sea. (After Dewey et al., *G.S.A. Bulletin,* vol. 84, Oct. 1973.)

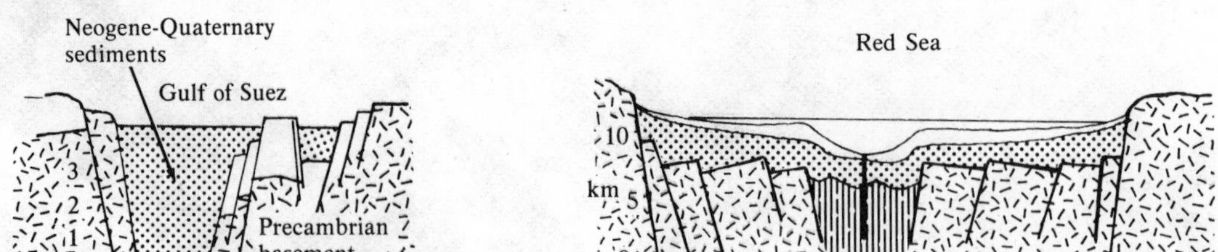

Figure 20.33. *Cross Sections of the Red Sea Today Showing Its Rift Zone Character.* **A.** *Gulf of Suez, an incipient rift;* **B.** *Red Sea proper, a mini-ocean basin.* (From Dewey and Bird, *Journal of Geophysical Research,* vol. 75, no. 14, May 10, 1970. By permission of authors and AGU.)

addition, high mountains all over the earth either had much larger glaciers than now or had glaciers where none exist today.

Other indirect evidences of glaciations and interglaciations were found. These include the pluvial lakes that existed during the early phases of glaciations in present arid regions such as the American southwest and the African Sahara (Figs. 20.38–39), large valleys and canyons cut by great amounts of glacial meltwater but now occupied by puny streams, and the rebounding of land areas after removal of the ice load (Figs. 20.40–41). The history of the Great Lakes of the midwestern United States is intimately involved with the glaciation in that region.

Figure 20.34. *The Himalayas.* These mountains, the highest on earth today, were formed by Late Cenozoic continent-continent collision. (Courtesy Sierra Club.)

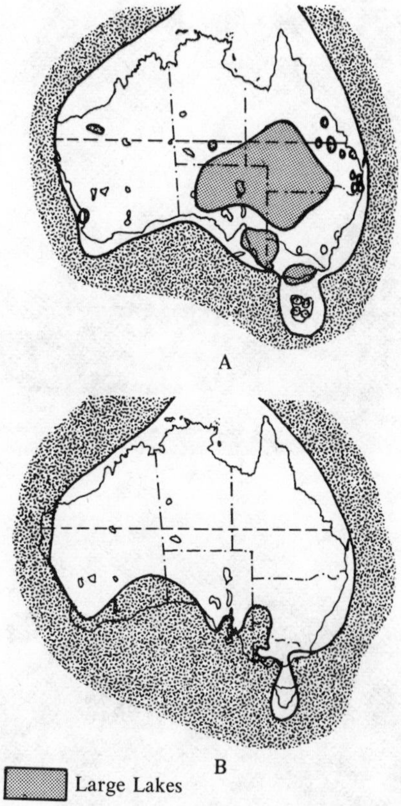

Figure 20.35. *Paleogeographic Maps of Australia.* **A.** *Oligocene,* **B.** *Miocene.* (Adapted from David, 1950.)

Geologists also have discovered that there was not a single glaciation, but several, separated by intervals that were warmer than present and during which the ice largely disappeared. The evidence (Fig. 20.42) includes multiple layers of glacial deposits separated by zones of deep weathering, soils, loess (wind-blown dust), and fossiliferous sediments bearing the remains of animals and plants of warm climates. In the oceans, layers of alternately warm and cold water faunas succeed one another in the sedimentary record (Fig. 20.43), offering further proof.

The traditional pattern (Fig. 20.44) was four glaciations called Nebraskan, Kansan, Illinoisan, and Wisconsinan from oldest to youngest in North America, and three interglacials called Aftonian, Yarmouthian, and Sangamonian from oldest to youngest in North America. Europeans call the glacial states Gunzian, Mindelian, Rissian, and Wurmian from oldest to youngest and have no separate names for the interglacials. (More recently, an earlier fifth or Donau Glacial has been added.) The system broke down, however, when the Wisconsinan tills proved to be multiple, and the stage was further subdivided into smaller glacial advances and retreats. Work in the ocean basins using temperature-sensitive foraminifera, geochemical techniques, and radiometric dating has shown that rather than four glacials and three interglacials, there has been a continuous succession of fluctuating glacial and interglacial intervals back into the Pleistocene (Fig. 20.45). Some of these appear to have been more or less glacial than others and perhaps correspond to the original seven subdivisions.

Debates rage about whether we are still in the Pleistocene and whether we are still in an interglacial epoch. The answer to the first question is no because Lyell originally established the Pleistocene on the basis of 90 percent living species. When we reach 100 percent living species, we are in his Recent (or Gervais' Holocene) Epoch. The different question about whether the present is or is not an interglacial is

A

B

Figure 20.36. *The Southern Alps of New Zealand.* **A.** *Mt. Sefton, Mt. Cook National Park;* **B.** *Milford Sound, Fiordlands National Park.* (Courtesy of New Zealand Tourist Office.)

intimately involved in understanding the causes of glaciation.

Causes of Glaciation

Glaciers have probably always existed in higher mountains, at least in high latitudes. But continental glaciation, in which large, thick masses of ice override vast lowland areas of continents, has not always existed. It occurred at least twice in the Precambrian, once in the Early Proterozoic (Aphebian) of Canada, and again in the Middle Proterozoic (Helikian). There was very widespread continental glaciation in the Late Precambrian-Eocambrian of nearly all the present continents. Glaciation was more or less continuous in Gondwanaland throughout the Paleozoic, but best known in the Ordovician of Africa and the Permocarboniferous of much of the supercontinent. (There was also a small Siberian icecap in the Late Paleozoic.) And finally, the great Quaternary glaciation occurred in the Late Cenozoic.

Various hypotheses have been advanced to explain the phenomenon of continental glaciation, but most have been either non-uniformitarian or lacking in supporting evidence. The lack of any periodicity in the occurrence of continental glaciations and the alternation of glacial and warm interglacial states have always been the biggest stumbling blocks to a coherent theory, if indeed all glaciations had the same cause. Clearly, some cyclical phenomenon must be responsible for the alternations of glacial and interglacial stages. This cannot be simply the starving of glaciers for lack of precipitation because the interglacials were warm intervals, not just cold times without glaciers.

There are three astronomical phenomena that do repeat themselves regularly and may influence temperatures. These are variations in the ellipticity of the earth's orbit, the obliquity of the earth's axis, and the wobble or precession of the earth's axis. The Yugoslavian meteorologist *Milankovitch* calculated the variations and their periods, and the German climatologist *Köppen* suggested that they might correpond with the glacial-interglaciation sequence. The combined effect of the three factors produces a greater cooling every 40,000 years and corresponds well with the Late Pleistocene cold and warm spells (Fig. 20.46).

Kukla noted that the earth's albedo ratio, that is, the ratio of the light reflected to the light received at its surface, varies in high latitudes according to the tilt of the earth's axis. Hence,

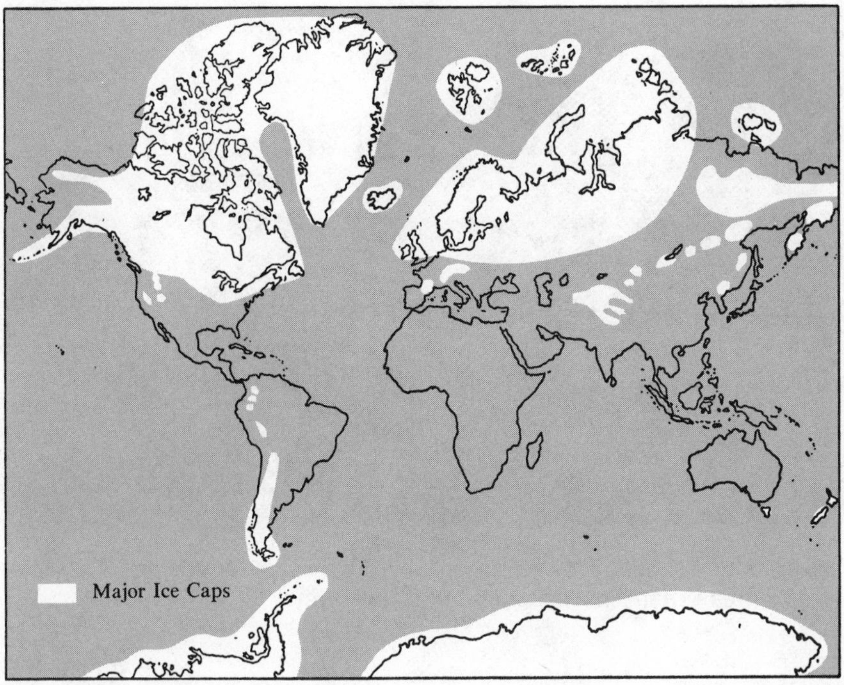

Figure 20.37. *Glaciated Regions During the Pleistocene.* Maximum extent of ice caps is shown. (After Antevs, Flint, Wills.)

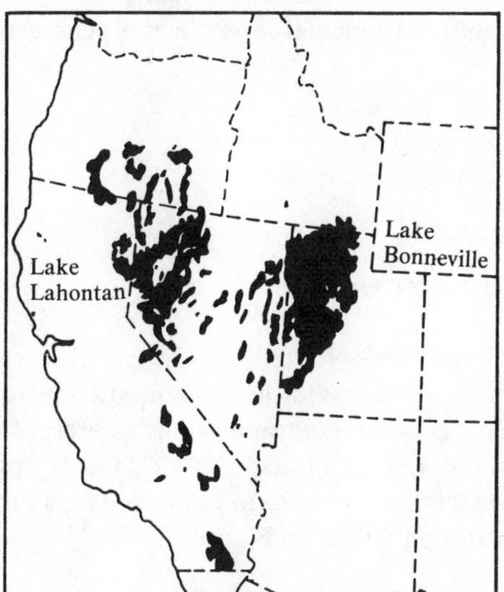

Figure 20.38. *Pleistocene Pluvial Lakes of the Southwest.* (After Thornbury.)

as the earth wobbles, its albedo ratio in high altitudes changes and the amount of heat received at the earth's surface there varies. In Kukla's view, the precession of the axis provides the most plausible cause for the rhythmic alternation of glaciations and interglacials. Glaciers form when in that part of the wobble when the albedo ratio is high, and they melt when it is low. A recent study of dated, sea-cut terraces in New Guinea has indicated that the timing of the glacial-interglacial sequences coincides with albedo fluctuations predicted astronomically.

But the precession of the axis and the other astronomical phenomena have occurred continuously since the earth's formation; why haven't glaciations? Perhaps paleogeography offers the answer, a suggestion made in the 1800s by *Lyell* and *Andrew Ramsay.* When the circulation of the oceans is upset by the situation of the continents, the earth's entire heat balance can be disturbed. Furthermore, high continents

Figure 20.39. *Pluvial Lake Shoreline.* Lake Bonneville terrace near Salt Lake City, Utah. (Courtesy Utah Travel Council and Hal Rumel.)

increase the albedo ratio and extensive ones lower temperatures. Certainly during the Paleozoic Gondwana glaciations and the Late Cenozoic glaciations at both poles, the continents were high, very extensive, and clustered near the poles. In the Paleozoic and Mesozoic, the Pacific Pole was in the Pacific basin, and there was generally no glaciation near it. In the Mesozoic, the Gondwana Pole was on the margin of Gondwanaland and the vast Pacific Ocean. Again, there was no glaciation. So we can say that perhaps glaciations occur when the three astronomical phenomena noted earlier coincide to reduce the earth's heat budget or during the high albedo ratio parts of the wobble cycle provided these conditions also exist when large, high continental masses cover or surround the polar regions.

Much further testing and observation are necessary, however, before we can confidently state that we have found the explanation for continental glaciations. If we are right though, we can say that the present is only an interglacial, because all the conditions are right for glaciation except the current astronomical positions which will return to glacial levels in the near geologic future.

The Future?

We close this final chapter on the earth's physical history with another prediction for the future. Based on extrapolations of present plate motions and amounts, the world fifty million years from now will probably look something like that portrayed in Figure 20.47.

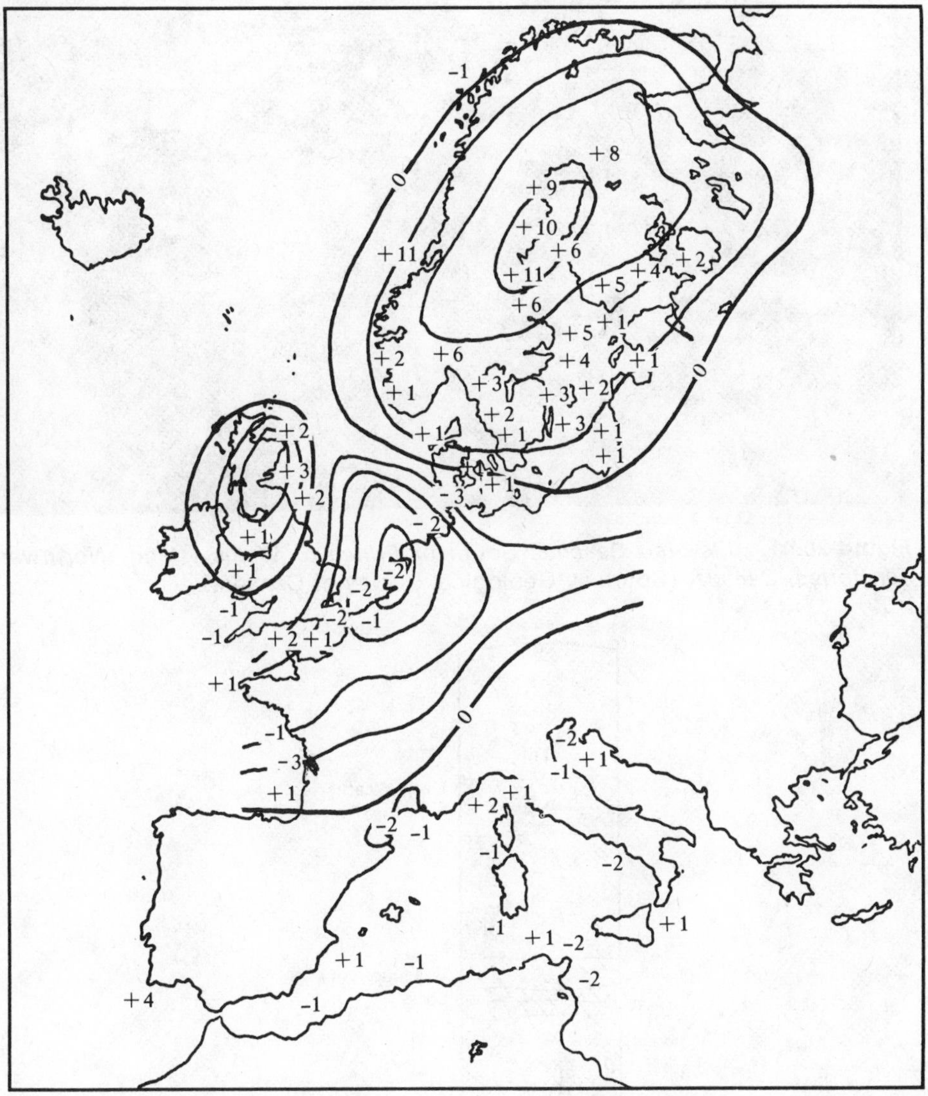

Figure 20.40. *Crustal Rebound in Scandinavia Following Deglaciation.* Positive numbers indicate uplift in millimeters per year; negative numbers indicate subsidence. (After Fairbridge.)

Figure 20.41. *Elevated Beaches South of Prince of Wales Island, Northwest Territories, Canada.* (Courtesy Geological Survey of Canada.)

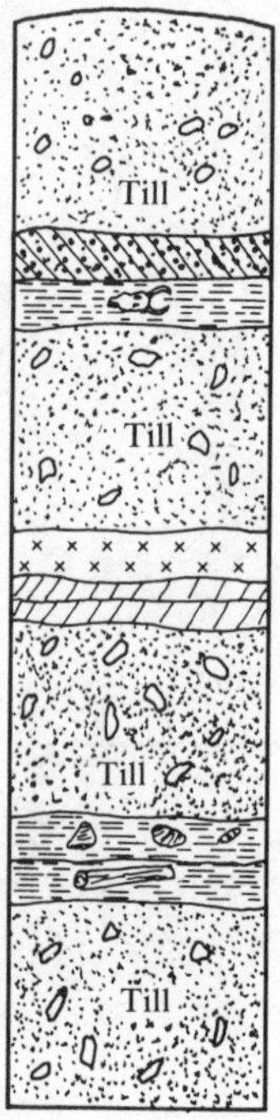

Figure 20.42. *Evidences of Multiple Glaciations in Pleistocene.* The materials in the layers indicated along the right margin could have formed only if the glacial ice melted and warmer conditions prevailed.

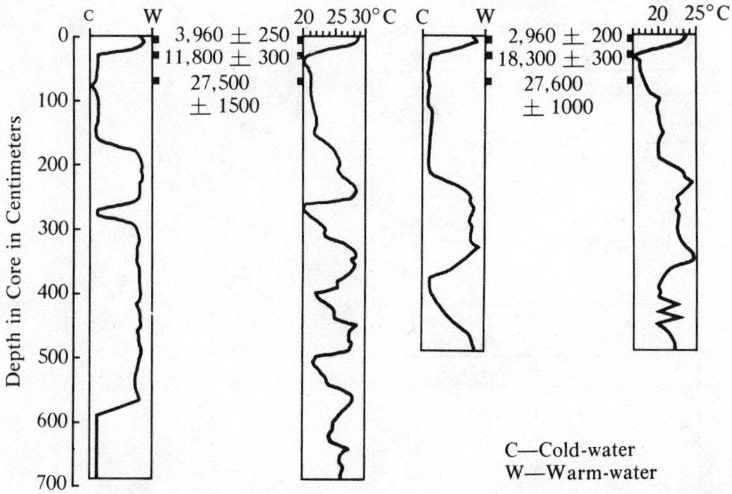

Figure 20.43. *Oceanic Evidence of Alternating Glacials and Interglacials in Sedimentary Cores.* Climatic curves were developed on foraminiferal assemblages and oxygen isotope ratios, indicating alternating warm-water and cold-water conditions. (From Leo F. Laporte, *Ancient Environments.* © 1968, Prentice-Hall, Inc., Englewood Cliffs, N.J.)

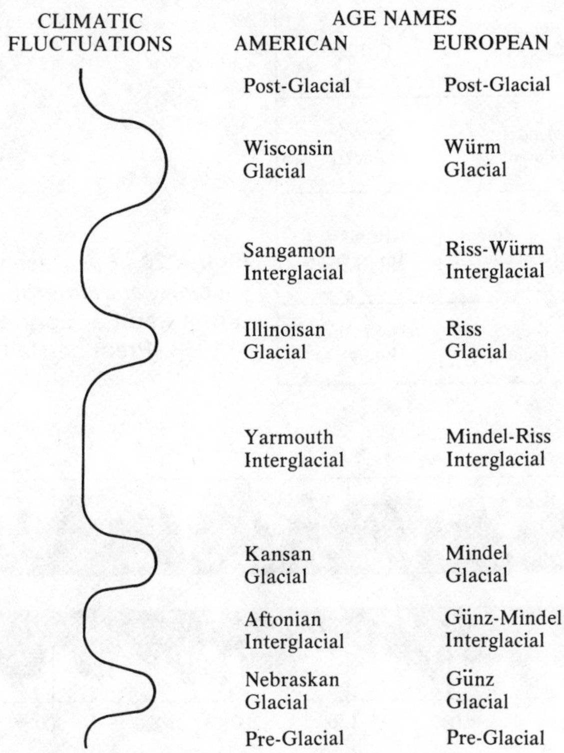

Figure 20.44. *Traditional Pattern of Four Glacials and Interglacials.*

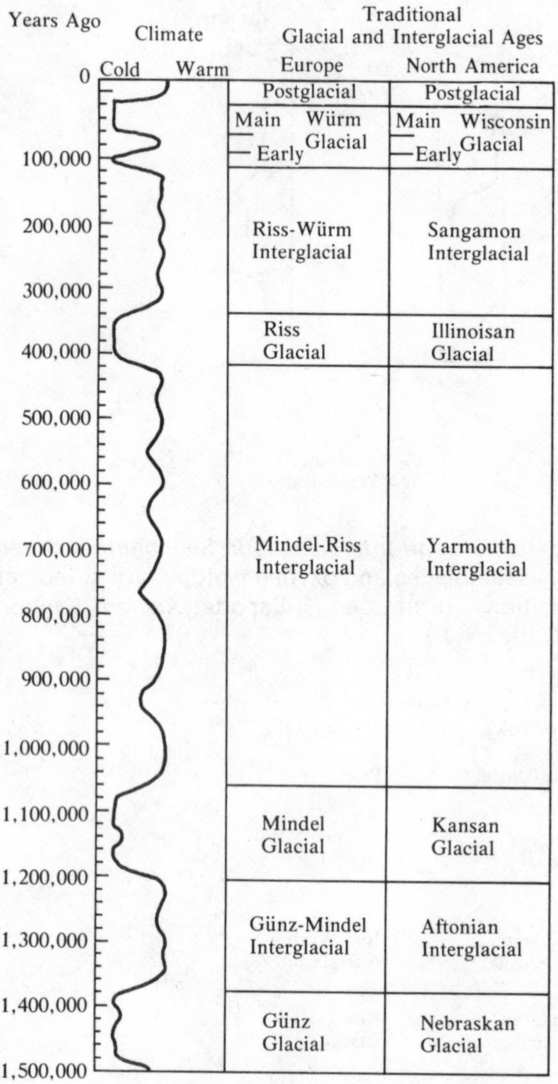

Figure 20.45. *Recent Evidence of Numerous Glacials and Interglacials in Oceanic Cores.* (From Leo F. Laporte, *Ancient Environments,* © 1968, Prentice-Hall, Inc., Englewood Cliffs, N.J.)

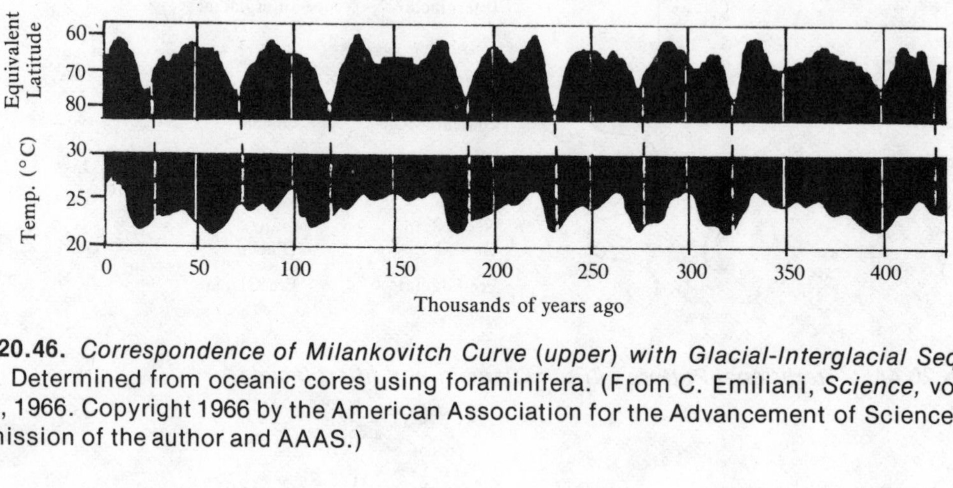

Figure 20.46. *Correspondence of Milankovitch Curve (upper) with Glacial-Interglacial Sequence (lower).* Determined from oceanic cores using foraminifera. (From C. Emiliani, *Science,* vol. 154, Nov. 18, 1966. Copyright 1966 by the American Association for the Advancement of Science. Used by permission of the author and AAAS.)

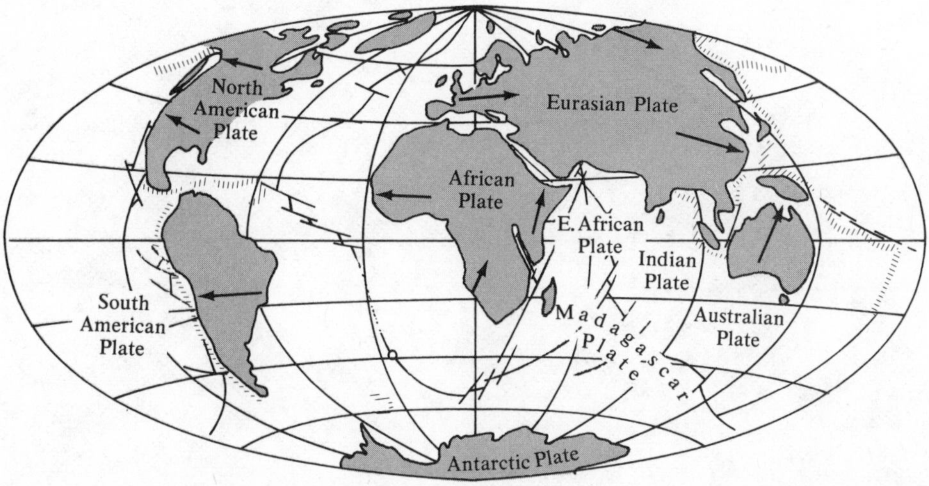

Figure 20.47. *The Surface of the Earth as It May Appear in Fifty Million Years.* Epeiric seas not shown. The present direction and rate of plate movements are assumed to remain constant. (Data from Dietz and Holden.)

REVIEW

1. What are the two major ways in which the Cenozoic epochs are distributed into periods?
2. What has been the main motion of the North American Plate in the Cenozoic Era?
3. What events produced the major deformation of the Tethyan region?
4. What is the relationship of the **Fort Union Clastic Wedge** to the **Tejas Sea?**
5. What is the origin of the spectacular canyons of the Rocky Mountains?
6. What is the historical significance of the **San Andreas Fault** of California? How might it be related to the Cenozoic deformation of the western United States?
7. How and when did the present **Panamanian land bridge** develop?
8. What major orogeny affected Europe in the Cenozoic?
9. How does the original definition of the Pleistocene differ from the concept of it as the great Ice Age?
10. What is the evidence for multiple Pleistocene glaciations?
11. Discuss at least one plausible theory for the origin of continental glaciations.
12. What is the future of California in fifty million years?

21

The Rise of the Modern Biota: Cenozoic Life

21

In the Cenozoic Era, the story of present-day life unfolds. On land, the mammals were the dominant large animal group, giving rise to the common name for the era, the **Age of Mammals.** Insects and angiosperms, so vital to the support of all terrestrial life, were just as significant, however. In the seas, the same groups that dominated the Mesozoic, the protistans and the mollusks, continued to expand and diversity with the significant exception of the extinct ammonoids.

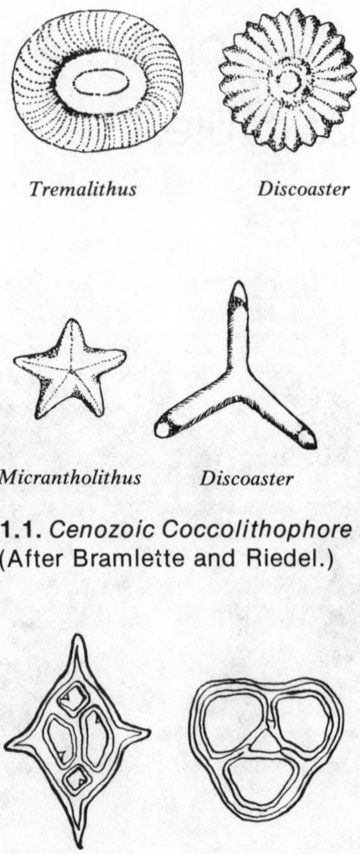

Figure 21.1. *Cenozoic Coccolithophore Plates.* X 1000. (After Bramlette and Riedel.)

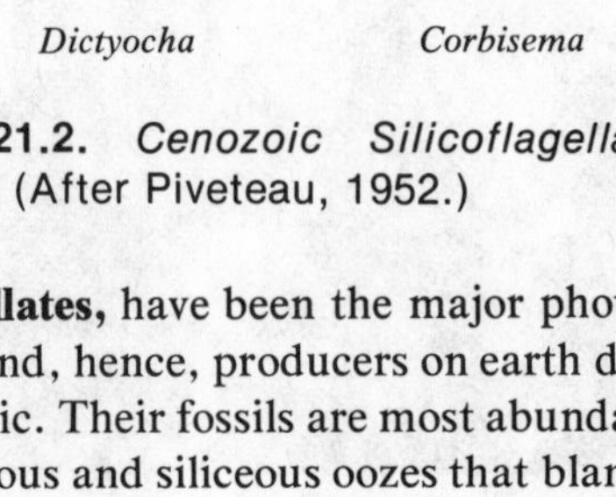

Figure 21.2. *Cenozoic Silicoflagellates.* X 500. (After Piveteau, 1952.)

Marine Life

The major marine fossil organisms of the Cenozoic are the **coccolithophores, diatoms, dinoflagellates, red algae, foraminifera, radiolaria, hexacorals, bryozoans, gastropods, pelecypods, annelids, crustaceans, echinoids,** and **fish.** The foraminifera, gastropods, and pelecypods are the major fossils for correlation of sediments formed in shallower water environments, while the coccolithophores, planktonic forams, and radiolaria are important in dating ocean basin deposits. The increasing provincialism that occurs in the Cenozoic because of the fragmented oceans and continents and the strong latitudinal gradients of the final ice age, make correlation from region to region more difficult than for any other interval of the geologic past since the Precambrian.

Protistans

The pelagic golden algae (Figs. 21.1–2), particularly the **coccolithophores,** but also the **silicoflagellates,** have been the major photosynthesizers and, hence, producers on earth during the Cenozoic. Their fossils are most abundant in the calcareous and siliceous oozes that blanket large areas of the ocean basins. One important variety of plate or coccolith is the star-shaped **discoaster,** which is important for correlation and disappeared in the Pleistocene.

The siliceous **diatoms** (Fig. 21.3) are second only to the golden algae as producers in the Cenozoic. Their remains are abundant in oceanic oozes, near-shore basin deposits, and they even invaded fresh water in the Neogene. Several formations are composed chiefly of diatom frustules forming the economically important rock diatomite, which is used in filters, insulation, etc.

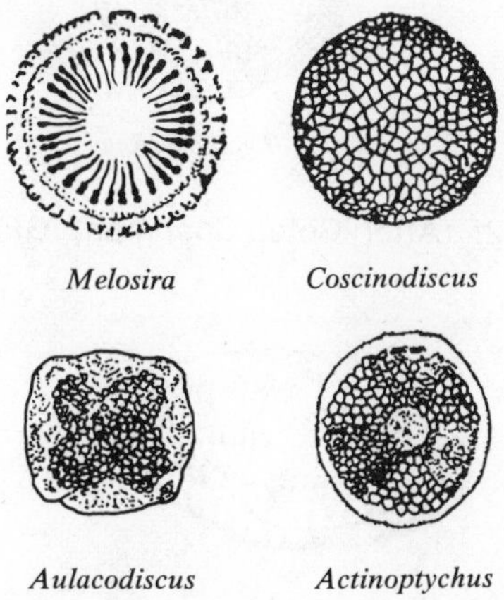

Figure 21.3. *Cenozoic Marine Diatoms.* X 200. (After Cupp, Lohman.)

The cellulose cysts of **dinoflagellates** (Fig. 21.4) including the **hystrichospheres** are another common Cenozoic fossil group. These pelagic producers are very important in the marine ecosystem also.

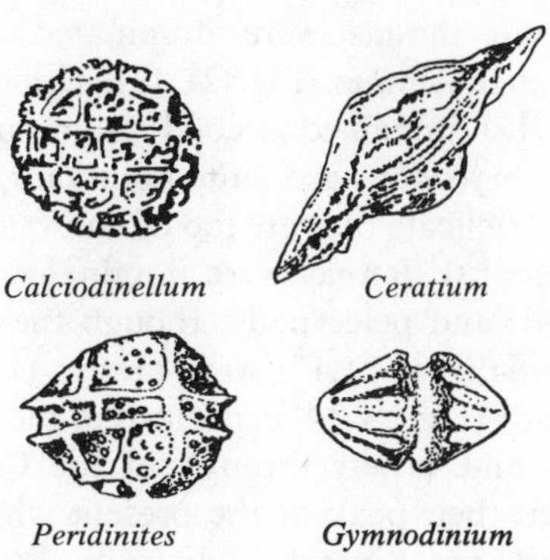

Figure 21.4. *Cenozoic Marine Dinoflagellates.* X 500. (After Deflandre.)

The **red algae,** like the green and the brown, include numerous seaweed types ill-suited for preservation and therefore probably much more abundant in the past than our record shows. More so than the green or brown, however, the red algae have evolved lineages with calcareous remains. The encrusting sheets of *Lithothamnion* and the delicate, jointed branches of other coralline red algae are significant, but usually overlooked, Cenozoic fossils (Fig. 21.5). They are major elements in coral reefs of the Cenozoic, where they served as cementing agents.

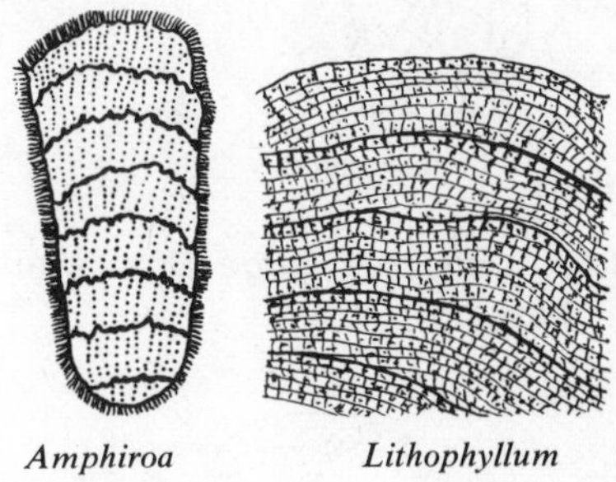

Figure 21.5. *Cenozoic Red Algae* (cross sections). X 50. (After Johnson and Ferris.)

The **foraminifera,** both benthic and planktonic, continued their Mesozoic expansion unabated in the Cenozoic. The most notable forms were the large, disk-shaped **nummulites** and **orbitoids** (Fig. 21.6) that especially characterized the Paleogene or Nummulitic Period. These forams were very numerous in the shallow, tropical seas of that time, particularly Tethys. Their remains are quite abundant in the limestone used to face the Egyptian pyramids, and the Greek historian *Heredotus* mistook them for food lentils carried by the slaves who toiled on the tombs. These large foraminifera, and the similarly adapted but elongate, wheat-grain-shaped **alveolinids** (Fig. 21.7), were drastically reduced and geographically restricted by the cooling climates of the Late Cenozoic. The abundant small forams (Fig. 21.8) include both benthonic and planktonic genera.

The siliceous **radiolaria** (Fig. 21.9) reach a peak in Cenozoic deposits. They are abundant in both the deposits of marine basins marginal to continents and the oceanic oozes.

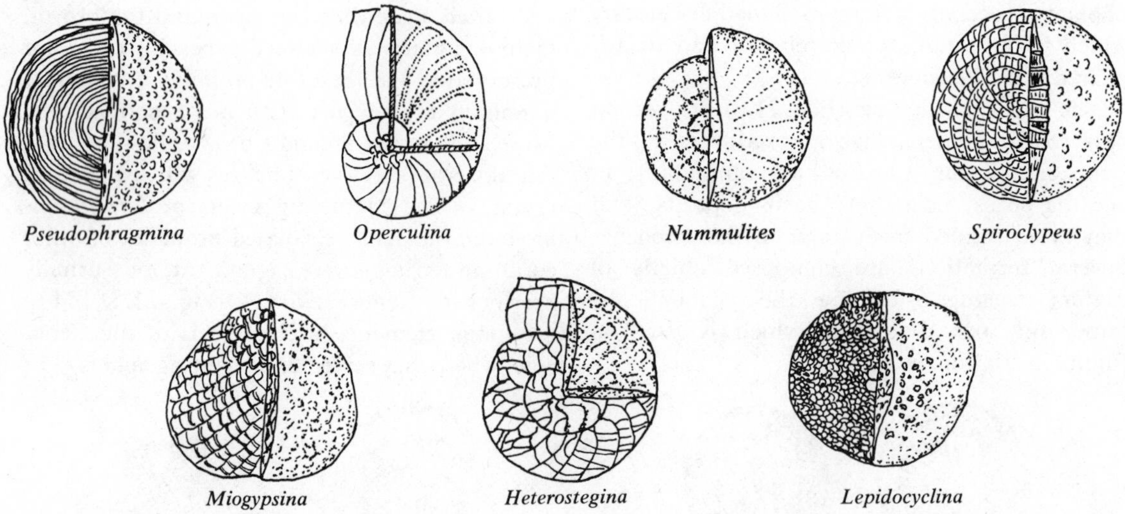

Figure 21.6. *Cenozoic Large Disc-Shaped Forams.* X 12. (After Cole, Cushman, Glaessner, Jones, and Vaughn.)

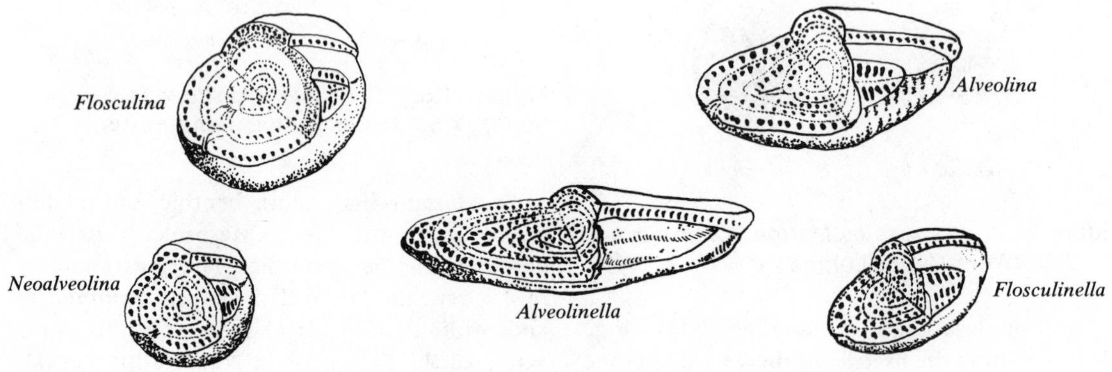

Figure 21.7. *Cenozoic Alveolinid Forams.* X 15. (After Sigal.)

Animals

Hexacorals, particularly the reef formers, were abundant and widespread in the Early Cenozoic epochs when shallow seas were more widespread and climates were milder. They became reduced and more restricted geographically with the cooling trend of the Late Cenozoic. Also, the mountain building that severed the worldwide east-west tropical oceanic connections in the Middle Cenozoic led to provincialism of the coral fauna. Both solitary and colonial genera are common (Fig. 21.10).

Bryozoan faunas were dominated by the delicate **cheilostomes** (Fig. 21.11). Cyclostomes (Fig. 21.12) remained second in importance. Cenozoic bryozoans are quite abundant, but so small and delicate they are too often overlooked.

The mollusk faunas were dominated by the gastropods and pelecypods, though the conical **scaphopods** (Fig. 21.13) were always common. **Gastropods** (Fig. 21.14) continued to increase in numbers and variety throughout the Cenozoic and are at their peak at the present when they are second only to insects in diversity. Worthy of special mention is the high-spired genus *Tur-*

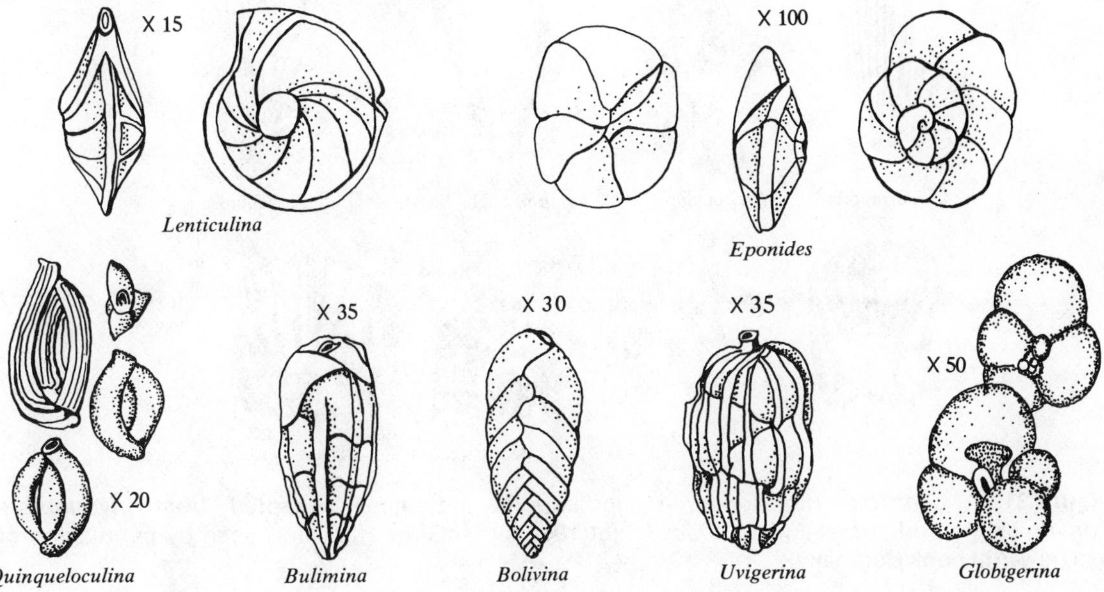

Figure 21.8. *Cenozoic Smaller Foraminifera.* Magnifications indicated. (After Bandy.)

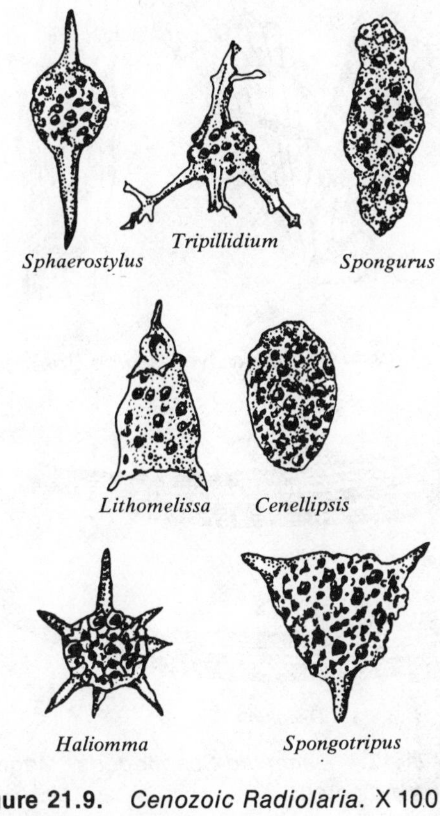

Figure 21.9. *Cenozoic Radiolaria.* X 100. (After Campbell and Clark.)

ritella and its relatives. This subtropical to tropical genus is very abundant and widespread through the Miocene when climates were milder than at present in current temperate regions and is widely used for correlation. Like most other warm-water marine forms, its range and numbers decreased in the cooling Late Cenozoic. **Pelecypods** (Fig. 21.15) also continued their great expansion in the Cenozoic. The modern burrowing types dominated, but the arks, the scallops, and the oysters continued to be important. A significant Paleogene group of large bivalves is represented by the genus *Venericardia.* In the Neogene, many of the scallops and oysters reached very large size.

Annelid worm tubes (Fig. 21.16), including coiled forms, are very common Cenozoic fossils, particularly as encrustations on mollusks. They are often overlooked, however, because of their small size and general simplicity.

The **crustacean** arthropods include several notable Cenozoic groups. The marine **ostracods** (Fig. 21.17) continue in abundance. The **barnacles** (Fig. 21.18) became important in the Cenozoic. Some species of *Balanus* reached truly

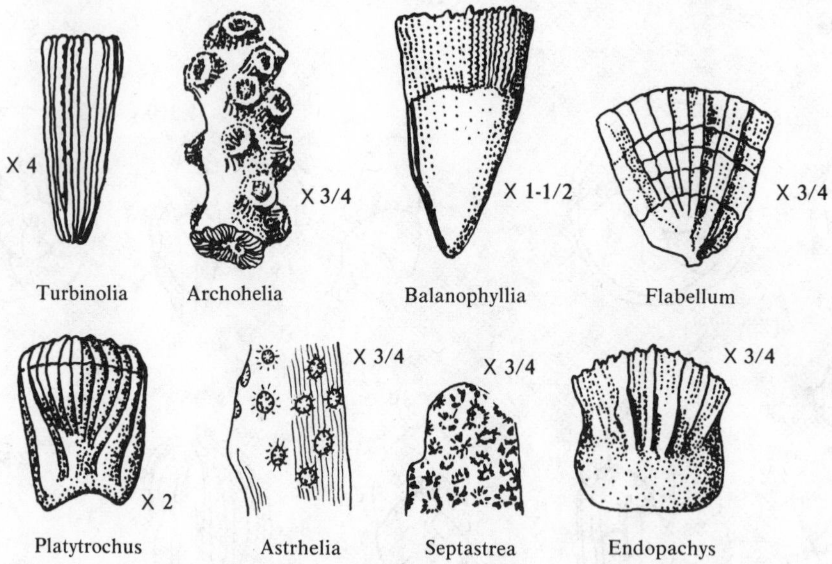

Figure 21.10. *Cenozoic Hexacorals.* Magnifications indicated. (Adapted from *Invertebrate Fossils* by Moore-Lalicker-Fischer, Copyright 1952 by McGraw-Hill, Inc., used by permission of McGraw-Hill book Company.)

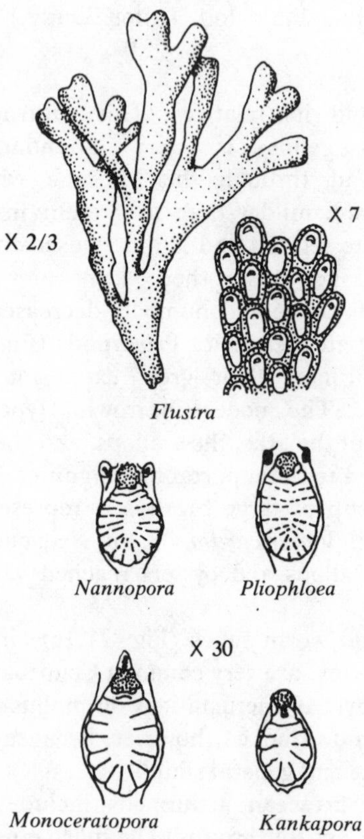

Figure 21.11. *Cenozoic Cheilostome Bryozoa.* Magnifications indicated. (After Bassler.)

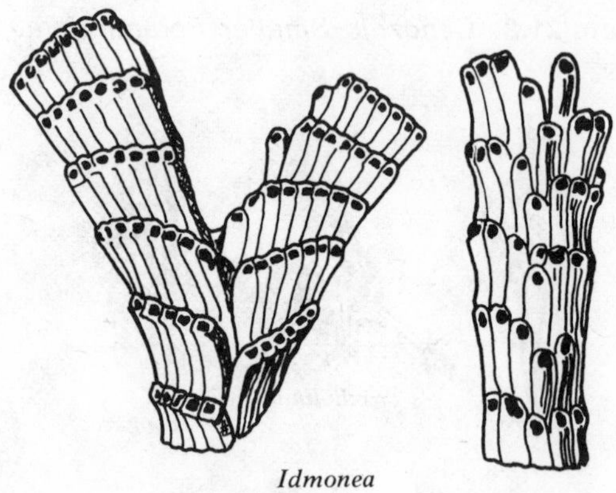

Figure 21.12. *Cenozoic Cyclostome Bryozoa.* X 8.

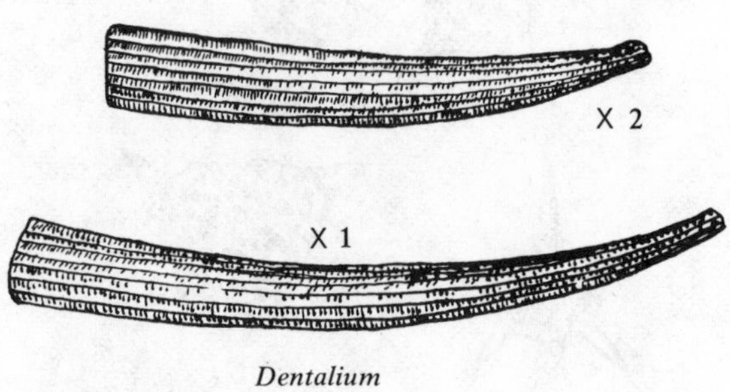

Figure 21.13. *Cenozoic Scaphopods.* Magnifications indicated.

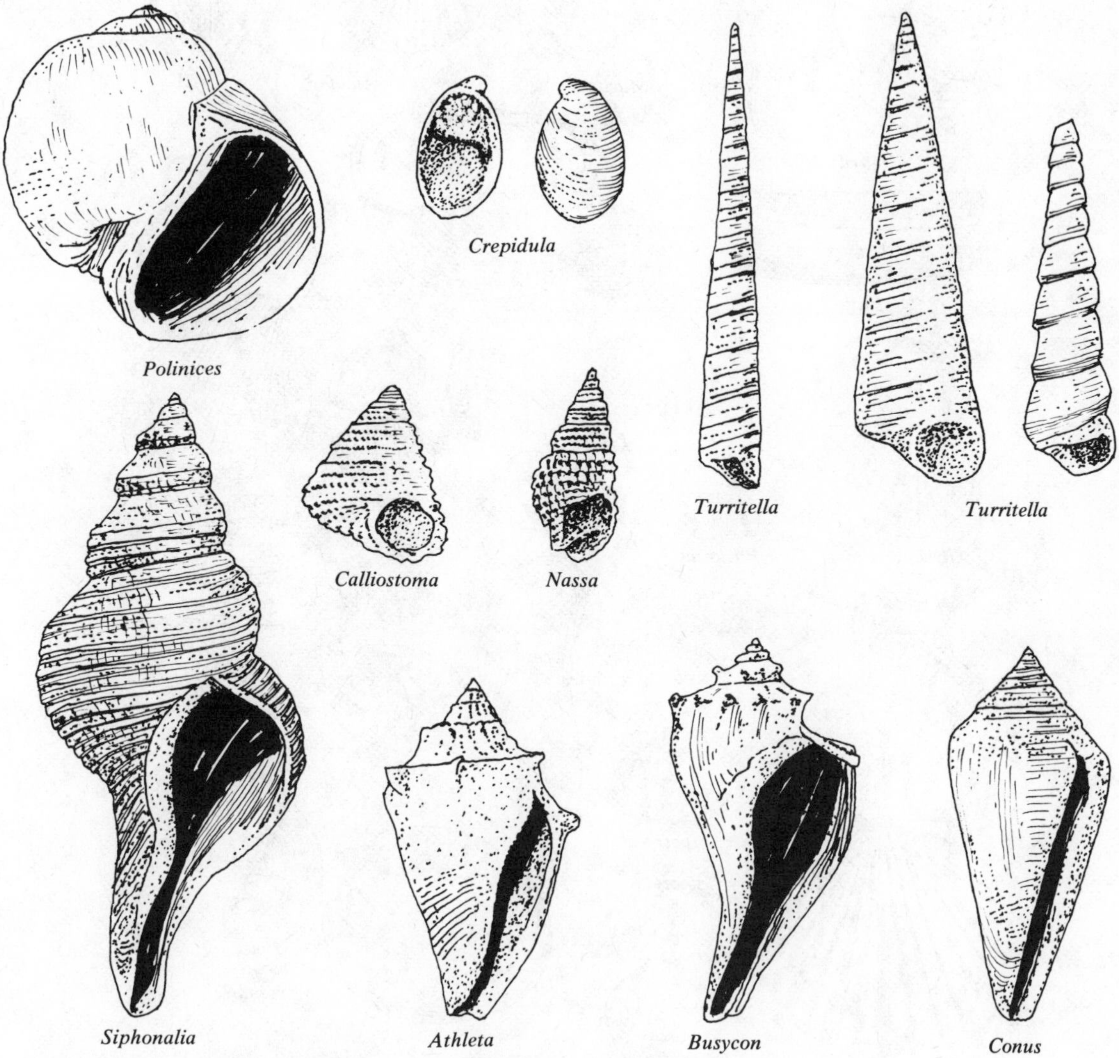

Figure 21.14. *Cenozoic Gastropods.* Natural size.

gigantic size in the Neogene. Finally, the **malacostracan** crustaceans are well represented by broken bits of appendages and carapaces and occasionally more complete fossils.

The **echinoids** (Fig. 21.19) are another group that reaches its peak in Cenozoic strata. The irregular burrowing forms dominate, but this is probably a function of their living in a favorable environment for fossilization. The **sand dollars** became the most successful Cenozoic echinoids, and their internally supported tests are well suited for preservation. They were common both in the tropical Early Cenozoic faunas and the cooler ones of the Late Cenozoic. The **cassiduloids** are chiefly a warm-water group that declined after an Eocene peak. The thin-shelled **heart urchins** of the modern Order Spatangoida are common in Cenozoic deposits.

Marine invertebrates have probably been more diverse in the Cenozoic than at any time in the geologic past because of the extreme fragmentation of continental shelves and ocean

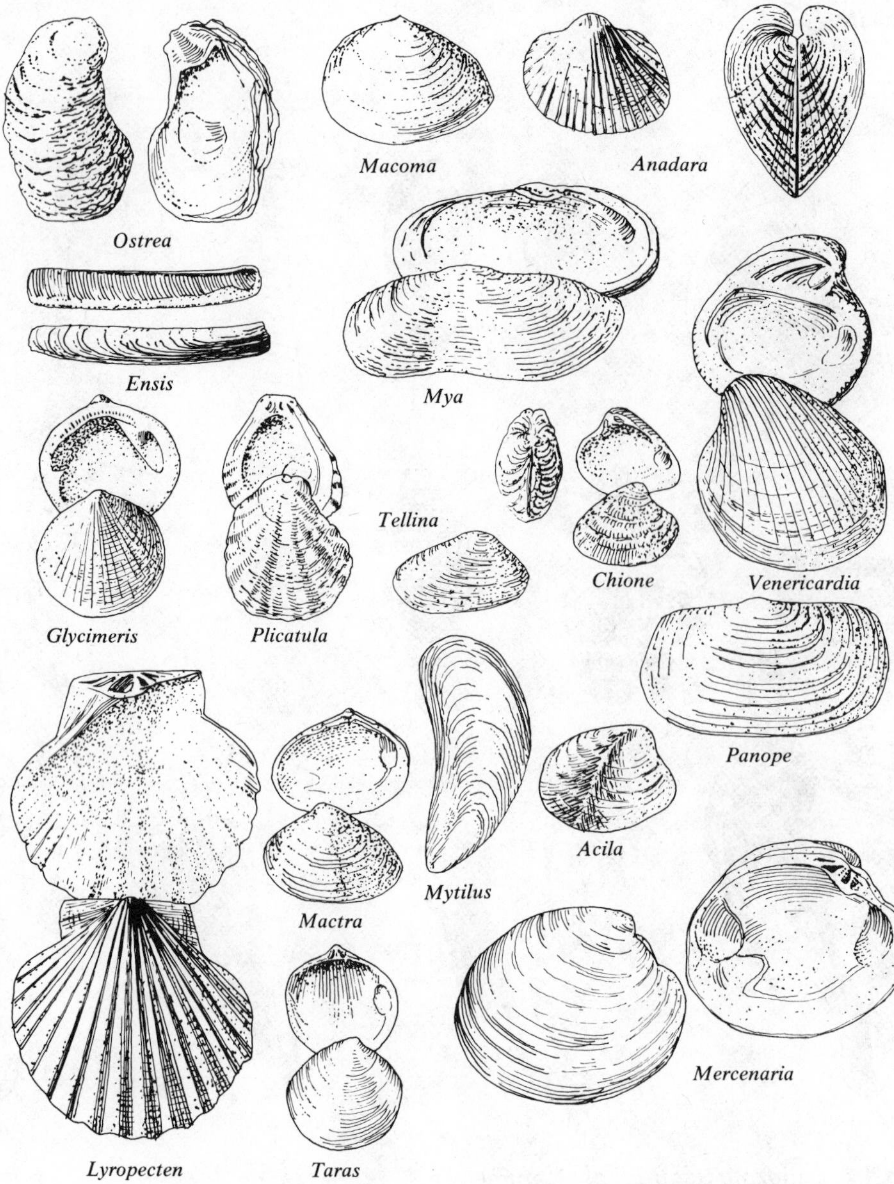

Figure 21.15. *Cenozoic Bivalves.* Natural size.

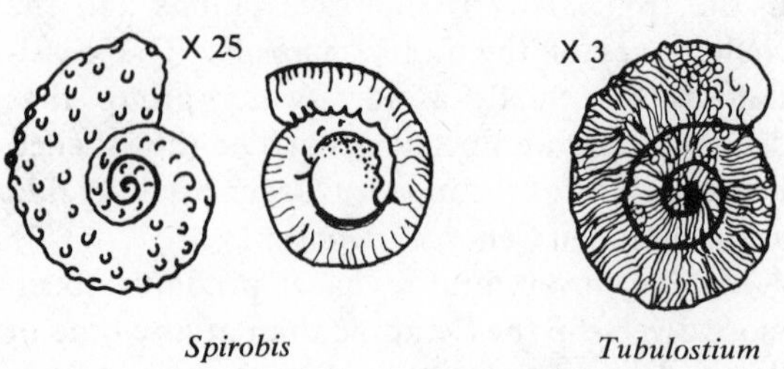

Figure 21.16. *Cenozoic Annelid Worm Tubes.* Magnifications indicated.

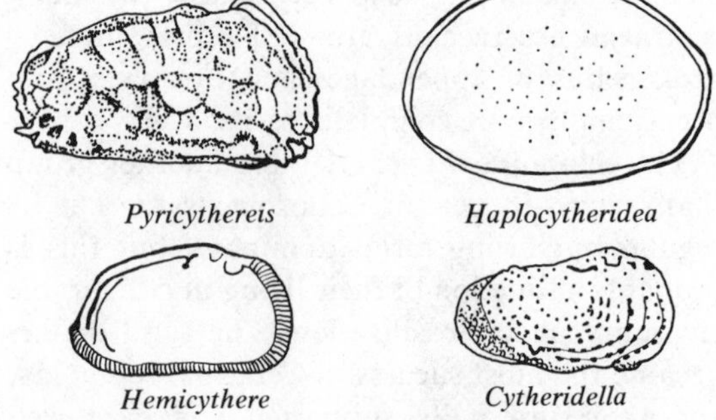

Figure 21.17. *Cenozoic Ostracods.* X 30.

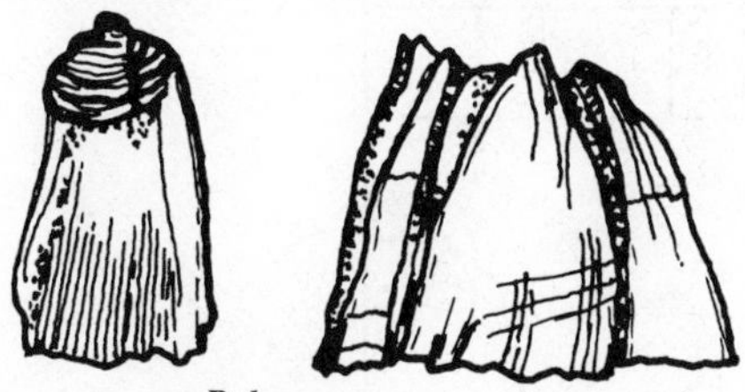

Figure 21.18. *Cenozoic Barnacles.* Individual plate is shown on left. X1/2.

basins and the strong north-south climatic gradients that developed in the Neogene (Figs. 3.10, 21.20).

The **fish** (Fig. 21.21), mostly modern ray-finned bony types, continue their success in the sea. Their scales, otoliths (ear bones), and disarticulated skeletons are abundant fossils (Fig. 21.22A-B), but complete, articulated remains are only locally common. The cartilaginous fish are represented by numerous shark, skate, and ray teeth (Fig. 21.22C). In some areas, these teeth were heaped together in such amounts that they now compose major elements of the sediment.

Mammals succeeded in the sea with the evolution of the **whales** (Fig. 21.23) in Africa in the Eocene and in the **pinniped carnivores** (Fig. 21.24) in the North Pacific in the Miocene.

Marine Ecosystem

The marine ecosystem was rejuvenated again in the Paleocene after the extinctions at the end of the Mesozoic. The pelagic coccolithophores, diatoms, and dinoflagellates form the basis of the modern food chains which attained maturity in the Eocene and the climax (fully mature) stage in the Oligocene. This state persists to the present. (See also Chapter 3.)

Fresh-Water Life

The fresh-water environment is abundantly represented by Cenozoic lake and stream deposits. The fossilizable denizens of this environment, **charophytes, diatoms, gastropods, pelecypods, ostracods,** and **fish,** are preserved in great abun-

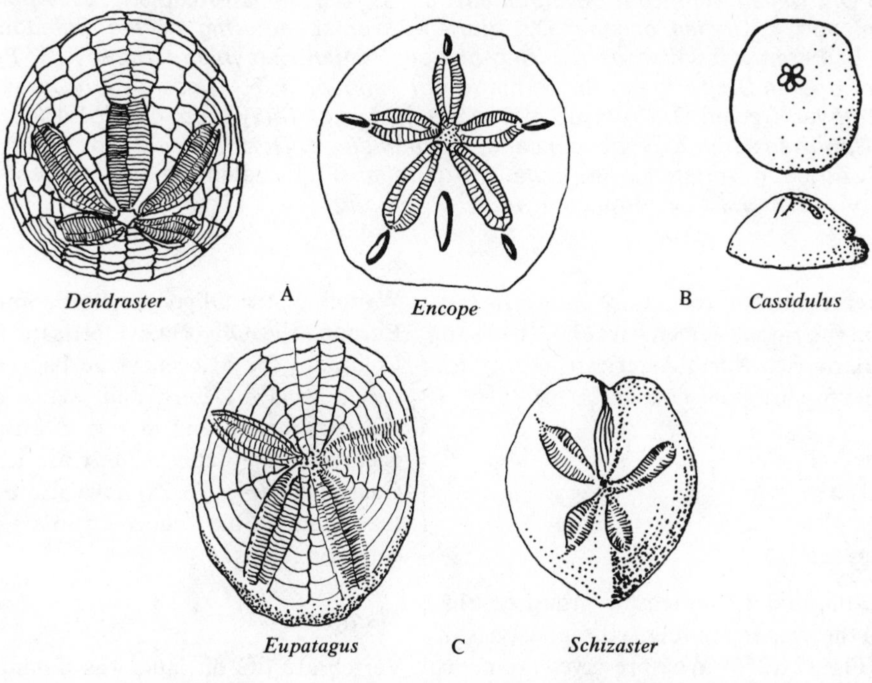

Figure 21.19. *Cenozoic Echinoids.* **A.** *Sand dollars,* **B.** *Cassiduloid,* **C.** *Heart urchins.* Three-fourths natural size.

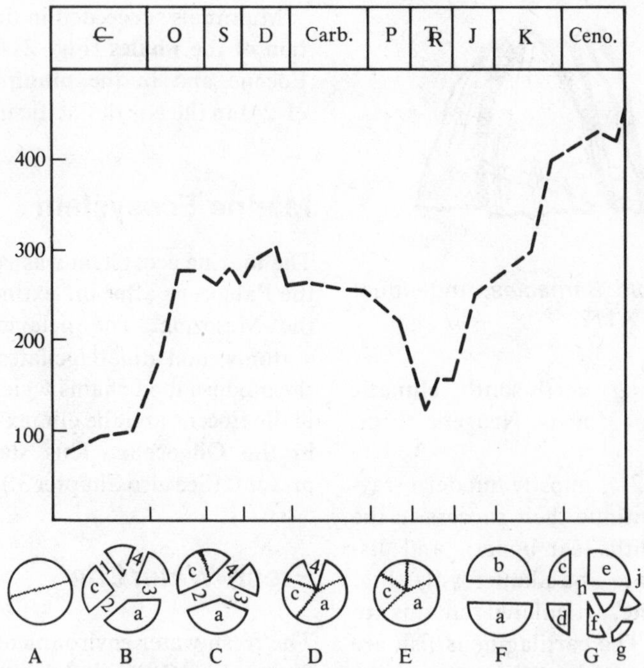

Figure 21.20. *Correlation of Organic Diversity with Continental Assembly and Fragmentation.* Figures along left margin of graph indicate numbers of families. **A.** *Eocambrian suturing of Pan-African-Baikalian Orogeny and the formation of Pangaea I;* **B.** *Fragmentation of Pangaea in Cambrian-Ordovician and the development of* **1.** *Caledonian-Acadian,* **2.** *Appalachian,* **3.** *Hercynian, and* **4.** *Uralian oceans;* **C.** *Silurian-Devonian suturing of the Caledonian-Acadian Orogeny;* **D.** *Permocarboniferous suturing of Appalachian-Hercynian Orogeny;* **E.** *Permotriassic suturing of Uralian Orogeny and the formation of Pangaea II;* **F.** *Triassic-Early Jurassic development of Tethyan Ocean;* **G.** *Cretaceous-Recent closing of Tethys, opening of Atlantic, fragmentation of Gondwanaland.* **a.** *Gondwanaland,* **b.** *Laurasia,* **c.** *North America,* **d.** *South America,* **e.** *Eurasia,* **f.** *Africa,* **g.** *Antarctica,* **h.** *India,* **j.** *Australia.* (From Valentine and Moores, *Nature,* vol. 228, Nov. 14, 1970. Used by permission.)

dance. Representative genera are shown in Figure 21.25. The Eocene Green River lake beds are a particularly rich North American locality for fresh-water fossil organisms (Fig. 21.26).

Land Life

Invertebrates

On land, the major invertebrate group continued to be the ever-expanding arthropod class of **insects** (Fig. 21.27). Well-preserved remains occur in the Eocene Green River lake beds of Wyoming, the Oligocene Baltic amber beds of Europe, the Oligocene Florissant lake beds of Colorado, the Miocene lake beds of the California Mojave Desert, and several other localities. Arachnean and myriapod arthropods also occur in these deposits, but are less common. Gastropods (Fig. 21.28) have also been successful on land in the Cenozoic and are occasionally common fossils.

Mammals

Vertebrate life on land was dominated by the **mammals.** Amphibians, reptiles, and birds oc-

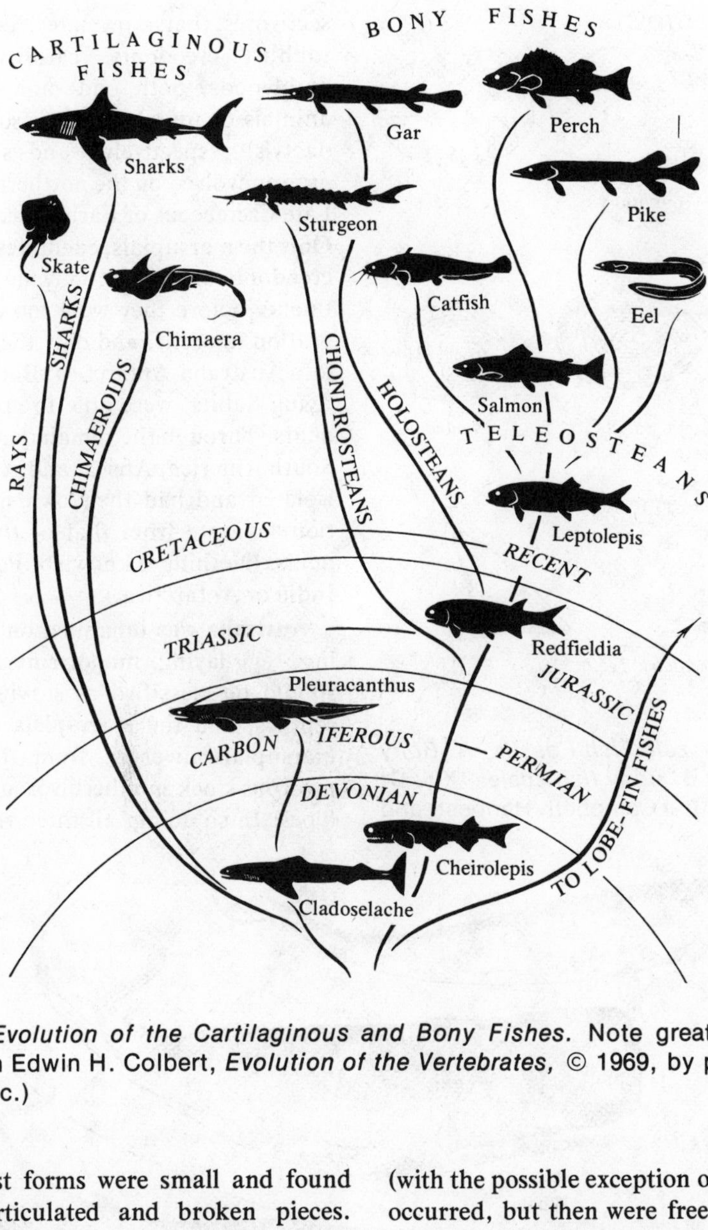

Figure 21.21. *Evolution of the Cartilaginous and Bony Fishes.* Note great expansion in the Cenozoic. (From Edwin H. Colbert, *Evolution of the Vertebrates,* © 1969, by permission of John Wiley & Sons, Inc.)

curred, but most forms were small and found chiefly as disarticulated and broken pieces. Mammalian evolution was strongly influenced by the shifting pattern of the continents in the Cenozoic (Fig. 21.29). In the Late Cretaceous, you may recall, the old Laurasia and Gondwanaland supercontinents had fragmented into most of their present pieces, and even some of these were separated into smaller entities by epeiric seas. Early marsupial and/or placental mammals invaded all the major continental blocks (with the possible exception of India) before this occurred, but then were free to evolve in isolation as the pieces moved apart.

North America and **Eurasia** remained in intermittent contact throughout the Cenozoic, and there was much interchange between them. In the Early Cenozoic, the Greenland-Norway rift was still narrow and permitted communication. In the Late Cenozoic, North America and Eurasia came into contact in the Alaska-East Siberia area. The carnivorous marsupials, in-

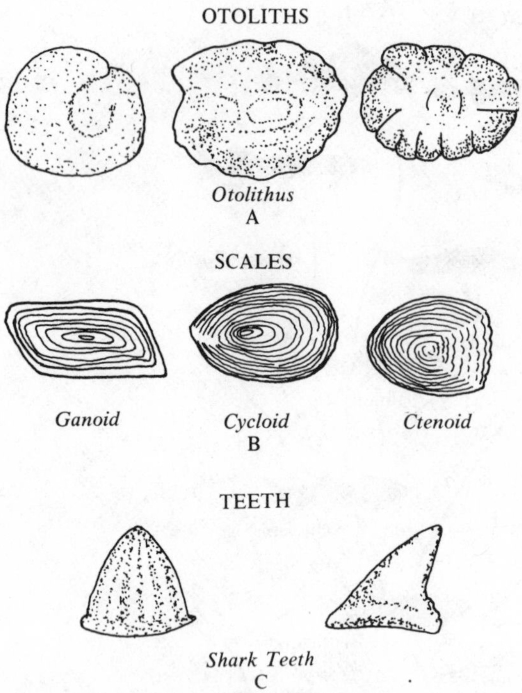

Figure 21.22. *Cenozoic Fish Fossils.* **A.** *Bony fish otoliths, X 10;* **B.** *Bony fish scales, X 5;* **C.** *Shark teeth, X 4.* (After Campbell, Henbest, and Jones.)

sectivores, bats, primates, edentates, rodents, rabbits, creodonts, carnivores, condylarths, amblypods, both odd- and even-toed hoofed animals or ungulates (perissodactyls and artiodactyls, respectively), and some other minor groups evolved on the northern continents in the Late Cretaceous or Early Paleogene (Fig. 21.29). Only the marsupials, edentates, condylarths, and creodonts were able to invade the southern continents before they were too isolated for immigration to occur, and only the marsupials made it to Australia-Antarctica. Bats, because of their flying habits, were able to spread to all continents. Through the remainder of the Paleogene, South America, Africa, and Australia were each isolated and had their own mammalian evolution separate from that of the northern continents. (Nothing is known of Paleogene faunas in India or Antarctica.)

Australia was inhabited only by some surviving, egg-laying **monotremes,** which perhaps should be classified as surviving mammal-like reptiles, and the **marsupials** (Fig. 21.29). The marsupials diverged from their original carnivorous stock into herbivorous and omnivorous types. In so doing, all three varieties converged

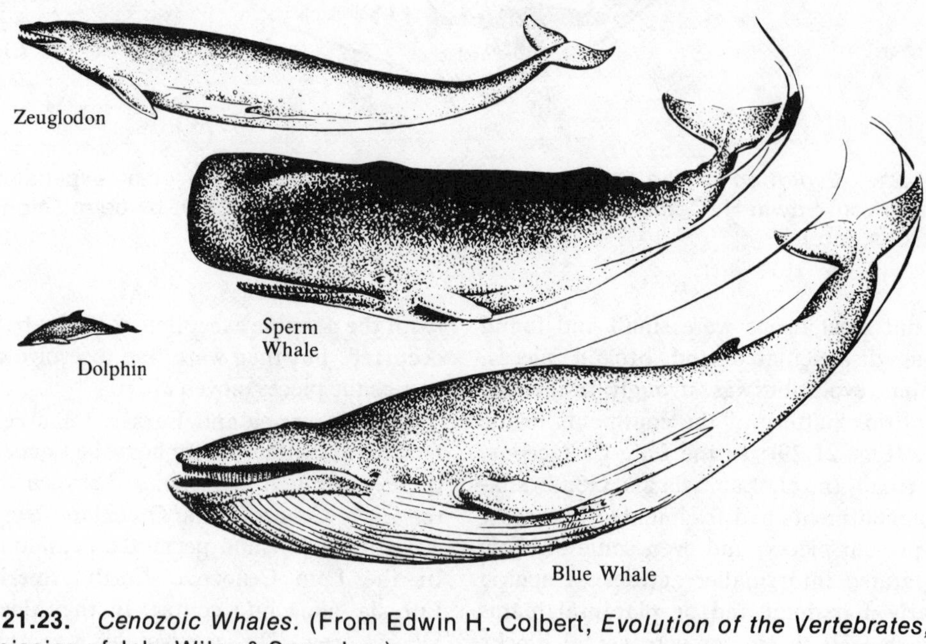

Figure 21.23. *Cenozoic Whales.* (From Edwin H. Colbert, *Evolution of the Vertebrates,* © 1969, by permission of John Wiley & Sons, Inc.)

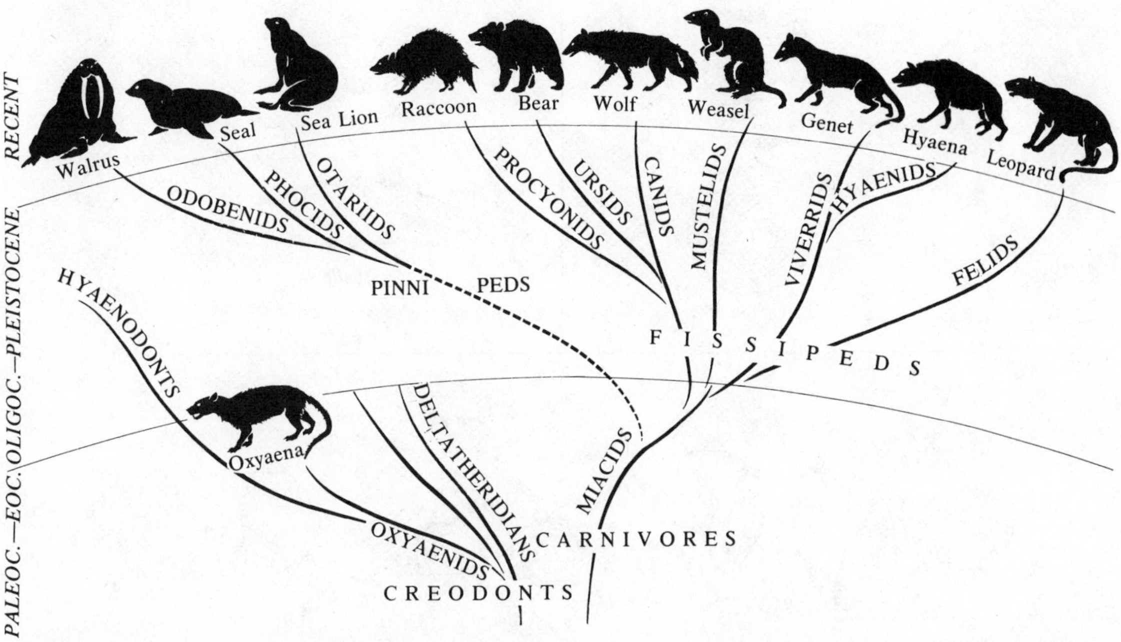

Figure 21.24. *Evolution of the Carnivorous Mammals.* Note marine pinniped carnivores at upper left. Creodonts and true carnivores may not be closely related. (From Edwin H. Colbert, *Evolution of the Vertebrates,* © 1969, by permission of John Wiley & Sons, Inc.)

in many ways on the placental orders filling the same niches elsewhere in the world (Figs. 4.12, 7.118). For example, there are Australian marsupials that closely resemble moles, squirrels, groundhogs, mice, and wolves, and others which do not closely resemble any northern continent animal, but are the ecologic equivalents of mammals found there. These include the kangaroo, which is the ecological equivalent of the large herbivores of the northern continents.

Only as Australia approached Southeast Asia in the Late Pleistocene were a few northern mammals such as the ubiquitous rats and the ingenious primate man able to invade the continent. The recent arrival of Western Man has, unfortunately, flooded the continent with northern mammals that are superior competitors to the native marsupials which they are rapidly exterminating.

South America has, as original mammalian inhabitants, the carnivorous marsupials, the edentates, and the herbivorous condylarths. The marsupials filled the carnivore role, as no creodont or true carnivore had appeared before its isolation, and also evolved into some rat-like forms (Fig. 21.29). The carnivorous forms were eliminated by competition from the northern continents' carnivores, such as cats, dogs, weasels, and raccoons, after the Mio-Pliocene reunion of North and South America.

The **edentates** (Fig. 21.30) had their center of evolution in South America, for the few ancestral Early Cenozoic forms of North America soon became extinct. (Some authorities remove the North American members from the order, restricting it to forms which evolved in South America.) The two main lineages, the armadillos and the sloths, underwent spectacular evolution, particularly in the Neogene. The giant armadillo-like glyptodonts of the Pleistocene reached 9-foot lengths, and the great ground sloths such as *Megatherium* reached 20-foot lengths. After the Neogene rejoining of North and South America, both the armadillos and the ground sloths successfully invaded North America, the latter, along with the glyptodonts,

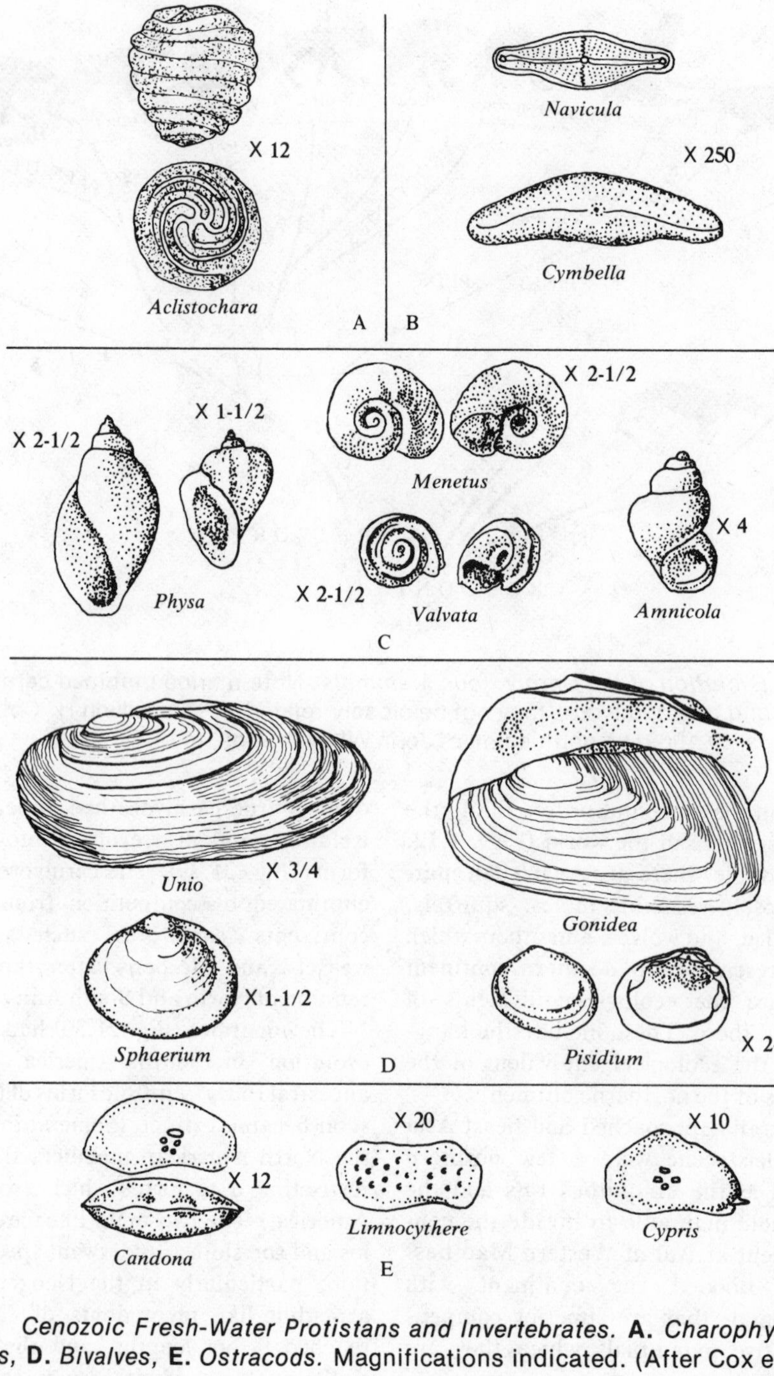

Figure 21.25. *Cenozoic Fresh-Water Protistans and Invertebrates.* **A.** *Charophyte,* **B.** *Diatoms,* **C.** *Gastropods,* **D.** *Bivalves,* **E.** *Ostracods.* Magnifications indicated. (After Cox et al., Jones.)

becoming extinct in the Early Holocene, perhaps by the hand of man.

The South American **condylarths** diversified and gave rise to three very successful orders of South American ungulates (Fig. 21.31): the bulky rhino-like or hippo-like **toxodonts** (a few northern forms may belong here, indicating the group arose before isolation but did not survive

Figure 21.26. *Diorama of Life in the Eocene Green River Lake.* Bony fish predominate in this scene. (Courtesy Field Museum of Natural History.)

Figure 21.27. *Oligocene Fly in Amber. Sphegina carpenteri* from the Baltic amber beds. Highly magnified. (Courtesy of Frank M. Carpenter.)

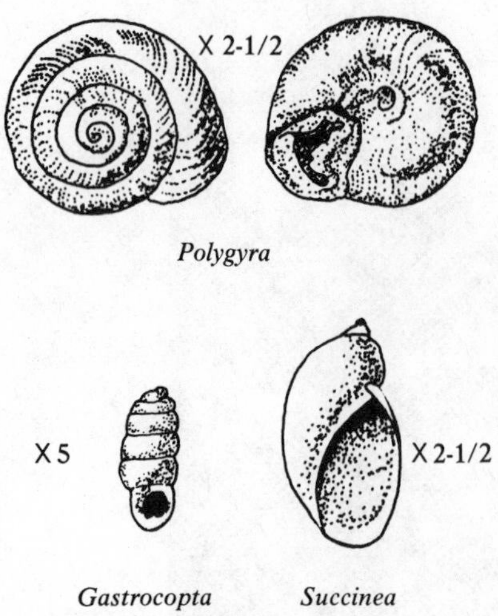

Figure 21.28. *Cenozoic Terrestrial Gastropods.* Magnifications indicated.

in the north), the elephant-like **astrapotheres,** and the horse- and camel-like **litopterns.** Once again, we see convergent evolution of isolated forms in the absence of the typical placental ungulate orders. These South American groups reached the Oligocene-Early Miocene peak, but when the Panama land bridge formed in the Late Miocene-Early Pliocene interval, the competition of the invading northern herbivores such as the perissodactyls, artiodactyls, and proboscideans drove them all to extinction before the close of the Pleistocene.

In the Oligocene, there was apparently close enough proximity of North America or Africa to South America, probably by way of volcanic islands, for a few small mammals to get into the South American continent. Some **rodents** entered and, in their isolation, evolved a whole host of unique forms, such as guinea pigs, chinchillas, pacas, and the New World porcupine (Fig. 21.32). They did not suffer attrition after the two continents joined, and a few, such as the porcupine, successfully invaded the north. The other Oligocene entrants were the small **primates** that gave rise to the unique South American monkey assemblage (Fig. 21.33). There being no present-day equatorial jungles in North America, the group has never migrated out of South America or come in contact with Old World monkeys.

Africa was isolated for much less time than the other southern continents. Its original complement of pre-isolation mammals consisted of the **creodonts** and **condylarths**. The carnivorous creodonts were not abundant, but are the apparent ancestors of the **whales,** whose first carnivorous members occur in the Eocene of North Africa (Fig. 21.23). Likewise, the condylarths amounted to very little themselves in Africa, but gave rise to several orders of herbivore subungulates, some of which converged on the mammals of the other continents. These groups (Figs. 21.29, 21.34) were the large herbivorous **proboscideans,** including the elephants and mastodons, the small rabbit-like **hyraxes** or true conies, the **embrithopods** which resembled huge rhinos, the **aardvarks** which resemble the South American edentate anteaters, and the **sirenians** (or **sea cows**) which took to the shallow seas. When Africa and Eurasia came into contact in the Oligocene and Miocene, the proboscideans not only survived, but eventually spread to all the continents except Australia; the hyraxes stayed at home save for one short-lived invader of Europe; the embrithopods became extinct, probably from competition with invading northern ungulates such as rhinos; the aardvarks remained in South Africa and no edentate ever has reached that area to compete with them; and the sea cows, being marine, rapidly spread through the world's oceans.

Another order of mammals, the **desmostyloids,** were Miocene and Pliocene mollusk-eating, amphibious forms of hippo-like proportions inhabiting the shallow seas of east Asia and western North America. Their closely packed cylindrical teeth (Fig. 21.35) are relatively common fossils in some California near-shore marine beds along the large oysters, pectens, and barnacles on which they fed. Their structure indicates they are related to the other subungulates, but their African origin and their route to the North Pacific are unknown, though the migration is not surprising because the creatures were amphibious.

The major northern invaders of Africa after the junction with Eurasia were the **primates,** which underwent a considerable African radiation discussed in the next chapter (Fig. 21.33); **rodents** (Fig. 21.32); the true **carnivores** including the hyaenas, dogs, and cats (Fig. 21.24); the **perissodactyls** including the claw-bearing chalicotheres and the rhinoceroses (Fig. 21.36); and the **artiodactyls** including pigs, hippos, camels, giraffes, and antelopes (Fig. 21.37). Africa has served as a Neogene refuge for many of the large members of these groups that did not survive the cooling climates of the northern continents.

The interconnected northern land masses of **North America** and **Eurasia** were the main center of evolution of the insectivore, primitive primate, rodent, rabbit, creodont, carnivore, condylarth, amblypod, perissodactyl, and artiodactyl mammalian groups (Figs. 21.38–43).

In addition, the flying **bats** originated there from insectivores, but have since spread to all continents. **Insectivores** include the Cenozoic shrews, moles, and hedgehogs, all rather small forms, but the nearest living relatives of the ancestral placental mammals. The primitive **primates,** such as the lemurs and tarsiers (Figs. 21.33, 22.2-4), were worldwide in the northern continents with their abundant, moist forests of the Paleogene. These prosimians retreated to isolated jungle refuges in Africa, Madagascar, and Southeast Asia in the Neogene. **Rodents** (Fig. 21.32) have become the most diverse of all mammalian orders and had spread to all continents by Late Neogene. The squirrel, rat-mouse, and beaver groups have centered their evolution in the northern continents, however. The **rabbits** (Fig. 21.44) were only slightly less successful, not reaching Australia until their disastrous recent introduction by man. The small bones and teeth of both rodents and rabbits are among the commonest of mammalian fossils.

The carnivorous **creodonts** and **carnivores** (Fig. 21.24) were almost exclusively northern continent forms until the Neogene when the true carnivores spread widely in Africa, India, and South America. The creodonts of the Paleogene were largely eliminated by competition with the superior true carnivores in the Neogene. The most spectacular true carnivores were the great stabbing cats (Fig. 21.45) and the cave bears of the Pleistocene. The fossil record is so complete for some lineages that we can trace the common ancestry of such groups as bears and dogs from bear-dogs. The **condylarths** and the **amblypods** (including **uintatheres** and **pantodonts**), like the creodonts, radiated in the Paleogene (Fig. 21.34), but were then gradually eliminated by competition from the more advanced herbivorous ungulates, the perissodactyls and artiodactyls. The odd-toed ungulates, or perissodactyls, reached an Early Neogene peak and have since declined nearly to extinction, while the even-toed artiodactyls have continued to expand, indicating they may be superior competitors in the same environments.

Among the **perissodactyls** (Fig. 21.36) are five main lineages: the horses, brontotheres, chalicotheres (or ancylopods), tapirs, and rhinos. All were common in North America; indeed, North America was the center for much of their evolution. Yet none have survived here until the Recent, and only the tapirs still live in the New World (excluding horses, which were reintroduced by man). The horses form the best documented evolutionary lineages known in the fossil record (Fig. 4.18). The early small, many-toed, browsing forms lived in the subtropical forests of the Paleogene; but in the Neogene, with the expansion of grasslands, one lineage invaded that environment and eventually evolved into the large, one-toed grazing horses of today. The extinct chalicotheres were of horse-like build, but had large claws on their feet for digging plant roots and bulbs. The extinct, bulky brontotheres probably occupied the niche filled in the Late Cenozoic by the bison, an artiodactyl which evolved in Eurasia and did not reach America until the Pleistocene. (The grazing horses may have replaced the grazing brontotheres and may in turn have themselves been replaced by superior competitors, the bison.) The rhinoceroses underwent a significant radiation into many more adaptive types than their few Old World survivors. These types included the agile running rhinos and the huge Asiatic Oligo-Miocene forms *Baluchitherium* (Fig. 21.46) and *Paraceratherium,* the largest land mammals known.

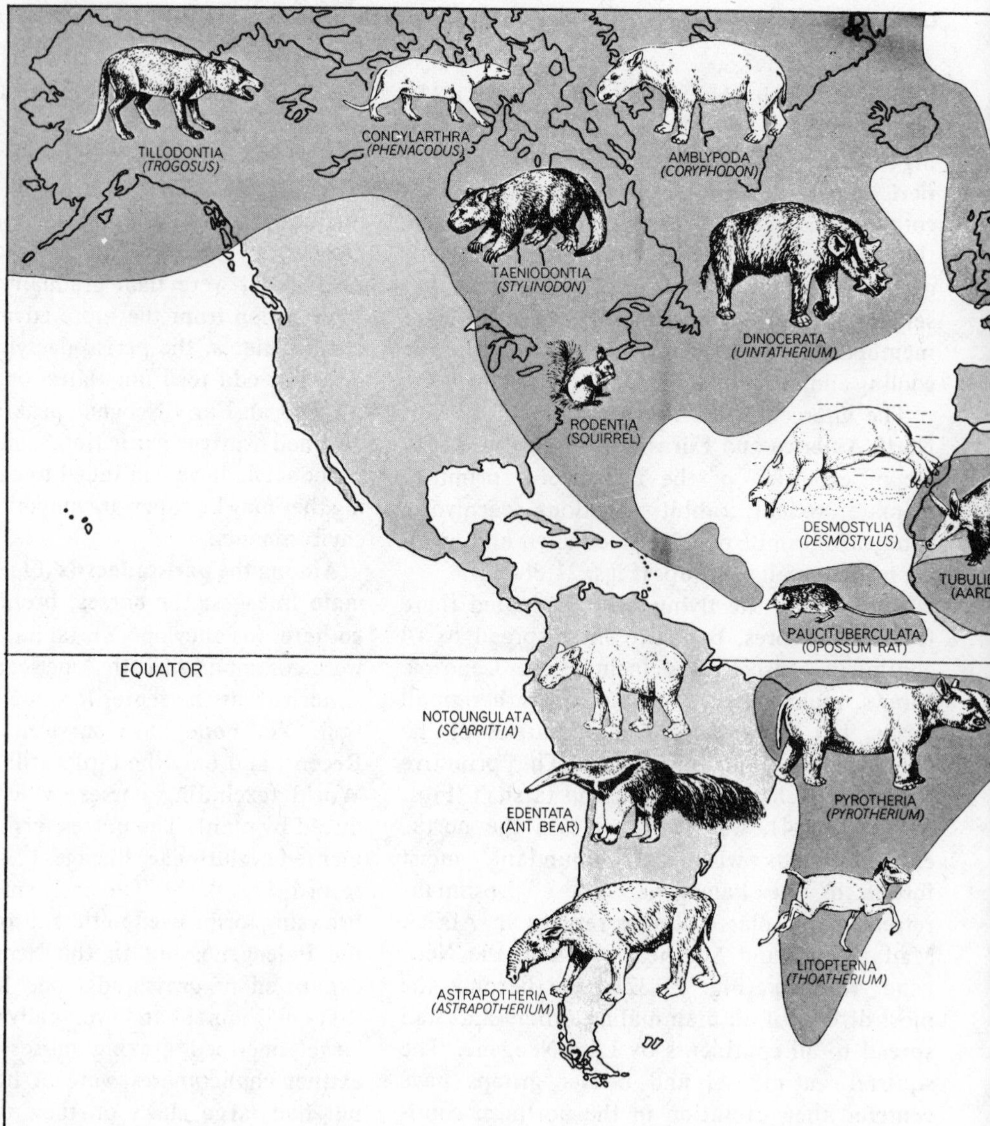

Figure 21.29. *Continental Drift and Mammalian Evolution.* See text for explanation. (From *Continental Drift and Evolution* by Björn Kurten, © 1969 by Scientific American, Inc. All rights reserved.)

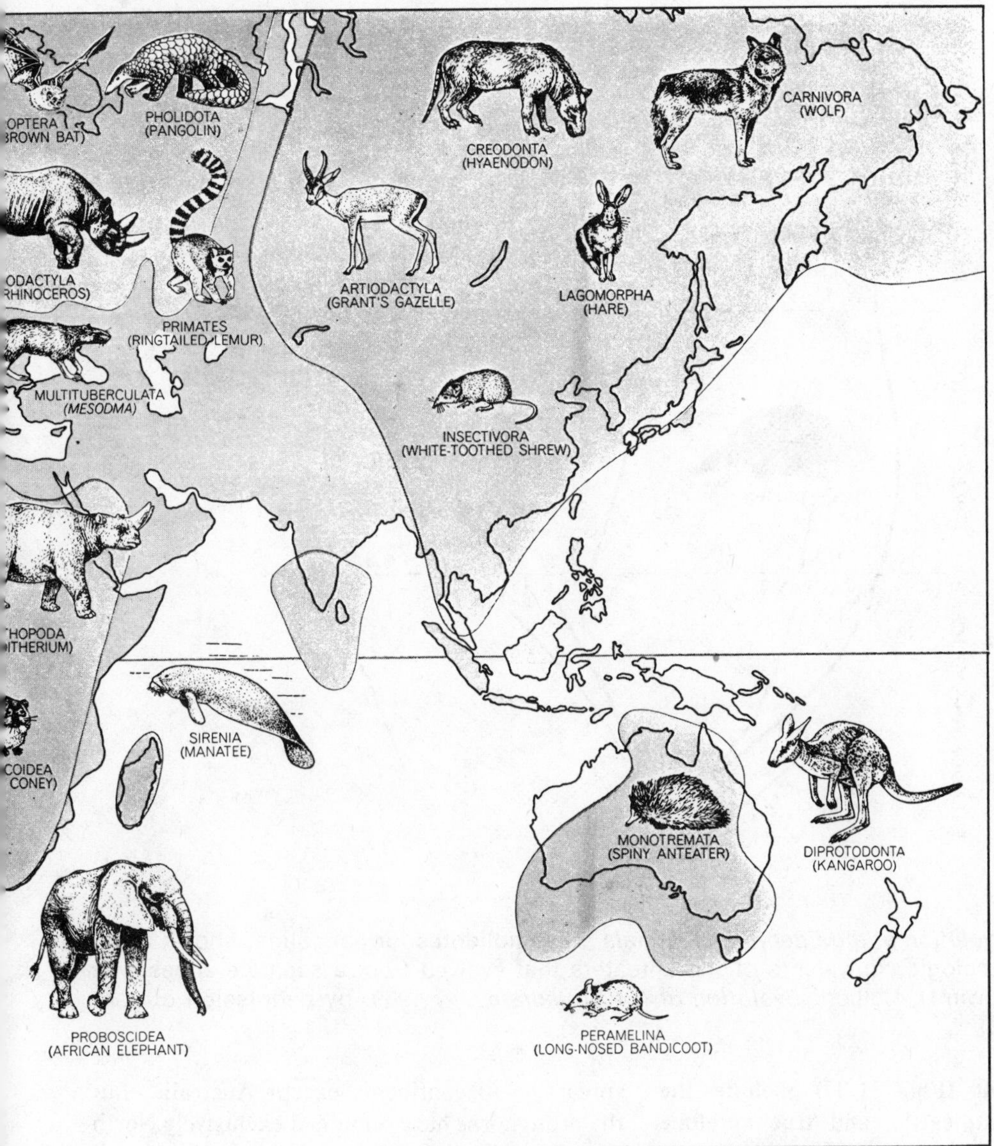

OPTERA
BROWN BAT)
PHOLIDOTA
(PANGOLIN)
CREODONTA
(HYAENODON)
CARNIVORA
(WOLF)
ODACTYLA
RHINOCEROS)
ARTIODACTYLA
(GRANT'S GAZELLE)
LAGOMORPHA
(HARE)
PRIMATES
(RINGTAILED LEMUR)
MULTITUBERCULATA
(MESODMA)
INSECTIVORA
(WHITE-TOOTHED SHREW)
THOPODA
OTHERIUM)
SIRENIA
(MANATEE)
COIDEA
CONEY)
MONOTREMATA
(SPINY ANTEATER)
DIPROTODONTA
(KANGAROO)
PROBOSCIDEA
(AFRICAN ELEPHANT)
PERAMELINA
(LONG-NOSED BANDICOOT)

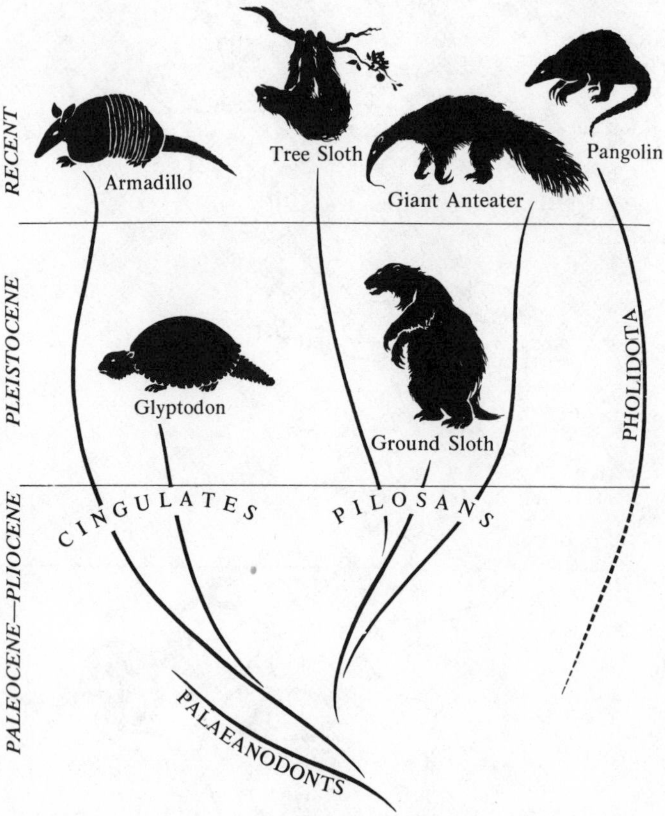

Figure 21.30. *Evolution of the Edentate Mammals.* The pholidotes, or pangolins, shown at the right margin are ecologic equivalents of the anteaters that evolved from a separate ancestry in Eurasia. (From Edwin H. Colbert, *Evolution of the Vertebrates,* © 1969, by permission of John Wiley & Sons, Inc.)

The **artiodactyls** (Fig. 21.37) include the swine, sheep, goat, cattle, and true antelope groups which underwent most of their evolution in the Old World and sent only a few, but very successful, invaders such as the peccaries, bighorn sheep, mountain goats, and bison into the New World in the Pleistocene. Other Old World artiodactyl groups such as the hippos, chevrotains, and giraffes never invaded the Americas. On the other hand, the camel, deer, and pronghorns underwent virtually all of their evolution in North America. The camels have since died out in North America, but survive in the Old World and South America which they invaded in the Late Cenozoic. The deer have spread to all continents except Australia, but the pronghorns have remained exclusively North American. The pig-like entelodonts, the sheep-like oreodonts, and the deer-like protoceratids, with their numerous cephalic horns, are extinct artiodactyl groups very abundant in North America in the Middle Cenozoic. The Oligo-Miocene rejoining of Africa with the northern continents permitted the successful invasion of the north by the proboscideans (Fig. 21.34), higher primates, and possibly the desmostylids.

Other notable features of mammalian evolution are the large members produced by almost every order living in the Pleistocene. These include giant marsupials, men, ground sloths,

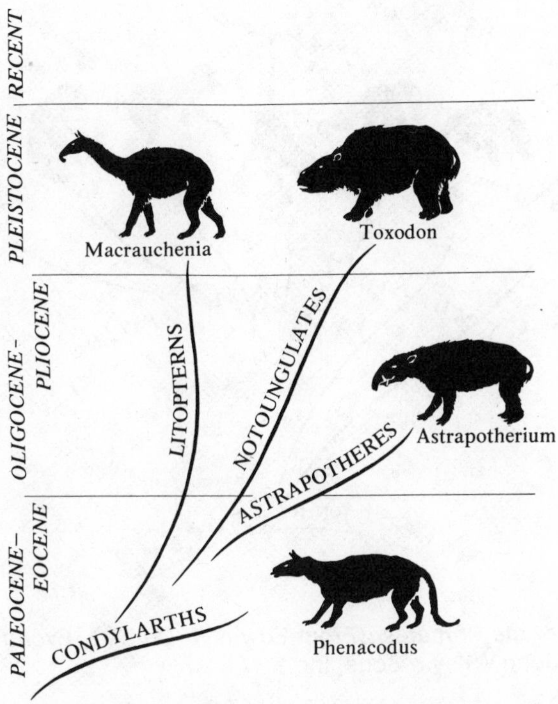

Figure 21.31. *Evolution of the South American Ungulate Mammals.* All of these groups have now become extinct. (From Edwin H. Colbert, *Evolution of the Vertebrates,* © 1969, by permission of John Wiley & Sons, Inc.)

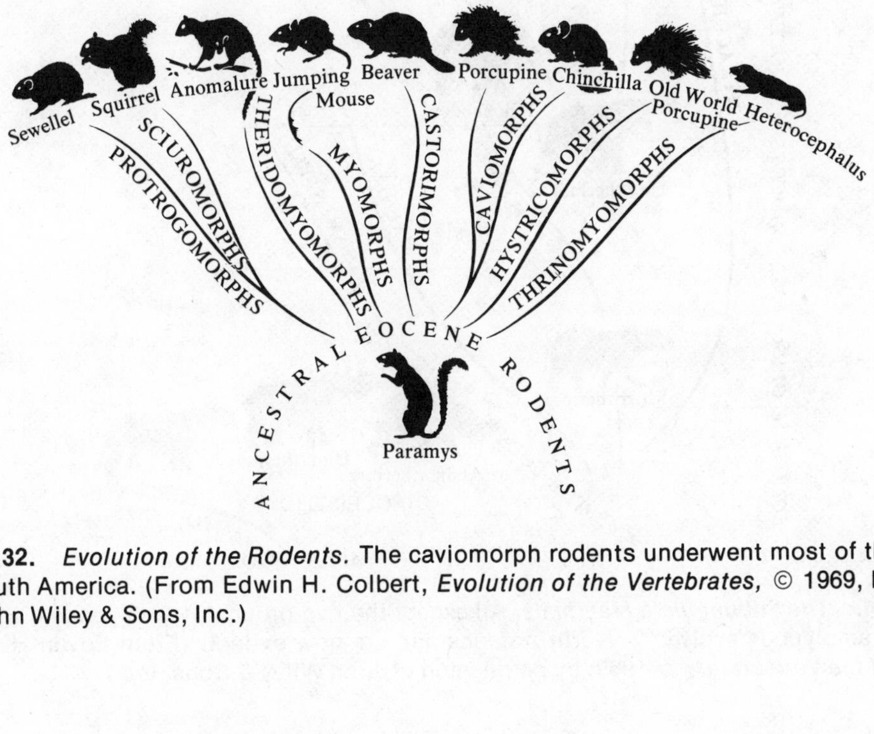

Figure 21.32. *Evolution of the Rodents.* The caviomorph rodents underwent most of their evolution in South America. (From Edwin H. Colbert, *Evolution of the Vertebrates,* © 1969, by permission of John Wiley & Sons, Inc.)

Figure 21.33. *Evolution of the Primates.* (From Edwin H. Colbert, *Evolution of the Vertebrates,* © 1969, by permission of John Wiley & Sons, Inc.)

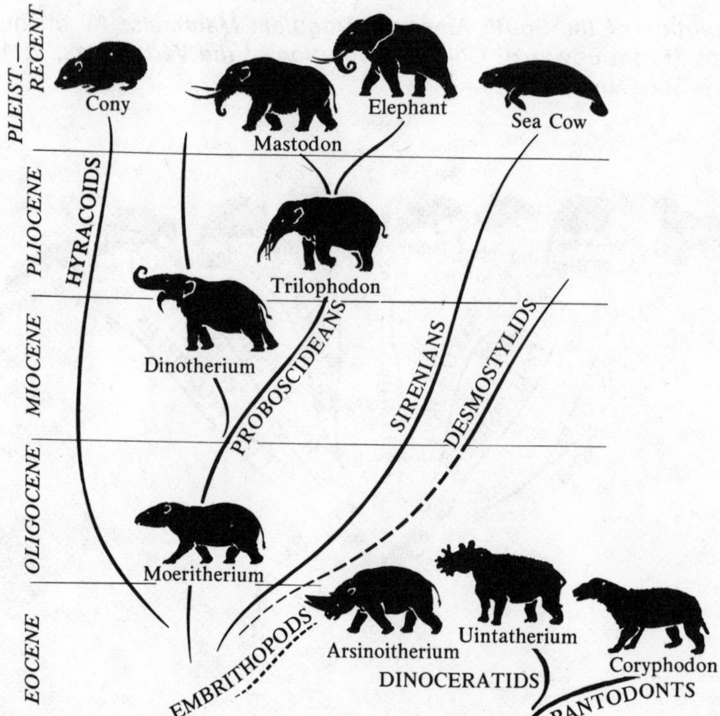

Figure 21.34. *The Subungulate Mammals.* All except the two groups of amblypods originated in Africa. The amblypods evolved in North America and are now extinct. (From Edwin H. Colbert, *Evolution of the Vertebrates,* © 1969, by permission of John Wiley & Sons, Inc.)

Figure 21.35. *Restoration and Teeth of a Desmostylid.* Top and side views of closely packed cylindrical teeth are shown. (After Stirton.)

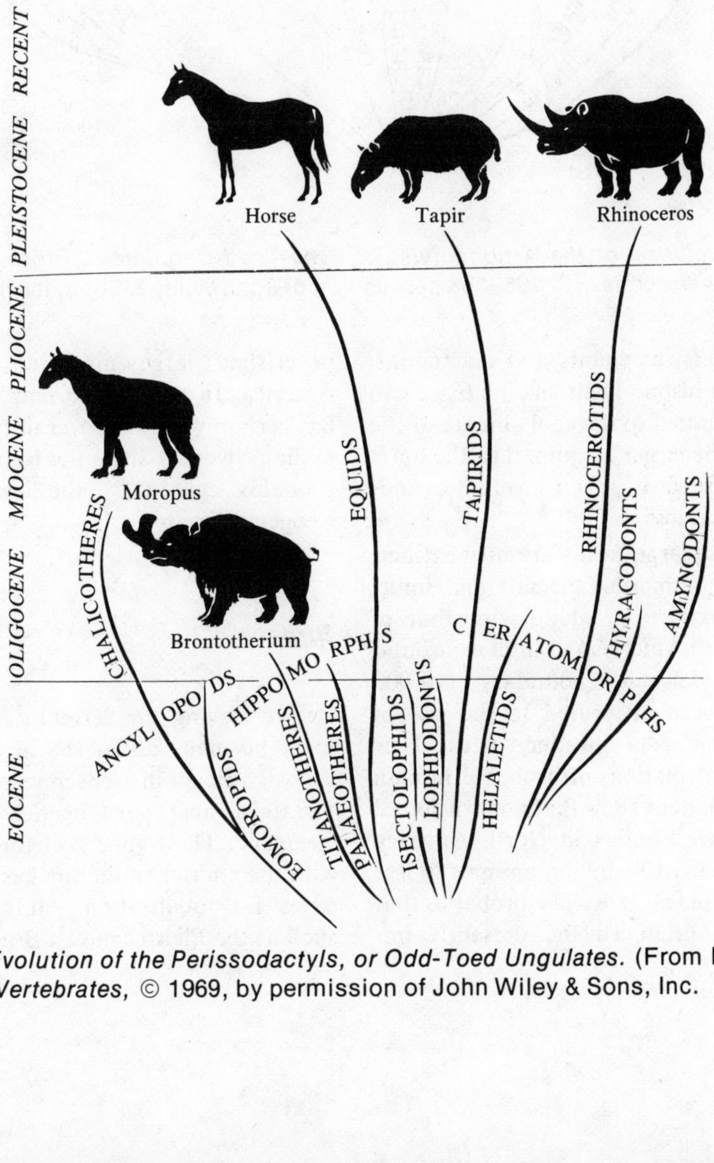

Figure 21.36. *Evolution of the Perissodactyls, or Odd-Toed Ungulates.* (From Edwin H. Colbert, *Evolution of the Vertebrates,* © 1969, by permission of John Wiley & Sons, Inc.

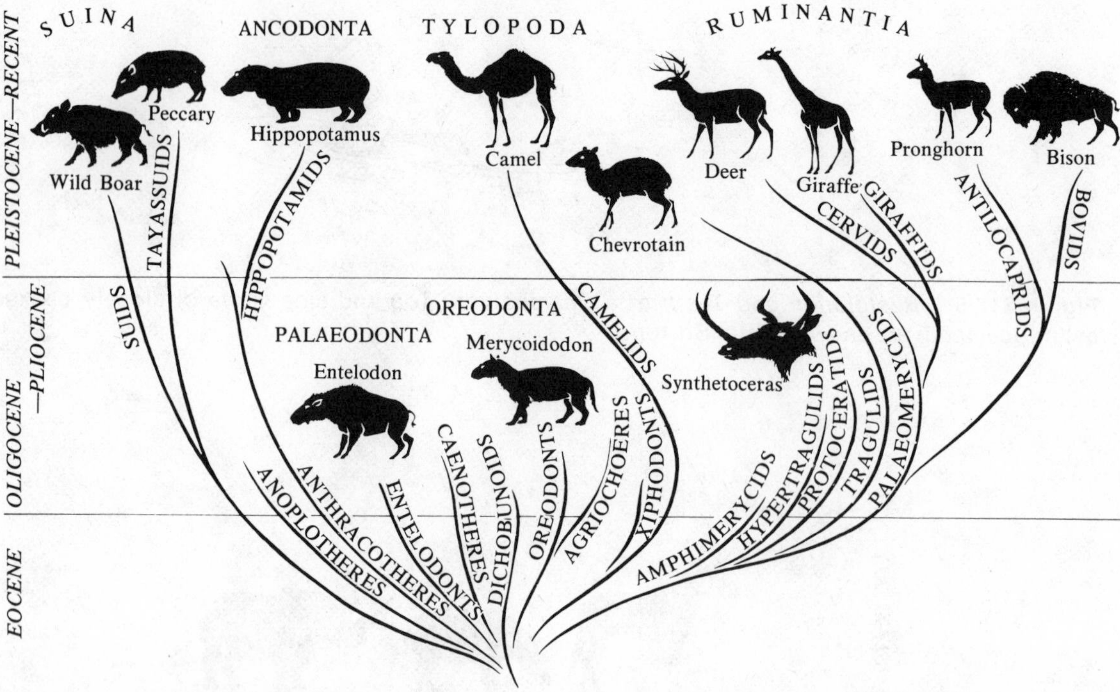

Figure 21.37. *Evolution of the Artiodactyls, or Even-Toed Ungulates.* (From Edwin H. Colbert, *Evolution of the Vertebrates,* © 1969, by permission of John Wiley & Sons, Inc.)

armadillos, beavers, elephants and mastodons, rhinos, deer, and bison. Their size increase can probably be attributed to the cool climate of the Pleistocene, for the larger an animal is, the lower the ratio of its surface area to volume, and, hence, the less heat loss.

Almost all these large forms are now extinct, despite surviving numerous glacials and interglacials. Their extinction, along with that of many typical North American mammals around the Pleistocene-Holocene boundary (10,000 years ago), has been attributed to the rise of man, a Late Pleistocene invader, in the New World. Recent calculations of probable human and animal populations show that an invasion of 300,000 prehistoric hunters in North America could wipe out nearly 100 million animals in less than 300 years. Hence, it is very probable that our own species produced the presently impoverished large mammal fauna of North America. In addition to man's influence, there has certainly been an overall reduction in mammalian diversity since the Miocene, a condition probably caused by the less favorable Late Neogene climate.

Birds

Before leaving the vertebrates, a few remarks must be made about the **birds.** This group is very significant in modern terrestrial faunas and certainly must have been so throughout the Cenozoic. The fragile skeleton of birds coupled with their aerial habitat makes them poor candidates for fossilization. Only scattered finds, such as the Pleistocene La Brea Tar Pit of Cali-

Figure 21.38. *Paleocene Mammals.* Rocky Mountain area. (Zallinger mural, courtesy of Yale Peabody Museum.)

fornia (Fig. 21.45), give us a glimpse of the magnitude of bird evolution. One significant development in bird history was the evolution in the Early Cenozoic of large, flightless carnivores, such as *Diatryma* and *Phororhacos* (Fig. 21.47). The former genus evolved in North America in the Eocene, but did not last long, probably because of the presence of carnivorous mammals. *Phororhacos* and a host of relatives survived much longer in South America, probably because placental carnivores were lacking until the mid-Neogene. Other, less ferocious, large flightless birds have survived on isolated islands and in some areas of continents down to the present (Fig. 21.47C). Their possible relationships with dinosaurs were noted in Chapter 19.

Figure 21.39. *Eocene Mammals.* Middle Eocene life of Wyoming. (Mural by J.H. Matternes, courtesy U.S. National Museum.)

Figure 21.40. *Oligocene Mammals.* Early Oligocene life of South Dakota-Nebraska area. (Mural by J.H. Matternes, courtesy U.S. National Museum.)

Figure 21.41. *Miocene Mammals.* Early Miocene life of Nebraska. (Mural by J.H. Matternes, courtesy U.S. National Museum.)

Figure 21.42. *Pliocene Mammals.* Early Pliocene life of the southern High Plains. (Mural by J.H. Matternes, courtesy U.S. National Museum.)

Figure 21.43. *Pleistocene Mammals.* At the upper left are several ungulates, including a horse and camels. Below these are two carnivores, a bear and a wolf, with a giant beaver to their right. In the center are a mastodon, a mammoth, and a large bison. At the right are a stabbing cat, a wolf, horses, a giant ground sloth, and a glyptodont. (Zallinger mural, courtesy Yale Peabody Museum.)

Figure 21.44. *Lagomorph Mammals.* (From Edwin H. Colbert, *Evolution of the Vertebrates,* ©1969, by permission of John Wiley & Sons, Inc.)

Plants

The plant life of the Cenozoic Era was dominated by the **angiosperms,** or flowering plants (Fig. 21.48). Their leaves, stems, roots, and particularly their pollen are abundantly preserved in continental strata (Fig. 21.49). The gymnosperms, such as the conifers and gingkos, declined to a few species common only in cooler and drier areas. Only the low-growing ferns remained diverse among the spore-bearing plants.

In the warmer and wetter conditions of the Early Cenozoic, most angiosperms had large, entire-margined leaves (Fig. 21.50), such as those of modern subtropical to tropical forms. In the early Cenozoic, a warm, wide equatorial Pacific Ocean contained currents which diverted considerable warm water to high latitudes, causing an expansion of the tropical and semitropical belts. There were no high-latitude, circumpolar currents because of the continental arrangement. In the late Cenozoic, the size of the Pacific decreased appreciably and a cold, circum-Antarctic current developed, thus cooling temperatures. As the climate became cooler and drier, the warm-weather plants contracted their ranges to low latitudes. Furthermore, smaller leaves with serrate margins (Fig 21.50), such as those characterizing present temperate to boreal angiosperms, became prevalent in mid-latitudes. In the dry climates of the Middle and Late Cenozoic, the **grasses** (Figs. 21.41, 21.42) became the dominant angiosperm group and carpeted vast plains areas on all the continents. Although their soft remains are uncom-

Figure 21.45. *La Brea Tar Pit Fauna.* A stabbing cat, *Smilodon*, is contesting a dead camel with a giant buzzard. (Mural by C.R. Knight, courtesy American Museum of Natural History.)

Figure 21.46. *Restoration of Baluchitherium.* This giant rhinoceros lived in southern Asia. (C.R. Knight painting, courtesy of the American Museum of Natural History.)

Figure 21.47. *Large Flightless Ground Birds of the Cenozoic.* **A.** *Diatryma, North America;* **B.** *Phororhacos, South America;* **C.** *Moa, New Zealand.* (After Lucas, Stirton.)

Figure 21.48. *Tropical Angiosperm Flora of England in the Eocene.* (After Seward.)

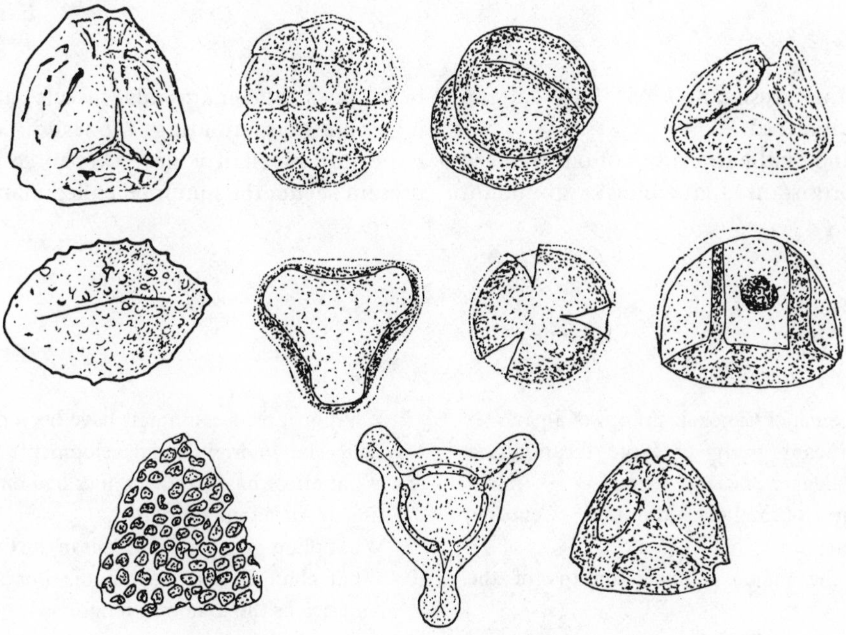

Figure 21.49. *Cenozoic Pollen Grains.* Highly magnified. (After Kuyl, Muller, and Waterbolk.)

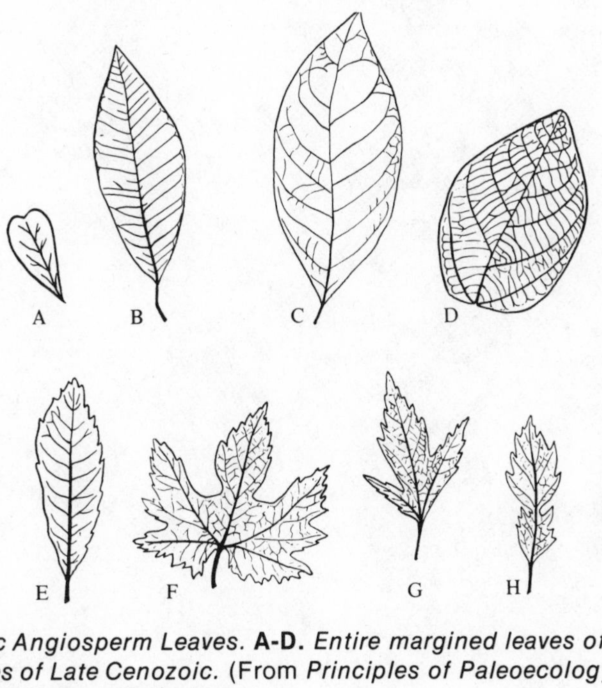

Figure 21.50. *Cenozoic Angiosperm Leaves.* **A-D.** *Entire margined leaves of Early Cenozoic.* **E-H.** *Dentate margined leaves of Late Cenozoic.* (From *Principles of Paleoecology* by D. Ager. © 1963. Used by permission of McGraw-Hill Book Co.)

mon fossils, their resistant seeds have left an abundant fossil record.

We now stand on the threshold of our modern world, with protistans and mollusks abundant in the seas, and mammals, insects, and flowering plants dominating the land. Only one important element is necessary to complete the present scene, the mammal called man.

REVIEW

1. What are the major Cenozoic groups of algae?
2. What significant group of large foraminifera reached a Paleogene peak?
3. What groups of mollusks dominate Cenozoic marine faunas?
4. What are the major echinoid groups of the Cenozoic?
5. Why is the Cenozoic Era known as the **Age of Mammals?**
6. Discuss the evolution of the faunas of South America and Australia in terms of plate tectonic history.
7. What groups of mammals have been dominantly Laurasian in origin and development?
8. What effect has plate tectonics had on the evolution of African mammals?
9. What plant group is dominant in the Cenozoic?
10. What changes occurred in the floras of North America as the Cenozoic progressed?

22

The Psychozoic "Era": The Rise of Man

22

The rise of man, a very recent event in the long history of the earth, deserves special attention not only because the author and readers of this text belong to the lineage, but also because it has had such a dramatic effect on the earth's entire ecosystem. Whether or not the human species survives, its span of existence on this planet will leave significant geologic reminders. Because of this, some scientists have suggested setting aside the last 5,000 years of earth history as a significant new chapter called the **Psychozoic Era,** characterized by man and his works. In this chapter we will use the term in an informal, expanded sense to discuss the entire rise of man.

Primate Characteristics

Man is a member of the mammal order **Primatoida** (Fig. 21.33), meaning the prime animals and clearly reflecting man's opinion of himself. The primate order arose from the insectivores in the Late Cretaceous. The probable ancestral group is that of tropical arboreal **tree shrews** (Fig. 22.1A), but fossils of this family are not

Figure 22.1. *Representatives of the Major Living Groups of Nonhuman Primates.* **A.** *Tree Shrew,* **B.** *Lemur,* **C.** *Tarsier,* **D.** *Old World Monkey,* **E.** *Apes.*

known before the Paleocene. Tree shrews have typical insectivore teeth and claws, but they have large brains with a diminished olfactory (sense of smell) region, the thumb is partially opposable, and the skull is lemur-like, all features of the primates. Some workers classify the tree shrews as insectivores; some, as primates; others, as a separate order. However one classifies them, they appear to be related to the lineage from which the typical primates arose.

Plesiadapidines

A primitive, extinct group of primates with chisel-like front teeth, the **plesiadapidines** (Fig. 22.2) are known from Upper Cretaceous, Paleocene, and Lower Eocene rocks of North America and Europe, indicating that the primates arose on the northern continents. The oldest-known tree shrew is also known from Europe, supporting this idea. The plesiadapidines already displayed the characteristic primate features of the grasping hand made possible by an opposable thumb, the large brain, and the shift of the eyes to the front of the face to permit overlapping stereoscopic vision. These features are all correlated with adaptation to a tree-living or arboreal existence, the mode of life of most primates. The teeth of most of the known plesiadapidines are different from those found in the main line of primates and indicate that they were not the immediate ancestors of the more advanced primates.

Lemurs

In the Late Paleocene and Eocene, the next step in primate evolution occurred with the appearance of the **lemurs** (Figs. 22.3–4). Lemurs are also

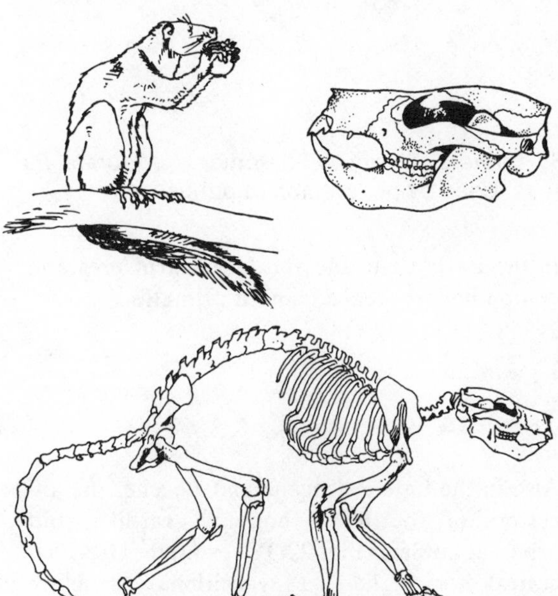

Figure 22.2. *Plesiadapis, A Paleocene Primate.* Restoration, skull, and skeleton. (Adapted from Simons, "The Early Relatives of Man." *Scientific American,* July 1964. ©1964 by Scientific American, Inc. All rights reserved.)

Figure 22.3. *Restoration of an Eocene Lemur. Notharctus osborni* of Middle Eocene age from North America. (Painting by Ferguson, courtesy of the American Museum of Natural History.)

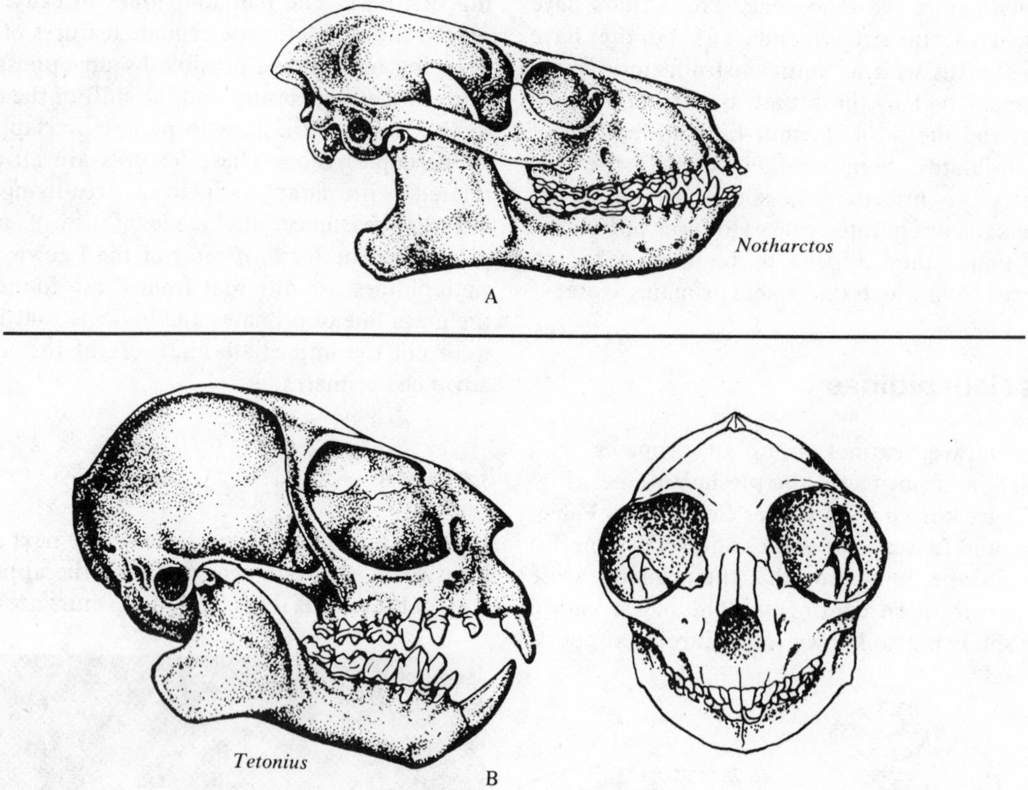

Figure 22.4. *Eocene Primate Skulls.* **A.** *Lemur,* **B.** *Tarsiers.* (From A.S. Romer, *Vertebrate Paleontology,* © 1966 by the University of Chicago Press. Used by permission of publisher.)

rather small quadrupedal forms and resemble squirrels. They retain a muzzle, and the eyes are not completely rotated to the front of the head. Most still have claw-like nails on at least some digits. The tail is usually long, but not prehensile for grasping. The early lemurs were widespread in the Eocene, occurring in the subtropical to tropical forests of North America, Europe, and Asia. As these forests disappeared in the Oligocene because of continental movements and the widespread mountain building that cooled climates and produced rain shadows, the range of lemurs contracted. Today they survive in southern Asia, eastern Africa, and Madagascar (Fig. 22.1B). The island of Madagascar is their chief refuge because it split off from Africa in the Early Cenozoic and lacks carnivores and competing arboreal advanced primates.

Tarsiers

Also in the Late Paleocene and Eocene, the ancestors of the living hopping, rat-like, but arboreal **tarsiers** (Fig. 22.1C) evolved. These ancestral forms had a transitional structure between the lemurs and the monkeys (Fig. 22.4). In particular, they and the living tarsiers have large, forward-directed eyes and have lost the muzzle. The brain, especially in the cerebral hemisphere area, is larger than that of lemurs.

The tarsiers, like the lemurs, were widespread on the northern continents in the Eocene, but have since become restricted, probably for the same reasons, to isolated regions, in this case, the East Indies.

Monkeys

The **South American** or **platyrrhine monkeys** (Fig. 21.33) probably evolved from a Paleogene lemur or tarsier that found its way down the volcanic island arcs between North and South America or migrated across the narrow early Atlantic from Africa via volcanic islands. Platyrrhine monkeys are primitive in being small and possessing claw-like nails, an only slightly opposable thumb, a small brain, and a long tail that is prehensile in most. They are apparently related to the Old World monkeys only through their common lemur or tarsier ancestor, and the present similarities of the two groups appear to be caused by convergent evolution in similar environments.

The **Old World** or **catarrhine monkeys** (Fig. 22.1D) appeared in the Oligocene of Egypt in northeast Africa. Africa was united with Eurasia in that region at the time, allowing entry of the tarsier ancestor of the Old World monkeys. Africa was to become the main center of evolution for the remaining groups of primates which spread from there to the other continents in the Late Neogene. The Oligocene Fayum fauna of Egypt contains fossils of several Old World monkeys. These are larger, bigger-brained, shorter-tailed and with a more opposable thumb and more typical nails than the New World monkeys. Several of these Old World monkeys show adaptations to ground-dwelling, a trait that culminates in the completely terrestrial Late Neogene baboons. The Fayum fauna also contains the oldest representatives of the **Superfamily Hominoidea,** the group to which man and the living great apes are assigned.

Hominoids

The Oligocene plexus of hominoids includes *Aegyptopithecus* (Fig. 22.5), the oldest known member, at 28 million years, of the specific lineage leading to man and the great apes. In addition to the original well-preserved skull, an arm bone of *Aegyptopithecus* has recently been found, and it indicates the creature was arboreal. The presence of mid-Oligocene apes and monkeys indicates that the divergence of the groups occurred earlier, perhaps at the beginning of that epoch nearly 40 million years ago.

Dryopithecines

Following this Late Oligocene genus, from the Early through the Late Miocene (20 to 8 million years ago), lived a group of hominoids called the **dryopithecines** (Fig. 22.6). The dryopithecines were near the point of divergence of men and the

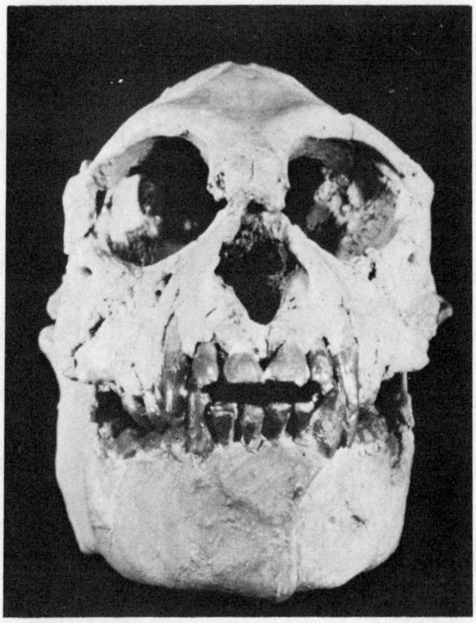

Figure 22.5. *Aegyptopithecus zeuxis Skull.* Oligocene, Fayum, Egypt. (Courtesy of Elwyn L. Simons, Yale University and *American Scientist.*)

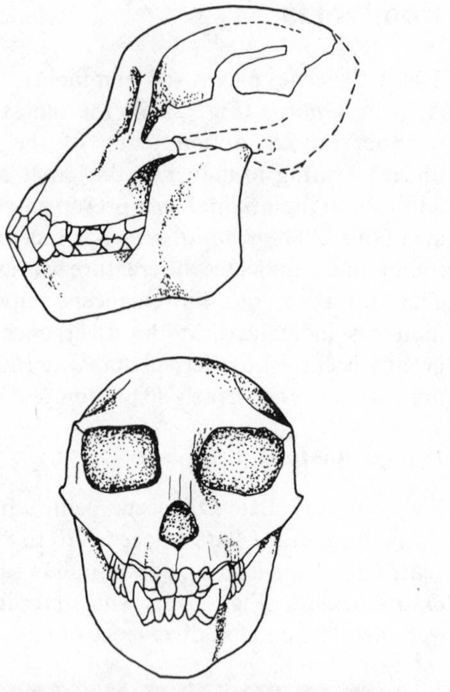

Figure 22.6. *Dryopithecine Skull.* (After Le Gros Clark.)

surviving ground-dwelling apes, the gorilla and the chimpanzee. At least the more advanced types, however, are more ape-like because the two sides of the lower jaw are nearly parallel as in that group, rather than slightly divergent as in man (Fig. 22.7). Dryopithecines are known from essentially complete skulls which display a low, rounded vault, moderate brow ridges, and a moderate forward protrusion of the jaws and face.

Unfortunately the limbs are poorly known, but they do suggest that the creature was not highly adapted to brachiation, and the foot is somewhat indicative of a bipedal posture. Perhaps *Dryopithecus* lived on the ground like the living baboon monkeys and occasionally reared up on its hind legs. The drying trend of the Neogene was restricting African forests at this time, and the ape-man lineage was evolving in response to the spreading open grassland country. Ground-dwelling attributes were being selected for, and one of these was surely the

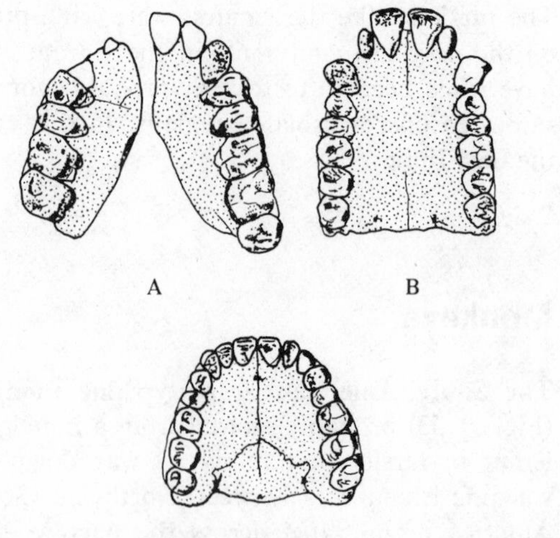

Figure 22.7. *Comparisons of* **A** *Ramapithecus Jaw with Jaws of* **B** *Ape (Orangutan) and* **C** *Man.* Note slight divergence of sides of jaw in human lineage as opposed to the parallel sides in apes. (After Hetrick, Simons.)

ability to rear up on the hind legs in order to view more of the surrounding countryside. The dryopithecines were most common in east Africa, but spread to the north and through the Near East region to southern Asia including India, now joined to Eurasia and Europe (Fig. 22.8).

Ramapithecus and Kenyapithecus

In the Middle and Late Miocene of east Africa and southern Asia have been found some fragments of jaws and teeth of a genus related to the dryopithecines. The genus from India, **Ramapithecus** (Figs. 22.7–10), has slightly divergent lower jaw bones as in man and is thus the oldest known fossil, about 10 million years old, unequivocally assigned to the **Family Hominidae,** the family of man. The east African form is known as *Kenyapithecus* (Fig. 22.8). It ranges in age from Early to Late Miocene and may be synonymous with *Ramapithecus* or closer to its dryopithecine ancestors. The 20-million-year-old species *K. africanus* evolved into the 12–14-million-year-old species *K. wickeri.* This latter

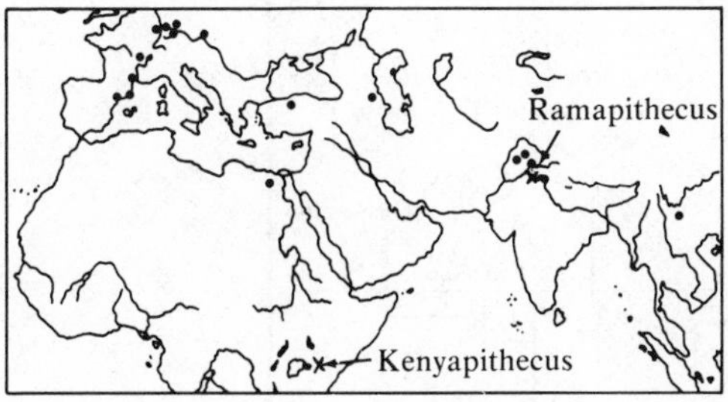

Figure 22.8. *Distribution of Dryopithecines and Ramapithecus-Kenyapithecus Complex.* Dots represent dryopithecine sites, X's represent *Ramapithecus-Kenyapithecus* sites. (After Simons.)

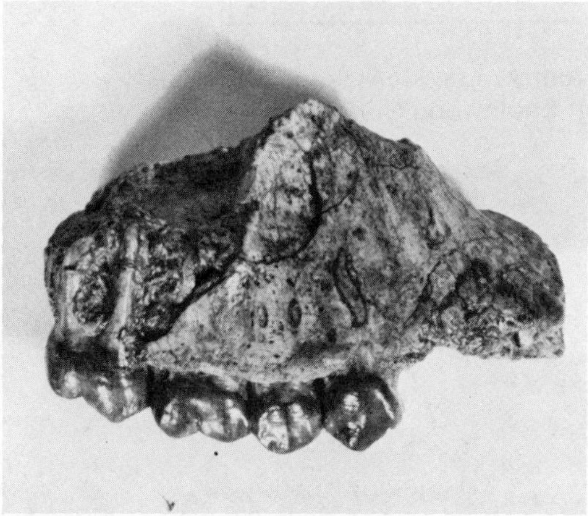

Figure 22.9. *Ramapithecus brevirostris.* (Courtesy of Elwyn L. Simons, Yale University.)

species may be the oldest known tool-using hominid. Blunt stones, not native to the site, were found associated with broken bones and crushed skulls of animals at Fort Ternan, Kenya. The use of tools indicates that the arms were in the process of being freed from walking and were being used for other purposes, perhaps indicating the onset of bipedalism.

Australopithecines

After the *Ramapithecus-Kenyapithecus* level of human evolution is that of the **australopithecines** (Figs. 22.10–14), chiefly found in Pleistocene rocks of 2 to .5 million years in age, but recently unearthed in 3–5-million-year-old Pliocene strata at Lake Rudolf, Africa. *Australopithecus* was a long-lived genus consisting of several species, some of which tended to a more carnivorous diet (gracile or africanus type) and others of which were more vegetarian (robustus or paranthropus type).

Australopithecus is best known from South Africa (Fig. 22.12), where skulls, jaws, teeth, and limbs have been found (Fig. 22.14). More recently, good east African material and fragments from southern and eastern Asia have been unearthed (Fig. 22.12). The three generally accepted species form a lineage whose members increased from chimpanzee to gorilla size through time, but all walked erect with limb bones essentially like those of modern man. The skull differed from modern man by its smaller brain case (400–500 cc, less than half that of modern man), more prominent brow ridges, and slightly more protruding face. The bipedal posture freed the front limbs for other functions such as tool-making and using. Simple pebble tools called hand axes, probably used for cutting, scraping, and chopping, have been found with the australopithecine fossils (Figs. 22.13, 22.23). These are assigned to the **Oldowayan** cultural level, the oldest unequivocal one yet found. Broken brain cases indicate that at least some australopithecines were cannibals. However, the teeth of australopithecines indicate their diet was basically vegetarian.

In 1974, a fairly complete fossil skeleton of a 3-million-year-old hominoid was unearthed in northeast Ethiopia, Africa. This specimen is of an erect but very small species, apparently more primitive than the australopithecines. This fossil, nicknamed "Lucy," may represent a late survivor of the group linking *Ramapithecus* and *Australopithecus.*

About 1.75 million years ago, in east Africa, a form of man appeared that was transitional, in some respects, between the advanced australopithecines and the next step in human evolution. This creature, unearthed by *Leaky* and called *Homo habilis* (Fig. 22.14), is considered by

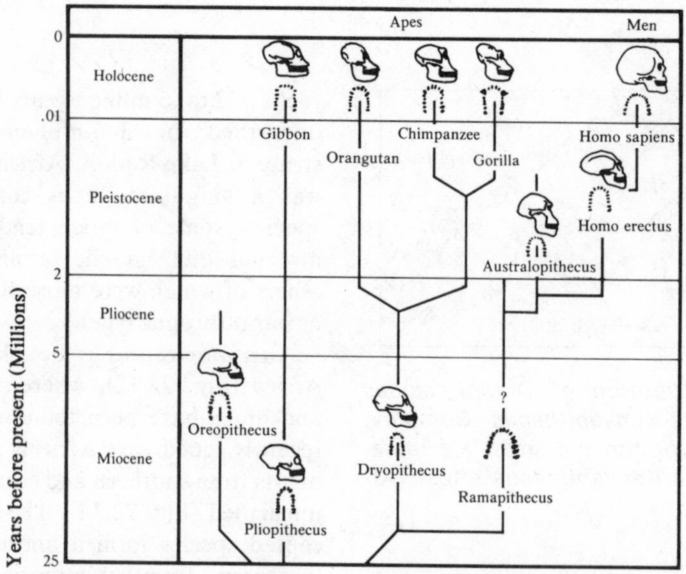

Figure 22.10. *Evolution of the Apes and Men.* (From A. Lee McAlester, *The History of Life,* © 1968. Reprinted by permission of Prentice-Hall, Inc., Englewood Cliffs, N.J.)

Figure 22.11. *Australopithecus Restoration.* (Augusta-Burian mural, courtesy of American Museum of Natural History.)

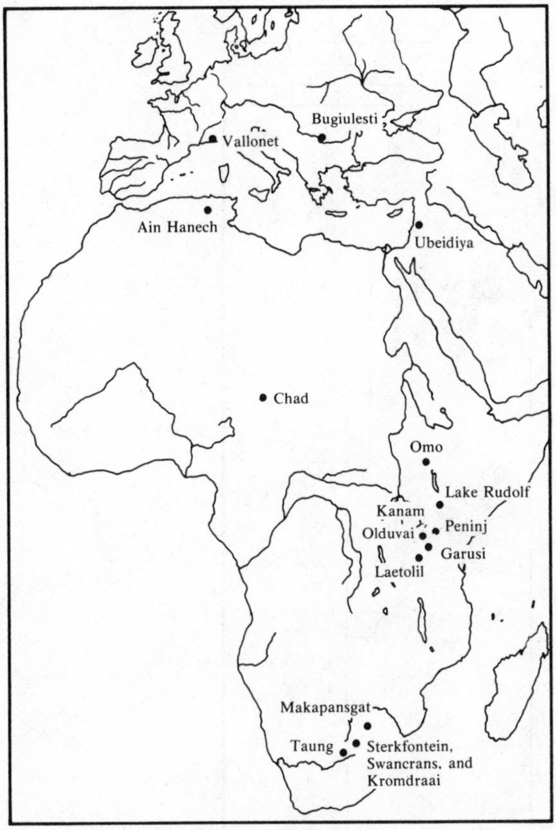

Figure 22.12. *Localities Where Australopithecus Remains or Artifacts Have Been Found.* (After Chard, Tobias.)

some to be nothing more than an advanced australopithecine and not the direct ancestor of modern man. The brain size, 600–700 cc, is intermediate between that of *Australopithecus* and *Homo erectus,* the oldest human species.

In 1972, a skull with a capacity of 750 cc, clearly belonging to the genus *Homo* was unearthed near Lake Rudolf in Africa. This find indicates that either *Homo habilis* or *Homo erectus* evolved earlier than originally thought and was contemporaneous with many australopithecines.

Homo erectus

The oldest unequivocal member of our genus, *Homo erectus* (Figs. 22.10, 22.13–15), probably evolved from *Australopithecus,* perhaps as early as 3 million years ago and certainly by 1.5 million years ago, was contemporaneous with it for a long period, and eventually supplanted it. The youngest representatives of *H. erectus* lived as recently as 200,000 years ago. *Homo erectus* was first known from Java where it was called *Pithecanthropus* and China where it was known as *Sinanthropus.* Experts now agree that it is sufficiently like modern man to be assigned to the same genus. The skull of *Homo erectus* has a brain capacity (750–1300 cc) bridging the gap between that of the australopithecines and modern man, but has a lower dome, more prominent brow ridges, and a more protruding face than that of modern man. It also lacks a chin, which suggests to some that language had not yet evolved. Remains of *Homo erectus* are now known in eastern and northern Africa, Asia, and Europe.

The tools of *Homo erectus* consisted of large, shaped hand axes of flint. Their quality of workmanship improved as the Middle Pleistocene progressed. The older hand axes, chipped on only one side, are assigned to the **Abbevillian** or **Chellean** cultural level (Fig. 22.16A), while the younger, completely worked, hand axes are called **Acheulian** (Fig. 22.16B). Scrapers and knives made from flakes of pebbles are called **Clactonian** (Fig. 22.16C). The somewhat different chopper tools of Asia are assigned to the **Soan** culture (Fig. 22.16D). The tools of these cultures are widely found in Europe, Asia, and Africa (Fig. 22.17). *Homo erectus* had also learned the use of fire by at least 800,000 years ago and was building crude shelters by 400,000 years ago.

Neanderthal Man

The earliest records of our species, *Homo sapiens,* are scattered fragments from Morocco, France, Germany, and England of approximately 300,000 to 500,000 years in age. These apparently represent primitive subspecies, as most have slightly lower brain capacities than modern man. The earliest complete skeletons of members of our species are those of the so-called

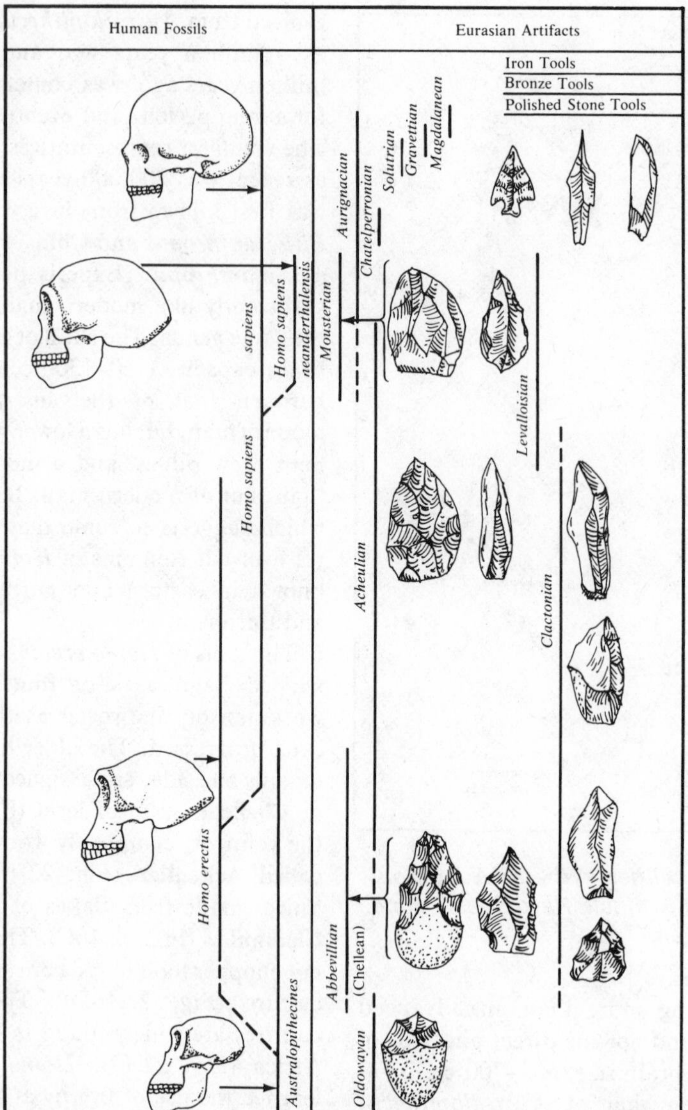

Figure 22.13. *Skulls, Ranges, and Tools of Major Pleistocene Hominoids.* (After Oakley.)

Neanderthal man (Figs. 22.13–14, 22.18), who lived from about 100,000 to 50,000 years ago during the Würm glacial stage in Europe, north Africa, and western Asia. He differed from typical *Homo sapiens* by possessing a lower forehead, wider skull, more prominent brow ridges and by lacking a chin. His brain capacity was slightly lower than the 1400 cc average of modern man. His body was modern in build, though many restorations picture him as a squat, slouching brute, an erroneous impression based on the skeleton of a diseased individual.

He developed the cultures known as **Levalloisian** and **Mousterian** (Figs. 22.13, 22.19) with their well-worked stone axes, scrapers, and points. The last were attached to shafts which

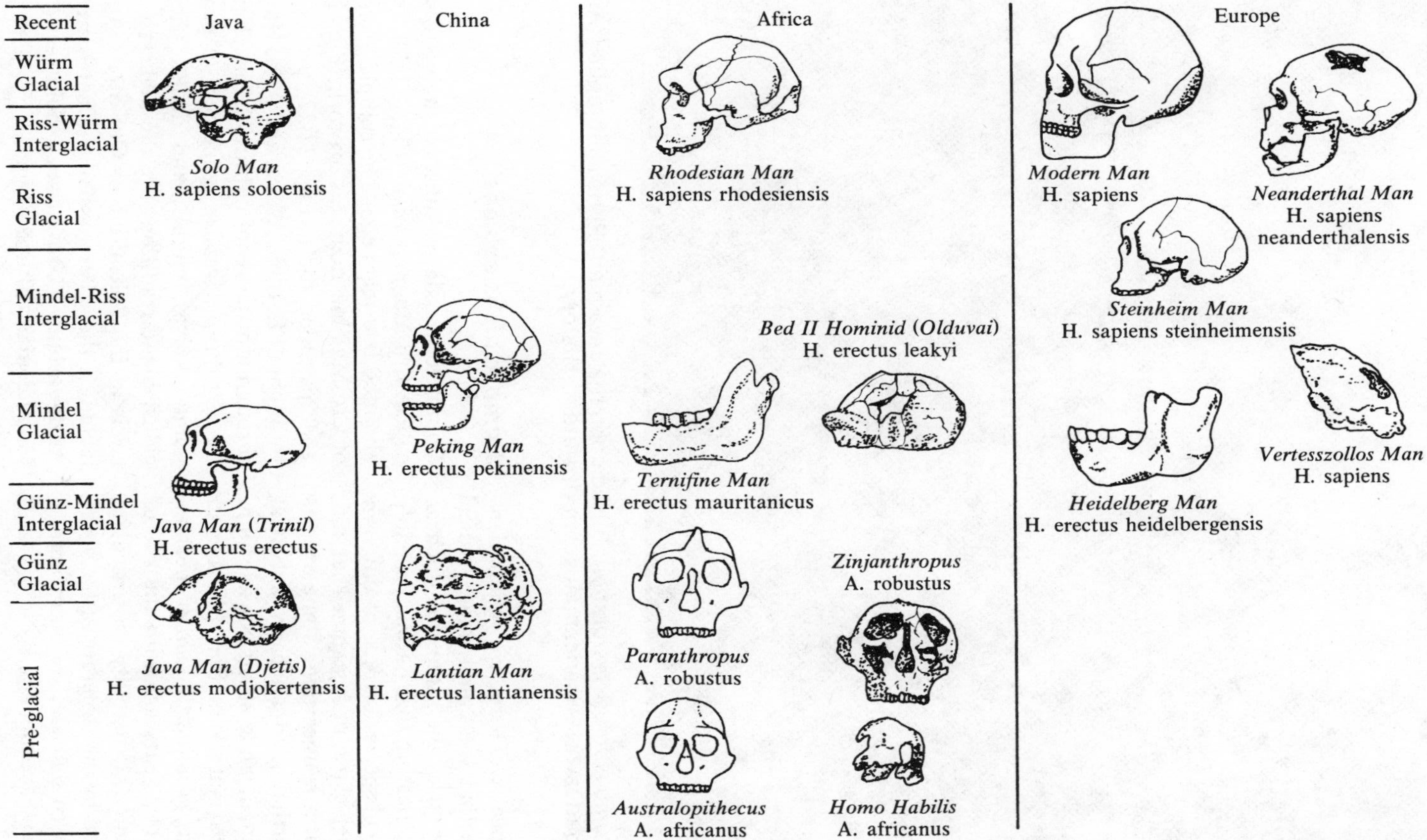

Figure 22.14. *Major Fossil Evidence of Human Evolution Known in 1966.* Italicized names are those originally assigned. Modern assignment appears below in roman type. (Adapted from *History of the Earth,* 2nd ed. by Bernhard Kummel, W.H. Freeman & Company. © 1970, by Scientific American, Inc. All rights reserved.)

Figure 22.15. *Homo erectus Restoration.* Based upon the Java man, Pithecanthropus. (Augusta-Burian mural, courtesy of American Museum of Natural History.)

then became spears. There are gradiations between Neanderthal man and more typical *Homo sapiens,* even at the same site, such as Carmel, Israel, indicating that he was probably no more than a subspecies or race of our own species. The exact cause of his somewhat different skull features is uncertain. Some experts consider Neanderthal man an early ancestral form of modern man, while others consider him a geographical variant. Whatever his role in human evolution, Neanderthal man became extinct about 50,000 years ago with the rise of Cro-Magnon man. The inability to articulate vowels as indicated by the structure of the vocal tract may have been the selective disadvantage leading to Neanderthal extinction.

Homo sapiens sapiens

Completely modern skulls, belonging to what is called **Cro-Magnon man** (Fig. 22.20), appear about 50,000 years ago. The complete skeletons of Cro-Magnon man, his artwork (Fig. 22.21), and his stone tools (Fig. 22.22) are common throughout Europe, Africa, and Asia. He passed through a diversity of cultural levels characterized by the progressive refinement of projectile points, knives, chisels, and drills. These so-called blade traditions are the highest cultural levels of the **Paleolithic** or Old Stone Age and include the **Perigordian, Chatelperronian, Aurignacian, Gravettian, Solutrean,** and **Magdalenian** cultures (Fig. 22.13). The people

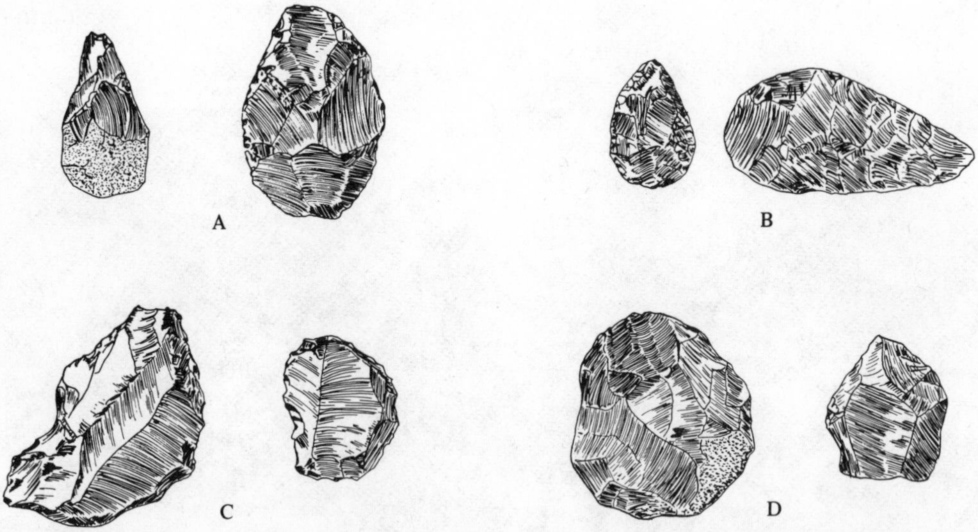

Figure 22.16. *Artifacts of Homo erectus.* **A.** *Early hand-axe, or Abbevillian;* **B.** *Late hand-axe, or Acheulian;* **C.** *European flake-tool, or Clactonian;* **D.** *Asian chopper-tool, or Soan.* X 1/3. (After Oakley.)

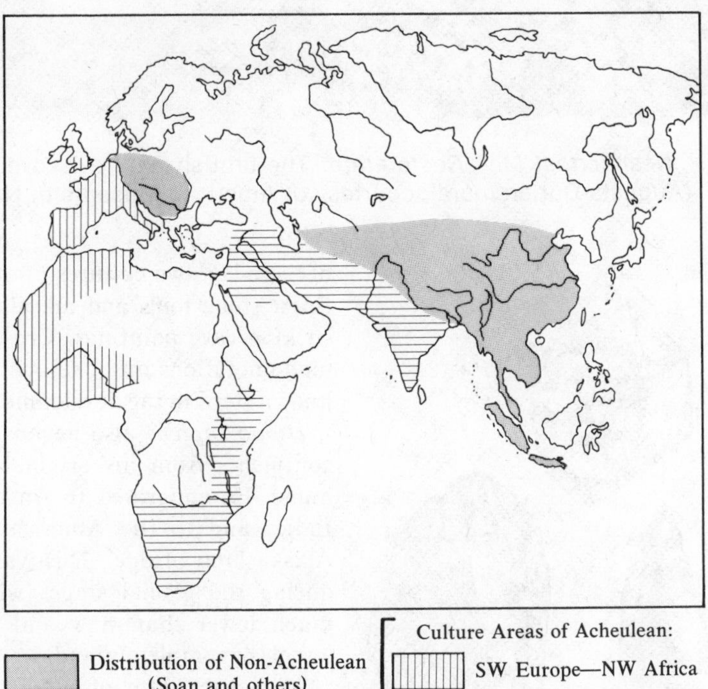

Figure 22.17. *Distribution of Artifacts of Homo erectus.* (From C.S. Chard, *Man in Prehistory,* © 1969, by permission of McGraw-Hill Book Co.)

Figure 22.18. *Neanderthal Man Restoration.* The brutish nature is overemphasized in this painting. (Augusta-Burian mural, courtesy of American Museum of Natural History.)

Figure 22.19. *Artifacts of Neanderthal Man.* Mousterian culture. (After Oakley.)

of these last two cultures produced probably the finest stone tools and are also known for their striking cave paintings. Cro-Magnon man also made notations on bone, which are the earliest known steps in the development of writing.

Homo sapiens also began to spread to other continents from his original African-Eurasian home. He emigrated to Australia via the East Indies and to the Americas via the Siberia-Alaska land bridge. This was particularly easy during the glacial stages when sea level was much lower than now and those areas were largely or completely land.

The exact date of man's entry into North America is still uncertain. The possibility existed at least as early as 50,000 years ago when typical *Homo sapiens* had evolved and the shelves were exposed. Until recently, however, only some possible charred wood, bones, and

Figure 22.20. *Cro-Magnon Man Restoration.* (Augusta-Burian mural, courtesy of American Museum of Natural History.)

Figure 22.21 *Artwork of Cro-Magnon Man.* Reproductions of late Paleolithic cave art from France. (After Oakley.)

ground provided evidence until near the end of the Pleistocene around 13,000 years ago when fluted spear points appeared. The extent of the glaciers at that time would have prevented man from entering North America, so it is probable he arrived during the preceding interglacial. Recently, some skulls from the San Diego area of California have been dated at 40,000–50,000 years in age, at last providing direct evidence of man's early entry into the New World.

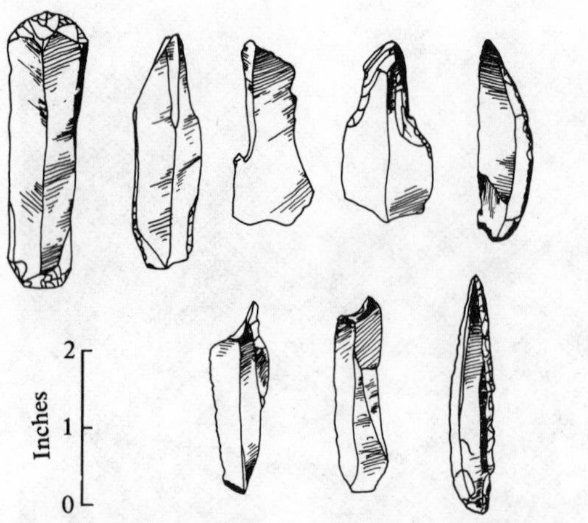

Figure 22.22. *Artifacts of Cro-Magnon Man.* (After Oakley.)

About 10,000 years ago, near the Pleistocene-Holocene boundary, *Homo sapiens* man began to make polished tools by grinding, and he moved into the **Neolithic** or New Stone Age cultures (Fig. 22.23). The neolithic people also began to make pottery, domesticate animals, and cultivate plants. Man could then settle down in permanent sites which became the first towns about 10,000 years ago. In these towns, some people could perform functions other than simply raising food. About 5,000 years ago the metal cultures replaced the stone age cultures, and man began to write his history. With this event, we pass from prehistory to history and thus from Historical Geology to History.

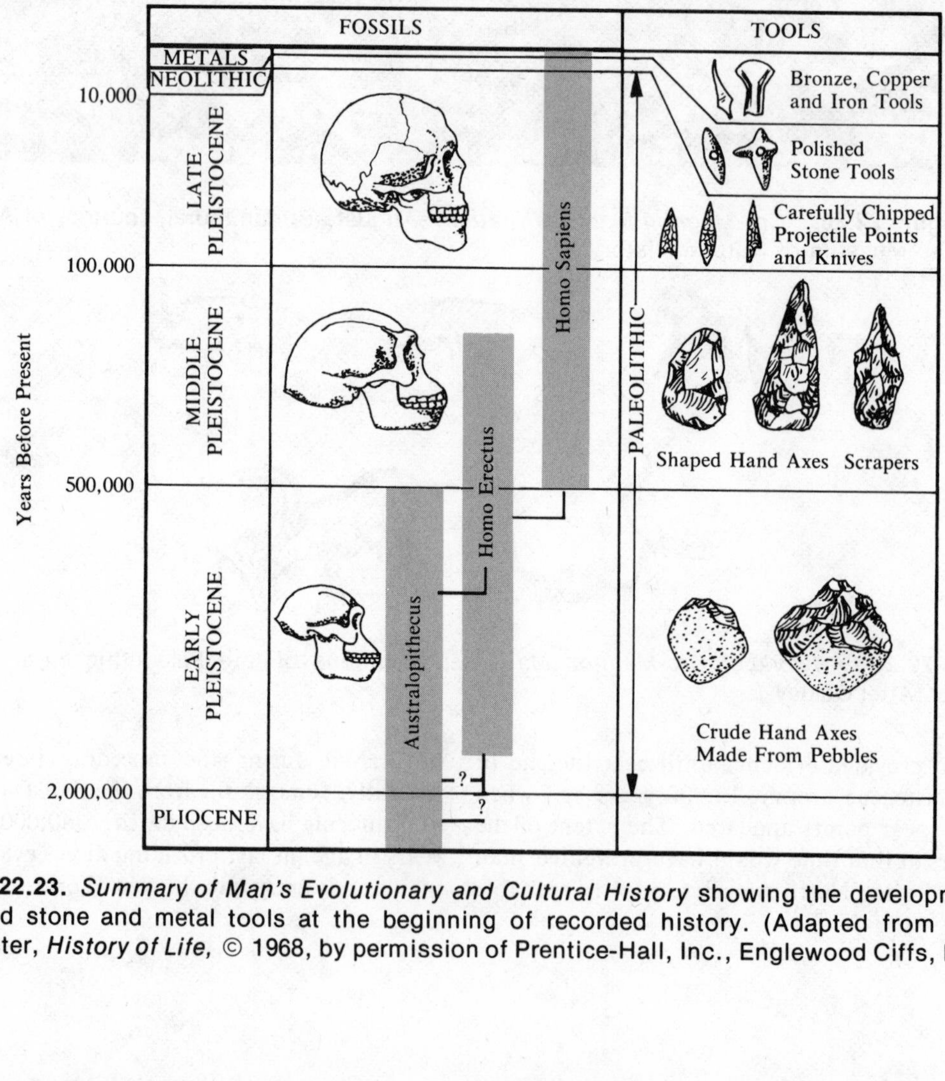

Figure 22.23. *Summary of Man's Evolutionary and Cultural History* showing the development of polished stone and metal tools at the beginning of recorded history. (Adapted from A. Lee McAlester, *History of Life,* © 1968, by permission of Prentice-Hall, Inc., Englewood Ciffs, N.J.)

REVIEW

1. To what order of mammals does man belong, and what other groups are also members of this order?
2. What is the oldest known group of fossil primates?
3. How do **lemurs** and **tarsiers** differ from more specialized primates?
4. Discuss the changes throughout the Cenozoic in the distribution of lemurs and tarsiers.
5. What is the oldest known member of the **ape-man group** of primates; when did it live; and what were its habits?
6. What is the significance of *Dryopithecus* and *Ramapithecus?* When did they live?
7. When did the **Australopithecines** live, and what was their general physical appearance?
8. What is *Homo habilis,* and what might be its significance?
9. How did *Homo erectus* differ from the Australopithecines and from modern man?
10. How did **Neanderthal man** differ from **Cro-Magnon man**? What might be their relationships?
11. Which type of man was responsible for the following types of implements: **Mousterian, Chellean, Aurignacian, Oldowayan**?
12. What are three recent fossil discoveries that have altered earlier ideas about the antiquity of man and his migrations?

Bibliography

Historical Geology

Clark, T.H., and Stearn, C.W. *The Geological Evolution of North America.* New York: Ronald, 1968.

Cloud, P., ed. *Adventures in Earth History.* San Francisco: Freeman, 1970.

Dott, R.H., and Batten, R.L. *Evolution of the Earth.* New York: McGraw-Hill, 1976.

Dunbar, C.O., and Waage, K.M. *Historical Geology.* New York: Wiley, 1969.

Kay, M., and Colbert, E.H. *Stratigraphy and Life History.* New York: Wiley, 1965.

Kummel, B. *History of the Earth.* San Francisco: Freeman, 1970.

Moore, R.C. *Introduction to Historical Geology.* New York: McGraw-Hill, 1958.

Peterson, M.S.; Rigby, J.K.; and Hintze, L.F. *Historical Geology of North America.* Dubuque, Iowa: Brown, 1973.

Seyfert, C.K., and Sirkin, L.A. *Earth History and Plate Tectonics.* New York: Harper & Row, 1973.

Spencer, E.W., *Basic Concepts of Historical Geology.* New York: Crowell, 1962.

Stirton, R.A. *Time, Life, and Man.* New York: Wiley, 1959.

Stokes, W.L. *Essentials of Earth History.* Englewood Cliffs, N.J.: Prentice-Hall, 1973.

Wells, A.K., and Kirkaldy, J.F. *Outline of Historical Geology.* London: Murby, 1966.

Woodford, A.O. *Historical Geology.* San Francisco: Freeman, 1965.

Paleontology

Ager, D.V. *Principles of Paleoecology.* New York: McGraw-Hill, 1963.

Andrews, H.N. *Studies in Paleobotany.* New York: Wiley, 1961.

Arnold C.A. *An Introduction to Paleobotany.* New York: McGraw-Hill, 1947.

Banks, H.P. *Evolution and Plants of the Past.* Belmont, Calif.: Wadsworth, 1970.

Beerbower, J.R. *Search for the Past.* Englewood Cliffs, N.J.: Prentice-Hall, 1968.

Black, R.M. *The Elements of Paleontology.* Norwich, England: Cambridge University, 1970.

Brouwer, A. *General Paleontology,* Chicago: University of Chicago Press, 1966.

Clark, D.L. *Fossils, Paleontology and Evolution.* Dubuque, Iowa: Brown, 1968.

Colbert, E.H. *Evolution of the Vertebrates,* New York: Wiley, 1969.

Darrah, W.C. *Paleobotany.* New York: Ronald, 1960.

Delevoryas, T. *Morphology and Evolution of Fossil Plants.* New York: Holt, Rinehart and Winston, 1962.

Easton, W.H. *Invertebrate Paleontology.* New York: Harper, 1960.

Fenton, C.L., and Fenton, M.A. *The Fossil Book.* Garden City, N.Y.: Doubleday, 1958.

Halstead, L.B. *The Pattern of Vertebrate Evolution.* San Francisco: Freeman, 1968.

Imbrie, J., and Newell, N., eds. *Approaches to Paleoecology.* New York: Wiley, 1964.

Jones, D.L. *Introduction to Microfossils.* New York: Harper, 1956.

Kurten, B. *The Age of Dinosaurs.* New York: McGraw-Hill, 1968.

Laporte, L. *Ancient Environments,* Englewood Cliffs, N.J.: Prentice-Hall, 1968.

Levin, H.L. *Life Through Time.* Dubuque, Iowa: Brown, 1975.

McAlester, A.L. *The History of Life.* Englewood Cliffs, N.J.: Prentice-Hall, 1968.,

Mathews, W.H. *Fossils.* New York: Barnes & Noble, 1962.

Moore, R.C., ed. *Treatise on Invertebrate Paleontology.* Lawrence, Kansas: University of Kansas, 1953–1972.

Moore, R.C.; Lalicker, C.G.; and Fischer, A.G. *Invertebrate Fossils.* New York: McGraw-Hill, 1952.

Olson, E.C. *Vertebrate Paleozoology.* New York: Wiley, 1971.

Raup, D.M., and Stanley, S.M. *Principles of Paleontology.* San Francisco: Freeman, 1971.

Rhodes, F.H.T.; Zim, H.S.; and Shaffer, P.R. *Fossils.* New York: Golden, 1962.

Romer, A.S. *The Procession of Life.* Cleveland: World, 1968.

Romer, A.S. *Vertebrate Paleontology.* Chicago: University of Chicago, 1966.

Seward, A.C. *Plant Life Through the Ages.* New York: Hafner, 1959.

Shrock, R.R., and Twenhofel, W.H. *Principles of Invertebrate Paleontology.* New York: McGraw-Hill, 1953.

Tasch, P. *Paleontology of the Invertebrates.* New York: Wiley, 1973.

Thenius, E. *Fossils and the Life of the Past.* New York: Springer-Verlag, 1973.

Weller, J.M. *The Course of Evolution.* New York: McGraw-Hill, 1969.

Stratigraphy

Berry, W.B.N. *Growth of a Prehistoric Time Scale.* San Francisco: Freeman, 1968.

Dunbar, C.O., and Rodgers, J. *Principles of Stratigraphy.* New York: Wiley, 1957.

Eicher, D.L. *Geologic Time.* Englewood Cliffs, N.J.: Prentice-Hall, 1976.

Garrels, R.M., and MacKenzie, F.T. *Evolution of Sedimentary Rocks.* New York: Norton, 1971.

Gignoux, M. *Stratigraphic Geology.* San Francisco: Freeman, 1955.

Harbaugh, J.W. *Stratigraphy and Geologic Time.* Dubuque, Iowa: Brown, 1968.

Krumbein, W.C., and Sloss, L.L. *Stratigraphy and Sedimentation.* San Francisco: Freeman, 1964.

Matthews, R.K. *Dynamic Stratigraphy.* Englewood Cliffs, N.J., Prentice-Hall, 1974.

Shaw, A. *Time in Stratigraphy.* New York: McGraw-Hill, 1964.

Weller, J.M. *Stratigraphic Principles and Practice.* New York: Harper, 1960.

Plate Tectonics

Bird, J.M., and Isacks, B., eds. *Plate Tectonics.* Washington, D.C.: American Geophysical Union, 1972.

Calder, N. *Restless Earth.* London: British Broadcasting Corporation, 1972.

Cox, A., ed. *Plate Tectonics and Geomagnetic Reversals.* San Francisco: Freeman: 1973.

Glen, W. *Continental Drift and Plate Tectonics.* Columbus, Ohio: Charles E. Merrill, 1975.

Hallam, A. *A Revolution in the Earth Sciences.* Oxford: Claredon Press, 1973.

Takeuchi, H.; Uyeda S.; and Kanamori, H. *Debate About the Earth.* San Francisco: Freeman, Cooper & Co., revised 1970.

Tarling, D., and Tarling, M. *Continental Drift.* Garden City, N.Y.: Doubleday & Co., Anchor Books, 1971.

Wilson, J.T., ed. *Continents Adrift.* San Francisco: Freeman, 1970.

Wilson, J.T., ed. *Continents Adrift and Continents Aground.* San Francisco: Freeman, 1976.

Index

Index

GEOLOGIC TIME SCALE

Eras	Periods	Alternate Periods	Epochs (North America)	Duration (in millions of years)		Abundant or Characteristic Fossil groups	
Cenozoic Era	Neogene Period	Quaternary Period	Holocene Epoch	23–24 M.Y.	3 M.Y.	Charaphytes Cocolithophores Diatoms Dinoflagellates Red Algae Rotaliine Forams Radiolaria Hexacorals Cheilostone Bryozoa Terebratuloid Brachs Scaphopods Gastropods Pelecypods	Ostracods Barnacles Malacostracan Crustaceans Insects Starfish Cassiduloid, Clypeasteroid, and Spatangoid Echinoids Fish Birds Mammals Angiosperms
			Pleistocene Epoch				
		Tertiary Period	Pliocene Epoch		63 M.Y.		
			Miocene Epoch				
	Paleogene Period		Oligocene Epoch	42 M. Y.			
			Eocene Epoch				
			Paleocene Epoch				
65 M. Y. Ago							
Mesozoic Era	Cretaceous Period		Gulfian Epoch	60–71 M. Y.		Charaphytes Coccolithophores Diatoms Dinoflagellates Rotaliine Forams Radiolaria Tintinnoids Sponges Hexacorals Cyclostome and Cheilostome Bryozoa	Terebratuloid Brachs Gastropods Pelecypods Ammonitic Ammonoids Belemnites Ostracods Holectypoid, Perischoechinoid, and Euechnoid Echinoids Spatangoid Echinoids Archosaur Reptiles
			Comanchean Epoch				
			Coahuilan Epoch				
125–136 M.Y. Ago							
	Jurassic Period		Upper Jurassic Epoch	45–60 M. Y.		Dasycladaceous Algae Dinoflagellates Radiolaria Tintinnoids Rhynchonelloid Brachs Ammonitic Ammonoids Belemnites	Disasteroid Echinoids Archosaur Reptiles Ferns Cycadeoids Conifers Gingkos
			Middle Jurassic Epoch				
			Lower Jurassic Epoch				
180–195 M.Y. Ago							
	Triassic Period		Upper Triassic Epoch	35–45 M. Y.		Dasycladaceous Algae Spongiomorphoids Ceratite Ammonoids Aulacocerids	Branchiopods Mammal-Like Reptiles Conifers Pectinoid Pelecypods
			Middle Triassic Epoch				
			Lower Triassic Epoch				
220–230 M.Y. Ago							
Paleozoic Era (continued)	Permian Period		Upper Permian Epoch	35–55 M. Y.		Fusuline Forams Cryptosome Bryozoa Chonetoid Brachs Productoid Brachs Davidsonioid Brachs	Goniatite Ammonoids Insects Amphibians Mammal-Like Reptiles Glossopterids
			Lower Permian Epoch				
265–280 M.Y. Ago							